Social Computing:
Concepts, Methodologies, Tools, and Applications

Subhasish Dasgupta
George Washington University, USA

Volume IV

INFORMATION SCIENCE REFERENCE
Hershey · New York

Director of Editorial Content: Kristin Klinger
Development Editor Julia Mosemann
Senior Managing Editor: Jamie Snavely
Assistant Managing Editor, MVB: Michael Brehm
Typesetters: Michael Brehm, Carole Coulson, Elizabeth Duke-Wilson, Devvin Earnest, Christopher Hrobak, Kurt Smith, Jamie Snavely, Sean Woznicki
Cover Design: Lisa Tosheff
Printed at: Yurchak Printing Inc.

Published in the United States of America by
Information Science Reference (an imprint of IGI Global)
701 E. Chocolate Avenue
Hershey PA 17033
Tel: 717-533-8845
Fax: 717-533-8661
E-mail: cust@igi-global.com
Web site: http://www.igi-global.com/reference

and in the United Kingdom by
Information Science Reference (an imprint of IGI Global)
3 Henrietta Street
Covent Garden
London WC2E 8LU
Tel: 44 20 7240 0856
Fax: 44 20 7379 0609
Web site: http://www.eurospanbookstore.com

Library of Congress Cataloging-in-Publication Data

Social computing : concepts, methodologies, tools and applications / Subhasish Dasgupta, editor.
p. cm.
Includes bibliographical references and index.
Summary: "This book uncovers the growing and expanding phenomenon of human behavior, social constructs, and communication in online environments, reflecting on social change, evolving networks, media, and interaction with technology and more"--Provided by publisher.
ISBN 978-1-60566-984-7 (hardcover) -- ISBN 978-1-60566-985-4 (ebook) 1. Information technology--Social aspects. 2. Technological innovations--Social aspects. 3. Internet--Social aspects. 4.
Technology--Social aspects. I. Dasgupta, Subhasish, 1966-
HM851.S62996 2010
303.48'34--dc22

2009047176

British Cataloguing in Publication Data
A Cataloguing in Publication record for this book is available from the British Library.

List of Contributors

Contents

Volume 1

Section I. Fundamental Concepts and Theories

This section serves as the foundation for this exhaustive reference source by addressing crucial theories essential to the understanding of social computing. Chapters found within this section provide a framework in which to position social computing within the field of information science and technology. Individual contributions provide overviews of computer-mediated communication, social networking, and social software. Within this introductory section, the reader can learn and choose from a compendium of expert research on the elemental theories underscoring the research and application of social computing.

Volume II

Section III. Tools and Technologies

This section presents extensive coverage of the tools and specific technologies that change the way we interact with and respond to our environments. These chapters contain in-depth analyses of the use and development of innumerable devices and also provide insight into new and upcoming technologies, theories, and instruments that will soon be commonplace. Within these rigorously researched chapters, readers are presented with examples of specific tools, such as social television, wikis, mobile photo galleries, and personal digital libraries. In addition, the successful implementation and resulting impact of these various tools and technologies are discussed within this collection of chapters.

Section IV. Utilization and Application

This section introduces and discusses the utilization and application of social computing technologies. These particular selections highlight, among other topics, the many applications of social networking technology, social software, and social marketing strategies. Contributions included in this section provide coverage of the ways in which technology increasingly becomes part of our daily lives as it enables the creation of new forms of interaction among individuals and across organizations.

Volume III

Section VI. Managerial Impact

This section presents contemporary coverage of the managerial implications of social computing. Particular contributions address the importance of guarding corporate data from social engineering attacks and explore a professional social network system. The managerial research provided in this section allows executives, practitioners, and researchers to gain a better sense of how social computing can impact and inform practices and behavior.

Section VII. Critical Issues

This section addresses conceptual and theoretical issues related to the field of social computing. Within these chapters, the reader is presented with analysis of the most current and relevant conceptual inquires within this growing field of study. Particular chapters discuss ethical issues in social networking, security concerns that arise when individuals share personal data in social networks, and reducing social engineering impact. Overall, contributions within this section ask unique, often theoretical questions related to the study of social computing and, more often than not, conclude that solutions are both numerous and contradictory.

Volume IV

Section VIII. Emerging Trends

This section highlights research potential within the field of social computing while exploring uncharted areas of study for the advancement of the discipline. Chapters within this section discuss social bookmarking, the emergence of virtual social networks, and new trends in educational social software. These contributions, which conclude this exhaustive, multi-volume set, provide emerging trends and suggestions for future research within this rapidly expanding discipline.

Preface

With increased access to social networking tools, the development of Web 2.0, and the emergence of virtual worlds, social computing crosses cultural boundaries to join people in the digital landscape.

As the world moves closer and closer to the integration of technology into traditional social behaviors, there is a greater need for innovative research and development into the various aspects of social computing. Information Science Reference is pleased to offer a three-volume reference source on this rapidly growing discipline, in order to empower students, researchers, academicians, and practitioners with a wide-ranging understanding of the most critical areas within this field of study. This publication uncovers the growing and expanding phenomenon of human behavior, social constructs, and communication in online environments and provides the most comprehensive, in-depth, and recent coverage of all issues related to the development of cutting-edge social computing technologies. This reference work presents the latest research on social change, evolving networks, media, and interaction with technology to offer audiences a comprehensive view of the impact of social computing on the way cultures think, act, and exchange information.

This collection entitled, "**Social Computing: Concepts, Methodologies, Tools, and Applications**" is organized in eight (8) distinct sections, providing the most wide-ranging coverage of topics such as: 1) Fundamental Concepts and Theories; 2) Development and Design Methodologies; 3) Tools and Technologies; 4) Utilization and Application; 5) Organizational and Social Implications; 6) Managerial Impact; 7) Critical Issues; and 8) Emerging Trends. The following provides a summary of what is covered in each section of this multi-volume reference collection:

Section 1, **Fundamental Concepts and Theories**, serves as a foundation for this extensive reference tool by addressing crucial theories essential to the understanding of social computing. Chapters such as, "Computer-Mediated Communication Learning Environments: The Social Dimension" by Stefania Manca, as well as "Online Communities and Social Networking" by Abhijit Roy, provide foundational overviews of how individuals interact with social computing tools and the impact these tools have on shaping and influencing behavior. "Mobile Social Networks: A New Locus of Innovation" by Nina D. Ziv and "Mobile Social Networks and Services" by Lee Humphreys offer investigations into the recent emergence of mobile social networks, reviewing current trends and technologies and offering suggestions for future research. As this section continues, authors explore the many uses of social software and its implications in contributions such as "Social Software (and Web 2.0)" by Jürgen Dorn, "Self-Organization in Social Software for Learning" by Jon Dron, and "Living, Working, Teaching and Learning by Social Software" by Helen Keegan and Bernard Lisewski. These and several other foundational chapters provide a wealth of expert research on the elemental concepts and ideas which surround investigations of social computing technologies.

Section 2, **Development and Design Methodologies**, presents in-depth coverage of design and architectures to provide the reader with a comprehensive understanding of the emerging technological developments within the field of social computing. A number of contributions, including "Distributed Learning Environments and Social Software: In Search for a Framework of Design" by Sebastian Fiedler and Kai Pata, "Electronic Classroom, Electronic Community: Designing eLearning Environments to Foster Virtual Social Networks and Student Learning" by Lisa Harris, and "A Methodology for Integrating the Social Web Environment in Software Engineering Education" by Pankaj Kamthan highlight the many methods for the effective design of social software and networks that support education. With contributions from leading international researchers, this section offers developmental approaches and methodologies for social computing.

Section 3, **Tools and Technologies**, presents extensive coverage of the various tools and technologies that define and continue to redefine social computing research and implementation. This section begins with "A Modern Socio-Technical View on ERP-Systems" by Jos Benders, Ronald Batenburg, Paul Hoeken, and Roel Schouteten, which discusses a specific approach to creating socio-technical systems. This section continues with an in-depth investigation of social television, contained within the selections "In Search of Social Television" by Gunnar Harboe, "Asynchronous Communication: Fostering Social Interaction with CollaboraTV" by Brian Amento, Chris Harrison, Mukesh Nathan, and Loren Terveen, "Examining the Roles of Mobility in Social TV" by Konstantinos Chorianopoulos, and "From 2BeOn Results to New Media Challenges for Social (i)TV" by Konstantinos Chorianopoulos and Pedro Almeida. With more than a dozen additional contributions, this section provides coverage of a variety of tools and technologies under development and in use in social computing and social networking communities.

Section 4, **Utilization and Application**, describes the implementation and use of an assortment of social computing tools and technologies. Including 20 chapters such as "Social Networking Sites and Critical Language Learning" by Andy Halvorsen, "Using Social Networking to Enhance Sense of Community in E-Learning Courses" by Steve Chi-Yin Yuen and Harrison Hao Yang, "Publishing with Friends: Exploring Social Networks to Support Photo Publishing Practices" by Paula Roush and Ruth Brown, this section provides insight into the utilization of social computing tools and technologies for both personal and professional initiatives. "Social Software Use in Public Libraries" by June Abbas offers suggestions for applying social software techniques such as tagging and cataloguing in library settings. "Designing for Disaster: Social Software Use in Times of Crisis" by Liza Potts presents another interesting application of social software, illustrating the need for sociotechnical interventions in systems design. Contributions found in this section provide comprehensive coverage of the practicality and present use of social computing by organizations and individuals.

Section 5, **Organizational and Social Implications**, includes chapters discussing the impact of social computing on organizational and individual behavior, knowledge, and communication. This section begins with an examination of organizational knowledge, investigating its foundations and management in chapters such as "Managing Organizational Knowledge in the Age of Social Computing" by V. P. Kochikar, "Social Software for Bottom-Up Knowledge Networking and Community Building" by Mohamed Amine Chatti and Matthias Jarke, and "The Essence of Organizational Knowledge: A Social Epistemology Perspective" by Fei Gao. "The Usability of Social Software" by Lorna Uden and Alan Eardley argues that despite the prevalence of Web 2.0 tools and technologies, there is little research on usability evaluation. Successful virtual communication and collaboration are explored in chapters including "Building Social Relationships in a Virtual Community of Gamers" Shafiz Affendi Mohd Yusof and "Entering the Virtual Teachers' Lounge: Social Connectedness among Professional Educators in Virtual Environments" by Randall Dunn. While these two chapters present very different applications of virtual collaboration, they both offer definitions of virtual communities and offer depictions of how virtual environments both differ from and resemble face-to-face communities.

Section 6, **Managerial Impact**, presents focused coverage of social computing in the workplace. Fernando Garrigos' chapter "Interrelationships Between Professional Virtual Communities and Social Networks, and the Importance of Virtual Communities in Creating and Sharing Knowledge" analyzes, as the title suggests, the relationship between professional virtual communities and social networks, and describes how these communities create and share knowledge. "Managing Relationships in Virtual Team Socialization," by Shawn D. Long, Gaelle Picherit-Duthler, and Kirk W. Duthler provides an overview of the emergence of virtual teams in the workplace and explores the specific issues virtual employees must overcome in order be efficient and productive. Also included in this section are chapters addressing topics related to social engineering attacks and enterprise social software, presenting an empirical view of managerial considerations for social computing.

Section 7, **Critical Issues**, addresses vital, conceptual issues related to social computing such as ethical considerations, security, and privacy. Chapters such as "Security and Privacy in Social Networks" by Barbara Carminati, Elena Ferrari, and Andrea Perego and "Emerging Cybercrime Variants in the Socio-Technical Space," by Wilson Huang and Shun-Yung Kevin Wang tackle the difficult question of privacy and data security in online environments. In "The Emergence of Agency in Online Social Networks," by Jillianne R. Code and Nicholas E. Zaparyniuk, the authors explores how agency emerges from social interactions, how this emergence influences the development of social networks, and the role of social software's potential as a powerful tool for educational purposes. "Social Network Structures for Explicit, Tacit and Potential Knowledge" by Anssi Smedlund highlights the role of knowledge, asserting that it is embedded in relationships between individuals rather than possessed by these individuals. These and other chapters in this section combine to provide a review of those issues which are the subject of critical inquiry in social computing research.

The concluding section of this authoritative reference tool, **Emerging Trends**, highlights areas for future research within the field of social computing, while exploring new avenues for the advancement of the discipline. "Legal Issues Associated with Emerging Social Interaction Technologies" by Robert D. Sprague depicts potential legal issues that can arise from social interaction technology use, such as employee behavior online impacting the ability to get or maintain a job. Similarly, "Public Intimacy and the New Face (Book) of Surveillance; The Role of Social Media in Shaping Contemporary Dataveillance" by Lemi Baruh and Levent Soysal investigates privacy implications of sharing personal data in a public environment. Other issues, such as codes of conduct in social networking sites, are explored in chapters such as "Conceptualizing Codes of Conduct in Social Networking Communities" by Ann Dutton Ewbank, Adam G. Kay, Teresa S. Foulger, and Heather L. Carter. New opportunities for using technology to maintain a healthy social network are demonstrated in "Using Ambient Social Reminders to Stay in Touch with Friends" by Ross Shannon, Eugene Kenny, and Aaron Quigley. These and several other emerging trends and suggestions for future research can be found within the final section of this exhaustive multi-volume set.

Although the primary organization of the contents in this multi-volume work is based on its eight sections, offering a progression of coverage of the important concepts, methodologies, technologies, applications, social issues, and emerging trends, the reader can also identify specific contents by utilizing the extensive indexing system listed at the end of each volume. Furthermore to ensure that the scholar, researcher and educator have access to the entire contents of this multi volume set as well as additional coverage that could not be included in the print version of this publication, the publisher will provide unlimited multi-user electronic access to the online aggregated database of this collection for the life of the edition, free of charge when a library purchases a print copy. This aggregated database provides far more contents than what can be included in the print version in addition to continual updates. This unlimited access, coupled with the continuous updates to the database ensures that the most current research is accessible to knowledge seekers.

The diverse and comprehensive coverage of social computing in this three-volume authoritative publication will contribute to a better understanding of all topics, research, and discoveries in this developing, significant field of study. Furthermore, the contributions included in this multi-volume collection series will be instrumental in the expansion of the body of knowledge in this enormous field, resulting in a greater understanding of the fundamental concepts and technologies while fueling the research initiatives in emerging fields. We at Information Science Reference, along with the editor of this collection and the publisher, hope that this multi-volume collection will become instrumental in the expansion of the discipline and will promote the continued growth of all aspects of social computing research.

Chapter 7.7
Effects of Digital Convergence on Social Engineering Attack Channels

Bogdan Hoanca
University of Alaska Anchorage, USA

Kenrick Mock
University of Alaska Anchorage, USA

ABSTRACT

Social engineering refers to the practice of manipulating people to divulge confidential information that can then be used to compromise an information system. In many cases, people, not technology, form the weakest link in the security of an information system. This chapter discusses the problem of social engineering and then examines new social engineering threats that arise as voice, data, and video networks converge. In particular, converged networks give the social engineer multiple channels of attack to influence a user and compromise a system. On the other hand, these networks also support new tools that can help combat social engineering. However, no tool can substitute for educational efforts that make users aware of the problem of social engineering and policies that must be followed to prevent social engineering from occurring.

DOI: 10.4018/978-1-60566-036-3.ch009

INTRODUCTION

Businesses spend billions of dollars annually on expensive technology for information systems security, while overlooking one of the most glaring vulnerabilities—their employees and customers (Orgill, 2004; Schneier, 2000). Advances in technology have led to a proliferation of devices and techniques that allow information filtering and encryption to protect valuable information from attackers. At the same time, the proliferation of information systems usage is extending access to more and more of the employees and customers of every organization. The old techniques of social engineering have evolved to embrace the newest technologies, and are increasingly used against this growing pool of users. Because of the widespread use of information systems by users of all technical levels, it is more difficult to ensure that all users are educated about the dangers of social engineering. Moreover, as digital convergence integrates previously separated communications channels, social engineers are taking advantage of these blended channels to reach new victims in new ways.

Social engineering is a term used to describe attacks on information systems using vulnerabilities that involve people. Information systems include hardware, software, data, policies, and people (Kroenke, 2007). Most information security solutions emphasize technology as the key element, in the hope that technological barriers will be able to override weaknesses in the human element. Instead, in most cases, social engineering attacks succeed despite layers of technological protection around information systems.

As technology has evolved, the channels of social engineering remain relatively unchanged. Attackers continue to strike in person, via postal mail, and via telephone, in addition to attacking via e-mail and online. Even though they arrive over the same attack channels, new threats have emerged from the convergence of voice, data, and video. On one hand, attacks can more easily combine several media in a converged environment, as access to the converged network allows access to all media types. On the other hand, attackers can also convert one information channel into another to make it difficult to locate the source of an attack.

As we review these new threats, we will also describe the latest countermeasures and assess their effectiveness. Convergence of voice, data, and video can also help in combating social engineering attacks. One of the most effective countermeasures to social engineering is the continued education of all information systems users, supplemented by policies that enforce good security practices. Another powerful countermeasure is penetration testing, which can be used to evaluate the organization's readiness, but also to motivate users to guard against social engineering attacks (see for example Argonne, 2006).

Throughout this chapter we will mainly use masculine gender pronouns and references to maleness when referring to attackers, because statistically most social engineering attackers tend to be men. As more women have become proficient and interested in using computers, some of the hackers are now female, but the numbers are still small. Nonetheless, there are some striking implications of gender differences in social engineering attacks, and we discuss those differences as appropriate.

SOCIAL ENGINEERING

Social engineering includes any type of attack that exploits the vulnerabilities of human nature. A recent example is the threat of social engineers taking advantage of doors propped open by smokers, in areas where smoking is banned indoors (Jaques, 2007). Social engineers understand human psychology (sometimes only instinctively) sufficiently well to determine what reactions they need to provoke in a potential victim to elicit the information they need. In a recent survey of black hat hackers (hackers inclined to commit computer crimes), social engineering ranked as the third most widely used technique (Wilson, 2007). The survey results indicate that 63% of hackers use social engineering, while 67% use sniffers, 64% use SQL injection, and 53% use cross site scripting.

Social engineering is used so widely because it works well despite the technological barriers deployed by organizations. Social engineers operate in person, over the phone, online, or through a combination of these channels. A report on the Australian banking industry in *ComputerWorld* claims that social engineering leads to larger losses to the banking industry than armed robbery. Experts estimate these losses to be 2-5% of the revenue, although industry officials decline to comment (Crawford, 2006). Social engineering is also used in corporate and military espionage, and no organization is safe from such attacks. A good overview of social engineering attacks and possible countermeasures can be found on the

Microsoft TechNET Web site (TechNET, 2006).

According to Gragg (2003), there are some basic techniques common to most social engineering attacks. Attackers tend to spend time building trust in the target person. They do that by asking or pretending to deliver small favors, sometimes over an extended period of time. Sometimes, the trust building is in fact only familiarity, where no favors are exchanged, but the victim and attacker establish a relationship. Social engineering attacks especially target people or departments whose job descriptions include building trust and relationships (help desks, customer service, etc). In addition to asking for favors, sometimes social engineers pretend to extend favors by first creating a problem, or the appearance of a problem. Next, the social engineers can appear to solve the problem, thus creating in a potential victim both trust and a sense of obligation to reciprocate. They then use this bond to extract confidential information from the victim. Finally, social engineers are experts at data aggregation, often picking disparate bits of data from different sources and integrating the data into a comprehensive, coherent picture that matches their information gathering needs (Stasiukonis, 2006b; Mitnick & Simon, 2002).

Although the description might seem complex, social engineering can be as simple as just asking for information, with a smile. A 2007 survey (Kelly, 2007) showed that 64% of respondents were willing to disclose their password in exchange for chocolate (and a smile). Using "good looking" survey takers at an IT conference led 40% of non-technical attendees and 22% of the technical attendees to reveal their password. Follow up questions, drilling down to whether the password included a pet name or the name of a loved one elicited passwords from another 42% of the technical attendees and 22% of the non-technical ones. While the survey respondents might have felt secure in only giving out passwords, user names were easier to obtain, because the full name and company affiliation of each survey respondent was clearly indicated on their conference badge. An earlier survey cited in the article reported similar statistics in 2004.

Another paper urging organizations to defend against social engineering illustrates the high levels of success of even simple social engineering attacks. Orgill (2004) describes a survey of 33 employees in an organization, where a "researcher" asked questions about user names and passwords. Only one employee of the 33 surveyed escorted the intruder to security. Of the 32 others that took the survey, 81% gave their user name and 60% gave their password. In some departments, all the employees surveyed were willing to give their passwords. In one instance, an employee was reluctant to complete the survey. A manager jokingly told the employee that he would not get a raise the next year unless she completes the survey. At that point, the employee sat down and completed the survey. This is a clear indication that management can have a critical role in the success or failure of social engineering attacks.

Statistically, an attacker needs only one gullible victim to be successful, but the high success rates mentioned above indicate that finding that one victim is very easy. If such "surveys" were to be conducted remotely, without a face to face dialog or even a human voice over the phone, success rates would likely be much lower, but the risks would also be lower for the attacker. Convergence of data, voice, and video allows attackers to take this alternative route, lower risk of detection at the expense of lower success rate. Given the ability to automate some of the attack avenues using converged media, the lower success rate is not much of a drawback. We will show how most social engineering attacks resort to casting a broad net, and making a profit even of extremely low success rates.

The basic tools of the social engineer include strong human emotions. Social engineers aim to create fear, anticipation, surprise, or anger in the victim, as a way to attenuating the victim's ability to think critically. Additionally, information overload is used to mix true and planted information to lead the victim to believe what the social engineer intends. Reciprocation is another strong emotion social engineers use, as we described earlier. Finally, social engineers combine using guilt (that something bad will happen unless the victim cooperates), transfer of responsibility (the social engineer offers to take the blame), and authority (where the social engineer poses as a supervisor or threatens to call in a supervisor). These are basic human emotions used in all social engineering attacks, whether using converged networks or not. This chapter will focus on how attackers use these emotions on a converged network, combining data, voice, and video.

SOCIAL ENGINEERING ON CONVERGED NETWORKS

Social engineering has seen a resurgence in recent years, partly due to the convergence of voice, data, and video, which makes it much easier to attack an organization remotely, using multiple media channels. The proliferation of computer peripherals and of mobile devices, also driven by network convergence, has further opened channels for attacks against organizations. In this section we discuss new attack vectors, combining some of the classical social engineering channels (in person, by phone, by e-mail) and show how they have changed on a converged network.

Social Engineering Attacks Involving Physical Presence

The classical social engineering attack involves a social engineer pretending to be a technical service person or a person in need of help. The attacker physically enters an organization's premises and finds a way to wander through the premises unattended. Once on the premises, the attacker searches for staff ID cards, passwords or confidential files.

Most of the in-person social engineering attacks rely on other information channels to support the in-person attack. Convergence of voice and data networks allows blended attacks once the attacker is within the victim's offices. Before showing up at the company premises, the attacker can forge an e-mail message to legitimize the purpose of the visit; for example, the e-mail might appear to have been sent by a supervisor to announce a pest control visit (applekid, 2007). Alternatively, the attacker might use the phone to call ahead for an appearance of legitimacy. When calling to announce the visit, the attacker can fake the telephone number displayed on the caller ID window (especially when using Voice over IP, Antonopoulos & Knape, 2002).

After entering the premises, an attacker will often try to connect to the organization's local area network to collect user names, passwords, or additional information that could facilitate subsequent stages of the attack. Convergence allows access to all media once the attacker is connected to the network; even copiers now have network connections that a "service technician" could exploit to reach into the organization's network (Stasiukonis, 2006c). Connecting to the company network using the port behind the copier is much less obvious than using a network port in the open. Finally, another powerful attack may involve a social engineer entering the premises just briefly, connecting a wireless access point to the organization's network, and then exploring the network from a safe distance (Stasiukonis, 2007). This way, the attacker can remain connected to the network, but at the same time minimize the risk of exposure while in the organization's building.

Social Engineering via Email, News, and Instant Messenger

One example of how convergence is changing the information security threats is the increased incidence of attacks using e-mail, HTML, and chat software. This is attractive to attackers, because it bypasses firewalls and allows the attacker to transfer files to and from the victim's computer (Cobb, 2006). The only requirement for such attacks is a good understanding of human weaknesses and the tools of social engineering. The attackers spend their time devising ways to entice the user to open an e-mail, to click on a link or to download a file, instead of spending time breaking through a firewall. One such attack vector propagates via IRC (Internet relay chat) and "chats" with the user, pretending to be a live person, assuring the downloader that it is not a virus, then downloading a shortcut to the client computer that allows the remote attacker to execute it locally (Parizo, 2005).

Because of the wide use of hyperlinked news stories, attacks are beginning to use these links to trigger attacks. In a recent news story (Naraine, 2006b), a brief "teaser" concludes with a link to "read more," which in fact downloads a keylogger by taking advantage of a vulnerability in the browser. This type of attack is in addition to the fully automated attacks that involve only "drive by" URL, where the malicious content is downloaded and executed without any intervention from the user (Naraine, 2006c). Analysis of the code of such automated attacks indicates a common source or a small number of sources, because the code is very similar across multiple different attack sites.

Convergence allows e-mail "bait" to use "hooks" in other applications. For example, an e-mail message with a Microsoft Word attachment may take advantage of vulnerability in Word and rely on e-mail as the attack channel (McMillan, 2006). Other vulnerabilities stem from more complex interactions between incompatible operating systems and applications. The recently released Microsoft Vista operating system has vulnerabilities related to the use of non-Microsoft e-mail clients, and requires user "cooperation" (Espiner, 2006). As such, Microsoft views this as a social engineering attack, rather than a bug in the operating system.

Other attacks are purely social engineering, as in the case of e-mail messages with sensationally sounding subject lines, for example, claiming that the USA or Israel have started World War III, or offering views of scantily clad celebrities. While the body of the message is empty, an attachment with a tempting name incites the users to open it. The name might be video.exe, clickhere.exe, readmore.exe, or something similar, and opening the attachment can run any number of dangerous applications on the user's computer (Gaudin, 2007). Other e-mail messages claim that the computer has been infected with a virus and instruct the user to download a "patch" to remove the virus (CERT, 2002). Instead, the "patch" is a Trojan that installs itself on the user's computer. The source of the message can be forged to make it appear that the sender is the IT department or another trusted source.

Finally, another way to exploit news using social engineering techniques is to send targeted messages following real news announcements. An article on silicon.com cites a phishing attack following news of an information leak at Nationwide Building Society, a UK financial institution. Soon after the organization announced the theft of a laptop containing account information for a large number of its customers, an e-mail began circulating, claiming to originate from the organization and directing recipients to verify their information for security reasons (Phishers raise their game, 2006). This is a much more pointed attack than the traditional phishing attacks (described next), where a threatening or cajoling email is sent to a large number of potential victims, in hope that some of them will react. Such targeted attacks are known as "spear phishing."

Phishing

Phishing is a special case of e-mail-based social engineering, which warrants its own section because of its widespread use (APWG, 2007). The first phishing attacks occurred in the mid 1990's and continue to morph as new technologies open new vulnerabilities. The classical phishing attack involves sending users an e-mail instructing them to go to a Web site and provide identifying information "for verification purposes."

Two key weaknesses of the user population make phishing a highly lucrative activity. As a larger percent of the population is using Web browsers to reach confidential information in their daily personal and professional activities, the pool of potential victims is greatly increased. At the same time, the users have an increased sense of confidence in the information systems they use, unmatched by their actual level of awareness and sophistication in recognizing threats.

A survey of computer users found that most users overestimate their ability to detect and combat online threats (Online Safety Study, 2004). A similar situation is probably the case for awareness of and ability to recognize phishing attacks. As phishers' sophistication increases, their ability to duplicate and disguise phishing sites increases, making it increasingly difficult to recognize fakes even by expert users.

More recently, pharming involves DNS attacks to lure users to a fake Web site, even when they enter a URL from a trusted source (from the keyboard or from a favorites list). To mount such an attack, a hacker modifies the local DNS database (a hosts file on the client computer) or one of the DNS servers the user is accessing. The original DNS entry for the IP address corresponding to a site like www.mybank.com is replaced by the IP address of a phishing site. When the user types www.mybank.com, her computer is directed to the phishing site, even though the browser URL indicates that she is accessing www.mybank.com. Such attacks are much more insidious, because the average user has no way of distinguishing between the fake and the real sites. Such an attack involves a minimal amount of social engineering, although, in many cases, the way the attacker gains access to the DNS database might be based on social engineering methods.

In particular, pharming attacks can rely on converged media, for example, using an "evil twin" access point on a public wireless network. By setting up a rogue access point at a public wireless hotspot and by using the same name as the public access point, an attacker is able to hijack some or all of the wireless traffic though the access point he controls. This way, the attacker can filter all user traffic through his own DNS servers, or more generally, is able to mount any type of man-in-the-middle attack. In general, man-in-the-middle attacks involve the attacker intercepting user credentials as the user is authenticating to a third party Web site and passing on those credentials from the user to the Web site. Having done this, the attacker can now disconnect the user and remain connected to the protected Web site. True to social engineering principles, these types of attacks are targeted at the rich. Evil twin access points are installed in first-class airport lounges, in repair shops specializing in expensive cars, and in other similar areas (Thomson, 2006).

Social Engineering Using Removable Media

In another type of social engineering, storage devices (in particular USB flash drives) might be "planted" with users to trick them into installing malicious software that is able to capture user names and passwords (Stasiukonis, 2006a). This type of attack is based on the fact that users are still gullible enough to use "found" storage devices or to connect to "promotional" storage devices purporting to contain games, news, or other entertainment media. Part of the vulnerability introduced by such removable storage devices is due to the option of modern operating systems to automatically open certain types of files. By simply inserting a storage device with auto run properties, the user can unleash attack vectors that might further compromise their system. In addition to USB flash drives, other memory cards, CDs, and DVDs can support the same type of attack.

Social Engineering via Telephone and Voice Over IP Networks

Using telephone networks has also changed. The basic attack is often still the same, involving a phone call asking for information. Convergence, in particular the widespread use of digitized voice channels, also allow an attacker to change his (usually) voice into a feminine voice (bernz, n.d.), which is more likely to convince a potential victim. Digitally altering one's voice will also allow one attacker to appear as different callers on subsequent telephone calls (Antonopoulos & Knape, 2002). This way, the attacker can gather information on multiple occasions, without raising as much concern as a repeat caller.

The wide availability of voice over IP and the low cost of generating and sending possibly large volumes of voice mail messages also enable new types of attacks. Vishing, or VoIP phishing (Vishing, 2006) is one such type of attack that combines the use of the telephone networks described with automatic data harvesting information systems. This type of attack relies on the fact that credit card companies now require users to enter credit card numbers and other identifying information. Taking advantage of user's acceptance of such practices, vishing attacks set up rogue answering systems that prompt the user for the identifying information. The call number might be located in a different location than the phone number might indicate.

The use of phone lines is also a way for attackers to bypass some of the remaining inhibitions users have in giving out confidential information on the World Wide Web. While many users are aware of the dangers of providing confidential information on Web sites (even those who appear genuine), telephone networks are more widely trusted than online channels. Taking advantage of this perception in conjunction with the widespread availability of automated voice menus has enabled some attackers to collect credit card information. Naraine (2006a) describes an attack where the victim is instructed via e-mail to verify a credit card number over the phone. The verification request claims to represent a Santa Barbara bank and directs users to call a phone number for verification. The automated answering system uses voice prompts similar to those of legitimate credit card validation, which are familiar to users. Interestingly, the phone system does not identify the bank name, making it possible to reuse the same answering system for simultaneous attacks against multiple financial institutions.

A vishing attack even more sophisticated than the Santa Barbara bank attack targeted users of PayPal (Ryst, 2006). PayPal users were sent an e-mail to verify their account information over the phone. The automated phone system instructed users to enter their credit card number on file with PayPal. The fraudulent system then attempted to verify the number; if an invalid credit card number was entered, the user was directed to enter their information again, bolstering the illusion of a legitimate operation. Although this type of multichannel attack is not limited to VoIP networks, such networks make the automated phone systems much easier to set up.

As VoIP becomes more prevalent we may begin to see Internet-based attacks previously limited to computers impact our telephone systems (Plewes, 2007). Denial of service attacks can flood the network with spurious traffic, bringing legitimate data and voice traffic to a halt. Spit (spam over Internet telephony) is the VoIP version of e-mail spam which instead clogs up a voice mailbox with unwanted advertisements (perhaps generated by text to speech systems) or vishing attacks. Vulnerabilities in the SIP protocol for VoIP may allow social engineers to intercept, reroute calls, and tamper with calls. Finally, VoIP telephones are Internet devices that may run a variety of services such as HTTP, TFTP, or telnet servers, which may be vulnerable to hacking. Since all of the VoIP phones in an organization are likely identical, a single vulnerability can compromise every phone in the organization.

SOLUTIONS AND COUNTERMEASURES TO SOCIAL ENGINEERING ATTACKS

Following the description of attacks, the chapter now turns to solutions. The first and most important level of defense against social engineering are organizational policies (Gragg, 2003). Setting up and enforcing information security policies gives clear indications to employees on what information can be communicated, under what conditions, and to whom. In a converged network, such policies need to specify appropriate information channels, appropriate means to identify the requester, and appropriate means to document the information transfer. As the attacker is ratcheting up the strong emotions that cajole or threaten the victim into cooperating, strong policies can make an employee more likely to resist threats, feelings of guilt, or a dangerous desire to help.

In addition to deploying strong policies, organizations can use the converged network to search for threats across multiple information channels in real time. In a converged environment, the strong emotions associated with social engineering could be detected over the phone or in an e-mail, and adverse actions could be tracked and stopped before an attack can succeed. In other words, convergence has the potential to help not just the social engineer, but also the staff in charge of countering such attacks.

Anti-Phishing Techniques

A number of anti-phishing techniques have been proposed to address the growing threat of phishing attacks. Most anti-phishing techniques involve hashing the password either in the user's head (Sobrado & Birget, 2005; Wiedenbeck, Waters, Sobrado, & Birget, 2006), using special browser plugins (Ross, Jackson, Miyake, Boneh, & Mitchell, 2005), using trusted hardware (for example on tokens) or using a combination of special hardware and software (e.g., a cell phone, Parno, 2006).

All the technological solutions mentioned involve a way to hash passwords so that they are not reusable if captured on a phishing site or with a network sniffer. The downfall of all of these schemes is that the user can always be tricked into giving out a password through a different, unhashed channel, allowing the attacker to use the password later on. A good social engineer would be able to just call the victim and ask for the password over the phone. Additionally, even though all these solutions are becoming increasingly user friendly and powerful, they all require additional costs.

Voice Analytics

We discussed earlier the negative implications of VoIP and its associated attacks (vishing). A positive outcome of data and voice convergence in the fight against social engineering is the ability to analyze voice on the fly, in real time as well as on stored digitized voice mail.

Voice analytics (Mohney, 2006) allows caller identification based on voice print, and can also search for keywords, can recognize emotions, and aggregate these information sources statistically with call date and time, duration, and origin. In particular, voice print can provide additional safeguards when caller ID is spoofed. At the same time, given that the social engineer has similar resources in digitally altering her voice, the voice analytics could employ more advanced techniques to thwart such attacks. For example, the caller could be asked to say a sentence in an angry voice or calm voice (to preempt attacks using recorded voice data). Attacks by people who know and avoid "hot" words can be preempted by using a thesaurus to include synonyms.

Blacklisting

Another common technological solution against social engineering is a blacklist of suspicious or unverified sites and persons. This might sound simple, especially given the ease of filtering Web sites, the ease of using voice recognition on digital phone lines, and the ease of using face recognition (for example) in video. However, maintenance of such a list can be problematic. Additionally, social engineers take precautions to disguise Web presence, as well as voice and physical appearance. Even though a converged network may allow an organization to aggregate several information sources to build a profile of an attack or an attacker, the same converged network will also help the social engineer to disperse the clues, to make detection more difficult.

Penetration Testing

Penetration testing is another very effective tool in identifying vulnerabilities, as well as a tool for motivating and educating users. As mentioned earlier, users tend to be overconfident in their ability to handle not just malware, but social engineering attacks as well. By mounting a penetration testing attack, the IT staff can test against an entire range of levels of sophistication in attack.

An exercise performed at Argonne National Labs (Argonne, 2006) involved sending 400 messages inviting employees to click on a link to view photos from an open house event. Such e-mail messages are easily spoofed and could be sent from outside the organization, yet they can be made to seem that they originate within the organization. Of the 400 recipients of the e-mail, 149 clicked on the link and were asked to enter their user name and password to access the photos, and 104 of these employees actually entered their credentials. Because this was an exercise, the employees who submitted credentials were directed to an internal Web site with information about phishing and social engineering.

A more complex and more memorable (for the victims) example of penetration testing was reported on the DarkReading site (Stasiukonis, 2006d). The attacker team used a shopping card to open the secure access door, found and used lab coats to blend in, and connected to the company network at a jack in a conference room. Several employees actually helped the attackers out by answering questions and pointing out directions. As part of the final report, the team made a presentation to the employees, which had a profound educational impact. Six months later, on a follow up penetration testing mission, the team was unable to enter the premises. An employee, who first allowed the attackers to pass through a door she had opened, realized her mistake as soon as she got to her car. She returned, alerted the security staff and confronted the attackers.

Palmer (2001) describes how an organization would locate "ethical hackers" to perform a penetration testing exercise. The penetration testing plan involves common sense questions, about what needs to be protected, against what threats, and using what level of resources (time, money, and effort). A "get out of jail free card" is the contract between the organization initiating the testing and the "ethical hackers" performing the testing. The contract specifies limits to what the testers can do and requirements for confidentiality of information gathered. An important point, often forgotten, is that even if an organization performs a penetration testing exercise and then fixes all the vulnerabilities identified, follow up exercises will be required to assess newly introduced vulnerabilities, improperly fixed vulnerabilities, or additional ones not identified during a previous test. In particular, despite the powerful message penetration testing can convey to potential victims of social engineering, there is always an additional vulnerability a social engineer may exploit, and there is always an employee who has not fully learned the lesson after the previous exercise.

Additionally, social engineering software is available to plan and mount a self-test, similar to the Argonne one reported earlier (Jackson Higgins, 2006a). Intended mainly to test phishing vulnerabilities, the core impact penetration testing tool from Core Security (www.coresecurity.com) allows the IT staff to customize e-mails and to use social engineering considerations with a few mouse clicks (Core Impact, n.d.).

Data Filtering

One application that may address social engineering concerns at the boundary of the corporate network is that proposed by Provilla, Inc. A 2007 Cisco survey identified data leaks as the main concern of IT professionals (Leyden, 2007). Of the 100 professionals polled, 38% were most concerned with theft of information, 33% were most concerned about regulatory compliance, and only 27% were most concerned about virus attacks (down from 55% in 2006). Provilla (www.provilla-inc.com) claims that their DataDNA™ technology allows organizations to prevent information leaks, including identity theft and to maintain compliance. The product scans the network looking for document fingerprints, on "every device...at every port, for all data types," according to the company. The channels listed include USB, IM, Bluetooth, HTTP, FTP, outside email accounts (Hotmail, Gmail, etc). Conceivably, the technology could be extended to include voice over IP protocols, although these are not mentioned on the company Web site at this time.

Reverse Social Engineering

Another defensive weapon is to turn the tables and use social engineering against attackers (Holz & Raynal, 2006). This technique can be used against less sophisticated attackers, for example, by embedding "call back" code in "toolz" posted on hacker sites. This can alert organizations about the use of such code and about the location of the prospective attacker. Alternatively, the embedded code could erase the hard drive of the person using it—with the understanding that only malevolent hackers would know where to find the code and would attempt to use it.

User Education

Among the tools available against social engineering, we saved arguably the most effective tool for last: educating users. Some of the technologies mentioned in this section have the potential to stop some of the social engineering attacks. Clearly, social engineers also learn about these technologies, and they either find ways to defeat the technologies or ways to circumvent them. Some experts go as far as to say that any "no holds barred" social engineering attack is bound to succeed, given the wide array of tools and the range of vulnerabilities waiting to be exploited. Still, educating users can patch many of these vulnerabilities and is likely to be one of the most cost-effective means to prevent attacks.

We cannot stress enough that user education is only effective when users understand that they can be victims of attacks, no matter how technologically aware they might be. Incidentally, penetration testing may be one of the most powerful learning mechanisms for employees, both during and after the attack. Stasiukonis (2006c) confesses that in 90% of the cases where he and his penetration testing team get caught is when a user decides to make a call to verify the identity of the attackers. The positive feelings of the person "catching the bad guys" and the impact of the news of the attack on the organization are guaranteed to make it a memorable lesson.

Educating the users at all levels is critical. The receptionist of a company is often the first target of social engineering attacks (to get an internal phone directory, to forward a fax or just to chat about who might be on vacation, Mitnick & Simon, 2002). On the other hand, the information security officers are also targeted because of their critical access privileges. An attacker posing as a client of a bank crafted a spoofed e-mail message supposedly to report a phishing attack. When the security officer opened the e-mail he launched an application that took control of the officer's computer (Jackson Higgins, 2006b). A social engineering attack may succeed by taking the path of least resistance, using the least trained user; at the same time, an attack might fail because one of the best trained users happened to notice something suspicious and alerted the IT staff.

Educational efforts often achieve only limited success and education must be an ongoing process. A series of studies by the Treasury Inspector General for Tax Administration (2007) used penetration testing to identify and assess risks, then to evaluate the effectiveness of education. The study found that IRS employees were vulnerable to social engineering even after training in social engineering had been conducted. In 2001, the penetration testers posed as computer support helpdesk representatives in a telephone call to IRS employees and asked the employees to temporarily change their password to one given over the phone. Seventy-one percent of employees complied. Due to this alarming rate, efforts were made to educate employees about the dangers of social engineering. To assess the effectiveness of the training, a similar test was conducted in 2004, and resulted in a response rate of 35%. However, another test in 2007 successfully convinced 60% of employees to change their password. One bright spot is that of the 40% of employees who were not duped by the social engineers, 50% cited awareness training and e-mail advisories as the reason for protecting their passwords, indicating that user education has the potential for success. In response to the latest study, the IRS is elevating the awareness training and is even emphasizing the need to discipline employees for security violations resulting from negligence or carelessness.

Clearly, user education is not limited to social engineering attacks that take advantage of converged networks. Any social engineering attack is less likely to succeed in an organization where employees are well informed and empowered by well-designed security policies. Education becomes more important on converged networks, to account for the heightened threat level and to allow users to take advantage of the available converged tools that may help prevent attacks.

CONCLUSION

Despite the negative press and despite the negative trends we discussed in this chapter, the good news is that the outlook is positive (Top myths, 2006a). The media is often portraying the situation as "dire" and reporting on a seemingly alarming increase in the number of attacks. For one, the number of users and the number and usage of information systems is increasing steadily. That in itself accounts for a staggering increase in the number of incidents reported. Additionally, the awareness of the general population with respect to information security issues in general and with respect to social engineering in particular is also increasing. The media is responding to this increased interest by focusing more attention on such topics. Surveys indicate that in fact the rate of occurrence of computer crime is actually steady or even decreasing, and that only the public perception and increased usage make computer crime seem to increase. A typical analogy is the seemingly daunting vulnerabilities in Microsoft operating systems, which are in fact only a perceived outcome of the increased usage base and increased attractiveness for attackers (Top myths, 2006b).

Whether the rate of computer crime is increasing or not, social engineering remains a real problem that needs to be continually addressed. Convergence in telecommunications makes it easier for users to access several information channels through a unified interface on one or a small number of productivity devices. This same trend makes it easier for attackers to deploy blended attacks using several information channels to a potential victim, and makes it easier to reach the user through the same converged interface or productivity device. By its nature, convergence means putting all one's eggs in one basket. The only rational security response is to guard the basket really well.

If there is one point we have tried hard to make painfully clear in this chapter, education, rather than technological solutions, appears to be the best answer to the social engineering problem. Users who are aware of attack techniques, who are trained in following safe usage policies, and who are supported by adequate technological safeguards are much more likely to recognize and deflect social engineering attacks than users who rely only on technology for protection.

REFERENCES

Antonopoulos, A. M., & Knape, J. D. (2002). Security in converged networks (featured article). *Internet Telephony, August*. Retrieved April 15, 2007, from http://www.tmcnet.com/it/0802/0802gr.htm

Applekid. (author's name). (2007). *The life of a social engineer*. Retrieved April 15, 2007, from http://www.protokulture.net/?p=79

APWG. (2007). *Phishing activity trends report for the month of February, 2007*. Retrieved April 15, 2007, from http://www.antiphishing.org/reports/apwg_report_february_2007.pdf

Argonne (2006). *Simulated 'social engineering' attack shows need for awareness*. Retrieved April 15, 2007, from http://www.anl.gov/Media_Center/Argonne_News/2006/an061113.htm#story4

bernz (author's name). (n.d.). *The complete social engineering FAQ*. Retrieved April 15, 2007, from http://www.morehouse.org/hin/blckcrwl/hack/soceng.txt

CERT. (2002). *Social engineering attacks via IRC and instant messaging* (CERT® Incident Note IN-2002-03). Retrieved April 15, 2007, from http://www.cert.org/incident_notes/IN-2002-03.html

Cobb, M. (2006). *Latest IM attacks still rely on social engineering*. Retrieved April 15, 2007, from http://searchsecurity.techtarget.com/tip/0,289483,sid14_gci1220612,00.html

Crawford, M. (2006). Social engineering replaces guns in today's biggest bank heists. *ComputerWorld (Australia), May*. Retrieved April 15, 2007, from http://www.computerworld.com.au/index.php/id;736453614

Damle, P. (2002). Social engineering: A tip of the iceberg. *Information Systems Control Journal, 2*. Retrieved April 15, 2007, from http://www.isaca.org/Template.cfm?Section=Home&CONTENTID=17032&TEMPLATE=/ContentManagement/ContentDisplay.cfm

Espiner, T. (2006). Microsoft denies flaw in Vista. *ZDNet UK, December 5*. Retrieved April 15, 2007, from http://www.builderau.com.au/news/soa/Microsoft denies flaw in Vista/0,339028227,339272533,00.htm?feed=rss

Gaudin, S. (2007). Hackers use Middle East fears to push Trojan attack. *Information Week, April 9*. Retrieved April 15, 2007, from http://www.informationweek.com/windows/showArticle.jhtml?articleID=198900155

Gragg, D. (2003). A multi-level defense against social engineering. *SANS Institute Information Security Reading Room*. Retrieved April 15, 2007, from http://www.sans.org/reading_room/papers/51/920.pdf

Hollows, P. (2005). Hackers are real-time. Are you? *Sarbanes-Oxley Compliance Journal, February 28*. Retrieved April 15, 2007, from http://www.s-ox.com/Feature/detail.cfm?ArticleID=623

Holz, T., & Raynal, F. (2006). Malicious malware: attacking the attackers (part 1), *Security Focus, January 31*. Retrieved April 15, 2007, from http://www.securityfocus.com/print/infocus/1856

Impact, C. (n.d.). *Core impact overview*. Retrieved April 15, 2007, from http://www.coresecurity.com/?module=ContentMod&action=item&id=32

Jackson Higgins, K. (2006a). *Phishing your own users*. Retrieved April 26, 2007, from http://www.darkreading.com/document.asp?doc_id=113055

Jackson Higgins, K. (2006b). *Social engineering gets smarter*. Retrieved April 26, 2007, from http://www.darkreading.com/document.asp?doc_id=97382

Jaques, R. (2007). *UK smoking ban opens doors for hackers*. Retrieved April 26, 2007, from http://www.vnunet.com/vnunet/news/2183215/uk-smoking-ban-opens-doors

Kelly, M. (2007). Chocolate the key to uncovering PC passwords. *The Register, April 17*. Retrieved April 26, 2007, from http://www.theregister.co.uk/2007/04/17/chocolate_password_survey/

Leyden, J. (2007). Data theft replaces malware as top security concern. *The Register, April 19*. Retrieved April 19, 2007, from http://www.theregister.co.uk/2007/04/19/security_fears_poll/

McMillan, R. (2006). *Third word exploit released, IDG news service*. Retrieved April 15, 2007, from http://www.techworld.com/applications/news/index.cfm?newsID=7577&pagtype=samechan

Mitnick, K. D., & Simon, W. L. (2002). *The art of deception: Controlling the human element of security*. Indiana: Wiley Publishing, Inc.

Mohney, D. (2006). Defeating social engineering with voice analytics. *Black Hat Briefings*, Las Vegas, August 2-3, 2006. Retrieved April 25, 2007, from http://www.blackhat.com/presentations/bh-usa-06/BH-US-06-Mohney.pdf

Naraine, R. (2006a). Voice phishers dialing for PayPal dollars. *eWeek, July 7*. Retrieved April 15, 2007, from http://www.eweek.com/article2/0,1895,1985966,00.asp

Naraine, R. (2006b). Hackers use BBC news as IE attack lure. *eWeek, March 30*. Retrieved April 15, 2007, from http://www.eweek.com/article2/0,1895,1944579,00.asp

Naraine, R. (2006c). Drive-by IE attacks subside; threat remains. *eWeek, March 27*. Retrieved April 15, 2007, from http://www.eweek.com/article2/0,1895,1943450,00.asp

Online Safety Study. (2004, October). *AOL/NCSA online safety study, conducted by America Online and the National Cyber Security Alliance*. Retrieved April 15, 2007, from http://www.staysafeonline.info/pdf/safety_study_v04.pdf

Orgill, G. L., Romney, G. W., Bailey, M. G., & Orgill, P. M. (2004, October 28-30). The urgency for effective user privacy-education to counter social engineering attacks on secure computer systems. In *Proceedings of the 5th Conference on Information Technology Education CITC5 '04*, Salt Lake City, UT, USA, (pp. 177-181). New York: ACM Press.

Palmer, C. C. (2001). Ethical hacking. *IBM Systems Journal, 40*(3). Retrieved April 15, 2007, from http://www.research.ibm.com/journal/sj/403/palmer.html

Parizo, E. B. (2005). *New bots, worm threaten AIM network*. Retrieved April 25, 2007, from http://searchsecurity.techtarget.com/originalContent/0,289142,sid14_gci1150477,00.html

Parno, B., Kuo, C., & Perrig, A. (2006, February 27-March 2). Phoolproof phishing prevention. In *Proceedings of the 10th International Conference on Financial Cryptography and Data Security*. Anguilla, British West Indies.

Phishers raise their game. (2006). Retrieved April 25, 2007, from http://software.silicon.com/security/0,39024655,39164058,00.htm

Plewes, A. (2007, March). *VoIP threats to watch out for—a primer for all IP telephony users*. Retrieved April 18, 2007, from http://www.silicon.com/silicon/networks/telecoms/0,39024659,39166244,00.htm

Ross, B., Jackson, C., Miyake, N., Boneh, D., & Mitchell, J. C. (2005). Stronger password authentication using browser extensions. In *Proceedings of the 14th Usenix Security Symposium, 2005.*

Ryst, S. (2006, July 11). The phone is the latest phishing rod. *BusinessWeek*.

Schneier, B. (2000). *Secrets and lies*. John Wiley and Sons.

Sobrado, L., & Birget, J.-C. (2005). *Shoulder surfing resistant graphical passwords*. Retrieved April 15, 2007, from http://clam.rutgers.edu/~birget/grPssw/srgp.pdf

Stasiukonis, S. (2006a). *Social engineering, the USB way*. Retrieved April 15, 2007, from http://www.darkreading.com/document.asp?doc_id=95556&WT.svl=column1_1

Stasiukonis, S. (2006b). *How identity theft works*. Retrieved April 15, 2007, from http://www.darkreading.com/document.asp?doc_id=102595

Stasiukonis, S. (2006c). *Banking on security*. Retrieved April 15, 2007, from http://www.darkreading.com/document.asp?doc_id=111503

Stasiukonis, S. (2006d). *Social engineering, the shoppers' way*. Retrieved April 15, 2007, from http://www.darkreading.com/document.asp?doc_id=99347

Stasiukonis, S. (2007). *By hook or by crook*. Retrieved April 15, 2007, from http://www.darkreading.com/document.asp?doc_id=119938

TechNET. (2006). *How to protect insiders from social engineering threats*. Retrieved April 15, 2007, from http://www.microsoft.com/technet/security/midsizebusiness/topics/complianceandpolicies/socialengineeringthreats.mspx

Thomson, I. (2006). 'Evil twin' Wi-Fi hacks target the rich. *iTnews.com.au, November*. Retrieved April 15, 2007, from http://www.itnews.com.au/newsstory.aspx?CIaNID=42673&r=rss

Top myths. (2006a). *The 10 biggest myths of IT security: Myth #1: 'Epidemic' data losses*. Retrieved April 15, 2007, from http://www.darkreading.com/document.asp?doc_id=99291&page_number=2

Top myths. (2006b). *The 10 biggest myths of IT security: Myth #2: Anything but Microsoft*. Retrieved April 15, 2007, from http://www.darkreading.com/document.asp?doc_id=99291&page_number=3

Treasury Inspector General for Tax Administration. (2007). *Employees continue to be susceptible to social engineering attempts that could be used by hackers* (TR 2007-20-107). Retrieved August 18, 2007, from http://www.ustreas.gov/tigta/auditreports/2007reports/200720107fr.pdf

Vishing. (2006). *Secure computing warns of vishing*. Retrieved April 15, 2007, from http://www.darkreading.com/document.asp?doc_id=98732

Wiedenbeck, S., Waters, J., Sobrado, L., & Birget, J. (2006, May 23-26). Design and evaluation of a shoulder-surfing resistant graphical password scheme. In *Proceedings of the Working Conference on Advanced Visual interfaces AVI '06*, Venezia, Italy,(pp. 177-184). ACM Press, New York: ACM Press. http://doi.acm.org/10.1145/1133265.1133303

Wilson, T. (2007). *Five myths about black hats*. Retrieved April 15, 2007, from http://www.darkreading.com/document.asp?doc_id=118169

This work was previously published in Social and Human Elements of Information Security: Emerging Trends and Countermeasures, edited by M. Gupta; R. Sharman, pp. 133-147, copyright 2009 by Information Science Reference (an imprint of IGI Global).

Chapter 7.8
On the Effectiveness of Social Tagging for Resource Discovery

Dion Hoe-Lian Goh
Nanyang Technological University, Singapore

Khasfariyati Razikin
Nanyang Technological University, Singapore

Alton Y. K. Chua
Nanyang Technological University, Singapore

Chei Sian Lee
Nanyang Technological University, Singapore

Schubert Foo
Nanyang Technological University, Singapore

Jin-Cheon Na
Nanyang Technological University, Singapore

Yin-Leng Theng
Nanyang Technological University, Singapore

ABSTRACT

Social tagging is the process of assigning and sharing among users freely selected terms of resources. This approach enables users to annotate/describe resources, and also allows users to locate new resources through the collective intelligence of other users. Social tagging offers a new avenue for resource discovery as compared to taxonomies and subject directories created by experts. This chapter investigates the effectiveness of tags as resource descriptors and is achieved using text categorization via support vector machines (SVM). Two text categorization experiments were done for this research, and tags and Web pages from del.icio.us were used. The first study concentrated on the use of terms as its features while the second used both terms and its tags as part of its feature set. The experiments yielded a macroaveraged precision, recall, and F-measure scores of 52.66%, 54.86%, and 52.05%, respectively. In terms of microaveraged values, the experiments obtained 64.76% for precision, 54.40% for recall, and 59.14% for F-measure.

DOI: 10.4018/978-1-59904-879-6.ch025

The results suggest that the tags were not always reliable indicators of the resource contents. At the same time, the results from the terms-only experiment were better compared to the experiment with both terms and tags. Implications of our work and opportunities for future work are also discussed.

INTRODUCTION

The increasing popularity of Web 2.0-based applications has empowered users to create, publish, and share resources on the Web. Such user-generated content may include text (e.g., blogs, wikis), multimedia (e.g., YouTube), and even organization/navigational structures providing personalized access to Web content. The latter includes social bookmarking/tagging systems such as del.icio.us and Connotea.

Social tagging systems allow users to annotate links to useful Web resources by assigning keywords (tags) and possibly other metadata, facilitating their future access (Macgregor & McCulloch, 2006). These tags may further be shared by other users of the social tagging system, in effect creating a community where users can create and share tags pointing to useful Web resources. Put differently, tags function both as content organizers and discoverers. Users create and assign tags to a useful resource they come across so that it would be easy for them to retrieve that resource at a later date. At the same time, other users can use one or more of these tags created to find the resource. The same tags may also be used to discover other related and relevant resources. In addition, through tags, a user can potentially locate like-minded users who hold interests in similarly-themed resources, leading to the creation of social networks (Marlow, Naaman, Boyd, & Davis, 2006).

Social tagging provides an alternative means of organizing resources when compared with conventional methods of categorization based on taxonomies, controlled vocabularies, faceted classification, and ontologies. Conventional methods require experts with domain knowledge and this often translates to a high cost of implementing such systems. They are also bound strictly by rules to ensure their classification schemes remain consistent (Morville, 2005). As the system becomes larger, the rules tend to be more complicated, leading to possible maintenance and accessibility issues. In contrast, the classification scheme in social tagging systems is deregulated. Instead of relying on (a few) experts, they are supported by a (possibly large) community of users. At the same time, tags are "flat," lacking a predefined taxonomic structure, and their use relies on shared, emergent social structures and behaviors, as well as a common conceptual and linguistic understanding within the community (Marlow et al., 2006). Tags are therefore also known as "folksonomies," short for "folk taxonomies," suggesting that they are created by lay users, as opposed to domain experts or information professionals such as librarians, and may in fact be more effective in describing the resource

While social tagging systems have become popular, it is not known if tags created by ordinary users (as opposed to experts) are useful for the discovery of information. A few studies have investigated the use of tags as resource descriptors. Examples include comparing the use of tags against author-assigned index terms in academic papers (Kipp, 2006; Lin, Beaudoin, Bui, & Desai, 2006), examining the ability of tags to classify blogs using text categorization methods (Sun, Suryanto, & Liu, 2007), and investigating the ability of del.icio.us tags to classify Web resources in a small scale study (Razikin, Goh, Cheong, & Ow, 2007). However, to be best of our knowledge, no large scale work has been conducted with del.icio.us, one of the earliest and more popular social tagging sites. The site has a diverse set of tags and Web resources, and its main function is to store, organize, and share bookmarks among a community of users.

The goal of this present chapter is to investigate if tags are useful in helping users to access relevant Web resources. Specifically, we obtain Web pages and their associated tags from del.icio.us and study whether the tags are good descriptors of these resources. Here, we adopt from techniques in text categorization (Sebastiani, 2002) and argue that an effective tag is one in which a classifier is able to assign documents to with high precision and recall. The rationale here is that if a classifier is able to accurately assign documents to their respective tags, then such tags are useful for organizing resources, implying that users would be able to utilize them for accessing information. The remainder of this chapter is organized as follows. In the next section, research related to this work is reviewed. A description of our experimental methodology and the results are then presented. We then provide a discussion of the implications of our findings and conclude with opportunities for future work in this area.

RELATED STUDIES

The use of tagging has become a popular way of organizing and accessing Web resources. Sites such as del.icio.us, Flickr, YouTube, and Last.fm offer this service for their users. Correspondingly, social tagging has also attracted much research, and work has mainly concentrated on the architecture and implementation of systems (e.g., Hammond, Hannay, Lund, & Scott, 2005; Puspitasari, Lim, Goh, Chang, Zhang, Sun, et al., 2007), usage patterns in tagging systems (e.g., Golder & Huberman, 2006; Marlow et al., 2006), user interfaces (e.g., Dubinko, Kumar, Magnani, Novak, Raghavan, & Tomkins, 2006), and the use of social tagging in search systems (e.g., Yanbe, Jatowt, Nakamura, & Tanaka, 2007) among others.

In particular, as tagging becomes an accepted practice among Web users, there is growing interest in investigating whether tags are a useful means for organizing and accessing content. For example, comparing tags with controlled vocabularies provide a basis for evaluating how tags differ from keywords assigned by experts. Lin et al. (2006) compared tags from Connotea and medical subject heading terms (MeSH terms) and found that there was only 11% similarity between MeSH terms and tags supplied by the users. The authors argued that this is because MeSH terms serve as descriptors while tags primarily focus on areas that are of interest to users. Kipp (2006) compared tags with author supplied keywords and indexing terms to determine the overlap in terms of usage. Results indicated that about 35% of the tags were related to the terms supplied by the authors and indexing terms. However, the relationship between tags and terms were not defined formally in thesauri.

An early work on automatic text categorization in social tagging systems was done by Brooks and Montanez (2006). Their study employed articles from the blogosphere. The authors used 350 popular tags from Technorati and 250 of the most recent articles of the collected tags. Using TF-IDF to cluster documents and pairwise cosine similarity to measure the similarity of all articles in each cluster, they found that tags categorize articles in the broad sense and users in a particular domain will not likely be able to find articles with a tag relating to a specific context. Sun et al. (2007) focused on classifying whole blogs with tags, and compared the classification results based on tags alone, tags together with blog descriptions (short abstract), and blog descriptions alone. It was found that tags together with descriptions had the best classification accuracy, while tags alone were more effective than blog descriptions alone for classification. Finally, in departure from the study of tags in blogs, Razikin et al. (2007) conducted a small scale study of the effectiveness of using tags to classify Web resources in del.icio.us. Using a support vector machine (SVM) classifier, relatively high precision and recall rates of 90.22% and 99.27%, respectively, were obtained.

Figure 1. The distribution of tags over the number of documents

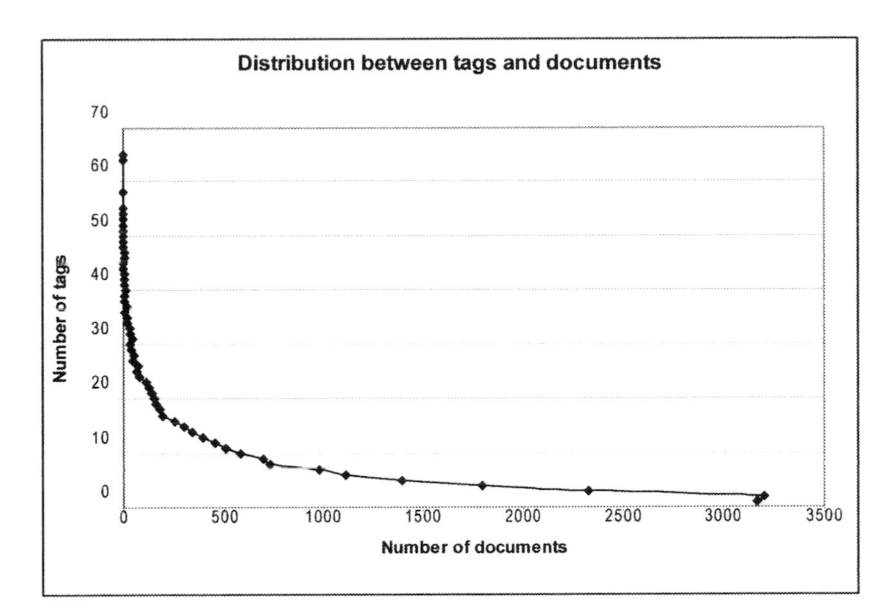

METHODOLOGY

Tags and Web pages from del.icio.us were mined from the site from August 2007 to October 2007. During this period, we randomly collected 100 tags and 20,210 pages that were in the English language. Pages that were primarily nontextual (e.g., images and video) were discarded. Consistent with the work of Brooks and Montanez (2006), we started mining the tags from the popular tags page. As such, our tags will be biased towards the more commonly used ones. The popularity of a tag indicates that there are a significant number of documents related to it, and therefore provide a sufficient size for the training and testing of the text classifier.

In our dataset, each tag was associated with an average of 1,331 documents, and each document was associated with an average of 6.66 tags. The minimum number of tags for a document was 1 while the largest number of tags for a single document was 65. Figure 1 shows the distribution of the tags for the number of documents. It clearly demonstrates that power law distribution applies. Interestingly, the same was observed for blogs (Sun et al., 2007). In the figure, there are 3,167 documents with one tag each while there is only a single document with the largest number of tags (65).

Two text categorization experiments were conducted in the present research. The SVM was the machine learning classifier selected for our work as it is a popular machine learning classifier used in Web-based text categorization studies with good performance. Specifically, the SVM^{light} package (Joachims, 1998) was used. Being a binary classifier, we created one classifier for each tag with the examples comprising Web pages belonging (positive examples) and not belonging (negative examples) to that tag. In total, 100 classifiers were trained with the default options of the package. The performance of the tags was evaluated based on the macroaveraged and microaveraged precision, recall, and F-measure.

The first experiment, which served as a baseline, used the terms from the Web pages as the features while the second experiment included tags, in addition to terms, as part of its feature set. The pages in our dataset were processed by removing the hypertext markup language (HTML)

Table 1. Top 15 tags with the highest F-measure values

	Experiment 1				Experiment 2			Diff
Tag	**Precision**	**Recall**	**F-measure**		**Precision**	**Recall**	**F-measure**	
reference	58.38	87.23	69.95		57.80	62.83	60.21	-9.74
howto	56.02	86.21	67.92		61.93	54.83	58.16	-9.76
politics	55.25	87.91	67.85		52.81	90.04	66.57	-1.28
imported	58.57	79.50	67.45		56.40	52.99	54.64	-12.81
fun	55.01	86.83	67.35		50.05	55.94	52.84	-14.51
blogs	55.07	85.74	67.06		59.14	73.92	65.71	-1.35
web	57.37	80.24	66.90		55.76	71.92	62.82	-4.08
web2.0	55.58	82.92	66.55		55.86	75.00	64.03	-2.52
inspiration	53.51	86.29	66.06		54.10	63.04	58.23	-7.83
Internet	54.90	82.18	65.83		55.17	66.22	60.19	-5.64
california	57.14	76.40	65.38		55.17	66.22	60.19	-5.64
restaurants	55.43	79.69	65.38		49.07	88.76	63.20	-2.18
osx	54.07	82.58	65.35		48.00	56.25	51.80	-13.58
recipe	56.83	73.79	64.21		54.92	69.30	61.28	-4.07
news	**54.93**	**76.52**	**63.96**		**58.19**	**88.24**	**70.13**	**6.17**

elements, JavaScript codes, and cascading style sheets elements. This is followed by the process of stop word removal and stemming of the remaining words. TF-IDF values of the terms were then obtained and these values were used as the feature vector for the SVM classifier. For each tag, we selected all the pages that were tagged with the keyword and these were grouped as the positive samples for the particular tag. An equal amount of pages, which were tagged with a different tag, were selected as negative samples. From this set of positive and negative samples, two-thirds of the pages were used as the training sample while the rest were part of the test set.

The second experiment augmented the first with additional features added with the aim of determining if these new features to the dataset would improve the results. The setup for the experiment was similar to that done for Experiment 1. The only difference is the addition of the document's tags to the feature set. The TF-IDF values for the tags were obtained and used as the feature values. Likewise in this experiment, the default parameters of the SVM package were used.

RESULTS FOR EXPERIMENT 1 – TERMS ONLY

Table 1 shows the top 15 tags which scored the highest F-measure obtained from Experiment 1. Table 2 shows the bottom 15 tags with the lowest F-measure obtained from the same experiment. In both tables, the extreme right column shows the difference in the F-measure values obtained in both. Entries in bold indicate an increase in the F-measure value for Experiment 2. The results are ranked in ascending order according to the tag's F-measure values obtained in this experiment.

As seen in Table 1, the top 15 tags that obtained the highest F-measure values had very broad meaning in that we were not able to determine a specific context with respect to the tag. Examples include "reference," "howto," and "politics." On the whole, the top 15 tags had better recall than precision values, indicating that the classi-

Table 2. Bottom 15 tags with the lowest F-measure values

	Experiment 1				Experiment 2			Diff
Tag	Precision	Recall	F-measure		Precision	Recall	F-measure	
templates	**49.63**	**31.60**	**38.62**		**63.27**	**43.87**	**51.81**	**13.19**
animation	46.99	31.97	38.05		52.43	22.13	31.12	-7.88
xml	47.03	31.52	37.74		51.30	28.42	36.57	-1.17
ajax	52.47	29.32	37.62		39.58	9.52	15.35	-22.27
economics	44.71	30.89	36.54		49.25	26.83	34.74	-1.80
windows	54.95	26.93	36.14		40.00	9.32	15.12	-21.02
accessories	**47.37**	**28.42**	**35.53**		**52.63**	**52.08**	**52.36**	**16.83**
cms	45.28	27.80	34.45		45.59	23.85	31.31	-3.14
journal	**51.32**	**25.83**	**34.36**		**42.74**	**35.10**	**38.55**	**4.19**
ruby	**55.56**	**24.15**	**33.67**		**55.64**	**35.75**	**43.53**	**9.86**
actionscript	43.36	26.34	32.78		49.38	21.51	29.96	-2.82
parts	**50.00**	**22.50**	**31.03**		**57.89**	**27.50**	**37.29**	**6.26**
self-improvement	43.55	23.28	30.34		44.00	18.97	26.51	-3.83
icons	**45.45**	**14.93**	**22.47**		**55.84**	**32.09**	**40.76**	**18.29**
adobe	**45.10**	**13.29**	**20.54**		**42.86**	**13.87**	**20.96**	**0.42**

fier was able to correctly assign the documents which actually belong to the tag more than 75% of the time. However, the bottom 15 tags paint a different picture (see Table 2). These tags appear to have a narrower definition in contrast to the top 15 tags. For example, the term "adobe" has a more objective and precise meaning than "fun." The recall values for these tags were lower than its precision values, implying the classifier tended to predict more true negatives than true positives. In other words, the number of pages that did not belong to the category was higher than the pages belonging to it.

Table 3 presents the macroaveraged and microaveraged values for precision, recall, and F-measure. The macroaveraged precision and recall were 52.66% and 54.86%, respectively, while the standard deviation for precision was 4.21 and 19.05 for recall. In contrast with the standard deviation for precision, the standard deviation for recall varied greatly and a reason could be contributed by the classifier's tendency to misclassify a page which actually belongs to the tag. The macroaveraged F-measure was 52.05% and suggests that the classifier managed to predict at least half of the test data correctly with a 10.99 standard deviation across tags.

Microaveraged values show how the classi-

Table 3. Macroaveraged and microaveraged values for Experiments 1 and 2

	Experiment 1			Experiment 2		
	Precision	Recall	F-measure	Precision	Recall	F-measure
Macroaveraged	52.66 (s = 4.21)	54.86 (s = 19.05)	52.05 (s = 10.99)	50.77 (s = 6.06)	45.24 (s = 20.75)	45.77 (s = 13.21)
Microaveraged	64.76	54.40	59.14	56.47	54.93	55.69

fier performed based on each document. Here, the microaveraged precision was 64.76% while recall was 54.4%. From the F-measure value, the document had a 59.14% chance of being correctly classified. As shown in Table 3, both the macroaveraged and microaverage F-measure values were quite close. However, the F-measure value suggests that the users are not exactly good tag creators in the sense that other users would not be able to locate related resources using these tags. It was discussed previously that a large group of users would provide more reliable tags for resource description in contrast to expert individuals, yet this is not the case as indicated by our results. We surmise that the underlying reason is that tags can have multiple meanings and that there is no agreement on their usage in del.icio.us. As a result, the Web documents that are associated with a tag are not semantically related to each other, which in turn reduces the classifier's precision. The vocabulary problem (Furnas, Landauer, Gomez, & Dumais, 1987) is another reason that could contribute to our results. In other words, a tag would be associated with a diverse set of pages, and most of these pages may have been tagged only once with a particular tag.

RESULTS FOR EXPERIMENT 2 – TERMS AND TAGS

The results obtained in this Experiment 2 are shown in the right columns of Tables 1 and 2. The same tags that were selected in Experiment 1 are again shown in the table. In addition, the difference between the F-measures obtained in both experiments for the selected tags are shown, and entries in bold show an increase in values from those obtained in Experiment 1. Here, only eight of the selected tags have an increase in their F-measure values. The tag "icons" has the largest gain with 18.29. Documents belonging to this tag have an increased chance of 18.29% to be selected than before. On the other hand, the tag "ajax" suffered the largest drop in F-measure value with a decrease of 22.27%.

Table 3 shows the macroaveraged and microaveraged values for precision, recall, and F-measure obtained for this experiment. On average, all the categories had precision and recall values of 50.77% and 45.24%, respectively. The standard deviation for precision was 6.06, smaller than that for recall at 20.75. This was similar to the values obtained in the previous experiment. The F-measure score suggests that the classifier only managed to predict 45.77% of the pages correctly for each category on average, with a standard deviation of 13.21. For microaveraged values, the classifier managed to predict the relevance of each document with a precision and recall of 56.47% and 54.93%, respectively. The microaveraged value for F-measure was 55.69%.

It can be seen from Table 3 that the macroaveraged and microaveraged values obtained in Experiment 2 were lower than those obtained in Experiment 1, implying that the addition of tags as part of the feature set did not help in improving the precision, recall, and F-measure values. This is an interesting outcome, as making use of the terms only resulted in better performance than having terms together with tags. Although tags are words themselves, they may degrade the performance of the classifier because they appear in every document associated with these tags, causing them to have smaller TF-IDF weights in the document collection.

DISCUSSION AND CONCLUSION

In contrast to taxonomies and subject directories, social tagging is an alternative means of organizing Web resources. It is growing in popularity and is being used in a number of Web sites. In this chapter, we have investigated the effectiveness of tags in assisting users to discover relevant content by employing a text categorization approach. Here, we considered tags as categories, and examined the performance of an SVM classifier in assigning a dataset of Web resources from del.icio.us to their respective tags. Two experiments were conducted. The first examined the use of document terms only as features while the second added the document's tags in addition to the previous feature set. Surprisingly, results from the first experiment were better than the second. The macroaveraged F-measure obtained from Experiment 1 was 52.05% while the F-measure from Experiment 2 was 45.77%. The microaveraged values from both experiments were 59.14% and 55.69%. Put differently, the use of terms only yielded slightly better results than using terms and tags.

The relatively low values for precision, recall, and F-measure suggest that not all tags are reliable descriptors of the document's content. Our findings are similar to Sun et al.'s (2007) work. In that study, the range of macroaveraged F-measure values obtained for the description-only experiments ranged from 32% to 41%. Perhaps the much lower values were a result of using a shorter length of text as descriptions. It was reported that the description contained an average of only 14.8 terms for each blog. While the work of Razikin et al. (2007) was similar to the present work, the results obtained in that study were better. A reason for this could be the fact that the tags chosen were not from the popular list, and thus the Web pages that were associated with these tags tended to be more specifically related to the tags themselves. As the present work used the more popular tags, the likelihood of pages being incorrectly assigned to a tag could have been much higher because of greater usage by a diversity of users, thereby contributing to the poorer performance scores.

Some implications can be drawn from our present study. First, proponents of social tagging argue that the knowledge from a group of users could be much better than those provided by an expert (Suroweicki, 2004). However, our results have shown that this may not be applicable for all the tags, as some tags were found to be good descriptors and some were not. Future research could investigate the specific characteristics that make tags good descriptors for resource discovery. For example, one could examine whether tags with objective meanings might yield better performance than those with subjective meanings. Next, because not all tags are created equally for resource discovery, a social tagging system could assist users by suggesting tags in addition to supporting free keyword assignment. For example, after identifying characteristics of tags that serve as good resource descriptors, such tags could be used as recommendations to a tag creator for a given Web resource. In addition, a social tagging system might recommend different tags depending on whether the tag would be for public access, in which case, the recommendations would focus only on "good" tags, or for private use, in which case the recommendations could include tags that have meaning only to the creator.

In conclusion, our findings have shown different levels of effectiveness of tags as resource descriptors. There are however, some limitations to our study that provide opportunities for future work. The first concerns the use of terms and tags of the documents. These might not be the only features that could be used. Additional features like the document's title and the anchor text could prove useful for classification. Different weight schemes could be attempted for different features as well. Further, our corpus had an average of 1,331 documents per tag and future work could look into increasing the number of documents per tag, and utilize tags other than those that were popular.

ACKNOWLEDGMENT

This work is partly funded by A*STAR grant 062 130 0057.

REFERENCES

Brooks, C. H., & Montanez, N. (2006). Improved annotation of the blogosphere via autotagging and hierarchical clustering. In *Proceedings of the 15th International Conference on the World Wide Web* (pp. 625-632).

Dubinko, M., Kumar, R., Magnani, J., Novak, J., Raghavan, P., & Tomkins, A. (2006). Visualizing tags over time. In *Proceedings of the 15th International Conference on the World Wide Web* (pp. 193-202).

Furnas, G. W., Landauer, T. K., Gomez, L. M., & Dumais, S. T. (1987). The vocabulary problem in human-system communication. *Communications of the ACM, 30*(11), 964–971. doi:10.1145/32206.32212

Golder, S. A., & Huberman, B. A. (2006). Usage patterns of collaborative tagging systems. *Journal of Information Science, 32*(2), 198–208. doi:10.1177/0165551506062337

Hammond, T., Hannay, T., Lund, B., & Scott, J. (2005). Social bookmarking tools (I). *D-Lib Magazine, 11*(4). Retrieved August 23, 2008, from http://www.dlib.org/dlib/april05/hammond/04hammond.html

Joachims, T. (1998). Text categorization with support vector machines: Learning with many relevant features. In *Proceedings of the 10th European Conference on Machine Learning* (pp. 137-142).

Kipp, M. E. (2006). *Exploring the context of user, creator and intermediate tagging.* Paper presented at the 7th Information Architecture Summit. Retrieved August 23, 2008, from http://www.iasummit.org/2006/files/109_Presentation_Desc.pdf

Lin, X., Beaudoin, J. E., Bui, Y., & Desai, K. (2006). *Exploring characteristics of social classification.* Paper presented at the 17th Workshop of the American Society for Information Science and Technology Special Interest Group in Classification Research. Retrieved August 23, 2008, from http://dlist.sir.arizona.edu/1790/

Macgregor, G., & McCulloch, E. (2006). Collaborative tagging as a knowledge organization and resource discovery tool. *Library Review, 55*(5), 291–300. doi:10.1108/00242530610667558

Marlow, C., Naaman, M., Boyd, D., & Davis, M. (2006). HT06, tagging paper, taxonomy, Flickr, academic article, to read. In *Proceedings of the Seventeenth Conference on Hypertext and Hypermedia* (pp. 31-40).

Morville, P. (2005). *Ambient findability.* Sebastopol, CA: O'Reilly Media.

Puspitasari, F., Lim, E. P., Goh, D. H., Chang, C. H., Zhang, J., Sun, A., et al. (2007). Social navigation in digital libraries by bookmarking. In *Proceedings of the 10th International Conference on Asian Digital Libraries, ICADL 2007* (LNCS 4822, pp. 297-306).

Razikin, K., Goh, D. H., Cheong, E. K. C., & Ow, Y. F. (2007). The efficacy of tags in social tagging systems. In *Proceedings of the 10th International Conference on Asian Digital Libraries, ICADL 2007* (LNCS 4822, pp. 506-507).

Sebastiani, F. (2002). Machine learning in automated text categorization. *ACM Computing Surveys, 34*(1), 1–47. doi:10.1145/505282.505283

Sun, A., Suryanto, M. A., & Liu, Y. (2007). Blog classfication using tags: An empirical study. In *Proceedings of the 10th International Conference on Asian Digital Libraries, ICADL 2007* (LNCS 4822, pp. 307-316).

Suroweicki, J. (2004). *The wisdom of crowds: Why the many are smarter than the few and how collective wisdom shapes business, economics, societies, and nations.* New York: Doubleday.

Yanbe, Y., Jatowt, A., Nakamura, S., & Tanaka, K. (2007). Can social bookmarking enhance search in the Web? In *Proceedings of the 2007 Conference on Digital Libraries* (pp. 107-116).

KEY TERMS

Social Tagging/Social Bookmarking: The process of sharing and associating a resource, such as Web pages and multimedia, with tags.

Support Vector Machines: A vector-based machine learning technique which makes use of the maximum distance between vector classes as a decision boundary.

Text Categorization: The process of assigning documents to predefined labels using various techniques such as statistical and vector space models.

Web 2.0: Web applications that enable the sharing or creation of resources by a group of users. Some of these applications include Weblogs, wikis, and social bookmarking sites.

This work was previously published in Handbook of Research on Digital Libraries: Design, Development, and Impact, edited by Y. Theng, S. Foo, D. Goh, J. Na, pp. 251-260, copyright 2009 by Information Science Reference (an imprint of IGI Global).

Chapter 7.9
Representing and Sharing Tagging Data Using the Social Semantic Cloud of Tags

Hak-Lae Kim
National University of Ireland, Galway, Ireland

John G. Breslin
National University of Ireland, Galway, Ireland

Stefan Decker
National University of Ireland, Galway, Ireland

Hong-Gee Kim
Seoul National University, South Korea

ABSTRACT

Social tagging has become an essential element for Web 2.0 and the emerging Semantic Web applications. With the rise of Web 2.0, websites that provide content creation and sharing features have become extremely popular. These sites allow users to categorize and browse content using tags (i.e., free-text keyword topics). However, the tagging structures or folksonomies created by users and communities are often interlocked with a particular site and cannot be reused in a different system or by a different client. This chapter presents a model for expressing the structure, features, and relations among tags in different Web 2.0 sites. The model, termed the social semantic cloud of tags (SCOT), allows for the exchange of semantic tag metadata and reuse of tags in various social software applications.

DOI: 10.4018/978-1-60566-368-5.ch046

INTRODUCTION

With the rise of Web 2.0, websites which provide content creation and sharing features have become extremely popular. Many users have become actively involved in adding specific metadata in the form of *tags* and content annotations in various social software applications. While the initial purpose of tagging is to help users organize and manage their own resources, collective tagging of common resources can be used to organize information via informal distributed classification systems called *folksonomies* (Mathes, 2004; Merholz, 2004).

Studies of tagging and folksonomies can be divided into two main approaches: (a) semantic tagging concentrates on folksonomies that are inconsistent and even inaccurate because a large group of untrained users assign free-form terms to resources without guidance. Since this approach aims to resolve tag ambiguities, a wealth of ideas and efforts is emerging regarding how to use and combine ontologies with folksonomies (Weller, 2007); (b) social networking focuses on a community of users interested in a specific topic that may emerge over time because of their use of tags (Mika, 2005). The power of social tagging lies in the aggregation of information (Quintarelli, 2005). Aggregation of information involves social reinforcement by reinforcing social connections and providing social search mechanisms. Thus, a community built around tagging activities can be considered a social network with an insight into relations between topics and users.

Using freely determined vocabularies by a participant is less costly than employing an expert (Sinclair & Cardew-Hall, 2007) and a cognitive load of tagging in comparison with taxonomies or ontology is relatively low (Merholz, 2004). However, tagging the data from social media sites without a social exchange is regarded as an individual set of metadata rather than a social one. Although tagging captures individual conceptual associations, the tagging system itself does not promote a social transmission that unites both creators and consumers. To create social transmission environments for tagging, one needs a consistent way of exchanging and sharing tagging data across various applications or sources. In this sense, a formal conceptual model to represent tagging data plays a critical role in encouraging its exchange and interoperation. Semantic Web techniques and approaches help social tagging systems to eliminate tagging ambiguities.

BACKGROUND

Social Tagging

Social tagging and folksonomies have received much attention from the Semantic Web and Web 2.0 communities as a new way of information categorization and indexing. Among the most popular websites that employ folksonomies are Del.icio.us[1] (social bookmarking system) and Flickr[2] (photo-sharing network). CiteULike, using a similar approach to Del.icio.us, focuses on academic articles.[3] There are a number of multimedia sites that support tagging, such as Last.fm[4] for music and YouTube[5] for video.

Although the idea of a *tag* is not new, most people agree that a tag is no longer just a keyword. There is semantic information associated with a tag (Weller, 2007). A tag represents a type of metadata used for items such as resources, links, web pages, pictures, blog posts, and so on. Tagging can be defined as a way of representing concepts through keywords and cognitive association techniques without enforcing a categorization. The term *folksonomy* is a fusion of the two words *folk* and *taxonomy* (Vander Wal, 2004); it became especially popular with the proliferation of web-based social software applications, such as social bookmarking or annotating photographs. Building on the above definitions, folksonomy can be considered as a collaborative practice and method of creating and managing tags for the purpose of annotating and categorizing content (Mathes, 2004).

Advantages and disadvantages of social tagging present an issue for discussion. Although social tagging and folksonomies have much to offer users who utilize tags in various social media sites, there are important drawbacks inherent within the current tagging systems: for example, there is no formal conceptualization to represent tagging data in a consistent way and no interoperability support for exchanging tagging data among different applications or people (Marlow et al., 2006; Kim et al., 2007). The simplicity and accessibility of tags may lead to a lack of precision resulting in keyword ambiguity caused by misspelling certain words, as well as using synonyms, morphologies, or over-personalized tags. Since there are many different ways of using tags, it may be easy to misunderstand the meaning of a given tag.

Aside from these problems, social tagging systems do not provide a uniform way to share and reuse tagging data amongst users or communities. There is no consistent method for transferring tags between the desktop and the web or for reusing one's personal set of tags between either web-based systems or desktop applications. Although some folksonomy systems support an export functionality using their Open APIs (Application Programming Interfaces) and share their data with a closed agreement among sites, these systems do not offer a uniform and consistent way to share, exchange, and reuse tagging data for leveraging social interoperability. Therefore, it may be difficult to meaningfully search, compare, or merge similar tagging data from different applications.

With the use of tagging systems increasing daily, these limitations will become critical. The limitations come from lack of standards for tag structure and the semantics for specifying the exact meaning. To overcome the current limitations of tagging systems, it may be beneficial to take into account not only standards for representing tagging data but also develop interoperable methods to support tag sharing across heterogeneous applications.

SEMANTIC TAGGING APPROACHES

Folksonomy vs. Ontology

In general, a *taxonomy* is the organization of a set of information for a particular purpose in a hierarchical structure. An *ontology* is set of well-defined concepts describing a specific domain. It has strict and formal rules for describing relationships among concepts and for defining properties. The distinction between an ontology and a taxonomy is sometimes vague. A simple ontology without properties and constraints (i.e., concept hierarchy) could be considered a taxonomy, but a heavyweight ontology should clearly specify its capabilities.

From a classification perspective, folksonomies and ontologies can be placed at the two opposite ends of the spectrum. When compared to a traditional classification system, a folksonomy can be seen as a set of terms forming part of a flat namespace; that is, a folksonomy is a completely uncontrolled and flat system (Tonkin, 2006). To its disadvantage, folksonomy has no hierarchy and there are no directly specified parent-child relationships between the varying descriptions of the same object. Despite these limitations, the usefulness of folksonomies has been acknowledged: a folksonomy is a user-generated classification created through a bottom-up consensus.

Tag Ontology

There are certain disagreements on the merits of folksonomies and traditional classifications (see, for example, Hendler, 2007; O'Reilly, 2005; Shirky, 2004; Spivack, 2005). Shirky (2004) makes an argument that ontological classification or categorization is overrated in terms of its value. Shirky views folksonomies as emergent patterns in users' collective intelligence and claims that they can be harnessed to create a bottom-up consensus view of the world. According to Shirky, traditional classification systems have been structured using hierarchical taxonomies by experts studying a particular domain. Therefore these systems do not satisfy user-specific ways of thinking and organizing information. Meanwhile, Gruber (2005) criticizes Shirky's approach in that he fails to point out that a folksonomy has limitations to represent, share, exchange, and reuse tags and confuses "ontology-as-specified-conceptualization" with a very narrow form of specification. Hendler

(2007) also argues that Shirky misunderstood how ontologies could be built on the principles of the Semantic Web. Spivack (2005) asserts that folksonomies are just specific, highly simplistic cases of ontologies with minimal semantics.

Despite conflicting differences between folksonomies and ontologies, the Semantic Web and ontologies can be seen as a complement to folksonomies. Gruber (2005) and Spivack (2005) emphasize the importance of folksonomies and ontologies working together. In particular, Gruber (2005) proposes the "Tag Ontology." This aims at identifying and formalizing a conceptualization of the activity of tagging, and building technology that commits to the ontology at a semantic level. This approach is a good starting point to bridge Web 2.0 and the Semantic Web:

- **Gruber's Conceptual Model:** Suggests a model that defines a tagging activity including an object, a tag, a tagging, and a source.
- **Richard Newman's Tag Ontology:** Defines the three core concepts of Tagger, Tagging, Tag for representing the tagging activity (Newman, 2005) and is based on a tripartite tagging (i.e., user, resource, and tag).

The two approaches are focused on tagging activities or events that people used to tag in resources using terms. Therefore the core concept is Tagging. The concept of tagging has a relationship, as a concept, with Tagger and Resource to describe people who participate in a tagging event and objects to where a tag is assigned. However, there are no ways to describe the frequency of tags in these ontologies.

SCOT: Social Semantic Cloud of Tags

SCOT (Social Semantic Cloud of Tags) is an ontology for sharing and reusing tag data and representing social relations among individuals. It aims to describe the structure and the semantics of tagged data as well as offer interoperability of data among different sources.

Figure 1 illustrates the relations among the elements as well as a tagging activity. The vocabularies can be used to make explicit a collection of users, tags, and resources; they are represented by

Figure 1. Terms and relations in SCOT that can be used to describe tagging activities

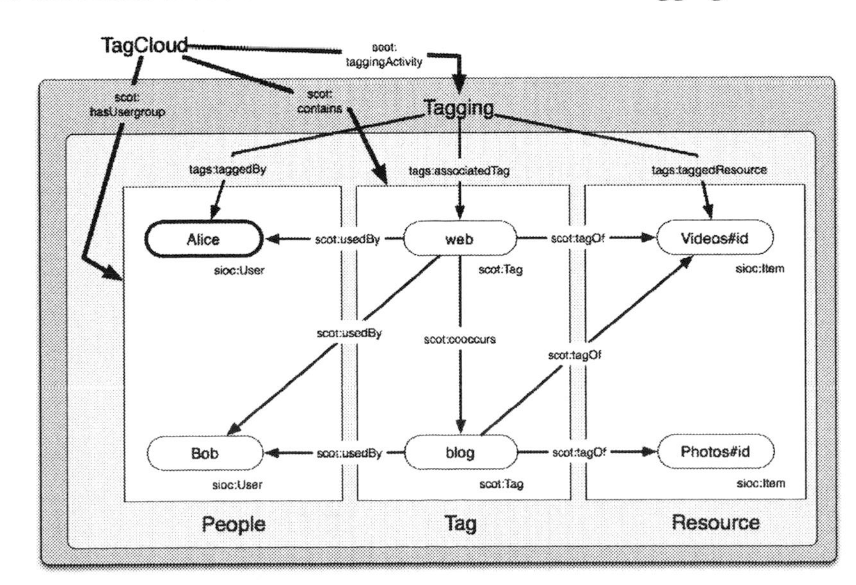

a set of RDF classes and properties that can be used to express the content and structure of the tagging activity as an RDF graph.

The TagCloud, which represents a conceptual model of a folksonomy, has a connection point to combine other concepts (i.e., tag, tagging, tagger, and so on.) and can be connected with other tag cloud with a unique namespace. The scot:TagCloud is a subclass of sioc:Container. All information consisting of relationships among taggers, items, and all tags is connected to this class. This class has the scot:contains and the scot:hasUsergroup properties. The former represents tags in a given domain or community while the latter describes taggers who participate in a tagging activity. A tagger may not be a single person according to contexts. For instance, if multiple taggers in a certain community generate a tag cloud, this tag cloud should contain all taggers. The scot:hasUsergroup property represents a person with a container. The scot:composedOf property describes a part of a TagCloud. In particular, if a TagCloud has more than two tag clouds, the property identifies each tag cloud. The scot:taggingActivity property present a relationship between a TagCloud and a Tagging.

The scot:Tag class, a subclass of tags:Tag (Tag class of Tag Ontology), allows users to assert that a tag is an atomic conceptual resource. A tag is a concept associated with a piece of information. The concept has many different variations according to taggers' cognitive patterns. Tag ambiguities, one of the most critical problems, result from this reason. The SCOT ontology provides several properties such as scot:spellingVariant, scot:synonym to solve this problem. It is called the "linguistic property" since these properties focus on representing the meaning of a tag and the relationships between each tag. In addition, the ontology has properties to describe occurrence of a tag (i.e., scot:frequency). A tag itself has its own frequency. The frequency is not unique, but it is an important feature to distinguish or compare with other tags. We called it a "numerical property." The properties have their own numerical values by computing. The properties in Figure 1 show high-level properties in the SCOT ontology.

In addition to representing the structure and the semantics of tags, the model allows the exchange of semantic tag metadata for reuse in social applications and interoperation amongst data sources, services, or agents in a tag space. These features are a cornerstone to being able to identify, formalize, and interoperate a common conceptualization of tagging activity at a semantic level.

SCOT aims to incorporate and reuse existing vocabularies as much as possible to avoid redundancies and to enable the use of richer metadata descriptions for specific domains. The ontology has a number of properties to represent social tagging activity and relationships among elements occurring in an online community.

- DC, or Dublin Core, provides a basic set of properties and types for annotating documents. In SCOT, we use the properties dc:title for the title of a TagCloud, dc:description to give a summary of the TagCloud, dc:publisher to define what system is generating the TagCloud, dc:creator to link to the person who created this set of tags. dcterms:created, from the Dublin Core refinements vocabulary, is used to define when a TagCloud was first created.
- FOAF (Brickley & Miller, 2004), or Friend of a Friend, specifies the most important features related to people acting in online communities. The vocabulary allows us to specify properties about people commonly appearing on personal homepages, and to describe links between people who know each other. foaf:Person is used to define the creator of a particular TagCloud. foaf:Group can be used to define a group of people who have created a group TagCloud.
- SIOC (Breslin et al., 2005), Semantically-Interlinked Online Communities, provides

the main concepts and properties required to describe information from online communities (e.g., message boards, wikis, blogs, etc.) on the Semantic Web. In the context of SCOT, sioc:Usergroup can be used to represent a set of sioc:User who have created the tags contained within a particular group TagCloud. A TagCloud is also a type of sioc:Container, in that it contains a set of Tags (subclass of sioc:Item).

- SKOS (Simple Knowledge Organization Systems) provides specifications and standards to support the use of vocabularies, such as thesauri, classification schemes, subject heading lists, taxonomies, other types of controlled vocabulary as well as terminologies and glossaries (Miles & Bechhofer, 2008). Tag is a subclass of skos:Concept, and a number of SKOS properties are used to define the relationships between Tags: broader, narrower, and so on.

Int.ere.st: Platform for Tag Sharing

Int.ere.st is a website where people can manage their tagging data from various sources, search resources based on their tags which were created and used by themselves, and leverage a sharing and exchanging of tagging data among people or various online communities.[6] The site is a platform for providing structure and semantics to previously unstructured tagging data via various mashups. The tagging data from distributed environments (such as blogs) can be stored in a repository, such as SCOT, via the Mashup Wrapper, which extracts tagging data using Open APIs from host sites. For instance, the site allows users to dump tagging data from Del.icio.us, Flickr, and YouTube; these tagging data are transformed into SCOT instances on a semantic level. Thus, all instances within Int.ere.st include different tagging contexts and connect various people and sources with the same tags. In addition, users can search people, tags, or resources and can bookmark some resources or integrate different instances. Through this iterative process, the tags reflect distributed human intelligence into the site.

Int.ere.st is the first OpenTagging Platform[7] of the Semantic Web, since users can manage a collection of tagging data in a smarter and more effective way as well as search, bookmark, and share their own as well as other's tagging data underlying the SCOT ontology. Those functionalities help users exchange and share their tagging data based on the Semantic Web standards. The site is compatible with other Semantic Web applications, and its information can be shared across applications. This means that the site enables users to create Semantic Web data, such as FOAF, SKOS, and SIOC automatically. RDF vocabularies can be interlinked with the URIs of SCOT instances that are generated in the site and shared in online communities.

FUTURE TRENDS

Social computing enables building social systems and software; it also allows for embedding social knowledge in applications rather than merely describing social information. Within social network analysis, traditional approaches have focused on static networks for small groups. As the technologies move forward, a major challenge for social network analysis is to design methods and tools for modeling and analyzing large-scale and dynamic networks. In particular, folksonomies are inherently dynamic and have different contexts among sources.

To facilitate the development of a social network for folksonomies, it is important to pay attention to social information. Although SNA allows analyzing phenomenon of social behavior at both individual and collective level, we do not have a solution that represents the relations among elements and reflects them to the objects. Tag ontologies are promising in providing the tools and formalism for representing social information including users, resources, tags, and their relationships.

Social search has become an active area in academic research as well as industry. Social search is a type of search engine that determines the relevance of search results by considering user interactions, contributions, or activities, such as bookmarking, tagging, and ranking. For instance, Del.icio.us and Spurl[8] (social bookmarking services) rely on user rankings, while Technorati[9] and Bloglines[10] (tag aggregators) analyze blogs and feed-based content. In particular, Swicki[11] and Rollyo[12] offer a community-based topic search as well as "searchles"[13] (these allow users to tag, group, and save links and create their own "SearchlesTV" channels through video mashups).

Most approaches, however, are limited to different types of resources and to show a comprehensive perspective on social relations across different applications. For instance, if users are involved in two different social spaces such as Del.icio.us and YouTube, one cannot build an integrated social network unless the two services have a mutual agreement. This issue, to some degree, is related to social information representation, since both websites have different aspects and events for building social connections. Thus, if one has a common conceptualization for social events, it is easy to build social connections among different spaces and to search them from different sources. Since tag ontologies are suitable to represent common conceptualization of social events, a social search can adopt Semantic Web technologies. We believe that a social search can benefit from a formal conceptualization of social knowledge, including tagging data based on the Semantic Web.

CONCLUSION

Tags have become an essential element for Web 2.0 and the Semantic Web applications. There is a vast collection of user-created content residing on the web. Tagging is a promising technological breakthrough offering new emerging opportunities for sharing and disseminating metadata. The critical issues discussed in this chapter offer many implications and challenges for representing tagging data semantically and exchanging them socially. With emphasis placed on tag ontologies and opportunities, these issues must be confronted without delay. Creators and consumers of folksonomies, as well as service providers, will profit from effective and efficient tagging methods that are socially and semantically enhanced.

REFERENCES

Breslin, J. G., Decker, S., Harth, A., & Bojars, U. (2005). SIOC: An approach to connect Web-based communities. *International Journal of Web-Based Communities*, *2*(2), 133–142.

Brickley, D., & Miller, L. (2004). *FOAF vocabulary specification*. Retrieved July 15, 2008, from http://xmlns.com/foaf/0.1

Golder, S. A., & Huberman, B. A. (2006). Usage patterns of collaborative tagging systems. *Journal of Information Science*, *32*(2), 198–208. doi:10.1177/0165551506062337

Gruber, T. (2005). Ontology of folksonomy: A mash-up of apples and oranges. *International Journal on Semantic Web and Information Systems*, *3*(2), 1–11.

Hendler, J. (2007). *Shirkyng my responsibility.* Retrieved July 15, 2008, from http://www.mindswap.org/blog/2007/11/21/shirkyng-my-responsibility

Kim, H. L., Breslin, J. G., Yang, S. K., & Kim, H. G. (2008). Social semantic cloud of tag: Semantic model for social tagging. In *Proceedings of the 2nd KES International Symposium on Agent and Multi-Agent Systems: Technologies and Applications* (pp. 83-92). Berlin, Germany: Springer.

Kim, H. L., Yang, S. K., Breslin, J. G., & Kim, H. G. (2007). Simple algorithms for representing tag frequencies in the SCOT Exporter. In *Proceedings of Intelligent Agent Technologies* (pp. 536-539).

Marlow, C., Naaman, M., Boyd, D., & Davis, M. (2006). HT06, tagging paper, taxonomy, Flickr, academic article, to read. In U. K. Wiil, et al. (Eds.), *Proceedings of the 17th Conference on Hypertext and Hypermedia* (pp. 31-39). New York: ACM Press.

Mathes, A. (2004). *Folksonomies: Cooperative classification and communication through shared metadata*. Retrieved July 15, 2008, from http://adammathes.com/academic/computer-mediated-communication/folksonomies.html

Merholz, P. (2004). *Metadata for the masses*. Retrieved July 15, 2008, from http://www.adaptivepath.com/ideas/essays/archives/000361.php

Mika, P. (2005). Ontologies are us: A unified model of social networks and semantics. *Web Semantics: Science . Services and Agents on the World Wide Web*, *5*(1), 5–15. doi:10.1016/j.websem.2006.11.002

Miles, A., & Bechhofer, S. (2008). *SKOS Simple knowledge organization system reference*. Retrieved July 15, 2008, from http://www.w3.org/TR/skos-reference

Newman, R. (2005). *Tag ontology design*. Retrieved July 15, 2008, from http://www.holygoat.co.uk/blog/entry/2005-03-23-2

O'Reilly, T. (2005). *What is Web 2.0: Design patterns and business models for the next generation of software*. Retrieved July 15, 2008, from http://www.oreillynet.com/pub/a/oreilly/tim/news/2005/09/30/what-is-web-20.html

Quintarelli, E. (2005). *Folksonomies: Power to the people*. Retrieved July 15, 2008, from http://www.iskoi.org/doc/folksonomies.htm

Shirky, C. (2004). *Ontology is overrated: Categories, links, and tags*. Retrieved July 15, 2008, from http://www.shirky.com/writings/ontology_overrated.html

Sinclair, J., & Cardew-Hall, M. (2008). The folksonomy tag cloud: When is it useful? *Journal of Information Science*, *34*(1), 15–29. doi:10.1177/0165551506078083

Spivak, N. (2005). *Folktologies--beyond the folksonomy vs. ontology distinction*. Retrieved July 15, 2008, from http://novaspivack.typepad.com/nova_spivacks_weblog/2005/01/whats_after_fol.html

Tonkin, E. (2006). *Folksonomies: The fall and rise of plain-text tagging*. Retrieved July 15, 2008, from www.ariadne.ac.uk/issue47/tonkin/

Vander Wal, T. (2005). *Explaining and showing broad and narrow folksonomies*. Retrieved July 15, 2008, from http://www.vanderwal.net/random/category.php?cat=153

Weller, K. (2007). Folksonomies and ontologies. 2007. Two new players in indexing and knowledge representation. In H. Jezzard (Ed.), *Applying Web 2.0. Innovation, impact and implementation: Online Information 2007 Conference Proceedings*, London (pp. 108-115). Retrieved July 15, 2008, from http://www.phil-fak.uni-duesseldorf.de/infowiss/admin/public_dateien/files/35/1204288118weller009_.htm

KEY TERMS AND DEFINITIONS

Folksonomy: A practice and method of collaboratively creating and managing tags for the purpose of annotating and categorizing content. The term *folksonomy* is a fusion of two words: *folk* and *taxonomy.* Folksonomies became popular with the introduction of web-based social software applications, for example, social bookmarking and photograph annotating.

Mashup: Involves web services or applications combining data from different websites. In general, mashup services are implemented by combining various functionalities with open APIs.

Ontology: Is set of well-defined concepts describing a specific domain.

Open API (Application Programming Interface): Is used to describe a set of methods for sharing data in Web 2.0 applications.

Semantic Web: Is an extension of the current World Wide Web that links information and services on the web through meaning and allows people and machines use web content in more intelligent and intuitive ways.

Social Computing: Is defined as any type of collaborative and social applications that offer the gathering, representation, processing, use, and dissemination of distributed social information.

Social Semantic Cloud of Tags (SCOT): Is an ontology for sharing and reusing tagged data and representing social relations among individuals. It aims to describe the structure and the semantics of data and to offer the interoperability of data among different sources.

Social Software: Can be defined as a range of web-based software programs that support group communication. Many of these programs share similar characteristics, for example, open APIs, customizable service orientation, and the capacity to upload data and media.

Social Tagging: Also known as *collaborative tagging*, refers to assigning specific keywords or *tags* to items and sharing the set of tags between communities of users.

Tag: A type of metadata used for items such as resources, links, web pages, pictures, blog posts, and so on.

Tagging: A way of representing concepts through tags and cognitive association techniques without enforcing a categorization.

Taxonomy: A method of organizing information in a hierarchical structure using a set of vocabulary terms.

ENDNOTES

1. http://del.icio.us
2. http://www.flickr.com
3. http://www.citeulike.org
4. http://www.lastfm.com
5. http://www.youtube.com
6. http://int.ere.st
7. http://opentagging.org
8. http://www.spurl.net
9. http://www.technorati.com
10. http://www.bloglines.com
11. http://www.swicki.com
12. http://rollyo.com
13. http://www.searchles.com

This work was previously published in Handbook of Research on Social Interaction Technologies and Collaboration Software; Concepts and Trends, edited by T. Dumova; R. Fiordo, pp. 519-527, copyright 2010 by Information Science Reference (an imprint of IGI Global).

Chapter 7.10
Anomaly Detection for Inferring Social Structure

Lisa Friedland
University of Massachusetts Amherst, USA

INTRODUCTION

In traditional data analysis, data points lie in a Cartesian space, and an analyst asks certain questions: (1) What distribution can I fit to the data? (2) Which points are outliers? (3) Are there distinct clusters or substructure? Today, data mining treats richer and richer types of data. Social networks encode information about people and their communities; relational data sets incorporate multiple types of entities and links; and temporal information describes the dynamics of these systems. With such semantically complex data sets, a greater variety of patterns can be described and views constructed of the data.

This article describes a specific social structure that may be present in such data sources and presents a framework for detecting it. The goal is to identify *tribes*, or small groups of individuals that intentionally coordinate their behavior—individuals with enough in common that they are unlikely to be acting independently.

While this task can only be conceived of in a domain of interacting entities, the solution techniques return to the traditional data analysis questions. In order to find hidden structure (3), we use an anomaly detection approach: develop a model to describe the data (1), then identify outliers (2).

BACKGROUND

This article refers throughout to the case study by Friedland and Jensen (2007) that introduced the tribes task. The National Association of Securities Dealers (NASD) regulates the securities industry in the United States. (Since the time of the study, NASD has been renamed the Financial Industry Regulatory Authority.) NASD monitors over 5000 securities firms, overseeing their approximately 170,000 branch offices and 600,000 employees that sell securities to the public. One of NASD's primary activities is to predict and prevent fraud among these employees, called registered representatives, or *reps*. Equipped with data about the reps' past employments, education, and "disclosable events," it must focus its investigatory resources on those reps most likely to engage in risky behavior. Publications by Neville et al. (2005) and Fast et al. (2007) describe the broader fraud detection problem within this data set.

DOI: 10.4018/978-1-60566-010-3.ch007

NASD investigators suspect that fraud risk depends on the social structure among reps and their employers. In particular, some cases of fraud appear to be committed by what we have termed *tribes*—groups of reps that move from job to job together over time. They hypothesized such coordinated movement among jobs could be predictive of future risk. To test this theory, we developed an algorithm to detect tribe behavior. The algorithm takes as input the employment dates of each rep at each branch office, and outputs small groups of reps who have been co-workers to a striking, or anomalous, extent.

This task draws upon several themes from data mining and machine learning:

Inferring latent structure in data. The data we observe may be a poor view of a system's underlying processes. It is often useful to reason about objects or categories we believe exist in real life, but that are not explicitly represented in the data. The hidden structures can be inferred (to the best of our ability) as a means to further analyses, or as an end in themselves. To do this, typically one assumes an underlying model of the full system. Then, a method such as the expectation-maximization algorithm recovers the best match between the observed data and the hypothesized unobserved structures. This type of approach is ubiquitous, appearing for instance in mixture models and clustering (MacKay, 2003), and applied to document and topic models (Hofmann, 1999; Steyvers, et al. 2004).

In relational domains, the latent structure most commonly searched for is clusters. Clusters (in graphs) can be described as groups of nodes densely connected by edges. Relational clustering algorithms hypothesize the existence of this underlying structure, then partition the data so as best to reflect the such groups (Newman, 2004; Kubica et al., 2002; Neville & Jensen, 2005). Such methods have analyzed community structures within, for instance, a dolphin social network (Lusseau & Newman, 2004) and within a company using its network of emails (Tyler et al., 2003). Other variations assume some alternative underlying structure. Gibson et al. (1998) use notions of hubs and authorities to reveal communities on the web, while a recent algorithm by Xu et al. (2007) segments data into three types—clusters, outliers, and hub nodes.

For datasets with links that change over time, a variety of algorithms have been developed to infer structure. Two projects are similar to tribe detection in that they search for specific scenarios of malicious activity, albeit in communication logs: Gerdes et al. (2006) look for evidence of chains of command, while Magdon-Ismail et al. (2003) look for hidden groups sending messages via a public forum.

For the tribes task, the underlying assumption is that most individuals act independently in choosing employments and transferring among jobs, but that certain small groups make their decisions jointly. These tribes consist of members who have worked together unusually much in some way. Identifying these unusual groups is an instance of anomaly detection.

Anomaly detection. Anomalies, or outliers, are examples that do not fit a model. In the literature, the term anomaly detection often refers to intrusion detection systems. Commonly, any deviations from normal computer usage patterns, patterns which are perhaps learned from the data as by Teng and Chen (1990), are viewed as signs of potential attacks or security breaches. More generally for anomaly detection, Eskin (2000) presents a mixture model framework in which, given a model (with unknown parameters) describing normal elements, a data set can be partitioned into normal versus anomalous elements. When the goal is fraud detection, anomaly detection approaches are often effective because, unlike supervised learning, they can highlight both rare patterns plus scenarios not seen in training data. Bolton and Hand (2002) review a number of applications and issues in this area.

MAIN FOCUS

As introduced above, the tribe-detection task begins with the assumption that most individuals make choices individually, but that certain small groups display anomalously coordinated behavior. Such groups leave traces that should allow us to recover them within large data sets, even though the data were not collected with them in mind.

In the problem's most general formulation, the input is a bipartite graph, understood as linking individuals to their affiliations. In place of reps working at branches, the data could take the form of students enrolled in classes, animals and the locations where they are sighted, or customers and the music albums they have rated. A tribe of individuals choosing their affiliations in coordination, in these cases, becomes a group enrolling in the same classes, a mother-child pair that travels together, or friends sharing each other's music. Not every tribe will leave a clear signature, but some groups will have sets of affiliations that are striking, either in that a large number of affiliations are shared, or in that the particular combination of affiliations is unusual.

Framework

We describe the algorithm using the concrete example of the NASD study. Each rep is employed at a series of branch offices of the industry's firms. The basic framework consists of three procedures:

1. For every pair of reps, identify which branches the reps share.
2. Assign a similarity score to each pair of reps, based on the branches they have in common.
3. Group the most similar pairs into tribes.

Step 1 is computationally expensive, but straightforward: For each branch, enumerate the pairs of reps who worked there simultaneously. Then for each pair of reps, compile the list of all branches they shared.

The similarity score of Step 2 depends on the choice of model, discussed in the following section. This is the key component determining what kind of groups the algorithm returns.

After each rep pair is assigned a similarity score, the modeler chooses a threshold, keeps only the most highly similar pairs, and creates a graph by placing an edge between the nodes of each remaining pair. The graph's connected components become the tribes. That is, a tribe begins with a similar pair of reps, and it expands by including all reps highly similar to those already in the tribe.

Models of "Normal"

The similarity score defines how close two reps are, given the set of branches they share. A pair of reps should be considered close if their set of shared jobs is unusual, i.e., shows signs of coordination. In deciding what makes a set of branches unusual, the scoring function implicitly or explicitly defines a model of normal movement.

Some options for similarity functions include:

Count the jobs. The simplest way to score the likelihood of a given set of branches is to count them: A pair of reps with three branches in common receives the score 3. This score can be seen as stemming from a naïve model of how reps choose employments: At each decision point, a rep either picks a new job, choosing among all branches with equal probability, or else stops working. Under this model, any given sequence of n jobs is equally likely and is more likely than a sequence of $n+1$ jobs.

Measure duration. Another potential scoring function is to measure how long the pair worked together. This score could arise from the following model: Each day, reps independently choose new jobs (which could be the same as their current jobs). Then, the more days a pair spends as co-workers, the larger the deviation from the model.

Evaluate likelihood according to a Markov process. Each branch can be seen as a state in a Markov process, and a rep's job trajectory can be seen as a sequence generated by this process. At each decision point, a rep either picks a new job, choosing among branches according to the transition probabilities, or else stops working.

This Markov model captures the idea that some job transitions are more common than others. For instance, employees of one firm may regularly transfer to another firm in the same city or the same market. Similarly, when a firm is acquired, the employment data records its workforce as "changing jobs" en masse to the new firm, which makes that job change appear popular. A model that accounts for common versus rare job transitions can judge, for instance, that a pair of independent colleagues in Topeka, Kansas (where the number of firms is limited) is more likely to share three jobs by chance, than a pair in New York City is (where there are more firms to choose from); and that both of these are more likely than for an independent pair to share a job in New York City, then a job in Wisconsin, then a job in Arizona.

The Markov model's parameters can be learned using the whole data set. The likelihood of a particular (ordered) sequence of jobs, P(Branch A → Branch B → Branch C → Branch D) is P(start at Branch A) • P(A → B) • P(B → C) • P(C → D)

The branch-branch transition probabilities and starting probabilities are estimated using the number of reps who worked at each branch and the number that left each branch for the next one. Details of this model, including needed modifications to allow for gaps between shared employments, can be found in the original paper (Friedland & Jensen, 2007).

Use any model that estimates a multivariate binary distribution. In the Markov model above, it is crucial that the jobs be temporally ordered: A rep works at one branch, then another. When the data comes from a domain without temporal information, such as customers owning music albums, an alternative model of "normal" is needed. If each rep's set of branch memberships is represented as a vector of 1's (memberships) and 0's (non-memberships), in a high-dimensional binary space, then the problem becomes estimation of the probability density in this space. Then, to score a particular set of branches shared by a pair of reps, the estimator computes the marginal probability of that set. A number of models, such as Markov random fields, may be suitable; determining which perform well, and which dependencies to model, remains ongoing research.

Evaluation

In the NASD data, the input consisted of the complete table of reps and their branch affiliations, both historical and current. Tribes were inferred using three of the models described above: counting jobs (JOBS), measuring duration (YEARS), and the Markov process (PROB). Because it was impossible to directly verify the tribe relationships, a number of indirect measures were used to validate the resulting groups, as summarized in Table 1.

Table 1. Desirable properties of tribes

Property	Why this is desirable
Tribes share rare combinations of jobs.	An ideal tribe should be fairly unique in its job-hopping behavior.
Tribes are more likely to traverse multiple zip codes.	Groups that travel long distances together are unlikely to be doing so by chance.
Tribes have much higher risk scores than average.	If fraud does tend to occur in tribe-like structures, then on average, reps in tribes should have worse histories.
Tribes are homogenous: reps in a tribe have similar risk scores.	Each tribe should either be innocuous or high-risk.

The first properties evaluate tribes with respect to their rarity and geographic movement (see table lines 1-2). The remaining properties confirm two joint hypotheses: that the algorithm succeeds at detecting the coordinated behavior of interest, and that this behavior is helpful in predicting fraud. Fraud was measured via a risk score, which described the severity of all reported events and infractions in a rep's work history. If tribes contain many reps known to have committed fraud, then they will be useful in predicting future fraud (line 3). And ideally, groups identified as tribes should fall into two categories. First is high-risk tribes, in which all or most of the members have known infractions. (In fact, an individual with a seemingly clean history in a tribe with several high-risk reps would be a prime candidate for future investigation.) But much more common will be the innocuous tribes, the result of harmless sets of friends recruiting each other from job to job. Within ideal tribes, reps are not necessarily high-risk, but they should match each other's risk scores (line 4).

Throughout the evaluations, the JOBS and PROB models performed well, whereas the YEARS model did not. JOBS and PROB selected different sets of tribes, but the tribes were fairly comparable under most evaluation measures: compared to random groups of reps, tribes had rare combinations of jobs, traveled geographically (particularly for PROB), had higher risk scores, and were homogenous. The tribes identified by YEARS poorly matched the desired properties: not only did these reps not commit fraud, but the tribes often consisted of large crowds of people who shared very typical job trajectories.

Informally, JOBS and PROB chose tribes that differed in ways one would expect. JOBS selected some tribes that shared six or more jobs but whose reps appeared to be caught up in a series of firm acquisitions: many other reps also had those same jobs. PROB selected some tribes that shared only three jobs, yet clearly stood out: Of thousands of colleagues at each branch, only this pair had made any of the job transitions in the series. One explanation why PROB did not perform conclusively better is its weakness at small branches. If a pair of reps works together at a two-person branch, then transfers elsewhere together, the model judges this transfer to be utterly unremarkable, because it is what 100% of their colleagues at that branch (i.e., just the two of them) did. For reasons like this, the model seems to miss potential tribes that work at multiple small branches together. Correcting for this situation, and understanding other such effects, remain as future work.

FUTURE TRENDS

One future direction is to explore the utility of the tribe structure to other domains. For instance, an online bookstore could use the tribes algorithm to infer book clubs—individuals that order the same books at the same times. More generally, customers with unusually similar tastes might want to be introduced; the similarity scores could become a basis for matchmaking on dating websites, or for connecting researchers who read or publish similar papers. In animal biology, there is a closely related problem of determining family ties, based on which animals repeatedly appear together in herds (Cairns & Schwager, 1987). These "association patterns" might benefit from being formulated as tribes, or even vice versa.

Work to evaluate other choices of scoring models, particularly those that can describe affiliation patterns in non-temporal domains, is ongoing. Additional research will expand our understanding of tribe detection by examining performance across different domains and by comparing properties of the different models, such as tractability and simplicity.

CONCLUSION

The domains discussed here (stock brokers, online customers, etc.) are rich in that they report the interactions of multiple entity types over time. They embed signatures of countless not-yet-formulated behaviors in addition to those demonstrated by tribes.

The tribes framework may serve as a guide to detecting any new behavior that a modeler describes. Key aspects of this approach include searching for occurrences of the pattern, developing a model to describe "normal" or chance occurrences, and marking outliers as entities of interest.

The compelling motivation behind identifying tribes or similar patterns is in detecting hidden, but very real, relationships. For the most part, individuals in large data sets appear to behave independently, subject to forces that affect everyone in their community. However, in certain cases, there is enough information to rule out independence and to highlight coordinated behavior.

REFERENCES

Bolton, R., & Hand, D. (2002). Statistical fraud detection: A review. *Statistical Science*, *17*(3), 235–255. doi:10.1214/ss/1042727940

Cairns, S. J., & Schwager, S. J. (1987). A comparison of association indices. *Animal Behaviour*, *35*(5), 1454–1469. doi:10.1016/S0003-3472(87)80018-0

Eskin, E. (2000). Anomaly detection over noisy data using learned probability distributions. In *Proc. 17th International Conf. on Machine Learning* (pp. 255-262).

Fast, A., Friedland, L., Maier, M., Taylor, B., & Jensen, D. (2007). Data pre-processing for improved detection of securities fraud in relational domains. In *Proc. 13th ACM SIGKDD International Conference on Knowledge Discovery and Data Mining* (pp. 941-949).

Friedland, L., & Jensen, D. (2007). Finding tribes: Identifying close-knit individuals from employment patterns. In *Proc. 13th ACM SIGKDD International Conference on Knowledge Discovery and Data Mining* (pp. 290-299).

Gerdes, D., Glymour, C., & Ramsey, J. (2006). Who's calling? Deriving organization structure from communication records. In A. Kott (Ed.), *Information Warfare and Organizational Structure*. Artech House.

Gibson, D., Kleinberg, J., & Raghavan, P. (1998). Inferring Web communities from link topology. In *Proc. 9th ACM Conference on Hypertext and Hypermedia* (pp. 225-234).

Hofmann, T. (1999). Probabilistic latent semantic analysis. In *Proc. 15th Conference on Uncertainty in AI* (pp. 289-296).

Kubica, J., Moore, A., Schneider, J., & Yang, Y. (2002). Stochastic link and group detection. In *Proc. 18th Nat. Conf. on Artificial Intelligence* (pp. 798-804).

Lusseau, D., & Newman, M. (2004). Identifying the role that individual animals play in their social network. *Proceedings. Biological Sciences*, *271*(Suppl.), S477–S481. doi:10.1098/rsbl.2004.0225

MacKay, D. (2003). *Information Theory, Inference, and Learning Algorithms*. Cambridge University Press.

Magdon-Ismail, M., Goldberg, M., Wallace, W., & Siebecker, D. (2003). Locating hidden groups in communication networks using hidden Markov models. In *Proc. NSF/NIJ Symposium on Intelligence and Security Informatics* (pp.126-137).

Neville, J., & Jensen, D. (2005). Leveraging relational autocorrelation with latent group models. In *Proc. 5th IEEE Int. Conf. on Data Mining* (pp. 322-329).

Neville, J., Şimşek, Ö., Jensen, D., Komoroske, J., Palmer, K., & Goldberg, H. (2005). Using relational knowledge discovery to prevent securities fraud. In *Proc. 11th ACM Int. Conf. on Knowledge Discovery and Data Mining* (pp. 449-458).

Newman, M. (2004). Fast algorithm for detecting community structure in networks. *Physical Review E: Statistical, Nonlinear, and Soft Matter Physics*, *69*, 066133. doi:10.1103/PhysRevE.69.066133

Steyvers, M., Smyth, P., Rosen-Zvi, M., & Griffiths, T. (2004). Probabilistic author-topic models for information discovery. In *Proc. 10th ACM SIGKDD International Conference on Knowledge Discovery and Data Mining* (pp. 306-315).

Teng, H. S., & Chen, K. (1990) Adaptive real-time anomaly detection using inductively generated sequential patterns. In *Proc. IEEE Symposium on Security and Privacy*, (pp. 278-284).

Tyler, J. R., Wilkinson, D. M., & Huberman, B. A. (2003). Email as spectroscopy: Automated discovery of community structure within organizations. In *Communities and Technologies* (pp. 81-96).

Xu, X., Yuruk, N., Feng, Z., & Schweiger, T. (2007). SCAN: A structural clustering algorithm for networks. In *Proc. 13th ACM SIGKDD International Conference on Knowledge Discovery and Data Mining* (pp. 824-833).

KEY TERMS AND DEFINITIONS

Anomaly Detection: Discovering anomalies, or outliers, in data.

Branch: In the NASD schema, a branch is the smallest organizational unit recorded: every firm has one or more branch offices.

Branch Transition: The NASD study examined patterns of job changes. If employees who work at Branch A often work at Branch B next, we say the (branch) transition between Branches A and B is common.

Latent Structure: In data, a structure or pattern that is not explicit. Recovering such structures can make data more understandable, and can be a first step in further analyses.

Markov Process: Model that stochastically chooses a sequence of states. The probability of selecting any state depends only on the previous state.

Registered Representative (rep): Term for individual in the NASD data.

Tribe: Small group of individuals acting in a coordinated manner, e.g., moving from job to job together.

This work was previously published in Encyclopedia of Data Warehousing and Mining, Second Edition, edited by J. Wang, pp. 39-44, copyright 2009 by Information Science Reference (an imprint of IGI Global).

Chapter 7.11
Emerging Online Democracy:
The Dynamics of Formal and Informal Control in Digitally Mediated Social Structures

Todd Kelshaw
Montclair State University, USA

Christine A. Lemesianou
Montclair State University, USA

ABSTRACT

The emergence and development of Web 2.0 has enabled new modes of social interaction that are potentially democratic, both within and across digitally mediated venues. Web-based interaction offers unlimited opportunities for organizing across geographic, demographic, and contextual boundaries, with ramifications in professional networking, political action, friendships, romances, learning, recreation, and entertainment. The authors conceptualize the democratization of Web-based social structures, defining online democracy as an imperfect balance of formal and informal modes of discursive control. The wrangling between formal and informal modes of discursive control ensures perpetual dynamism and innovation; the wrangling also offers the promise that diverse voices are not only welcome but also potentially responsive and responsible. The conclusion advocated is the importance of paying attention to these tendencies since they demonstrate that the Web's proclivities for decentralization and pluralism do not necessarily lead to relativistic and nihilistic hypertextuality but to potentially novel forms of shared social control.

DOI: 10.4018/978-1-60566-368-5.ch036

INTRODUCTION

With the advent of web-based social interaction technologies, new opportunities have arisen for user control and interactivity. These opportunities range widely in their relational complexities, spanning information gathering and opinion sharing, the formation of interpersonal relationships and online communities, and the development and maintenance of sophisticated organizational and global networks. Across these varied modes of interactivity, control of the technology, the media, and the communicative content is becoming increasingly decentralized and populistic. It is necessary, therefore, to address what is commonly called the "democratization" of the web.

Democracy, as conceived here, is not characterized by wholly unregulated chaos, despite the relativistic potential of hypertextual communication. Whereas interactivity in digitally mediated venues may range in quality from "anything goes" anarchy to rigid authoritarianism, this chapter addresses the emergence of democratic moderation, in which online participants "concertively" regulate their communication. Here we identify some past research and thoughts on the Internet's democratic qualities;[1] describe and illustrate online contexts as potentially democratic social structures that experience interplay between formal and informal communicative forces; and anticipate future trends of theory, empirical research, and practice.

BACKGROUND

The introduction of a new generation of social interaction technologies opens a possibility of the transformation of structural and social reality (McLuhan & Fiore, 1967; Ong, 1982). Since its emergence, the Internet[1] has been approached, theoretically and empirically, as a momentous and consequential social, cultural, economic, and political force (DiMaggio, Hargittai, Neuman, & Robinson, 2001; Wilson & Peterson, 2002). Some have theorized that the Internet might drastically transform the self, interaction, and social order and serve as a catalyst for social justice, empowering individuals to find spaces within which their voices may count (Negroponte, 1995). Others have cautioned that the Internet constrains and disempowers individuals within structured routines and cultural norms. They have argued that in the new virtual world some would emerge as winners (e.g., transnational corporations and interests) and others as losers (Beniger, 1996). Castells (1996), for instance, proposed that the Internet would follow the commercial path of its media predecessors and predicted a web "populated by two essentially distinct populations, *the interacting* and *the interacted*" (p. 371): the first group exemplifying the web's fragmentation potential and, the second, its reproduction of traditional media's massification patterns; and both groups reflecting the divide between the information rich and poor.

Studies on the Internet's potential to rearrange social, cultural, economic, and political life have focused on such issues as access to open information flows across national and global systems (Bimber, 2000; Norris, 2001; Schiller, 1995), identity construction (Cutler, 1996; Morse, 1998; Turkle, 1995), community formation and mobilization (Foster, 1997; Rheingold, 1993; Zappen, Gurak, & Doheny-Farina, 1997), and civic and political participation and deliberation (Putnam, 2000). The potential of the Internet to promote civic engagement and political democracy have also gained attention (e.g., Agre, 2002), especially since the 1990s when United States' congressional, state, and presidential candidates began deploying campaign websites (Hurwiz, 1999). The increased opportunities for interactivity among citizenries through blogs, chat rooms, and Internet forums are now being investigated (e.g., Best & Krueger, 2005; Dahlgreen, 2000; Endres & Warnick, 2004) as are the perceived public risks posed by such online participation (Andrejevic, 2006; Best, Krueger, & Ladewig, 2007). While some suggest that the Internet has the potential to mobilize "netizens" in new ways and to support democratic processes (Carpini, 2000; Deuze, 2006; Min, 2007), at times lending authorial voice to marginalized constituents and concerns, there is also increased support for the polarization of the public and the unfulfilled potential of deliberative democracy (Noveck, 2000; Selnow, 1998; Streck, 1998; White, 1997; Wilhelm, 2000). Still others point out that online deliberative democracy can be actualized, but its impact is greatly diminished when positioned within a dominant commercialized and individualized culture (Dahlberg, 2001).

On a more micro-analytic level, various studies have explored language, communication practices, and social interaction and relations on the Internet (e.g., Crystal, 2001). These inquiries address the emergence and negotiation of "netiquette" rules that socialize online participants, as well as more subtle self-imposed or community-imposed levels of informal control that undermine the Internet's democratic potential. McLaughlin, Osborne, & Smith (1995) identified early misconduct on Usenet that included misuse of technology, bandwidth waste, factual errors, inappropriate violation of language guidelines, newsgroup-specific conventions, and ethical codes. The extent to which the Internet liberates or simply reproduces patterns of expression and participation, in relation to gender, race, and sexual politics, continues to interest participants, observers, and researchers (Bromseth, 2001; Dibbell, 1993; Herring, 1996; Herring & Paolillo, 2006; Kendall, 2000; Lessig, 1999; Miller, 1995; Stivale, 1997) as do the Internet's descriptive and prescriptive rules and various interactive forms, where behaviors range from disruption and hostility (e.g., flaming) to disciplinary control and normalization (Dutton, 1996; Lee, 2005; Phillips, 1996; Thompsen, 1996). Much of the research has examined online contexts such as MUDs, MOOs, and Usenet communities that are now being replaced in prominence by new Internet venues.

Janack (2006) studied how participants of *Blog for America*—a feature of the 2000 U. S. presidential candidate Howard Dean's campaign website—discursively disciplined themselves. That study echoes this chapter's focus: While new social media feature pluralistic participation, interactivity, and ungrounded hypertextuality, they operate within both formally structured and emergent frameworks of values, practices, and expectations that significantly reaffirm and rewrite authority and control. In Janack's study, for instance, blog participants behaved as gatekeepers, a role traditionally enacted by campaign staffers, to silence Dean's critics through various rhetorical strategies (e.g., ignoring or minimizing critical comments, *ad hominem* attacks). If Internet-enabled discourse may be described in Bakhtin's (1984) language as "carnivalesque," we can see how it potentially brings people together as equals to revel in liberation from exogenous constraints (legal, governmental, religious, and so on) and social stratifications. At once, though, it is necessary to recognize that the carnival, no matter how liberating, is a social order nonetheless. Its playfulness, unpredictability, populism, and multi-voicedness are inevitably bounded and regulated. This moderation is a hallmark of online democracy, and an important potential characteristic of web-based interaction.

ONLINE DEMOCRACY'S MODES OF SOCIAL CONTROL

Online communication occurs within and across diverse contexts and forms, ranging from relatively simple information posts to highly sophisticated social and political commentary. Accordingly, discursive qualities vary greatly throughout the web. Some utterances are noisily and chaotically ungrounded whereas others are kept orderly by the moderator's iron fists. Of interest is what happens somewhere between these extremes, when administrative and lay users participate together to negotiate and maintain their shared social orders.

To understand such potentially democratic qualities, it is necessary to define online democracy and to address how it may be enacted. This section begins with the general idea that all online interaction is governed by tensional interplay between formal and informal modes of control. Then we define online democracy as a particular mode of dialectical tension, and illustrate this dynamic—and the general potential for online democracy in Internet-mediated venues—by exploring some actual online interaction.

Online Social Structures and Democracy

Any social structure is continuously rebuilt through participants' communication (Giddens, 1984). This applies not only to face-to-face and other traditional social contexts but also to emerging venues that are digitally mediated, such as online forums, discussion boards, blogs, wikis, and social networking sites. These online settings may be thought of as both nouns and verbs; they are as much solid frameworks of norms as they are fluid interactions among people. In Giddens' terms, such simultaneously solid and fluid structures comprise socially made rules (implicit and explicit formulas for action) and resources (participants' commoditized abilities, knowledge, designated roles, etc.) that at once permit and limit interaction. The constant "reinscription" of rules and resources occurs not just within a given social structure as a demarcated and unitary body but also *across* overlapping structures that may mediate and contradict each other.

One way to understand this complex nature of social structures is to address how they are enabled and constrained by a dialectical interplay of social forces—what Bakhtin (1981) characterizes as "centripetal" and "centrifugal." Centripetal forces are authoritative, stabilizing, decisive, and preserving of traditions. Directives, punishments, and behaviors that concretize stratified roles are instances of centripetally leaning talk. Conversely, centrifugal forces are insurgent, destabilizing, equivocally open-ended, and change-minded. Questions, evasions, and behaviors that challenge authority are centrifugally leaning.

When examining online interaction, each discrete utterance (e.g., a blog post) manifests a quality that is roughly mappable along the *centripetal—centrifugal* continuum. Directives and pronouncements, for instance, tug the discourse centripetally whereas questions and invitations for response pull things centrifugally. In looking at a communicative context's utterances in aggregate, it is possible to recognize a particular meta-dynamic, the cumulative tugging of which defines the social structure's general character. How imbalanced are its forces? How do they tend?

If a given online setting experiences a preponderance of explicit rules, web forum moderator-enforced censorship, gate-keeping, and other centripetally-leaning discourses, the social structure tends toward autocracy or authoritarianism as an organizational form. Its social order is maintained by verbal and nonverbal talk that enforces centralized control while stifling insurgence. Leaders are typically designated and recognizable as established roles, and control mechanisms tend to be formally explicit and enforceable as codified rules. At the other end of the sliding scale—the centrifugal one—a given online venue with few (or no) explicit regulations, disciplinary procedures, and plenty of cacophonous interaction leans toward anomie or anarchy as the basis of social (dis)order. Discursive struggles occur through decentralization of authority and destabilization of public order. Leadership is emergent and fluid, and social norms—which are likely implicit—are negotiated among participants rather than imposed from atop or outside of the social structure.

Online Democracy

Whereas social forms such as anarchy and autocracy inhabit the continuum's two dialectical extremes, online democracy hovers (however imprecisely) near the middle. Online democracy may be defined as the quasi-counterbalance of normalizing and destabilizing discursive forces; a subtle and messy negotiation of control in which centripetal and centrifugal pushes and pulls more-or-less even out. The "more-or-less" characteristic of this equilibrium reflects the general idea that social structures, which are constantly remade in participants' interaction, are always dynamic and never static. For this discussion of online venues—which are potentially democratic but certainly not always—this definition is important. Given the seemingly pluralistic decentralization of discourse in web-based settings, the Internet appears to come to life as a noisy "marketplace of ideas" (Mill, 1956); a panoply of public "sphericules" (Gitlin, 1998) in which interaction is most easily described as ungrounded. The risk of these characterizations is that they may overemphasize centrifugal forces while denying or trivializing the emergence and functions of social control.

Control within and among online social structures is ubiquitous, and it comes in two generally different modes that are dialectically engaged. The first mode is what Spitzer (1982) describes as "formal," embodying centripetal qualities with their hierarchical stratification. The second, "informal" mode advances centrifugal qualities that disrupt hierarchical structures in socially emergent ways. As formal and informal modes of control interact within a democratic social system, discourses that "seem to weaken hierarchies of power may actually establish new channels through which those hierarchies can be strengthened, extended, and made more responsive to the complexities of modern social administration" (Spitzer, 1982, pp. 187-188). Accordingly, in addressing the democratization of online social spheres, it would be a mistake to highlight a presumed absence of discursive control. Instead it is important to acknowledge how control functions, both formally and informally.

Formal Control in Online Democracy

Democratic online venues are, in part, maintained by formal discursive control. There are many common instances of formal control in contemporary Internet life. Registration forms that are created and processed by centralized (usually institutional) website managers, for instance, require prospective users to provide their names, contact information, and other identifiers to gain access to the venue. Often, registrants are required to accept explicit terms and conditions. The contents of posts are filtered, either by moderators or automatically. Rules for conduct are stated, as are enforcement mechanisms, which range from message flagging to censorship to banishment. In short, formal control is usually obvious to participants, manifesting in "power-down" policies and actions that demand and enforce respect while constraining behavior.

Informal Control in Online Democracy

In roughly equal measure, informal control mechanisms also fulfill important functions in web-based democracies. Unlike formal control, which maintains social order authoritatively, informal enactments of control are essentially disruptive. They may directly undermine explicit authority, as do acts of spamming, flaming, hacking, and impersonation. They may also rewrite, reframe, or appropriate the online venue's content and norms to shift power away from the institution or privileged users. Informal control may be recognized in tagging and other XML-enabled tools that users apply to customize and order online content according to their own preferences.

But this disruption is not as chaotic, unregulated, and destructive as one might assume. If formal control is "top-down" then informal control is "bottom-up," manifesting what Foucault (1995) addresses in terms of a "panopticon"—an invisible mode of omniscience. Fostered online by the blurring of private and public identities within contexts that are rife with mutual voyeurism, panoptic omniscience compels participants to discipline not only each other, but also themselves. Barker (1999) describes the effect in organizational settings as "concertive control," through which team members in a "supervisorless culture" develop communicative patterns such as "informal hierarchies, particular power relationships, and team norms" (p. 13). In this negotiation there is both disruption and discipline—as well as the (re) construction of order. So informal control may be understood as emergent and participant-centered co-regulation.

THE DYNAMICS OF FORMAL AND INFORMAL CONTROL IN ONLINE DEMOCRACY: A CASE OF JUICYCAMPUS.COM

One specific case that offers an insightful view of how formal and informal kinds of control play out on the Internet is JuicyCampus.com, a website that aims to enable "online anonymous free speech on college campuses" (JuicyCampus.com, About Us). The website claims to provide a forum "where college students discuss the topics that interest them most, and in the manner that they deem most appropriate" (JuicyCampus.com, About Us) and promotes itself as the "world's premier college gossip site" that attracts "nearly one million unique visitors per month, while serving 500 campuses across the country" (JuicyCampus.com, Official Blog Announcements). Critics view the website as a gossip mill and compare it to "a dorm bathroom wall writ large, one that anyone with Internet access can read from and post to" (Morgan, 2008). However controversial it may occur to an outside observer, JuicyCampus.com represents a growing number of unmoderated online forums that promise anonymity to the users. The provision of general communicative guidelines coupled with the absence of gatekeeping and censorship-minded monitoring results in rich meta-discussions among participants about what kinds and styles of talk are appropriate. In the course of this site's interaction and meta-interaction, an illustration emerges of the carnival that is web-based democracy.

The Web as a Borderless System of Social Structures

JuicyCampus.com was founded in 2007 by Matt Ivester, a Duke University graduate who characterized the site as an attempt to cultivate "gossip 2.0" (cited in Morgan, 2008). The site does not require participants to register, and posts are anonymous. The site has spread to 500 college campuses in the U.S. and has been the center of a number of controversies with regard to its function and nature of participation. For the purposes of this analysis we focus on one of the most discussed JuicyCampus.com's threads, *The Yale Women's Center is Genius*, which as of March 2008 had generated 153 responses.

The Internet's borderlessness and carnivalesque qualities become apparent when one traces the emergence of this forum: a group of Yale students, members of the Zeta Psi fraternity, posed in front of the Yale Women's Center holding a sign reading, "We love Yale sluts" (Abrahamson, 2008). The photograph was uploaded on Facebook.com on January 16, 2008 and came to the attention of the Yale Women's Center members and the university community. On January 21 the Center declared an intention to pursue legal action (Abrahamson, 2008). On the same day, a blog posting appeared on IvyGateblog.com (O'Connor, 2008) that generated 216 responses from January 21 to February 28. On February 29 the forum's thread, entitled *The Yale Women's Center is Genius*, was posted on JuicyCampus.com with entries followed by the authors until March 16. The list of responses that have emerged or directed attention to this case is not meant to be exhaustive but rather an indication of the dynamic and hypertextual way in which discourse unfolds.

The discursive positions and moves of participation in all these social networks simultaneously invoke, reflect, co-opt, and control the Internet's carnivalesque potential and offer a macro-analytic glimpse of the centripetal and centrifugal forces that are generally at work in public online interaction. Participants' posts on JuicyCampus.com exemplify the range of potential discursive positions from which democratic talk can unfurl—with appeals to legal issues and ramifications, celebrations of freedom of expression and feminist ideology, references to sexual and gender politics, concerns about advertising and the economies of cyberspace, and, of course, frequent mention of netiquette rules and what constitutes appropriate discourse. It is to this exploration of the formal and informal enactments of control that we now turn.

Enactments of Formal Control

JuicyCampus.com's official policies straddle the legal and contextual. The site has an official policy outlining terms and conditions of use, an intellectual property policy, and a privacy policy. Although the site's *Privacy & Tracking Policy* page claims, "we do not track any information that can be used by us to identify you" (JuicyCampus.com, Privacy & Tracking Policy), the site managers have assisted police in identifying individuals who have made explicit threats (see, for example, Morgan, 2008). The site also has frequent announcements by the site managers that contextualize much of the more formal legal language, at times in contradictory ways. For example, the site outlines specific *Terms & Conditions* with regard to user conduct that also address defamation. However, on the *Frequently Asked Questions* page the site creators note: "Facts can be untrue. Opinions can be stupid, or ignorant, or mean-spirited, but they can't be untrue. And we believe everyone is entitled to their opinion" (JuicyCampus.com, Frequently Asked Questions). Site managers' specific announcements have tackled the issue of defamation (December 9, 2007; December 11, 2007); anonymity (December 9, 2007); use of real names of people being discussed (January 29, 2008); copyrighted material; the posting of contact information; spamming; and what constitutes "juicy," which the site founder says is *not hate* (February 29, 2008).

What is notable in reviewing these announcements is the celebration of the carnival—the recognition that online democracy's nature and form are contested and emergent. What is even more noteworthy is that it is the *institution* that enacts this celebration of carnival. As JuicyCampus.com (Official Blog Announcements, February 29, 2008) states, "Ultimately, JuicyCampus is created by our users, and we ask that you please take this responsibility seriously." So, at once the site managers are doing two things: advancing and maintaining formal control by inviting and permitting emergent and user-controlled discourse (invitation and permission being particular modes of control); and delimiting discourse in a way that protects themselves legally while defining the parameters of "juiciness."

Enactments of Informal Control

Anonymous users enact emergent and decentralized control in various ways throughout JuicyCampus.com's *The Yale Women's Center is Genius* discussion. Most basically, informal control is typified in users' ability to initiate discussion threads and reply without fear of monitoring, filtering, censorship, or expulsion. The important point about this is that, whereas users are liberated from formal oversight, they are subject to one another's responses. This cultivates a mode of discipline that emerges among participants, akin to Foucault's (1995) "panopticon" and Barker's (1999) "concertive control." Participants exercise discipline on issues ranging from who can participate, how they may appropriately do so, and why. Informal enactments of control are manifest in various strategies such as name-calling, threats, irony, caution, silencing, confrontation, and othering, among others (see Table 1):

Disciplining potentially defamatory talk (relating mostly to the Women's Center coordinator), participants argue from various positions that engender much more than just the legal perspec-

Table 1. Types of informal control

Participation issue	Strategy	Illustration of informal control
Who?	Name-calling	you a * get off the site. "impartial observer?" NEWSFLASH: everyone here is * invested. stop slowing the dialogue down. (3/4/08)
What?	Threat	the author of the other thread – "someone needs to * chase o-m till she cries" and the post on this thread of the same name - should be found, shamed, and lethally injected. how dare you speak of her? (2/29/08)
Where?	Caution	as a fan of the women's center: can we not talk about their strategies online, let alone on this site? FOR BLARINGLY OBVIOUS REASONS. (3/6/08)
When?	Irony	i just hate feminists and wrongly assumed it was safe to post at 5 a.m. because I wrongly assumed that they would be asleep. but now i see my mistake: they didn't have any sex so they couldn't fall asleep. (3/4/08)
Why?	Othering	i just wanted to express my opinion. I am not like any of you crazies. (3/4/08)

tive witnessed at the formal level of administrative control of the site (see Table 2):

Here we see the carnival's explosively disruptive potential as competing languages and perspectives struggle to frame the legitimacy of participation. We also see the attempts to control, rein in, redirect, and reframe the discussion's boundaries. The tension between formal (centripetal) and informal (centrifugal) modes of control is evident throughout the JuicyCampus.com site and the Yale Women's Center forum thread, and point to web-based democracy's inherent messiness. In these matters, there will be no "resolution." And, in these matters, as one participant observes, the carnival comes to life meta-discursively: "i love that the most discussed thread on juicycampus is about the women's center's attempt to destroy juicycampus" (3/4/08).

FUTURE TRENDS

As the Internet has emerged in popular use and evolved into Web 2.0, it has become less a medium for the sheer expression and transference of *information* and more a setting for *relationship-making*. Web-based interaction offers unlimited opportunities for organizing across geographic, demographic, and contextual boundaries, with ramifications for professional networking, po-

Table 2. Illustrations of informal control

What counts as defamation?	Illustration of informal control
Moral perspective	The * making these nasty comments about people should be thoroughly ashamed of themselves! (2/29/08)
Democratic perspective	these big anonymous sites make everybody forget that the people we are talking about are not public property, that their lives are not performed for our benefit, that the warriors are people too. * is as good a person as she is a warrior. let's make sure we remember that. (2/29/08)
Relational Perspective	WHY ARE ALL YOU PEOPLE SICKOS???? SHE IS MY * SISTER. I HAVE HELPED HER FIND TEDDY BEARS THAT SHE HAS LOST FOR YEARS. SHE IS * AWESOME. (3/2/08)
Public Domain Perspective	i am totally in awe of * and i became a feminist since arriving at yale because of two talks i went to at the women's center that she moderated... but she is a celebrity here. you shouldn't nipe at kids for talking about her because she is a public person, like a politician. we do have some right to talk about her as a person in a way that we dont have a right to talk about other people. (3/2/08)
Legal Perspective	dude, why are you mouthing off about girls on this thread? ITS LIKE SUICIDE. dumb *. everyone knows that they are planning to sue the site. EVERYONE knows that they have lawyers cuz of the frat stuff. EVERYONE knows that your * will get subpeonaed and EVERYONE will know who you are. (3/5/08)

litical action, friendships, romances, education, recreation, healthcare, and so on. One example of real but previously unforeseeable use of online interaction is *Naughtie Auties*, a resource center in the Second Life virtual venue where those with autism spectrum disorders can practice social interaction (Saidi, 2008). Theorists, researchers, and practitioners of web-based communication will increasingly have opportunities to engage these new relationship structures' discursive negotiations of control, and to recognize democracy when it occurs and assess its functions and sociologic consequences.

Internet scholars and users will also have increasing opportunities to consider how technological populism contributes to open and democratic discourse. It is noteworthy that means for producing and distributing visual, literary, musical, etc. creations are increasingly accessible. The result is that "authorship," broadly understood, is becoming less elite and more decentralized. It is now quite easy for amateurs to self-produce and distribute their art, ideas, and home videos, and to redefine "celebrity" in populistic terms via so-called "first-person media." As well, through Web 2.0-based technologies, lay people may co-opt, remix, and redistribute canonic pieces of art in ways that redefine authorship and ownership, problematizing notions of intellectual property. This wrangling between formal control mechanisms (e.g., copyright law) and insurgent, populistic inter-activities has everything to do with democracy, and will, in the future, be an important locus of concern.

Generally, the emergent democratization of online discourse is important to study and reflect upon since its social practices both influence and are influenced by life beyond "virtual" space. Democracy as a social order and even as a moral ideal is a historical phenomenon that has yet to be fully realized. It may be enacted in daily life, however imperfectly, throughout many contexts, spanning family, labor, education, entertainment, community, politics, etc. In a sense, participants in online venues that enable democratic interaction have opportunities to learn how to "do" democracy in effective and satisfying ways. There are particular procedures (e.g., deliberative decision making), responsibilities (e.g., shared leadership), and expectations of decorum (e.g., mutual respect, even in light of disagreement) that support effective democratic organizing. How democracy is practiced in online environment has great consequence for many dimensions of 21st century life, including citizenship (national, global, corporate, etc.), community building, resource management, innovation, and so on. In short, the Internet's new opportunities for experiencing and practicing democracy both reflect and contribute to the emerging democratization of broader social life.

CONCLUSION

The authors have approached online interaction as potentially democratic. The discussion's emphasis has not been upon democratic political interaction *per se*, but upon communicative practices of online democracy as they may be enacted in (and across) a wide range of online settings, such as blogs, wikis, and social networks. The definition of online democracy that we advance—a subtle and messy negotiation of control in which centripetal and centrifugal pushes and pulls more-or-less even out—is important for understanding how online contexts, potentially, are neither strictly controlled by their institutional creators/managers nor entirely disrupted by their "mobs" of users. This imperfect balance between formal and informal modes of discursive control ensures perpetual dynamism and innovation, as well as the promise that diverse voices are not only welcome but also potentially responsive and responsible.

Our intentions have been to recognize that, first, as in "real" life, so-called "virtual" interactions and relationships may take various organizational forms that range from anarchy to absolute control; and, second, that the Internet provides some new opportunities that are democratic in quality. It is important to pay attention to these opportunities since they demonstrate that the web's proclivities for decentralization and pluralism do not necessarily lead to relativistic and nihilistic hypertextuality but to potentially novel forms of shared social control. When formal and informal regulatory forces temper each other, there are consequences for the kinds of relationships and communities that interactants may forge. As participants in online venues increasingly engage in online democratic social structures, they may learn to relate on both substantive and meta-discursive levels in order to negotiate mutually recognized rules and resources. Such practice might have great consequence for broader social systems, and for the character of democracy throughout 21st century life.

REFERENCES

Abrahamson, Z. (2008, January 22). Misogyny claim leveled at frat. *Yale Daily News*. Retrieved from http://www.yaledailynews.com/articles/comments/23045

Agre, P. E. (2002). Real-time politics: The Internet and the political process. *The Information Society*, *18*, 311–331. doi:10.1080/01972240290075174

Andrejevic, M. (2006). Interactive (in)security. *Cultural Studies*, *20*(4-5), 441–458. doi:10.1080/09502380600708838

Bakhtin, M. M. (1981). *The dialogic imagination* (C. Emerson & M. Holquist, Trans.). Austin, TX: University of Texas Press.

Bakhtin, M. M. (1984). *Rabelais and his world* (H. Iswolsky, Trans.). Bloomington, IN: Indiana University Press.

Barker, J. R. (1999). *The discipline of teamwork: Participation and concertive control*. Thousand Oaks, CA: Sage Publications.

Beniger, J. R. (1996). Who shall control cyberspace? In L. Strate, R. Jacobson, & S. B. Gibson (Eds.), *Communication and cyberspace* (pp. 49-58). Creskill, NJ: Hampton Press.

Best, S. J., & Krueger, B. S. (2005). Analyzing the representativeness of Internet political participation. *Political Behavior*, *27*(2), 183–216. doi:10.1007/s11109-005-3242-y

Best, S. J., Krueger, B. S., & Ladewig, J. (2007). The effects of risk perceptions on online political participation decisions. *Journal of Information Technology & Politics*, *4*(1), 5–17. doi:10.1300/J516v04n01_02

Bimber, B. (2000). The gender gap on the Internet. *Social Science Quarterly*, *81*, 868–876.

Bromberg, H. (1996). Are MUDs communities? In R. Shields (Ed.), *Cultures of the Internet: Virtual spaces, real histories, living bodies* (pp. 143-152). Thousand Oaks, CA: Sage.

Bromseth, J. C. H. (2001). Constructions of and negotiations on interaction norms and gender on electronic discussion lists in Norway. *NORA*, *9*(2), 80–88.

Carpini, M. X. D. (2000). Gen.com: Youth, civic engagement, and the new information environment. *Political Communication*, *17*, 341–349. doi:10.1080/10584600050178942

Castells, M. (1996). *The rise of the network society*. Cambridge, MA: Blackwell.

Crystal, D. (2001). *Language and the Internet*. Cambridge, UK: Cambridge University Press.

Cutler, R. H. (1996). Technologies, relations, and selves. In L. Strate, R. Jacobson, & S. B. Gibson (Eds.), *Communication and cyberspace* (pp. 317-334). Creskill, NJ: Hampton Press.

Dahlberg, L. (2001). The Internet and democratic discourse. *Information Communication and Society*, *4*(4), 615–633. doi:10.1080/13691180110097030

Dahlgreen, P. (2000). The Internet and the democratization of civic culture. *Political Communication*, *17*, 335–340. doi:10.1080/10584600050178933

Deuze, M. (2006). Participation, remediation, bricolage: Considering principal components of a digital culture. *The Information Society*, *22*, 63–75. doi:10.1080/01972240600567170

Dibbell, J. (1993). *A rape in cyberspace, or, how an evil clown, a Haitian trickster spirit, two wizards, and a cast of dozens turned a database into a society*. Retrieved from http://edie.cprost.sfu.ca/253/readings/village-voice

DiMaggio, P., Hargittai, E., Neuman, R. W., & Robinson, J. P. (2001). Social implications of the Internet. *Annual Review of Sociology*, *27*, 307–336. doi:10.1146/annurev.soc.27.1.307

Dutton, W. H. (1996). Network rules of order: Regulating speech in public electronic fora. *Media Culture & Society*, *18*, 269–290. doi:10.1177/016344396018002006

Endres, D., & Warnick, B. (2004). Text-based interactivity in candidate campaign Web sites: A case study from the 2002 elections. *Western Journal of Communication*, *68*(3), 322–342.

Foster, D. (1997). Community and identity in the electronic village. In D. Porter (Ed.), *Internet culture* (pp. 23-37). New York: Routledge.

Foucault, M. (1995). *Discipline and punish: The birth of the prison* (2nd ed., A. Sheridan, Trans.). New York: Vintage Books.

Giddens, A. (1984). *The constitution of society: Outline of the theory of structuration*. Berkeley, CA: The University of California Press.

Gitlin, T. (1998). Public sphere or public sphericules? In T. Liebes & J. Curran (Eds.), *Media, ritual and identity*. London: Routledge Press.

Herring, S. C. (1996). Posting in a different voice: Gender and ethics in computer-mediated communication. In C. Ess (Ed.), *Philosophical perspectives on computer-mediated communication* (pp. 115-145). Albany: SUNY Press.

Herring, S. C., & Paolillo, J. C. (2006). Gender and genre variation in Weblogs. *Journal of Sociolinguistics*, *10*(4), 439–459. doi:10.1111/j.1467-9841.2006.00287.x

Hurwirz, R. (1999). Who needs politics? Who needs people? The ironies of democracy in cyberspace. *Contemporary Sociology*, *28*(6), 655–661. doi:10.2307/2655536

Janack, J. A. (2006). Mediated citizenship and digital discipline: A rhetoric of control in a campaign Blog. *Social Semiotics*, *16*(2), 283–301. doi:10.1080/10350330600664862

JuicyCampus.com. (n.d.). *About us*. Available at http://www.juicycampus.com/posts/about-us

JuicyCampus.com. (n.d.). *Frequently asked questions*. Available at http://juicycampus.com/faqs.php

JuicyCampus.com. (n.d.). *Official blog announcements*. Available at http://juicycampus.blogspot.com/search/label/Announcements

JuicyCampus.com. (n.d.). *Privacy & tracking policy*. Available at http://juicycampus.com/privacy_policy.php

JuicyCampus.com. (n.d.). *Terms & conditions*. Available at http://www.juicycampus.com/posts/terms-condition

Kendall, L. (2000). 'Oh, no: I'm a nerd!': Hegemonic masculinity on an online forum. *Gender & Society*, *14*(2), 256–274. doi:10.1177/089124300014002003

Lee, H. (2005). Behavioral strategies for dealing with flaming in an online forum. *The Sociological Quarterly*, *46*, 385–403. doi:10.1111/j.1533-8525.2005.00017.x

Lessig, L. (1999). *Code and other laws of cyberspace*. New York: Basic Books.

McLaughlin, M. L., Osborne, K. K., & Smith, C. B. (1995). Standards of conduct on Usenet. In S. G. Jones (Ed.), *Cybersociety: Computer-mediated communication and community* (pp. 90-111). Thousand Oaks, CA: Sage.

McLuhan, M., & Fiore, Q. (1967). *The medium is the massage*. New York: Random House.

Mill, J. S. (1956). *On liberty*. New York: Liberal Arts Press.

Miller, L. (1995). Women and children first: Gender and the settling of the electronic frontier. In J. Brook & I. A. Boal (Eds.), *Resisting the virtual life: The culture and politics of information* (pp. 49-57). San Francisco: City Lights.

Min, S. J. (2007). Online vs. face-to-face deliberation: Effects on civic engagement. *Journal of Computer-Mediated Communication*, *12*(4). doi:10.1111/j.1083-6101.2007.00377.x

Morgan, R. (2008, March 16). A crash course in online gossip. *The New York Times*, p. ST7.

Morse, M. (1998). *Virtualities: Television, media art, and cyberculture*. Bloomington, IN: Indiana University Press.

Negroponte, N. (1995). *Being digital*. New York: Vintage Books.

Norris, P. (2001). *Digital divide? Civic engagement, information poverty and the Internet in democratic societies*. New York: Cambridge University Press.

Noveck, B. S. (2000). Paradoxical partners: Electronic communication and electronic democracy. In P. Ferdinand (Ed.), *The Internet, democracy, and democratization* (pp. 18-35). London: Frank Cass.

O'Connor, M. (2008, January 21). *Zeta Psi pledges "Love Yale sluts," Women's Center pledges to sue.* Messages posted to http://www.ivygateblog.com/2008/01/zeta-psi-pledges-love-yale-sluts-womens-center-pledges-to-sue/

Ong, W. (1982). *Orality and literacy*. London: Methuen Press.

Phillips, D. J. (1996). Defending the boundaries: Identifying and countering threats in a Usenet newsgroup. *The Information Society*, *12*, 39–62. doi:10.1080/019722496129693

Putnam, R. D. (2000). *Bowling alone: The collapse and revival of American community*. New York: Simon & Schuster.

Rheingold, H. (1993). *Virtual communities*. New York: Addison-Wesley.

Saidi, N. (2008, March 28). iReport: 'Naught Auties' battle autism with virtual interaction. Retrieved from http://www.cnn.com/2008/HEALTH/conditions/03/28/sl.autism.irpt

Schiller, H. I. (1995). The global information highway: Project for an ungovernable world. In J. Brook & I. A. Boal (Eds.), *Resisting the virtual life: The culture and politics of information* (pp. 17-33). San Francisco: City Lights.

Selnow, G. W. (1998). *Electronic whistle stops: The impact of the Internet on American politics*. Westport, CT: Praeger.

Spitzer, S. (1982). The dialectics of formal and informal control. In R. L. Abel (Ed.), *The politics of informal justice*. New York: Academic Press.

Stivale, C. J. (1997). Spam: Heteroglossia and harassment in cyberspace. In D. Porter (Ed.), *Internet culture* (pp. 133-144). New York: Routledge.

Streck, J. M. (1998). Pulling the plug on electronic town meetings: Participatory democracy and the reality of Usenet. In C. Toulouse & T. W. Luke (Eds.), *The politics of cyberspace* (pp. 18-47). New York: Routledge.

Thompsen, P. A. (1996). What's fueling the flames in cyberspace? A social influence model. In L. Strate, R. Jacobson, & S. B. Gibson (Eds.), *Communication and cyberspace* (pp. 297-315). Creskill, NJ: Hampton Press.

Turkle, S. (1995). *Life on the screen: Identity in the age of Internet*. New York: Basic Books.

White, C. S. (1997). Citizen participation and the Internet: Prospects for civic deliberation in the information age. *Social Studies*, *88*, 23–28.

Wilhelm, A. G. (2000). *Democracy in the digital age: Challenges to political life in cyberspace*. New York: Routledge.

Wilson, S. M., & Peterson, L. C. (2002). The anthropology of online communities. *Annual Review of Anthropology*, *31*, 449–467. doi:10.1146/annurev.anthro.31.040402.085436

Zappen, J. P., Gurak, L. J., & Doheny-Farina, S. (1997). Rhetoric, community, and cyberspace. *Rhetoric Review*, *15*(2), 400–419.

KEY TERMS AND DEFINITIONS

Carnival: A concept developed by Bakhtin (1984) that illustrates how people come together as a collective of equals and interact in a way that defies exogenous sociologic divisions. In the experience of carnival, there is an air of playfulness and multi-voicedness that invigorates participants' understandings of self and community.

Formal Discursive Control: Communicative currents within a social structure that are institutionally centralizing, and which impose and enforce regulations in a "top-down" manner that is typically explicit (e.g., codes of conduct). Formal control is manifested in what Bakhtin (1981) terms "centripetal" forces, which are authoritative, stabilizing, decisive, and preserving of traditions.

Informal Discursive Control: Communicative currents within a social structure that disrupt institutional authority and enable the negotiation of emergent norms in a "bottom-up" manner that is often implicit and subtle. Informal control is manifested in what Bakhtin (1981) terms "centrifugal" forces, which are insurgent, destabilizing, equivocally open-ended, and change-minded.

Online Democracy: The quasi-counterbalance of normalizing and destabilizing discursive forces; a subtle and messy negotiation of control in which centripetal and centrifugal pushes and pulls more-or-less even out.

Social Structure: A relational system that is both constrained and enabled by norms (beliefs, values, rules, and roles) that are made and remade in participants' interaction. Social structure is partly stable and partly dynamic; at once a thing and a process.

ENDNOTE

1. We use the Internet here to reference mostly the various examples of mediated networks including the World Wide Web and electronic mail, social and interactive spaces such as weblogs and wikis, and sharing media such as podcasts and social bookmarking tools rather than the technological infrastructure itself.

This work was previously published in Handbook of Research on Social Interaction Technologies and Collaboration Software; Concepts and Trends, edited by T. Dumova; R. Fiordo, pp. 404-416, copyright 2010 by Information Science Reference (an imprint of IGI Global).

Chapter 7.12
Agent Cognitive Capabilities and Orders of Social Emergence

Christopher Goldspink
Incept Labs, Australia

Robert Kay
Incept Labs, Australia,
University of Technology, Sydney, Australia

ABSTRACT

This chapter critically examines our theoretical understanding of the dialectical relationship between emergent social structures and agent behaviors. While much has been written about emergence individually as a concept, and the use of simulation methods are being increasingly applied to the exploration of social behavior, the concept of "social emergence" remains ill defined. Furthermore, there has been little theoretical treatment or practical explorations of how both the range and type of emergent structures observed may change as agents are endowed with increasingly sophisticated cognitive abilities. While we are still a very long way from being able to build artificial agents with human-like cognitive capabilities, it would be timely to revisit the extent of the challenge and to see where recent advances in our understanding of higher order cognition leave us. This chapter provides a brief recount of the theory of emergence, considers recent contributions to thinking about orders of emergence, and unpacks these in terms of implied agent characteristics. Observations are made about the implications of alternative cognitive paradigms and the position is proposed that an enactivist view provides the most logical pathway to advancing our understanding. The chapter concludes by presenting an account of reflexive and non-reflexive modes of emergence, which incorporates this view.

INTRODUCTION

Building and working with artificial societies using the methods of multi-agent social simulation serves us in several ways: 1) It allows us to operationalize social theories and to compare simulated behaviors with those observed in the real world; and 2) it allows us to build new theory by exploring the minimal mechanisms that might explain observed social behavior. Most importantly 3) it provides a unique ability to explore the interplay between levels of phenomena and to understand dynamic

DOI: 10.4018/978-1-60566-236-7.ch002

properties of systems. A great deal can and has been achieved in both these areas with even the simple methods we currently have available. However, Keith Sawyer (2003) has recently reminded us that, to date, we have worked with agents with very limited cognitive capability and that this necessarily limits the range and type of behavior which can be explored. This echoes a sentiment made a decade ago by Christiano Castelfranchi (1998a) that social simulation is not really *social* until it can provide an adequate account of the implication of feedback between macro and micro which becomes possible with higher cognitive functioning of social agents.

In many respects, developments in our capacity to simulate artificial societies have led us to confront anew a long-standing issue within social theory. This is a problem that social science conducted within traditional disciplinary boundaries has become quite adept at avoiding. Indeed it can be argued that the particular form disciplinary fragmentation takes in social science is a primary strategy for avoiding it. The problem is referred to in a number of ways depending on the disciplinary tradition. This chapter begins by revisiting this most important of problems. In terms of the challenge it poses to artificial societies it can be expressed in the following three questions:

1. What are the fundamental cognitive characteristics which distinguish human agents from animal or automaton?
2. How do these characteristics influence the range and type of behaviors agents may generate and the emergent structures which they may give rise to?
3. How can we theorize about the relationship between cognitive capability and categories of emergent form?

These questions form the focus for this chapter. We begin to address them by revisiting the contribution of alternative schools of thought to our understanding of the nature and origins of emergent structure and alternative concepts of orders of emergence. We then discuss the implications of the two competing cognitive paradigms within AI—that of cognitivism and the enactive view. Finally we turn to current research on the development of human cognition and examine its implications for anticipating different orders of emergent structure—proposing what we call reflexive and non-reflexive classes of emergence. Finally a research program for the advancement of understanding in this area is proposed.

This work has its origins in two strands of research with which the authors are currently involved. The first addresses the relationship between micro and macro levels of social behavior and organization directly. Over the past decade we have explored the characteristics of the micro-macro problem (see Chris Goldspink & Kay, 2003, 2004) in pursuit of a coherent and consistent account of the interpenetration (circular causality) between micro and macro phenomena. Our aim is to develop a theory which can provide a substantive account of fundamental social generative mechanisms. To date no such social theory exists that satisfactorily explains this dynamic.

The other strand is one author's involvement with the Centre for Research in Social Simulation and though it the European Union funded project titled Emergence in the Loop (EMIL). The aim of EMIL is to: a) provide a theoretical account of the mechanisms of normative self-regulation in a number of computer mediated communities b) specify the minimum cognitive processes agents require to behave in normative ways c) develop a simulator which can replicate the range and type of normative behavior identified by the empirical research so as to further deepen our understanding of how and under what conditions normative self-regulation is possible and the range and type of environmental factors which influence it.

A BRIEF RECOUNT OF THE THEORY OF EMERGENCE

The notion of emergence has a long history, having been invoked in a number of disciplines with varying degrees of centrality to the theoretical and methodological development of associated fields. Unfortunately the concept has largely remained opaque and ambiguous in its conceptualization, leading to the criticism that it stands as little more than a covering concept – used when no adequate account or explanation exists for some unexpected phenomena. Clayton has argued that the concept covers:

...a wide spectrum of ontological commitments. According to some the emergents are no more than patterns, with no causal powers of their own; for others they are substances in their own right... (Clayton, 2006: 14).

The origin of the concept has been attributed to George Henry Lewes who coined the term in 1875 (Ablowitz, 1939). It subsequently found wide adoption within the philosophy of science but more recently has been advanced within three distinct streams: *philosophy*, particularly of science and mind; *systems theory*, in particular complex systems; and *social science* where it has largely been referred to under the heading of the micro-macro link and/or the problem of structure and agency. Interestingly there has been relatively little cross fertilization of thinking between these streams.

The Contribution from Philosophy of Science

The philosophy of science and philosophy of mind stream is arguably the oldest – some date it back to Plato (Peterson, 2006) but the debate is widely seen as having come to focus with the British Emergentists (Eronen, 2004; Shrader, 2005; Stanford Encyclopaedia of Philosophy, 2006). This school sought to deal with the apparent qualitatively distinct properties associated with different phenomena (physical, chemical, biological, mental) in the context of the debate between mechanism and vitalism: the former being committed to Laplacian causal determinism and hence reductionism and the latter invoking 'non-physical' elements in order to explain the qualitative difference between organic and inorganic matter. This stream remains focused on explaining different properties of classes of natural phenomena and with the relationship between brains and minds (See Clayton & Davies, 2006 for a recent summary of the positions). As a consequence this has been the dominant stream within artificial intelligence. Peterson (2006: 695) summarizes the widely agreed characteristics of emergent phenomena within this stream as follows. Emergent entities:

1. Are characterized by higher-order descriptions (i.e. form a *hierarchy*).
2. Obey higher order *laws*.
3. Are characterized by *unpredictable novelty*.
4. Are *composed of* lower level entities, but lower level entities are *insufficient* to fully account for emergent entities (*irreducibility*).
5. May be capable of *top-down causation*.
6. Are characterized by *multiple realization or wild disjunction* (Fodor, 1974) (alternative micro-states may generate the same macro states).

A key concept within these discussions is that of *supervenience*: a specification of the 'loose' determinisms held to apply between levels such that '*...an entity cannot change at a higher level without also changing at a lower level*' (Sawyer, 2001: 556). Within this stream prominence of place is given to both downward and upward causation. Clayton and Davies (2006) specify downward causation as involving macro structures placing *constraint* on lower level processes hence *'Emergent entities provide the context in which local, bottom up causation takes place and is made possible'* (Peterson, 2006: 697). Davies (2006) argues that the mechanism of downward causation can usefully be considered in terms of boundaries. Novelty, he argues, may have its origin in a system being 'open'. If novel order emerges it must do so within the constraints of physics. He concludes:

...top-down talk refers not to vitalistic augmentation of known forces, but rather to the system harnessing existing forces for its own ends. The problem is to understand how this harnessing happens, not at the level of individual intermolecular interactions, but overall – as a coherent project. It appears that once a system is sufficiently complex, then new top down rules of causation emerge. (Davies 2006: 48).

For Davies then, top-down causation is associated with self-organization and may undergo qualitative transitions in form with increasing system complexity. For Davies also it is the 'openness' of some systems that 'provides room' for self-organizing process to arise, but he concludes, *'openness to the environment merely explains why there may be room for top-down causation; it tells us nothing about how that causation works.'* The devil then, is in the detail of the mechanisms specific to particular processes in particular contexts and particular phenomenal domains. Perhaps then a part of the problem with the concept is that it has been approached at too abstract a level.

The Contribution from Social Science

The micro-macro problem—the relationship between the actions of individuals and resulting social structures and the reciprocal constraint those structures place on individual agency—has long standing in social science as well as in philosophy. The problem is central to many social theories developed throughout the 19th and 20th century. Examples include: Marxian dialectical materialism (Engels, 1934) built upon by, among others, Vygotsky (1962) and Lyont'ev (1978); the social constructionism of Berger and Luckmann (1972); Gidden's structuration theory (1984); and the recent work of critical realists (Archer, 1998; Archer, Bhaskar, Ciollier, Lawson, & Norrie, 1998; Bhaskar, 1997, 1998). These alternative theories are frequently founded on differing assumptions, extending from the essentially objectivist/rationalist theory of Coleman (1994), through the critical theories of Habermas and to the radical constructivism of Luhmann (1990; 1995).

Fuchs & Hofkirchner (2005: 33) have recently suggested a four category schema for classifying social theory according to the ontological position adopted with respect to the micro-macro relationship. The majority of existing social theories, they argue, fall into one or other of two categories which they label *individualism* and *sociologism*. Neither of these 'paradigms' provides a theoretical foundation which supports exploration let alone the possibility of advancing understanding of the interplay between agency and structure, rather the problem is avoided by restricting analysis to one level or the other. A third category, *dualism*, while considering both aspects, insists on the adoption of a dichotomous stance and as a consequence does not support any understanding of the interplay between levels. Only those theories categorized as *dialectical* therefore have relevance. Even here, it is reasonable to conclude that little practical advance has been achieved, as most positions result in a straddling of bottom up and top-down arguments and/or suffer from excessively vague conceptualization. These theories also quickly break down into a dichotomy the moment an attempt is made to make them operational.

What has been largely agreed, despite the very different theoretical and often inadequate handling of this problem, is that structure and agency come together in *activity* or in *body-hood* – the specific psycho-motor state at the instant of enaction. Both Vygotsky and Giddens, for example, focus on action as the point of intersection between human agency and social structures and it is implicit in Bourdieu's *habitus* also.

The Contribution from Systems Theory

Systems language was clearly evident in the work of the early Emergentists and in a great deal of sociology and anthropology which took seriously the structure/agency problem – notably that of Margaret Mead and Gregory Bateson. However, 'systems' as a focus of systematic research arguably took form with von Bertalanffy's attempt to establish a General Systems Theory in 1950 (Bertalanffy, 1950; Bertalanffy_von, 1968). As the science of 'wholes' systems theory stands in contrast to reductionism's concern with parts. Systems theory was put forward as a counter to what was perceived as excessive reductionism dominating scientific discourse during much of the 20th century.

In the early stages of development of the theory systems tended to be modeled as 'black boxes' effectively masking the relationship between micro and macro elements. The application of the concept to social science, in particular through the development by Ernst von Glasersfeld and Heinz von Foerster (Keeney, 1987) of social cybernetics along with soft systems approaches (Checkland, 1988) provided a theoretical lens and methods useful for describing the systemic behavior of social systems. So while the aspiration of GSM to establish a general science of systems is generally regarded to have failed (Jackson, 2000), systems approaches have contributed valuable methods for the study of the interplay between levels in a social system. The Systems view of emergence was founded on:

- Holism; the whole is greater than the sum of its parts.
- A concern with *feedback both positive and negative*.
- A concern with boundaries and boundary conditions.

More recently the development of complex systems theory and its application to natural, social and cognitive phenomena has provided additional concepts upon which much current debate about emergence draws. Many of these concepts and methods have become widely used within the multi-agent modeling community (Castelfranchi, 1998b; Conte, Hegselmann, & Terna, 1997; Gilbert, 1995; Holland, 1998).

Within contemporary debate, and in contrast to the position taken by the British Emergentists who argued that irreducibility was the *exception* (Eronen, 2004), most real world systems are now argued to be non-linear (S. Kauffman, 2000; S. A. Kauffman, 1993, 1996; Stewart, 1990) and hence irreducible. It is non-linearity which contributes to these system's capacity for novelty and unpredictability through the presence of deterministic Chaos (Lorenz, 2001; Williams, 1997) and/or equifinality. Equifinality as it is known within systems theory, or the principle of 'wild disjunction' as it is known in philosophy, refers to a system where a single high level property may be realized by more than one set of micro-states which have no lawful relationship between them (Richardson, 2002a, 2002b; Sawyer, 2001). As there is no a-priori basis by which the likely micro state can be determined, such systems are irreducible and unpredictable in principle.

Observations

The concept of emergence has led to the establishment of a number of general principles which describe the relationship between micro and macro phenomena, as well as some methods and techniques for identifying and exploring it. Specifically, we can conclude that there are systems which are:

- Inherently analytically reducible (to which the concept of emergence does not apply);
- Analytically reducible in principle but difficult to reduce in practice and/or where an advance in science/knowledge is needed for reduction to be possible because the results were 'unexpected' (Chalmers, 2006) (to which the concept of 'weak' emergence can be applied);
- Not reducible in principle (to which the principle of 'strong' emergence is relevant).

We argue that all living systems and all social systems belong to the latter class. Accordingly we agree with McKelvey (1997) that a great deal of social order may be attributable to complex organization involving non-liner relations between elements. It is for this reason that simulation methods are regarded as important but only to the extent that we can construct artificial societies which are reasonable analogues of the social systems we want to understand and this implies agent architectures which are capable of generating the range of social behaviors/structures of interest. The problem here is that we still have a very rudimentary understanding of what cognitive capabilities support or are necessary for what range and types of social structures.

In the following section we draw on the limited prior attention given to this problem and attempt to clarify what is currently known. Throughout the discussion, pointers are provided to where the mechanisms being outlined have, at least in part, been incorporated into computer simulations of artificial intelligence or artificial societies.

ORDERS OF EMERGENCE

A number of authors have identified what they refer to as orders of emergence. Gilbert, for example distinguishes between first and second order emergence. First order emergence includes macro structures which arise from local interactions between agents of limited cognitive range (particles, fluids, reflex action). By contrast, second order emergence is argued to arise *'where agents recognise emergent phenomena, such as societies, clubs, formal organizations, institutions, localities and so on where the fact that you are a member or a non-member, changes the rules of interaction between you and other agents.'* (Gilbert, 2002). This reflects high order cognition on the part of the agent. In particular it reflects a range of capabilities including but not limited to the ability to distinguish class characteristics; assess 'self' for conformity with class characteristics and/or signals from other agents which suggest acceptance or belonging; the ability to change rule associations and behavior as a function of these changes. First and second order emergence then each imply qualitatively distinct mechanisms and suggest a continuum of orders of emergence linked, in biological entities at least, to cognitive capability.

In a similar vein, Castelfranchi (1998a: 27) has distinguished what he refers to as cognitive emergence. *'Cognitive emergence occurs where agents become aware, through a given 'conceptualization' of a certain 'objective' pre-cognitive (unknown and non deliberated) phenomenon that is influencing their results and outcomes, and then, indirectly, their actions.'* This approach is based on a first generation AI (Franklin, 1998) approach to conceptualizing agents: agent cognition is assumed to involve acting on beliefs desires and intentions (BDI). Thus Castelfranchi conceives of a feedback path from macro pattern to micro behavior in much the same way as Gilbert, except that here a cognitive mechanism is specified. Castelfranchi argues that this mechanism has a significant effect on emergence and indeed *'characterises the theory of social dynamics'* – that is, it gives rise to a distinct class of emergent phenomena. In this account, the representations agents have about the beliefs, desires and intentions of other agents plays a causal role in their subsequent behavior and therefore shapes the structures they participate in generating. In this same chapter Castelfranchi argues that understanding this process is fundamental to social simulation: it is where social simulation can make its greatest contribution.

These ideas are more comprehensively reflected in the five orders of emergence suggested by Ellis (2006:99-101). These are:

1. Bottom up leading to higher level generic properties (examples include the properties of gases, liquids and solids)
2. Bottom up action plus boundary conditions leading to higher level structures (e.g. convection cells, sand piles, cellular automata)
3. Bottom up action leading to feedback and control at various levels leading to meaningful top down action - teleonomy (e.g. living cells, multi-cellular organisms with 'instinctive' – phylogenetically determined reactive capability)
4. as per 3 but with the addition of explicit goals related to memory, influence by specific events in the individuals history (i.e. learning)
5. In addition to 4 some goals are explicitly expressed in language (humans).

Ellis's framework makes clear that the range and type of emergence possible in a system depends fundamentally on the range and class of behavior agents are able to generate and that this varies depending on the properties of the agent.

If we consider Ellis' category one emergence, it is apparent that particles have fixed properties and are able to enter into a limited range of interactions (specified by physical laws) based on those properties. Swarms of particles can nevertheless demonstrate some rudimentary self-organization and hence emergence (Kennedy & Eberhart, 2001). Physics has furnished good accounts of many specific examples (Gell-Mann, 1995) but they have limited implication for our understanding of social behavior.

Category two has also recently been well explored as it is the focus of complexity theorists. Examples include the work of Per Bak (1996) on sand piles and earthquakes, Lorenz (2001) on weather systems and Prigogine (1997; 1985) on far from equilibrium systems. Many so called social simulations also belonging here– specifically those which incorporate agents which have fixed behaviors and no capacity for learning (individual or social). These include classic simulations based on swarms (Boids) and/or involving fixed decision criteria or rules such Schelling's segregation model, the cooperation models of Axelrod (1984) or the Sugarscape models of Epstein and Axtell (1996). Some may argue that these models involve agents with goals and therefore represent examples of fourth order emergence. The transition between 3rd order and fourth, as will be argued below, involves a move to agent autonomy that is missing in these models: their goals are designed in and not a result of their own operation it is for this reason that we argue they belong to order two.

It is significant that Ellis provides primarily biological examples for his category three order of emergence. The paradigmatic biological entity which illustrates the processes of reciprocal micro-macro causality and for which we have an excellent description which has been made operational both in vitro and in silico (see for example McMullin & Grob, 2001; F. Varela, Maturana, & Uribe, 1974) is the cell. While the mechanisms of autocatalysis and the metabolic pathways of cell self-production are well known, well documented and closely studied, the most concise articulation of the fundamental processes involved come with the theory of autopoiesis developed by the theoretical biologists Humberto Maturana and Francisco Varela (H. Maturana & Varela, 1980; H. R. Maturana & Varela, 1992; F. Varela, 1979; F. Varela et al., 1974). Unfortunately this account is not widely appreciated even within biology itself[1]. Varela (1997: 78) states:

Autopoiesis is a prime example of a ...dialectics between the local component levels and the global whole, linked together in reciprocal relation through the requirement of constitution of an entity that self-separates from its background.

The theory of autopoiesis provides a foundation for understanding other emergent processes, particularly those associated with biological entities. The originating authors themselves extended it to cover multi-cellular entities and to provide a more general theory of cognition. Others have gone so far as to argue that it furnishes a theory of society and/or organization (Niklas Luhmann, 1995; von_Krogh & Roos, 1995; Zeleny, 1991) although this remains controversial (Bednarz, 1988; Mingers, 2002, 2004) and we specifically reject it as incompatible with the original concept and as unnecessary (Goldspink, 2000; Kay, 1999).

Unlike the self-organizing processes which characterize the second order, the defining characteristic of biological self-organization is the attainment of 'strong autonomy' (Rocha, 1998). While Ellis does not say so directly, it would appear that it is the advent of a self-referential operational closure which demarcates third and higher orders of emergence from the lower orders.

Maturana and Varela argue that cognition is associated with this operational closure or autonomy. Autonomy is used here to refer to a *constitutive* process rather than as a *categorical* distinction and cognition is defined as the range of behaviors the agents can generate to remain *viable* or *to retain its identity* as a self-constituting agent (Froesea, Virgo, & Izquierdo, 2007; Thompson & Varela, 2001). For those immersed in symbolic AI it may come as a surprise that a biological cell may thus be described as a cognitive entity. This theme will be developed further in a following section as it is central to the idea of enactive cognition finding increasing uptake within second generation AI, artificial life and robotics (Barandiaran, 2005; Di Paolo & Lizuka, 2007; Di Paolo, Rohde, & De Jaegher, 2007; Moreno & Etxeberria, 1995; Moreno, Umerez, & Ibanes, 1997).

In his third order category Ellis includes a range of capabilities of biological entities up to and including 'instinctive' action. These suggest that this category would pertain to single and multi-cellular organisms including those with a central nervous system. It may be that this order is too broadly cast. Multi-cellularity is arguably another threshold point as differentiated aggregates of cells display greater capacity to respond to their environment, even where they do not possess a central nervous system, than do individual cells. Furthermore those with a central nervous system enjoy even greater behavioral flexibility. As a consequence each probably originates a distinct macro phenomenology different from that of the cells that constitute them (H. R. Maturana & Varela, 1992).

The primary point of distinction between order three and order four would appear to be between (phylogenetically) fixed individual characteristics and a capacity for an individual agent to have goals and to learn. The mechanisms by which these characteristics are acquired and fixed at the level of individuals (sexual transmission and natural selection) are ignored by Ellis or seen as unimportant from the perspective of emergence. This is reasonable if our concern is with social behavior which manifests over relatively short time cycles in geological terms. When does a capacity to adjust structure in response to an environment as implied by the characteristics of Ellis' third order become the learning ability associated with the forth order?

Ellis explicitly demarcates the goal directedness of the fourth order from apparent goals implied in the teleonomic operation of living things implicit in the third. We must therefore assume he means active goal-setting: the exercise of what we commonly refer to as agency or free will. Agency results from the vastly expanded behavioral plasticity available when an organism develops an advanced nervous system. Also, to learn an agent must have some form of memory. Memory too is generally associated with the existence of a central nervous system and is often seen as involving stored representations. But the idea of 'representations' is highly problematic from a biological point of view. What is it that is represented and how? We consider this problem in the next section.

Ellis would seem to be pointing to a category here which deals with non-human animals but the transition points are not well defined from the perspective of mechanisms of emergence. Learning in animals can stretch from simple operant conditioning to complex evaluative processes involving logical reflexion. Different stages along this continuum would appear to support significantly different forms of emergent structure. Ellis makes no distinction, for example, between individual and social learning.

Ellis marks his final transition from category four to category five by moving from simple learning capability to the capacity for language. Animals such as apes have rudimentary language ability – are they included in here or is this category the human catch-all category? Unfortunately the more closely we look at the jump between fourth and fifth order the more it resembles an abyss.

There has been a considerable research effort directed at understanding the origins and developmental phases associated with the attainment of the distinctive human cognitive capabilities. These are the capabilities which seem to relate to the transition between Ellis' category four and five orders of emergence. Much of this has drawn on comparative neurology, and sociological and psychological study of non-human animals, in particular apes. Insights are available also from developmental psychology and neurology directed at understanding human ontogeny: the phases of development from infant to adult. Note that these may overlap as phylogenetically determined capabilities characteristic of some animals may correspond to early stages of human ontogenetic development. This corpus offers those of us involved with AI two opportunities a) a capacity to aim to better stage the development of agent specifications - aiming to provide a reasonable model for simple intelligence before the more complex and b) a capacity, even before we can effectively model or simulate more advanced intelligence, to theorize about the implications it may have for emergence of social structure.

Some work has already been undertaken in this area, most notably in the area of robotics rather than computer simulation of social phenomena (although robots can be regarded as physical simulations and multi-agent software simulations as simulated robotics). Of particular note here is the work of Dautenhahn (2001; 2002), Bryson (2007; n.d) and Steels (1997; 2005; 1999) in the area of language.

Gardenfors (2006) identifies the following as needing to be explained (presented in order of their apparent evolution).

- Sensations
- Attention
- Emotions
- Memory
- Thought and imagination
- Planning
- Self-consciousness/theory of mind
- Free-will
- Language

These are present to varying degrees in different organisms and develop at different stages in humans as they develop from infancy to adulthood. The degree of interrelatedness is not, however, straight forward. Apes for example demonstrate self-awareness and theory of mind but do both without language whereas in humans language appears to play a significant role in both. For the time being then too little is known about these transitions.

It is perhaps in understanding these transitions that we find the greatest challenges for advancing artificial societies and it is here that we find philosophy may have dealt us an unhelpful turn. The advent of the central nervous system and the observation that cognitive function is correlated with brain size has contributed to a distinctive account of the function of brain and its relationship to mind (Johnson, 1990; Lakoff & Johnson, 1999). In this convention, mind and hence cognition has been argued to originate in brains and to involve symbol manipulation. As we consider the literature on what makes human cognition distinctive, we need to be mindful of the effect of this and alternative paradigms. What are these alternatives and what difference do they make to our understanding of orders of emergence in general and social emergence in particular?

TWO PARADIGMS: TWO POSSIBLE APPROACHES

Within AI there are two alternative and some argue antithetical paradigms of cognition – symbolic and connectionist. Symbolic AI assumes that it is possible to model every general intelligence using a suitable symbol system and that intelligence involves symbol manipulation (Franklin, 1998).

In their book *The Embodied Mind*, Varela & Rosch (1992) state:

The central intuition ... is that intelligence—human intelligence included—so resembles computation in its essential characteristics that cognition can actually be defined as computations of symbolic representations (F. Varela, Thompson, & Rosch, 1992: 40).

The symbolic approach inevitably constructs a duality. The environment is experienced as a facticity and acted upon directly, but is also conceived and symbolically represented in the mind. Mind and behaviour are linked as hypothesis and experiment. The mind looks for patterns in representations and tests the degree to which these accord with the outside world.

More recently, this tradition has been challenged. The advent of complexity theory has given greater impetus to connectionist models of mind such as neural networks. Here emergent structure or pattern arises from massively interconnected webs of active agents. Applied to the brain, Varela et al state:

The brain is thus a highly cooperative system: the dense interconnections amongst its components entail that eventually everything going on will be a function of what all the other components are doing (1992: 94).

It is important to note that no symbols are invoked or required by this model. Meaning is embodied in fine-grained structure and pattern throughout the network. Unlike symbolic systems, connectionist approaches can derive pattern and meaning by mapping a referent situation in many different (and context dependent) ways. Meaning in connectionist models is embodied by the overall state of the system in its context. It is implicit in the overall 'performance in some domainre'. Herein lays its major problem from the perspective of multi-agent simulation. In connectionist models the micro-states which support a given macro state is opaque – relatively inaccessible to an observer and difficult to interpret – indeed, there will often be several or many micro configurations compatible with a given macro-state (Richardson, 2002b). Several attempts have been made to address this problem. The first was to consider hybrid systems in an attempt to gain the advantage of each (Khosla & Dillon, 1998). The second has been to find a middle ground. This is apparent for example in Gardenfors' theory of conceptual spaces (Gardenfors, 2004). At the same time the practical value of connectionist systems – their capacity to categorize contexts or situations in a non-brittle way– has been seen as a significant advantage in robotics (Brooks, 1991).

Back in 1992 Varela et al noted that:

...an important and pervasive shift is beginning to take place in cognitive science under the very influence of its own research. This shift requires that we move away from the idea of the world as independent and extrinsic to the idea of a world as inseparable from the structure of [mental] processes of self modification. This change in stance does not express a mere philosophical preference; it reflects the necessity of understanding cognitive systems not on the basis of their input and output relationships but by their operational closure (1992: 139).

They go on to argue that connectionist approaches, while an advance on cognitivism are not consistent with an approach which views biological agents as operationally closed in that '*...the results of its processes are those processes themselves*' (1992, p. 139). They assert:

Such systems do not operate by representation. Instead of representing an independent world, they enact a world as a domain of distinctions that is inseparable from the structure embodied by the cognitive system (1992: 140).

These authors argue for an approach of cognition as 'enaction', an intertwining of experience and conceptualization which results from the structural coupling of an autonomous organism and its environment. Autopoietic theory provided a concrete and operationalizable account of the intertwining of micro and macro at the level of the cell. The enactive theory of cognition goes some way towards providing a basis for understanding this process in multi-cellular animals. Enactive cognition is currently enjoying significant attention and hence conceptual extension as well as experimental grounding in the field of robotics (see for example De Jaegher & Di Paolo, 2007; Di Paolo et al., 2007; Metta, Vernon, & Sandini, 2005). The attraction here is pragmatic – it helps to address longstanding problems within robotics, in particular the problem of symbol grounding (Harnad, 1990). To date it has seen little uptake within social simulation. The implications of enaction go well beyond pragmatics however.

The enactive turn in AI has as an explicit target a resolution of the micro-macro problem. While symbolic AI assumes the existence of an objective independent world and a mental model with some correspondence to the real world, enaction dispenses with this dichotomy. As an autonomous entity, the cognizing agent is concerned only to maintain its viability in an environment. It adjusts its structure to accommodate perturbation from the environment (which includes other cognitive agents) in order to do so. Advanced nervous systems and capabilities such as language simply extend the requisite variety available to the agent extending the range and type of environmental perturbations it can survive. As agents and environments structurally couple they co-determine one another to 'satisfice' the conditions for mutual viability. From this perspective, the importance of environment recedes from determinant to constraint. Intelligence moves from problem solving capacity to flexibility to enter into and engage with a shared world. However, McGee (2005a; 2005b) has recently argued that despite its promise, enactive cognition is not yet sufficiently well articulated to 'speak of hypothetical mechanisms'. The limiting factor here would appear to be as much one of insufficient application as theoretical difficulty. In the final section we attempt a definition of two classes of emergence which we call reflexive and non-reflexive. These draw on the enactive paradigm and attempt to provide a concrete specification of the mechanisms which underlay each.

TOWARDS AN ENACTIVE SPECIFICATION OF ASPECTS OF COGNITION AND THEIR ASSOCIATED ORDERS OF EMERGENCE

How then do we advance our understanding of the effect of different cognitive capability on orders of emergence? A useful strategy may be to simplify the problem. By way of a mental exercise we will take simple extremes and recast the problem in terms of an enactive view. From an enactive position the critical phases of cognitive development appear to be as follows:

- Autonomy (operational closure)
- Structural Coupling
- Reflexivity/self consciousness
- Language/consensual domains

All living beings (from amoeba to humans) are distinguished by autonomy and as autonomous entities they necessarily enter into structural coupling with their environment. We take this as one pole of the continuum and identify the class of emergence which it can support as non-reflexive. This is the enactive equivalent to social order which is a product of emergence *without* the feedback loop from macro to micro which Castelfranchi (1998a) refers to as immergence. The mechanisms are, however, more sophisticated than are currently modeled in Artificial Societies as they involve autonomous agents – these are essentially what Ellis refers to in his category four – i.e. biological agents which can change their structure (learn) in response to environmental perturbation. It should be feasible to simulate this type of agent with current technology or at least to achieve a close proxy although we have not yet managed to do so beyond the most basic chemical system analogues of cell autopoiesis. If we were to achieve it how might we describe the system operation?

Non-Reflexive Social Emergence

Non-reflexive emergence arises from the mechanism of structural coupling between operationally closed (autonomous) agents. Structural coupling will arise between such agents which have sufficient cognitive range (behavioral repertoire) when they are located in a common environment. Assuming that their phylogeny and ontogeny is such that they can co-exist, through the process of recurrent mutual perturbation, each will adjust its structure so as to accommodate the other – their structures will become mutually aligned or structurally coupled. This process has been approximated in a simulation by Stoica-Kluver and Kluver (2006).

An observer may notice regularities in the resulting patterns of interaction and these may be labeled as 'norms' for example although Castelfranchi would refer to them as social functions as they 'work without being understood'. These patterns represent mutual accommodations, and an observer might attribute to those accommodations some social 'function'. The accommodations an agent makes to remain viable in one domain of interaction will need to be reconciled (within its body-hood) against accommodations being made (simultaneously) as it also participates with different agents in other domain/s in which it is simultaneously participating – agency and structure converge and are both instantiated at the point of enaction. The accommodations made will be those that allow the agent to remain viable and to maintain its organization (i.e. which 'satisfice' the constraints and allow conservation of identity) based on its unique ontogeny (structure resulting from its history of interactions in a variety of domains including the current one).

Here the emergent structure can be seen to be 'in' (i.e. internalized within its own cognitive structure) each agent to the extent that each has had to make structural adjustments to operate in the shared domain. The structural adjustment each needs to make in order to persist will, however, be unique. In other words the structural accommodations each has made in order to contribute to the patterns, will *not* be the same. The structure, then, can also be regarded as 'in' the network, as it is the intersection of these disparate agent structures which gives it its particular form at a particular time. As any agent could leave the domain and have minimal effect on the resulting pattern, each agent's 'contribution' will be relatively small. The pattern can be thought about as like a hologram. The whole is in every part (agent) such that removal of parts (agents) reduces the resolution (coherence) but does not constitute loss of overall pattern. However, the loss of too many components may reduce the coupling to the point that the existing pattern de-coheres and transforms into something different. Each agent contributes to the pattern formation, so it is conceivable that the pattern will only be realized with some critical minimal number of agents present which have had a sufficient mutual history to have aligned their structures.

In natural systems, the local level interactions between agents are constrained by the existing structures of the agents and the state of their environment. With biological agents the system is open in that any emergent structure is possible as long as it remains consistent with the biological viability of the agents as living (autopoietic) entities. This biological constraint includes limits to environmental conditions conducive to life (i.e. not too hot or too cold, the need for energy, limitations to sensory channels, channel bandwidths and affective/psychomotor response capabilities etc). These are primarily a product of phylogeny (the evolutionary history of the organism at the level of the species) rather than ontogeny (the history of development at the level of the individual), and are therefore slow to change and not under the control of the emergent social system. As a consequence the basic dimensionality of the phase space of the social system does not change over the time frame of interest for understanding social systems. The dimensionality of the phase space is determined by the dimensions of variability possible by individuals – i.e. the plasticity of their nervous systems and by higher order dimensions which emerge from their interaction.

Reflexive Social Emergence

What changes if we now jump to the opposite pole on our hypothetical continuum? Here we attempt to outline the difference made by agents which are self aware and which can interact in language.

Biological agent's sensory surfaces are selected to be sensitive to difference in dimension of their world relevant to their survival and their cognitive apparatus is thus geared to make distinctions relevant to maintaining their viability in past environments. Once cognitive complexity exceeds a critical threshold (Gardenfors, 2006) these distinctions can be represented in language. Maturana and Varela (1980) describe language as involving the co-ordination of the co-ordination of actions – i.e. language provides a meta process by which agents orientate themselves within a world. Structural coupling can arise purely through behavioral coordination of action (as discussed above), but it can also take place in and through linguistic exchange – the mutual co-ordination of co-ordination of behaviors. This gives rise to a consensual linguistic domain characterized by a more or less shared lexicon. This process has been simulated using both shared referents and simple structural coupling in the absence of objective referents (Gong, Ke, Minett, & Wang, 2004; Hutchins & Hazlehurst, 1995; Steels, 1997, 1998; Steels, 2005; Steels & Kaplan, 1998; Steels & Kaplan, 1999), as has the emergence of a rudimentary grammar (Howell & Becker, n.d; Vogt, n.d).

The advent of language radically increases the behavioral plasticity of agents and has significant implications for the dimensionality of the phase space and of the resulting higher order structures it can generate and support. This is because language makes possible the emergence of domains of interaction which can themselves become the target for further linguistic distinction and hence new domains. In other words, language allows the agent to make distinctions on prior distinctions (to language about its prior language or to build further abstractions on prior abstractions). This supports the possibility of infinite recursion and infinite branching (there are no doubt biological constraints on this in humans). This is an intrinsically social process. Furthermore, a capacity to distinguish (label or categorize) processes supports reification and this simplifies the cognitive handling of processual phenomena and allows the resulting reifications to be treated by the agent in the same manner as material objects.

These capabilities greatly expand the structural flexibility of the agents: they can now invent shared epistemic worlds. The phase space of agent cognition is now based primarily on constraints of ontogeny rather than phylogeny and is hence under the influence of the agent/s.

Language makes possible a further major qualitative difference in natural and human social emergence. Humans (and possibly some other primates, cetaceans and elephants)[2] have developed sufficient cognitive capacity to become self-aware and as such exhibit reflexive behavior. This occurs when the agent is capable of distinguishing 'self' and 'other' i.e. the agent can entertain the notion of 'I' as a concept and treat that concept as an object. The advent of this capacity for reflexive identity also supposes the existence of a range of conceptual operators that act on identity – identity construction and maintenance becomes a part of the agent's world creation. Exploration of this process is proceeding under the title of Neurophenomenology (Rudrauf, Lutz, Cosmelli, Lachaux, & Le Van Quyen, 2003; Thompson & Varela, 2001).

In other words, agents can now notice the patterns that arise as they interact with others and distinguish those patterns in language. Such a mechanism would be the enactive equivalent to Castelfranchi's (1998a) Cognitive Emergence. Here a reflexive agent can notice an emergent pattern of social behavior and explicitly denote it as a 'norm' for example. While this denotation may be idiosyncratic (i.e. based on the necessarily limited perception of the individual agent), the agent can nonetheless act on the basis of this denotation. Once distinguished and reified within a domain, agents can decide (on the basis of rational as well as value based or emotional criteria) how to respond – they can choose to ignore the norm or to behave in ways they believe will limit the reoccurrence of the behaviors that are outside the agreed/shared patterns of the group. Once a pattern has been distinguished in language it can make the transition to a rule: a formally stated, linguistically explicit requirement with stated conditionals and possible resources to maintain it. This suggests that an agent can form hypotheses about the relationship between a macro structural aspect of the social system in which it is a participant and then act on that hypothesis, potentially changing the structure which it participates in generating. This gives rise to a feedback path between macro and micro phenomena that is not present in any other natural phenomena.

Consistent with Castelfranchi's claim, agents possessing this cognitive complexity form the components of a social system which would exhibit a distinct class of emergence. From the emergent perspective this is argued on the basis that reflexive agents will display qualitatively different behaviors from non-reflexive through the ability to modify their own sets of behavioral change triggers. For agents which have linguistic capability, the two processes (linguistic and non-linguistic) intertwine or even become one and would not be able to be empirically disentangled. Their respective influences will only be able to be examined through simulations or by comparing agents with different (phylogenetic) capabilities (i.e. different species) and this sets some interesting methodological challenges.

The Role of the Observer

Another significant implication of the relationships described above is the observer dependant nature of emergence in social systems. In human social systems every agent is an observer and it is the process of observation and the associated distinction-making which is the reflexive engine of emergence. In natural systems, the agents of the system are unable to observe and distinguish linguistically or to distinguish external structures as separate from themselves hence the process of observation has no impact on the dynamics of the system or the way in which emergence takes place. To some extent we can see an acknowledgement of this effect in methodological discussions within ethnography, action research (Carr & Kemmis, 1986) and grounded theory (Corbin & Strauss, 1990). In each of these methodologies the impact of the researcher on the social system under study is acknowledged and seen as part of the process. The view being proposed here is that any agent that becomes a part of the system being observed has the potential to influence that system. An agent can become a part of the system simply by being itself observed or conceived as observing by those who constitute the system. In other words, the effect of the entry of a new observing agent is to change the system boundary so as to include that agent. The boundary is itself an entity of ambiguous status – it is an epistemic distinction albeit one based on potentially ontological markers. In most social theory, positing the observer as a necessary part of the system removes any ontological privilege and threatens either infinite recursion or paradox. Based on the position advocated here, a degree of both may well be fundamental to the type of system being described (Hofstadter, 2007).

Implications for Emergence

Complex systems of all kinds demonstrate a capacity to give rise to complex macro patterns as a result of local interactions between agents in highly connected webs. This local interaction can often be characterized as involving some signaling between agents. As we have seen above, in human social systems, this signaling behavior takes on a qualitatively different form. This has three key implications for our understanding of emergence that to date have largely been ignored by the literature.

1. **Social systems will display an increased range of emergent possibilities:** The reflexive nature of social systems implies that a greater range of emergent structures should be expected and they will be subject to more rapid change.
2. **Dimensions of phase space are non-constant:** As the agents in the social system define and redefine the phase space as a function of their reflexive distinctions they will create and change the dimensions of that phase space, in order to support their own viability in that space.
3. **Phase space comes under control of the system and is dynamic:** The dimensionality of the phase space associated with ontogenetic parameters is derived through the self-distinguishing characteristics of the agents and can be influenced by their situated behavior. Significantly the feedback path between macro and micro would add significant non-linearity to the system and it becomes important to identify and explain order producing mechanisms within the network.

CONCLUSION AND FUTURE DIRECTIONS

In this chapter we have attempted to provide an operational specification of the gap implicit in Ellis' fourth and fifth order emergence. In a sense we have demarcated the extremes using the lens of enactive cognition. Enactive cognition was selected as it provides a theoretical underpinning which avoids the dualism inherent in symbolic systems and the confusion of fundamental processes which results from this. It has been argued to be both theoretically better capable of capturing the essential mechanisms and of providing a practical way of avoiding the now well documented pitfalls of symbolic AI. From this perspective the first challenge that must be addressed to advance social simulation is to achieve some form or proxy of constitutive autonomy in our multi-agent models. Significant work is currently underway on this problem in robotics but there have been few systematic attempts within social simulation.

Once this has been achieved we then need to model autonomous closure in linguistic systems. We would seem to be a very long way from this at present. It may be possible however to achieve this first in some abstract domain – simulating perhaps Luhmann's self-referential systems of communicative acts. This is probably unlikely however.

In our sketching out the extremes many questions remain about what might lay in the middle. This middle includes very significant phases of human cognitive development – including theory of mind and narrative intelligence. There can be no doubt that these will support qualitatively distinct classes of emergent social phenomena. There is evidence from the study of apes that forms of these cognitive capabilities do not require language. These may be much more accessible to our still limited capacity to simulate than the human equivalents which appear to intertwine with linguistic capability. We probably have much to learn then from the study of primate communities and from research into cognition in species other than humans. At present these attract considerably less attention within the social simulation community and perhaps this is a mistake. We have learned a lot from ants – how much more from apes? Robotics also appears well equipped to incorporate the insights coming from situated, embodied and enactive cognition. It s more difficult to see how embodied proxies may be incorporated into multi-agent simulations but no doubt there are ways. Such systems will doubtless need to be able to bootstrap some level of operational closure and it will be behavior within the self-determining boundary that – free from the inevitable teleological hand of the designer can reveal insights into how we humans do what we seem to do so effortlessly – construct social worlds in which we can live viable and interesting lives.

REFERENCES

Ablowitz, R. (1939). The Theory of emergence. *Philosophy of Science*, *6*(1), 16. doi:10.1086/286529

Archer, M. (1998). Realism in the social sciences. In M. Archer, R. Bhaskar, A. Collier, T. Lawson & A. Norrie (Eds.), *Critical realism: Essential readings*. London: Routledge.

Archer, M., Bhaskar, R., Ciollier, A., Lawson, T., & Norrie, A. (1998). *Critical Realism: Essential Readings*. London: Routledge.

Axelrod, R. (1984). *The evolution of cooperation*. New York: Basic Books.

Bak, P. (1996). *How nature works: The science of self-organized criticality*. New York: Copurnicus.

Barandiaran, X. (2005). Behavioral adaptive autonomy. A milestone on the ALife roue to AI? San-sebastian, Spain: Department of Logic and Philosophy of Science, University of the Basque Country.

Bednarz, J. (1988). Autopoesis: The organizational closure of social systems. *Systems Research*, *5*(1), 57–64.

Berger, P. L., & Luckman, T. (1972). *The social construction of reality*. Penguin.

Bertalanffy, L. v. (1950). An outline of general systems theory. *British Journal for the Philosophy of Science, 1*(2). Bertalanffy_von, L. (1968). *General systems theory*. New York: Braziller.

Bhaskar, R. (1997). A realist theory of science. London: Verso.

Bhaskar, R. (1998). *The possibility of naturalism*. London: Routledge.

Brooks, R. A. (1991). Intelligence without representation. *Intelligence without . Reason*, (47): 569–595.

Bryson, J. J. (2007). Embodiment vs. memetics. Bath: Artificial Models of Natural Intelligence, University of Bath.

Bryson, J. J. (n.d). Representational requirements for evolving cultural evolution, *Interdiciplines*.

Carr, W., & Kemmis, S. (1986). *Becoming critical: Knowing through action research*. Deakin University.

Castelfranchi, C. (1998a). Simulating with cognitive agents: The importance of cognitive emergence. In J. S. Sichman, R. Conte & N. Gilbert (Eds.), *Multi-agent systems and agent based simulation*. Berlin: Springer.

Castelfranchi, C. (1998b). Simulating with cognitive agents: The importance of cognitive emergence. In J. S. Sichman, R. Conte & N. Gilbert (Eds.), *Lecture Notes in Artificial Intelligence*. Berlin: Springer Verlag.

Chalmers, D. J. (2006). Strong and weak emergence. Canberra: Research School of Social Sciences, Australian National University.

Checkland, P. (1988). *Systems thinking systems practice*. G.B.: John Wiley.

Clayton, P. (2006). Conceptual foundations of emergence theory. In P. Clayton & P. Davies (Eds.), *The re-emergence of emergence: The emergentist hypothesis from science to religion*. Oxford: Oxford University Press.

Clayton, P., & Davies, P. (2006). *The re-emergence of emergence: The emergentist hypothesis from science to religion*. Oxford: Oxford University Press.

Coleman, J. S. (1994). *Foundations of social theory*. Cambridge: Belknap.

Conte, R., Hegselmann, R., & Terna, P. (1997). *Simulating social phenomena*. Berlin: Springer.

Corbin, J. M., & Strauss, A. (1990). Grounded theory research: Procedures, canons, and evaluative criteria. *Qualitative Sociology, 13*(1), 18. doi:10.1007/BF00988593

Dautenhahn, K. (2001). The narrative intelligence hypothesis: In search of the transactional format of narratives in humans and other social animals. In *Cognitive Technology*, (pp. 248-266). Hiedelberg, Germany: Springer-Verlag.

Dautenhahn, K. (2002). The origins of narrative. *International Journal of Cognitive Technology, 1*(1), 97–123. doi:10.1075/ijct.1.1.07dau

Davies, P. (2006). The physics of downward causation. In P. Clayton & P. Davies (Eds.), *The Re-Emergence of Emergence: The Emergentist Hypothesis from Science to Religion*. Oxford: Oxford University Press.

De Jaegher, H., & Di Paolo, E. A. (2007). (forthcoming). Participatory Sense-making: An enactive approach to Social Cognition. *Phenomenology and thec ognitive . The Sciences*.

Di Paolo, E. A., & Lizuka, H. (2007). How (not) to Model Autonomous Behaviour. *Biosystems*.

Di Paolo, E. A., Rohde, M., & De Jaegher, H. (2007). Horizons for The Enactive Mind: Values, Social Interaction and Play. In J. Stewart, O. Gapenne & E. A. Di Paolo (Eds.), *Enaction: Towards a New Paradigm for Cognitive Science*. Cambridge MA: MIT Press.

Ellis, G. F. R. (2006). On the Nature of Emergent Reality. In P. Clayton & P. Davies (Eds.), *The Re-Emergence of Emergence: The Emergentist Hypothesis from Science to Religion*. Oxford: Oxford University Press.

Engels, F. (1934). *Dialectics of Nature*. Moscow: Progress Publishers.

Epstein, J. M., & Axtel, R. (1996). *Growing Artificial Societies*. Cambridge, Ma.: MIT Press.

Eronen, M. (2004). *Emergence in the Philosophy of Mind.* University of Helsinki, Helsinki.

Fodor, J. A. (1974). Special; Sciences or The Disunity of Science as a Working Hypothesis. *Synthese*, *28*, 18. doi:10.1007/BF00485230

Franklin, S. (1998). *Artificial Minds*. London: MIT press.

Froesea, T., Virgo, N., & Izquierdo, E. (2007). Autonomy: a review and a reappraisal. Brighton Uk: University of Sussex.

Fuchs, C., & Hofkirchner, W. (2005). The Dialectic of Bottom-up and Top-down Emergence in Social Systems. *tripleC 1*(1), 22.

Gardenfors, P. (2004). *Conceptual Spaces*. London: The MIT Press.

Gardenfors, P. (2006). *How Homo became Sapiens: On the evolution of Thinking*. Oxford: Oxford University Press.

Gell-Mann, M. (1995). *The Quark and the Jaguar: Adventures in the simple and the complex*. Great Britain: Abacus.

Giddens, A. (1984). *The Constitution of society: Outline of the theory of structuration*. Berkeley: University of California Press.

Gilbert, N. (1995). Emergence in Social Simulation. In N. Gilbert & R. Conte (Eds.), *Artificial Societies*. London: UCL Press.

Gilbert, N. (2002). *Varieties of Emergence.* Paper presented at the Social Agents: Ecology, Exchange, and Evolution Conference Chicago.

Goldspink, C. (2000). *Social Attractors: An Examination of the Applicability of Complexity theory to Social and Organisational Analysis.* Unpublished PhD, University Western Sydney, Richmond.

Goldspink, C., & Kay, R. (2003). Organizations as Self Organizing and Sustaining Systems: A Complex and Autopoietic Systems Perspective. *International Journal of General Systems*, *32*(5), 459–474. doi:10.1080/0308107031000135017

Goldspink, C., & Kay, R. (2004). Bridging the Micro-Macro Divide: a new basis for social science. *Human Relations*, *57*(5), 597–618. doi:10.1177/0018726704044311

Gong, T., Ke, J., Minett, J. W., & Wang, W. S. (2004). A Computational Framework to Simulate the co-evolution of language and social structure.

Harnad, S. (1990). The Symbol Grounding Problem. *Physica*, *42*, 335–346.

Hofstadter, D. R. (2007). I am a Strange Loop. In: Basic Books. Holland, J. H. (1998). *Emergence: from chaos to order*. Ma.: Addison Wesley.

Howell, S. R., & Becker, S. (n.d). Modelling Language Aquisition: Grammar from the Lexicon? Hutchins, E., & Hazlehurst, B. (1995). How to invent a lexicon: the development of shared symbols. In N. Gilbert & R. Conte (Eds.), *Artificial Societies*. London: UCL Press.

Jackson, M. C. (2000). *Systems Approaches to Management*. London: Kluwer Academic.

Johnson, M. (1990). *The Body in the Mind: The Bodily Basis of Meaning, Imagination and Reason*. Chicago: The University of Chicago Press.

Kauffman, S. (2000). *Investigations*. New York: Oxford.

Kauffman, S. A. (1993). *The Origins of Order: Self Organization and Selection in Evolution*: Oxford University Press.

Kauffman, S. A. (1996). *At home in the Universe: The Search for Laws of Complexity*. London: Penguin.

Kay, R. (1999). *Towards an autopoietic perspective on knowledge and organisation*. Unpublished PhD, University of Western Sydney, Richmond.

Keeney, B. P. (1987). *Aesthetics of change*: Guilford.

Kennedy, J., & Eberhart, R. C. (2001). *Swarm Intelligence* (1 ed.). London: Academic Press.

Khosla, R., & Dillon, T. S. (1998). Welding Symbolic AI with Neural Networks and their applications. *IEEE Transactions on Evolutionary Computation*.

Lakoff, G., & Johnson, M. (1999). *Philosophy in the flesh: The embodied mind and its challenge to Western thought*. New York: Basic Books.

Leont'ev, A. N. (1978). *Activity, Consciousness and Personality*. Engelwood Cliffs: Prentice Hall.

Lorenz, E. N. (2001). *The Essence of Chaos* (4 ed.). Seattle: University of Washington Press.

Luhmann, N. (1990). *Essays on Self Reference*. New York: Columbia University Press.

Luhmann, N. (1995). *Social Systems*. Stanford: Stanford University Press.

Maturana, H., & Varela, F. (1980). *Autopoiesis and Cognition: The Realization of the Living* (Vol. 42). Boston: D. Reidel.

Maturana, H. R., & Varela, F. J. (1992). *The Tree of Knowledge: The Biological Roots of Human Understanding*. Boston: Shambhala.

McGee, K. (2005a). Enactive Cognitive Science. Part 1: Background and Research Themes. *Constructivist Foundations*, *1*(1), 15.

McGee, K. (2005b). Enactive Cognitive Science. Part 2: Methods, Insights, and Potential. *Constructivist Foundations*, *1*(2), 9.

McKelvey. (1997). Quasi-Natural Organisation Science. *Organization Science*, *8*, 351–380. doi:10.1287/orsc.8.4.351

McMullin, B., & Grob, D. (2001). Towards the Implementation of Evolving Autopoietic Artificial Agents, *6th European Conference on Artificial Life ECAL 2001*. University of Economics, Prague.

Metta, G., Vernon, D., & Sandini, G. (2005). *The Robotcup Approach to the Development of Cognition*. Paper presented at the Fifth International Workshop on Epigenetic Robotics: Modeling Cognitive Development in Robotic Systems Lund University Cognitive Studies.

Mingers, J. (2002). Are Social Systems Autopoietic? Assessing Luhmanns Social Theory. *The Sociological Review*, *50*(2). doi:10.1111/1467-954X.00367

Mingers, J. (2004). Can Social Systems be Autopoietic? Bhaskar's and Giddens' Social Theories. *Journal for the Theory of Social Behaviour*, *34*(4), 25. doi:10.1111/j.1468-5914.2004.00256.x

Moreno, A., & Etxeberria, A. (1995). Agency in natural and artificial systems. San Sabastian, Spain: Department of Logic and Philosophy of Science University of the Basque Country.

Moreno, A., Umerez, J., & Ibanes, J. (1997). Cognition and Life. *Brain and Cognition, 34*, 107–129. doi:10.1006/brcg.1997.0909

Oyama, S. (2000). *The Ontogeny of Information: Developmental Systems and Evolution*: Duke University Press.

Peterson, G. R. (2006). Species of Emergence. *Zygon, 41*(3), 22. doi:10.1111/j.1467-9744.2005.00769.x

Prigogine, I. (1997). *The End of Certainty: Time, Chaos and the New Laws of Nature*. New York: The Free Press.

Prigogine, I., & Stengers, I. (1985). *Order out of Chaos: Man's New Dialogue with Nature*: Flamingo.

Richardson, K. A. (2002a). Methodological Implications of a Complex Systems Approach to Sociality: Some further remarks. *Journal of Artificial Societies and Social Simulation, 5*(2).

Richardson, K. A. (2002b). *On the Limits of Bottom Up Computer Simulation: Towards a Non-linear Modeling Culture.* Paper presented at the 36th Hawaii International Conference on Systems Science, Hawaii.

Rocha, L. M. (1998). Selected Self-Organization: and the semiotics of evolutionary systems In S. Salthe, G. Van de Vijver & M. Delpos (Eds.), *Evolutionary Systems: Biological and Epistemological Perspectives on Selection and Self-Organization* (pp. 341-358): Kluwer Academic Publishers.

Rudrauf, D., Lutz, A., Cosmelli, D., Lachaux, J.-P., & Le Van Quyen, M. (2003). From Autopoiesis to Neurophenomenology: Francisco Varela's exploration of the biophysics of being. *Biological Research, 36*, 27–65.

Sawyer, K. R. (2001). Emergence in Sociology: Contemporary Philosophy of Mind and Some Implications for Sociology Theory. *American Journal of Sociology, 107*(3), 551–585. doi:10.1086/338780

Sawyer, K. R. (2003). Artificial Societies: Multiagent Systems and the Micro-macro Link in Sociological Theory. *Sociological Methods & Research, 31*, 38. doi:10.1177/0049124102239079

Shrader, W. E. (2005). *The Metapysics of Ontological Emergence.* University of Notre Dame.

Stanford Encyclopedia of Philosophy. (2006). Emergent Properties, *Stanford Encyclopedia of Philosophy.*

Steels, L. (1997). *Constructing and Sharing Perceptual Distinctions.* Paper presented at the European Conference on Machine Learning, Berlin.

Steels, L. (1998). *Structural coupling of cognitive memories through adaptive language games.* Paper presented at the The fifth international conference on simulation of adaptive behavior on From animals to animats 5, Univ. of Zurich, Zurich, Switzerland.

Steels, L. (2005). The emergence and evolution of linguistic structure: from lexical to grammatical communication systems. *Connection Science, 17*(3 & 4), 17.

Steels, L., & Kaplan, F. (1998). Stochasticity as a Source of Innovation in Kanguage Games. In C. Adami, R. K. Belew, H. Kitano & C. Taylor (Eds.), *Artificial Life VI.* Cambridge, MA: MIT Press.

Steels, L., & Kaplan, F. (1999). Bootstrapping grounded word Semantics. In T. Briscoe (Ed.), *Linguistic evolution through language acquisition: formal and computational models*. Cambridge, UK: Cambridge University Press.

Stewart, I. (1990). *Does God Play Dice - The New Mathematics of Chaos*: Penguin.

Stioica-Kluver, C., & Kluver, J. (2006). Interacting Neural Networks and ther Emergence of Social Structure. *Complexity, 12*(3), 11.

Thompson, E., & Varela, F. J. (2001). Radical Embodiment: neural dynamics and consciousness. *Trends in Cognitive Sciences, 5*(10), 418–425. doi:10.1016/S1364-6613(00)01750-2

Varela, F. (1979). *Principles of Biological Autonomy*. New York: Elsevier-North Holland.

Varela, F. (1997). Patterns of Life: Intertwining Identity and Cognition. *Brain and Cognition, 34*, 72–87. doi:10.1006/brcg.1997.0907

Varela, F., Maturana, H., & Uribe, R. (1974). Autopoiesis: The Organization of Living Systems, Its Characterization and a Model. *Bio Systems, 5*, 187–196. doi:10.1016/0303-2647(74)90031-8

Varela, F., Thompson, E., & Rosch, E. (1992). *The Embodied Mind*. Cambridge: MIT Press.

Vogt, P. (n.d). Group Size Effects on the Emergence of Compositional Structures in Language. Tilburg, Netherlands: Tilburg University. von_Krogh, G., & Roos, J. (1995). *Organizational Epistemology*. London: St Martins Press.

Vygotsky, L. S. (1962). *Thought and Language*. Cambridge, Mass: MIT Press.

Williams, G. P. (1997). *Chaos Theory Tamed*. Washington D.C: Joseph Henry Press.

Zeleny, M. (1991). *Autopoiesis: A Theory of Living Organization*. New York: North Holland.

ENDNOTES

1. Quite why this should be the case is not clear. It does challenge the dominant paradigm within molecular biology and may have been displaced by the apparent potential offered by genomics (Oyama, 2000). It may also be that its implications are most significant outside of the biology discipline.
2. It is important to note that we can infer the existence of threshold effects here but cannot precisely specify the critical points of complexity at which self-awareness and language becomes possible. The ability for language is of course evident in species other than humans, but the degree to which their linguistic plasticity involves or enables reflexivity in the system is a subject for further research.

This work was previously published in Handbook of Research on Agent-Based Societies: Social and Cultural Interactions, edited by G. Trajkovski; S. Collins, pp. 17-34, copyright 2009 by Information Science Reference (an imprint of IGI Global).

Chapter 7.13
Social and Distributed Cognition in Collaborative Learning Contexts

Jeffrey Mok
Miyazaki International College, Japan

ABSTRACT

Technological artifacts such as computers and mobile electronic devices have dramatically increased our learning interactions with machines. Coupled with the increasingly different forms of collaborative learning situations, our contemporary learning environments have become more complex and interconnected in today's information age. How do we understand the learning and collaborative processes in such environments? How do members receive, analyze, synthesize, and propagate information in crowded systems? How do we investigate the collaborative processes in an increasingly sophisticated learning environment? What is collaboration in the current technological age? This chapter, using the conceptual framework of distributed and social cognition, will seek to answer these questions. It will describe the current perspectives on social and distributed cognition in the context of learning, and examine how these theories can inform the processes of collaborative learning with computers. The chapter will conclude with implications to our learning environments today.

DOI: 10.4018/978-1-60566-106-3.ch020

INTRODUCTION

At the heart of educational psychology, is the search for a deeper and broader understanding on how learners acquire knowledge that is realistic and ecologically valid. The pervasiveness of, and increasing reliance on, electronic devices is challenging and transforming the way learners obtain, store and share information. Collaborative learning has also taken new levels of meaning and practice with these ubiquitous digital devices. Snapshots of typical learning situations see a learner accessing a personal digital assistant while listening to a lecture; another sees a learner sending text messages or surfing the Internet while talking to a peer. Collaborative learning is no longer content with just face-to-face group discussion confined within four walls or supported by the computer only. Contemporary collaborative learning environments are becoming more complex.

Evidently, today's collaborative learning environments are vastly different from the past and there is a need to understand them for classroom design, as well as to enrich educational psychology. How do we understand the learning processes and cognitive activity in such environments? How do learners collaborate in an ever crowded cognitive system? Is there a theoretical framework where we can begin to appreciate and study this increasingly sophisticated learning environment? How do the current perspectives on social cognition and educational psychology inform us in our understanding of this phenomenon? This chapter will attempt to answer these questions by discussing the current perspectives on social cognition, describing distributed cognition as a framework and drawing some implications for studying today's learning environments.

WHAT ARE LEARNING ENVIRONMENTS LIKE TODAY?

The continuing emergence of more sophisticated technology is radically challenging and changing the way students think and learn. The reliance on increasingly powerful computational artifacts has made technology ubiquitous in most classrooms and student life. This sophistication has also been taken to higher levels with the increasing availability of all types of digital information and the myriad of networked and integrated infrastructures. Our Internet and information age has given us tools and resources for engaging in learning that we never had before.

Take any typical learning situation in developed countries. In classrooms or outside schools, you will invariably see students using handheld electronic devices to enter data or check information. They can text message, surf the Internet and "google" what the teacher is saying in class. In study rooms, cafeteria, or homes, students engaging in learning will be seen using cell phones, laptops and other electronic devices. An example of today's (and tomorrow's) learning environments is the Technology Enabled Active Learning (TEAL) project at MIT (Dori, Belcher, Bessette, Danziger, McKinney, & Hult, 2003), where a studio-based learning session takes place with students engaging in and solving projects. The classroom scene is full of students discussing in groups, consulting their computer laptops, running tests with electronic equipment and communicating through electronic devices. The teacher roves from table to table, offering feedback and asking questions. Increasingly integral to these learning environments are collaborative activities involving synchronous (occurring at the same time) and asynchronous (not occurring at the same time) communication to mediate learning and knowledge building. We see students consulting each other in class groups, through e-mails, forums, and blog discussions. Learning projects and papers are written with feedback and proofreading from others. More sophisticated learning environments such as online learning, virtual learning and learning with artificial intelligence (AI) are enabling different forms of collaboration. The Internet and digital age have made our generation characteristic of sharing and learning from one another. Solo learning is increasingly difficult to accomplish in today's commonplace tasks.

WHAT ARE THE EDUCATIONAL ISSUES FACING OUR DIGITAL AGE?

Several issues confront our current understanding of learning environments. First, the multiple interactions of human and electronic devices are posing challenges to the traditional scientific method of investigation. These interactions are raising questions about the reductionist approach and ecological validity. They are also questioning how we analyze, identify and exclude variables in this complex learning process. Most empirical studies deal with the unit of analysis comprising of a single discrete task analysis without external aids

(Williamson, 2004). This reductionist approach to experiments may illuminate the single cause cognitive relationships, but in reality, contemporary settings are more complex. Perret-Clermont, Perret, and Bell (1991) are right to say that "the causality of social and cognitive processes (in a system) is, at the very least, circular and is perhaps even more complex" (p. 50). For any single effect, there are multiple causes and influences.

Second, the advent of social cognition into cognitive psychology has introduced many other considerations such as social aspect (Vygotsky, 1981), culture (Bruner, 2005), and emotions (Hatano, Okada, & Tanabe, 2001) in the study of affect and cognition. However, most of them are studied as a singular influence, rather than in a holistic or interdisciplinary manner. Rarely do we see a consideration of two or more influences at the same time in a study. While there are attempts, such as Newell (1994), to study cognition as a unified whole, Newell still regards the mind as a unit, in spite of "enlarging" it to a whole in seeking explanations for experiences and multiple influences to cognition. Is there a larger perspective of cognition or a bigger paradigm to study multiple influences to cognition?

Third, the increased complexity of collaborative activity serves as a basis from which to question the preoccupation with the individual as the unit of analysis. Also, where are the boundaries and what is included in the collaborative activity? In the study of collaborative workplaces, Kling (1991) is concerned with the problem created by the loaded concept of "collaboration" in computer supported collaborative work (CSCW). The complexity and associated issues of conflict and interpersonal dynamics was proving too much of a minefield to study. The crux of collaboration is the joint activity of "coordination, cooperation and communication" (Engelstrom, 1992, p. 64). The joint activities and interplay of the coordinating (organizing), cooperation (sharing), and communication (discourse) of knowledge, present challenging mental representations which have yet to be accounted. Members not only share the objects in the cognitive system, but cocreate a shared script of joint activities. This knowledge building (Stahl, 2002) in terms of integrating, synthesizing and creating of knowledge needs accountability. Englestrom (1992) sees communication as the higher form of collaboration and the mediated activity as the key. Do we also include this mediation in our study of cognition in a social setting? Is there a theory on cognition to help educators understand these interactive and mediated joint activities in a seemingly difficult study of a collaborative setting?

Fourth, this digital phenomenon requires a framework to provide a coherent and comprehensive paradigm to make sense of the complexity. The emergence in the 1990s of CSCW and subsequently computer supported collaborative learning (CSCL) as paradigms addressing the emergence of computer use in the workplace and classroom served the needs of that time. The interest in collaborative activity began in workplaces (CSCW) and extended into educational settings (CSCL). Where CSCW is concerned with how groups collaborate in performing tasks, CSCL looks at how groups learn in educational settings. However, as Lipponen (2002) questioned the state of CSCL in 2002 as a paradigm, this chapter is asking the same question in the light of proliferation of other electronic devices, besides computers, that aid learning. While collaborative learning in CSCL recognizes the interdisciplinary approach to such studies and several frameworks have been proposed to provide a comprehensive account of the learning contexts, none exists to incorporate the pervasive use of digital devices. Designs on not only the technology in support of learning, but the learning environment and the artifacts in the cognitive system, would yield more in terms of our understanding of learning in today's classrooms.

WHAT DO CURRENT PERSPECTIVES TELL US?

Cognitive science has kept strictly focused on the brain and its law of singular causes (Popper, 1999) while ignoring other social and cultural factors (Gardner, 1985). At the turn of the 20th century, the challenge and confluence of ideas in the epistemology of cognitive science and cognitive psychology changed the functionalist view of cognitive scientists towards cognition. The notion of "causation" was even challenged, and replaced by "relation" (Mach, 1976), which in turn led to qualitative causally interpreted Bayesian nets (Williamson, 2004) and the introduction of the notion of "probability" (Popper, 1959). The positivistic reductionism of the sciences was also challenged as the only means to understand the world (Putnam, 1981). In the 1970s, the introduction of "deterministic chaos" (Goodwin, 2003) into scientific studies began to acknowledge the recognition of the indeterminateness of scientific and objectivity of values.

At the same time, there was a movement to view cognition beyond the confines of the skull (Clark, 2002; Salomon, 1993). The analogous comparison of the brain to the computer, led to studies into the computational representations of how the mind works (Turing, 1950). This computational approach recognized that mental phenomena arose from the operation of multiple distinct processes rather than a single undifferentiated one. Connectionists, who are also concerned about learning, such as Rumelhart and McClelland (1986), used the "Parallel Distributed Processing" model to study cognition that is distributed in a network of computers, believing it to be similar to the neural networks of the brain. This was one important early work that explored the distribution of cognition. Connectionists focus on learning from environmental stimuli and storing this information in a form of connections between computers (neurons). This was an early attempt to see cognition as occurring outside the skull.

In cognitive psychology, the influences of human and social sciences, in particular, anthropology and sociology have been instrumental in the emergence of social cognition in the late 1960s. This is now the dominant model and approach in mainstream social psychology. While the cognitive aspect of learning focuses on the effects of external stimuli on individual cognition, the social aspect of learning looks at social relationships that influence human cognition. The external stimuli included interactions with other humans but it was the effects of the influences that were being studied rather than the relationships. Social cognition, on the other hand, considers the social aspects and roles of the individuals: how people process social information, the encoding, storage, retrieval, and application. The advent of social cognition and its related movements has challenged and freed cognitive studies, shifting it from outcome-oriented to process-oriented (Fiske & Taylor, 1991), recognizing cognition and learning as socially influenced. An antithesis to cognitive processes, social cognition advocates continue the debate till this day about how learning is to be studied. This resulted in most researchers' focus on either the cognitive or social processes in studying learning and cognition, such as systems supported by computers (Kreijns, Kirschner, & Jochems, 2003). It would appear that Perret-Clermont et al. (1991) are right to allude that research paradigms stressing on what is social and what is cognitive will fail because "the causality of social and cognitive processes is [...] perhaps even more complex" (p. 50).

Social cognition began influencing educational psychology and ushered in the current constructivist learning theory. The rise of social constructivism via Vygotsky's social development theory and the emerging social and cultural theory of language and thought forced reconsideration about how people learn in educational psychology (Wood, 1998, p. 39). Within a sociocultural constructivist framework, the notion of learning is seen as a coconstruction of knowledge between individuals.

Seen as dialogical interaction of a community, the social cognition in education can range from a simple joint learning activity between two individuals to an extended and complex network of multiparty interaction of knowledge building. Through collaboration, learning now extends to participation in a community of learners (Brown & Campione, 1990) and community of practices (Wenger, 1999).

Thus, the dissatisfaction with the reductionist thought, singular cause method, and the belief that cognition resides only in the head, led to the developments in social cognition spawning several popular movements: situated cognition (also known as situated learning) (Lave & Wenger, 1991), activity theory (Leont'ev, 1978), embodied cognition (Varela, Thompson, & Rosch, 1992), distributed cognition (Hutchins, 1995), and the recent enactivism (Cowart, 2004). While each of the movements attend to the concerns peculiar to their areas, there appears to be none that offer a comprehensive framework, embracing both the cognitive and social processes, such as the theory of distributed cognition that we shall now turn to.

WHAT IS DISTRIBUTED COGNITION?

Consensus and acceptance of distributed cognition is still inconclusive (Salomon, 1993). However, as evident in the growing literature, distributed cognition is becoming a recognized theory. The definition of distributed cognition varies from the radical view to a loose position. Hutchins' (1995) distributed cognition theory is a study of cognition distributed across individuals and artifacts in a social-cultural and technical system as defined by the members and artifacts in a context. He challenged cognitive science's traditional preoccupation with the individual and the brain as the boundary of the unit of analysis. As such, he also challenges the "range of mechanisms" (Hutchins, 1995, p. 373) that participates in the cognitive process. For Hutchins, the study of cognition erred in confining the study within the skull of the individual and ignoring the context and the individual's interaction with others and artifacts. External elements should not be only treated as stimuli or aids to cognition but rather as equal partners in exhibiting, distributing and creating cognition. Any study into cognition should include all the elements that are directly, and even indirectly, working towards the accomplishment of a cognitive activity instead of the singularity approach. An individual's memory by itself is insufficient to understand how a memory system works (Hollan, Hutchins, & Kirsh, 2000), citing the rich and complex cognitive interactions in a cockpit or a ship's bridge involving the manipulation of artifacts. Pea (1993, p. 69) refers this as "off-loading"—when humans rely on artifacts to help them remember or compute cognitive tasks. The classic description of how a person requires an external representation by writing the multiplication on a piece of paper when called upon to solve a mathematical problem, is evidential to the use of the artifact (pen and paper) to facilitate the multiplication process which was mentally difficult to do.

Engeström (1992) also argues that computer supported collaborative work (CSCW) suffered from the Cartesian focus on the mind as the unit of analysis while relegating the collaboration to efforts to harmonize with the individual. Sociocultural aspects should be included in the study of CSCL (Kling, 1991; Reason, 1990). Reason (1990) differentiates latent human error from active error, attributing the former to the collective and suggests studying the group's interrelationships to help understand thinking better. Similarly, Cole (1991) sees cognition as a jointly and socially mediated activity. Perkins' (1993, pp. 93–95) views knowledge as "represented," "retrieved," and "constructed" jointly by the "person plus." This is a radical departure from the traditional view of cognition. Currently, there is a growing consensus that the concept of intelligence should not be confined as a property of the mind (Pea, 1993).

The unit of analysis consists of human agents and nonhuman artifacts in the environment and the unit varies with each different context (Hutchins, 1995). This focuses on whole environments as a unit of analysis. So, instead of "keeping" cognition inside the skull, cognition is now seen as external and being distributed in order to accomplish the cognitive task at hand. Lave (1988) and Saxe (1988) observe behavior and cognition in a social (or/and technological) context in their work. Hollan et al. (2000) opine that cognition can be effectively observed as occurring in a distributed manner. Some may argue that cognition is non-symbolic (Dreyfus & Dreyfus, 1986) and therefore cannot be studied. While others like Glaser and Chi (1988) believe that thinking is represented and can be studied.

DISTRIBUTED OR NOT?

Distribution is the spreading or circulating of things over an area as opposed to a single locus. Distribution considers the sharing, transformation and propagation of any form of information processing in the system. Hutchins (1995) postulates that the cognition process is distributed across members and artifacts of a social group involving coordination between internal and external structures. The locus of cognition is no longer centered on *one* individual. Rather, there are several loci of cognition in a system, each one contributing to the distribution as well as processing the cognitive activity. Cognition is also distributed across time with the earlier events affecting later ones. This means that the manner of distribution is time sensitive. The timing and aging of the cognition affects the cognitive process and system. The cognitive system is also seen as a whole rather than its discrete parts and the boundaries of the unit of analysis are now extended. Halverson (2002) sees distributed cognition focusing on the organization and operation of the cognitive system where its mechanisms make up the cognitive process and seek cognitive accomplishment. Pea regards intelligence as distributed (Pea, 1993, p. 50) to the artifacts alleviating the tedious and burdensome cognitive tasks that humans have to undertake. For him, computer tools and programs are the natural artifacts enabling distributed intelligence to occur and it is preferable for humans to partner them than go solo in any given cognitive task.

As a cognitive science anthropologist, Hutchins (1995) sees all cognition as being distributed to both individuals and artifacts. Salomon (1993), a psychologist and an educator, is more guarded and acknowledged that cognition was distributed but keeps the individual cognition as separate while operating together with others in the system. Fearing that distributed cognition may be seen as the only explanation that ignores the other aspects, Salomon (1993) is careful not to attribute cognitive powers to nonhuman artifacts. Because of the overemphasis on "what's outside" the brain, he feels the extreme position was truncated conceptually. While espousing the overall concept of distributed cognition, he points out that not all cognition is distributed and suggested the middle road: recognizing some distribution of cognition while affirming the individual plays a significant cognitive role in the system. Salomon (1993) maintains that in any given distributed system, there are "sources" of cognition (p. 111) which he attributes to human minds. So, for Salomon (1993), he also sees the interconnectedness between what was distributed versus the internal solo cognition (p. 113) of the individuals.

However, following Vygotsky's notion of internalization, "any higher mental function necessarily goes through an external stage in its development because it is initially a social function" (1981, p. 162), cognition can be viewed as distributed because of its social origins. Individual cognition is even argued to be socially mediated where the individual thought (and action) is shaped by the social context of social relationships, self identities and group associations (Clancey, 1997). For cognition to be functionally meaningful, it has to be socially mediated whether by the individual or by others. The classic example of using a pen and paper to externally represent the cognition process during solving a complex math problem clearly suggests solo cognition is distributed between the mind and external representations. In a more complicated cognitive context like negotiating a ship into a harbor, there are some subsystems of cognitive activity where solo cognition exists, which may seem to be not distributed, such as an in-situ reflection. However, even personal reflection or any other forms of solo cognition are a result, and also a consequence, of a social interaction. Subsequently, the cognition is manifested later in the distribution; even though it was not distributed initially.

Cognition can be categorized from a range of lower-order to higher-order: from comprehension, recall to analysis, synthesis and problem solving. Pea (1993) argues that higher-order thinking belongs to solo thinking and cannot be distributed. Perkins (1993), like Pea, feels too that higher-order knowledge cannot be distributed. They argue that such complex activity occurs in the head and what is distributed is knowledge resulting from that activity. However, if the system is considered as a *whole* unit of analysis, consisting of the different sources of cognition (humans, artifacts and environment), then the cognitive system as a whole is capable of higher order cognition and can be considered as such. The distribution of cognition is within this whole system and any higher-order thinking would occur *within* the system. In reality, higher-order thinking begins with lower-order thinking and as the organizing, integrating and synthesizing (higher-order thinking) of knowledge begins, the social and mental representations are distributed. This organized knowledge may be observed as visually presented or verbally described. Although this may sound like a technical justification for higher-order cognition to be seen as distributed, the fact remains that cognition, whether it is higher or lower-order is distributed.

One of the foreseeable difficulties (but a liberating aspect) in studying distributed cognition is the indeterminateness of the system boundaries. Unlike the traditional cognitive studies where the constructs are clearly stated, ethnographic studies into cognitive behavior and patterns allow undetermined influences to be considered during the study, including new and emerging influences that interact with previous ones in the cognitive system. These recursive and emerging influences on cognition can be very exciting. While this is the nature of the study and characteristic of the analysis, the questions of limit and termination of the cognitive activity are left open. Theoretically, the cognitive system and process is limited by the cognitive task and time taken to accomplish the task, but the extended boundaries that contributed to the task and duration may be difficult to ascertain, due to the dynamic nature of the distribution. Giere (2002) went, as far as to consider the coalmines in Montana as the boundary of the distribution of his science laboratory task. But certainly, any objective researcher will not risk such an irrational stretching of the theory to its limits.

COLLABORATING ALL THE TIME?

Collaborative learning is a process of interaction of knowledge and the joint working of two or more people in an intellectual undertaking of a task or goal. Forms of collaboration range from common task completion, joint decision making to complex problem solving. Implicit in its understanding is the interaction of human members. However, collaboration can also include other intelligent entities. These other intelligent entities may come in the form of computers or highly sophisticated AI machines like robots. If the focus of collaboration is in the "joint working" aspect, then would it not be too preposterous to say that humans might collaborate with a robot or even a computer? To stretch this further, we may even be working jointly with less intelligent (but nonetheless intelligent) artifacts such as a personal digital assistant or a cell phone. Take an intellectual endeavor, for example, writing a paper. To write a paper in today's context, I will have to use a computer writing software program. The program is "intelligent" as it picks out my spelling and grammatical errors. And if I need to refer to types of format and style, it offers an array of choices. It has indeed "worked jointly" with me on my paper, although not exactly in the conceptual domain. Certainly, if I used the computer to surf the Internet for ideas and discussions on the topic I am writing, it would certainly have contributed, as a conduit, to my intellectual endeavor. Clearly, I am not equating a computer to a human, but increasingly, technology is advancing at a rate that in the near future, we may consult fairly intelligent devices for original thoughts.

COGNITIVE ARTIFACTS ARE SOURCES OF COGNITION?

In the framework of distributed cognition, artifacts are considered cognitive. This may be a radical idea to some but it may not be a far fetched notion. Take the common practice of using a personal digital assistant (PDA) to aid our memory by storing the information into its database. Did the PDA help our memory? Did it amplify our recall ability such that we are able to remember it the next time? Although it did not really change our memory, it organized the information we entered in the system so that we can retrieve it at an incredible speed, which humans are cognitively incapable of. The artifact was involved in the cognitive function of organizing the information in a way that we can search for it easily and quickly. So, the artifact performed a cognitive task: "organizing" the input data, and "searching and gathering" the required data. Technological devices that aid our memory and computation are known as cognitive artifacts (Norman, 1993).

A distinction needs to be made between cognition and semantics. Searle (1980), using the classic Chinese room experiment where a non-Chinese speaker had to use a rule book to construct a response to a question in Chinese, argues that the machine does not have the semantics of symbols it is manipulating compared to the human mind. So, the machine may act as if it is "thinking," but in reality, it had no clue to the semantic meaning even though it is able to successfully construct a "meaningful" answer to the response. The argument is that the system therefore does not understand the meaning attached to the symbols but merely processes it due to its programming. Searle (1980), therefore, argues that semantic cognition is not distributed between artifacts and humans. So the issue is, does the system really learn as compared to the human mind? In considering distributed cognition, should it include semantics in cognition?

Nardi (1996) argues that the theory of distributed cognition devalues or restricts the meaning of cognition when there is no distinction made between people and things as cognitive agents. Her contention is that for an artifact to exhibit cognition, it must possess the quality of having the "act of or process of knowing, including awareness and judgment." On this definition of cognition, she feels that artifacts are incapable of consciousness and therefore, should not be put on the same level of consideration in a cognitive system. Technically, cognition is any activity that involves the act of recall, comprehension, critical thinking (organizing, sorting, sequencing, comparing and contrasting, etc.) or creative thinking (brainstorming, predicting, synthesizing, etc.), that involves information processing. As such, any artifact that is capable of this action is performing some form of cognition. Recalling and generating knowledge is not the sole prerogative of consciousness. With huge strides in AI, we are seeing robotic machines that are capable of initiating interaction and performing complex cognitive activities but devoid of consciousness. The issue of awareness, consciousness and emotions may yet be elusive to the most advanced or powerful machine at present but in the future, who can tell?

SOCIAL, CULTURAL, HISTORICAL, AND EMOTIONAL INFLUENCES

Halverson (2002) points out that distributed cognition explores the broader sociocultural-technical system of the cognitive system. Clark (1998, p. 258) submits that the mind is best understood as the activity of "an essentially situated brain" in its bodily, cultural and environment context. Hatch and Gardner (1993) feel that the reason why cognitive scientists stayed away from the sociocultural elements is because of their unquantifiable nature and they first needed to understand the brain on its own before considering other aspects. Epistemologically, when a learner engages new learning materials, he does not interact with the material solely on a linear basis, detached from his or her surroundings. The people in the zone of proximity, the artifacts that the learner uses, the physical surroundings and context contribute to the learning process. Socially, the social role of a learner with peers affects learning. If he is held in high esteem by the peers or considered by the teacher to be a favorite student, the learning experience will be different from one who is not. Culturally, those from bigger families and are more outspoken at home, will find group activity more familiar and learning easier than those from a single child family. Learners with different histories with the teacher and classroom environments will differ in the processing of information. Personal histories with each other, with the artifacts, and with the environment will affect the learning. Learning with an unfamiliar face, machine or place compared with the familiar will yield different cognitive results. The emotional state of the learners also provides different learning experiences even when going through the same program and in the same context. Evidence has shown that emotions (Hatano et al., 2001) affect cognition and as a result, affects both individual as well as group performances too.

All these influences: social (Vygotsky, 1981), cultural (Bruner, 2005), personal histories, and emotions (Hatano et al., 2001), should be considered in the cognitive system, at the beginning, during and ending in distributed cognition. This description of learning challenges the idea that knowledge can be transmitted in an absolute and linear relationship. It also challenges the assumption that objective knowledge can be acquired in individuals. Whatever it is, the consideration of all possible influences in a cognitive system clearly seeks to give a holistic and comprehensive picture on how learning and cognition happens.

HIGHLY CONTEXTUAL AND MULTILAYERED COGNITIVE ACTIVITY

In distributed cognition, a system is observed to re-configure itself with subsystems enjoining in the interactions in the system while accomplishing the cognitive functions and task. The cognitive process is bordered by the functional relationships among elements that are participating in the process and not by the spatial distance relationship (Hollan et al., 2000). This suggests an emerging character of the cognitive processes in the system. Hutchins (2005) uses the term, "conceptual blending" (p. 1556), which involves the interactions between the mental spaces of the people, artifacts and environment. These include the body language, coordinating mechanisms, various forms of communication and how tacit knowledge is shared and accessed. Salomon (1993, p. 112) also points to the "joint nature" of the distribution rather to one agent. These layers of cognition and interplay of mental spaces clearly present a multilayered cognitive activity to behold.

Contextually, Bruner (2005) clearly believes that any social cognition is highly situated in its local context and culture. This places the cognitive system as highly contextualized in its own setting: the human members and artifacts situated in the environment. The interplay and interconnection of each member's histories and culture clearly make the study immensely rich with many layers of relationships.

Zhang and Patel (2006) consider affordances as "allowable actions" offered by the environment coupled by the properties of the agent. Affordances are the functions that can be carried out (afforded) by the properties in the environment (Gibson, 1977), including the human agents. Simply put, art studios afford drawing, computer rooms afford computer work. Affordances are another key element in considering the environment as part of the cognitive system. This means that anything that affords an executive function that contributes to the accomplishment of the cognitive task are considered in the study of the cognition distributed in the system.

While all these may paint a rather complex and seemingly incomprehensible picture of what and how learning takes place, the consideration of these factors will not only open a wider and perhaps deeper understanding of learning but in doing so, offers a more holistic and authentic picture of what learning really is.

OLD WINE IN NEW WINE BOTTLES

Cole and Engestrom (1993) cite Wilhelm Wundt and Hugo Munsterberg as the early psychologists who were the forerunners in recognizing a different form of psychology that regards cognition as requiring interaction outside the brain. Unfortunately, their writings were not picked up and developed to any recognizable cognitive psychological strands. Subsequently, Leont'ev, Luria, and Vygotsky, the progenitors of cultural-historical psychology, sought to mediate basic cognitive tasks to more complex ones with cultural tools, including the use of language (Cole & Engestrom, 1993). This means that in order to perform higher cognitive tasks, more than just the brain alone is involved, and the mediation of other cultural artifacts is also required. Hutchins (1995) alludes to Vygotsky's "Mind in Society" (1978) where his notions of treating the society as having mind like properties. By this, he is using language of the mind to describe the activities of the group. Also, for Vygotsky, every high level cognitive function appears as an interpsychological process first before the intrapsychological process occurs. Conversely, Hutchins draws on Minsky's (1986) work, "Society of the Mind" where the language of the group can describe what is inside the mind. Minsky (1986) regards the higher level of cognition as composed of several lower level agencies and are interconnected.

However, this was not the case in the Soviet Union in the early 1900's where Lev Vygotsky's social-historical school, now known as activity theory, began (Rogoff & Wertsch, 1984, pp. 1–6). Vygotsky (1978) postulates that mental functioning occurs first between people in social interaction and later within the child's mind. Similar schools of thought also arose in Scandinavia and Germany, under the banner of activity theory, action theory, and situated action.

In educational psychology, Dewey (1963) warns against treating experience (learning and development) as something going on inside one's head. He recognizes that there are "sources outside an individual which give rise to experience" (Dewey, 1963, p. 39). Evidently, distributed cognition was not entirely new in its concept.

IMPLICATIONS OF USING DISTRIBUTED COGNITION IN COLLABORATIVE LEARNING ENVIRONMENTS

The first implication in using distributed cognition in the study of collaborative learning environments is that it is an authentic and naturalistic study, rather than a de-contextualized one where the results are not tenable when put to real life situations. In terms of research validity, using the holistic approach to study cognition and learning will ensure ecological validity. Using distributed cognition as a theory presupposes a qualitative case study approach, and mixed method to understand the various influences and relationships within the cognitive system. This snapshot of the cognitive activity legitimizes the findings while respecting the sensitivity of time. Such a naturalistic study too, when trustworthiness is ensured, has translatability value.

Second, distributed cognition demands that any cognitive system be studied as a whole environment. This holistic approach means that every human and nonhuman artifact, their embedded cultural symbolism, historical data, emotional state and social relationships are considered. At the same time, the source, transformation, propagation and emerging of cognition through time, including the subsystems of cognitive relationships are duly considered. On top of this, together with the single and multiple relationships involved within the cognitive system, the system is, as a whole, also looked at. Collaboration is thus seen as a whole, together with its subcollaborations (subsystems) and the relationship between the whole and its parts. This three-tiered matrix relational study seeks to capture as much data as possible in order to holistically and comprehensively understand the collaborative and cognitive system. This will give a more accurate picture of the learning environment.

Third, the recognition of cognition as a whole unit, allows a holistic understanding of learning in a given context: seeing the cognitive actions as a whole culminating in the aggregate performance and allowing the researcher to see how each cognitive action and relationship contribute to the performance. This bigger picture of cognition will also better inform our studies into collective intelligence (Levy, 1997) or groupthink (Janis, 1997). Salomon, Perkins, and Globerson (1991), on ways of evaluating intelligence between people and technology partnerships, cite both "systemic" and "analytic" when considering both aggregate performance and specific contribution by each member.

Fourth, this whole-environment approach radically regards the individual as a member rather than central to the study, allowing an unbiased treatment of each member and artifact in the system. This means that no one member is prejudicially seen to contribute more or less to the process and performance. This will lead to greater integration of both technological and nontechnological artifacts, human actions and the environment. At the same time, this may reveal hitherto unconsidered elements that may surprise the research with potentially significant impact, due to the impartial treatment of all members and artifacts as equal partners in the cognitive system. Educational goals would then, shift from individual mastery to jointly accomplished performance (Salomon, 1993).

Fifth, the inclusive nature of this framework, in considering all observable representations of cognition in a cognitive system allows an unbridled approach to the study of cognition. This opens up a wider sphere of possible influences that affect the performance of the cognitive tasks. In effect, we can apply these theoretical constructs to a much wider range of considerations in the study of cognitive phenomenon.

With these macro implications delineated, clearly this perspective will render learning with a larger holistic feel: no longer restricted to the linearity of cognitive relationships, but the ability to see the whole. This includes the specific contributions of each element, unbiased analysis for all members, and a broader understanding of the cognitive system.

At the micro level, the equal treatment of human and nonhuman artifacts as cognitive agents is significant in the collection and analysis of data. First, the distribution pattern and learning process will allow insights to what and how information is gathered, analyzed, synthesized, transformed, stored and created. This will help us see the flow of information, identifying both convergence and divergence points. Second, it also involves the study of the coordination between the internal and external structures in and outside the cognitive system, i.e. looking at sub systems. This will identify critical points of influences and effects to the relationships and the cognitive system as a whole. Third, scrutinizing the effects of time that each mental representation has on the cognitive processes in the system will reveal the dynamics and time sensitive nature of the human, artifacts and environment. This is central to the distributed cognition theory, the interplay and emergent qualities among the three entities: human agents, artifacts and environment. This will inform not only the design of future learning environments and its members within, but also reveal how humans actually think and learn. Critical learning incidents can be identified and enhancements be made. Fourth, the notion of cognitive artifacts opens up the perspective that intelligent artifacts are capable of cognitive abilities and of joint activity. The PDA example highlighted earlier is just one of the million ways an electronic device can contribute and collaborate in the cognitive systems to accomplish higher and more complex cognitive tasks. AI or cybernetic systems are examples of higher intelligent artifacts that may one day play a major role in learning. Fifth, the notion of affordances by artifacts and the environment gives us a dimension to consider in our studies. This study into affordances will greatly inform the future designs of learning environments.

Finally, the holistic approach stemming from distributed cognition will yield insights into the characteristics and influences of the coherent and emergent wholes (Goodwin, 2003) that make up much of the naturalistic learning environments. The inclusion of culture of the members in the analysis allows examination of the symbols and meaning attached to each visual, audio, feeling and verbal expression found in the cognitive system, and will give us insights into the reasons for human actions and behavior. Cultural considerations will force the researcher to consider the cultural perceptions of the human members towards each other, as well as the artifacts. The inclusion of the historical aspect of the members and artifacts in the unit of analysis allows the researcher to see why certain actions are taken and behavior is manifested. The historical aspect will mean looking into the histories of the cognitive systems as well as personal histories of both human and nonhuman artifacts. The technological experience or academic history of the learner will affect the interactions with other humans (with different histories) and artifacts (again with different histories). The inclusion of social structure informs the analysis of the hierarchical structure of human relationships. The inclusion of emotions into the study of cognition will let us see the behavioral and attitudinal dispositions of the interactions. The further inclusion of discourse analysis will allow researchers to examine the language used and communication that affects the distribution of the cognition. Potentially, with insights into how culture, history, emotions, communication and social structures affect cognition and its distribution, these will greatly inform the design learning environments, artifacts and learning strategies. It is believed that this framework will help to advance the studies of "effects with technology" (Salomon et al., 1991, p. 3), the partnerships with machines, which will lead to redefinition and enhancement of learners' performance with technology.

CONCLUSION

With the general dissatisfaction of reductionism and restrictive "within the skull" cognition, together with the prevalent exploratory social learning theories and social cognition movements, the emergence of distributed cognition is timely, especially since learning environments are constantly changing. The theory that cognition can be studied as distributed across human members and artifacts is fascinating as it offers a holistic view involving not just the social, historical and cultural aspects of a cognitive system but also the very idea that the cognitive system has an entity of its own is very intriguing. The theory also liberates the idea that artifacts can "do" cognition which opens up a world of possibilities in dealing with artifacts of the future, where they will certainly be more capable and more powerful as cognitive entities in their own right. Thus, distributed cognition sees cognition as one unit of analysis of the cognitive system, which was traditionally bounded by the skull, but is now extended to the elements outside the skull and bounded by whatever artifact and human agents that play a part in the cognitive system situated in a learning environment.

This chapter began with the nature of today's complex collaborative learning environments confronting our educational studies. Learning is currently seen as social, as well as situated in a context: involving other human agents as well as intelligent electronic devices. Understanding learning via distributed cognition as an extended cognitive system addresses the challenges highlighted by being nonreductionist, inclusive of various influences, and allowing reconciliation of multiple and emergent joint mediated activities. With increasing sophisticated cognitive tasks being introduced into our world today, the learning environment has become not only more crowded but filled with more sources and artifacts for cognition to be distributed. Distributed cognition can open up a new vista in understanding how cognition and learning can take place. This vista will allow us to examine the layers of cognitive processes and interplay of the internal and external structures resulting in insights that will assist us in human learning and cognition as never before. Crucially, because of its whole environment approach, influences can be identified from a broader and even deeper perspective and this will help inform technological design issues as well and human learning strategies.

As we advance in our thinking and research on learning *with* technology, distributed cognition gives us that breadth and depth to study, in detail as well as holistically, today's complex learning environments. And if learning is to be an enculturation of the practice of life-long learning and personal and professional development, then, the sooner we begin to understand the influences from as many disciplinary aspects as possible, the better we will be able to design and facilitate learning environments for our students of tomorrow.

Perhaps a more fundamental question and issue to address is how do our modern young learn. How do they learn and what makes them want to learn? What gives them meaning in learning? Answering these questions will begin to help us design learning environments that are suited for them and address the challenges of today's emerging technology. And we need a framework that can adequately address these questions in a comprehensive, holistic and ecologically valid manner.

This chapter may be in part, advancing what Kuhn (1970) advocates in his book, *The Structure of Scientific Revolutions*: the notion that science has become overly specialized with each succeeding paradigm, thus losing sight of the forest for the trees in investigating cognitive phenomenon. Nonetheless, this chapter is mindful that this paradigm of distributed cognition will not take that route.

REFERENCES

Brown, A. L., & Campione, J. C. (1990). Communities of learning and thinking, or a context by any other name. *Contributions to Human Development, 21*, 108–126.

Bruner, J. (2005). Homo sapiens, a localized species. *The Behavioral and Brain Sciences, 28*, 675–735. doi:10.1017/S0140525X05250124

Clancey, W. J. (1997). *Situated cognition: On human knowledge and computer representations*. New York: Cambridge University Press.

Clark, A. (1998). Where brain, body and world collide. [Special issue on the brain]. *Daedalus: Journal of the American Academy of Arts and Sciences, 127*(2), 257–280.

Clark, A. (2002, November). *Cognition beyond the flesh: Singing burrows and surrogate situations*. Paper presented at International interdisciplinary seminar on new robotics, evolution and embodied cognition (IISREEC). Lisbon, Portugal.

Cole, M. (1991). Conclusion. In L. Resnick, J. Levine, & S. Teasley (Eds.), *Perspectives on Socially Shared Cognition* (pp. 398–417). Pittsburgh: APA and LRDC.

Cole, M., & Engeström, Y. (1993). A cultural-historical approach to distributed cognition. In G. Salomon (Ed.), *Distributed Cognitions: Psychological and Educational Considerations* (pp. 1–46). Cambridge: Cambridge University Press.

Cowart, M. (2004). Embodied cognition. *Internet Encyclopedia of Philosophy*. Retrieved July 6, 2008, from http://www.iep.utm.edu/e/embodcog.htm

Dewey, J. (1963). *Experience an education*. New York: McMillan.

Dori, J., Belcher, J., Bessette, M., Danziger, M., McKinney, A., & Hult, E. D. (2003, December). Technology for active learning. *Materials Today*, 44–49. doi:10.1016/S1369-7021(03)01225-2

Dreyfus, H. L., & Dreyfus, S. E. (1986). *Mind over machine. The power of human intuition and expertise in the era of the computer*. Oxford: Basil Blackwell.

Engeström, Y. (1992). Interactive expertise: Studies in distributed working intelligence. [Department of Education, University of Helsinki.]. *Research Bulletin (Sun Chiwawitthaya Thang Thale Phuket)*, 83.

Fiske, S. T., & Taylor, S. E. (1991). *Social cognition* (2nd ed.). New York: McGraw Hill.

Gardner, H. (1985). *The mind's new science*. New York: Basic Books.

Gibson, J. (1977). The theory of affordances. In R.E. Shaw, & J. Bransford (Eds.), *Perceiving, Acting, and Knowing* (pp. 67–82). Hillsdale, NJ: Lawrence Erlbaum Associates.

Giere, N. (2002). Scientific cognition as distributed cognition. In P. Carruthers, S. Stich, & M. Siegal (Eds.), *The cognitive basis of science*. Cambridge, UK: Cambridge University Press.

Glaser, R., & Chi, M. (1988). Overview. In M. Chi, R. Glaser, & M. Farr (Eds.), *The nature of expertise* (pp. xv–xxvii). Hillsdale, NJ: Erlbaum.

Goodwin, B. (2003). Patterns of wholeness: Introducing holistic science. *Resurgence, 216*. Retrieved July 6, 2008 from http://www.resurgence.org/resurgence/issues/goodwin216.htm

Halverson, C. (2002). Activity theory and distributed cognition: Or what does CSCW need to do with theories? *Computer Supported Cooperative Work, 11*(1–2), 243–267. doi:10.1023/A:1015298005381

Hatano, G., Okada, N., & Tanabe, H. (Eds.). (2001). *Affective minds*. Amsterdam: Elsevier.

Hatch, T., & Gardner, H. (1993). Finding cognition in the classroom: An expanded view of human intelligence. In G. Salomon (Ed.), *Distributed Cognitions: Psychological and Educational Considerations* (pp. 164–187). Cambridge: Cambridge University Press.

Hollan, J., Hutchins, E., & Kirsh, D. (2000). Distributed cognition: Towards a new foundation for human computer interaction research. [TOCHI]. *ACM Transactions on Human-Computer Interaction*, *7*(2), 174–196. doi:10.1145/353485.353487

Hutchins, E. (1995). *Cognition in the wild*. Cambridge, MA: MIT Press.

Hutchins, E. (2005). Material anchors for conceptual blends. *Journal of Pragmatics*, *37*, 1555–1577. doi:10.1016/j.pragma.2004.06.008

Janis, I. L. (1997). Groupthink. In R.P. Vecchio (Ed.), *Leadership: Understanding the Dynamics of Power and Influence in Organizations* (pp. 163–176). U.S.: University of Notre Dame Press.

Kling, R. (1991). Cooperation, coordination and control in computer supported work . *Communications of the ACM*, *34*(12), 83–88. doi:10.1145/125319.125396

Kreijns, K., Kirschner, P. A., & Jochems, W. (2003). Identifying the pitfalls for social interaction in computer-supported collaborative learning environments: A review of the research. *Computers in Human Behavior*, *19*(3), 335–353. doi:10.1016/S0747-5632(02)00057-2

Kuhn, T. S. (1970). *The Structure of scientific revolutions* (2nd ed.). Chicago: University of Chicago Press.

Lave, J. (1988). *Cognition in practice*. Cambridge: Cambridge University Press.

Lave, J., & Wenger, E. (1991). *Situated learning: Legitimate peripheral participation*. Cambridge: Cambridge University Press.

Leont'ev, A. (1978). *Activity, consciousness, and personality*. Retrieved July 6, 2008, from http://marxists.org/archive/leontev/works/1978/index.htm

Levy, P. (1997). *Collective intelligence: Mankind's emerging world in cyberspace* (R. Bononno, Trans.). Cambridge, MA: Perseus Books.

Lipponen, L. (2002). *Exploring foundations for computer-supported collaborative learning*. Retrieved July 6, 2008, from http://www.helsinki.fi/science/networkedlearning/publications/publ-2002main.html

Mach, E. (1976). *Knowledge and error*. Netherlands: Reidel Publishing Company.

Minsky, M. (1986). *The society of mind*. New York: Simon and Schuster.

Nardi, B. (1996). Concepts of cognition and consciousness: Four Voices. *Australian Journal of Information Systems*, *4*(1), 64–79.

Newell, A. (1994). *Unified theories of cognition*. MA: Harvard University Press.

Norman, D. A. (1993). *Things that make us smart: Defending human attributes in the age of the machine*. New York: Addison-Wesley.

Pea, R. (1993). Practices of distributed intelligence and designs for education. In G. Salomon (Ed.), *Distributed Cognitions: Psychological and Educational Considerations* (pp. 47–87). Cambridge: Cambridge University Press.

Perkins, D. (1993). Person-plus: A distributed view of thinking and learning. In G. Salomon (Ed.), *Distributed Cognitions: Psychological and Educational Considerations* (pp. 88–110). Cambridge: Cambridge University Press

Perret-Clermont, A. N., Perret, J. F., & Bell, N. (1991). The social construction of meaning and cognitive activity in elementary school children. In L. Resnick, J. Levine, & S. Teasley (Eds.), *Perspectives on Socially Shared Cognition* (pp. 41–62). Hyattsville, MD: American Psychological Association.

Popper, K. (1959). The propensity interpretation of probability. *The British Journal for the Philosophy of Science*, *10*, 25–42. doi:10.1093/bjps/X.37.25

Popper, K. (1999). *The logic of scientific discovery*. London: Routledge.

Putnam, H. (1981). *Reason, truth and history*. Cambridge: Cambridge University Press.

Reason, J. (1990). *Human error*. Cambridge: Cambridge University Press.

Rogoff, B., & Wertsch, V. J. (1984). *Children's learning in the zone of proximal development*. San Francisco: Jossey-Bass Inc. Publishers.

Rumelhart, D., & McClelland, J. (Eds.). (1986). *Parallel distributed processing*. The MIT Press/Bradford Books.

Salomon, G. (1993). No distribution without individual's cognition: A dynamic interactional view. In G. Salomon (Ed.), *Distributed Cognitions: Psychological and Educational Considerations* (pp. 111–138). Cambridge: Cambridge University Press.

Salomon, G., Perkins, D., & Globerson, T. (1991). Partners in cognition: Extending human intelligence with intelligent technologies. *Educational Researcher*, *20*(3), 2–9.

Saxe, B. (1988). Candy selling and math learning. *Educational Researcher*, *17*, 14–21.

Searle, J. (1980). Minds, brains, and programs. *The Behavioral and Brain Sciences*, *3*(3), 417–457.

Stahl, G. (2002). Contributions to a theoretical framework for CSCL. In G. Stahl (Ed.), *Computer Support for Collaborative Learning: Foundations for a CSCL Community*. Mahwah, NJ: Lawrence Erlbaum Associates.

Turing, A. M. (1950). Computing machinery and intelligence. *Mind*, *49*, 433–460. doi:10.1093/mind/LIX.236.433

Varela, F. J., Thompson, E. T., & Rosch, E. (1992). *The embodied mind: Cognitive science and human experience*. Cambridge, MA: The MIT Press.

Vygotsky, L. S. (1978). *Mind in society: The development of higher psychological process*. MA: Harvard University Press.

Vygotsky, L. S. (1981). The genesis of higher mental functions. In J.V. Wertsch (Ed.), *The Concept of Activity in Soviet Psychology* (pp. 144–188). Armonk, NY: Sharpe.

Wenger, E. (1999). *Communities of practice: learning, meaning, and identity*. Cambridge: Cambridge University Press.

Williamson, J. (2004). A dynamic interaction between machine learning and the philosophy of science. *Minds and Machines*, *14*(4), 539–549. doi:10.1023/B:MIND.0000045990.57744.2b

Wood, D. (1998). *How children think and learn* (2nd ed.). Oxford: Blackwell Publishers.

Zhang, J., & Patel, V. (2006). Distributed cognition, representation, and affordance. *Pragmatics & Cognition*, *14*(2), 333–341. doi:10.1075/pc.14.2.12zha

KEY TERMS AND DEFINITIONS

Artifact: An object or document created by humans.

Cognition: An act of information processing pertaining to memory, attention, perception, action, problem solving and mental imagery.

Cognitive System: An area or space where interconnected items of knowledge and representations of human cognitive processes are studied.

Collaborative Learning Environment: A situated area or space, networked or otherwise where there is sharing, coordinating, and cocreating of knowledge between two or more persons aided by artifacts to achieve outcomes they could not accomplish independently.

Computer-Supported Collaborative Learning: A process of increasing in knowledge through joint intellectual effort with the help of computers.

Distributed Cognition: A framework of understanding how information processing is circulated across individuals and artifacts in an environment.

Human and Computer Interaction: A study on interaction between people and computers.

Reductionist: An idea that all complex systems can be completely understood in terms of their components.

Social Cognition: A study on how people process information socially in encoding, storage, retrieval, and application to social situations.

Socially Mediated: How information and knowledge are exchanged and negotiated between humans.

This work was previously published in Handbook of Research on Electronic Collaboration and Organizational Synergy, edited by J. Salmons; L. Wilson, pp. 295-311, copyright 2009 by Information Science Reference (an imprint of IGI Global).

Chapter 7.14
Social Networking Sites (SNS) and the 'Narcissistic Turn': The Politics of Self-Exposure

Yasmin Ibrahim
University of Brighton, UK

ABSTRACT

The advent of the Internet hailed the ability of users to transform their identity and expression and articulation of the 'self' through their digital interactions. The Internet in its early days enabled the user to re-define identity through the text-based environment of the internet without declaring their offline persona or identity. In comparison new social software like *Facebook* have brought about a narcissistic turn where private details are placed on a global arena for public spectacle creating new ways of connecting and gazing into the lives of the others. It raises new social issues for societies including the rise of identity fraud, infringement of privacy, the seeking of private pleasures through public spectacle as well as the validation of one's identity through peer recognition and consumption.

DOI: 10.4018/978-1-60566-727-0.ch006

INTRODUCTION

The Internet in its early days signified the rebirthing of the individual and most prominently the 'self' as technology enabled the user to remediate identity through a text-based environment. Anonymity and virtuality constituted a form of 'avatarism' where individuals could re-invent their presence online without declaring their offline persona or identity (See Donath 1998; Froomkin 1995). In comparison, new social networking sites (SNS), such as Facebook, signify a 'narcissistic turn' where offline identities are publicized online and constructed through a multimedia platform to create new forms of self-expression, gaze, spectacle, and sociabilities. Equally, social networking is embedded within a new economy of sharing and exchanging personal information between friends and strangers. The sharing and communication of personal details have reached unprecedented levels with the proliferation of e-commerce and social networking sites in recent years (See Szomsor et al. 2008; Geyer et al. 2008; Strater & Richter 2007; Stefanone 2008; Lampe et al. 2006; Joinson 2008).

This marks a shift from the earlier 'virtuality' discourses of the Internet which perceived anonymity and the ability to transform identities online as a form of empowerment whilst raising the tenuous issues of trust, intimacy and deception. The increasing popularity of social networking sites, on the other hand, emphasizes the narcissistic tendency in the human condition manifested through an exhibition of the self through photos and other multimedia content. The publicizing of personal details on a global arena for public spectacle creates new ways of connecting and gazing into the lives of others. It raises new social issues for societies, including the rise of identity fraud, infringement of privacy, the seeking of private pleasures through public spectacle, as well as the validation of one's identity through peer recognition, connection and consumption online. The ability to connect with offline networks through online self-profiles and content and additionally the possibility of inviting audiences to be part of the 'friends' list celebrates the declaration of offline identities.

The politics of self-revelation on the Internet creates the need to understand new forms of computer-mediated behavior which are emerging and may have implications for the ways in which users construct and express their identities. The creation of profiles and the ability to make connections through these constructs indicate how these become a form of social capital in forming connections and communion with a wider imagined community offline and online. This chapter examines the phenomenon of self-exposure through social networking sites on the Internet and discusses how the emergence and popularity of these sites reflects a shift in debates about identity discourses on the Internet on a theoretical and societal level. The chapter also delves into the social and legal implications of self-revelation and, more specifically, how social networking sites create risk communities where an awareness of risks exists along with the urge to reveal in order to make contact and connections with others. Social networking sites function through complicit risk communities which highlights both the narcissistic strand as well as the postmodern hazards that lurk in the online environment.

THE EARLY DISCOURSES OF THE INTERNET

The term *cyberspace* was coined by science fiction writer William Gibson in 1982 to capture the nature of a space both real and illusory. This duality is one of the fundamental reasons why investigations of online spaces are complex and multi-dimensional. Early writings on the Internet portrayed the new medium as constituting a virtual space which was divorced from offline existence (Miller and Slater, 2000: 4). Miller and Slater (2000) define *virtuality* as the capacity of communicative technologies to constitute, rather than mediate, realities and to form relatively bounded spheres of interaction. These discourses often portrayed the emergence of new forms of society and identity (Rheingold, 2000) in which the 'virtual was often disembodied from the real' (Miller and Slater, 2000: 4). This disembodiment represented a form of escapism from real society where individuals could invent, deconstruct, and re-invent their identities. As such, cyberspace created fluidity in terms of identity as well as a form of release from the confines of the real world.

From this perspective Computer Mediated Communication (CMC) represents an unusual form of communication, as it does not fit into the conventional distinctions between public and private, and direct and mediated communication. (Diani, 2000: 386). CMC stands in a somewhat ambiguous relationship to other forms of communication. Its private and public nature is unclear. In line with the nature of communications on the Internet, there is also the question of how people establish identities in cyberspace. Because of the fact that we are not physically present on the Internet and because we can present many different personas there, the individual voices that make up a cyber community are often referred to as 'avatars' (Jordan, 1999: 59, 67). The irrelevance of geographical location with regard to CMC contributes to the phenomenon known as *disembedding*: that is, enabling users to transcend their immediate surroundings and communicate on a global platform (Giddens, 1990). Besides this feature of 'disembedding', the online environment prior to Web 2.0 was defined by the use of discursive form (textuality) and the ability to communicate anonymously. These features are seen as empowering since users are not constrained by both time and space and are at liberty to recreate and deconstruct their identities online.

The 'avatar' culture may offer people a degree of freedom to conceal identity and to approach taboo topics without the constraints of the real world. It can help forge new identities as well as new relationships. As the 'virtual' world is mediated through technology, the user may be bound by a new set of rules to negotiate the virtual world. Nevertheless, this hybrid form of communication does not exist in a vacuum. Discourses in the Internet can interact with happenings in other media and reflect one's physical context of existence. As such various online interactions can be embedded in disparate ways into larger social structures, such as professions and social movements. The dynamics of online interactions may be difficult to comprehend except through this physical embedding (Friedland, 1996; Miller and Slater, 2000; Slevin, 2000; Wynn & Katz, 1997; Slater, 2002). In this sense, the culture of the physical world can invariably transcend into cyberspace, thus further altering the pattern of mediated communications.

Early discourses of the Internet and present debates about the Web indicate that identity is a contentious and fragmented construct in view of the absence of physical cues in a discursive and subsequently a multimedia environment (Stefanone et al. 2008:107). Compared to the earlier Internet environment which leveraged on experimentation with identity, today's computer-mediated communication aligns the users closer to their offline selves. The increasing emphasis on existing offline identities and relationships as well as physical and non-verbal communication cues and manipulation defines the nature of computer-mediated communication today (Stefanone et al. 2008).

THE PROLIFERATION OF NETWORKING SITES

The shift in the discourses of empowerment and increasing need to declare and share identities may also be attributed to the technological advancements inherent in the new social Web. Web 2.0 encapsulates a plethora of tools - including wikis, blogs, and folksonomies - which promote creativity, collaboration, and sharing between users (Szomszor et al. 2008:33). The multimedia experience and social communication platforms thus equally characterise the Web 2.0 user culture. From a technical and social perspective, Web 2.0, in comparison to its earlier manifestation, refers to improved applications, increased utilization of applications by users, and the incorporation of content generative technologies into everyday life by those who can afford and access such technologies. Anderson (2007) identifies the Web 2.0 environment as a new and improved second version and particularly a user-generated Web which is characterised by blogs, video sharing, social networking, and podcasting - delineating both the production and consumption of the Web environment where both activities can be seamless. Beyond its technical capacities, the term is a convenient social construct to analyze new forms of processes, activities, and behaviors - both individual and collective as well as public and commercial - that have emerged from the Internet environment.

Unlike earlier Websites which thrived on the notion of anonymity and virtuality, these new platforms for social communion emphasise the declaration of real offline identities to participate in the networking phenomenon. While the forerunners of social networking sites in the 1990s included sites (such as Classmates.com), the advent of the new millennium heralded a new generation of Websites which celebrated the creation of self-profiles with the launch of Friendster in 2002 which attracted over 5 million registered users in a span of a few months (See Rosen, 2007; Mika 2004; Mislove et al. 2007; DiMicco & Millen 2007). Friendster was soon followed by MySpace, Livejournal.com, and Facebook; these sites convened around existing offline communities such as college students. In the case of MySpace the site was originally launched by musicians to upload and share videos while Facebook initially catered to college students but is presently open to anyone who wants to network socially online. Some of these sites have witnessed phenomenal growth since their inception. MySpace, for example, has grown from 1 million accounts to 250 million in 2008 (Caverlee et al. 2008: 1163).

Additionally, several of the top ten most visited sites on the Web are social networking sites (cf. Golbeck & Wasser 2007: 2381).The creation and exhibition of self-profiles can be historically located and is not unique to the new media environment. Christine Rosen (2007) points out that historically the rich and powerful documented their existence and their status through painted portraits. In contemporary culture, using a social networking site is akin to having one's portrait painted, although the comparative costs make social networking sites much more egalitarian. She contends that these digital 'self-portraits' signify both the need to re-create identity through the online platform as well as to form social connections. Invariably images play a role in the representation of self and in fostering communication (See Froehlich et al.2002; Schiano et al. 2004Bentley et al. 2006). For Rosen (2007: 15), the resonant strand that emerges is the 'timeless human desire for attention'.

Undoubtedly, users join such sites with their friends and use the various messaging tools to socialize, share cultural artefacts and ideas, and communicate with one another (Boyd, 2007). As such, these sites thrive on a sense of immediacy and community (Barnes, 2006). With social networking sites there is a shift in the re-making of identity. While social connective sites in the 1990s illuminated the sense of place with home pages, global villages, and cities, with social networking sites there has been an emphasis on the creation of the 'self' through hobbies, interests, interactions, and a display of users' contacts through multimedia formats (Rosen, 2007). According to Boyd and Heer (2006: 2), 'the performance of social identity and relationships through profiles has shifted these from being a static representation of self to a communicative body in conversation with other represented bodies. The emphasis of self-expression, through the creation of profiles, anchors publicity, play, and performance at the core of identity formation and communication. As such, identity is mutable online and not embodied by the body, and often the need to disclose real-life identities is intimately tied to this community's code of authenticity in making identity claims where friends and peers can verify claims made in the profiles (Donath & Boyd, 2004).

Social networking sites can support a variety of shared multimedia content beyond photos to include video and music, can be constitutive of self-identity and representation, and become a playground for the creativity of millions (Geyer et al. 2008:1545). Geyer et al. (2008) point out that while such sites connect people with each other through content and profile sharing, some sites focus on a single content type as in the case of Flickr and YouTube when communities form through the sharing of photos or videos. Other sites may entail sharing of many types of content.

In assessing SNS's, Boyd and Ellison (2007) highlight three distinctive features: the user's ability to construct a public profile, articulate a list of other users with whom they share a connection, and view and traverse their list of connections and those made by others within the system. According to Ralph Gross and Alessandro Acquisti (2005), such sites, through the emphasis on personal profiles, offer a representation of users for others to peruse with the intention of contacting or being contacted by others to meet new friends or dates, find new jobs, receive or provide recommendations, and much more.

Dana Boyd (2006) postulates that while the meanings of practices and features can differ across sites and individuals the notion of sharing is intrinsic to these sites. Personal information and private comments on a public platform then become a form of social capital which people trade and exchange to build new ties and to invite different types of gaze and spectatorship. Chapman and Lahav (2008) point out that while there is a novelty surrounding social networking behavior from the perspective of researchers, this behavior will become increasingly integrated with other forms of communication as social networking becomes increasingly incorporated into one's everyday routines. This means that social networking behavior will function in conjunction with other communication options including email, instant messaging, and mobile devices.

The need to attract public attention in some way through daily interactions and to seek familiar and unknown audiences characterizes social networking sites. Stefanone et al. (2008) maintain that this behavior is linked to the 'celebrity culture' that is evident in mainstream media and particularly in television genres such as reality TV (See Stefanone et al 2008). With user-generated content and the ability to host profiles on interactive sites, the Web 2.0 environment enables users to participate in celebrity culture by constructing themselves as active personas online. Stefanone et al. (2008: 107) contend that new multimedia technologies erode 'the behavioural and normative distinctions between the celebrity world and the mundane everyday lives of the users.' They argue that the dissolution of this boundary is discernible in two resonant strands: the popularization of the reality television genre and the proliferation of social networking sites which hinge on the revelation of offline identities. They identify these two trends as reconfiguring the media environment where audiences are more than the recipients of media messages. Audiences as users and consumers can become 'protagonists of media narratives and can integrate themselves into a complex media ecosystem' (Stefanone et al. 2008: 107). They argue that platforms, such as social networking sites, emphasize aspects of human interaction that have been traditionally associated with celebrities, including the primacy of image and appearance in social interaction. This may have social implications, such as 'promiscuous friending,' where the network is both a collection of known relationships as well as people with whom users may have never met. Beyond enabling social connections, this could lead to fame seeking or the desire to be 'popular' through the social imaginary of the multimedia environment.

The popularity of such sites may also be explained by the need of some people to look into other peoples' lives or to increase awareness of others within their physical and virtual communities (Strater & Richter 2007: 157). Inherent to such landscape is the ability to track other members of the community where the 'surveillance' functions allow an individual to track the actions and beliefs of the larger groups to which they belong (Lampe et al. 2006: 167). Lampe et al. (2006) define this as social searching or social browsing where it enables users to investigate specific people with whom they share an offline connection. Lampe et al. (2006:167) take the relationship between social networking and social browsing further by asserting that 'users largely use social networking to learn more about people they meet offline and are less likely to use the site to initiate new connections.'

IDENTITY AND SOCIAL NETWORKS

The identities established in social networking sites function to enable offline and online networks. Often, the identity that is constructed online reflects the complex entwining of computer-mediated communications on the one hand and offline social networks on the other. Jon Kleinberg (2006: 5) contends that 'distributed computing systems have incessantly been entwined with social networks that fuse user populations' in the online and offline environments. The growth of activities, such as blogging, social network services, and other forms of social media on the Internet has made large-scale networks more evident and visible to the general public. As Adam Joinson (2008: 1027) points out, Websites, such as Facebook, were originally built around existing geographical networks of student communities. This meant that offline communities were reflected in the online environment; such communities function in a number of ways: for example, as a means of sustaining relationships by providing social and emotional support; as an information repository; and, as offering the potential to expand one's offline and online networks. In view of this, on-line settings beyond being rich data on the construction of identity by users have also become rich sources of data for large-scale studies of social networks (See Backstorm et al. 2007; Caverlee et al. 2008; Mislove 2007; Mori 2005; Goussevskaia 2007; Joinson 2007).

Wellman (cf. Lange, 2007: 2) defines *social networks* as 'relations among people who deem other network members to be important or relevant to them in some way, with media often used to maintain such networks.' Another essential component of such sites is that user profile information involves some element of 'publicness' (Preibusch et al,. 2007) and it is the consumption of private details which sustains the culture of gaze and the curiosity of the invisible audience. Communication technologies, such as the Internet with its global platform where data can be endlessly circulated and anyone can leave electronic footprints, 'erode the boundaries between 'publicity' and 'privacy'(Weintraub & Kumar, 1997). Lange (2007), explains that social network sites are Websites that allow users to create a public or semi-public profile within the system and one that explicitly displays their relationship to other users in a way that is visible to anyone who can access their profile. Consequently, Boyd (2007) considers SNS's as the latest generation of 'mediated publics' where people can gather publicly through mediated technology. She points out that features (such as, persistence - i.e. the permanence of a profile and its circulation in cyberspace, searchability, replicability, and invisible audiences) constitute the key elements of this environment. Users' behavior may be mediated by these features without necessarily integrating the underlying immediate and future consequences or risks embedded in these technologies or their actions.

The ability to tag photos to profiles and the presence of photo recognition software means that there is a loss of visual anonymity which can be complemented by new forms of gaze (Montgomery, 2007). Equally, the semantic nature of the Web can simplify or reduce 'Web-based communications to the descriptive and may unconsciously ascribe social values by describing these relationships as 'friends' or 'acquaintances.' According to a study done by Sheffield Hallam University, while the number of friends people can have on such sites is massive, the actual number of close friends is approximately the same in the face-to-face real world (Randerson, 2007). Mori et al. (2005: 82) point out that Semantic Web-based ontologies deliberately simplify such relationships. Danah Boyd (2004:1280), in her enthnographic study of the Friendster site, found 'people are indicated as friends even though the user does not particularly know or trust the person.' In this sense, the Semantic Web flattens the complexity of relationships and falsely assumes that publicly visible or articulated social networks and relationships can be conflated with private relationships.

As such, publicity (and public labels such as 'friends'), exchange, and sharing are integral and definitive parts of the SNS culture where the emphasis is not entirely on the authenticity of the information but the elements of connection and connectivity it can create (Nardi, 2005). Social networking sites can also capture the shift in identities of users when they transition from one life phase into another. Such transitions can include progress from school to work place where the social connections and identity of the user can shift. In tandem with this, DiMicco and Millen (2007: 383) argue that these platforms are complex as such sites can reflect the fact that users have 'transitioned between life stages and expanded their number of social connections, and these sites can assist users in maintaining social networks and diverse social relations.' This then entails a degree of managing self-identity on such spaces and perhaps the creation of multiple identities through multiple profiles to delineate a distinction between corporate and non-coporate identites, for example, or between formal and professional relationships compared to long-term friendships. DiMicco and Millen's (2007: 386) research reveals that multiple identities can nevertheless be burdensome where users may require more technical knowledge to navigate access control mechanisms which may deny access to one set of users while allowing a target audience to view selected profiles.

Gross & Acquisti (2005), point out that the most common model of the SNS is the presentation of the participant's profile and the visualization of his or her network of relations to others. Such sites can encourage the presentation of a member's profile (including their hobbies and interests and the publication of personal and identifiable photos) to the rest of the community through technical specifications, while visibility of information can be highly variable amongst such sites. Most networking sites make it easy for third parties, from hackers to government agencies, to access participants' data without the site's direct collaboration, thereby exposing users to risks ranging from identity theft to online and physical stalking and blackmailing (Gross & Acquisti, 2005).

Additionally users browse neighbouring regions of their social network as they are likely to find content that is of interest to them. Thus search engines may use social networks to rank Internet search results relative to the interests of users' neighbourhoods in the social network.

RISKS

The casualness with which people reveal personal details online is related to the different norms people apply to online and offline situations where these variant norms have implications for notions of privacy, authenticity, community, and identity. Research conducted by Mislove et al. (2007:41) on sites (such as Flickr, LiveJournal, Orkut and YouTube) reveals that these sites are the portals of entry into the Internet for millions of users. Equally, they invite advertisements as well as the pursuit of commercial interest; this means users in these networks tend to trust each other having been brought together through common interests. *Trust* has been defined in various ways in sociological literature. Golbeck et al. (2003) define it as credibility or reliability in human interaction where it can entail the according of a degree of credence to a person through interpersonal communication. With specific reference to information sharing, Mori et al. (2005:83) infer that trust could relate to reliability with regard to how a person handles information that has been shared or reciprocated. Gips et al. 2005 argue that the 'social' aspect of these networks is self-reinforcing; this means to be trusted one must make many 'friends' and create many links that will slowly pull the user into the core of activities.

Barnes (2006), in citing Benniger, postulates that electronic forms of communication are gradually replacing traditional modes of interpersonal communication as a socializing force, mediating and at times displacing social norms in different contexts. In the interactive spaces of the Internet, there may be a disconnect between the way users say they feel about the privacy settings of their blogs and how they react once they have experienced the unanticipated consequences of a breach of privacy (cf. Barnes, 2006; Gibson 2007; Mannan & Oorschot 2008). The issue of privacy setting can be problematic on social networking sites for users since a default privacy setting can constrain a user's need to meet and network with more people beyond their offline network (Joinson 2008: 1035). Mannan & Oorschot (2008: 487) concur that there is a tendency to overlook privacy implications in the current rush to join others in 'lifecasting' and users may work on the false impression that only friends and family are consuming the personal content.

Gross & Acquisti's (2005) anecdotal evidence suggests that participants are happy to disclose as much information as possible to as many people as possible, thus highlighting the design and architecture of sites which hinge on the ease with which personal information is volunteered and the willingness of users to disclose such information. The perceived benefits of selectively revealing data to strangers may appear larger than the perceived costs of possible privacy invasions. Other factors (such as, peer pressure and herding behavior, relaxed attitudes towards or lack of interest in, personal privacy, incomplete information about the possible privacy implications of information revelation, faith in the networking service or trust in its members, and the myopic evaluation of privacy risks of the service's own user interface) may drive the unchallenged acceptance by users of compromises to their safety (Gross & Acquisti 2005; Strater & Richter 2007; Gibson 2007), thus sealing the role of SNS's as complicit risk communities.[this sentence was long enough to confuse and the puncuation added to the confusion; I hope these changes clarify the author's meaning.] Strater & Richter (2007) point out that large-scale analyses of Facebook have revealed that a majority (87% on average) of students have default or permissive settings. While a significant majority have an awareness of privacy options, less than half ever alter their default setting. This means that while users do not underestimate the privacy threats of online disclosure, they can nevertheless misjudge the 'extent, activity and accessibility of their social networks' (Strater & Richter 2007: 158).

According to a 2008 study by Ofcom, over one-fifth of UK adults have at least one online community profile (cf. Szomszor et al. 2008). Caverlee at al (2008: 1163) nevertheless point out that the growth of social networking sites has come with a huge cost as these sites have been subject to threats: such as, specialised phishing attacks, the impersonation of profiles, spam, and targeted malware dissemination. Unanticipated new threats, they state, are also bound to emerge. They identify three resonant vulnerabilities which plague social network users: malicious infiltration, nearby threats, and limited network view. Malicious infiltration covers the illusion of such networks being secure through the provision of requiring a valid email address or a registration form when in effect malicious participants can still gain access. Similarly, nearby threats allude to the nearness of malicious users who can be a 'few hops away' despite users believing they have a tight control over their direct friends. Lastly, a limited network view describes the fact that users have a myopic perspective on the entire network as they may not be privy to information about the vast majority of participants in the entire network. The Facebook site for example maintains over 18 million user profiles with 80% to 90% of college undergraduates as users where users are allowed to disclose more varied information fields on the site (cf. Strater & Richter 2007: 157). Strater and Richter's (2007) research on Facebook also reveals that users were unaware of the ability of others to remove, delete, and in other ways control tagged photographs and wall posts from their profiles, thereby consigning such personal images and information to a life of permanent circulation and consumption on the Web.

Barnes (2006), in citing Katz and Rice (2000), describes the Internet as a 'Panopticon where surveillance is part of the architecture'. There are a myriad of risks lurking in the trails of data people leave in SNS sites and in the ways it is mined for commercial, legal, and criminal purposes. SNS's (such as, LiveJournal.com, Facebook, Myspace, Friendster, and Google's Orkut.com) have been a source of concern in the US, initiating federal laws that require most schools and libraries to render such Web spaces inaccessible to minors in order to protect them from harm (McCullagh, 2006). Similarly, in the UK, the House of Lords Science and Technology select committee has suggested that both private and public sectors need more effective ways to deal with the rise of online fraud and hacking and have recommended the formation of a new national police squad charged with reducing online crime (Johnson, 2007). The Information Commissioner's Office (IOC) in the UK has also drawn up official guidelines for the millions of people who use such sites, offering warnings such as 'a blog is for life' and 'reputation is everything'. People are also advised that entries can leave an 'electronic footprint' and that the lives of people can be put at risk by the reckless disclosure of information (Hough, 2007). The notion of data and profiles having a permanence and circulation in unexpected ways is something the ICO wants to impress on people in terms of potential harm and transgression of privacy.

Early discourses of the Internet celebrated not only the ability to re-invent identity online but also the concept of 'avatarism' where a user can have multiple identities. But although this can certainly be empowering, it can also enable new forms of deception. New forms of narcissism enabled by SNS's, however, celebrate the notion of constructing one's offline profile online and inviting others to start friendships through such representations of self. Users may not then think beyond the cultural ethos of these spaces. Additionally, in tandem with the declaration of real identities online, deception and faking are also part of the terrain. Dana Boyd (2004), in observing the Fakesters in Friendsters Website, notes that users' appropriation of well-known celebrity and media profiles, or the invention of their own, 'exercises a certain creativity and introduces playful expression' which draws an audience that wishes to engage with these users. She asserts that 'fakesters' were means of 'hacking the system to introduce missing social texture'. Boyd's argument is that the phenomenon of the Fakesters reflects the fundamental weakness of trust in the network (in this particular instance the reference was to Friendsters) where there is an ambiguity between the real and the parody.

Beyond the personal information posted by social networkers, there are also worries about privacy after Facebook's secret operational code was published on the Internet. The Facebook site in the UK has 3.5 million users and about 30 million users worldwide. The company blamed the leaked code on a 'bug' which meant that it was published accidentally (Johnson, 2007). While such glitches may not necessarily allow hackers to access private information directly, they could nevertheless help criminals close in on personal data. While some personal information listed on the site is semi-private, government and quasi-government agencies, such as Get Safe Online in the UK, are worried that criminals who become friends with other users have the potential to find out much more information about them (Johnson, 2007). Research by Websense supports the idea that criminals 'work as an underground community, sharing information on what tools and methods work when it comes to tricking consumers on SNS and hackers have realized that they need to become discreet when it comes to social networking since they need to blend in with the crowd where links can be added to sites, such as Wikipedia, to lure users onto corrupt sites' (Vassou, 2006; Newman

2006). There have also been numerous incidents of spyware and spamming being employed by such sites (Rosen, 2007).

The constant demand to make these sites attractive to advertisers means that privacy of users can be compromised in other ways. Wendlandt (2007) notes that online advertising is the fastest growing segment of the advertising industry, currently accounting for more than 25% of advertising growth per year, translating to more than five times the recent average annual growth of other types of media with about 6-7% spent on Internet advertising globally. Recently, 13,000 Facebook users signed a petition protesting against the networking site's new advertising system which alerts members of friends' purchases online. Some Facebook members have even threatened to leave due to the fact that the new system allowed their friends to find out what they were planning to give them for Christmas (Wendlandt, 2007). Preibusch et al. (2007) point out that popular SNS sites, such as MySpace.com, collect data for e-commerce purposes. User profiles are important for data mining in such Websites. Data that accrues on the Web is not only used for communicating but also for secondary purposes that may be covered in the SNS's terms of use. Such data can be acquired by marketing agencies for targeted marketing or by law enforcement agencies and secret services, etc (Preibusch et al, 2007).

FUTURE TRENDS

With the increasing popularity of social networking sites, the incorporation of various multimedia formats and functions in these platforms, the supplanting of actual offline networks through social networks on the internet construct social networking sites as viable spaces for the movement of new forms of both social and financial capital (i.e. advertising, e-commerce and data mining). Here the act of connecting with larger user communities present challenges and risks for users, social software designers, commercial organizations and government bodies. The increasing appropriation of social networking sites into our everyday lives (through mobile technologies) and everyday engagements mean that visibility and non-visibility of social and personal networks will construct online identities as a vital part of a data economy. The need to reveal and to limit information flows and to enact a secure environment for users whilst enforcing users to comply with data management protocols on such sites will enact these as a contested space of new forms of sociability and social deviance. The users' notion of security, privacy and the human need for communion will continue to temper the social networks as complex and complicit risk communities.

CONCLUSION

The narcissistic streak in social networking sites that is evident through the creation of self- profiles hinges on the disclosure of offline identities where public spectacle and gaze repoliticize the construction of self. The notion of self in social networking sites is both imagined through self-description and crafted through textual and multimedia environments but equally through its articulation and display of contacts and its ability to invite or deny communion with other users. In this sense, the concept of self is anchored through both individual agency and imagination as well as other users' gaze and consumption of these profiles. This explicit ethos of exposure, display, and spectacle define the cultural ethos of social networking sites. This phenomenon again ignites debates about the issues of identity formation on the Internet where identity can be created and defined in multiple ways and is amenable to deception and inauthenticity. In the process, it highlights the complex nature of the Internet environment which can demand different cultural responses from different online spaces and communities of users. Self-exposure and narcissism gives a platform for re-definition of offline identities and new sociabilities which can in turn reconfigure and redefine the notion of friendship and community in these spaces. SNS's also herald the emergence of complicit risk communities where personal information becomes social capital which is traded and exchanged and where the concept of public or private can be defined through the nature of users' access, gaze, and the transactions and interactions they permit.

The culture of social networking sites thrives on the narcissistic and the performative, on one hand, and reciprocity and exchange, on the other. Hence the potential dangers and risks of willingly disclosing and displaying personal details become part of the architecture or code of these sites. The appropriation of new technologies by individuals in order to communicate, form new communities, and maintain existing relationships signifies new ways in which risk becomes embedded and encoded into our social practices, posing new ethical and legal challenges which inadvertently expand the landscape of risk.

REFERENCES

Anderson, P. (2007). *What is Web 2.0? ideas, technologies and implications for education,' JISC Technology & Standards Watch*. Retrieved, July 19, 2008, from http://www.jisc.ac.uk/media/documents/techwatch/tsw0701b.pdf

Backstrom, L., Cynthia, D., & Kleinberg, J. (2007). Wherefore art thou R3579X? Anonymized social networks, hidden patterns, and structural steganography. *Proceedings of the WWW 2007 conference*, May 8-12, 2007, Alberta, Canada.

Barnes, S. B. (2006). A privacy paradox: Social networking in the United States. *First Monday, 1*(9). Retrieved July 19, 2008, from http://firstmonday.org/issues/issue11_9/barnes/index.html

Bentley, F., Metcalf, C., & Harboe, G. (2006). Personal vs. commercial content: The similarities between consumer use of photos and music. *Proceedings of the SIGCHI conference on Human Factors in computing systems*, April 22-27, 2006, Montréal, Québec, Canada

Boyd, D. (2004). Friendster and publicly articulated social networking. *CHI 2004,* April 24-29, 2004, Vienna, Austria.

Boyd, D. (2007). Social network sites: public, private or what? *Knowledge Tree*, 13, Retrieved Dec 12, 2007, from http://kt.flexibelearning.net.au/tkt2007/?page_id=28

Boyd, D., & Ellison, N. B. (2007). Social networking sites: Definition, history and scholarship. *Journal of Computer Mediated Communication, 13*(1), 11. Retrieved Dec 12, 2007, from http://jcmc.indiana.edu/vol13/issue1/boyd.ellison.html

Boyd, D., & Heer, J. (2006). Profiles as conversation: Networked identity performance on friendster. *Proceedings of the Hawaii International Conference on System Sciences (HICSS-39)*, January 4-7, Persistent Conversation Track, Kauai, Hawaii.

Caverlee, J., Liu, L., & Webb, S. (2008). Towards robust trust establishment in Web-based social networks with social trust. [Beijing, China.]. *WWW, 2008*(April).

Chapman, C., & Lahav, M. (2008). International ethnographic observation of social networking sites. *Proceedings of the CHI 2008 conference on Human Factors in Computing Systems*, April 5-10, 2008, Florence, Italy.

Diani, M. (2000). Social movement networks: Virtual and real. *Information Communication and Society, 3*(3), 386–401. doi:10.1080/13691180051033333

DiMicco, J., & Millen, D. (2007). Identity management: multiple presentations of self in facebook. *Group '07, Conference on Supporting Group Work*, November 4-7, 2007, Sannibel Island, Florida, USA.

Donath, J. (1998). Identity and deception in the virtual community. In P. Kollock, & M. Smith (Eds.), *Communities in cyberspace* (29-59). London: Routledge.

Donath, J., & Boyd, D. (2004). Public displays of connection. *BT Technology Journal*, *22*(4), 71–81. doi:10.1023/B:BTTJ.0000047585.06264.cc

Friedland, L. A. (1996). Electronic democracy and the new citizenship. *Media Culture & Society*, *18*, 185–212. doi:10.1177/016344396018002002

Froehlich, D., Kuchinsky, A., Pering, C., Don, A., & Ariss, S. (2002). Requirements for photoware. *Proceedings from the Conference on Computer Supported Cooperative Work: CSCW 2002*. New Orleans, LA.

Froomkin, A. (1995). Anonymity and its enmities. *Journal of Online Law*, 4. Retrieved Dec 12, 2007 from http://www.wm.edu/law/publications/jol/95_96/froomkin.html

Geyer, W., Dugan, C., DiMicco, J., Millen., D., et al. (2008). Use and reuse of shared Lists as a social Content Type. *Proceedings of the CHI 2008, conference on Human Factors in Computing Systems,* April 5-10, 2008, Florence, Italy.

Gibs, J., Fields, N., Liang, P., & Plipre, A. (2005). SNIF: Social networking in Fur. *Proceedings of the CHI 2005 conference on Human Factors in Computing Systems*, April 2-7, 2005, Portland, Oregon, USA.

Gibson, R. (2007). Who's really in your top 8: Networking security in the age of social networking. *Proceedings of the SIGUCCS' 07, conference on user services*, October 7-10, 2007, Orlando, FL.

Gibson, W. (1984). *Neuromancer*. New York: Ace Books.

Giddens, A. (1990). *The Consequences of modernity*. Stanford, CA: Stanford University Press.

Golbeck, J., Hendler, J., & Parsia, B. (2003). Trust networks on the Semantic Web. *Proceedings from Agents 2003, conference on cooperative information agents*, August 27-29, Helsinki, Finland.

Golbeck, J., & Wasser, M. (2007). Social browsing: Integrating social networks and Web browsing. *Proceedings from CHI 2007, conference on Human Factors in Computing Systems,* April 28 – May 34, San Jose, CA.

Goussevskaia, O. Kuhn. M., & Wattenhofer, R. (2007). Layers and hierarchies in real virtual networks. In *proceedings of IEEE/WIC/ACM International Conference on Web Intelligence* (pp 89-94). IEEE Computer Society: Washington, DC.

Gross, R., & Acquisti, A. (2005). Information revelation and privacy in online social networks. *Workshop on Privacy in the Electronic Society* (WPES). Retrieved from Dec 4, 2007 from http://privacy.cs.cmu.edu/dataprivacy/projects/facebook/facebook1.html

Horst, S. A., & Miller, D. (2006). *The Cell Phone: An Anthropology of Communication*. Oxform, UK: Berg.

Hough, A. (2007, November 23). Fraud warning for users of social networking sites. *Reuters*. Retrieved Dec 4, 2007, from http:today.reuters.co.uk/misc/

Johnson, B. (2007, Aug 13). Facebook's Code Leak Raises Fears of Fraud. *Guardian Unlimited*. Retrieved April 12, 2007, from http://www.guardian.co.uk/technology/2007/aug/13/internet

Joinson, A. (2008). 'Looking at,' 'looking up' or 'keeping up with people?' motives and uses of facebook. *Proceedings from CHI 2008, conference on human factors in computing systems*, April 5-10, Florence, Italy.

Jordan, T. (1999). *Cyberpower: the culture and politics of cyberspace and the internet*. London: Routledge.

Kendall, L. (2002). *Hanging out in the virtual pub: identity, masculinities, and relationships online*. Berkeley, CA: University of California Press.

Kleinberg, J. (2006). Distributed social systems. *Proceedings from PODC'06, conference on principles of distributed computing*, July 22-26, Denver, CO.

Lampe, C., Ellison, N., & Steinfield, C. A. (2006). Face(book) in the crowd: Social searching vs. social browsing. *Proceedings of the CSCW'06, Conference on Computer Supported Cooperative Work*, November 4-8, Banff, Alberta, Canada.

Lange, P. (2007). Publicly private and privately public: Social networking on youtube. *Journal of Computer-Mediated Communication, 13*(1), 18. Retrieved December 4, 2007, from http://jcmc.indianna.edu/vol113/issue1/lange/html

Mannan, M., & Oorschot, P. (2008). Privacy-enhanced sharing of personal content on the Web. *WWW '08*, April 21-25, 2008, Beijing, China.

Mika, P. (2004). Social networks and the Semantic Web. *Proceedings of the IEEE/WIC/ACM International Conference on Web Intelligence*, 20-24 September, Beijing, China.

Miller, D., & Slater, D. (2000). *Internet: An Ethnographic Approach*. London: Berg.

Mislove, A., Marcon, M., & Gummadi, K. (2007). Measurement and analysis of online social networks. *Internet Measurement Conference '07*, October 24-26, San Diego, CA.

Montgomery, E. (2007). Facebook: Fraudsters' Paradise? *Money.UK.MSN.com*, November 20, 2007. Retrieved December 4, from http://money.uk.msn.com/banking/id-fraud/article.aspx?cp-documentid=5481130

Mori, J., Sugiyama, T., & Matsuo, Y. (2007). Real-world oriented information sharing using social Networks. *Proceedings from SIGGROUP '07 conference on supporting group work*, Sanibel Island, FL.

Nardi, B. A. (2005). Beyond bandwidth: Dimensions of connection in interpersonal communitation. *Computer Supported Cooperative Work, 14*, 91–130. doi:10.1007/s10606-004-8127-9

Newman, R. (2006). Cybercrime, identity theft and fraud: practicing safe internet – Nework security threats and vulnerabilities. *InfoSecCD Conference '06*, September 22-23, 2006, Kennesaw, GA.

Preibusch, S., Hoser, B., Gurses, S., & Berebdt, B. (2007, June). Ubiquitous social networks – opportunities and challenges for Privacy-aware user modelling. *Proceedings of the Data Modelling Workshop*, Corfu, Greece. Retrieved December 4, 2007, from http://vasarely.wiwi.hu-berlin.de/DM.UM07/Proceedings/05-Preibusch.pdf

Randerson, J. (2007). Social network sites do not deepen friendships. *The Guardian*, September 10. Retrieved December 4, 2007, from http://guardian.co.uk/science/2007/sep/10/socialnetwork/print

Rheingold, H. (2000). *The Virtual Community*. New York: Harper Collins.

Rosen, C. (2007). Virtual friendship and the new narcissism. *New Atlantis (Washington, D.C.), 17*, 15–31.

Schiano, D., Nardi, B., Gumbrecht, M., & Swartz, L. (2004). Blogging by the rest of us. *Proceedings in CHI 2004, Conference on Human Factors in Computing Systems*, April 2004, Vienna, Austria.

Slater, D. (2002). Social relationships and identity on/off-line. In L. Lievrouw & S. Livingstone (Eds.), *Handbook of New Media: Social Shaping and Consequences of ICTs.* London: Sage.

Slevin, J. (2000). *The Internet and Society*. Cambridge, UK: Polity Press.

Stefanone, M., Lackoff, D., & Rosen, D. (2008). We're all stars now: Reality television, Web 2.0, and mediated identities. *HT*'08, June 19-21, Pittsburgh, PA.

Strater, K., & Richter, H. (2007). Examining Privacy and Disclosure in a Social Networking Community. *Symposium on Usable Privacy and Security* (SOUPS) 2007, July 18-20, 2007, Pittsburgh, PA.

Szomszor, M. (2008). Correlating user profiles from multiple folksonomies. *HT '08 Conference on Hypertext and Hypermedia,* June 19-21, 2008, Pittsburgh, PA.

Vassou, A. (2006). Social networking sites driving new wave of security. *Computeractive*, December 13, 2006. Retrieved Dec 4, 2007, from http://www.computeractive.co.uk/articles/print2170872

Weintraub, J., & Kumar, K. (Eds.). (1997). *Public and private in thought and practice*. Chicago: University of Chicago Press.

Wendlandt, A. (2007). Web advertising to come under EU scrutiny. *Reuters*, November 23. Retrieved Dec 4, 2007, from http://www.reuters.com/articlePrint?articleId=USL229260820071123

Wynn, E., & Katz, J. E. (1997). Hyperbole over cyberspace: Self-presentation and social boundaries in internet home pages and discourse. *The Information Society, 13*(4), 297–327. doi:10.1080/019722497129043

This work was previously published in Collaborative Technologies and Applications for Interactive Information Design: Emerging Trends in User Experiences, edited by S. Rummler & K.B. Ng, pp. 82-95, copyright 2010 by Information Science Reference (an imprint of IGI Global).

Chapter 7.15
Audience Replies to Character Blogs as Parasocial Relationships

James D. Robinson
University of Dayton, USA

Robert R. Agne
Auburn University, USA

ABSTRACT

News anchors, talk show hosts, and soap opera characters often become objects of parasocial affection because of the nature of these program genres. This chapter explores the concept of parasocial interaction by focusing on audience replies to blog posts made on behalf of a TV character, Jessica Buchanan of ABC Television Network's *One Life to Live* show. The authors employ communication accommodation theory to illuminate the concept and to identify specific communicative behaviors that occur during parasocial interaction. The chapter presents evidence of parasocial interaction within the blog replies and audience accommodation to the blog posts. Analysis suggests that parasocial interaction is the mediated manifestation of the relationship dimension inherent in television messages and used by audience members in much the same way it is used during face-to-face interaction.

DOI: 10.4018/978-1-60566-368-5.ch027

INTRODUCTION

It is estimated that in the U. S. 12 million adults "blog" or keep online journals and 57 million adults or 39% of all adult Internet users report reading blogs (Lenhart & Fox, 2006). A worldwide total of 175,000 new blogs are created every day, and the web search engine Technorati (2008) reports tracking 112.8 million blogs worldwide. Blogs are used as a vehicle for providing commentary to the public. The critical differences in blogs and diaries are the opportunities for reaching a mass audience and the opportunity for that mass audience to respond to the commentary found within the blog. Because of the interactive nature of the blogs and blogging software, readers are able to add comments, links, pictures, video, or any other media format to the blog for the edification and entertainment of other denizens of the Internet.

Popular television characters - such as, Dwight Schrute (*The Office*), Joe the Bartender (*Grey's Anatomy*), and Jessica Buchanan (*One Life to Live*) - have blogs that allow audience members' additional insight into the character's identity and additional information about the story or plotline. These blogs are different from the blogs maintained by actors since they are written from the perspective of a fictional character. More importantly these character blogs allow audience members the perception that they can interact with the character – even though this interaction is parasocial. While audience members have always had some opportunity to interact, or more often, parasocially interact with characters through fan mail, the messages they send have not been readily available to scholars for study. Blog messages are more plentiful and easier to access, and provide communication scholars an invitation for studying parasocial interaction in depth.

In this chapter we first address what is known about parasocial relationships between the audience and TV characters. We then introduce communication accommodation theory as a framework for identifying specific communicative behaviors that are likely to occur during parasocial interaction. An analysis of a TV character blog determines whether parasocial behaviors occur in blog replies and whether there is evidence of audiences accommodating the communicative behavior of the character. Finally we offer some suggestions for future research and future trends in this line of research.

BACKGROUND

Parasocial Interaction

The term *parasocial interaction* was used by Horton and Wohl (1956) to explain feelings of closeness audience members feel toward television characters. This closeness is believed to arise when TV characters behave in ways that resemble face-to-face interaction. This feeling of intimacy can be enhanced by production characteristics, such as the selection of shots and the format of the program. Bell (1991) suggests audience members may also feel as if they are engaged in an interaction when the characters seem to be adapting their behavior to the anticipated reaction of the audience. An example may help illustrate this notion. Imagine a scene where a talk show host is performing a monologue. On a small scale, a pseudo-interaction sequence might look something like this:

TV Host: Tells a bad joke.

Audience: Groans, boos, or merely does not laugh.

TV Host: Does a double take and makes a face.

In this example the audience members may feel as if the character told them the joke and then responded to their reaction. (Note that this is not a real interaction and the audience is aware of that.) Such interaction may seem more dynamic than a simple monologue because the character appears to be reacting to the audience.

Audience members are limited in their ability to reply or interact with their favorite TV characters. The audience member may "reply" by making commentary or talking back to the TV, laughing, or nodding their heads in agreement. Rather than sending fan mail, viewers may imitate face-to-face interaction. Again it is critical to acknowledge that the audience members understand that they are not actually interacting with the character. In a sense, the audience member is also acting like he or she is interacting with the character. More often, however, the audience member will do nothing more than think about the character's message and generate a reply. These parasocial interactions only occur in the minds of audience members but are nonetheless similar in some ways to actual interactions.

Since these faux interactions occur largely in the mind of the audience member, their responses to the character's messages can be viewed as cognitive. Greenwald (1968) recognized that people are often influenced more by their thoughts or cognitive responses to a message than by the message itself. More importantly audience members often recall their cognitive responses more accurately than they can remember the actual messages. These cognitive responses, then, may be considered as the cognitive enactment of parasocial interaction.

Audience replies to the blog posts of a TV character can be viewed as cognitive responses. In the cognitive response literature, the thought listing procedure is used by researchers to ascertain the thoughts of study participants. Applying this procedure researchers ask respondents to write down each individual thought that they go through while they are listening to a message (see Cacioppo, von Hippel, & Ernst, 1997 for an excellent overview of the technique). Audience replies to character blogs are similar to the thought listing procedure in that the replies identify the thoughts or issues of importance to them in response to the program and/or the character's blog postings. It is not unreasonable to expect that audience members in a close parasocial relationship will respond differently from an audience member who is less involved with the character.

Research investigating factors that increase the likelihood of parasocial interaction suggests audience members are more likely to report that they are in a parasocial relationship with a character when:

- A program presents a character in ways that resemble face-to-face interaction (Meyrowitz, 1986; Nordlund, 1978; Rubin, Perse, & Powell, 1985);
- The character engages in a conversational style of speaking and provides the audience an opportunity to respond (Rubin, Perse, & Powell, 1985); and
- Viewer involvement in a program is high (Rubin & Perse, 1987). Parasocial interaction is also more likely to occur with media personae that appear frequently on television (Levy, 1979).

Audience members are more likely to initiate face-to-face contact with characters when they report high levels of parasocial interaction (Gans, 1977; Horton & Wohl, 1956; McGuire & LeRoy, 1977). Talk show hosts, soap opera and TV shows characters, and news anchors have been most often examined as objects of parasocial affection because of the nature of those program genres.

Meyrowitz (1994) investigated the impact of losing parasocial relationships on audience members by examining audience reactions to the deaths of celebrities, such as John Lennon and Elvis Presley. He found that such a loss is not unlike the loss of a close friend and characterized these relationships as being very warm and caring. Cohen (2004) and others have examined the impact of characters being lost to show cancellations or cast restructuring and found a correlation between levels of parasocial interaction and expected breakup distress. This is consistent with Koenig and Lessan (1985) who found that viewers reported feeling closer to a favorite TV personality than to mere acquaintances (but not as close as a good friend). Audience members report feeling sorry for characters, missing characters, looking forward to seeing characters, seeking out information about the characters, and desiring to meet them in person (Rubin, Perse, & Powell, 1985).

While researchers have long suggested parasocial relationships can serve as a substitute for interpersonal relationships, the research has generally not supported this claim (Finn & Gorr, 1988; Rubin Perse, & Powell, 1985). Rather, there is little reason to believe that parasocial interaction can be predicted by social deficits, such as chronic loneliness (Ashe & McCutcheon, 2001; Perse & Rubin, 1989; Rubin, Perse, & Powell, 1985), neuroticism (Tsao, 1996), and low self-esteem (Tsao, 1996; Turner, 1993). In fact, people who have difficulty developing interpersonal relationships also often have trouble developing parasocial relationships (Cohen, 2004). These scholars have begun to suggest that parasocial relationships provide company for audience members (Isotalus, 1995) and are complementary to interpersonal relationships (Kanazawa, 2002, Perse & Rubin, 1990; Taso, 1996).

Given the increasing acceptance among researchers that parasocial relationships should be examined in much the same way that interpersonal relationships are studied, it may behoove scholars to employ communication accommodation theory in their efforts. Horton and Wohl (1956) suggest as much: "The more a performer seems to adjust his performance to the supposed response of the audience, the more the audience tends to make the response anticipated." Furthermore, the "simulacrum of conversational give and take may be called para-social interaction" (p. 215). Bell (1991) concurs and identifies accommodation theory as an excellent candidate for navigating the nexus between mass and interpersonal communication. They argue that audience perceptions of character accommodation – as manifest by the appearance of characters' adjusting their communicative performance to expected audience responses – should increase the likelihood of parasocial interaction.

Communication Accommodation Theory

Communication accommodation theory (Giles, 1973; Giles, Coupland, & Coupland, 1991) is an explanation of why people modify their communicative behavior to match the communicative behavior of others during face-to-face interaction. The theory is an extension of social identity theory (Tajfel & Turner, 1979; Tajfel, 1982) and suggests people behave in ways that will result in being seen as socially desirable. Proponents of social identity theory believe our conceptions of self are based on the social status ascribed to our membership groups. So from the perspective of social identity theory, an individual's social status is determined by social status attributed by themselves and others to their social groups. Similarly, when an individual is evaluating the social status of someone else, the social status of their group memberships plays a large role in determining that social calculus. Who we are is determined, in part, by the social groups that include and exclude us as members.

Accommodation theory suggests that when we interact with socially desirable others we adapt our communicative behavior to more closely approximate their behavior. Through this process of imitation or convergence, we are trying to ingratiate ourselves to be socially desirable and be viewed by others as being a member of that socially desirable group. We hope that through accommodation, the high status individual will like us and invite us into his or her social group. Strategically speaking we adopt the behavior of the socially desirable individual to reduce the communicative differences between us. On the most basic level, this adaptation of communicative style and content may be little more than imitating the behavior and language of the individual we are trying to accommodate so that we are accepted and liked. Of course, more skilled social interlocutors are able to converge using far less obvious and far more sophisticated techniques. Borrowing a phrase, adopting the cadence or speech patterns of the other, employing similar literary references, or even adopting the same metaphorical world view are all ways of converging with a conversational partner.

When we find ourselves interacting with someone less socially desirable, we maintain or increase communicative differences between us. In fact we behave in ways that signal to all that we are not like the person we are currently interacting. Behaving in ways that maintain our differences is called *divergence*; it is a strategy we use to maintain or increase the social distance between us.

Giles, Coupland, and Coupland (1991) suggest that sometimes people accommodate or adjust their linguistic and their nonverbal behavior in face-to-face conversations as a conscious strategy to gain approval from or to influence the other communicant. At other times we accommodate without being consciously aware of our behavior change. Whether the convergence is mindful or not, a growing body of research suggests people are influenced by people who accommodate or imitate their language (Giles, Coupland, & Coupland, 1991; van Baaren, Holland, Steenaert, & van Knippenberg, 2003) or their gestures (Chartrand & Baragh, 1999; Mauer & Tindall, 1983).

Research into the accommodation process has identified a number of behaviors including: being attentive (Ng, Liu, Weatherall, & Loong, 1997), offering compliments (Williams & Giles, 1996), head nodding, facial affect and smiling (Hale & Burgoon, 1984), pause lengths (Giles et al., 1987), posture (Condon & Ogston, 1967), self disclosure (Ehrlich & Graeven, 1971; Henwood, Giles, Coupland, & Coupland, 1993; Giles & Harwood, 1997), speech rate (Street, 1983), speech volume (Ryan, Hummert, & Boich, 1995), being supportive, (Ng et al., 1997), utterance length (Matarazzo et al., 1968; Giles et al., 1987), vocabulary (Giles, Mulac, Bradac, & Johnson, 1987), and vocal intensity (Natale, 1975; Welkowitz, Feldstein, Finkelstein, & Aylesworth, 1972).

While it could be argued that all audience members' responses to character blogs are parasocial interactions, other motivations, uses, or gratifications may also account for this behavior. If parasocial interaction is indeed motivated identification with or affinity toward the character, it is reasonable to expect audience replies to blog posts to prominently feature accommodation behavior. Accepting the premise that parasocial interaction should be studied in much the same way that interpersonal interaction should be studied, audience members should enact convergence behaviors when they reply to character blogs.

In an effort to test these ideas, character blogs and the replies to those blog posts were examined. Since the previous research suggests daytime serials have been one of the most widely researched program genres within the parasocial interaction literature and because the nature of "soap operas" encourage the development of parasocial relationships, the subsequent analyses focus on audience replies to the blog posts of Jessica Buchanan. Jessica Buchanan has been a character on the daytime program *One Life to Live* since its beginning. Taking place in Llanview – a fictitious suburb of Philadelphia, *One Life to Live* was created by Agnes Nixon and premiered on ABC in July 1968. Jessica Buchanan suffers a multiple personality disorder. Jess's second and quite distinct personality is Tess.

AN ANALYSIS OF AUDIENCE BLOG REPLIES

All of the blog posts penned by the character Jessica Buchanan between October 15, 2007 and January 11, 2008 were analyzed. Jessica made 20 entries to her blog during that time and a total of 56 audience members responded by sending 117 replies. A content analysis of the blog posts and the audience replies yields several interesting findings. None of the blogging was done by Jessica's other personality Tess. In two instances an audience member asked Jessica about the reappearance of Tess – most often in response to problems that Tess would remedy through violence. The typical blog post by Jessica was 75 words long – not including a heading and a date. These posts focus on her feelings about what has been happening within the plotline of the program. Jessica makes no mention of subplots or other characters unless they directly affect her somehow. In a sense, Jessica writes in her blog as if it were a journal or a diary that is being shared with the audience. There is no acknowledgement of the audience in her blog posts nor are there ever replies to questions or comments made by audience members.

Examination of audience replies to the blog posts indicates great disparity in their communicative behavior. Of the 56 individuals posting replies, only 9 wrote more than one time to Jessica. That is to say 83.9% of the replies were written by people who only wrote one message to Jessica over the three month period. Four of those 9 audience members wrote between 2 and 4 messages to Jessica and the remaining five audience members replied to blog posts 9 or more times during the period. This disparity is both statistically significant and theoretically significant since it provides evidence that even among fans of the show motivated enough to reply to a blog post, there is a great deal of variation in the amount of character contact audience members desire. In fact only one of the 117 blog replies contains a request for actual interaction. The fan wrote: "email me – it's the address above." That same fan also wrote the only blog replies indicating they had actually seen the character on the street. On these two different occasions the audience member wrote: "I saw you on Friday and you looked great – cool shirt," "I saw you today and you looked pretty cool," and "I just wanted to tell you that. Ok?" This particular fan replied to blog posts on 9 different occasions and repeatedly tells Jessica, "I'm your biggest fan." This last comment still brings to mind John Hinkley's last words to John Lennon even 28 years after the fact.

Audience members desiring contact are clearly interested in contact with Jessica and not the other audience members. Only three blog replies ask other audience members for information about a character or situation occurring on the soap. All three of these messages oriented toward other fans were written by individuals who only replied to one blog post during the entire period. Additionally, these requests for information from other fans only yielded one reply during the entire three month period. This seems to reduce the viability of the interactional starvation explanation for audience behavior.

The audience comments suggest clear evidence of parasocial interaction. Those audience members replying frequently tend to reply in much the same way they would reply to a friend via e-mail or a letter. In addition, the audience members making multiple posts to the blog often offer sympathy (e.g., "I hope you…" "I think you're better off…," and "Good for you Jess …") or advice (e.g., "you should …," "you're better off if you …," and "I think you can …"). It is also quite common for audience members to ask questions of Jessica – as though they were literally interacting (e.g., "Is Tess gone for good?" "What is the matter with Sarah lately?" "Why don't you just tell the family the truth?" "Is Vicki gonna be found in Paris, Texas?"). Audience members also use the "xoxo" convention to extend hugs and kisses to the character in their e-mail. Thus there does appear to be some evidence that some, if not many, audience members are behaving in ways that are consistent with parasocial interaction.

From the perspective of accommodation theory, fans should enact convergence behaviors or adopt behaviors consistent with Jessica's behavior if they perceive a character to be highly socially desirable. Analysis of the blog replies provides evidence of characters engaging in accommodation behaviors in their replies to Jessica's blog posts. For example, there is evidence that offering compliments constitutes convergence behavior (Williams & Giles, 1996) and can be found in the audience replies (e.g., "You pretending to be Jess is a great idea …," "I love you guys …," "Dressing her up as Jared was a stroke of genius…," "Of course you're fiercely protective of your twin and everyone else you love. That's one of your best qualities…," "You go girl …," "Applauds Jessica...," "You are a neat family …").

Self disclosure is another communicative behavior that has been associated with convergence and again can be found, although much less frequently, in the audience replies (Ehrlich & Graeven, 1971; Henwood, Giles, Coupland, & Coupland, 1993; Giles & Harwood, 1997). Responses that illustrate this say: for example, "Get a dog, I have two dogs…," "I would miss him too …," "I was heartbroken when you chose to be with Nash – I actually cried," and "I seen a lot of snakes since I been (Sic) dating…."

Other more common convergence behaviors found in audience replies include being attentive (Ng et al., 1997) and supportive (Ng et al., 1997). In nearly every blog reply, there is clearly some evidence of audience members expressing support for Jessica. This support includes informational support, such as telling Jessica things that are going on in the program as well as being empathetic and providing her emotional support. Examples of informational support include: "You might want to get that dizziness checked out …," "There might be a slight chance you are pregnant…," "I see romance between Antonio and the lady cop anyway …," and "Please don't let Dorian get Charlie, Vicki found him first and she deserves him a lot more than you know."

Examples of empathy and the provision of emotional social support are also very common in audience replies to Jessica: "You are right. You've been through hell & back. I can understand why you would be afraid to marry again. But Nash is not Antonio or Tico…," "If I was in your shoes, I would miss him too," "Jessica you're doing the right thing for you, Bree and Nash …," "Don't feel guilty for hurting Antonio. Eventually he will understand that things are better this way. It wasn't fair to you or Antonio living the lie of your marriage when you were in love with Nash. You were dying and you had to tell Antonio the truth once and for all. Eventually, he will move on and he can be happy with Jamie and whoever comes into his love life next…. Enjoy your new life and take it easy. Just because you've been released from the hospital doesn't mean you're invincible."

Other examples of convergence, including the borrowing of vocabulary from another person in conversation, can be found in the language used by audience members. For example, in one blog entry, Jessica refers to Jared Banks as a "snake." In two of the nine replies to that post, the fans described Jared as being a snake and a third reply suggested Jessica throw Jared out on his "reptilian ass." Obviously this type of language convergence is quite difficult to code unless the vocabulary is idiosyncratic, but it is clear from going through all of the blog replies that there is a great deal of mirroring of language and style in blog replies.

We suggest parasocial interaction can be studied by examining the audience replies to character blogs. Certainly the analysis of blog posts is not the only way to study parasocial interaction, but it represents a relatively unobtrusive way for researchers to gain insight. In addition, communication accommodation theory appears to be useful for identifying specific behaviors that demonstrate convergence and divergence. Examining parasocial relationships in much the same way interpersonal relationships are studied should yield important insights into the potential uses and effects of the mass media. Perhaps more importantly, the use of accommodation theory may also help us better understand the differences and similarities between interpersonal and mass communication.

FUTURE TRENDS

In the future, researchers will examine more closely convergence and divergence behaviors that occur in parasocial relationships. A growing body of research on the notion of parasocial breakups (e.g., the consequences of shows being taken off the air or characters being written out of shows) will undoubtedly help in this endeavor (Cohen, 2004). New studies that use attachment theory (Bowlby, 1973) may also shed some light on the ways audience members become involved in parasocial relationships. Borrowing from attachment theory, once again, points to the increased use of research on interpersonal communication and interpersonal relationships to understand parasocial interaction. It makes sense that the reasons children form attachments to their parents and that adults form attachments to other adults should also apply to the reasons they form attachments to mediated characters.

Future researchers will undoubtedly begin focusing their efforts on explaining the communicative processes that occur during parasocial interaction. We suggest the interactional view

(Watzlawick, Beavin & Jackson, 1967) as a starting point for such efforts. This theoretical structure may identify the underlying mechanism for parasocial relationships as well as a mechanism for explaining how individuals can use the same media content and characters to fulfill their needs.

The key axiom applicable here is that messages have both a content and relationship dimension. The content component is the message or the words within a message while the relationship component tells communicants how the content/message should be interpreted. How the message should be interpreted is based on a number of factors – the most important being the nature of the relationship between the communicants. The relationship between communicants is believed to be the single most important contextual factor and can obviously alter the meaning of message content.

If the interactional perspective is adapted to mediated communication, audience members establish relationships with all TV characters so that they can contextualize the messages they receive from television. Certainly not all relationships are particularly close – just as most of our interpersonal contact is not particularly intimate. Thus parasocial relationships are an example of close relationships between audience members and characters. These relationships are interesting because close relationships have idiosyncratic relational rules or rules that are specific to a particular relationship. That audience members' can develop relationships – albeit parasocial relationships – with characters allows audience members to shape a message to better use the media for gratification.

Future research also needs to compare the communicative behavior of fans writing to TV characters with those replying to blogs written actors. The teen actress Kristen Aldersen who plays Starr Manning on *One Life to Live* also has a blog. Kristen writes her blog not in character but instead as herself (a high school aged actress). Audience members behave quite differently when replying to her blog, just as they behave differently to other TV characters (e.g., Kendall Hart-Slater of *All My Children* and Dr. Robin Scorpio of *General Hospital*).

The ultimate value of such research may be the explication and clarification of the parasocial interaction concept. It seems less useful if it is a loss of touch with reality as it has been written about in the past. If parasocial interaction is more akin to being highly involved and a key determinant in the way audiences contextualize mediated messages, then this should become clear under careful scrutiny.

CONCLUSION

A life-long fan of General Hospital, Mary Ann Gayonski (personal communication, September 13, 2008) summed up audience perceptions of parasocial interaction accurately when she said, "I feel like I know them." "If I saw a character on the street I would want to talk with them about what is happening on the show." "I don't know that much about the actor, but I do know the character." "It really is like knowing someone – it's not the same as knowing someone – but it is like knowing someone." "Lots of time I know what a character is going to do before they do it." "They are not really friends but they do seem like friends sometimes." This testimony reveals that audience members would treat characters as if they were friends or acquaintances – even though they recognize them for what they are in most cases – characters on a fictional program.

As friends and acquaintances, audience members are likely to accommodate or converge given the affect they feel toward the character when given a chance to interact. If audience members feel that characters are "like friends," they are likely to treat them as if they are friends – just as dog owners may treat their pets as if they have language skills they do not possess. Like using heuristics in the evaluation of information, these behavioral heuristics occur because it is easier to behave toward characters as if they were actually people.

With the addition of an audience of other fans able to view character blog posts and replies to blog posts, it is again reasonable to expect some effort at accommodation to demonstrate their similarity with the character as well as some divergence behaviors to indicate to others that they indeed recognize the characters are not real. If audience members are behaving toward the other audience members, we would expect more divergence and less convergence to occur. This preliminary research certainly does not support that hypothesis. Very little audience-to-audience interaction can be observed within these blog posts and replies. Audience members appear to be writing to the characters with little regard for others "cavesdropping" into their conversation.

REFERENCES

Ashe, D. D., & McCutcheon, L. E. (2001). Shyness, loneliness, and attitude toward celebrities. *Current Research in Social Psychology, 6*, 1–7.

Bell, A. (1991). Audience accommodation in the mass media. In H. Giles, J. Coupland, & N. Coupland (Eds.), *Contexts of accommodation: Developments in applied sociolinguistics* (pp. 69- 102). Cambridge, UK: Cambridge University Press.

Bowlby, J. (1973). *Attachment and loss: Vol. II. Separation.* New York: Basic Books.

Cacioppo, J. T., von Hippel, W., & Ernst, J. M. (1997). Mapping cognitive structures and processes through verbal content: The thought-listing technique. *Journal of Consulting and Clinical Psychology, 65*, 928–940. doi:10.1037/0022-006X.65.6.928

Chartrand, T. L., & Bargh, J. A. (1999). The chameleon effect: The perception-behavior link and social interaction. *Journal of Personality and Social Psychology, 76*, 893–910. doi:10.1037/0022-3514.76.6.893

Cohen, J. (2004). Parasocial break-up from favorite television characters: The role of attachment styles and relationship intensity. *Journal of Social and Personal Relationships, 21*(2), 187–202. doi:10.1177/0265407504041374

Condon, W. S., & Ogston, W. D. (1967). A segmentation of behavior. *Journal of Psychiatric Research, 5*, 221–235. doi:10.1016/0022-3956(67)90004-0

Ehrlich, H. J., & Graeven, D. B. (1971). Reciprocal self disclosure in a dyad. *Journal of Experimental Social Psychology, 7*, 389–400. doi:10.1016/0022-1031(71)90073-4

Finn, S., & Gorr, M. B. (1988). Social isolation and social support as correlates of television viewing motivations. *Communication Research, 15*, 135–158. doi:10.1177/009365088015002002

Gans, H. J. (1977). Audience mail: Letters to an anchorman. *The Journal of Communication, 27*(3), 86–91. doi:10.1111/j.1460-2466.1977.tb02130.x

Giles, H. (1973). Accent mobility: A model and some data. *Anthropological Linguistics, 15*, 87–105.

Giles, H., Coupland, N., & Coupland, J. (1991). Accommodation theory: Communication, context, and consequence. In H. Giles, J. Coupland, & N. Coupland (Eds.), *Contexts of accommodation: Developments in applied sociolinguistics* (pp. 1-69). Cambridge, UK: Cambridge University Press.

Giles, H., & Harwood, J. (1997). Managing intergroup communication: Lifespan issues and consequences. In S. Eliasson & E. H. Jahr (Eds.), *Language and its ecology: Essays in memory of Einar Haugen* (pp. 105-130). New York: Elsevier/ North Holland.

Giles, H., Mulac, A., Bradac, J., & Johnson, P. (1987). Speech accommodation theory: The first decade and beyond. In M. L. McLaughlin (Ed.), *Communication yearbook* (Vol. 10, pp. 13-48). Beverly Hills, CA: Sage.

Greenwald, A. G. (1968). Cogntive learning, cognitive response to persuasion, and attitude change. In A. G. Greenwald, T. C. Brock, & T. C. Ostrom (Eds.), *Psychological foundations of attitudes* (pp. 63-102). New York: Academic Press.

Hale, J. L., & Burgoon, J. K. (1984). Models of reactions to changes in nonverbal immediacy. *Journal of Nonverbal Behavior, 8*, 287–314. doi:10.1007/BF00985984

Henwood, K., Giles, H., Coupland, J., & Coupland, N. (1993). Stereotyping and affect in discourse: Interpreting the meaning of elderly painful self-disclosure. In D. M. Mackie & D. L. Hamilton (Eds.), *Affect, cognition, and stereotyping: Interactive processes in group perception* (pp. 269-296). New York: Academic Press.

Horton, D., & Wohl, R. (1956). Mass communication and para-social interaction. *Psychiatry, 19*, 215–229.

Isotalus, P. (1995). Friendship through screen: Review of parasocial relationship. *Nordicom Review, 1*, 59–64.

Kanazawa, S. (2002). Bowling with our imaginary friends. *Evolution and Human Behavior, 23*, 167–171. doi:10.1016/S1090-5138(01)00098-8

Koenig, F., & Lessan, G. (1985). Viewers' relationship to television personalities. *Psychological Reports, 57*, 263–266.

Lenhart, A., & Fox, S. (2006). *Bloggers: A portrait of the Internet's new storytellers. Pew Internet & American Life*. Retrieved December 28, 2007, from http://www.pewinternet.org/pdfs/PIP%20Bloggers%20Report%20July%2019%202006.pdf

Levy, M. R. (1979). Watching TV news as parasocial interaction. *Journal of Broadcasting, 23*, 69–80.

Matarazzo, J. D., Weins, A. N., Matarazzo, R. G., & Saslow, G. (1968). Speech and silence behavior in clinical psychotherapy and its laboratory correlates. In J. Schlier, H. Hunt, J. D. Matarazzo, & C. Savage (Eds.), *Research in psychotherapy, Vol. 3* (pp. 347-394). Washington, DC: American Psychological Association.

Mauer, R. E., & Tindall, J. H. (1983). Effects of postural congruence on client's perceptions of counselor empathy. *Journal of Counseling Psychology, 30*, 158–163. doi:10.1037/0022-0167.30.2.158

McGuire, B., & LeRoy, D. J. (1977). Audience mail: Letters to the broadcaster. *The Journal of Communication, 27*(3), 79–85. doi:10.1111/j.1460-2466.1977.tb02129.x

Meyrowitz, J. (1986). Television and interpersonal behavior: Codes of perception and response. In G. Gumpert & R. Cathcart (Eds.), *Inter/media: Interpersonal communication in a media world* (pp. 253-272). New York: Oxford University Press.

Natale, M. (1975). Convergence of mean vocal intensity in dyadic communications as a function of social desirability. *Journal of Personality and Social Psychology, 32*, 790–804. doi:10.1037/0022-3514.32.5.790

Ng, S. H., Liu, J. H., Weatherall, A., & Loong, C. S. F. (1997). Younger adults, communication experiences and contact with elders and peers. *Human Communication Research*, *24*, 82–108. doi:10.1111/j.1468-2958.1997.tb00588.x

Nordlund, J. (1978). Media interaction. *Communication Research*, *5*, 150–175. doi:10.1177/009365027800500202

Rubin, A. M., & Perse, E. M. (1987). Audience activity and soap opera involvement: A uses and effects investigation. *Human Communication Research*, *14*(2), 246–268. doi:10.1111/j.1468-2958.1987.tb00129.x

Rubin, A. M., Perse, E. M., & Powell, R. A. (1985). Loneliness, parasocial interaction and local television news viewing. *Human Communication Research*, *12*(2), 155–180. doi:10.1111/j.1468-2958.1985.tb00071.x

Ryan, E. B., Hummert, M. L., & Boich, L. H. (1995). Communication predicaments of aging: Patronizing behavior toward older adults. *Journal of Language and Social Psychology*, *13*, 144–166. doi:10.1177/0261927X95141008

Shepard, C. A., Giles, H., & Le Poire, B. A. (2001). Communication accommodation theory. In W. P. Robinson & H. Giles, *The new handbook of language and social psychology* (pp. 33-56). Bristol, UK: John Wiley & Sons.

Street, R. L., Jr. (1983). Noncontent speech convergence in adult – child interactions. In R. N. Bostrom (Ed.), *Communication yearbook* (Vol. 7, pp. 369-395). Beverly Hills, CA: Sage.

Tajfel, H. (Ed.). (1982). *Social identity and intergroup relations*. Cambridge, UK: Cambridge University Press.

Tajfel, H., & Turner, J. C. (1979). An integrative theory of intergroup conflict. In W. G. Austin & S. Worchel (Eds.), *The social psychology of intergroup relations* (pp. 33-47). Monterey, CA: Brooks/Cole.

Technorati. (2008). *About us*. Retrieved June 29, 2008, from http://technorati.com/about

Tsao, J. (1996). Compensatory media use: An exploration of two paradigms. *Communication Studies*, *47*, 89–109.

Turner, J. (1993). Interpersonal and psychological predictors of parasocial interaction with different television performers. *Communication Quarterly*, *41*, 443–453.

van Baaren, R. B., Holland, R. W., Steenaert, B., & van Knippenberg, A. (2003). Mimicry for money: Behavioral consequences of imitation. *Journal of Experimental Social Psychology*, *39*(4), 393–398. doi:10.1016/S0022-1031(03)00014-3

Watzlawick, P., Beavin, J. H., & Jackson, D. (1967). *Pragmatics of human communication: A study of interactional patterns, pathologies, and paradoxes*. New York: Norton.

Welkowitz, J., Feldstein, S., Finkelstein, M., & Aylesworth, L. (1972). Changes in vocal intensity as a function of interspeaker influence. *Perceptual and Motor Skills*, *35*, 715 718.

Williams, A., & Giles, H. (1996). Intergenerational conversations: Young adults' retrospective accounts. *Human Communication Research*, *23*, 220–250. doi:10.1111/j.1468-2958.1996.tb00393.x

KEY TERMS AND DEFINITIONS

Accommodation: The modifications in communicative behavior made by individuals during interaction. Accommodation may include changes to verbal, vocal and non-verbal behaviors. The process of accommodation may occur as an intentional communicative strategy or may occur without the conscious awareness of the individual; it is motivated by the desire to be liked. Accommodation can be manifest as convergence (adopting the communicative behavior of another) or divergence (behaving stylistically different from another to maintain our differences).

Cognitive Response: Thoughts that occur while we are listening to someone talk are called *cognitive responses*. Cognitive response is not a synonym for decoding a message. "Decoding" refers to a completely separate process. In decoding, sound or visual stimuli are translated back into language. Once we have decoded the message, our idiosyncratic responses or thoughts to those messages are described as our cognitive responses. If we are very interested in the topic, our cognitive responses may be message relevant. Message relevant responses focus on counter-arguments or additional evidence supporting a particular position. If we are not interested in the topic, our cognitive responses may not be particularly message relevant (e.g., "I "I need to get gas on my way home"). In short, our cognitive responses are the things we think of while listening to the messages of others. Cognitive responses occur while reading, watching television, listening to the radio, or surfing the Internet.

Communication Accommodation Theory: Proposed by Howard Giles, professor of communication at the University of California, Santa Barbara, to explain the adaptations people make in their communicative behavior during conversation. The theory assumes people adapt their communicative behaviors and message content in an effort to be perceived favorably by high social status individuals. When interacting with individuals of low social status, we are motivated to maintain our distance or be perceived as being different from the low status individuals. This theory is based on many of the same tenets as social identity theory.

Convergence: When an individual imitates or adopts the communicative behaviors of another in conversation, we say they are converging or becoming more like the other communicatively. For example a person may accommodate the communicative behavior of another by talking louder or adopting an accent (vocal accommodation), by appropriating the language of another in conversation (verbal accommodation), or by imitating kinesic or facial behaviors (nonverbal accommodation). These adaptations are efforts by a communicant to be viewed favorably by the high social status other during conversation.

Divergence: When we communicate with someone we perceive to be socially unattractive, we diverge or behave in such a way that we will be viewed as being different from that person. Motivated by the desire to be seen as socially desirable, in the presence of an undesirable communicative partner, we fail to accommodate and actually behave in ways that will distance us from another. For example, if someone uses coarse language or slang, we might diverge by employing formal or more precise language. If they talk in a loud voice, we might talk in a quiet voice; and, if they wave their hands, we might maintain a more still communicative style.

Parasocial Interaction: The term *parasocial interaction* is often used as a synonym for parasocial relationship. When the two terms are differentiated, parasocial interaction is used to describe the specific audience and/or character behaviors. One character winking at an audience and another giving a soliloquy are examples of parasocial interaction. Similarly audience members talking to their TV or generating verbal replies that go unexpressed are examples of parasocial action too. It is generally believed that the way a show is designed and shot contributes to the likelihood of audience members engaging in parasocial interaction and/or establishing parasocial relationships.

Parasocial Relationship: The term *parasocial relationship* was first coined by Horton & Wohl (1956) to describe the pseudo-friendships that occur between audience members and TV characters and other media personae. The notion of relationship is used here to describe faux interpersonal relationships that typically share some commonalities with actual interpersonal relationships. For example, an audience member can feel affect toward a character, "know" or understand the character, or relate to a character as if the character was an actual acquaintance. These relationships can represent little more than the liking of a character; they can also extend into the realm of delusion. In such extreme cases, audience members may actually believe they have a relationship with the character. The term is often used to identify the similarities between interpersonal relationships and mediated relationships.

Thought Listing Procedure: A social psychological procedure for cognitive response evaluation technique used by researchers to gather the cognitive responses of individuals. After exposing subjects to a message, the researcher asks subjects to list thc thoughts that ran through their heads during message presentation. Each thought or cognitive response can then be examined to see whether respondent thoughts are consistent or inconsistent with the message and ultimately how effective a particular message may be.

Web Logging or Blogging: Blogs are online diaries or journals used by their authors as vehicles for providing commentary. They are updated on a regular basis and tend to focus on the personal experiences of the author. The critical differences in blogs and diaries are the opportunities for reaching a mass audience and the opportunity for that mass audience to respond to the commentary found within the blog. Because of the interactive nature of the blogs and blogging, software readers are able to add comments, links, pictures, video, or any other media format to the blog for the edification and entertainment of other denizens of the Internet. Blogging, then, is the act of updating a blog or an online diary.

This work was previously published in Handbook of Research on Social Interaction Technologies and Collaboration Software; Concepts and Trends, edited by T. Dumova & R. Fiordo, pp. 302-314, copyright 2010 by Information Science Reference (an imprint of IGI Global).

Chapter 7.16
The Generative Potential of Appreciative Inquiry as an Essential Social Dimension of the Semantic Web

Kam Hou Vat
Faculty of Science and Technology, University of Macau, Macau

ABSTRACT

The mission of this chapter is to present a framework of ideas concerning the expected form of knowledge sharing over the emerging Semantic Web. Of specific interest is the perspective of appreciative inquiry, which should accommodate the creation of some appreciative knowledge environments (AKE) based on the peculiar organizational concerns that would encourage or better institutionalize knowledge work among people of interest in an organization. The AKE idea is extensible to the building of virtual communities of practice (CoP) whose meta-data requirements have been so much facilitated in today's Web technologies including the ideas of data ownership, software as services, and the socialization and co-creation of content, and it is increasingly visible that the AKE model of knowledge sharing is compatible for the need of virtual collaboration in today's knowledge-centric organizations. The author's investigation should provide a basis to think about the social dimension of today's Semantic Web, in view of the generative potential of various appreciative processes of knowledge sharing among communities of practice distributed throughout an organization.

DOI: 10.4018/978-1-60566-650-1.ch021

INTRODUCTION

In the late 20th century, Tim Berners-Lee (1999) had the idea of providing rapid, electronic access to the online technical documents created by the world's high-energy physics laboratories. He sought to make it easier for physicists to access their distributed literature from a range of research centers scattered around the world. In the process, he laid the foundation for the World Wide Web. Berners-Lee has a two-part vision for the working of the World Wide Web (http://public.web.cern.ch/Public/Welcome.html). The first is to make the Web a more collaborative medium. The second is to make the Web understandable and thus serviceable by machines. Yet, it was not his intention that someday his idea to link technical reports via hypertext then has actually revolutionized essential aspects of human communication and social interaction. Today, the Web provides a dazzling array of information services designed for use by human, and has become an ingrained part of our lives. There is another Web coming, however, where online information will be accessed by intelligent agents that will be able to reason about that information and communicate their conclusions in ways that we can only begin to dream about. This is the Semantic Web (Berners-Lee, Hendler, & Lassila, 2001; Berners-Lee, 1998a, 1998b, 1998c; http://www.SemanticWeb.org), representing the next stage in the evolution of communication of human knowledge. The developers of this new technology have no way of envisioning the ultimate ramifications of their work. Still, they are convinced that "creating the ability to capture knowledge in machine understandable form, to publish that knowledge online, to develop agents that can integrate that knowledge and reason about it, and to communicate the results both to people and to other agents, will do nothing short of revolutionize the way people disseminate and utilize information" (Musen, 2006, pp. xii). This article is meant to provide a strategic view and understanding of the Semantic Web, including its attendant technologies. In particular, our discussion situates on an organization's concerns as to how to take advantages of the Semantic Web technologies, by focusing on such specific areas as: diagnosing the problems of information management, providing an architectural vision for the organization, and steering an organization to reap the rewards of the Semantic Web technologies. Of interest here is the introduction of the appreciative context of organizational systems development based on the philosophy of appreciative inquiry (Cooperrider, 1986; Gregen, 1990), a methodology that takes the idea of social construction of reality to its positive extreme especially with its relational ways of knowing.

THE TECHNOLOGICAL BACKGROUND OF SEMANTIC WEB

Most of today's Web content is suitable for human understanding. Typical uses of the Web involve people's seeking and making use of information, searching for and getting in touch with other people, reviewing catalogs of online stores and ordering products by filling out forms, as well as viewing the confirmation. The main tool of concerns is the search engine (Belew, 2000), with its key-word search capability. Interestingly, despite much improvement in search engine technology, the difficulty remains; namely, it is the person who must browse selected documents to extract the information he or she is looking for. That is, there is not much support for retrieving the information, which is a very time-consuming activity. The main obstacle to providing better support to Web users is the non-machine-serviceable nature of Web content (Antoniou & van Harmelen, 2004); namely, when it comes to interpreting sentences and extracting useful information for users, the capabilities of current software are still very limited. One possible solution to this problem is to represent Web content in a form that is more readily machine-processable and to use intelligent techniques (Hendler, 2001) to take advantage of these representations. In other words, it is not necessary for intelligent agents to understand information; it is sufficient for them to process information effectively. This plan of Web revolution is exactly the initiative behind the Semantic Web, recommended by Tim Berners-Lee (1999), the very person who invented the World Wide Web in the late 1980s. Tim expects from this initiative the realization of his original vision of the Web, i.e. the meaning of information should play a far more important role than it does in today's Web. Still, how do we create a Web of data that machines can process? According to Daconta and others (2003), the first step is a paradigm shift in the way we think about data. Traditionally, data has been locked away in proprietary applications, and it was seen as secondary to the act of processing data. The path to machine-processable data is to make the data progressively smarter, through explicit metadata support (Tozer, 1999). Roughly, there are four stages in this smart data continuum (Daconta, Obrst, & Smith, 2003), comprising the pre-XML stage, the XML stage, the taxonomies stage, and the ontologies stage. In the pre-XML stage where most data in the form of texts and databases, is often proprietary to an application, there is not much smartness that can be added to the data. In the XML stage where data is enabled to be application independent in a specific domain, we start to see data moving smartly between applications. In the third stage, data expected to be composed from multiple domains is classified in a hierarchical taxonomy. Simple relationships between categories in the taxonomy can be used to relate and combine data, which can then be discovered and sensibly combined with other data. In the fourth stage based on ontologies which mean some explicit and formal specifications of a conceptualization (Gruber, 1993), new data can be inferred from existing data by following logical rules. This should allow combination and recombination of data at a more atomic level and very fine-grained analysis of the same. In this stage, data no longer exists as a blob but as a part of a sophisticated microcosm. Thereby, a Semantic Web implies a machine-processable Web of smart data, which refers to the data that is application-independent, composable, classified, and part of a larger information ecosystem (ontology).

Understanding Semantic Web Technologies

Today, XML (extensible markup language; http://www.xml.com) is the syntactic foundation of the Semantic Web. It is derived from SGML (standard generalized markup language), an international standard (ISO8879) for the definition of device- and system-independent methods of representing information, both human- and machine-readable. The development of XML is driven by the shortcomings of HTML (hypertext markup language), the standard language also derived from SGML, in which Web pages are written. XML is equipped with explicit metadata support to identify and extract information from Web sources. Currently, many other technologies providing features for the Semantic Web are built on top of XML, to guarantee a base level of interoperability, which is important to enable effective communication, thus supporting technological progress and business collaboration. For brevity, the technologies that XML is built upon are Unicode characters and Uniform Resource Identifiers (URI). The former allows XML to be authored using international characters, whereas the URI's are used as unique identifiers for concepts in the Semantic Web. Essentially, at the heart of all Semantic Web applications is the use of ontologies. An ontology is often considered as an explicit and formal specification of a conceptualization of a domain of interest (Gruber, 1993). This definition stresses two key points: that the conceptualization is formal and hence permits reasoning by computer; and that a practical ontology is designed for some particular domain of interest. In general, an ontology describes formally a domain of discourse. It consists of a finite list of terms and the relationships between these terms. The terms denote important concepts (classes of objects) of the domain. The relationships include hierarchies of classes. In the context of the Web, ontologies provide a shared understanding of a domain, which is necessary to overcome differences in terminology. The search engine can look for pages that refer to a precise concept in an ontology instead of collecting all pages in which certain, generally ambiguous, keywords occur. Hence, differences in terminology between Web pages and the queries can be overcome. At present, the most important ontology languages for the Web include (Antoniou & Harmelen, 2004): XML (http://www.w3.org/XML/), which provides a surface syntax for structured documents but imposes no semantic constraints on the meaning of these documents; XML Schema (http://www.w3.org/XML/Schema), which is a language for restricting the structure of XML documents; RDF (Resource Description Framework) (http://www.w3.org/RDF/), which is a data model for objects ("resources") and relations between them; it provides a simple semantics for this data model; and these data models can be represented in an XML syntax; RDF Schema, (http://www.w3.org/TR/rdf-schema/) which is a vocabulary description language for describing properties and classes of RDF resources, with a semantics for generalization hierarchies of such properties and classes; OWL (http://www.w3.org/TR/owl-guide/), which is a richer vocabulary description language for describing properties and classes, such as relations between classes, cardinality, equality, richer typing of properties, characteristics of properties, and enumerated classes.

Clarifying the Meta-Data Context of Semantic Web

It is hard to deny the profound impact that the Internet has had on the world of information over the last decade. The ability to access data on a variety of subjects has clearly been improved by the resources of the Web. However, as more data becomes available, the process of finding specific information becomes more complex. The sheer amount of data available to the Web user is seen as both the happy strength and also the pity weakness of the World Wide Web. Undoubtedly, the single feature that has transformed the Web into a common, universal medium for information exchange is this: using standard search engines, anyone can search through a vast number of Web pages and obtain listings of relevant sources of information. Still, we have all experienced such irritation (Tozer, 1999; Belew, 2000) as: the search results returned are incomplete, owing to the inability of the search engine to interpret the match criteria in a context sensitive fashion; too much information is returned; lack of intelligence exists in the search engine in constructing the criteria for selection. Likewise, what is the Semantic Web good for? Perhaps, a simple example in the area of knowledge management could help clarify the situation. The field of organizational knowledge management typically concerns itself with acquiring, accessing, and maintaining knowledge as the key activity of large businesses (Liebowitz, 2000; Liebowitz & Beckman, 1998). However, the internal knowledge from which many businesses today presumably can draw greater productivity, create new value, and increase their competitiveness, is available in a weakly structured form, say, text, audio and video, owing to some limitations of current technology (Antoniou & Harmelen, 2004, p.4) in such areas as: *searching information*, where companies usually depend on keyword-based search engines, the limitation of which is that even though a search is successful, it is the person who must browse selected documents to extract the information he or she is looking for; *extracting information*, where human time and effort are required to browse the retrieved documents for relevant information, and current intelligent agents are unable to carry out this task in a satisfactory manner; *maintaining information*, where there are current problems such as inconsistencies in terminology and failure to remove outdated information; *uncovering information*, where new knowledge implicitly existing in corporate databases is extracted using data mining, but this task is still difficult for distributed, weakly structured collections of documents; and *viewing information*, where it is often desirable to restrict access to certain information to certain groups of employees, and views which hide certain information, are known from the area of databases but are hard to realize over an intranet or the Web. The aim of the Semantic Web is to allow much more adaptable technologies in handling the scattered knowledge of an organization (Swartz & Hendler, 2001) such as: knowledge will be organized in conceptual spaces according to its intended meaning; automated tools will support maintenance by checking for inconsistencies and extracting new knowledge; keyword-based search will be replaced by query answering—requested knowledge being retrieved, extracted, and presented in a human-friendly manner; query over several documents will be supported; and defining who may view certain parts of information will also be made possible.

CRAFTING THE KNOWLEDGE-CENTRIC ORGANIZATION

It is not uncommon to hear any Chief Executive Officer (CEO) respond to the question, "What distinguishes your company from its competitors?" with the emphatic "Our knowledge." Yet, it is also not surprised to see the same CEO become somewhat puzzled when the follow-up question, "What comprises your knowledge assets and value on this knowledge?" is presented. Many leading organizations nowadays are discovering they need to do a better job of capturing, distributing, sharing, preserving, securing, and valuing their precious knowledge in order to stay ahead of their competition, or at least survive (Liebowitz, 1999). By the term knowledge-centric (Daconta, Obrst, & Smith, 2003; Liebowitz & Beckman, 1998), we mean the process of managing knowledge in organizations with the focus to provide mechanisms for building the knowledge base of the firm to better apply, share, and manage knowledge across various components in the company. The use of Semantic Web technologies is a means to achieving the knowledge-centric organization by weaving the underlying technologies into every part of the organization's work life cycle, including production, presentation, analysis, dissemination, archiving, reuse, annotation, searches, and versioning of the knowledge work. To situate our discussion on the Semantic Web context, it is helpful to investigate what a typical non-knowledge-centric organization scenario is like in its daily operations.

Making Sense of Information Overload

To remain competitive, many an enterprise today accrue numerous information resources to use in their problem solving, decision making and creative thinking for improving products, processes, and services. Yet, the critical problem for the typical organization is the sheer volume of information coming in, from a wide variety of sources, in various formats (papers, emails, and different electronic media), and it is difficult to manage such resources and turn them into knowledge, which according to Tom Davenport (1997), is a synthesis of information. The knowledge process in a non-knowledge-centric organization typically comprises five stages of information management. The first stage is often characterized by a capture process, in which a human being in the organization takes information from somewhere (newspaper, radio, Internet, database, phone call, or email), and brings it to the organization, via some means such as vocally by mentioning the information to someone, or electronically by sending it through email to someone. If the data is not lost in the process, the recipient writes a paper or presentation, or even a status report. The second stage is often characterized by a securing process, in which the data is put into a database, recorded to a digital file, or indexed into a search engine. Now that entering information is always the first step, but the potential problem is this: each division, group, or project in the organization may enter the information into different systems. Assuming there is only one database per project, and assuming a division has only ten projects, there may be ten different databases containing data in a division. What if there is a different database system for each project? There then will be ten different software systems containing data. What if there are five divisions in the company, with similar systems in use? We now have many data sources that might be individual stovepipes in the organization, each of which perform a specific task at the expense of trapping the data and robbing the organization of business agility in adapting such data to new systems of interest. The third stage in the knowledge process is often characterized by integration, depending on the complexity of the organization's information architecture, a blueprint based on which different information systems services are rendered. Perhaps, since most of the information systems are stove-piped (namely, information cannot be shared by other systems that need it), there is usually no good way to combine different information systems into a coherent picture. In other words, any attempt to combine the information must involve data conversions across incompatible software systems, in which each database and software system is designed differently and has different interfaces to talk to them. As a result, there is usually little or no integration of these databases, because it is prohibitively difficult and expensive. Even if there is an integration solution, the result is often another stove-piped system. The fourth stage of the knowledge process is often characterized by searching, or discovery of an organization's internal resources. This is a haphazard and time-consuming activity because it involves so many different systems. Imagine we have to login to different databases and search engines, and manually compare and contrast the information we find into a coherent picture or thought. This is the most wasteful part of the knowledge process in person-hours. Finally, the fifth stage is concerned with the application of the search results (if we succeed in the last stage). After the tedious search process, the result is usually a presentation or paper report. Many times, this process of creating the report involves several people. The approval process is done by manual reviews and is often slow. After the new product is created, the information is supposed to be filed, say, onto a Web server that may or may not be indexed by one of the organization's search engines. The issues with this approach of knowledge process are many: How are we to know what version of the document we have? There is no way to tell if the information has been superseded once this new document is integrated into one of the organization's stove-piped databases. How are we to reuse the information, in terms of the ability to discover, refine, annotate, and incorporate past knowledge?

Making Use of Semantic Web Technologies

The knowledge process in a knowledge-centric organization starts with the discovery and production phase where an individual member of the organization receives an information item and would like to turn that into a knowledge item. It is intended as a process that could be repeated by many others in the same organization. With Semantic Web technologies, any new piece of information must be marked up with XML using a relevant organizational schema. Once this is done, the individual should digitally sign the XML document using the XML signature specification to provide strong assurance that the individual verified the validity of the information. The next step is the annotation process, in which the individual may want to use RDF to annotate the new information with his or her notes or comments, adding to the XML document, but without breaking the digital signature seal of the original material. At the end of the annotation process, the author should digitally sign the annotation with XML signature. Then, the annotated information must be mapped to topics in the taxonomy and entities in the corporate ontology so that pieces of the information can be compared to other pieces of information in the organization's knowledge base. Example annotations include: Who is the person that authored this document? What department does he or she work in? Is the individual an expert on this topic? Is this topic in the organization's taxonomy? Once this is completed, it is time to store the information in an application with a Web service interface. If that is a new Web service, the Web service should be registered in the organization's registry, along with its taxonomic classifications. The result of the discovery and production process is that the information coming into the organization has been marked up with standard XML, digitally signed to show assurance of trust, annotated with an author's comments, mapped to the organization's ontology, and published to a Web service and registered in a Web service registry. Consequently, because the Web service is registered in a registry, people and programs in the organization can discover the Web service based on its name or taxonomic classification. Besides, now that any incoming information is stored in an easily accessible format (Web services) and is associated with the organization's ontology and taxonomy, retrieval of information is much facilitated.

Preparing for Change via the Semantic Web

It follows from our previous discussion that in order to take advantage of Semantic Web technologies, most organizations need to change the way they manage information resources (Van den Hoven, 2001) such as: encouraging the sharing of information resources by using common terminology, definitions, and identifiers across the enterprise; establishing an enterprise-wide information architecture, which show the relationships between information held in various parts of the enterprise; ensuring information integrity through procedures to ensure accuracy and consistency; improving information accessibility and usability by putting it in useful formats to make it accessible in any way that makes business sense; and enforcing security to protect the information resources from accidental or deliberate modification, destruction, or unauthorized access. Fortunately, these changes can mostly be implemented evolutionarily over time so as to realize the vision of a knowledge-centric organization. In fact, the most challenging aspect may not be the technology, but the cultural transformation of the mind-set of employees because the use of Semantic Web represents a whole system change of the behavior in accessing, integrating, and leveraging knowledge throughout the organization. So, how do we get started? Our learning indicates that the IDEAL model (Gremba & Myers, 1997) originally conceived as a life cycle model for software process improvement based on the capability maturity model (CMM) for software at the CMU-SEI (Paulk, Weber, Curtis & Chrissis, 1994), has been found helpful in the change management process. IDEAL suggests a useable and understandable approach to continuous improvement by outlining the steps required to establish a sustainable improvement program, through five different stages of work. Initiating (I) is to lay the groundwork for a workable improvement effort. Diagnosing (D) is to determine where we are relative to where we want to be. Establishing (E) is to plan the specifics of how we will reach our destination. Activating (A) is to do the work according to the plan. Learning (L) is to learn from the experience and improve our ability to adopt new technologies in the immediate future. In the context of the knowledge-centric organization using Semantic Web, *Initiating* involves developing a clear vision for changing the information management process in the organization. What is the clear and compelling business case for change? How will the Semantic Web technologies enable the organization to achieve its business goal? How does this change link to other, broader corporate goals? If these issues are not well elaborated, it is very hard for members of the organization to buy into the change. A clear, concise, and simple mission statement may help. *Diagnosing* involves setting clear goals and milestones specific to the organization, based on the vision (or mission) communicated in the Initiating stage. Often, visionary goals (not technical goals) are what are needed. An example is: "Be able to look up all project information across the organization by spring 2009." *Establishing* involves identifying critical stakeholders who will be impacted by the change. Oftentimes, it helps to divide stakeholders into different groups to assess the unique impact on each group and develop targeted plans to help them work through change. For example, what kind of resources or tools can help each group manage the change? It might also help if some change facilitators are made available to address the cultural and organizational change issues identified in the process. *Activating* involves picking a core team to spread the vision throughout the organization. This team preferably composed of both technical and management people, is charged with the mission to mobilize the change efforts among members of the organization. It is also important to identify a change champion to help lead the effort of organizational and cultural transformation to ensure that the company embraces the new technology. At this point, *learning* is the most important because the core team will need to understand the high-level concepts of the Semantic Web, the purpose behind it, and the core business benefits it brings. Once the management and the technical staff are on board the core team, it is time to determine the technical goals to implement the plan. Example technical goals could include (Daconta, Obrst, & Smith, 2003, pp. 252-254): *Mark up documents in XML*—After this step, all new document development in the organization should have XML formats, to enable data content to be separate from presentation, and style sheets can be used to add different presentations to content later. *Expose applications as Web services*—so as to publish the application's interfaces as self-describing knowledge objects, with a goal of delivering small, modular building blocks that can be assembled by the intended users. *Establish an organizational registry*—so as to register different applications and provide query for Web services. *Build ontologies*—so as to overlay higher-level semantic constructs on the documents marked up with XML which provides facilities and syntax for specifying a data structure that can be semantically processed. *Integrate search tools*—so as to allow members of the organization to do searches of documents based on specific ontology. *Provide an enterprise portal*—so as to provide some aggregation points to integrate knowledge management into the organization through specific user-interfaces of search engines.

ORGANIZATIONAL CHALLENGES FACING THE SEMANTIC WEB

Based on our earlier discussion, it is not difficult to see that in an organization with Semantic Web technologies, because any incoming information has been marked up with XML, standard techniques and technologies can be used to store it and style its presentation. Still, because the information has been mapped to the organization's ontology, any new information can be easily associated and compared with other information in the organization. Also, because the original information has been digitally signed, anyone looking at the information will have assurance of its validity. Besides, because author annotations are added and also digitally signed, there is convenient tracking of who found the information and their comments. Furthermore, because it is stored in a Web service, any software program can communicate with it using open standards. Nonetheless, what do all these technology-made conveniences mean for the social dimension of the Semantic Web installed inside an organization? It is no denial that organizational knowledge synthesis (or creation and transfer) is a social as well as an individual process (Nonaka, 2002). Sharing tacit knowledge requires individuals to share their personal beliefs about a situation with others. At that point of sharing, justification often becomes public. Each individual is faced with the tremendous challenge of justifying his or her beliefs in front of others—and it is this need for justification, explanation, persuasion and human connection that makes knowledge synthesis a highly dynamic as well as fragile process (Markova & Foppa, 1990; Vat, 2003). To bring personal knowledge into an organization, within which it can be amplified or further synthesized, it is necessary to have a field (Ichijo & Nonaka, 2007; Von Krogh, Ichijo, & Nonaka, 2000; Nonaka & Takeuchi, 1995) that provides a place in which individual perspectives are articulated, and conflicts are resolved in the formation of higher-level concepts. In the specific context of Semantic Web, this field of interaction is yet to be defined and engineered by the organization architect of the company, or of the organizational change management behind the Semantic Web. Principally, this field should facilitate the building of mutual trust among members of the organization, and accelerate the creation of some implicit perspective shared by members as a form of tacit knowledge. Then, this shared implicit perspective is conceptualized through continuous dialogue among members. It is a process in which one builds concepts in cooperation with others. It provides the opportunity for one's hypothesis or assumption to be tested. Typically, one has to justify the truthfulness of his or her beliefs based on his or her unique viewpoint, personal sensibility, and individual experience, sized up from the observations of any situation of interest. In fact, the creation of knowledge, from this angle, is not simply a compilation of facts but a uniquely human process that can hardly be reduced or easily replicated. Yet, justification must involve the evaluation standards for judging truthfulness, and there might also be value premises that transcend factual or pragmatic considerations before we arrive at the stage of cross-leveling any knowledge (Von Krogh, Ichijo, & Nonaka, 2000); namely, the concept that has been created and justified is integrated into the knowledge base of the organization. The key to understand the social dimension of the Semantic Web is to ask how it could support or facilitate knowledge sharing among individuals. Putting knowledge sharing (or rather conversation among individuals) to work means bringing the right people with the requisite knowledge together and motivating their online interaction. That way, they could work collaboratively to solve real and immediate problems for the organization. To reach that level of practical impact, there must be trust and commitment among the participants apart from software and online connectivity. In light of our discussion, that means leading and fostering the kind of organizational culture that motivates people to share what they know with their peers (co-workers) without a fear of being questioned, critiqued or put on the defense. In the specific context of this article, this culture of knowledge sharing which should be in the driver's seat for selecting and configuring the Semantic Web technologies for an organization, could be developed from the idea of appreciative inquiry (AI) (Cooperrider & Whitney, 2005).

THE GENERATIVE POTENTIAL OF APPRECIATIVE INQUIRY

The contributions behind the work of appreciative inquiry (AI), is mainly attributed to David L. Cooperrider's (1986) doctoral research at Case Western Reserve University. The context of AI is about the co-evolutionary search for the best in people, their organizations, and the relevant world around them. In its broadest focus, it involves systematic discovery of what gives life to a living system when it is most alive, most effective, and most constructively capable in economic, ecological, and human terms. Principally, AI involves the art and practice of asking questions that strengthen a system's capacity to apprehend, anticipate, and heighten positive potential. AI has been described in different ways since its publication: as a paradigm of conscious evolution geared for the realities of the new century (Hubbard, 1998); as a methodology that takes the idea of the social construction of reality to its positive extreme especially with its relational ways of knowing (Gergen, 1990); as the most important advance in action research in the last decade of the 20th century (Bushe, 1995); as offspring to Abraham Maslow's vision of a positive social science (Chin, 1998; Curran, 1991); as a powerful second generation practice of organizational development (Watkins & Cooperrider, 1996); as model of a much needed participatory science (Harman, 1990); as a radically affirmative approach to change which completely lets go of problem solving mode of management (White, 1996), and others as an approach to leadership and human development (Cooperrider & Whitney, 2005). In essence, AI is an attempt to determine the organization's core values (or life giving forces). It seeks to generate a collective image of a future by exploring the best of what is in order to provide an impetus for imagining what might be (Cooperrider & Srivastva, 1987). Positively, Thatchenkery and Chowdhry (2007, p.33) says it well, "To be appreciative, we must experience a situation, accept the situation, make sense of the situation (pros/cons), and do a bit of mental gymnastics to understand the situation, with an appreciative lens. Not only that, the appreciative lens that we put on the situation impacts our next experience as well." Indeed, the interpretive scheme we bring to a situation significantly influences what we will find. Seeing the world is always an act of judgment. We can take an appreciative judgment or a critical or deficit oriented judgment. AI takes the former. Geoffrey Vickers (1965, 1968, 1972), a professional manager turned social scientist, was the first to talk about appreciation in a systematic way. Vickers' main contribution is that of appreciation and the appreciative process which constitutes a system. An appreciative system may be that of an individual, group, or an organization. In explaining appreciation, Vickers used systems thinking (Checkland & Casar, 1986), which provided basic concepts to describe the circular human processes of perceiving, judging, and acting. Specifically, Vickers focused on five key elements of appreciation, including respectively: the experience of day-to-day life as a flux of interacting events and ideas; reality judgments about what goes in the present or moment and a value judgment about what ought to be good or bad, both of which are historically influenced; an insistence on relationship maintaining (or norm seeking) as a richer concept of human action than the popular notion of goal seeking; a concept of action judgments stemming from both reality and value judgments; and action, as a result of appreciation, contributing to the flux of events and ideas, as does the mental act of appreciation itself. This leads to the notion that the cycle of judgments and actions is organized as a system. Simply put, as humans, we are in a state of flux. We judge the events we experience based on our individual history. We make meaning based on the interactions with other humans to enrich our lives. Our judgments, relationships, and values dictate how we act in subsequent events. By framing our perceptions and judgments on appreciation, we can change our behavior. In the context of fostering a knowledge-centric culture for an organization including possibly various communities of practice (Wenger, 1998), we can change the way we hoard knowledge to a philosophy of sharing knowledge. Indeed, the basic rationale of AI is to begin with a grounded observation of the best of what is, articulate what might be, ensure the consent of those in the system to what should be, and collectively experiment with what can be.

VIRTUAL ORGANIZING IN SUPPORT OF APPRECIATIVE INQUIRY

The idea of virtual organizing, attributed to Venkatraman and Henderson (1998), can be considered as a method to operationalize the context of appreciative inquiry, dynamically assembling and disassembling nodes on a network of people or groups of people in an organization, to meet the demands of a particular business context. This term emerged in response to the concept of *virtual organization*, which appeared in the literature around the late twentieth century (Byrne, Brandt, & Port 1993; Cheng 1996; Davidow, & Malone 1992; Goldman, Nagel, & Preiss 1995; Hedberg,

Dahlgren, Hansson, & Olve 1997; Mowshowitz, 1997). There are two main assertions associated with virtual organizing. First, virtual organization should not be considered as a distinct structure such as a network organization in an extreme and far-reaching form (Jagers, Jansen, & Steenbakkers 1998), but virtuality is a strategic characteristic applicable to every organization. Second, information technology (IT) (not excluding Semantic Web technologies) is a powerful enabler of the critical requirements for effective virtual organizing. In practice, virtual organizing helps emphasize the ongoing process nature of the organization, and it presents a framework of achieving virtuality in terms of three distinct yet interdependent vectors: virtual encounter for organization-wide interactions, virtual sourcing for asset configuration, and virtual expertise for knowledge leverage. The challenge of virtual organizing is to integrate the three hitherto separate vectors into an interoperable IT platform that supports and shapes any new organizational initiative, paying attention to the internal consistency across the three vectors.

Understanding the Three-Vector Framework

The first of the three vectors of virtual organizing deals with the new challenges and opportunities for interacting with the members of an organization. The second focuses on the organization's requirements to be virtually integrated in a network of interdependent (business) partners, so as to manage a dynamic portfolio of relationships to assemble and coordinate the necessary assets for delivering value for the organization. The third is concerned with the opportunities for leveraging diverse sources of expertise within and across organizational boundaries to become drivers of value creation and organizational effectiveness. All these three vectors are accomplished by the provision of suitable information system (IS) support, whose ongoing design represents the IS challenge of every organization in the Internet age.

- **Virtual Encounter:** This idea of providing remote interaction with the organization is not new, but has indeed been redefined since the introduction of the Internet, and particularly, the World Wide Web. Many an organization feels compelled to assess how its products and services can be experienced virtually in the new medium of the Internet. The issue of customization is important. It requires a continuous information exchange with parties of interest, which in turn requires an organizational design that is fundamentally committed to operating in this direction. Practically, organizations need to change from an inside-out perspective to an outside-in perspective. This is often characterized by the emergence of online customer communities, with the capacity to influence the organization's directions with a distinct focus. It is believed that with virtual organizing becoming widespread, organizations are increasingly recognizing communities as part of their value system and must respond appropriately in their strategies.
- **Virtual Sourcing:** This vector focuses on creating and deploying intellectual and intangible assets for the organization in the form of a continuous reconfiguration of critical capabilities assembled through different relationships in the business network. The mission is to set up a resource network, in which the organization is part of a vibrant, dynamic network of complementary capabilities. The strategic leadership challenge is to orchestrate an organization's position in a dynamic, fast-changing resource network where the organization can carefully analyze her relative dependence on other players in the resource coalition and ensure her unique capabilities.
- **Virtual Expertise:** This vector focuses on the possibilities for leveraging expertise at different levels of the organization. In

today's organizations, many tasks are being redefined and decomposed so that they can be done at different locations and time periods. However, the real challenge in maximizing work-unit expertise often rests not so much in designing the technological platform to support group work but in designing the organization structure and processes. The message is clear: knowledge lives in the human act of knowing, and it is an accumulation of experience that is more a living process than a static body of information; so, knowledge must be systematically nurtured and managed. In effect, organizations are increasingly leveraging the expertise not only from the domain of a local organization but also from the extended network (Figallo & Rhine, 2002) of broader professional community.

Adapting the Three-Vectors to an Appreciative Knowledge Environment

What makes managing knowledge through the Semantic Web a challenge is that knowledge comes often not as an object that can be stored, owned, and moved around like a piece of equipment or a document. It resides in the skills, understanding, and relationships of its members as well as in the tools and processes that embody aspects of this knowledge. In order for knowledge sharing within an organization to be successful, it is convinced that the people involved must be excited about the process of sharing knowledge. For many people, the primary reason for knowledge sharing is not that they expect to be repaid in the form of other knowledge, but the conviction that their individual knowledge is worth knowing, and that sharing this knowledge with others will be beneficial to their reputation (van den Hoof et al., 2004, p.1). There is some psychological benefit to sharing knowledge as the sharer may be held in higher esteem by the receiver(s) of the knowledge and may gain status as a result. Thereby, an appreciative sharing of knowledge must be viewed as the non-threatening and accepting approach that makes people realize what they do can make a difference. One common example is the communities of practice (CoP) (Wenger, McDermott, & Snyder, 2002) (be it physical or online) mentioned earlier. Many organizations today are comprised of a network of interconnected communities of practice each dealing with specific aspects such as the uniqueness of a long-standing client, or technical inventions. Knowledge is created, shared, organized, revised, and passed on within and among these communities. In a deep sense, it is by these communities that knowledge is owned in practice. Yet, knowledge exists not just at the core of an organization, but on its peripheries as well (as part of the knowledge network) (Tsoukas, 1996; Figallo & Rhine, 2002). So, communities of practice truly become organizational assets when their core and their boundaries are active in complementary ways, to generate an intentionally appreciative climate for organizational knowledge synthesis. In response to the knowledge challenge in a knowledge-centric organization, it is useful to conceive of an appreciative knowledge environment (AKE) based on virtual organizing, and experiment with how the ideas of its three vectors can be applied to nurture online the growth of different communities of practice (Wenger, 1998) scattered throughout an organization.

- **Virtual Encountering the AKE:** From a management perspective, it is important to identify what CoP's currently exist in the organization, and how, if they are not already online, to enable them to be online in order to provide more chances of virtual encounter of such communities, to the organizational members. For those communities already online, it is also important to design opportunities of interaction among different online communities, to activate their knowledge sharing. Since it is not a

CoP's practice to reduce knowledge to an object, what counts as knowledge is often produced through a process of communal involvement, which includes all the controversies, debate and accommodations. This collective character of knowledge construction is best supported online with individuals given suitable IS support to participate and contribute their own ideas. An IS subsystem, operated through virtual encounter, must help achieve many of the primary tasks of a community of practice, such as establishing a common baseline of knowledge and standardizing what is well understood so that people in a specific community can focus their creative energies on the more advanced issues.

- **Virtual Sourcing the AKE:** From the discussion built up in the first vector, it is not difficult to visualize the importance of identifying the specific expertise of each potential CoP in the organization, and if not yet available, planning for its acquisition through a purposeful nurture of expertise in various specific CoP's. In order to enable an organization to be part of a vibrant, dynamic network of complementary capabilities, in which the same organization could claim others' dependence and ensure her unique capabilities, an IS subsystem, operated through virtual sourcing, must help the organization understand precisely what knowledge will give it the competitive edge. The organization then needs to acquire this knowledge, keep it on the cutting edge, deploy it, leverage it in operations, and steward it across the organization.
- **Virtual Expertizing the AKE:** It is important to understand that not everything we know can be codified as documents and tools. Sharing tacit knowledge requires interaction and informal learning processes such as storytelling, conversation, coaching, and apprenticeship. The tacit aspects of knowledge often consist of embodied expertise—a deep understanding of complex, interdependent elements that enables dynamic responses to context-specific problems. This type of knowledge is very difficult to replicate. In order to leverage such knowledge, an IS subsystem, operated through virtual expertise, must help hooking people with related expertise into various networks of specialists, to facilitate stewarding such knowledge to the rest of the organization.

FUTURE TREND OF THE SEMANTIC WEB

The future of the Semantic Web must not be seen only from its technological possibilities, but also from its social dimension to operationalize knowledge sharing among members of the organization (Argyris, 1993). In order to facilitate the stewarding of knowledge through the various online communities of practice in an organization, it is important to have a vision that orients the kind of knowledge an organization must acquire, and wins spontaneous commitment by the individuals and groups involved in knowledge creation (Dierkes, Marz, and Teele, 2001; Kim, 1993; Stopford, 2001). This knowledge vision should not only define what kind of knowledge the organization should create in what domains, but also help determine how an organization and its knowledge base will evolve in the long run (Leonard-Barton, 1995; Nonaka & Takeuchi, 1995). The central requirement for organizational knowledge synthesis (or sharing) is to provide the organization with a strategic ability to acquire, create, exploit, and accumulate new knowledge continuously and repeatedly. To meet this requirement, we need an interpretation framework, which could facilitate the development of this strategic ability through the various communities. It is believed that there are at least three major appreciative processes constituting the interpretation framework of a knowledge-centric organization, including the personal process, the social process, and the organizational process. What follows is our appreciation of these three important processes (Checkland & Holwell, 1998, pp.98-109; Checkland, & Casar, 1986) considered as indispensable in the daily operations of the organization with the Semantic Web capability. Of particular interest here is the idea of providing meta-data support for various appreciative settings, which according to Vickers (1972, p.98), refer to the body of linked connotations of personal interest, discrimination and valuation which we bring to the exercise of judgment and which tacitly determine what we shall notice, how we shall discriminate situations from the general confusion of ongoing event, and how we shall regard them.

- **The Personal Process:** Consider us as individuals each conscious of the world outside our physical boundaries. This consciousness means that we can think about the world in different ways, relate these concepts to our experience of the world and so form judgments which can affect our intentions and, ultimately, our actions. This line of thought suggests a basic model for the active human agent in the world. In this model we are able to perceive parts of the world, attribute meanings to what we perceive, make judgments about our perceptions, form intentions to take particular actions, and carry out those actions. These change the perceived world, however slightly, so that the process begins again, becoming a cycle. In fact, this simple model requires some elaborations. First, we always selectively perceive parts of the world, as a result of our interests and previous history. Secondly, the act of attributing meaning and making judgments implies the existence of standards against which comparisons can be made. Thirdly, the source of standards, for which there is normally no ultimate authority, can only be the previous history of the very process we are describing, and the standards will themselves often change over time as new experience accumulates. This is the process model for the active human agents in the world of individual learning, through their individual appreciative settings. This model has to allow for the visions and actions, which ultimately belong to an autonomous individual, even though there may be great pressure to conform to the perceptions, meaning attributions and judgments, which belong to the social environment, which, in our discussion, is the community of practice.
- **The Social Process:** Although each human being retains at least the potential selectively to perceive and interpret the world in their own unique way, the norm for a social being is that our perceptions of the world, our meaning attributions and our judgments of it will all be strongly conditioned by our exchanges with others. The most obvious characteristic of group life is the never-ending dialogue, discussion, debate and discourse in which we all try to affect one another's perceptions, judgments, intentions and actions. This means that we can assume that while the personal process model continues to apply to the individual, the social situation will be that much of the process will be carried out inter-subjectively in discourse among individuals, the purpose of which is to affect the thinking and actions of at least one other party. As a result of the discourse that ensues, accommodations may be reached which lead to action being taken. Consequently, this model of the social process which leads to purposeful or intentional action, then, is one in which appreciative settings lead to

particular features of situations as well as the situations themselves, being noticed and judged in specific ways by standards built up from previous experience. Meanwhile, the standards by which judgments are made may well be changed through time as our personal and social history unfolds. There is no permanent social reality except at the broadest possible level, immune from the events and ideas, which, in the normal social process, continually change it.

- **The Organizational Process:** Our personal appreciative settings may well be unique since we all have a unique experience of the world, but oftentimes these settings will overlap with those of people with whom we are closely associated or who have had similar experiences. Tellingly, appreciative settings may be attributed to a group of people, including members of a community, or the larger organization as a whole, even though we must remember that there will hardly be complete congruence between the individual and the group settings. It would also be naïve to assume that all members of an organization share the same settings, those that lead them unambiguously to collaborate together in pursuit of collective goals. The reality is that though the idea of the attributed appreciative settings of an organization as a whole is a usable concept, the content of those settings, whatever attributions are made, will never be completely static. Changes both internal and external to the organization will change individual and group perceptions and judgments, leading to new accommodations related to evolving intentions and purposes. Subsequently, the organizational process will be one in which the data-rich world outside is perceived selectively by individuals and by groups of individuals. The selectivity will be the result of our predispositions to "select, amplify, reject, attenuate or distort" (Land, 1985, p.212) because of previous experience, and individuals will interact with the world not only as individuals but also through their simultaneous membership of multiple groups, some formally organized, some informal. Perceptions will be exchanged, shared, challenged, and argued over, in a discourse, which will consist of the inter-subjective creation of selected data and meanings. Those meanings will create information and knowledge which will lead to accommodations being made, intentions being formed and purposeful action undertaken. Both the thinking and the action will change the perceived world, and may change the appreciative settings that filter our perceptions. This organizational process is a cyclic one and it is a process of continuous learning, and should be richer if more people take part in it. And it should fit into the context of the appreciative knowledge environment scenario.

REMARKS OF CHALLENGE FOR KNOWLEDGE-CENTRIC ORGANIZATIONS

Earlier in the manuscript, we have associated the social context of Semantic Web to that of a knowledge-centric organization, and the appreciative importance of communities of practice (CoP) online. In this regard, there is an active role such communities can play in enabling the organization to learn from the experience of its members. Traditional organization (hierarchical) structures are designed to control activities and often discourage the easy sharing of knowledge and learning. Communities, nonetheless, help to foster relationships based on mutual trust, which are the unspoken and often unrecognized channels through which knowledge is shared. In fact, CoPs have profound implications for the management of knowledge work. They highlight the limits of management control in that CoPs are voluntary entities, depending entirely on the interest and commitment of their members. They cannot be designed or imposed in a top-down manner. Knowledge does not circulate through them in any officially prescribed form or procedures. Rather knowledge exchange through suitable means such as stories, jokes and anecdotes which serve to enliven and enhance a shared learning experience, has become important under the following contexts:

- **Perceiving the importance of story-telling:** It is not difficult to understand why story-telling has become a more important way of communicating knowledge than codifying it using specific IS/IT systems (Brown & Duguid, 1991): Firstly, stories present information in an interesting way with a beginning, a body, and an end, as well as people behaving goodly or badly. Secondly, stories present information in a way people can empathize with—recounting a situation which each of us might face, so it has greater perceived relevance. Thirdly, stories personalize the information—instead of talking about the situations in the abstract, we hear about the doings of individuals whom we might know or have heard of. Fourthly, stories bring people together, emphasizing a shared social identity and interests—we share knowledge rather than transfer it. More, stories express values—they often contain a moral about certain kinds of behavior leading to either positive or negative outcomes. In this way, stories link information with interest, values and relevance, giving us a sense of the context in which experience has been developed and helping us to grasp the tacit nature of some of the knowledge being communicated.
- **Understanding the nature of community knowing:** Perceptively, the importance of story-telling also provides an insight into the limits of technology for managing knowledge. Often, the design of IS/IT systems is based on a cognitive model of seeing knowledge as a "thing" (Malhortra, 2000) which is possessed by individuals, whereas the CoPs see it as the product of social interaction and learning among members of the same. By being a member of a community, individuals are able to develop their practice, sharing experience and ideas with others involved in the same pursuit. In light of this, the essence of understanding the social dimensions of managing knowledge work through the Semantic Web comes down to a few key points about the nature of knowing (Nonaka and Takeuchi, 1995; O'Leary, 1998; Wenger, 1998; Wenger et al., 2002):
    - **Knowledge lives in the human act of knowing:** In many instances of our daily living, our knowledge can hardly be reduced to an object that can be packaged for storage and retrieval. Our knowledge is often an accumulation of experience—a kind of residue of our actions, thinking, and conversations—that remains a dynamic part of our ongoing experience. This type of knowledge is much more a living process than a static body of information.
    - **Knowledge is tacit as well as explicit:** Not everything we know can be codified as explicit knowledge such as documents or tools. Sharing tacit knowledge requires interaction and informal learning processes which often involve a deep understanding of complex, interdependent elements that enables dynamic responses to context-specific problems, even though it is very difficult to document such knowledge in whatever manner

serves the needs of practitioners.

- ◦ **Knowledge is dynamic, social as well as individual:** It is important to accept that though our experience of knowing is individual, knowledge is not. Appreciating the collective nature of knowledge is especially important in an age when almost every field changes too much, too fast for individuals to master. Today's complex problems solving requires multiple perspectives. We need others to complement and develop our own expertise. In fact, our collective knowledge of any field is changing at an accelerating rate. What was true yesterday must be adapted to accommodate new factors, new data, new inventions, and new problems.

- **Positioning an appropriate appreciation for the Semantic Web:** The move to Semantic Web has been developing rapidly over the last decade, and has attracted a lot of attention in the development of different demonstration projects (Davies, Studer, & Warren, 2006) that can serve as reference implementations for future developers. Yet, what makes managing knowledge work through the Semantic Web a challenge is that today many an organization has come to the realization that unless knowledge is owned by people to whom it matters, it will not be developed, used, and kept up to date optimally. Knowledge is not a thing that can be managed at a distance like in an inventory. It is part of the shared practice of communities that need it, create it, use it, debate it, distribute it, adapt it, and transform it. As the property of a community, knowledge is not static; it involves interactions, conversations, actions, and inventions. Thereby, networking knowledge in a virtual community of practice is not primarily a technological challenge, but one of community development. Addressing the kind of dynamic knowing that makes a difference in practice requires the participation of people who are fully engaged in the process of creating, refining, communicating, and using knowledge. The thrust to develop, organize, and communicate knowledge must come from those who will use it. What matters is not how much knowledge can be captured, but how documenting can support people's abilities to know and to learn when the community itself becomes the living repository of people's knowledge. The Semantic Web works best when it is used to connect communities, not just to capture or transfer knowledge. Because much knowledge is embedded in particular communities, developing a shared understanding and a degree of trust is often the most critical step towards knowledge sharing in an organization. The use of Semantic Web technologies can complement but not replace the importance of social networks in this aspect (DiSessa & Minstrell, 1998). Indeed, the Semantic Web can support the development of new communities of practice through problem-solving interactions that allow individuals to appreciate the different perspectives which others bring to their work. Specifically, the Semantic Web can sustain the development of communities by allowing them to develop and exchange shared cultural objects of interest, such as texts, stories, and images, which help reinforce the meaning and purpose of the communities (Bodker, 1991). From a knowledge-building perspective (Bajjaly, 1999; Cohill & Kavanaugh, 1997), the design of Semantic Web must be based on understanding such concerns as: communities must be viewed as supporting networks of personal relationships in which people can collaboratively construct understanding to enable the exchange of resources and the development of a common framework for the analysis and evaluation of such resources. Thereby, it is important to consider how different strategies of the Semantic Web implementation can progressively involve individual members by helping them become resources for other community members.

- **Managing the knowledge-centric resources:** In 1969, Peter Drucker emphasized that knowledge had become the crucial resource of the economy. He claims the credit for coining the notion of 'knowledge work', which he contrasted with more traditional forms of work such as service work and manual work. Today, the term 'knowledge work' tends to refer to specific occupations which are "characterized by an emphasis on theoretical knowledge, creativity and use of analytical and social skills" (Frenkel et al., 1995, p.773). Knowledge work, interpreted this way, encompasses both what is traditionally referred to as professional work, such as accountancy, scientific and legal work, and more contemporary types of work, such as consultancy, software development, advertising and public relations. Understandably, these types of knowledge work are not susceptible to be easily imitated because there is a significant application of both tacit and explicit knowledge (Nonaka, 1994). Those engaged in these types of work are often individuals with high levels of education and specialist skills, who demand autonomy over their work processes to get the job done; namely, to demonstrate their ability to apply those skills to identify and solve problems. What is significant about these types of knowledge workers is that they own the organization's primary means of production—that is, knowledge. Nowadays, with the advent of the Semantic Web, we are ready to construct knowledge portfolio (Birchall & Tovstiga, 2002; Dove, 1999) for the organization, to track the knowledge contributions of individual knowledge workers, and different grouping of the same in the form of group-based project work. The management of knowledge workers assumes greater importance for sustaining productivity than the management of machines, technologies, or work processes. Like musicians, Drucker (1988) sees such employees exploring outlets for their creative abilities, seeking interesting challenges, enjoying the stimulation of working with other specialists. This, he argues, poses new management challenges in knowledge-centric organizations: developing rewards, recognition and career opportunities; giving an organization of specialists a common vision; devising a management structure for coordinating tasks and task teams; and ensuring the supply and skills of top management people.

CONCLUSION

Finally, in closing our discussion, it is essential to articulate the promise of appreciative inquiry (AI) (Reed, 2007; Lewis, Passmore, & Cantore, 2008) for a knowledge-centric organization. In the broadest sense, the major theme of appreciative knowledge sharing in and among virtual communities of practice (Hoadley & Pea, 2002) could be understood from the perspective of effectively applying information and communications technologies, ICT (including the Semantic Web technologies) to improve the lives of people (organizational members), in terms of getting knowledge to those of a community who need it in the right time. Of much concern here is an effort to theorize the social dimensions of this ICT-based knowledge sharing. In the words of David Hakken (2002, p.362), we have to ask "what kinds of theorizations make sense in analyzing what happens when a concerted effort is made to introduce a technology supportive of knowledge sharing in a 'holistic' way—that is, to try to anticipate and address the social context/consequences of the interventions." In simpler terms, we can describe AI as an exciting philosophy for change. The major assumption of AI is that in every organization something works and change can be managed through the identification of what works, and the analysis of how to do more of what works. A key characteristic of AI is that it is a generative process. That means it is a moving target, and is created and constantly re-created by the people who use it. While the electronic stewarding of knowledge in an online community is based upon the Semantic Web technologies, its success rests with its people (Linn, 2000)—organizers, information and knowledge providers, sponsors, users, volunteers—who support the organization (comprising various CoPs) in a variety of ways. Therefore, when attempting to design technology in support of a knowledge-centric organization, it is important to remember "what is working around here?" in the organization. The tangible result of the appreciative inquiry process should be a series of vision statements that describe where the organization wants to be, based on the high moments of where they have been. Because the statements are grounded in real experience and history, it is convinced that people in the organization know how to repeat their success. In retrospect, think about a time when you shared something that you knew that enabled you or your company to do something better or achieve success. What happened? Share your story. Such activities include not only information capture and transmission, but also the establishment of social relationships in which people can collaboratively construct understanding. It is this energy that distinguishes AI's generative potential that presumably has no end because it is a living process. And it is quite promising for any knowledge-centric organization pursing the Semantic Web technologies.

REFERENCES

Antoniou, G., & van Harmelen, F. (2004). *A Semantic Web primer.* Cambridge, MA: The MIT Press.

Argyris, C. (1993). *Knowledge for action: A guide to overcoming barriers to organizational change.* San Francisco: Jossey-Bass.

Bajjaly, S. T. (1999). *The community networking handbook.* Chicago: American Library Association.

Belew, R. K. (2000). *Finding out about: A cognitive perspective on search engine technology and the WWW.* Cambridge, UK: Cambridge University Press.

Berners-Lee, T. (1998a). *Semantic Web road map.* Retrieved July 30, 2008, from http://www.w3c.org/DesignIssues/Semantic.html

Berners-Lee, T. (1998b). *Evolvability.* Retrieved July 30, 2008, from http://www.w3c.org/DesignIssues/Evolution.html

Berners-Lee, T. (1998c). *What the Semantic Web can represent.* Retrieved July 30, 2008, from http://www.w3c.org/DesignIssues/RDFnot.html

Berners-Lee, T. (1999). *Weaving the Web.* San Francisco, CA: Harper San Francisco.

Berners-Lee, T., Hendler, J., & Lassila, O. (2001, May). The Semantic Web. *The Scientific American.* Retrieved July 30, 2008, from http://www.sciam.com/article.cfm?id=the-semantic-web

Birchall, D. W., & Tovstiga, G. (2002). Assessing the firm's strategic knowledge portfolio: A framework and methodology. *International Journal of Technology Management, 24*(4), 419–434. doi:10.1504/IJTM.2002.003063

Bodker, S. (1991). *Through the interface: A human activity approach to user interface design.* Hillsdale, NJ: Lawrence Erlbaum.

Brown, J., & Duguid, P. (1991). Organizational learning and communities-of-practice: Towards a unified view of working, learning and innovation. *Organization Science, 2*, 40–57. doi:10.1287/orsc.2.1.40

Browne, J., Sacket, P. J., & Wortmann, J. C. (1995). Future manufacturing systems—towards the extended enterprise. *Computers in Industry, 25*, 235–254. doi:10.1016/0166-3615(94)00035-O

Bushe, G. R. (1995). Advances in appreciative inquiry as an organization development intervention. *Organization Development Journal, 13*(3), 14–22.

Byrne, J. A., Brandt, R., & Port, O. (1993, February 8). The virtual corporation. *Business Week*, 36–41.

Checkland, P., & Holwell, S. (1998). *Information, systems and information systems: Making sense of the field.* New York: John Wiley & Sons Ltd.

Checkland, P. B., & Casar, A. (1986). Vicker's concept of an appreciative system: A systematic account. *Journal of Applied Systems Analysis, 3*, 3–17.

Cheng, W. (1996, February 5-7). *The virtual enterprise: Beyond time, place and form* (Economic Bulletin). Singapore International Chamber of Commerce.

Chin, A. (1998). Future visions. *Journal of Organizational Change Management, 11*(1).

Cohill, A. M., & Kavanaugh, A. L. (1997). *Community networks: Lessons from Blacksberg, Virginia.* Norwood, MA: Artech House.

Cooperrider, D. (1986). *Appreciative inquiry: Toward a methodology for understanding and enhancing organizational innovation.* Unpublished doctoral dissertation, Case Western Reserve University, Cleveland, Ohio.

Cooperrider, D. L., & Srivastva, S. (1987). Appreciative inquiry in organizational life. In W. Pasmore & R. Woodman (Eds.), *Research in organization change and development* (Vol. 1, pp. 129-169). Greenwich, CT: JAI Press.

Cooperrider, D. L., & Whitney, D. (2005). *Appreciative inquiry: A positive revolution in change.* San Francisco: Berrett-Koehler.

Curran, M. (1991). Appreciative inquiry: A third wave approach to organization development. *Vision/Action,* (December), 12-14.

Daconta, M. C., Obrst, L. J., & Smith, K. T. (2003). *The Semantic Web: A guide to the future of XML, Web services, and knowledge management.* Indianapolis, IN: Wiley Publishing, Inc.

Davenport, T. H. (1997). *Information ecology: Mastering the information and knowledge environment.* Oxford, England: Oxford University Press.

Davidow, W. H., & Malone, M. S. (1992). *The virtual corporation—structuring and revitalizing the corporation for the 21st century.* New York: HarperCollins.

Davies, J., Studer, R., & Warren, P. (2006). *Semantic Web technologies: Trends and research in ontology-based systems.* Chichester, England: John Wiley & Sons Ltd.

Davis, B. H., & Brewer, J. (1997). *Electronic discourse: Linguistic individuals in virtual space.* Albany, NY: State University of New York Press.

Dierkes, M., Marz, L., & Teele, C. (2001). Technological visions, technological development, and organizational learning. In M. Dierkes, A. B. Antal, et al. (Eds.), *Handbook of organizational learning and knowledge* (pp. 282-304). Oxford, UK: Oxford University Press.

DiSessa, A. A., & Minstrell, J. (1998). Cultivating conceptual change with benchmark lessons. In J. G. Greeno & S. Goldman (Eds.), *Thinking practices* (pp. 155-187). Mahwah, NJ: Lawrence Erlbaum.

Dove, R. (1999). Managing the knowledge portfolio. *Automotive Manufacturing & Production, April*(52). Retrieved July 30, 2008 from http://www.parshift.com/Essays/essay052.htm

Drucker, P. F. (1988). The coming of the new organization. *Harvard Business Review*, (Summer): 53–65.

Figallo, C., & Rhine, N. (2002). *Building the knowledge management network*. New York: John Wiley & Sons.

Frenkel, S., Korczynski, M., Donoghue, L., & Shire, K. (1995). Re-constituting work: Trends towards knowledge work and info-normative control. *Work, Employment and Society, 9*(4), 773–796.

Gergen, K. J. (1990). Affect and organization in postmodern society. In S. Srivastva, D. L. Cooperrider, et al. (Eds.), *Appreciative management and leadership: The power of positive thought and action in organizations* (1st ed., pp. 289-322). San Francisco, CA: Jossey-Bass Inc.

Goldman, S., Nagel, R., & Preiss, K. (1995). *Agile competitors and virtual organizations: Strategies for enriching the customer*. New York: van Nostrand Reinhold.

Gremba, J., & Myers, C. (1997). *The IDEAL model: A practical guide for improvement*. Pittsburgh, PA: CMU-SEI. Retrieved July 30, 2008, from http://www.sei.cmu.edu/ideal/ideal.bridge.html

Gruber, T. (1993). A translation approach to portable ontologies. *Knowledge Acquisition, 5*(2), 199-220. Retrieved July 30, 2008, from http://ksl-web.stanford.edu/KSL_Abstracts/KSL-92-71.html

Hakken, D. (2002). Building our knowledge of virtual community: Some responses. In K. A. Renninger & W. Shumar (Eds.), *Building virtual communities: Learning and change in cyberspace* (pp. 355-367). Cambridge, UK: Cambridge University Press.

Harman, W. W. (1990). Shifting context for executive behavior: Signs of change and re-evaluation. In S. Srivastva, D. L. Cooperrider, et al. (Eds.), *Appreciative management and leardership: The power of positive thought and action in organizations* (1st ed., pp. 37-54). San Francisco, CA: Jossey-Bass Inc.

Hedberg, B., Dahlgren, G., Hansson, J., & Olve, N. (1997). *Virtual organizations and beyond: Discover imaginary systems*. New York: John Wiley & Sons Ltd.

Hemlin, S., Allwood, C. M., & Martin, B. R. (2004). *Creative knowledge environments: The influences on creativity in research and innovation*. Northampton, MA: Edward Elgar.

Hendler, J. (2001). Agents and the Semantic Web. *IEEE Intelligent Systems, 16*(March-April), 30–37. doi:10.1109/5254.920597

Hoadley, C., & Pea, R. D. (2002). Finding the ties that bind: Tools in support of a knowledge-building community. In K. A. Renninger & W. Shumar (Eds.), *Building virtual communities: Learning and change in cyberspace* (pp. 321-354). Cambridge, UK: Cambridge University Press.

Hubbard, B. M. (1998). *Conscious evolution: Awakening the power of our social potential*. Novato, CA: New World Library.

Ichijo, K., & Nonaka, I. (Eds.). (2007). *Knowledge creation and management: New challenges for managers.* New York: Oxford University Press.

Jagers, H., Jansen, W., & Steenbakkers, W. (1998, April 27-28). Characteristics of virtual organizations. In P. Sieber & J. Griese (Eds.), *Organizational virtualness, Proceedings of the VoNet-Workshop*. Bern, Switzerland: Simowa Verlag.

Kim, D. (1993). The link between individual and organizational learning. *Sloan Management Review*, (Fall): 37–50.

Land, F. (1985). Is an information theory enough? *The Computer Journal*, *28*(3), 211–215. doi:10.1093/comjnl/28.3.211

Leonard-Barton, D. (1995). *Wellsprings of knowledge: Building and sustaining the sources of innovation*. Boston: Harvard Business School Press.

Lewis, S., Passmore, J., & Cantore, S. (2008). *Appreciative inquiry for change management: Using AI to facilitate organization development.* London: Kogan Page.

Liebowitz, J. (1999). *Knowledge management handbook.* Boca Raton, FL: CRC Press.

Liebowitz, J. (2000). *Building organizational intelligence: A knowledge management primer.* Boca Raton, FL: CRC Press.

Liebowitz, J., & Beckman, T. (1998). *Knowledge organizations: What every manager should know.* Boca Raton, FL: CRC Press.

Linn, M. C. (2000). Designing the knowledge integration environment: The partnership inquiry process. *International Journal of Science Education*, *22*(8), 781–796. doi:10.1080/095006900412275

Malhotra, Y. (2000). Knowledge management and new organization forms: A framework for business model innovation. In Y. Malhotra (Ed.), *Knowledge management and virtual organizations* (pp. 2-19). Hershey, PA: Idea Group Publishing.

Markova, I., & Foppa, K. (Eds.). (1990). *The dynamic of dialogue*. New York: Harvester Wheatsheaf.

Mowshowitz, A. (1997). Virtual organization. *Communications of the ACM*, *40*(9), 30–37. doi:10.1145/260750.260759

Musen, M. A. (2006). Foreword. In J. Davies, R. Studer, & P. Warren (Eds.), *Semantic Web technologies: Trends and research in ontology-based systems* (pp. xi-xiii). Chichester, England: John Wiley & Sons, Ltd.

Nonaka, I. (2002). A dynamic theory of organizational knowledge creation. In C. W. Choo & N. Bontis (Eds.), *The strategic management of intellectual capital and organizational knowledge* (pp. 437-462). Oxford, UK: Oxford University Press.

Nonaka, I., & Takeuchi, H. (1995). *The knowledge creating company: How Japanese companies create the dynamics of innovation*. Oxford, UK: Oxford University Press.

O'Leary, D. E. (1998). Enterprise knowledge management. *IEEE Computer*, *31*(3), 54–61.

Paulk, M. C., Weber, C. V., Curtis, B., & Chrissis, M. B. (1994). *The capability maturity model: Guidelines for improving the software process.* Reading, Ma: Addison Wesley.

Reed, J. (2007). *Appreciative inquiry: Research for change*. London: Sage Publications.

Stopford, J. M. (2001). Organizational learning as guided responses to market signals. In M. Dierkes, A. B. Antal, et al. (Eds.), *Handbook of organizational learning and knowledge* (pp. 264-281). New York: Oxford University Press.

Swartz, A., & Hendler, J. (2001). The Semantic Web: A network of content for the digital city. In *Proceedings of the Second Annual Digital Cities Workshop*, Kyoto, Japan. Retrieved July 30, 2008, from http://blogspace.com/rdf/SwartzHendler.html

Thatchenkery, T. (2005). *Appreciative sharing of knowledge: Leveraging knowledge management for strategic change*. Chagrin Falls, OH: Taos Institute Publishing.

Thatchenkery, T., & Chowdhry, D. (2007). *Appreciative inquiry and knowledge management.* Northampton, MA: Edward Elgar.

The Complete Oxford English Dictionary. (1971). Oxford: Oxford University Press.

Tozer, G. (1999). *Metadata management for information control and business success.* Norwood, MA: Artech House, Inc.

Tsoukas, H. (1996). The firm as a distributed knowledge system: A social constructionist approach. *Strategic Management Journal, 17*(Winter Special Issue), 11-25.

van den Hoof, B., de Ridder, J., & Aukema, E. (2004). *The eagerness to share: Knowledge sharing, ICT and social capital* (Working Paper). Amsterdam, The Netherlands: Amsterdam School of Communication Research, University of Amsterdam.

Van den Hoven, J. (2001). Information resource management: Foundation for knowledge management. *Information Systems Management, 18*(2), 80–83.

Vat, K. H. (2003, June 24-27). Toward an actionable framework of knowledge synthesis in the pursuit of learning organization. In *Proceedings of the 2003 Informing Science + IT Education Conference (IsITE2003)*, Pori, Finland.

Venkatraman, N., & Henderson, J. C. (1998). Real strategies for virtual organizing. *Sloan Management Review, 40*(1), 33–48.

Vickers, G. (1965). *The art of judgment.* New York: Basic Books.

Vickers, G. (1968). *Value systems and social process*. New York: Basic Books.

Vickers, G. (1972). Communication and appreciation. In G. B. Adams et al. (Eds.), *Policymaking, communication and social learning: Essays of Sir Geoffrey Vickers*. New Brunswick, NJ: Transaction Books.

Von Krogh, G., Ichijo, K., & Nonaka, I. (2000). *Enabling knowledge creation: How to unlock the mystery of tacit knowledge and release the power of innovation.* New York: Oxford University Press.

Watkins, J. M., & Cooperrider, D. L. (1996). Organization inquiry model for global social change organizations. *Organization Development Journal, 14*(4), 97–112.

Wenger, E. (1998). *Communities of practice: Learning, meaning, and identity*. Cambridge, UK: Cambridge University Press.

Wenger, E., McDermott, R., & Snyder, W. M. (2002). *Cultivating communities of practice: A guide to managing knowledge.* Cambridge, MA: Harvard Business School Press.

White, T. W. (1996). Working in interesting times. *Vital Speeches of the Day, LXII*(15), 472–474.

KEY TERMS AND DEFINITIONS

Appreciative Inquiry (AI): Appreciative Inquiry is about the co-evolutionary search for the best in people, their organizations, and the relevant world around them. In its broadest focus, it involves systematic discovery of what gives "life" to a living system when it is most alive, most effective, and most constructively capable in economic, ecological, and human terms.

Appreciative Processes: These are processes to leverage the collective individual learning of an organization such as a group of people, to produce a higher-level organization-wide intellectual asset. This is supposed to be a continuous process of creating, acquiring, and transferring knowledge accompanied by a possible modification of behavior to reflect new knowledge and insight, and to produce a higher-level intellectual content.

Appreciative Settings: A body of linked connotations of personal or collective interest, discrimination and valuation which we bring to the exercise of judgment and which tacitly determine what we shall notice, how we shall discriminate situations of concern from the general confusion of ongoing event, and how we shall regard them.

Appreciative Knowledge Environment (AKE): A work, research or learning environment to incorporate the philosophy of appreciative inquiry in support of a cultural practice of knowledge sharing among organizational members.

Community of Practice (CoP): These are people who come together around common interests and expertise. They create, share, and apply knowledge within and across the boundaries of teams, business units, and even entire organizations—providing a concrete path toward creating a true knowledge organization.

Knowledge-Centric Organization: Any organization whose knowledge focus is to provide mechanisms for building the knowledge base of the firm to better apply, share, and manage knowledge resources across various components in the company.

Semantic Web: The Semantic Web is an evolving extension of the World Wide Web in which the semantics of information and services on the web is defined, making it possible for the web to understand and satisfy the requests of people and machines to use the Web content. It derives from W3C director Tim Berners-Lee's vision of the Web as a universal medium for data, information, and knowledge exchange.

Virtual Organizing: A method to operationlize the context of appreciative inquiry, with the technology-enabled capability to assemble and disassemble nodes on a network of people or groups of people in an organization, to meet the demands of a particular business context. In virtual organizing, virtuality is a strategic characteristic applicable to every organization.

This work was previously published in Handbook of Research on Social Dimensions of Semantic Technologies and Web Services, edited by M. M. Cruz-Cunha; E. F. Oliveira; A. J. Tavares & L. G. Ferreira, pp. 411-434, copyright 2009 by Information Science Reference (an imprint of IGI Global).

Chapter 7.17
The Emergence of Agency in Online Social Networks

Jillianne R. Code
Simon Fraser University, Canada

Nicholas E. Zaparyniuk
Simon Fraser University, Canada

ABSTRACT

Social and group interactions in online and virtual communities develop and evolve from expressions of human agency. The exploration of the emergence of agency in social situations is of critical importance to understanding the psychology of agency and group interactions in social networks. This chapter explores how agency emerges from social interactions, how this emergence influences the development of social networks, and the role of social software's potential as a powerful tool for educational purposes. Practical implications of agency as an emergent property within social networks provide a psychological framework that forms the basis for pedagogy of social interactivity. This chapter identifies and discusses the psychological processes necessary for the development of agency and to further understanding of individual's engagement in online interactions for socialization and learning.

DOI: 10.4018/978-1-60566-208-4.ch008

INTRODUCTION

Social and group interactions in online and virtual communities develop and evolve from expressions of human agency. Agency is the capability of individuals to consciously choose, influence, and structure their actions (Emirbayer & Mische, 1998; Gecas, 2003) and is an active exercise of ability and will. The ways in which individuals express agency are associated with their motivational orientation, intentionality, and choice (volition), and relates to their ability to engage these characteristics in social contexts to achieve their goals. As agents, individuals formulate intentions, execute decisions, and produce motivation in an effort to communicate. Understanding how agency develops and emerges within social networks is a key factor in identifying *why* online social networks develop and *how* they influence individual processes such as cognition, motivation, behavior, and ultimately learning.

The exploration of the emergence of agency in social situations is of critical importance to understanding the psychology of agency and group interactions in social networks. Research in social psychology provides a context in which to investigate the psychological effects of online social software as it relates to motivation (see Ryan & Deci, 2000), interactions within the social networks (see Thompson & Fine, 1999), and how individuals vary in their ability to express agency (see Martin, 2003, 2004).

Agency emerges out of interactions and goal directed activities within social networks. Similarly, social networks emerge through the interactions and characteristics of agents support their formation, development, and evolution. Socially situated emergent properties of agency and social networks connect them as a dynamic complex system. Social software is software that "supports, extends, or derives added value from human social behavior" (for a review see boyd, 2007; Coates, 2005). Online friendship websites, massively multiplayer online games, and social groupware, such as Facebook (2008), MySpace (2008), Bebo (2008), and Second Life (Linden Research Inc., 2008) provide frameworks in which social dynamics can mediate the development of agency within social networks.

The purpose of this chapter is to introduce the concept of agency as it relates to the formation, development, and evolution of social networks. This chapter explores how agency emerges from social interactions, how this emergence influences the development of social networks, and the potential role of social software as a tool with educational applications. Practical implications of agency as an ability to engage within social networks provides a psychological framework that forms the basis for a pedagogy of social interactivity. This chapter discusses the psychological processes necessary for the development of agency, how these processes affect an individual's engagement in online interactions for both socialization and learning, and how social software such as Facebook (2008), MySpace (2008), Bebo (2008), and Second Life (Linden Research Inc., 2008) can be used in educational contexts. As agency directly affects how an individual understands their various roles, beliefs, and decisions in social contexts, there are far reaching implications for social software as an educational tool.

AGENCY

Agency is an ability developed through social means and human experience (Mead, 1932, 1934). As an ability to act independently despite the immediate situation, agency engages habit, imagination, and judgment (Emirbayer & Mische, 1998, p. 970). Agency also involves the knowledge, experience, and the ability to achieve one's goals (Little, Hawley, Henrich, & Marsland, 2002). Within the social framework, agency abilities develop through the interaction of social processes, the dynamics of which can be explained using action theory.

For action theorists (e.g. Parsons, 1968), agency is captured in the notion of *effort*. In this view, agency acts as the force that achieves, where conditions for achievement are at one end of a spectrum and the normative rules are at the other (Emirbayer & Mische, 1998). Agency ability is ultimately a *temporal* continuum through which an individual exercises personal influence and in return affects environmental processes that ultimately affect other personal self-processes. Thus, personal influence becomes a reciprocal collective determinant even though it also determines the individual (Martin, 2003). Agency remains a strong dynamic and causal force underlying individual action.

As a dynamic process, agency is a motivating force of action. The ways in which individuals express and develop agency are associated with their motivational orientation, intentionality, and choice, and speaks to their ability to engage these characteristics in social contexts to achieve their particular goals. Internal personal factors, behavioral patterns, and environmental influences require agency ability to facilitate social processes. Agency-related constructs associated with social interaction include self-efficacy, locus of control, and volition.

Self-Efficacy

A belief in one's capability to succeed is an essential condition of human functioning. Whether one believes that they can produce a certain action is as important as having the skills available to succeed (Bandura, 1997). Self-efficacy is a generative property, meaning it is a capacity that originates within the *Self*. It is generative in that it is a belief that an individual holds to be true. Self-efficacy is also an evaluative capacity in which one perceives their ability to perform a particular action. When an individual then deems themselves effective enough to complete a task, they anticipate the result to be positive. Thus, an efficacy belief propagates from the belief in one's own ability and that they have the skill necessary to complete a task successfully. In relation to social networks and social interaction, self-efficacy for socialization is an important part in determining whether an individual feels they can successfully communicate within a social setting. Social software enables the development of self-efficacy for socialization as it removes social barriers that may otherwise inhibit individuals from interacting in certain ways. For example, massive multiplayer online games (MMOG) and social software such as MySpace (2008), Facebook (2008), Bebo (2008), and Second Life (Linden Research Inc., 2008), provide opportunities for interaction where individuals can socialize and develop confidence in their socialization skills without the awkwardness individuals may encounter in a face-to-face setting. Individuals come to believe that they can communicate relatively successfully, and ultimately develop a higher self-efficacy for socialization and feel that they have control over their actions within this particular context. As a result, individuals who have high self-efficacy also have a tendency to believe that they control their actions and the outcomes that result.

Locus of Control

The causal relationship between's one own behavior and that of an outcome affects a range of choices an individual makes (Lefcourt, 1966; Rotter, 1966). Social Learning Theory (Rotter, 1954, 1966), not to be confused with Social Cognitive Theory, posits that control is considered a *generalized expectancy* which operates across a large number of situations and relates to whether or not an individual believes they possess or lack power over what happens to them. How an individual attributes causal beliefs to outcomes is a central argument of Social Learning Theory and the locus of control concept.

Individuals often believe that they have the power to control the outcome of any given situation. If one believes that a cause of an outcome is a result of personal skill, one has an internal control expectancy, or an internal locus of control. Within a MMOG or other online environment, if an individual perceives a threat or is in a situation in which they are required to make a decision, they must first recognize that they have choices available. The individual can then engage the situation or leave it. How an individual reacts to any situation requires self-control and self-regulation in the form of volition.

Volition

Volition incorporates factors of self-control and self-regulation. Contemporary ideas of volition from an information processing perspective were adapted by Kuhl (1985) are based on a theory of motivation and action originally developed by Ach (1910), (as in Corno, 2001). According to Kuhl (1985), self-control and self-regulation are *modes* of volition coordinated through a central executive. *Self-control* is the mode of volition that supports the maintenance of an active goal, whereas *self-regulation* involves the maintenance of one's actions in line with an integrated *Self*. Volition is a "post-decisional, self-regulatory process that energizes the maintenance and enactment of intended actions" (Kuhl, 1985, p. 90). Volition is a self-regulatory process that provides the means for maintaining the commitment and motivation of an individual to their actions. Within social software, volition enables individuals to persist in achieving a desired outcome, such as meeting new friends or getting a date. Volition becomes particularly important in social networks because intentions are fragile and people often waver on commitments especially when they are faced with challenging problems to solve (Corno, 2001). As an agency-related construct, volition ensures that individuals persist in their motivation to achieve their goals.

Volition, self-efficacy, and locus of control are agency-related constructs that demonstrate agency ability. Each of these constructs interacts within the self-system and collectively enables the expression agency. Assumptions within dynamical systems theory assist in interpreting agency as a relationship between several self-processes *emergent* of that relationship.

EMERGENCE OF AGENCY

Agency develops through a socially mediated ability exercised through human interaction (Mead, 1934). As an emergent entity (Martin, 2003; O'Connor & Wong, 2002), agency develops out of the fundamental characteristics of ability (physical, psychological, and behavioral components), will (i.e. volition, locus of control), and action. Emergence, from an ontological perspective is a non-reducible phenomenon, meaning, that if a construct is emergent it has several component parts but is irreducible with respect to them (Martin, 2003; O'Connor & Wong, 2002). For example, water has its own properties that are complex and novel and are not just a collection of the properties of its components, oxygen, and hydrogen. Oxygen and hydrogen are necessary for the creation of water, however, water also has properties that are uniquely its own; in this way the properties of water are *emergent* (Martin, Sugarman, & Thompson, 2003). Martin, Sugarman and Thompson (2003) propose that agency possesses emergence, and that agency itself contains emergent properties generated by a combination of mental and social events. As water has its own properties, when a heat source is applied to water, it boils, as the water molecules act in response to this external force, the property of the water changes from liquid into a gas. Similarly, the properties of agency change when external forces interact with the different *Self*-factors such as self-efficacy, locus of control, and volition. Thus, agency is a systemic construct, a dynamic interaction among a number of associated agency factors. Changes in the properties of agency are a result of mental (internal) and environmental (external) relationships. Similarly, social networks emerge as a construction of the individuals who interact (internal) and the groups they form (external), however, social networks are not simply sum of their parts.

EMERGENCE OF SOCIAL NETWORKS

Individuals are both a product of and producer of their socio-cultural world (Martin, 2003; Martin et al., 2003). Environmental and social factors through interactions with people are both producers and are a products of social systems (Bandura, 1997) and involve the coordination and interdependence of personal and situational forces (Markus & Nurius, 1984). The dynamics of a social network are a function of both informal and formal factors and affect the emergence of social roles, specifically informally self-generated social roles referred to as virtual identities (see Code & Zaparyniuk, this volume). The emergence of informal social roles have variable effects on the patterns of interaction and connection among individuals in the network and ultimately on the performance, productivity, evolution, and sustainability of the social network (Jeffrey C Johnson, Palinkas, & Boster, 2003). Critical aspects of these emergent properties are the adaptability of the social network to internal and external patterns of change.

The adaptability of a social network is dependent upon the cohesion of individuals within the network. As individuals utilize, model, and emulate behaviors (cognitive and otherwise) projected by their peers and other agents, they effectively co-regulate their development of social competence. Through this process, individuals exploit the abilities of others to enhance their own capabilities, but also to facilitate their achievement of social outcomes. In this context, individuals co-regulate in social networks to achieve personal social goals. This *co-regulation* is a result of an individual resolution to utilize the abilities and efforts of others to achieve personal, social, or other goals. As co-regulation is an on-going collaborative process, the cooperative relationship between individuals within a social network enables adaptation.

During co-regulation, individuals become agents or 'causal contributors' (Bandura, 1997) of their own social experience. Seeking the meditative efforts of others, helps develop the competence to self-regulate. Self-regulatory competence is a skill, an instrument of agency, that is acquired through collaboration (Bruner, 1997). Ultimately, agency is expressed by the capability of individuals to consciously choose, influence, and structure their actions (Emirbayer & Mische, 1998; Gecas, 2003), and in the context of social networks enables them to formulate intentions, execute decisions, and produce motivation in their effort to interact and communicate within the network.

AGENCY AND SOCIAL NETWORKS AS DYNAMIC COMPLEX SYSTEMS

The emergence of agency and social networks involves particular mental and social causations. On the assumption that agency develops as a result of the interaction of these mental causations (Martin, 2003; Martin et al., 2003), and social networks emerge as result of informal social roles and cohesion, agency and social networks can be described as dynamic systems. As agency and social networks are dynamic, they also contribute to their own creation and evolution (see Code & Zaparyniuk, this volume); however, the system is irreducible with respect to them. Just as one individual or group does not embody the dynamics of the social network, the interaction of the group entities brings about group formation. Any one of the agentic factors alone cannot measure agency, but they may be able to indicate it collectively through their interaction. Similarly, each of the individuals within a social network can only create a network (system) through their interactions. Dynamic systems theorists describe general functions of a nonlinear dynamical system (e.g. Carver & Schier, 2002; Neil F Johnson, 2007; Vallacher, Nowak, Froehlich, & Rockloff, 2002), however relative to a discussion of agency and social networks as dynamic systems, four of these factors are briefly outlined.

Factor 1. The System Cannot be Decomposed into Separate Additive Influences.

As agency is an emergent abstraction of the relationship among factors such as self-efficacy, locus of control, and volition, social networks are similarly an emergent abstraction of the relationships between the individuals within the system. From an ontological perspective, an emergent entity cannot be broken down into its constituent parts. Agency is not merely a sum of self-efficacy, locus of control, and volitional measures. Social networks do not exist without its members; however, each individual member on their own does not characterize it.

Factor 2. The System has Memory or Includes Feedback.

As agentic variables attributed to self-efficacy, locus of control, and volition, are interdependent; they affect and influence each other in both positive and negative ways. From the perspective of agency, self-regulatory competence (e.g. Bouffard, Bouchard, Goulet, Cenoncourt, & Couture, 2005; Wolters, 1999) is affected by a student's self-efficacy for the task (e.g. Loedewyk & Winne, 2005; Schunk & Ertmer, 2001) and motivation to complete the task (e.g. Wolters, 2003; Wolters, Yu, & Pintrich, 1996). Similarly, social networks and the individuals within them are interdependent and influence each other. Within social networks, the presence or absence of factors such as informal social roles has an impact on a network's emergent properties such as stability, adaptability, and robustness (Neil F Johnson, 2007).

Factor 3. TheSystem can Adapt itself According to its History, Feedback, and Environment.

A system's properties and their patterns of change emerge from 'rules' specifying how the system's elements interact. Emergent or 'macro-level' properties can be understood as features (usually in the case of events and processes) that supervene on, and are thus realized in, 'lower-level' features (Henderson, 1994). Related to agency, characteristics of metacognition are identified through personal awareness and cognitive control (Brown, 1987; Flavell, 1979), and provide feedback for such high-order functions as planning, strategy selection, and monitoring (Sternberg, 1999). Alternatively, social networks have 'group' level properties such as cohesion and coherence that supervene on, and are realized in the members of the network.

Factor 4. The System is Self-Organizing and Non-Linear.

A system can only exist if it has autonomous organizing capacities (Gergen, 1984). As a system, agency is self-organizing as it is an emergent function of its constituents, as are social networks an emergent function of its membership. Self-organization of cognitive and affective elements into higher order structures have been revealed in experimental work on social judgment and action identification (Vallacher, Nowak, Markus, & Strauss, 1998) and in computer simulations of self-reflection processes (Nowak, Vallacher, Tesser, & Borkowski, 2000).

Agency and social networks as dynamic systems have particular explanatory value on the causes of human action, but also on the formation of particular social groups. Within education, this explanatory value provides new opportunities for teaching and learning.

IMPLICATIONS FOR EDUCATION

Agency and social networks emerge within social contexts. Social software as a tool for facilitating the development of social networks and agency development has far-reaching implications for educational practice. As individuals interact and groups form the purpose of learning, agency and social networks will also emerge within educational contexts. Conceptualizing social software as a cultural tool, using social networks to represents multiple ways of knowing, knowledge-building, promoting communities of practice, and enabling self-regulated learning, enables a clear application of social software as an educational tool.

Social Software as a Cultural Tool

Cultural tools mediate communication within social settings. Social experience involves the interactions between individuals, and involves the tools, symbols, and values that influence the action (Gauvain, 2001). Vygotsky's sociocultural theory of development posits that the transformation and development of cognitive and social skills occurs within social interactions. Vygotsky (1962, 1978) believed that children (and individuals) learn using *cultural tools* which mediate higher-order mental processes such as reasoning and problem solving. Cultural tools include both *technical tools* such as books, media, and computers, and *psychological tools* such as language, signs, writing, and symbols.

"By being included in the process of behavior, the psychological tool alters the entire flow and structure of mental functions" (Vygotsky, 1981, p. 137).

Online social networks and social software changes the way we perceive and act within social settings. As social software is both a social and psychological tool, it provides a computing environment in which actions are mediated through the appropriation of language, writing, signs, and symbols. As a result, online social networks and social software are cultural tools, and are carriers of social, cultural, and historical formations that amplify certain social actions (Jones & Norris, 2005). The enactment of social software as a cultural tool promotes the development of a unique and particular social language that mediates agentic expression.

Mediated Agency

Mediated agency aids individuals in interpreting the meaning of a situation. As a social construction, agency is mediated within a social setting by psychological tools such as language, and technical tools such as computers. Wertsch, Tulviste, and Hagstrom (1993) refer to agency within a sociocultural situation as *mediated agency* as agency is 'mediated' by the available cultural tools. Psychological and technical tools mediate agency as individuals usually "operat[e] within [these] meditational means" (Wertsch et al., 1993, p. 342). Meditation of thought and action through social software enables the generation of social structures, histories, and ideologies (Jones & Norris, 2005). Individuals and groups use these cultural tools to understand their social world and to draw meaning from their interactions.

Multiple Ways of Knowing in Social Software

Meaning making is entirely situational. Individual construction of meaning is dependent upon the active interpretation of the situation by the participants which is referred to as a *situation definition* (Park & Moro, 2006). Situation definitions are of interest in a discussion of online social networks because how one interprets a situation includes how to use particular mediational means such as cultural tools (social software) and social genres (social language) within a given social context. Understanding how individuals actively interpret situations and construct meaning involves 'multiple ways of knowing' in which individuals use situation definitions to establish contexts for meaning making that are dynamic and only partially shared. Situation definitions enable a shift in authority structure that makes the individual the 'author' of any virtual situation (Rowe, 2005).

Situational definitions within online social software establish *contexts* for meaning making. Contexts describe circumstances that give meaning and form the setting for an event; a focal event set within its cultural setting (Duranti & Goodwin, 1992). Situation definitions activated within the context of online social software have four general attributes as outlined by Duranti and Goodwin (1992) and Gilbert (2006). Relative to online social software, these attributes provide a strong educational context in which learning can take place. First, the situation definition is within a social, spatial, and temporal framework in which individual encounters with events are situated. In other words, social software provides a social, spatial, and temporal framework in which social educational encounters can occur. Second, the situation definition provides a behavioral environment to frame the 'talk' that takes place. Social software provides a collaborative framework to formulate ideas and associate tasks and actions. Third, the use of specific language associated with the focal event enables a situational definition to develop within the socially networked environment. Social software provides a context in which *social languages* or *social genres* (Bakhtin, 1986/1978) are developed. Finally, situation definitions enable individuals to connect relationships between prior knowledge. Social software enables the connection between ideas and tasks situated within the network to be readily associated to background knowledge.

Situation definitions within are only partially shared between individuals and others within the network. Multiple situation definitions can exist simultaneously as individuals change their representation and understanding of events over time (Rowe, 2005). Interaction with others' competing understanding of an event or situation and the process of coming to share situation definitions is a crucial feature of learning and perspective taking (Wertsch, 1985). Thus, socio-cultural accounts of learning within social software are intersubjective, engaging participants in dynamic and discursive interactions shifting the authority of defining the situation from one individual to another.

Situation definitions within social software are associated with a shift in authority structure. "The authority to define situations is the authority to author one's own life and circumstances rather than simply to respond to what is given" (Rowe, 2005, p. 128). Within the context of social software, this enables individuals to become the authors of their own 'situations' or contexts. As a result, an individual could take on any particular identity by defining the 'rules' in which they choose to engage with others thus enacting on their agency. The *authority* of any given situation resides within the recognition of the power an individual has to create a context (enact a situation definition) rather than simply to engage within one, which by definition is an engagement of agency. The recognition that individuals have authority to enact and define a particular situation contributes to their perceived value in the online social community and for contributions to the community's growing knowledge.

Social Software as a Knowledge Building Environment

Knowledge building is a collective inquiry embedded in cultural practice. "Knowledge advancement is fundamentally a socio-cultural process, enhanced by cultures of innovation" (Scardamalia, 2004). A shift in authority structure encourages agency mediation, which makes online social software conducive to knowledge generation or *knowledge building*.

Knowledge building environments "enhance collaborative efforts to create and continually improve ideas" (Scardamalia, 2004). Embedded in practice, knowledge building characterizes learning as an evolutionary and creative process but also recognizes the importance of 'tools' necessary to facilitate learning as a collective endeavor making learning a complementary process (Bereiter & Scardamalia, 1996).

Social software shares similar characteristics of knowledge building environments. Scardamalia (2004) outlines several characteristics that distinguish knowledge building environments from other similar environments, such as those commonly discussed in the Computer Supported Collaborative Learning (CSCL) literature. Using Scardamalia's outline, each characteristic is adapted to the application of online social software as knowledge building environments for the purposes of education.

1. Online social software provides shared, user-configured spaces that represent the

collective contributions of group members. Providing users with shared, user-contributed spaces that represent the collective contributions of group members enables a sense of community 'authorship' in which each member develops a sense of accountability to the group. The value on their individual contributions increases as their motivation to contribute increases. From the perspective of education, collective ownership and accountability enhances collaborative inquiry and cooperation.

2. Online social software supports linking and referencing ideas so that the development of the ideas can be traced. Providing users with the ability to track idea formation and development, social software promotes the recognition of group participation in the idea creation process and acknowledges that ideas have a 'history of thought.' As an educational tool, social software provides a framework for the discovery of the origins of ideas within the collaborative setting, further enhancing a sense of community authorship, value in individual contribution, and accountability to the collective.
3. Online social software provides ways to represent higher-order organizations of ideas. Social software enables the visualization of idea organization which scaffolds individual schema formation and linkages to prior knowledge. Scaffolding the connection between a learner's prior knowledge is a critical component in aiding in the transformation of conceptual understanding, which is a foundation of learning and development.
4. Online social software provides ways for the same idea to be worked within varied and multiple contexts and to appear in different higher-order organizations of knowledge. As social software enables the visualization of idea organizations, linkages to prior knowledge within multiple contexts promotes cross 'disciplinary' innovation. Further to characteristic 3, social software that enables the connection between multiple contexts promotes cross organizational understanding and individual conceptual change.
5. Online social software has different kinds of systems of feedback to enhance self- and group-monitoring processes. Within the context of social software, users have the ability to provide feedback enabling the group to collectively 'self'-organize and 'self'-monitor ongoing knowledge creation processes. The ability of a learner to be self-reflective is a critical component in self-regulated learning and metacognitive development. As social software enables collective and individual feedback processes, individuals within the social network have frequent opportunities to give and receive feedback on collaborative idea development.
6. Online social software provides opportunistic linking of persons and groups—with the possibility of crossing traditional disciplinary, cultural, and age boundaries. As social software enables temporal relationships among members, opportunity for interactions among individuals is exponential. The possibilities of linking individuals from cross disciplinary fields, cultural, and age groups are more likely, thus encouraging alternative perspective taking; a critical component in higher-order thinking.
7. Online social software supports ways for different user groups to customize the environment and to explore within- and between community trajectories. As social software enables individual and group customization, identity formation at the individual generates a sense of ownership over the 'cultural' space. Further to a sense of community membership, individuals have the ability to construct their own 'situation definitions' and express different aspects of their personalities within the social network.

Characteristics of social software may vary with different kinds; however, if researchers and educators are to understand how to implement social software in education, then identifying the aforementioned characteristics in practice is critical for facilitating learning. Each kind of characteristic *in situ* has its own distinctive knowledge acquired through an individual's complete participation within the 'community of practice' (Bereiter & Scardamalia, 1996).

Social Software Promotes Communities of Practice

Members of an online social community are bound by what they do together and develop around a collective sense of meaning (Wenger, 1998). "Communities of practice are groups of people who share a concern, a set of problems, or a passion about a topic, and who deepen their knowledge and expertise in this area by interacting on an ongoing basis" (Wenger, McDermot, & Snyder, 2002, p. 4). Online social networks form 'communities of practice' as individuals within these environments share concerns, as in advocating for social justice for women (Pierce, 2007), collaborate on a solution to set of problems, as in solving school management issues by principals (Smith, 2007), and deepen their knowledge and expertise, as in contributing to open-source software development projects (Hemetsberger & Reinhardt, 2006). Ultimately, members of a 'community of practice' share a collective sense of purpose supported by the structural elements the social software provides.

Communities of practice arise out of a collective sense of need and intention. To achieve a particular outcome communities of practice develop initially because of a necessity to fulfill a particular purpose, perform a particular function, and produce a particular product or action. Key ideas Wenger (1998) identifies as the primary characteristics of a community of practice involve questions of 1) what is the network about, 2) how the network functions, and 3) what the network produces. Within the context of online social software, particularly in its use in education, these particular questions are essential in determining the context, tasks, and outcomes for learning. Online social software enables the expression of these intentions as it provides particular structural elements that aid in the development of the community.

Online social software establishes a social, spatial, and temporal framework for the development of communities of practice. Structural capabilities of online social software do not inhibit the size of a network (small or large), how long the network exists (long-lived or short-lived), where members of the network are situated (co-located or distributed), whether members are homogenous or heterogeneous, and whether networks develop within (inside) and across boundaries (organizational, corporate, educational, country) (Wenger et al., 2002). Although, traditional social networks and communities of practice also share many of these structural elements, online social software makes each of these capabilities more prevalent and strategically adaptable to the educational and classroom setting. Incorporating social software for the purposes of teaching and learning within and beyond the classroom context engages a student in multiple ways of knowing through knowledge building, which, in turn cultivates a community of practice in the classroom and enables self- and co- regulated learning.

Social Software Enables Self-Regulated Learning

Social and academic competence is a highly improbable occurrence without the ability to self-regulate. The conditions that promote the development of self-regulatory systems are fundamental considerations in understanding the emergence of complex dynamic systems. Agency, social networks, as well as self-regulation are not isolated occurrences. They require the presence of multiple interacting factors that influence their dynamic relationship.

Individuals manipulate and reframe on-going activity to situate meaning construction through self-regulation. Self-regulated learning (SRL) involves an active, effortful process in which learners set goals for their learning and then attempt to monitor, regulate, and control their cognition, motivation, and behaviour (Pintrich, 2000). Guiding and constraining SRL processes are contextual features of the social and learning environments, such as teachers and other students, learning outcomes, and acceptable 'norms and practices' as defined by culture. Thus, SRL is a personal and collective process that is both metacognitive and socially mediated by *others*. Social software facilitates SRL as an individual system exercised through socially mediated means, and enables forethought, performance control, and self-reflection (Bandura, 1986, 1997, 2001, 2006; Zimmerman, 1998, 2004).

Forethought. Influenced by personal motivational factors, the process of forethought involves the ability to anticipate the outcomes of a particular action and then strategically plan for a desired goal. In regards to academic learning, forethought involves goal setting, task analysis, motivational beliefs (goal orientation), and self-efficacy for the specified academic task. For example, a student believes that they are good at organic chemistry and is effacious about their upcoming midterm. The student then sets a goal to achieve an *A* on the midterm and begins to plan on how to study for the exam. In this example, the student believes they will do well, is confident in their abilities in the domain, sets a goal to achieve a particular grade on an assessment task, then makes a plan on how to achieve this academic outcome. Thus, in this manner, the student has "a forethoughtful perspective [that] provides direction, coherence, and meaning" to their achievement (Bandura, 2001, p. 7). Social software provides a supportive environment in which students are scaffolded through the analysis of a task, which in turn affects their motivation and self-efficacy for the task in question. In addition to forethought, learners must also be able to control and monitor the implementation of the plans they consider during forethought.

Performance control. Involving strength of will, self-control, and self-observation, learners engage in strategies to monitor and implement plans developed during forethought. In the development of academic competence, this stage of the SRL process is particularly important as the strategies learners employ, or develop, are of critical import if they are to achieve their desired outcome. For example, research on proactive learners suggests that a large percentage of them control environmental variables. Minimizing distractions in their study space by using earplugs while they study is a type of attention-focusing strategy (Corno, 1993; Corno & Kanfer, 1993), and just one example of the types of self-control strategies competent self-regulated learners use to achieve.

Using self-control strategies, proactive learners exercise self-observation processes to metacognitvely monitor their progress. Self-observation processes include self-monitoring, which refers to mentally tracking one's performance, and self-recording which involves a physical record of how one is doing (Zimmerman, 2004). For example, research on learners who use self-recording strategies demonstrated enhanced self-regulatory processes such as self-efficacy beliefs, which in turn improved goal attainment (Zimmerman & Kitsantas, 1997, 1999). Within social software, students have opportunities to practice regulatory strategies and observe other, potentially more effective strategies through other students in the network and have the control to monitor how they are performing and make adaptations as needed. To develop self-regulatory competence, in conjunction with performance control strategies, learners must also continuously reflect and evaluate on their progress on a task.

Self-reflection. A critical component in the self-regulatory process involves the continuous comparison of present levels of achievement with personal goals and standards. Self-regulatory comparisons or *judgments of success* involve comparisons along three different dimensions (Zimmerman, 2004). First, during self-improvement a learner evaluates their achievement based on progress over prior experience. Second, learners evaluate their achievement relative to the performance of their peers and thus, make comparative judgments along a *social* dimension. Finally, learners also use *mastery judgments* whereby they compare their achievement relative to a mastery source. Through *self* and *other* comparative processes reflective "actions give rise to self-reactive influence through performance comparison with goals and standards" (Bandura, 2001, p. 8). The comparison of goals and standards demonstrates judgments of success, decision-making, and is an ultimate expression of agency. Social software provides an environment in which self and other comparative processes are possible within the social dimension (network). Agency not only includes the ability to make choices, but also the ability to take appropriate courses of action and evaluate success based on standards set both by the individual learner and by social others.

CONCLUSION

Social and group interactions in online communities develop and evolve from expressions of human agency. Agency is a result of an emergent causal relationship between ability, will, and action. Social networks are a result of mediated expressions of agency that challenge the existing authority structure of classroom discourse. Social software provides students with opportunities to manipulate contexts and strategically interact with other students (agents) to achieve a desired outcome. Thus, agency ability links attributes of motivation to courses of action. Expressions of agency through online social networks, promotes the idea that an individual has authority over their *virtual* cultural space.

Understanding human interaction within online social networks begins with an appreciation of agency. Agency ability is a social construction that develops through mediation, the appropriation of cultural tools, and facilitates a novel means of community formation. Social software in education gives students the power (ability) to do what they want in the absence of internal or external constraints, to understand and reflectively evaluate their intentions, reasons, and motives, and to control their own behavior. The authors suggest that future research explore questions of:

1. The impact of the formal or informal nature of the educational context on agency;
2. The affordances different types of social software tools have relative to the emergence of human agency;
3. The role of the facilitator in orchestrating and encouraging agency emergence; and
4. The potential implications each of these aspects have on learning design.

Education has a responsibility to engage the use of social software to encourage student's development of agency and responsible social action. A shift in authority structure is required to utilize online social networks as a means for knowledge construction, meaning making, and building community within the classroom. Educators can encourage the emergence of agency in social networks through establishing contexts for meaning making, collective inquiry, and knowledge building that develop a community of practice. Recognizing social software's potential as a cultural tool is critical for education as cultural tools not only enhance human thinking, they transform it (Gauvain, 2001).

REFERENCES

Ach, N. (1910). *Uber den willensakt und das temperament. [On the will and temperament]*. Leipzig, Germany: Quelle & Meyer.

Bakhtin, M. M. (1986/1978). Speech genres and other late essays. In C. Emerson & M. Holquist (Eds.). Austin, TX: University of Texas Press.

Bandura, A. (1986). *Social foundations of thought and action*. Englewood Cliffs, NJ: Prentice-Hall.

Bandura, A. (1997). *Self-Efficacy*. New York: W. H. Freeman and Company.

Bandura, A. (2001). Social cognitive theory: An agentic perspective. *Annual Review of Psychology*, *52*, 1–26. doi:10.1146/annurev.psych.52.1.1

Bandura, A. (2006). Towards a psychology of human agency. *Perspectives on Psychological Science*, *1*(2). doi:10.1111/j.1745-6916.2006.00011.x

Bebo, Inc. (2008). About Bebo. Retrieved April 21, 2008, from http://www.bebo.com/StaticPage.jsp?StaticPageId=2517103831

Bereiter, C., & Scardamalia, M. (1996). Rethinking learning. In D. R. Olson & N. Torrance (Eds.), *The Handbook of Education and Human Development* (pp. 485-513). Cambridge, MA: Blackwell Publishers.

Bouffard, T., Bouchard, M., Goulet, G., Cenoncourt, I., & Couture, N. (2005). Influence of achievement goals and self-efficacy on students' self-regulation and performance. *International Journal of Psychology*, *40*(6), 373–384. doi:10.1080/00207590444000302

Boyd, D. M. (2007). The significance of social software. In T. N. Burg & H. Schmidt (Eds.), *BlogTalks reloaded. Social software: Research & cases* (pp. 15-30). Herstellung: Books on Demand GmbH, Norderstedt.

Brown, A. (1987). Metacognition, executive control, self-regulation, and other more mysterious mechanisms. In F. E. Weinert & R. H. Kluwe (Eds.), *Metacognition, motivation, and understanding* (pp. 65-116). Hillsdale, NJ: Lawrence Earlbaum Associates.

Bruner, J. (1997). *The culture of education*. Cambridge, MA: Harvard University Press.

Carver, C., & Schier, M. (2002). Control processes and self-organization as complementary principles underlying behavior. *Personality and Social Psychology Review*, *64*(4), 304–315. doi:10.1207/S15327957PSPR0604_05

Coates, T. (2005). An addendum to a definition of Social Software. *Plasticbag.org (blog)* Retrieved October 12, 2007, from http://www.plasticbag.org/archives/2005/01/an_addendum_to_a_definition_of_social_software/

Code, J., & Zaparyniuk, N. (this volume). Social identities, group formation, and the analysis of online communities. In S. Hatzipanagos & S. Warburton (Eds.), *Handbook of Research on Social Software and Developing Community Ontologies*. New York, NY: IGI Publishing.

Corno, L. (1993). The best laid plans: Modern conceptions of volition and educational research. *Educational Researcher*, *22*(2), 17–22.

Corno, L. (2001). Volitional aspects of self-regulated learning. In B. Zimmerman & D. Schunk (Eds.), *Self-regulated learning and academic achievement* (Vol. 191-225). Mahwah, NJ: Lawrence Earlbaum Associates.

Corno, L., & Kanfer, R. (1993). The role of volition in learning and performance. *Review of Research in Education*, *19*, 301–341.

Duranti, A., & Goodwin, C. (Eds.). (1992). *Rethinking context: Language as an interactive phenomenon*. Cambridge, UK: Cambridge University Press.

Emirbayer, M., & Mische, A. (1998). What is agency? *American Journal of Sociology*, *103*(4), 962–1023. doi:10.1086/231294

Facebook, Inc. (2008). About Facebook. Retrieved April 21, 2008, from http://www.facebook.com/about.php

Flavell, J. H. (1979). Metacognition and cognitive monitoring: A new era of cognitive developmental inquiry. *The American Psychologist*, *34*, 906–911. doi:10.1037/0003-066X.34.10.906

Gauvain, M. (2001). Cultural tools, social interaction, and the development of thinking. *Human Development*, *44*(2-3), 126–143. doi:10.1159/000057052

Gecas, V. (2003). Self-agency and the life course. In J. T. Mortimer & M. J. Shanahan (Eds.), *Handbook of the Life Course*. New York, NY: Kluwer Academic Publishing/Plenum Publishers.

Gergen, K. J. (1984). Theory of the self: Impasse and evolution. In L. Berkowitz (Ed.), *Advances in experimental psychology* (Vol. 17, pp. 49-115). New York, NY: Academic.

Gilbert, J. K. (2006). On the nature of "context" in chemical education. *International Journal of Science Education*, *28*(9), 957–976. doi:10.1080/09500690600702470

Hemetsberger, A., & Reinhardt, C. (2006). Learning and knowledge-building in open-source communities: A social-experiential approach. *Management Learning*, *37*(2), 187–214. doi:10.1177/1350507606063442

Henderson, D. K. (1994). Accounting for macro-level causation. *Synthese*, *101*(2), 129–156. doi:10.1007/BF01064014

Johnson, J. C., Palinkas, L. A., & Boster, J. S. (2003). Informal social roles and the evolution and stability of social networks. In R. Breiger, K. Carley & P. Pattison (Eds.), *Dynamic social network modeling and analysis: Workshop summary and papers* (pp. 121-132). Washington, D.C.: National Research Council.

Johnson, N. F. (2007). *Two's Company, Three is complexity: A simple guide to the science of all sciences*. Oxford, UK: Oneworld.

Jones, R. H., & Norris, S. (2005). Introducing mediational means / cultural tools. In S. Norris & R. H. Jones (Eds.), *Discourse in Action* (pp. 49-51). New York, NY: Routledge.

Kuhl, J. (1985). Volitional mediators of cognition-behavior consistency: Self-regulatory processes in action versus state orientation. In J. Kuhl & J. Beckmann (Eds.), *Action control: From cognition to behavior* (pp. 101-128). West-Berlin: Springer-Verlag.

Lefcourt, H. M. (1966). Internal versus external control of reinforcement: A review. *Psychological Bulletin*, *65*(4), 206–220. doi:10.1037/h0023116

Linden Research Inc. (2008). About Second Life. Retrieved April 21, 2008, from http://secondlife.com/

Little, T. D., Hawley, P. H., Henrich, C. C., & Marsland, K. W. (2002). Three views of the agentic self: A developmental synthesis. In E. Deci & R. Ryan (Eds.), *Handbook of self-determination research* (pp. 390-404). Rochester, NY: University of Rochester Press.

Loedewyk, K., & Winne, P. (2005). Relations among the structure of learning tasks, achievement, and changes in self-efficacy in secondary students. *Journal of Educational Psychology*, *97*(1), 3–12. doi:10.1037/0022-0663.97.1.3

Markus, H. J., & Nurius, P. S. (1984). Self-understanding and self-regulation in middle childhood. In W. A. Collins (Ed.), *Development during middle childhood: The years from six to twelve*. Washington D.C: National Academy Press.

Martin, J. (2003). Emergent persons. *New Ideas in Psychology, 21*, 85–99. doi:10.1016/S0732-118X(03)00013-8

Martin, J. (2004). Self-regulated learning, social cognitive theory, and agency. *Educational Psychologist, 39*(2), 135–145. doi:10.1207/s15326985ep3902_4

Martin, J., Sugarman, J., & Thompson, J. (2003). *Psychology and the question of agency*. New York, NY: State University of New York Press.

Mead, G. H. (1932). *The philosophy of the present*. Chicago, IL: University of Chicago Press.

Mead, G. H. (1934). *Mind, self and society from the standpoint of a social behaviorist.* Chicago, Il: Chicago University Press.

MySpace, Inc. (2008). About MySpace. Retrieved April 21, 2008, from http://www.myspace.com/index.cfm?fuseaction=misc.aboutus

Nowak, A., Vallacher, R. R., Tesser, A., & Borkowski, W. (2000). Society of self: The emergence of collective properties in self-structure. *Psychological Review, 39*, 39–61. doi:10.1037/0033-295X.107.1.39

O'Connor, T., & Wong, H. Y. (2002). Emergent properties. In E. N. Zalta (Ed.), *Stanford encyclopedia of philosophy*. Stanford, CA: The Metaphysics Research Lab, Stanford University.

Park, D., & Moro, Y. (2006). Dynamics of Situation Definition. *Mind, Culture, and Activity, 13*(2), 101–129. doi:10.1207/s15327884mca1302_3

Parsons, T. (1968). *The Structure of Social Action*. New York, NY: Free Press.

Pierce, T. (2007). *Women, weblogs, and war: Digital culture and gender performativity. three case studies of online discourse by muslim cyberconduits of Afghanistan, Iran, and Iraq.* ProQuest Information & Learning, US.

Pintrich, P. (2000). The role of goal orientation in self-regulated learning. In M. Boekaerts, P. Pintrich & M. Zeidner (Eds.), *Handbook of Self-Regulations* (pp. 451-502). San Diego, CA: Academic Press.

Rotter, J. B. (1954). *Social learning and clinical psychology*. Englewood Cliffs, NJ: Prentice-Hall.

Rotter, J. B. (1966). Generalized expectancies for internal versus external control of reinforcement. *Psychological Monographs, 80*(1), 1–28.

Rowe, S. (2005). Using multiple situation definitions to create hybrid activity space. In S. Norris & R. H. Jones (Eds.), *Discourse in action: Introducing mediated discourse analysis* (pp. 123-134). New York, NY: Routledge`.

Ryan, R., & Deci, E. (2000). Intrinsic and extrinsic motivations: Classic definitions and new directions. *Contemporary Educational Psychology, 25*(1), 54–67. doi:10.1006/ceps.1999.1020

Scardamalia, M. (2004). Knowledge building environments: Extending the limits of the possible in education and knowledge work. In A. Distefano, K. E. Rudestam & R. Silverman (Eds.), *Encyclopedia of distributed learning*. Thousand Oaks, CA: Sage Publications.

Schunk, D., & Ertmer, P. (2001). Self-regulation and academic learning: Self-efficacy enhancing interventions. In M. Boekaerts, P. Pintrich & M. Zeidner (Eds.), *Handbook of Self-Regulation* (pp. 630-649). New York: Academic Press.

Smith, A. A. (2007). Mentoring for experienced school principals: Professional learning in a safe place. *Mentoring & Tutoring: Partnership in Learning*, *15*(3), 277–291. doi:10.1080/13611260701202032

Sternberg, R. J. (1999). *Cognitive psychology*. Fort Worth, TX: Harcourt Brace.

Thompson, L., & Fine, G. A. (1999). Socially shared cognition, affect and behavior: A review and integration. *Personality and Social Psychology Review*, *3*(4), 278–302. doi:10.1207/s15327957pspr0304_1

Vallacher, R. R., Nowak, A., Froehlich, M., & Rockloff, M. (2002). The dynamics of self-evaluation. *Journal of Personality and Social Psychology Review*, *6*, 370–379. doi:10.1207/S15327957PSPR0604_11

Vallacher, R. R., Nowak, A., Markus, J., & Strauss, J. (1998). Dynamics in the coordination of mind and action. In M. Kofta, G. Weary & G. Seflek (Eds.), *Personal control in action: Cognitive and motivational mechanisms* (pp. 27-59). New York, NY: Plenum.

Vygotsky, L. S. (1962). *Thought and language*. Cambridge, MA: MIT Press.

Vygotsky, L. S. (1978). *Mind in society: The development of higher psychological processes*. Cambridge, MA: Harvard University Press.

Vygotsky, L. S. (1981). The instrumental method in psychology. In J. V. Wertsch (Ed.), *The concept of activity in Soviet psychology* (pp. 134-143). New York, NY: M. E. Sharpe.

Wenger, E. (1998). Communities of practice: Learning as a social system [Electronic Version]. *The Systems Thinker, 9*. Retrieved November 10, 2007, from http://www.ewenger.com/

Wenger, E., McDermot, R. M., & Snyder, W. M. (2002). *Cultivating communities of practice: A guide to managing knowledge*. Boston, MA: Harvard Business School Press.

Wertsch, J. V. (1985). *Culture, communication, and cognition: Vygotskian perspectives*. New York, NY: Cambridge University Press.

Wertsch, J. V., Tulviste, P., & Hagstrom, F. (1993). A sociocultural approach to agency. In E. A. Forman, N. Minick & C. A. Stone (Eds.), *Contexts for learning: Sociocultural dynamics in children's development* (pp. 336-356). New York, NY: Oxford University Press.

Wolters, C. (1999). The relation between high school students' motivational regulation and their use of learning strategies, effort, and classroom performance. *Learning and Individual Differences*, *11*(3), 281–299. doi:10.1016/S1041-6080(99)80004-1

Wolters, C. (2003). Regulation of motivation: Evaluating an underemphasized aspect of self-regulated learning. *Educational Psychologist*, *38*(4), 189–205. doi:10.1207/S15326985EP3804_1

Wolters, C., Yu, S., & Pintrich, P. (1996). The relation between goal orientation and students' motivational beliefs and self-regulated learning. *Learning and Individual Differences*, *8*(3). doi:10.1016/S1041-6080(96)90015-1

Zimmerman, B. (1998). Academic studying and the development of personal skill: A self-regulatory perspective. *Educational Psychologist*, *33*(2/3), 73–86. doi:10.1207/s15326985ep3302&3_3

Zimmerman, B. (2004). Sociocultural influence and students' development of academic self-regulation: A social-cognitive perspective. In D. M. McInerney & S. Van Etten (Eds.), *Big Theories Revisited* (Vol. 4 In: Research on Sociocultural Influences on Motivation and Learning, pp. 139-164). Greenwich, CT: Information Age Publishing.

Zimmerman, B., & Kitsantas, A. (1997). Developmental phases in self-regulation: Shifting from process goals to outcome goals. *Journal of Educational Psychology*, *89*(1), 29–36. doi:10.1037/0022-0663.89.1.29

Zimmerman, B., & Kitsantas, A. (1999). Acquiring writing revision skill: Shifting from process to outcome self-regulatory goals. *Journal of Educational Psychology*, *91*(2), 241–250. doi:10.1037/0022-0663.91.2.241

KEY TERMS

Agency: The capability of individuals to consciously choose, influence, and structure their actions (Emirbayer & Mische, 1998; Gecas, 2003) and is an exercise of ability and will through action.

Communities of Practice: Involve groups of people who share concerns, problems, and passions about a topic, and who choose to interact to deepen their knowledge and expertise in this area by interacting on an ongoing basis (Wenger et al., 2002).

Cultural Tools: Mediate higher-order mental processes such as reasoning and problem solving (Vygotsky, 1962, 1978). Cultural tools include both *technical tools* such as books, media, computers, and social software, and *psychological tools* such as language, signs, writing, and symbols.

Emergence: From an ontological perspective is a non-reducible phenomenon. Meaning, that if a construct is emergent it has several component parts but is irreducible with respect to them (Martin, 2003; O'Connor & Wong, 2002).

Knowledge-Building Environments (KBES): Environments that "enhance collaborative efforts to create and continually improve ideas" (Scardamalia, 2004).

Locus of Control: A belief in the causal relationship between's one own behavior and that of an outcome affects a range of choices an individual makes (Lefcourt, 1966; Rotter, 1966).

Self-Efficacy: A belief in one's capability to succeed at a given task (Bandura, 1997).

Social Software: Software which "supports, extends, or derives added value from human social behavior" (Coates, 2005).

Volition: A "post-decisional, self-regulatory processes that energize[s] the maintenance and enactment of intended actions" (Kuhl, 1985, p. 90).

This work was previously published in Handbook of Research on Social Software and Developing Community Ontologies, edited by S. Hatzipanagos & S. Warburton, pp. 102-118, copyright 2009 by Information Science Reference (an imprint of IGI Global).

Chapter 7.18
Two Informational Complexity Measures in Social Networks and Agent Communities

António Jorge Filipe Fonseca
ISCTE, Portugal

ABSTRACT

Several informational complexity measures rely on the notion of stochastic process in order to extract hidden structural properties behind the apparent randomness of information sources. Following an equivalence approach between dynamic relation evolution within a social network and a generic stochastic process two dynamic measures of network complexity are proposed.

INTRODUCTION

Most of the statistical social network analysis methods rely on a fixed network structure. Earlier methods of dyadic statistical analysis manage to provide quantification on degrees of mutuality between actors and triadic analysis does a step forward allowing validation of theories of balance and transitivity about specific components of a network. Each of these constitutes a subset of a more general *k-sub graph analysis* based on *k-sub graph census* extracted from the network architecture. Some sort of frequency analysis is done over these censuses from which probabilistic distributions can be evaluated. More recent single relational statistical analysis additionally allows the validation of statistic models through parametric estimation. Looking further for positional assumptions of groups of actors, stochastic block model analysis measure the statistical fitness of defined equivalent classes on the social network (Carrington, 2005) (Wasserman, 1994). All of these tools assume some sort a static network structure, a kind of snapshot of reality, over which statistical measuring is done and some degree of confidence is evaluated against real data. It is not of our knowledge any method of providing some sort of analysis on the dynamic structure of social networks in which the set of relations evolves over a certain amount of time. The method we propose has the purpose of allowing a probabilistic informational based evaluation of each actor's inner complex motivated behaviour on the process of relation change inside his network. The evaluation is supported on the measurement of the *Entropy Density* and after of the *Excess Entropy* over the stochastic relational changing. Measuring the evolution of this interplay complexity measure can provide some insights into each actor degree of inner structure pertaining to the specific kind of relation that the network is supposed to represent. *Entropy Density* and *Excess Entropy* are theoretical measures and for practical purposes only estimative can be obtained. The study of estimative computation for entropy is still subject to intense research in the Physics community. As the measurement only approaches the real relational entropy of each actor, it can be considered within the context of the set of members of the social network an absolute measure as the same estimate bias is applied to the community as an all.

NETWORK DYNAMICS AND INFORMATION

The complexity of the information that may be extracted from observations of a source, as a phenomenological observation of its behavior, without any prior knowledge of the source's internal structure, may reveal missed regularities and structural properties hidden behind the apparent randomness of the stochastic process that the source generates. This information can provide useful clues in order to predict the source future behavior or its characteristic properties. In social network dynamics each actor generally performs some sort of action, communication acts; exchange of capital or economic goods or any other possible social interplay or attribute for which a well defined time limit and intensity can be pinpointed in time. This definition can generally be extended and even applied to objects as for example a correlation of goods purchased by each client of a supermarket. The agent that determines the dynamic nature of a network is primarily Time. Topologically a network that evolves as the one depicted in *Figure 1* is completely defined as a set of graphs, each constituted by a set of nodes and link, eventually tagged with some intensity attribute, that have been established during a defined time slot T of the interval of observation. The sequence of the n time slot started at time t_0 pretend to report the dynamical relational evolution that the network is supposed to represent. There are some approaches to dynamic network that take into account possible resilience of links as an intensity attribute. In fact, the process of time partitioning can be tricky, dynamic features that appear relevant at some time slot durations can in fact be insignificant look at other time spans. There exist ways to circumvent this problem taking into account the kind of relation the network represents. For example normalizing the duration of the interplay through an inverse exponential function of time slot duration can avoid some dependence on value of T.

Having a dynamic network and coding the combination of all the established links that each node performed during each time-slot, we obtain a relational stochastic process for that particular node. This process will symbolically represent the node relational evolution within the community. The extraction of informational quantities from this process is straightforward.

INFORMATIONAL MEASURES OF COMPLEXITY

Given a symbolic process $\vec{S} \equiv \cdots S_{-2}, S_{-1}, S_0, S_1, S_2 \cdots$ of random variables S_{T_i} that range over an alphabet A, taking sequence values $\ldots s_{-2}, s_{-1}, s_0, s_1, s_2 \cdots \in A$, the total Shannon entropy of *length-L* sequences of $\vec{S}$ is defined as (Cover, 2006):

$$H(L) = \sum_{s^L \in A^L} Pr(s^L) \log_2 Pr(s^L)$$

Figure 1. A network that evolves over time. The arcs between the nodes represent all the interplay activity performed during the timeslot $nT, (1 \leq n \leq N)$

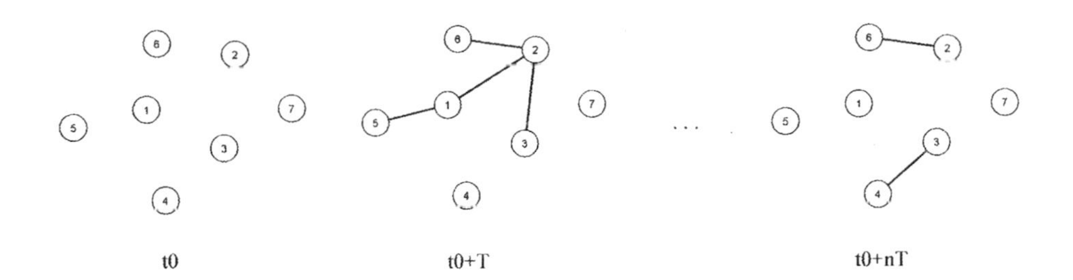

Where s^L are symbols belonging to these sequences of *L* length that represent combinations of symbols of A constituting the alphabet A^L. On a binary alphabet with $A^L = \{0,1\}$ and with *L*=3, $A^L = \{000, 001, 010, 011, 100, 101, 110, 111\}$. *H(L)* is calculated over all the extension of the process $\vec{S}$ for any possible consecutive combination of *L* symbols. It happens that *H(L)* is a non-decreasing function of *L*: $H(L) \geq H(L-1)$ and it is also a concave function: $H(L) - 2H(L-1) + H(L-2) \leq 0$. This fact can easily be understood if we notice that the size of A^L increases exponentially with *L* and so inversely does $Pr(s^L)$. Figure 2 depicts the growth of *H(L)* function of increasing *L* length:

$$h_\mu \equiv \lim_{L \to \infty} \frac{H(L)}{L}$$

REDUNDANCY

The entropy rate h_μ quantifies the amount of *irreducible randomness* that remains after all the correlations and patterns embedded in longer and longer *length-L* sequence blocks are extracted from the entropy computation (Crutchfield, 2001). It is the rate at which the source transmits *pure randomness* and it is measured as *bits/symbol*. Since each symbol belongs to an alphabet of size $|A|$ the *Redundancy* within the source, R, is given by the difference between this maximum theoretical entropy rate, the *channel capacity* that is needed to transmit any optimally coded message from an arbitrary source using an alphabet A, $C = \log_2 |A|$, and h_μ:

$$\mathrm{R} \equiv \log_2 |\mathrm{A}| - h_\mu$$

The redundancy R is a measure of the information that an observer gains after expecting a maximally entropic uniform probability distribution from the source and actually learns the correct distribution $Pr(\vec{S})$ of the sequence.

It is possible to define derivatives for *H(L)*. Having the first derivative we obtain the *apparent entropy rate* or *apparent metric entropy* at a given length *L*:

$$h_\mu(L) \equiv \Delta H(L) \\ = H(L) - H(L-1), L \geq 1$$

This derivative constitutes an *information gain*. The function $h_\mu(L)$ is an estimate of how random the source appears if only blocks of the process up to length *L* are considered. The difference between $h_\mu(L)$ and the true h_μ on the measure into the infinite *L* limit, give a related L-estimate, the *per-symbol L-redundancy*:

$$r(L) \equiv \Delta R(L) \equiv h_\mu(L) - h_\mu$$

Figure 2. Growth of H(L) function of L. The picture also shows the metric entropy coefficient h_μ as the rate of increase with respect to l of the total Shannon entropy in the large L limit

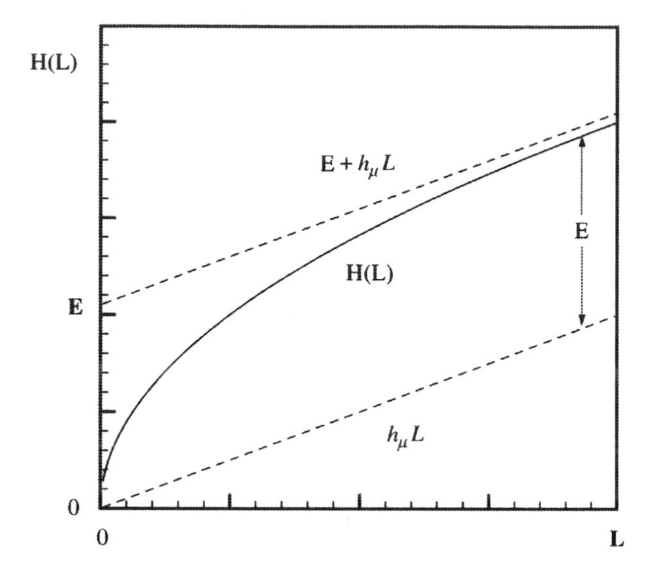

$r(L)$ measures how the *apparent entropy rate* computed at a finite L exceeds the actual *entropy rate*. Any difference between the two indicates there is redundant information in the L-blocks in the amount of $r(L)$ bits. Interpreting $h_\mu(L)$ as an estimate of the source's unpredictability we can look further at the rate of change of $h_\mu(L)$, the rate at which the unpredictability is lost. This is given by the second order derivative:

$$\Delta^2 H(L) \equiv \Delta h_\mu(L) = h_\mu(L) - h_\mu(L-1)$$

PREDICTABILITY GAIN

The previous second order derivative provides a measure of the change, for each increment on the size of increasing larger L blocks, of the *metric entropy* estimate $h_\mu(L)$. This measure constitutes a *predictability gain* (Crutchfield, 2001). From the previous equation we can see that the quantity $-\Delta^2 H(L)$ measures the reduction in per-symbol uncertainty in going from (L-1) to L block statistics. Computing how this derivative converges to the limit $L \to \infty$, which is to say that for every possible large combinations of symbols, we obtain the a *Total Predictability* G of the process.

G is defined as:

$$G \equiv \sum_{L=1}^{\infty} \Delta^2 H(L)$$

And as $h_\mu(L)$ is always positive it can be sown that $\Delta^2 H(L) \leq 0$ and also that:

$$G = -R$$

Like R the unit of G is *bits/symbol.* For a periodic process $G = \log_2 |A|$, $h_\mu = 0$ so it assumes its maximum value for a completely predictable process. G however does not tell us how difficult it is to carry out any prediction, nor how many symbols must be observed before the process can be optimally predicted, but in fact give us a measure "disequilibrium" between the actual entropy rate of the process h_μ and the maximum possible entropy rate of a periodic process with the same alphabet A.

A finite L estimate of G is given by:

$$G(L) = H(1) + \sum_{l=2}^{L}[H(l) - 2H(l-1) + H(l-2)]$$

FINITARY AND INFINITARY PROCESSES

If, for all i and j, the probabilistic distribution of each symbol obeys the equality $Pr(S_i) = Pr(S_j)$, the probability distribution of the stochastic process:

$$Pr(\overleftrightarrow{S}) = Pr(\cdots, S_i, S_{i+1}, S_{i+2}, S_{i+3}, \cdots)$$

is given by:

$$Pr(\overleftrightarrow{S}) = \cdots Pr(S_i)Pr(S_{i+1})Pr(S_{i+2})\cdots$$

and we say that the process is *independently and identically distributed*. If however the probability of the next symbol depends on the previous symbol, we then call the process *Markovian*:

$$Pr(\overleftrightarrow{S}) = \cdots Pr(S_{i+1} \mid S_i)Pr(S_{i+2} \mid S_{i+1})\cdots$$

In a general case, if the probability depends on previous R symbols of the sequence:

$$Pr(S_i \mid \cdots, S_{i-2}, S_{i-1}) = Pr(S_i \mid S_{i-R}, \cdots, S_{i-1})$$

Then we call this type of process an *order-R Markovian*. A *hidden Markov* process consists of an internal *order-R Markov* process that is observed by a function of its internal-state sequences. These are usually called *functions of a Markov chain* which we suppose is embedded inside the system. These kind of processes are considered *finitary* since any Markov Chain as a finite amount of memory.

Generally any stochastic process is said to be *finitary* if at the $L \to \infty$ limit:

$$H(L) \sim h_\mu L$$

$$\lim_{L\to\infty} \Delta H(L) = h_\mu,$$

$$\lim_{L\to\infty} \Delta^n H(L) = 0 \text{ for } \lim_{L\to\infty} \Delta^n H(L) = 0$$

$n \geq 2$ for $n \geq 2$

On a *finitary* process the *entropy rate* estimate $h_\mu(L)$ decays faster than $1/L$ to the actual entropy rate h_μ. Other way of defining a finitary process is to admit that the *Excess Entropy* E, the second measure we will examine, is finite on *finitary* processes and infinite otherwise.

EXCESS ENTROPY

In order to capture the structural properties of memory embedded into the system, we need to look at other entropy convergence integrals. Several authors (Packard,1982) (Grassberger,1986) (Li,1991) (Crutchfield,2001) defined a quantity named *Excess Entropy* E or *Complexity*, *Effective Measure of Complexity* or *Stored Information*, as measure of how $h_\mu(L)$ converges to h_μ. This quantity is expressed as:

$$E \equiv \sum_{L=1}^{\infty}\left[h_\mu(L) - h_\mu\right] = \sum_{L=1}^{\infty} r(L)$$

The unit of E is *bits*. Following the reasoning explained above this quantity gives us the difference between the per-symbol entropy conditioned on L measurements and the per-symbol entropy conditioned on an infinite number of measurements. As the source appears less random at length L by the amount $r(L)$ this constitutes a measure of information carrying capacity in the L-blocks that is not actually random but due instead to correlations. If one sums these individual per-symbol L-redundancy contributions we obtain the total amount of apparent memory in the source, which can be interpreted as an *intrinsic redundancy*. The entropy-rate convergence is controlled by this intrinsic redundancy as a property of the source. At each L we obtain additional information on the way about how $h_\mu(L)$ converges to h_μ, this information is not contained in $H(L)$ and $h_\mu(L)$ for smaller L. Thus each measure of $h_\mu(L)$ is an independent indicator of how $h_\mu(L)$ converges to h_μ. For *non-finitary* processes E does not converge at all. We will admit however, as we will deal with estimates for finite L relative within a definite context of similar estimates, that they provide an absolute characterization of *intrinsic redundancy* up to L-block estimates within the context. The structure of this quantified memory of the system cannot be analyzed within the framework of information theory, for this purpose complexity measures based on computation theory like *Kolmogorov Complexity* or *Logic Depth* must be used (Shannon, 1948)(Kolmogorov,1965).

The excess entropy can also be seen as the mutual information between the left and right (past and future) semi-infinite halves of the process $\vec{S}$:

$$\mathrm{E} = \lim_{L\to\infty} MI\left[S_0 S_1 \cdots S_{2L-L}; S_{L+1} S_{L+2} \cdots S_{2L}\right]$$

Whenever this limit exist.

A finite L estimate of E is given by:

$$\mathrm{E} = \sum_{l=1}^{L}\left[h_\mu(l) - h_\mu(L)\right]$$

$$= \sum_{l=1}^{L}\left[H(l) - H(l-1)\right] - L\left[H(L) - H(L-1)\right]$$

MEASURING COMMUNITY DYNAMICS

Having a social network or a community of actors which relations evolve over a specific time interval we now want to quantify each agent's degree of complex intentionality that determines dynamical structural change within the community network of interactions. Recalling the equivalence of relational change with a stochastic process examined in the above section, it is reasonable to establish a phenomenological equivalence between each agent's observed dynamical changing of relations process as a stochastic process eventually determined by each actor specific structure.

Assuming a community with N actors, we may divide the interval duration of our observation into K time slots of duration T. At each time slot k each agent or node $ni, (1 \le i \le \mathrm{N})$ has a domain $A = 2^{N-1}$ of possibly different relation configurations c_i^k directly with other agents. The L-block entropy of this stochastic process C_i as A^L possibly different configurations c_i^l for each of the blocks which is given by:

$$H_i(L) = -\sum_{k=1}^{K-L} Pr(c_i^{kL}) \log_2 Pr(c_i^{kL})$$

Where:

$$Pr\left(c_i^{kL}\right) = \frac{1}{A^L}\sum_{l=1}^{\left|A^L\right|} \delta_{c_i^{kL}, c_i^L}, \quad c_i^{kL} \in A^L,$$

$$Pr\left(c_i^{kL}\right) = \frac{1}{A^L}\sum_{l=1}^{\left|A^L\right|} \delta_{c_i^{kL}, c_i^L} \quad c_i^{kL} \in A^L \; c_i^L \in A^L,$$

$$c_i^{kL} \in A^L, \; c_i^L \in A^L$$

c_i^{kL} is the L-length configuration c_i^{kL} c_i^l is the L-length configuration c_i^l that starts at time slot k.

Recalling the above definition of G, an estimate of the total predictability $G_i(L)$ for a time interval of K time slots, with a sufficiently large L (L can be as large as K, but we should note that the computational effort grows exponentially with the size L) is given by:

$$G_i(L) = Hi(1) + \sum_{l=2}^{L} \left[H_i(l) - 2H_i(l-1) + H_i(l-2) \right]$$

And an estimate of the excess entropy $E_i(L)$ for each agent of the community is given by:

$$Ei(L) = \sum_{l=1}^{L} \left[H_i(l) - H_i(l-1) \right] - L\left[H_i(L) - H_i(L-1) \right]$$

With respect to the definition of *total predictability* we can interpret $G_i(L)$ at the end of the network evolution observation, as the node or actor n_i *predictability*, which is an equivalent of measuring the periodicity of the actor's interplay within the community. An actor with a large amount of *predictability* should stick to regular patterns of relations both in time and in interplay with other actors. On the other hand the quantity $E_i(L)$ provides an estimate of the n_i node or actor magnitude of *complex subjective commitment* to its relational choices within the network. This means to have some kind of memory that dictates his relational patterns, in other word to be socially complex.

At this point we should note that these two measures, although egocentric and also estimates, should be considered absolute within the complete set of relations between nodes of the network. In fact, when both measures are using the same L estimate and are applied to the set of nodes as an all, they can be considered as an absolute measurement with respect to all the nodes of a particular observation. Then, when we are referring to the *complex subjective commitment* or *predictability* on some node on the context of one observation, we are considering two properties of the particular situated role of each node in respect to the particular kind of relation that the network is supposed to describe.

As we have stated above, there is an equivalence of E_i with the mutual information between the past and the future of the series of connections each node performs. Thus, we can see why E_i is adequately defined as a *commitment* towards a subjective preference for some dynamic pattern of relations within the network as opposed to no preference at all. The excess entropy E_i quantifies the magnitude of correlation between the past observations of the set of relational patterns at each node with future observations and vice versa, thus reflecting the commitment of the node to enforce the observed set of patterns.

THE COMMUNITY OF AGENTS PARALLEL

Until now we have been considering normal human interactions, however these same two measures of complex social activity can also be applied to computer agent communities. Multi-agents systems normally implement some kind of protocol for message exchange that allows the agents to cooperate and coordinate (Weiss, 2000)(Shoham, 2009). The fine tuning of message exchanging and work distribution among agents can benefit from informational measuring of agent communication. The application of these two measures can provide useful tools to adjust agent communication in very sophisticated and complex agent environments.

AN EXAMPLE

In order to better illustrate the measures proposed on the previous section we will examine a hypothetical community of agents constituting an artificial social network with some given assumptions about the agent roles. A general human social network could be used instead. Let us consider the message exchange between agents during a defined period *T*. Admissibly we have a chronological log of all the messages which we partition at a fixed time intervals. Let us also assume that we want to restrict the focus of our study into a restrict subset of agents and that those agents have arbitrarily complex cognitive and communication capabilities:

1. One coordinator for a given task
2. A agent that performs part of the task
3. A directory facilitator

We want have some insight, having only logged timestamp; sender and receiver for each message, on the memory extent and role of each agent. Considering that we have *a posteriori* knowledge that these agents have the following roles:

1. The coordinator communicates only with the agents under his responsibility in order to perform his coordinated task. At fixed intervals he checks with a non coordinated agent the state of accomplishment of the task.
2. The coordinated agent periodically consults the coordinator in order to adjust his goal. Otherwise he just silently prosecutes his goal.
3. The directory facilitator randomly exchanges messages with every agent of the community at their request in order facilitate agent communication.

From this description we should expect to obtain at the end of our observation a great level of *predictability* for the working agent, a moderate level for the coordinator and vanishing levels for the directory facilitator. On the other hand the *subjective commitment* to patterns of communication should be greater for the coordinator as he should possess a greater amount of memory that should allows him to coordinate his well defined team of agents, a lesser one for the directory facilitator, as although communicating randomly within the community he should however reflect some patterns of communication behavior product of the system's global memory, and a lesser degree for the worker agent as he only communicates at regular interval with his single coordinator.

CONCLUSION

We strongly believe these two measures of ego-centric social complexity have great potential on the evaluation of social networks. Actor or agent *social predictability* constitutes an important indicator in many evaluations. Also the *commitment to the community* is an important factor on distinguishing and characterizing its members. We should recall that the nature of the social relation that is evaluated for each particular network as direct influence on the interpretation of these measures. Social network science abstracts the nature of the relation from the actors involved. A dynamic relation, as we stated above, could mean communicating but also buying and selling or just having some kind of affinity between actors or agents. Thus the nature of the relation that is studied has direct impact on the nature of the measurements. Also a correct choice of sampling rate of the network timeline constitutes a critical matter that directly influences the results. For each dynamics, different time slot duration will have direct impact on the results obtained. This impact should be subject of further investigation, it is more or less obvious however that each dynamic process has his own time scale. This scale has thus direct influence on the correct choice of time partition. A concrete application of these measures to a real case scenario will follow.

REFERENCES

Carrington, P. J., Scott, J., & Wasserman, S. (2005). *Models and Methods in Social Network Analysis.* UK: Cambridge University Press.

Cover, T. M., & Thomas, J. A. (2006). *Elements of Information Theory.* US: John Wiley and Sons.

Crutchfield, J. P., & Feldman, D. P. (2001). Regularities unseen, randomness observed: Levels of entropy convergence. *Santa Fe Institute Working Papers* 01-02-012.

Grassberger, P. (1986). Toward a Quantitative Theory of Self-Generated Complexity [Springer Netherlands.]. *International Journal of Theoretical Physics*, *25*, 907–938. doi:10.1007/BF00668821

Kolmogorov, A. N. (1965). Three approaches to the quantitative definition of information. *Problems of Information Transmission*, *1*, 1–7.

Li, W. (1991). On the Relationship between Complexity and Entropy for Markov Chains and Regular Languages. *Complex Systems . Complex Systems Publications*, *5*, 381–399.

Packard, N. H. (1982). Measurements of Chaos in the Presence of Noise. *Phd Thesis, University of California*

Shannon, C. (1948). A Mathematical Theory of Communication. *Bell System Technical Journal, Bell Laboratories, 27*, 379-423, 623-656.

Shoham, Y., & Leyton-Brown, K. (2009). *Multiagent Systems – Algorithmic, Game-theoretic, and Logical Foundations*. US: Cambridge University Press

Wasserman, S., & Faust, K. (1994). *Social Network Analysis: Methods and Applications*. UK: Cambridge University Press.

Weiss, G. (2000). *Multiagent systems a modern approach to distributed artificial intelligence.* Cambridge MA USA: The MIT Press.

This work was previously published in International Journal of Agent Technologies and Systems, Vol. 1, Issue 4, edited by G. Trajkovski, pp. 49-57, copyright 2009 by IGI Publishing (an imprint of IGI Global).

Chapter 7.19
Social Self-Regulation in Computer Mediated Communities:
The Case of Wikipedia

Christopher Goldspink
University of Surrey, UK

ABSTRACT

This article documents the findings of research into the governance mechanisms within the distributed on-line community known as Wikipedia. It focuses in particular on the role of normative mechanisms in achieving social self-regulation. A brief history of the Wikipedia is provided. This concentrates on the debate about governance and also considers characteristics of the wiki technology which can be expected to influence governance processes. The empirical findings are then presented. These focus on how Wikipedians use linguistic cues to influence one another on a sample of discussion pages drawn from both controversial and featured articles. Through this analysis a tentative account is provided of the agent-level cognitive mechanisms which appear necessary to explain the apparent behavioural coordination. The findings are to be used as a foundation for the simulation of 'normative' behaviour. The account identifies some of the challenges that need to be addressed in such an attempt including a mismatch between the case findings and assumptions used in past attempts to simulate normative behaviour.

INTRODUCTION

The research documented in this article is part of the EU funded project titled 'Emergence in the Loop: Simulating the two way dynamics of norm innovation' (EMIL) which aims to advance our understanding of emergent social self-organisation. The project involves conducting several empirical case studies the first of which is the Wikipedia.

When people encounter Wikipedia for the first time and learn how it works, they commonly express surprise. The expectation appears to be that an open collaborative process of such magnitude should not work. Yet the Wikipedia has been shown to produce credible encyclopaedic articles (Giles, 2005) without the hierarchical and credentialist controls typically employed for this type of production.

The research presented here is framed within the debate about governance mechanisms associated with Open Source production systems. This is not the only perspective which could be adopted but it does serve to provide some initial orientation. Consistent with the wider project focus, the relationship between these theories and the theory of social norms is examined.

In the empirical research we examine the extent to which communicative acts are employed by editors to influence the behaviour of others. Particular attention is given to the illocutionary force of utterances (Searle, 1969) and the effect of deontic commands linked to general social norms and Wikipedia specific rules. In the conclusion some observations are made about the agent-level cognitive mechanisms which appear necessary to explain the observed social order as well as the apparent influence of social artefacts, goals and the wiki technology.

The following questions are canvassed through this research.

- What processes appear to operate in computer mediated organizations which enable them to be, in effect, self-regulating?
- How consistent are the findings with established theories for understanding norms and governance, particularly in on-line environments?
- What alternative hypotheses are there which appear to explain the phenomena and which can provide the foundation for future research?

Governance Theory

According to the relevant Wikipedia article, the word 'governance' derives from the Latin that suggests the notion of "steering". The concept of governance is used in a number of disciplines and a wide range of contexts and the range and type of steering mechanisms differ depending on whether the focus is with states or institutions. While both have been applied to Open Source, it is most common (and arguably most appropriate) to use institutional concepts of governance. Institutional steering mechanisms may be: formal (designed rules and laws) or informal (emergent as with social norms); extrinsic (involving contracts and/or material incentives) or intrinsic (involving values and principles); and the mechanisms by which governance operates may be top down (imposed by authority) or bottom up (invented by the participants as a basis for regulating each other). Theories vary with respect to the mechanisms advanced and the emphasis placed on different mechanisms. Theory is also advanced for different purposes: to explain or to prescribe. In broad terms the debate is often dichotomised with economics derived theories (Agency and Transaction Cost) on one side and sociological theories (stewardship) on the other (see J. H. Davis, D. Schoorman, & L. Donaldson, 1997; Donaldson & Davis, 1991). Depending on the position of the advocate these may be presented as antithetical or as viable alternatives for different contexts.

Agency theory derives from neo-classical economics and shares the foundational assumption of agent utility maximization. Advocates argue that many productive transactions involve *principals* who delegate tasks to *agents* to perform on their behalf (Donaldson & Davis, 1991). This gives rise to what is known as the 'principal's dilemma'. Simply stated this dilemma asks *'how can the principal ensure that the agent will act in its interest rather than on the basis of self-interest?'* Note that this dilemma arises from the assumed self-interested nature of agents –it is a dilemma intrinsic to the assumptions upon which the theory is based even though this is argued to have empirical support. Two general solutions are offered: the use of formal contracts and sanctions and the use of material incentives.

Critics argue that not all human decisions are made on the basis of self-interest. Sociological and psychological models of governance posit various alternatives: some remain committed to assumptions of rational action and goal seeking, while others address issues of power or various forms of intrinsic motivation, including a desire to conform to social norms. These latter positions generally form the basis of theories of *stewardship* (J. H. Davis, D. F. Schoorman, & L. Donaldson, 1997).

While these two broad sets of ideas form the backdrop to most debates about governance in traditional institutions increased recourse has also been made to

(Coase, 1993, 1995; Williamson, 1996). TCE is concerned with the relative merit of alternative governance arrangements for differing production environments. Oliver Williamson (1985), a key contributor, states *'The choice of governance mode should be aligned with the characteristics of the transaction...'*. Principals are presented with a continuum of possible ways of trying to achieve effective regulation from open markets to hierarchy. Both of these are seen as imposing costs (agency costs for hierarchy and transaction costs for markets). The aim is to combine them to achieve an optimum balance between these costs. This 'balancing' implies a top down rational decision making role for institutional managers.

More recently two additional categories of governance have been added to the TCE family – 'networks' and 'bazaars'. Both have arisen to explain the emergence of production and exchange arrangements which do not seem to fit on the market-hierarchy continuum. Both Network Governance (Candace Jones, William S Hesterly, & Stephen P Borgatti, 1997) and Bazaar Governance (Demil & Lecocq, 2003) are argued to be particularly relevant to understanding the flexible structures associated with Open Source production. Demil & Lecocq (2003: 8) cite Jones et al (1997: 916) and argue that network governance:

> *...involves a select, persistent, and structured set of autonomous forms [agents] engaged in creating products or services based on implicit and open ended contracts to adapt to environmental contingencies and to coordinate and safeguard exchanges. These contracts are socially – not legally – binding.*

The final sentence highlights the key difference between network and more conventional TCE mechanisms. To achieve cooperation the network form of governance relies on social control, such as *'occupational socialization, collective sanctions, and reputations'* rather than on formal authority.

Bazaar governance is also argued to rely heavily on the mechanism of reputation (Demil & Lecocq, 2003: 13). Reputation is assumed to provide the incentive to become involved and to comply with group expectations and norms. Unlike network systems, however, agents are free to enter or leave the exchange process – there are no obligations to become or to remain engaged. Raymond (1999) states that *'contrary to network governance, free-riders or opportunistic agents cannot be formally excluded from the open-source community'*.

To summarise:

- Free markets are characterised by: a lack of obligation to engage in a transaction; low interdependence between parties involved with the exchange; and transactions regulated only by price. Within a pure market the individual identities of the transacting parties are not important.
- In Hierarchies, there are formal contracted obligations on all parties, these are maintained by fiat but may also be supported by wider formal institutions e.g. Courts. Obligations are associated with formal position making the official (role) identity of the parties the key determinant of the relationship.
- Within network structures, exchanges are

regulated using relational contracts – there is a formal obligation to remain engaged even though specific actions and operational responsibilities may not be included in a contract. There is also some reliance on social norms– the socialised position of actors becomes important. Exchange commitments may be relatively short lived and persist only so long as they offer mutual benefit.

- With bazaar governance there is no obligation on any party to perform particular duties or even to remain engaged: there are low entry and exit costs. There are few formal mechanisms for policing or sanction but sufficient regulation is achieved by means of shared task, reciprocity norms and/or informal group sanctioning with participants influenced by their desire to build reputation.

Understanding the Role of Norms

As can be seen, 'norms' are argued to play a role in a number of theories of governance, with their being particularly significant in Stewardship, Network and the Bazaar theories. Sociologists have long argued that norms are fundamental mechanisms for social regulation. What though is a 'norm'? How do norms emerge and how are they influenced and by what?

Gibbs (1981) argues that *'Sociologists use few technical terms more than norms and the notion of norms looms large in their attempt to answer a perennial question: How is social order possible?'*. Not surprisingly then the concept has been incorporated into a wide range of alternative and often competing bodies of theory.

The normative literature can be divided into two fundamentally distinct groups. In the social philosophical tradition (Lewis, 1969) norms are seen as a particular class of emergent social behaviour which spontaneously arise in a population. From this perspective, a 'norm' is a pattern identified by an observer ex-post. The defining characteristic of the pattern is the apparently prescriptive/proscriptive character: people behave 'as if' they were following a rule. By contrast, the view offered by the philosophy of law sees norms as a *source* of social order. This standpoint assumes the prior existence of (powerful) social institutions and posits them as the source of rules, which, when followed, lead to social patterns. These positions appear antithetical although following the work of Berger and Luckman (1972) each may be seen as a part of a dialectic whereby emergent social patterns become reintegrated and formalised in institutions.

Therborn argues (2002: 868) that people follow norms for different reasons. The extremes run from habit or routine to rational knowledge of consequences for self or the world. Between these lie:

- Identification with the norm or values – linking sense of self (identity) to the norm source (person, organization or doctrine) often leading to in-group-out-group.
- Deep internalization – self-respect – done independently to what others are doing.

Bicchieri (2006: 59) provides a rare hint at the cognitive process involved stating:

To 'activate' a norm means that the subjects involved recognise that the norm applies: They infer from some situational cues what the appropriate behaviour is, what they should expect others to do and what they are expected to do themselves, and act upon those cues.

This suggests a complex process of self-classification (how am 'I' situated with respect to this group and what is the nature of the situation in which 'I' find myself, does a norm pertain to 'me' in this situation and under what conditions and to what extent am I obliged to comply?).

To begin to identify which (if any) of these loosely defined mechanisms might be supported by evidence and to aid in the development of a theory of norms helpful for understanding the more general mechanisms at play in social self-regulation we selected the Wikipedia as a preliminary case study. Wikipedia belongs to the Open Source movement as it has adopted the Open Source License. It was originally designed to operate under the umbrella of a conventional hierarchical form of governance and its unanticipated success as a radical governance experiment makes it a particularly interesting case study. It was anticipated that findings in relation to the Wikipedia may have some wider relevance to understanding the open source phenomena but also serve to cast light on mechanisms which underpin human institutions– particularly those that are more normative in nature. In order to be able to judge the degree of generalisation that may be possible it is first important to identify the distinctive features of the Wikipedia.

The Wikipedia

Wikipedia grew out of an earlier Web encyclopaedia project called Nupedia founded by Jimmy Wales with Larry Sanger appointed as its first editor-in-chief. From its inception Nupedia was linked to a free information concept and thus the wider open source movement. Nupedia used traditional hierarchical methods from compiling content with contributors expected to be experts. The resulting complex and time consuming process and an associated lack of openness have been argued to explain the failure of the Nupedia. Sanger (2006; 2007), however, questions this view, arguing that the expert model was sound but needed to be simplified.

Sanger was introduced to the WikiWiki software platform in 2001 and saw in it a way to address the limitations hampering Nupedia. The inherent openness of the Wikiwiki environment was, however, seen as a problem so Wikipedia began as an experimental side project. Sanger notes that a majority of the Nupedia Advisory Board did not support the Wikipedia, being of the view '*...that a wiki could not resemble an encyclopaedia at all, that it would be too informal and unstructured'* (Sanger, 2007). However the intrinsic openness of Wikipedia attracted increasing numbers of contributors and quickly developed a life of its own. Almeida et al (2007) note that growth in articles, editors and users have all shown an exponential trajectory. From Sanger's earlier comments it is clear that he had been surprised at the rate of development and of the quality achieved by the relatively un-coordinated action of many editors.

The Debate Over Governance in Wikipedia

The use and enforcement of principles and rules has been an ongoing issue within the Wikipedia community with a division emerging between the founders and within the wider community about whether rules were necessary and if they were, how extensive they should be and how they should be policed. The power to police rules or impose sanctions has always been limited by the openness of the technology platform. Initially Sanger and Wales, were the only administrators with the power to exclude participants from the site. In 2004 this authority was passed to an Arbitration Committee which could delegate administrator status more widely. The Arbitration Committee is a mechanism of last resort in the dispute resolution process, only dealing with the most serious disputes. Recommendations for appointment to this committee are made by open elections with appointment the prerogative of Wales.

In the early stages Sanger argues the need was for participants more than rules and so the only rule was 'there is no rule'. The reason for this, he explains, was that they needed to gain experience of how wikis worked before over prescribing the mechanisms. However, *'As the project grew and the requirements of its success became increasingly obvious, I became ambivalent about this particular "rule" and then rejected it altogether'* (Sanger, 2007). However, in the minds of some members of the community, it had become 'the essence' of Wikipedia.'

In the beginning, complete openness was seen

as valuable to encourage all comers and to avoid them feeling intimidated. Radical collaboration – allowing everybody to edit everyone's (unsigned) articles – also avoided ownership and attendant defensiveness. Importantly it also removed bottle necks associated with 'expert' editing. That said the handpicking of a few core people is regarded by Sanger as having had an important and positive impact on the early development of Wikipedia. Sanger argues for example *'I think it was essential that we began the project with a core group of intelligent good writers who understood what an encyclopaedia should look like, and who were basically decent human beings'* (2005). In addition to 'seeding' the culture with a positive disposition, this statement highlights the potential importance of establishing a style consistent with the Encyclopaedia genre – a stylistic model which might shape the subsequent contributions of others.

Sanger argues that in the early stages 'force of personality' and 'shaming' were the only means used to control contributors and that no formal exclusion occurred for six months, despite there being difficult characters from the beginning. The aim was to live with this 'good natured anarchy' until the community itself could identify and posit a suitable rule-set. Within Wikipedia rules evolved and as new ones were needed they were added to the 'What Wikipedia is not' page'. Wales then added the 'Neutral Point of View' (NPOV) page which emphasised the need for contributions to be free of bias. The combination of clear purpose and the principle of neutrality provided a reference point against which all contributions could be easily judged. Sanger regards the many rules, principles and guidelines which have evolved since as secondary and not essential for success.

How do newcomers learn these (ever increasing) rules and do they actually influence behaviour? Bryant et al (2005) suggest that there is evidence of 'legitimate peripheral practice', a process whereby newcomers learn the relevant rules, norms and skills by serving a kind of apprenticeship. These authors argue that this is evident in new editors of Wikipedia initially undertaking minor editing tasks before moving to more significant contributions, and possibly, eventually, taking administrative roles. These authors tend to project a rather idealistic view of involvement, however, overlooking a key attribute of the wiki environment –newcomers have the same rights as long standing participants and experts and this mechanism for socialising newcomers can be effectively bypassed.

In some Open Source environments (such as Open Source Software) it is possible to gain reputation which may be usable in the wider world. The commitment to the community is often explained (for an excellent overview see Rossi, April, 2004) by arguing that a desire for reputation increases compliance. However, in the Wikipedia environment there is no list of contributors to which an editor can point as evidence of their contribution (although they can self-identify their contributions on their user page). Contributions are, in essence, non attributable. In the case of Wikipedia identification with product, community and values appears a more likely reason for remaining involved than does reputation.

In a study specifically designed to study the conflict and coordination costs of Wikipedia, Kittur, Suh, Pendleton, & Chi (2007: 453) note that there has a been a significant increase in regulatory costs over time. *'...direct work on articles is decreasing, while indirect work such as discussions, procedure, user coordination, and maintenance activity (such as reverts and anti-vandalism) is increasing'*. The proportion of indirect edits (i.e. those on discussion or support pages) has increased from 2% to 12%. Kittur et al cite an interview respondent as stating *'the degree of success that one meets in dealing with conflicts (especially conflicts with experienced editors) often depends on the efficiency with which one can quote policy and precedent.'* (Kittur et al., 2007: 454). This suggests that force of argument supported by the existence of the formal rules and etiquette are important to the governance process.

This is however based on ex post attributions.

Wiki Technology: The Artifact

Wiki technology has a very flat learning curve: contributing is extremely simple. There are few technical impediments confronting novice users. Wiki platforms are intrinsically open supporting decentralised action unless modified to control or restrict access. Division of labour emerges as editors choose which pages interest them and which they want to focus on contributing to or maintaining.

Wikipedia has added a number of facilities which support the ready detection and correction of vandalism. Watch lists support users in taking responsibility for the oversight and monitoring of particular topics. Changes made to a page are logged using a history list which supports comparison between versions as well as identifying the time and date of any change and the ID of who made that change. The reversion facility supports the rapid reinstatement of the page content. Lih (2004: 4) attributes significance to this feature noting that *'This crucial asymmetry tips the balance in favour of productive and cooperative members of the wiki community, allowing quality content to emerge'*. and Stvilia et all (2004: 13) note that *'By allowing the disputing sides to obliterate each others contributions easily, a wiki makes the sides interdependent in achieving their goals and perhaps surprisingly may encourage more consensus building rather than confrontation'*.

Stvilia, Twidale, Gasser, & Smith (2005) among others identify discussion pages as an important *'...coordination artefact which helps to negotiate and align members perspectives on the content and quality of the article.'* Discussion pages provide an opportunity for managing minor disputes about content or editing behaviour and for movement towards the agreement.

Ciffiolilli (2007) has argued that a significant consequence of these technical features is the way in which they alter transaction costs (Coase, 1993; Williamson & Winter, 1993). Transaction costs result from information overheads associated with complex coordination. However, the technology does not cancel other costs of coordination and control. These are commonly referred to as agency costs and the highly open nature of the wiki may increase them. In hierarchies, this cost is evident in the cost of command and control (management overhead) whereas in the Open Source environment they are borne by the participating community (and not necessarily equitably). The cost burden will be less where there is a high level of self-regulation and lower where a lack of goal alignment or low social commitment leads contributors to disregard others and act individualistically or opportunistically. The efficacy of cultural control will be influenced by factors such as the homogeneity of the user group and that group's propensity for self-organisation (endogenous norm formation), rates of turnover of the group, and the effect of external perturbation of the group or of the task on which they are working. This may also be subject to feedback effects: reduced norm compliance may lead to higher turnover and reduced commitment, further reducing norm compliance for example.

In conclusion then, Wikipedia is a volunteer open source project characterised by low ties between contributors, no formal obligations and very few means for the exercise of formal sanction. There is a low level of reciprocity with contributors under no obligation to maintain engagement. The wiki technology is open, inviting many to the task and imposing low costs to participation while reducing transaction costs. There is however high reliance on pro-social behaviour dominating if agency costs (borne by individuals) is not to lead to high turnover and possible governance failure. The anonymity of Wikipedia precludes any significant reputation effects outside of the small group of co-editors who maintain extended involvement with an article and to a very limited degree the wider Wikipedia community.

Wikipedians have produced a set of permissions, obligations, rules and norms which have been documented in guidelines and etiquettes as well as embedded in technical artefacts such as style bots. The need for and effect of these is however controversial. From a governance perspective there are relatively few means within Wikipedia by which formal control can be exercised using these rules and the community relies instead on the use of informal or 'soft' control. These mechanisms need to be effective in the face of perturbation from 'vandals' (task saboteurs), 'trolls' (social saboteurs), as well as turnover of contributors in the context of a task which can require the accommodation of emotionally charged and value based issues.

Analysis of Governance Micro-Mechanisms

In Wikipedia there are two classes of activity: editing; and conversation about editing. This article is not concerned with the editing activity (although this is to be considered in future research) but with the self-organising and self-regulating phenomena which make it possible. Insight into this can be gained by examining the Discussion pages which accompany many of the articles rather than the articles themselves. The activity on the Discussion pages comprises a series of 'utterances' or speech acts between contributors about editing activity and the quality of product. On the face of it then, these pages should provide a fertile source of data to support analysis of how governance operates in the Wikipedia, in particular informal or 'soft' governance.

Within these pages we expected to see attempts by editors to influence the behaviour of one another through the only means available to them – communicative acts. We anticipated that these may exhibit some regularity which would allow us to examine both the range and type of events that led to the explicit invocation of rules and norms and which revealed emergent influence patterns which were themselves normative. We wanted also to examine what conventions prevailed and how these compared and interacted with the goal of the community and its policies. A convention is defined here as a behavioural regularity widely observed by members of the community. Policies include explicit codes of conduct as well as guidelines (etiquettes) and principles.

Methodology

For the study we randomly selected a sample of Discussion pages associated with both Controversial and Featured articles. At the time of the study (May/June 2007) there were 583 articles identified by the Wikipedia community as controversial. The featured articles are more numerous. At the time of the study there were approximately 1900 of them. The analysis reported here is based on a sample of nineteen Controversial and eleven Featured articles. The most recent three pages of discussion were selected for analysis from each Discussion page associated with the article included in the sample.

These were subjected to detailed coding using the Open Source qualitative analysis software WeftQDA. Both qualitative and quantitative analysis was performed. The latter was undertaken by re-processing the coded utterances such that each utterance constituted a case and each applied code a variable associated with that case. This data set was then analysed using SPSS.

A number of coding schemes for natural speech were considered before choosing to use the Verbal Response Mode (VRM) taxonomy (Stiles, 1992). VRM has been developed over many years and used in a wide range of communication contexts. Stiles defines it as *'a conceptually based, general purpose system for coding speech acts. The taxonomic categories are mutually exclusive and they are exhaustive in the sense that every conceivable utterance can be classified.'* (Stiles, 1992: 15). The classification schema is attractive where there is a need (as here) to capture many of the subtleties of natural language use that derive from and rely on the intrinsic flexibility and ambiguity of natural language yet map them to a more formal system needed for computer simulation.

Additional codes were applied to identify: valence, subject of communication, explicit invocation or norms or rules and the associated deontic and trigger, whether the receiver/s accepted the illocutionary force of the utterance and the ID and registration status of the person making the utterance.

There were 3654 utterances coded in these thirty three documents.

FINDINGS

Style of Communication

There was a statistically significant correlation between the article group (Controversial vs Featured) and broad style of communication. This was however very small at -0.078 (p=.01 2-tailed). This difference was most apparent when examined at the level of specific styles. Both groups had approximately similar proportions of neutrally phrased utterances (approximately 64%). Nearly one quarter (22.5%) of all utterances in Featured articles were positive compared to only eleven percent in controversial sites. By comparison nearly one quarter (23.9%) of all utterances in controversial sites were negative compared to fourteen percent for featured. The positive styles of 'affirming', 'encouraging' and 'acknowledging' were significantly overrepresented in the featured articles but underrepresented in the controversial articles. The reverse was the case for the negative styles of 'aggressive', 'contemptuous' and 'dismissive'.

There was a statistically significant correlation between the broad style of communication and the editor status. The correlation was again very low at -.054 (p=.01 two tailed).

Overall, the most common positive utterance was affirming (4.7%) closely followed by encouraging (4.7%) and acknowledging (4.3%). The most common negative utterance was dismissive (8.2%) followed by defensive (6.4%) and contemptuous (3.5%).

All the Wikipedia discussions sampled reflected a strongly neutral-objective *style* (although from the qualitative observations it was apparent that the content was sometimes far from objective or balanced). The statistically significant difference between Controversial and Featured sites was in the relative balance of positive and negative utterance and was not so great as to explain the different status awarded the associated articles.

Validation

Within speech act theory (Habermas, 1976; Searle, 1969), validation refers to whether an utterance made by one speaker is accepted, rejected, ignored or let go unquestioned by the intended recipient/s.

In the Wikipedia sample half of all utterances were accepted without question. A further eighteen percent were explicitly accepted by at least one editor; eleven percent were explicitly rejected and a substantial twenty two percent were ignored. Twenty five percent of positive style utterances were accepted by at least one editor compared to eighteen percent of neutral and only nine percent of negative. By comparison only two percent of positive utterances were rejected compared to nine percent of neutral and twenty six percent of negative. Positive utterances were more likely to be accepted without question (61%) compared to negative (21.7%) and neutral (54.4%). Negative comments were more likely to be ignored (44.1%) compared to neutral (18.2%) and positive (11.4%).

From this we can conclude that positive utterances are more likely to be validated than negative, but that overall, a significant number of utterances are ignored or rejected.

Normative and Rule Invocation

Overall 5.2% of all utterances involved norm or rule invocation. This meant that Wikipedia rules were invoked 122 times and general social norms a further 77 times in 3654 utterances. This overall number was contributed to disproportionately by three (outlier) articles in the sample. Rules were most commonly invoked in response to neutral style communication (63.9%) followed by twenty seven percent in response to a negative style. Only nine percent of positive style utterances were responded to with a rule invocation. By comparison, norms were most commonly invoked in response to negative style utterances (53.2%) followed by neutral (44.2%) and then positive (2.6%). The difference in likelihood of invocation by style was statistically significant (p=.001).

A Wikipedia rule invocation was most likely to be triggered by the *form* of an article (44.9%) an *edit action* (22%); an *article fact* or a *person's behaviour* (both 16%). A norm was most likely to be triggered by a *person's behaviour* (35.6%), an *edit action* (23.3%), *article form* (21.9%), or *article fact* (19.2%). This pattern did not differ to a significant degree between the Featured and Controversial sites.

Nearly three quarters (73.6%) of rule invocations had the implicit deontic of 'it is obligatory' Norms also were most likely to carry this deontic (61.3%). The second most likely deontic was 'it is permissible that' (9.7%).

While there was no statistically significant difference in the degree to which either norms or rules were invoked between the Featured and Controversial articles, there was a qualitative difference in the role norm and rule invocation played. In Controversial discussions, social norms and rules were most likely to be invoked against the behaviour of an editor who was of a different view (group?) while in Featured sites, norms and rules were somewhat more often used by the editor as a reflection on their own contribution – i.e. involved a level of self-check. This might take the form of a statement such as 'I know this is not NPOV but…..'.

Registered vs Non-Registered Users

There was no statistically significant difference in the likelihood for either registered or non-registered users to invoke norms or rules. There was a statistically significant difference between registered and non-registered editors (p=.000) when it came to validation. Registered editors were more likely than non-registered to be explicitly accepted (18.7% of utterances compared to 13.9%), less likely to be rejected (9.9% compared to 13.7%), considerably less likely to be ignored (18.3% compared to 34.7%) or unquestioned (53.1% compared to 37.6%). Qualitatively, however, it was much more common that un-registered users would make suggestions before undertaking edits, particularly in the Features articles, so their behaviour was less likely to attract action or comment.

Non-registered editors were more likely to make negative style utterances (24.3% compared to 18.5%) and less likely to make positive utterances (9.5% compared to 17.4%). This difference was significant (p=.000).

Influence Through Illocutionary Force

The theory of speech acts distinguishes between the meaning of an utterance and its pragmatic intent. With the VRM coding frame used in this research each utterance is coded twice, once to capture the semantic form and again to capture the use of language to exert (illocutionary) force (Searle, 1969). A typical utterance may have a *form* which differs from the *intent*. The utterance 'could you close the door?', for example, has the form of a *question* but the intent of *advisement:* the speaker intends the listener to close the door. In VRM, the relationship of form to intent is expressed, using the statement "in service of" (Stiles, 1992). In this example the question 'could you close the door' is 'in service of' the advisement 'close the door'. In standard presentation this is recorded as (QA).

Edification in service of Edification (EE) is the

most frequent form of utterance in the Wikipedia sample – 37% of all utterances were of this mode. The Edification mode is defined as deriving from the speaker's frame of reference, making no presumption about the listener and using a neutral (objective) frame of reference shared by both speaker and listener. This mode is informative, unassuming and acquiescent. As a strategy for influencing others it reflects attempts to convince by neutral objective argument.

The second most common mode is that of Disclosure in service of Disclosure (DD). Disclosure is defined as being from the speaker's experience, making no presumption, but being framed using the speaker's frame of reference. This is summarised as informative, unassuming but directive. Unlike EE mode, DD mode represents an attempt by the speaker to impose or have the listener accept the speaker's frame. Twelve percent of all utterances adopted this form.

The third most common mode is Disclosure in service of Edification (DE). The DE mode represents an utterance which is from the speaker's frame of reference but as if it is neutral or from a shared frame. Eight percent of all utterances used this mode. This is a somewhat neutral mode where the speaker offers clearly labelled personal knowledge as information.

The fourth most common mode is Advisement in service of Advisement (AA). AA mode represents speech from the speaker's experience, which makes presumptions about the listener and adopts the speaker's frame of reference. It can be summarised as informative, presumptuous and directive. It commonly takes the form of 'you should....' Approximately 7% of utterances were in this mode. A further 12% of utterances have the directive pragmatic intent of advisement masked by using a less presumptuous form – that of Edification or Disclosure.

Significantly, utterances associated with politeness (such as acknowledgements 5%) and with discourse which aims at mutual understanding, such as confirmation (1.5%) and reflection (1%), were very rare in the Wikipedia sample.

Discussion of Findings

What is significant about the utterance strategies is that they typically involve an exchange of assertions delivered with a neutral – i.e. non-emotive style. There are very few explicit praises, or put downs, and few niceties like explicit acknowledgements of one another. Seldom do contributors refer to one another by name – the exchanges are rather impersonal. This does not tally with what one would expect if the Wikipedia etiquette (http://en.wikipedia.org/wiki/Wikipedia:Etiquette) had been institutionalised. The Featured articles conform a little more closely with what one would expect than do the Controversial, but if we assume that the etiquette captures the community's ideal, the emerged patterns do not conform to that 'ideal' to the extent that might be expected in either case. Similarly we see low levels of questioning or of reflection (i.e. feeding back the words of the speaker to check understanding or to come to better understand the other's intentions). This is arguably inconsistent with the task needs – to reach consensus on controversial topics. The frequency with which utterances were ignored also suggested low engagement by participants in the discussion. All of this would seem to need some explanation.

The absence of any expression of acknowledgement of emotions and/or similarity of attitude (homophilly) among many contributors suggests that Wikipedia lacks many of the qualities of verbal exchange that would identify it as strong community. It is more consistent with being a place to share coordination of a task. This could suggest that the goal is the primary orientating point. However, the lack of quality of discourse needed to achieve consensus is more indicative of a brief encounter between different and established milieux which struggle to find common understanding rather than of a community committed to a common goal (Becker & Mark, 1997). This might suggest that the shared goal may be subordinate to more personal goals by a considerable proportion of contributors. Or it may be that the technology and environment will support no more than this.

The Wikipedia environment supports saboteurs who can use the opportunity afforded by the open and anonymous platform to use identity deception i.e. to mimic the language and style of an 'expert' or to present as a genuine editor while trying to pursue a personal or political agenda hostile to the aims or interests of the Wikipedia. We found no direct evidence of this behaviour in the pages we sampled even though the discussions about controversial articles provide particularly fertile ground for such sabotage. Nevertheless the threat of it could have an overall influence on the type of communication conventions which arise. Editors may, for example, display reserve and suspicion, withholding trust and taking conventional signals of authority and identity (Donath, 1998) as unreliable. The first principle in the Wikipedia etiquette is 'assume good faith'. To do so would, however, leave the process more vulnerable to 'troll' activity.

Utterance strategies between registered and unregistered editors did not vary greatly, although unregistered editors were more likely to use disclosure intent and more likely to ask questions (possibly associated with the increased likelihood that they are relatively new to Wikipedia). They are also more likely to be negative – reflecting their potentially lower commitment to the article or the community.

Qualitatively there was considerable evidence of mind reading (theory of mind) – i.e. editors appeared to form judgements about the intent of others on relatively little information. There was, however, little evidence of the use of utterance strategies to better understand or check these theories of mind. Some editors, particularly in the Controversial discussions appeared quick to judge and then follow response patterns consistent with those judgements (e.g. ignoring or accepting utterances of others). There were also few instances of renegotiated patterns of communication style. Positions and styles stayed relatively constant over the period of the interaction. Only occasionally would an editor modify his/her style significantly if challenged. Of the rule invocations 26% were accepted, a similar proportion were rejected or ignored and the remainder went unquestioned (but generally had no affect on behaviour). This is consistent with norms being triggered by a limited range of cues which allow individuals to locate themselves and select identities appropriate to a context and which then remain essentially stable. The invocation of rules and norms appears to have little to no immediate effect on behaviour although it is not clear if it has an effect in subsequent behaviour as this cannot be ascertained from the available data.

CONCLUSION AND FUTURE WORK

In this study we set out to identify mechanisms which underpin the emergence of systemic self-organisation in a volunteer on-line global institution. The aim was to specify the mechanisms involved in order to support the design of a simulation architecture suitable for the wider study of normative mechanisms. The findings have challenged some of our assumptions and expectations, in particular:

- The more detailed and specific behavioural etiquette seems to have little influence on the overall character and style of interaction.
- The overall quality of interaction of editors falls short of the range and quality of communicative style characteristic of a community and that would be consistent with what one would expect, given the nature of the task.
- Most regulation is achieved without the need for frequent explicit invocation of rules or norms. Rather, behaviour seems to accord to a convention which editors quickly recognise and conform to (or bring to the Wikipedia)

and which minimally accommodates what needs to be done to satisfy the task in a context of potentially heterogeneous personal goals.

- There was a lack of evidence of active negotiation of expectations and standards and convergence of behaviour towards a norm. Within the discussion pages there appeared to be little obvious norm innovation, evolution, adaptation or extension. This suggests that on first encounter with Wikipedia, editors read a set of cues as to what constitutes appropriate or acceptable behaviour and then accommodate it. Alternatively the order observed may be largely attributable to the prior socialisation of participants with local norms and rules playing a very minor part in supporting task regulation.
- While there is a difference between controversial and featured sites this is minimal and the quality of the interaction cannot explain the difference in status. Similarly there appeared to be little in the subject matter of the two groups of articles which would explain the difference – both contained subject matter which was contestable and subject to significantly diverse opinion.

Wikipedia is not a market as there is no tradable product or price, either in a conventional sense or in the form of tradable reputation. Nor is Wikipedia a command hierarchy: the openness of the wiki platform and the low cost of joining and leaving precludes formal control as a primary means for governance. Neither is Wikipedia well described by the network theory of governance as there is no obligation to maintain involvement. While it might be expected that the Bazaar Governance would apply, the absence of a reputation mechanism suggests that it may be better considered through the more general lens of stewardship theory. Even here, there is no role for moral leadership but rather a diffused willingness to comply with certain minimum standards on the part of a sufficient majority.

There is no clear basis to argue that the apparent order is a direct result of the use of deontic commands associated with social norms and environment specific rules. Despite the fact that the community has been a prolific rule generator, they appear to play a minor role. Contributors demonstrate a style which is broadly inconsistent with these rules and not a good fit with the task.

Overall though there is order and it appears to be emergent. The mechanisms which underpin this emergence have not been revealed by the analysis undertaken to date although some hypotheses can be tentatively suggested. The neutral-objective style may be a consequence of the anonymity and open nature of the environment – leading to a suspension of trust. It may propagate as new comers copy the pattern through a process of behavioural cueing. It is possible also that the order is due to pro-social behaviour internalized and brought to the task. The volunteer nature of Wikipedia, and the level of commitment required, is likely to mean that long term editors reflect a pro-social disposition (Penner, Dovidio, Piliavin, & Schroeder, 2005). In this context a little norm/rule invocation may go a long way if not by influencing immediate behaviour then by encouraging future compliance and/or by giving incentive for non-compliers to leave. The relatively small difference in overall style apparent in relation to the diverse range of articles may have little to do with the specific communicative behaviours adopted in communication about that article but rather due to the chance association of individuals at a given point of time and how this subtle process of encouragement and dissuasion plays out over time. Such a view is quite different from that modelled in past attempts to simulate social norms.

A review of past approaches to the simulation of norms undertaken by EMIL partners at the University of Bayreuth concluded that the past research drew on the traditions of game theory and artificial intelligence. The latter were exclusively in the first generation AI tradition. Significantly, data drawn from real social situations was seldom used and there was a strong tendency to build on prior work with little questioning of assumptions about the nature of normative behaviour. Seldom was any mainstream theory of social behaviour employed as a part of the research program. The EMIL project is notable, therefore, for its insistence on the need to adopt an empirical orientation: for models to be designed in the light of and tested against observations drawn from real world cases of normative behaviour as well as in its avoidance of pre-commitment to particular simulation models or traditions.

While the findings of the research to date are far from conclusive they do challenge many of the assumptions incorporated into past simulations and suggest a range of alternative hypotheses. Some of these will be able to be critically examined by further analysis of the current data and/or by data currently being collected through a controlled wiki experiment as well as data proposed to be collected in a case study in Second Life. The EMIL simulator is being designed to support a range of alternative assumptions and so should allow us to test alternative hypotheses and contribute to our understanding of this increasingly significant phenomena.

ACKNOWLEDGMENT

The work on which this article is based has been supported by EMIL a three-year EU funded project (Sixth Framework Programme - Information Society and Technologies - Citizens and Governance in the Knowledge Based Society).EMIL comprises six Partner institutions: Institute of Cognitive Science and Technology, National Research Council, Italy; University of Bayreuth, Dept. of Philosophy, Germany; University of Surrey, Centre for Research on Social Simulation, Guildford, United Kingdom; Universität Koblenz – Landau, Germany; Manchester Metropolitan University, Centre for Policy Modelling, Manchester, United Kingdom; and AITIA International Informatics Inc. Budapest, Hungary.

I thank the partners for their contributions to the research.

REFERENCES

Almeida, R. B., Mozafari, B., & Cho, J. (2007). On the Evolution of Wikipedia, *International Conference on Weblogs and Social Media* Boulder Colorado.

Becker, B., & Mark, G. (1997). Constructing Social Systems through Computer Mediated Communication. Sankt Augustin, Germany: German National Research Center for Information Technology.

Berger, P. L., & Luckman, T. (1972). *The Social Construction of Reality*: Penguin.

Bicchieri, C. (2006). *The Grammar of Society*. Cambridge: Cambridge University Press.

Bryant, S. L., Forte, A., & Bruckman, A. (2005). Becoming Wikipedian: Transformation of Participation in a Collaborative Online Encyclopedia, *GROUP 05*. Sanibel Island Florida USA.

Ciffolilli, A. (2007). Phantom Authority, self-selective recruitement and retention of members in virtual communities: The case of Wikipedia, *FirstMonday* (Vol. 8).

Coase, R. H. (1993). The Nature of The Firm. In O. E. Williamson & S. G. Winter (Eds.), *The Nature of the Firm: Origins, Evolution and Development*. N.Y.: Oxford University Press.

Coase, R. H. (1995). *Essays on Economics and Economists*. USA: University of Chicago Press.

Davis, J. H., Schoorman, D., & Donaldson, L. (1997). Towards a Stewardship Theory of Management. *Academy of Management Review, 22*(1), 20-47.

Davis, J. H., Schoorman, D. F., & Donaldson, L. (1997). Towards a Stewardship Theory of Management. *The academy of management review, 22*(1), 20-47.

Demil, B., & Lecocq, X. (2003). Neither market or hierarchy or network: The emerging bazaar governance (pp. 36): Université Lille/Institut d'Administration des Entreprises.

Donaldson, L., & Davis, J. H. (1991). Stewardship Theory or Agency Theory: CEO Governance and Shareholder Returns. *Australian Journal of Management, 16*(1), 49-65.

Donath, J. S. (1998). Identity and deception in the virtual community. In P. Kollock & M. Smith (Eds.), *Communities in Cyberspace*. London: Routledge.

Gibbs, J. P. (1981). *Norms, Deviance and social control: Conceptual matters*. New York: Elsevier.

Giles, J. (2005). Internet Encyclopaedias go head to head, *Nature*.

Habermas, J. (1976). Some Distinctions in Universal Pragmatics: A working paper. *Theory and Society, 3*(2), 12.

Jones, C., Hesterly, W. S., & Borgatti, S. P. (1997). A General Theory of Network Governance: Exchange Conditions and Social Mechanisms. *Academy of Management review, 22*(4), 911-945.

Jones, C., Hesterly, W. S., & Borgatti, S. P. (1997). A General Theory of Network Governance: Exchange Conditions and Social Mechanisms. *Academy of Management Review 22*(4), 911-945.

Kittur, A., Suh, B., Pendleton, B. A., & Chi, E. H. (2007). *He Says, She says: Conflict and coordination in Wikipedia.* Paper presented at the Computer/Human Interaction 2007, San Jose USA.

Penner, L. A., Dovidio, J. F., Piliavin, J. A., & Schroeder, D. A. (2005). Prosocial behavior: Multilevel perspectives. *Annual Review of Psychology, 56*, 365-392.

Rossi, M. A. (April, 2004). Decoding the "Free/Open Source (F/OSS) Puzzle" - a Survey ofTheretical and Empirical Contributions (pp. 42): University of Sienna.

Sanger, L. (2005). The Early History of Nupedia and Wikipedia: A Memoir, *Slashdot*.

Sanger, L. (2006). The Nupedia myth, *ZDNet.com*.

Sanger, L. (2007). The Early History of Nupedia and Wikipedia: A Memoir, *Slashdot*.

Searle, J. R. (1969). *Speech Act: An Essay in the Philosophy of Language*. Cambridge: Cambridge University Press.

Stiles, W. B. (1992). *Describing Talk: A Taxonomy of Verbal Response Modes*: Sage.

Stvilia, B., Twidale, M. B., Gasser, L., & Smith, L. C. (2005). *Information Quality Discussions in Wikipedia*. Illinois: Graduate School of Library and Information Science.

Therborn, G. (2002). Back to Norms! On the Scope and Dynamics of Norms and Normative Action. *Current Sociology, 50*(6), 17.

Williamson, O. E. (1996). *The Mechanisms of Governance*. New York: Oxford University Press.

Williamson, O. E., & Winter, S. G. (1993). *The Nature of The Firm: Origins, Evolution and Development*. New York: Oxford University Press.

This work was previously published in International Journal of Agent Technologies and Systems, Vol. 1, Issue 1, edited by G. Trajkovski, pp. 19-33, copyright 2009 by IGI Publishing (an imprint of IGI Global).

Chapter 7.20
Social Network Structures for Explicit, Tacit and Potential Knowledge

Anssi Smedlund
Helsinki University of Technology, Finland and Tokyo Institute of Technology, Japan

ABSTRACT

The purpose of this conceptual article is to develop argumentation of the knowledge assets of a firm as consisting of three constructs, to extend the conventional explicit, tacit dichotomy by including potential knowledge. The article highlights the role of knowledge, which has so far not been utilized in value creation. The underlying assumption in the article is that knowledge assets can be thought of as embedded in the relationships between individuals in the firm, rather than possessed by single actors. The concept of potential knowledge is explained with selected social network and knowledge management literature. The findings suggest that the ideal social network structure for explicit knowledge is centralized, for tacit knowledge it is distributed, and for potential knowledge decentralized. Practically, the article provides a framework for understanding the connection between knowledge assets and social network structures, thus helping managers of firms in designing suitable social network structures for different types of knowledge.

INTRODUCTION

This article starts from the notion that knowledge is an asset for the firm in value creation (e.g., Spender, 1996). According to research in social networks and in the theory of the firm, value creation with knowledge can be considered as something that is embedded in the relationships between individuals, thus making the research on firms' social network structures important (Nelson & Winter, 1982; Granovetter, 1985; Winter, 1987; Kogut & Zander, 1992; Uzzi, 1996). A common saying in the social networks literature is "it's not what you know, it's who you know" (e.g., Cohen & Prusak, 2001).

The main message of this article is that there are fundamentally different types of knowledge assets that produce value with fundamentally different types of social network structures. Based on a short overview of knowledge management literature, an idea is proposed that there are three types of knowledge assets in a firm: explicit, tacit and potential, as well as corresponding three ideal types of social network structures: centralized, distributed and decentralized. The general purpose of this article is to develop convincing arguments to show that knowledge should be described with three constructs, to extend the conventional dichotomous view of knowledge. This line of thought makes it possible to start thinking of unrealized, not yet implemented, knowledge as a strategic asset, in addition to the knowledge assets already utilized by the firm.

The dichotomous view of knowledge as either explicit or tacit has been dominant in the theory of knowledge management after Nonaka and Takeuchi (1995) introduced their model of knowledge creation, the so-called SECI model. It has been claimed, however, that although the SECI model is excellent in describing a process after the initial idea has been developed for a new innovation, it does not necessarily explain the time before clarifying the idea (Engeström, 1999). One possible explanation for this is that the constructs of explicit and tacit knowledge alone are not sufficient to explain the varying nature of knowledge, and how knowledge should be utilized in the very early phases of innovation processes.

This article elaborates arguments about a third knowledge construct, potential knowledge. Potential knowledge is first explained through theory, and illustrated with social network structures. Potential knowledge is defined as a *knowledge asset either in codified or experience-based form that has not yet been utilized in value creation.*

A so-called Coleman-Burt debate on ideal social network structure appears in the social networks literature. This debate is about whether the most optimal network should be structurally sparse and decentralized (Burt, 1992; 2004) or dense and distributed (Coleman, 1988; Uzzi, 1996). There are empirical suggestions towards solving this debate, arguing that the optimal network structure is a combination of sparseness and density, including network ties among the actors that enable both closure and reach simultaneously (Uzzi & Spiro, 2005; Baum, van Liere, & Rowley, 2007; Schilling & Phelps, 2007).

As a result of this theoretical article, it is suggested that the type of knowledge asset—explicit, tacit or potential—is a contingency for the social network structure. It is suggested that there is no one ideal social network structure. Instead, the social network structure of a firm includes a centralized structure for explicit knowledge, a distributed structure for tacit knowledge, and a decentralized structure for potential knowledge. All the types of knowledge and the corresponding social network structures are needed, and individuals can belong to many types of networks simultaneously.

Besides categories of knowledge, another approach to the concept is to consider knowledge as a continuum. There, knowledge is never purely either tacit or explicit, but a combination of both (e.g., Jasimuddin, Klein, & Connell, 2005). Following this line of thought, knowledge that is utilized in the creation of value can be thought to include all three types, with the weighting of the different types changing from one situation to another. The role of potential knowledge is essential in the early phases of the innovation process, whereas tacit knowledge is important in the development phases, and explicit knowledge in the commercialization phases (c.f., Nonaka & Takeuchi, 1995). Based on the knowledge continuum insight, it is proposed in the discussion section that the weights of the different knowledge types, and also the social network structures are different in the idea, development and commercialization phases of the innovation process. Implications for managers are presented and further research issues suggested in the concluding section.

EXPLICIT, TACIT AND POTENTIAL KNOWLEDGE OF A FIRM

An epistemological definition suitable of describing the nature of potential knowledge is "knowing about the thought origins for doing things" (Scharmer, 2001, p. 6). Potential knowledge is knowledge whose value for the organization has not been discovered yet. To borrow from physics, potential energy is stored and available to call on when needed, while kinetic energy is in use, in motion. In the context of an expert's work at the individual level, potential knowledge has been defined as the total amount of knowledge the person has, in contrast to the "actual" knowledge that the individual uses in his or her work (Hollnagel, Hoc, & Cacciabue, 1995).

In the categorical approach to knowledge, knowledge is usually seen as either explicit or tacit. Explicit knowledge is knowledge that is codified, in the form of books, documents and written procedures, "knowledge about things." Tacit knowledge, on the other hand, can be defined as "knowledge about how to do things" (Scharmer, 2001, p. 6), and it is located in the routines of individuals and the organization, as well as in the ways of working between the individuals in the firm (Nelson & Winter, 1982).

Tacit knowledge, according to Polanyi's (1966) original definition, cannot be made explicit, but in the knowledge management theory, a fundamental insight of Nonaka's SECI model (1995) includes the transformation of tacit knowledge into explicit and back during the innovation process in a firm, as highlighted in the cases presented in Nonaka and Takeuchi's (1995) book.

Definitions for tacit knowledge vary in the literature. Hansen (1999) sees tacit knowledge in the firm as corresponding to knowledge that has a low level of codification, that is complex and hard, but not impossible to articulate, or can be acquired only through experience. According to Teece (1986), knowledge in the tacit form is transferable, but it has to be transferred by those who posses the knowledge, due to difficulties in the articulation of tacit knowledge.

The discussion on the definitions and types of knowledge has been guiding the knowledge management literature since the birth of the field. Snowden (2002) states that knowledge management as a discipline has gone through three phases since the early 1990s. The first phase considered the efficient use and storing of codified knowledge, the second phase was started by Nonaka and Takeuchi's (1995) book, and the attention was directed towards learning and conversion between tacit and explicit knowledge. The third phase deals with innovation, complexity and self-emergence of knowledge.

A shortcoming that Nonaka and Takeuchi's SECI model has, despite of its undisputed explanation power on the knowledge creation process, is that it does not take into account the emergence of knowledge in the very early phases of the innovation process (Engeström, 1999; Scharmer, 2001). Scharmer illustrates this with the well- known home bakery example of Nonaka and Takeuchi (1995, p. 100) by arguing that certain kinds of information about bread, such as weight, price and ingredients are explicit knowledge. The activities of baking and producing the bread are examples of tacit knowledge. Finally, the knowledge that enables a baker to create the baking process in the first place is self-transcending, emergent type of knowledge. This is the type of knowledge that could be labeled as potential, and that is what the SECI model lacks. Potential knowledge is the starting point for the knowledge spiral in the SECI model of Nonaka and Takeuchi (1995).

To position the concept of potential knowledge in conjunction with the value creation of the firm, it can be reflected through the explicit-tacit dichotomy (Table 1). Firm-level explicit knowledge, knowledge about things, is knowledge in a codified form that can be stored, managed and used electronically with data mining and document mining techniques in the organization. Firm-level tacit

Table 1. Potential knowledge defined through explicit and tacit types of knowledge

Tacit	Experience-based knowledge	Experience-based or codified knowledge that has not yet been utilized in value creation.
Explicit	Codified knowledge	
	Realized knowledge assets	**Potential knowledge assets**

knowledge, respectively, is knowledge that exists in the skills and perceptions of the employees or groups of employees in a given area, is stored in organizational routines (Nelson & Winter, 1982), and cannot be handled electronically. Potential knowledge is a knowledge asset either in a codified or experience-based form that has not yet been utilized in value creation.

Table 1 describes potential knowledge as an unrealized form of tacit or explicit knowledge. The three knowledge assets pose a challenge in the management of social network structures—what kinds of social network structures are needed, and what managerial action should be taken to create these structures?

SOCIAL NETWORK STRUCTURES FOR EXPLICIT, TACIT AND POTENTIAL KNOWLEDGE IN A FIRM

In this section, the ideal social network structures for explicit, tacit and potential knowledge are presented. By leveraging social network structures, a firm can produce most value with its knowledge assets. It has been argued that decentralized, distributed and centralized social network structures (see Barabási, 2002) are ideal for potential, tacit and explicit knowledge. These networks make it possible not only to search for new, non-redundant sources of potential knowledge, but also to transfer experience-based, tacit knowledge, and to implement explicit knowledge in the firm.

The management of potential knowledge requires scanning the firm's environment through social network ties and seizing the value creation potential in that network. Therefore, the management challenges related to building a social network structure for potential knowledge are twofold: 1) how to increase the span of the network in order to increase the knowledge potential, and 2) how to transform potential, unrealized knowledge into realized knowledge.

In the case of tacit knowledge, which is hard to transfer due to its experience-based, un-codified and complex nature, the management challenge is how to arrange the social network structure to allow close, personal and reciprocal social network relationships for the transfer of tacit knowledge across the organization.

Explicit knowledge, which is exact, codified knowledge about things, poses the challenge of how to implement that knowledge in practice. The main topic of interest in the implementation of explicit knowledge is creating a structure of accuracy and discipline to ensure flawless flow of explicit knowledge. Table 2 summarizes the three classifications of knowledge and the management challenges related to the social network structure in each type of knowledge.

From the point of view of the contingency theory (e.g., Burns & Stalker, 1961), it has been stated that environmental conditions affect the structures of networks. For example, Podolny and Baron (1994; 1997) argue that under uncertainty, and when facing authoritative power, social network ties are formed more likely with similar others, which leads to the formation of strong ties. Also, cultural traditions and institutions have been found to influence the social network structures in a way that in the environment with highly profound institutions, such as the lifetime

Table 2. Three classifications of knowledge and social network structure-related management challenges

Type of knowledge	Definition	Social network management challenge
Potential knowledge	Codified or experience-based knowledge that has not yet been utilized in value creation	How to build an extensive social network and how to scan and seize knowledge from this structure?
Tacit knowledge	Experience-based knowledge	How to transfer experience-based knowledge in the social network structure?
Explicit knowledge	Codified knowledge	How to implement codified knowledge in value creation with the social network structure?

employment and seniority-based promotion in Japan, taller hierarchies and greater formal centralization can be found (Lincoln, Hanada, & McBride, 1986).

The type of knowledge is a contingency for the social network structure. The explicit knowledge of a firm is used to reach efficiency in producing already designed products or services, because it includes well-codified rules, documents and procedures. This knowledge is, for example, knowledge related to stock levels or blueprints of products. Therefore, an ideal social network structure for explicit knowledge is the centralized structure (see Barabási, 2002), in the sense of the mechanic management system presented by Burns and Stalker (1961).

Tacit knowledge is a knowledge asset that has accumulated through past experiences in the firm. It is mainly used for gradual improvement of existing products, services or production methods and processes. Tacit knowledge can be, for example, the know-how of the employees, or past customer experiences. It is experiences that are difficult to transfer and require reciprocal and close relationships with the individuals. The distributed social network structure (see Barabási, 2002) is the most suitable one for tacit knowledge. In the traditional contingency theory, Burns and Stalker's (1961) organic management system resembles the distributed social network structure.

Potential knowledge is either codified or experience-based, an unrealized knowledge asset that has future value creation potential. It is used to initiate the innovation of something totally new in the very early phases of the innovation process. Potential knowledge is characterized with connections to many non-redundant sources of knowledge, which makes the ideal network structure for potential knowledge decentralized (see Barabási, 2002). There are no equivalent structures of decentralized networks in the contingency theory. Table 3 summarizes this chapter by presenting the social network structures for explicit, tacit and potential knowledge.

Decentralized, distributed and centralized social network structures have notable differences in their functioning mechanisms. The decentralized social network is built on individuals as hubs of knowledge who gather and broker knowledge from different sources. The distributed social network structure does not have brokers, because there the relationships are distributed evenly, and individuals are connected with a few links to a couple of others. In the centralized social network, the functioning mechanism is based on a focal individual who manages the flows of knowledge with disconnected others. In the next section, the types of knowledge and social network structures are connected to different tasks in the firm.

DISCUSSION

Based on a common presupposition in the theory of the firm, and in the social network literature, knowledge can be thought of as embedded in the interactions between individuals in the firm (i.e., Nelson & Winter, 1982; Kogut & Zander, 1992; Uzzi, 1996). Therefore, each of the three knowledge constructs poses different challenges for the management of social network relationships in a firm. In order to produce value with potential knowledge, the firm must be able to grow the reach of its network to include non-redundant, new sources of knowledge. In explicit knowledge, the social network should allow quick and flawless implementation, and in tacit knowledge comprehensive and reciprocal transfer.

There has been a debate among knowledge management scholars about whether the concept of knowledge should be treated as a categorical construct (e.g., Nonaka & Takeuchi, 1995), or as a continuum (e.g., Jasimuddin, Klein, & Connell, 2005). The starting point of this article has been the categorical perspective to knowledge, although the continuum perspective that sees knowledge as existing along a continuum of tacitness and explicitness is usable also in the context of the theory of the firm. The suggestion towards knowledge consisting of both tacit and explicit components at the same time is plausible for example in the context of absorptive capacity (Cohen & Levinthal, 1990). There, a firm must posses a certain knowledge base before it is able to learn anything new.

Figure 1 integrates the categorical and continuum perspectives of knowledge in the value creation of a firm. Here, a firm creates value basically in three ways: 1) invention, 2) development, and 3) commercialization. These three ways of value creation are present in the innovation process of product innovation. In the invention phase, the role of potential knowledge is emphasized, and the management of the firm should concentrate on building a decentralized social network structure to support free and fast flow of ideas from distant and non-redundant sources. However, based on the insight of absorptive capacity and the knowledge continuum view, the invention phase needs also

Table 3. Social network structures for explicit, tacit and potential knowledge in a firm

	Potential knowledge	**Tacit knowledge**	**Explicit knowledge**
Illustration			
Network structure	Decentralized	Distributed	Centralized
Functioning mechanism	There are hubs in the knowledge network that control the flows of knowledge and intermediate between different groups. Some of the actors are more connected than the others.	There is no specific actor who manages the knowledge flows. The knowledge flows horizontally from one actor to another. Every actor has knowledge links to a couple of other actors.	The focal node in the network manages the knowledge flows. The knowledge flows hierarchically from the top down and from the bottom to the top. There are no knowledge exchange links between the subordinates.

Figure 1. Potential, tacit and explicit knowledge in different phases of the innovation process

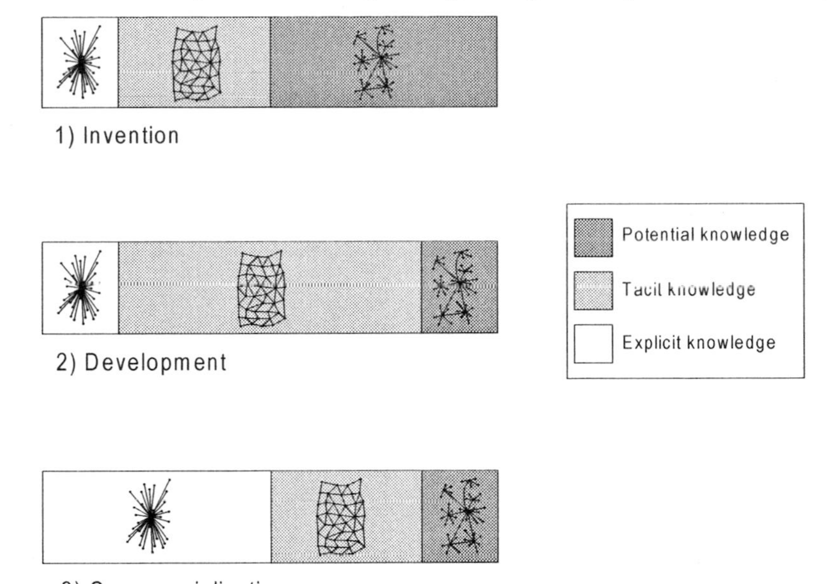

tacit knowledge and explicit knowledge, but to a lesser extent. This is because without certain explicit procedures and methods, or without some kind of individual's past experiences in ideation work, new invention is not possible.

Similarly with the invention phase, also the development and commercialization phases include all three types of knowledge, but with different weights (see Figure 1). In the development phase, an idea is gradually developed, based on the expertise of the individuals, with the reciprocal and distributed social network structure. Finally, in the commercialization phase, the developed product is produced as efficiently as possible along the unambiguous rules and procedures in the centralized social network structure.

Jasimuddin et al. (2005) present the paradoxes and difficulties related to the concept of knowledge convincingly in their literature review. There is confusion in the field of knowledge management about the concept, not only because different scholars have their own theoretical backgrounds, but also because it is not always clear whether knowledge is discussed from the point of view of the individual or the organization, from the epistemological or managerial standpoint.

Connecting potential, tacit and explicit knowledge constructs to the different phases in the innovation process makes it possible to see knowledge as both a categorical and a continual concept. Categorizations are needed to distinguish different aspects of knowledge assets that create value for the firm, and the knowledge continuum is needed when the categories are applied into practice in the value creation process.

CONCLUSIONS AND FURTHER RESEARCH

The cycles of the economy are becoming shorter, and firms are expected to bring new products and services to the market at an increasing pace. Rapid, centralized implementation of explicit knowledge is essential, because firms must be able to transform their product definitions, processes and production methods rapidly across a possibly globally distributed hierarchical demand-supply chain. Besides efficient production, firms must be capable of improving their products or services gradually to meet the needs of the customers. Gradual development can be achieved by allowing reciprocal and thorough transfer of tacit knowledge in a distributed social network. Last but not least, firms face a challenge of innovativeness, scanning and seizing their environment for new possible trends and ideas. There, the decentralized social network structure for potential knowledge comes into place.

It is clear that there is no one optimal social network structure for a firm but many, and each social network type requires unique management initiatives. Potential knowledge can be harnessed with managerial actions that aim to create a decentralized social network structure, for example, by investing in search capabilities or by increasing the pool of different types of talent in the firm, and emphasizing creativity and sharing of ideas in the leadership style. In order to create decentralized structures, individuals with knowledge broker capabilities (c.f., Burt, 2004) should be encouraged to improve the sharing, gathering and flow of ideas across the firm.

Tacit knowledge can be leveraged with managerial actions that aim for distributed social network structures. This can be achieved by emphasizing learning and trust building in the leadership style of managers. Investments in team building, team cohesion, and building of cross-functional teams with members from different parts of the organization should be made to ensure a timely transfer of tacit knowledge.

Finally, the management of explicit knowledge should focus on efficiency and a time-to-market mindset. There, a centralized social network structure can be achieved by investing in efficient, hierarchical management systems and systems engineering with industrial organization logic.

The divisions according to the type of knowledge presented in this article offer one theoretical framework for managing the complex whole of the social network ties in a firm. Decentralized, distributed and centralized social network structures all exist in the firm, and the same individuals can be a part of a social network for potential, tacit and explicit knowledge at the same time. The plethora of types of relationships between individuals in an organization is vast. The fundamental question for future research is which kinds of layers of relationships should be investigated more thoroughly, and how the social networks that the different types of relationships uncover should be operationalized. More work should be done in terms of defining the potential knowledge construct to find answers to why some firms are essentially better in sensing new opportunities in the market than others.

By influencing the social network structures in the firm with managerial action, the use of knowledge assets in value creation can be encouraged. The aim of further research based on this article is to connect different types of knowledge assets to different value-creating tasks in the firm, and to study empirically the inter-firm and intra-firm management initiatives suitable for each of the social network types. From the intra-firm social network perspective, possible interesting research questions would be related to the ideal structures in, for example, transferring ideas across the firm, business development, or in efficient production. Also individual-level research questions on, for example, the social network characteristics of highly innovative individuals should be studied from the knowledge management perspective.

In many cases, innovation occurs in the relationships between different firms (Powell, Kogut, & Smith-Doerr, 1996), and also development and production functions are increasingly operated across firm and industry borders. The three types of knowledge and the corresponding three social network types should be investigated from the point of view on how the three knowledge types can be separated from each other in inter-firm relationships. Creating awareness of different types of knowledge assets in the collaboration relationships between many firms, and different types of social networks to support them, would allow more efficient management of globally distributed innovation, development and production activities.

REFERENCES

Barabási, A.-L. (2002). *Linked: The new science of networks.* Cambridge, MA: Perseus Publishing.

Baum, J., van Liere, D., & Rowley, T. (2007). Between closure and holes: Hybrid network positions and firm performance. Working paper, Rotman School of Management, University of Toronto.

Burns, T., & Stalker, G. (1961). *The management of innovation.* London: Tavistock Publications Ltd.

Burt, R. (2004). Structural holes and good ideas. *American Journal of Sociology, 110*(2), 349-399.

Burt, R. (1992). *Structural holes: The social structure of competition.* Cambridge, MA: Harvard University Press.

Cohen, D., & Prusak, L. (2001). *In good company: How social capital makes organizations work.* Boston: Harvard Business School Press.

Cohen, W., & Levinthal, D. (1990). Absorptive capacity: A new perspective on learning and innovation. *Administrative Science Quarterly, 15*(1),128-152.

Coleman, J. (1988). Social capital in the creation of human capital. *The American Journal of Sociology, 94*(1), 95-120.

Engeström, Y. (1999). 23 Innovative learning in work teams: Analyzing cycles of knowledge creation in practice. In: Y. Engeström, R. Miettinen , and R.-L. Punamäki-Gitai (Eds.), *Perspectives on activity theory.* Cambridge: Cambridge University Press.

Granovetter, M. (1985). Economic action and social structure: The problem of embeddedness. *The American Journal of Sociology, 91*(3), 481-510.

Hansen, M. (1999). The search-transfer problem: The role of weak ties in sharing knowledge across organization subunits. *Administrative Science Quarterly, 44*(1), 82-111.

Hollnagel, E., Hoc, J., & Cacciabue, P. (1995). Expertise and technology: I have a feeling we are not in Kansas anymore. In: J. Hoc, P. Cacciabue, & E. Hollnagel (Eds.), *Expertise and technology,* (pp. 279-286). New Jersey: Lawrence Erlbaum Associates Publishers.

Jasimuddin, S., Klein, J., & Connell, C. (2005). The paradox of using tacit and explicit knowledge. Strategies to face dilemmas. *Management Decision, 43*(1), 102-112.

Kogut, B., & Zander, U. (1992). Knowledge of the firm, combinative capabilities, and the replication of technology. *Organization Science, 3*(3), 383.

Lincoln, J., Hanada, M., & McBride, K. (1986). Organizational structures in Japanese and U.S. manufacturing. *Administrative Science Quarterly, 31*(3), 338-364.

Nelson, R., & Winter, S. (1982). *An evolutionary theory of economic change.* Cambridge, MA: Belknap.

Nonaka, I., & Takeuchi, H. (1995). *The knowledge-creating company: How Japanese companies create the dynamics of innovation.* New York: Oxford University Press.

Podolny, J. (1994). Market uncertainty and the social character of economic exchange. *Administrative Science Quarterly, 39*(3), 458.

Podolny, J., & Baron, J. (1997). Resources and relationships: Social networks and mobility in the workplace. *American Sociological Review, 62*(5), 673-693.

Polanyi, M. (1966). *The tacit dimension.* London: Routledge & Kegan.

Powell, W., Kogut, K., & Smith-Doerr, L. (1996). Interorganizational collaboration and the locus of innovation: Networks of learning in biotechnology. *Administrative Science Quarterly, 41*(1), 116-145.

Scharmer, C. (2001). Self-transcending knowledge: Organizing around emerging realities. In: I. Nonaka & D. Teece (Eds.), *Managing industrial knowledge: Creation, transfer and utilization.* London: Sage Publications.

Schilling, M., & Phelps, C. (2007). Interfirm collaboration networks: The impact of large-scale network structure on firm innovation. *Management Science, 52*(11), 1113-1126.

Snowden, D. (2002). Complex acts of knowing: Paradox and descriptive self-awareness. *Journal of Knowledge Management, 6*(2), 100-111.

Spender, J.-C. (1996). Making knowledge the basis of a dynamic theory of the firm. *Strategic Management Journal, 17*(Winter Special Issue), 45-62.

Teece, D. (1986). Profiting from technological innovation: Implications for integration, collaboration, licensing and public policy. *Research Policy, 15*(6), 285-305.

Uzzi, B. (1996). The sources and consequences of embeddedness for the economic performance of organizations: The network effect. *American Sociological Review, 61*(4), 674-698.

Uzzi, B., & Spiro, J. (2005). Collaboration and creativity: The small world problem. *The American Journal of Sociology, 111*(2), 447-504.

Winter, S. (1987). Knowledge and competence as strategic assets. In: D. Teece (Ed.), *The competitive challenge: Strategies for industrial innovation and renewal.* Centre for Research Management.

This work was previously published in International Journal of Knowledge Management, Vol. 5, Issue 1, edited by M. Jennex, pp. 78-87, copyright 2009 by IGI Publishing (an imprint of IGI Global).

Chapter 7.21
Motif Analysis and the Periodic Structural Changes in an Organizational Email-Based Social Network

Krzysztof Juszczyszyn
Wrocław University of Technology, Poland

Katarzyna Musiał
Wrocław University of Technology, Poland

ABSTRACT

Network motifs are small subgraphs that reflect local network topology and were shown to be useful for creating profiles that reveal several properties of the network. In this work the motif analysis of the e-mail network of the Wroclaw University of Technology, consisting of over 4000 nodes was conducted. Temporal changes in the network structure during the period of 20 months were analysed and the correlations between global structural parameters of the network and motif distribution were found. These results are to be used in the development of methods dedicated for fast estimating of the properties of complex internet-based social networks

INTRODUCTION

Communication technologies enabled the emergence of complex, evolving social networks built on various services like e–mail, P2P and community portals. In general they are similar to the traditional social networks based on relations between humans, but there are also some significant differences. First, the information about the users in virtual communities and their activities is stored in electronic form which allows precise inference of the network structure and parameters. On the other hand, the networks created by means of communication technologies show incomparable size and dynamics (for more differences between regular social networks and virtual ones see Sec.3).

When investigating the topological properties and structure of complex networks we must face a number of complexity–related problems. In large social networks, tasks like evaluating the centrality measurements, finding cliques, etc. require significant computing overhead. In this context the methods, which proved to be useful for small and medium sized networks fail when applied to the huge structures. Moreover, our knowledge about the actual network structure may be incomplete especially due to its size and dynamics.

In this work we present the results of the analysis of local topology structure of large e–mail based organizational social network. The investigated e–mail network of the Wroclaw University of Technology (WUT), consisting of over 4000 nodes (e–mail addresses of the users) was generated basing on server logs from the period of February 2006 – September 2007, analyzed with standard methods – clustering and centrality assessment. Periodic changes in the network structure, connected with the business profile of the organization (university) were discovered and analyzed. In the series of experiments the global features of the network (clustering coefficients, centrality as well as the number of edges and nodes) were checked for correlations with the distribution of small network subgraphs, called network motifs. The existence of relations between motif profile of the network and its global structural properties may allow their fast estimation.

Dependencies between global network characteristics and the distribution of local topology features may have numerous applications. They can help to estimate the measures like centrality and clustering without the complete knowledge of the network structure. This feature appears very appealing especially when we deal with evolving networks consisting of millions of nodes (like social networks of mobile phone users, Web communities and so on) and the distribution of motifs may be determined by various sampling techniques which do not assume the exhausting processing of the entire structure of the network.

In the following sections we briefly introduce the networks motifs concept as well as the basics of online social networks. Furthermore, the process of social network extraction from WUT e–mail logs is presented. Finally, the results of motif detection and temporal changes in the structure of this network are discussed.

NETWORK MOTIFS

Complex networks, both biological and engineered, were analyzed with respect to so–called network motifs (Milo, 2002). They are small (usually 3 up to 7 nodes in size) subgraphs which occur in the given network far more (or less) often then in the equivalent random (in terms of the number of nodes and edges) networks. Despite all known structural and statistical similarities, networks from different fields have very different local topological structure. It was recently shown that concentration of network motifs may help to distinguish and classify complex biological, technical and social networks (Milo et al.,2004). We can define so–called superfamilies of networks, which correspond to the specific significance profiles (SPs). To create SP for the motifs in a given network, the concentration of individual motifs is measured and compared to their concentration in a number of random networks. The statistical significance of motif M is defined by its Z–score Z_M:

$$Z_M = \frac{n_M - \left\langle n_M^{rand} \right\rangle}{\sigma_M^{rand}} \qquad (1)$$

where

n_M – the frequency of motif M in the given network,

$\left\langle n_M^{rand} \right\rangle$ and σ_M^{rand} – the mean and standard deviation of M's occurrences in the set of random networks, respectively (Itzkowitz, 2003).

Most algorithms for detecting network motifs assume exhaust enumeration of all subgraphs with a given number of nodes in the network. Their computational cost dramatically increases with the network size. However, it was recently shown that it is possible to use random sampling to effectively estimate concentrations of network motifs. The algorithm presented in (Kashtan, 2004) is asymptotically independent of the network size and enables fast detection of motifs in very large networks with hundreds of thousands of nodes and larger. In result we do not have to process the entire network and the prohibitive computational cost may be avoided.

The existence of network motifs affects not only topological but also functional properties of the network. For biological networks, it was suggested that network motifs play key information processing roles (Shen-Orr, 2002). For example, so–called FFL motif – has been shown both theoretically and experimentally to perform tasks like sign–sensitive filtering, response acceleration and pulse–generation (Mangan, 2003). Such results reveal that, in general, we may conclude about function and properties of very large networks from their basic building blocks (Mangan, Zaslaver & Alon 2003).

In another work, motif analysis was proved to have ability of fast detection of the small–world and clustering properties of the large artificial Watts–Strogatz network (Chung-Yuan, 2007). This result open promising but still unexplored possibilities of reasoning about network's global properties with sampling of local topological structures.

Very little research has been done on motifs in computer science and sociology. SPs for small social networks (<100 nodes) were studied in (Milo, 2004). A web network counting 3.5×10^5 nodes (Barabasi, 1999) was used to show the usability of sampling algorithm (Kashtan, 2004).

The motif analysis of Internet–based social networks is also hardly represented in recent works. It was recently shown that the email–based social networks appear to have the SP very similar to traditional social networks, but there are also some distinctive features which define their unique SP (Juszczyszyn, 2008).

VIRTUAL SOCIAL NETWORKS

People who interact with one another, share common activities or even posses similar demographic profiles can form a social network. Overall, the concept of a social network is quite simple and can be described as a finite set of individuals, by sociologists called actors, who are the nodes of the network, and relationships that are the linkages between them (Adamic, 2003; Ehrlich, 2006; Garton, 1997; Hanneman, 2005; Weng, 2005). In other words, social network concept is utilized to describe the relationships between friends, co–workers, members of the particular society, relatives in the family, etc. Not only can the character of the relationships be analyzed, but also their strength and direction. Although social network analysis (SNA) emphasizes the connections between people, the results of SNA provide also much information about individuals themselves.

The concept of social network, first coined in 1954 by J. A. Barnes (1954), have been in a field of study of modern sociology, anthropology, geography, social psychology and organizational studies for last few decades. The regular social networks are based on the in person contact. However, new trends in social network have emerged with development of the Internet where there is no in person contact. The online social networks can be extracted from the data about users and their behaviors that are gathered in various types of services that are available in the computer network.

The social networks that can be identified in the real world such as network of co–workers, friends or family members differs from social networks existing in the Internet

The main features that distinguish social networks on the Internet from regular ones are as follows:

1. Lack of physical, in person contact – only by distance, even very long distances.
2. Usually the lack of unambiguous and reliable correlation between member's identity in the virtual community – internet identity and their identity in the real world.
3. The possibility of multimodal communication; simultaneously with many members but also the possibility of easy switches between different communication channels, especially online and offline, e.g. online VoIP and offline text communication.
4. The simplicity of a break up and suspension of contacts or relationships.
5. The relatively high ease of gathering the data about communication or common activities and its further processing.
6. The lower reliability of the data about users and their activities available on the Internet. Users of Internet services relatively frequently provide fake personal data due to privacy concerns.

Although social networks in the Internet have already been studied in many contexts and many definitions were created, they are not consistent. Also, different researchers name these networks differently. In consequence, these networks are called: computer–supported social networks, CSSNs (Wellman, 1996), online social networks (Garton, 1997), web–based social networks (Goldbeck, 2005), web communities (Gibson, 1998; Flake, 2000), or virtual communities (Adamic, 2003).

In the literature, the name web communities was firstly used to describe the set of web pages that deal with the same topic (Gibson, 1998; Flake, 2000). Adamic and Adar (2003) argue that a web page must be related to the physical individual in order to be treated as a node in the online social network. Thus, they analyze the links between users' homepages and form a virtual community based on this data. Additionally, the equivalent social network can also be created from an email communication system. On the other hand, a computer–supported social network introduced in (Wellman, 1996) appears when a computer network connects people or organizations. Finally, Golbeck (2005) affirms that a web–based social network must fulfill the following criteria: users must explicitly establish their relationships with others, the system must have explicit support for making connections, and relationships must be visible and browsable.

Based on the kind of service people use, many examples of the social networks in the Internet can be enumerated. To the most commonly known belong: a set of people who date using an online dating system (Boyd, 2004), a group of people who are linked to one another by hyperlinks on their homepages (Adamic, 2003), customers who buy similar stuffs in the same e–commerce (Weng, 2005), the company staff that communicates with one another via email (Adamic, 2005;, Kazienko, 2007; Shetty, 2005; Culotta, 2004; Zhu, 2006), people who share information by utilizing shared bookmarking systems (Zhu, 2006) such as del.icio.us.

E–mails that was mentioned above are bidirectional and asynchronous way of communication. They posses limited social presence but on the other hand enable to exchange the information between people who are in different places and on different schedules (Wellman, 1996; Musiał, 2008). The email service is an example of social network in the computer network where the users communicate with one another by exchanging messages. The relationship in such a network can be derived both from the communication between two users as well as from the address books available in the given email system. The email addresses can be derived from different domains, e.g. Gmail, Yahoo, etc. Note that during analysis of such networks the cleansing process is a very complex issue, e.g. the junk mail should be excluded or two email addresses can belong to one social entity.

EMAIL–BASED SOCIAL NETWORK OF WUT

The experiments were carried out on the logs from the Wroclaw University of Technology (WUT) mail server, which contained only the emails incoming to the staff members as well as organizational units registered at the university. The communication with the external addresses was not considered – only organizational social network was investigated in our experiments. Motif detection was performed with FANMOD tool (Wernicke, 2006; Wernicke & Rasche, 2006) dedicated for motif detection in large networks.

First, the data cleansing process was executed. The bad email addresses was removed from the analysis and the duplicates were unified. WUT social network consists of nodes and relations between these nodes. The email addresses are the nodes of this network and the relationship between two nodes exists if and only if there is any email communication between them. In order to tract the temporal changes in the considered network the information from the logs was extracted for every month in the period of February 2006 – September 2007 separately. This allowed us to create 20 networks reflecting the structure of organizational communication between the WUT employees. This network was used for motifs detection in order to check how the motif SP changes when different periods of time are considered. The size of the networks varied from 3 257 to 4465 nodes, reflecting fact that various numbers of employees were active in different months.

Figure 1 shows that the changes in network size are periodic – this effect is connected with the general business profile of the WUT employees' activities. Obviously – August (months nr 7 – 3257 nodes and 19 – 2905 nodes on Figure1), as a peak of summer holiday season, may be associated with minimal communication activity of the university personnel. This is reflected in

Figure 1. Temporal changes in the size of WUT email network [16]

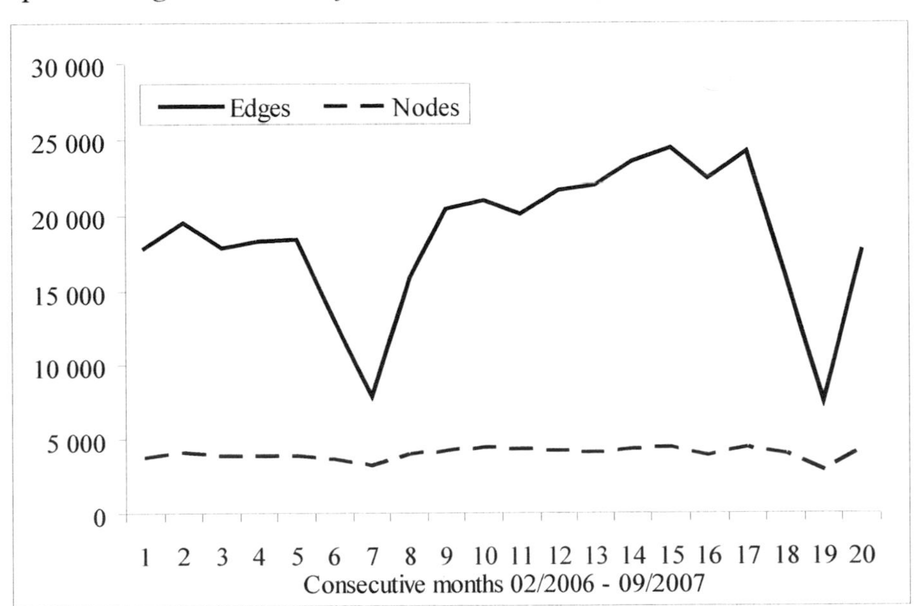

Figure 2. Average clustering coefficient CC1 and Betweenness centrality

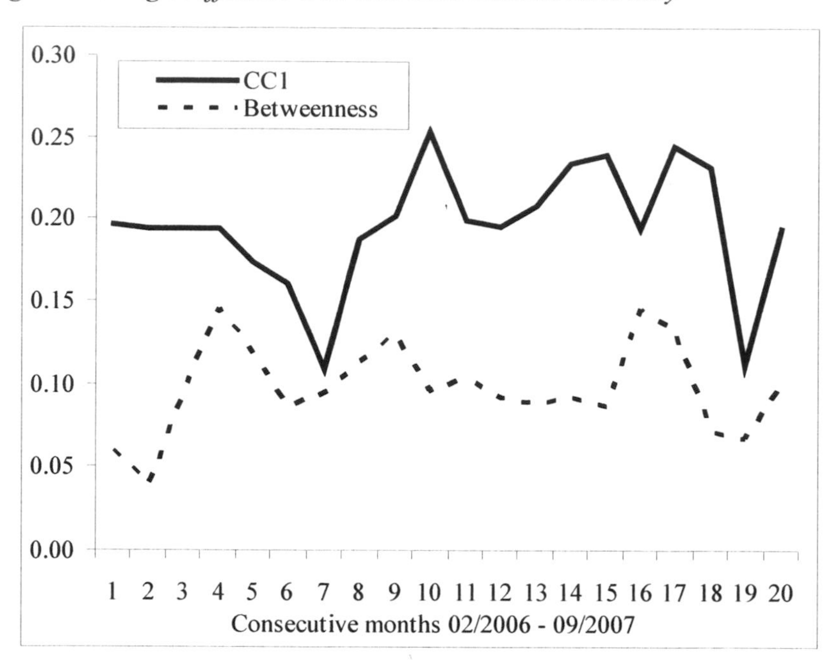

small decrease in the number of active e–mail users (network nodes) and rapid drop in the number of network edges – 7941 edges in August 2006 and 7555 edges in August 2007 (which stand for incoming/outcoming messages). Similar effect (although less in size) is typical also for July and September, when the holidays respectively start and end. The short winter university holidays (two weeks in the middle of February) are not so clearly visible but, as we will see below, detectable with the motif analysis.

Along with the size of the network (measured as the number of nodes and edges), three more structural measures were computed: two clustering coefficients (*CC1*, *CC2*) and the betweenness centrality. The clustering coefficients were defined in the following way:

$$CC1 = \frac{2\left|E\left(G1(n)\right)\right|}{\deg(n)\left(\deg(n)-1\right)} \quad (2)$$

$$CC2 = \frac{\left|E\left(G1(n)\right)\right|}{\left|E\left(G2(n)\right)\right|} \quad (3)$$

where:

deg(n) – denotes degree of node *n*,
|E(G1(n))| – number of lines among nodes in 1–neighborhood of node *n*,
|E(G2(n))| – number of lines among nodes in 1 and 2–neighborhood of node *v*.

The two coefficients considering 1– and 2–neighborhood are given by Eq. 2 and 3. We also assume that for a node *n* such that $deg(n) \leq 1$ all clustering coefficients are 0.

Betweenness centrality was defined in standard way (according to [18]) as a measure which returns for given node *n* the medium mediating between all pairs of nodes. Figure 2 presents the changes in the values of *CC1* and the network betweenness centrality. We may note that two moments of the most significant changes in the network structure (months no 7 and 19) are visible in the values of *CC1*, while fluctuations of betweenness are random–like and their interpretation is not straightforward.

Figure 3. Average clustering coefficient CC2

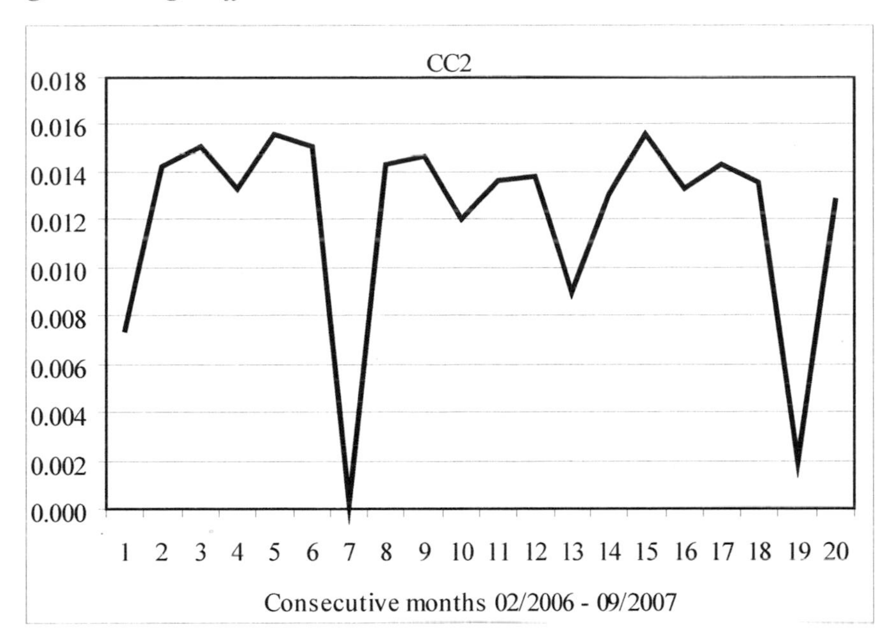

Figure 4. Network motifs sensitive to structural changes in an e-mail network (Juszczyszyn, 2008)

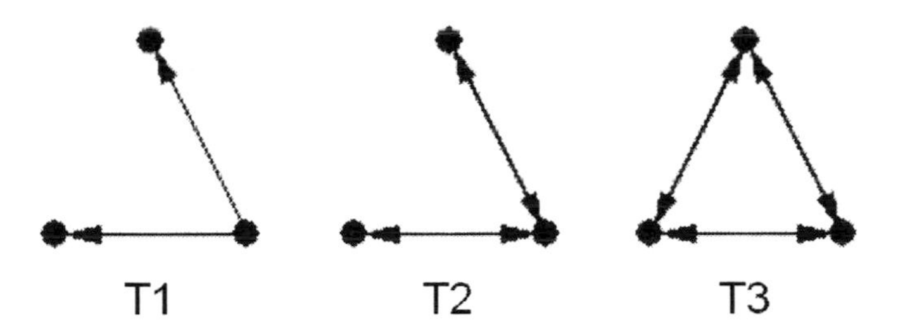

The temporal characteristic of *CC2* (which takes into account the density of the 2–neighbourhood of given node) is shown on the Figure 3. We can see that the two minima clearly denote August–the peak of summer holidays, associated with rapidly fading communication between the employees of WUT. Moreover, the next two smallest values (for February'2006 and February'2007) point to the two–week winter holidays (also: the end of winter semester on the university).

CC1, *CC2* and the betweenness centrality will be used in Sec. 5 to check for possible correlations with the motif distribution in the network.

Investigation of the local network topology with motif analysis, presented recently in (Juszczyszyn, 2008) have shown that the abovementioned periodic activity changes affect not only the scale but also the structure of communication. The distribution of motifs detected in (Juszczyszyn, 2008) (triads consisting of three network nodes and up to six directed edges) proved that holiday seasons are associated with the lack of the broadcast–type

Figure 5. Social network discovered from the email communication between employees of WUT (Juszczyszyn, Kazienko & Musiał, 2008)

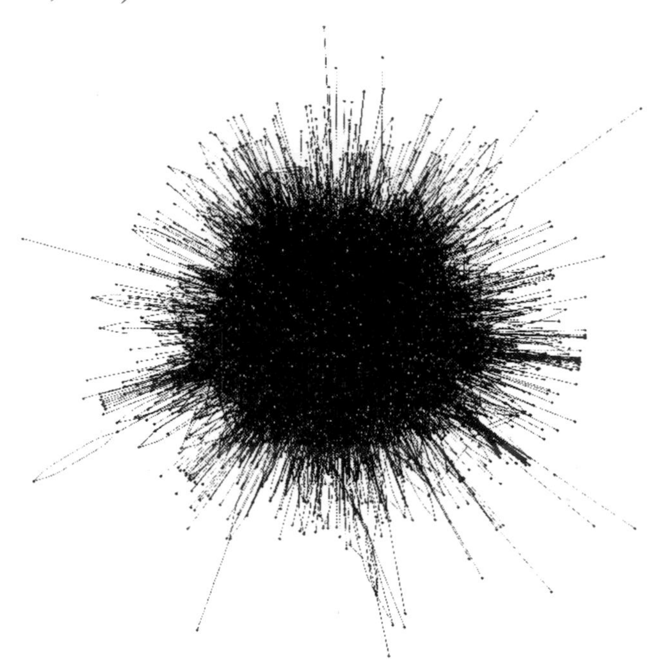

communication. Also, the users forming densely connected cliques tend to sustain their communication during holidays which was reflected by the large number of occurrences of fully connected triads in the investigated network.

In particular – three triads (motifs) were identified as good candidates for tracking changes in large, evolving networks. They are presented on the Figure 4.

The Z–scores of these three motifs were visibly changing (with the highest variance) according to periodic structural changes in the network. In result, for the further experiments, we chose to investigate the behaviour of fully connected subgraphs (like T3 on Figure2) along with those reflecting broadcast type communication directed towards targets which are not connected (like T1 and T2. Note that they may be also associated with the existence of weak links and connections between cliques). Also, in order to follow the conclusions from these first experiments we assumed that the distribution of the four–node subgraphs (quadruples) will be investigated, because they better reflect the nature of dense cliques and connections between then (this will be discussed in detail in the next section).

On the Figure 5 a visualization of the network created on the basis of e–mail communication in entire 20–months period is shown (Juszczyszyn, Kazienko & Musiał, 2008). All 20 networks created for our experiments described below were similar in size and complexity.

MOTIF DETECTION AND ANALYSIS

The motif analysis was performed in this paper based on four–node motifs. Six possible four–node connected subgraphs (a.k.a. quadruples) exist. As defined in Sec. 2 if their occurrence in the given network differs significantly from that of a random network that consists of the same number of nodes and edges as the investigated network, they become

Figure 6. Four-node network motifs (quadruples)

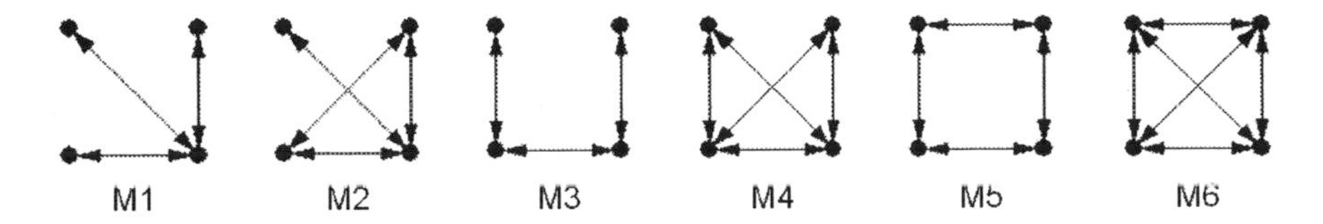

Figure 7. Temporal significance profile for motifs M1-M5

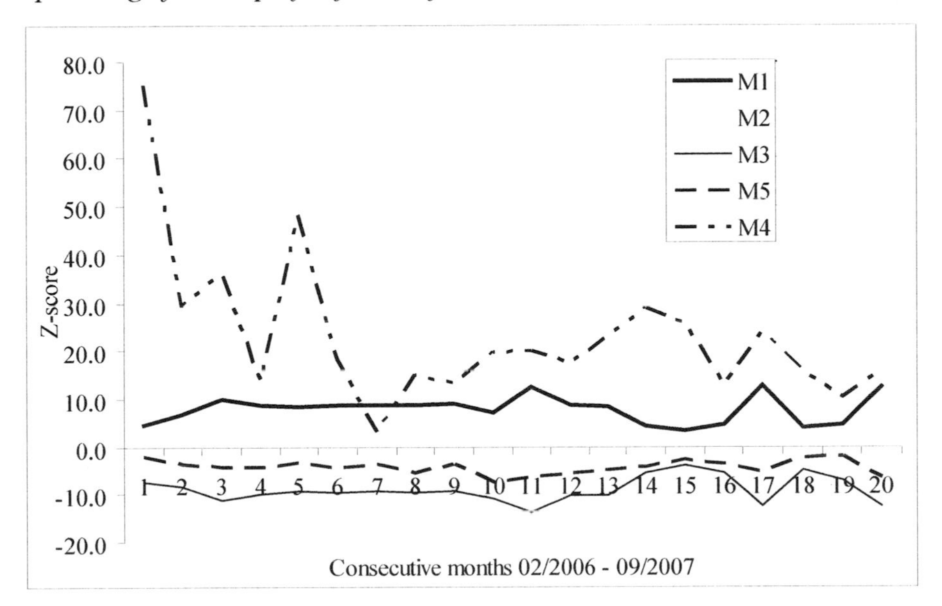

network motifs. Figure 6 introduces the subgraphs we used in our experiments. Note, that all arrows on the picture are bi–directional, in motif detection we chose to disregard the direction or the relations in the network – analysing over one hundred of possible directed quadruples would impose inevitable interpretation problems.

Detection of the above six subgraphs was carried in all 20 networks derived from e–mail communication. Figure 7 shows the temporal significance profiles for motifs M1 to M5 and their changes during 20 months.

Conducted analysis have shown no visible correlation between the Z–scores of motifs M1–M5 and the changes of the network connectivity illustrated on Figure1. The same result was obtained for centrality and clustering measures. For example, the Pearson correlation coefficients computed for M1–M5 and all the measures introduced in Sec. 3 do not exceed 0.25, showing that there is no linear relationship between them. Moreover, the Z–scores of M1–M5 (with the exception of M4, which may be explained but its close edit distance to M6 discussed below) are close to zero, in terms of motif analysis it means that their distribution is similar to the one met in random networks. They also do not change sign that sometimes may suggest a structural change in the network.

However, the situation with M6 is different (Figure 8). First of all, its Z–score is bigger by the order of magnitude, taking values from 69 to 539. It also shows the biggest variance in the set of 20 considered months.

The analysis of correlations between the

Figure 8. Temporal significance profile for motif M6

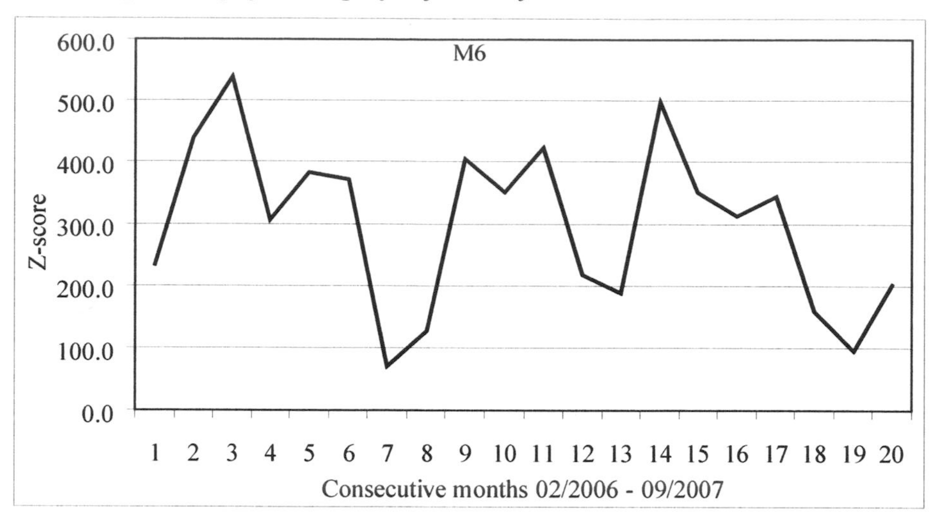

Z–score of M6 and the values of *CC1, CC2* and betweennes centrality have shown dependencies between them. The Pearson coefficient computed for Z–score of M6 equals:

- 0,5618 for number of edges in the network,
- 0,4993 for number of nodes in the network,
- 0,4503 for *CC1,*
- 0,6538 for *CC2,*
- 0,1231 for betweenness centrality.

According to discussion presented by Cohen (2007) the values obtained for *CC1* and *CC2* should be treated (when dealing with social networks) as significant. There is positive linear relationship between the frequency of M6 occurrences and *CC1* and *CC2.* Correlation with betweenness centrality was not confirmed – it will be the addressed during further experiments.

CONCLUSION

The authors investigated the local topology and its dynamics in the email based social network by utilizing the concept of network motifs. Additionally, we analyzed some characteristics of the global structure (such as centrality, clustering coefficient, number of nodes or edges) and presented how the values of these measures change over time. Finally, we uncovered the correlation between the network characteristics and some of the investigated four-nodes motifs.

We demonstrated that there is a relationship between the Z–score of M6 motif (fully connected quadruple) and the values of chosen structural network measures. Temporal changes in local connection patterns of the social network are correlated with changes in the intensity of communication and clustering coefficients and are detectable by means of motifs analysis (which, along with fast network sampling techniques, gives us a possibility to insight into the structure of large social network without significant computational overhead). The further investigations will aim to develop methods of estimating the values of global network characteristics by subgraph sampling techniques. We intend to use these methods for analysis of large, changing network structures (like internet communities, evolving networks of mobile phone users etc.) which require a new ways of detecting their properties with low computational overhead and minimal cost.

An interesting possibility of future temporal motif analysis will be also addressing relationship strengths in detail. Complementarities between results of (Juszczyszyn, Kazienko & Musiał, 2008) and the temporal TSP changes suggest that further experiments may bring interesting results.

REFERENCES

Adamic, L.A., & Adar, E. (2003), Friends and Neighbors on the Web, *Social Networks*, 25(3), pp. 211-230.

Adamic, L.A., & Adar, E. (2005), How to search a social network, Social Networks, 27(3), 187 – 203.

Barabasi A.-L. Albert R. (1999) Emergence of scaling in random networks. Science, 286, 509–512.

Barnes, J. A. (1954), Class and Committees in a Norwegian Island Parish, Human Relations.

Boyd, D.M. (2004), Friendster and Publicly Articulated Social Networking, CHI 2004, *ACM Press 2004*, 1279 – 1282.

Chung-Yuan H., Chuen-Tsai S., Chia-Ying C., & Ji-Lung H. (2007) Bridge and brick motifs in complex networks, Physica A, 377, 340–350.

Cohen, J., Cohen P., West, S.G., & Aiken, L.S. (2003). Applied multiple regression/correlation analysis for the behavioral sciences. (3rd ed.) Hillsdale, NJ: Lawrence Erlbaum Associates.

Culotta, A., Bekkerman, R., & McCallum, A. (2004), "Extracting social networks and contact information from email and the Web", CEAS 2004, First Conference on Email and Anti-Spam.

Ehrlich, D.M. (2006), "Social network survey paper," *International Journal of Learning and Intellectual Capital* 3 (2), pp. 167-177.

Flake, G., Lawrence, S., & Lee Giles, C. (2000), Efficient identification of web communities. In: Proceedings of the Sixth ACM SIGKDD International Conference on Knowledge Discovery and Data Mining. Boston, MA, pp. 150–160.

Garton, L., Haythorntwaite, C. & Wellman, B. (1997) 'Studying Online Social Networks', Journal of Computer-Mediated Communication, 3(1), retrieved 1-11-2008 from: http://jcmc.indiana.edu/vol3/issue1/garton.html.

Gibson, D., Kleinberg, J., & Raghavan, P. (1998). Inferring Web communities from link topology. In: Proceedings of the Ninth ACM Conference on Hypertext and Hypermedia.

Golbeck, J. (2005) 'Computing and Applying Trust in Web-Based Social Networks', Dissertation Submitted to the Faculty of the Graduate School of th Universtity of Maryland, College Park in partial fulfillment of the requirements for the degree of Doctor of Philosophy.

Garton, L., Haythorntwaite, C., & Wellman B. (1997), "Studying Online Social Networks," *Journal of Computer-Mediated Communication*, 3 (1), retrieved 1-11-2008 from: http://jcmc.indiana.edu/vol3/issue1/garton.html.

Hanneman, R., & Riddle, M. (2005), *Introduction to social network methods, Online textbook*, retrieved 1-11-2008 from: http://faculty.ucr.edu/~hanneman/nettext/.

Itzkovitz, S., Milo R., Kashtan N., Ziv G., & Alon U. (2003) Subgraphs in random networks. Physical Review E., 68, 026127.

Juszczyszyn, K., Kazienko P., Musiał K., & Gabrys B. (2008), Temporal Changes in Connection Patterns of an Email-based Social Network, IEEE/WIC/ACM Joint International Conference on Web Intelligence and Intelligent Agent Technology 2008, Sydney, Australia, IEEE Computer Society Press.

Juszczyszyn, K., Kazienko P., & Musiał K. (2008),

Local Topology of Social Network Based on Motif Analysis, 12th International Conference on Knowledge-Based and Intelligent Information & Engineering Systems, Springer Lecture Notes in Artificial Intelligence, Zagreb, Croatia.

Kashtan, N., S. Itzkovitz, S., Milo, R., & Alon U. (2004) Efficient sampling algorithm for estimating subgraph concentrations and detecting network motifs. Bioinformatics, 20 (11), 1746–1758.

Kazienko, P., Musiał K., & Zgrzywa A. (2008) Evaluation of Node Position Based on Email Communication. Control and Cybernetics, to appear.

Mangan, S., & Alon, U. (2003) Structure and function of the feedforward loop network motif. Proc. of the National Academy of Science, USA, 100 (21), 11980–11985.

Mangan S., Zaslaver, A. & Alon, U. (2003) The coherent feedforward loop serves as a sign-sensitive delay element in transcription networks. J. Molecular Biology, 334, 197–204.

Milo, R., Itzkovitz S., Kashtan, N., Levitt, R., Shen-Orr, S., Ayzenshtat, I., Sheffer, M., & Alon, U. (2004) Superfamilies of evolved and designed networks. Science 303(5663), 1538–42.

Milo, R., Shen-Orr, S., Itzkovitz, S., Kashtan, N., Chklovskii, D., & Alon, U. (2002) Network motifs: simple building blocks of complex networks. Science, 298, 824–827.

Millen, D., Feinberg, J., & Kerr B. (2005) Social bookmarking in the enterprise, Queue 3(9), ACM Press.

Musiał, K., Kazienko P., & Kajdanowicz, T. (2008) Multirelational Social Networks in Multimedia Sharing Systems. Chapter 18 in N.T.Nguyen, G.Kołaczek, & B.Gabryś (eds.) Knowledge Processing and Reasoning for Information Society, EXIT, Warsaw, 275-292.

Shen-Orr, S., Milo, R., Mangan, S., & Alon, U. (2002) Network motifs in the transciptional regualtion network of Escherichia coli. Nat. Genet., 31, 64–68.

Shetty, J., & Adibi, J. (2005), Discovering Important Nodes through Graph Entropy The Case of Enron Email Database, LinkKDD '05, 3rd International Workshop on Link Discovery, KDD 2005, ACM Press 2005, 74-81.

Valverde, S., Theraulaz, G., Gautrais, J., Fourcassie, V., & Sole R.V. (2006) Self-organization patterns in wasp and open source communities. IEEE Intelligent Systems 21 (2), 36–40.

Wernicke, S. (2006) Efficient detection of network motifs. IEEE/ACM Transactions on Computational Biology and Bioinformatics, 3 (4), 347–359.

Wernicke, S., & Rasche, F. (2006) FANMOD: a tool for fast network motif detection. Bioinformatics, 22 (9), 1152–1153.

Wasserman,S., & Faust,K. (1994) Social network analysis: Methods and applications, Cambridge University Press, New York.

Weng, L.T, Xu Y., & Li Y., "A Framework For E-commerce Oriented Recommendation Systems," *The 2005 International Conference on Active Media Technology, AMT05*, IEEE Press, 2005, pp. 309-314.

Wellman, B., Salaff, J., Dimitrova, D., Garton L., Gulia M., & Haythornthwaite C. (1996) 'Computer Networks as Social Networks: Collaborative Work, Telework, and Virtual Community', Annual Review Sociology, 22, 213 – 238.

Zhu W., Chen, C. & Allen, R.B. (2006), Visualizing an enterprise social network from email, 6th ACM/IEEE-CS joint Conference on Digital Libraries, ACM Press, 383.

This work was previously published in International Journal of Virtual Communities and Social Networking, Vol. 1, Issue 2, edited by S. Dasgupta, pp. 22-35, copyright 2009 by IGI Publishing (an imprint of IGI Global).

Chapter 7.22
"Social Potential" Models for Modeling Traffic and Transportation

Rex Oleson
University of Central Florida, USA

D. J. Kaup
University of Central Florida, USA

Thomas L. Clarke
University of Central Florida, USA

Linda C. Malone
University of Central Florida, USA

Ladislau Boloni
University of Central Florida, USA

ABSTRACT

The "Social Potential", which the authors will refer to as the SP, is the name given to a technique of implementing multi-agent movement in simulations by representing behaviors, goals, and motivations as artificial social forces. These forces then determine the movement of the individual agents. Several SP models, including the Flocking, Helbing-Molnar–Farkas-Visek (HMFV), and Lakoba-Kaup-Finkelstein (LKF) models, are commonly used to describe pedestrian movement. A systematic procedure is described here, whereby one can construct and use these and other SP models. The theories behind these models are discussed along with the application of the procedure. Through the use of these techniques, it has been possible to represent schools of fish swimming, flocks of birds flying, crowds exiting rooms, crowds walking through hallways, and individuals wandering in open fields. Once one has an understanding of these models, more complex and specific scenarios could be constructed by applying additional constraints and parameters. The models along with the procedure give a guideline for understanding and implementing simulations using SP techniques.

INTRODUCTION

Modeling traffic and transportation requires consideration of how individuals move in a given environment. There are three general aspects to consider when looking at movement: reactive behaviors, cognitive behaviors and constraints due to environmental factors. Individual drivers and pedestrians have a general way of dealing with certain situations, some of which comes from experience and some from personality. In this situation, there is generally only one specific response for any given agent. In other situations, one needs to allow an individual to choose from a set of various possible decisions based on how they affect movement and path planning. A final consideration is how the environment will constrain the general movement of the individual.

Much of an individual's movement, especially when driving a vehicle, is reactive. This is due to the fact that most actions are reactions to the conditions of the road and events which are occurring nearby. This is similar to pedestrian movement since walking becomes routine for people. Individuals do not think about every step that they are going to make and every possible outcome, they simply step forward and know the general outcomes they expect. When things deviate from the expected, then their movements are adjusted. Individuals transporting cargo, have a defined origin and destination which requires some decision making such as route planning. There is a goal they are trying to reach, and decisions are made along the way to achieve this goal. We will refer to these as cognitive behaviors, due to the fact that they take some conscious thought to achieve the goal. Techniques of path planning, seeking or organization can be used to represent these choices. The final aspect of movement is the definition of the environment. The individuals need to know where obstacles are and how they interact with them in order to avoid collisions and other unwanted contact.

In multi-agent systems there are numerous techniques which can be used to describe how each agent makes decisions and moves, such as Genetic Programming, Reinforced Learning, Case Based Reasoning, Rules Based Reasoning, Game Theory, Neural Network, Context Based Reasoning, Cellular Automata, and SP. The two primary techniques which are used to represent the decisions of individuals in pedestrian simulations are Cellular Automata and SP.

This chapter will focus on SP techniques for modeling and how to use it to represent individuals' desires and movements during a simulation. A description of the technique is given along with a detailed example of constructing a model from scratch. This will give some insight into the elements of the technique and the process which must be taken to use it effectively. There are a few commonly used models which represent pedestrian movement: Flocking (Reynolds, 1987), HMFV (Helbing, 2002), and LKF (Lakoba, 2005). A brief description of these models will be given along with the forces which are used in the model. Then cognitive behaviors will be discussed which can be added to any of the existing models to create specific desired movements in the individuals. Next, a description of different techniques used to interact with the environment is given. We then conclude by looking at how to apply this technique to more than individuals' movements.

BACKGROUND

Individuals tend to move in predictable manners due to the fact that walking in an environment becomes an automatic process where decisions are made instinctively (Helbing, 2005). People are familiar with walking and the paths they tend to follow. This fact allows for the construction of models which should represent the movement of individuals in reasonably simple terms. The same could be said for traffic and transportation movements, except that the possible movements for these are constrained more than for individuals. Nonetheless, the same techniques can be used for both systems.

One manner of looking at how an object moves is to relate it to the physical forces acting on the object, referred to as Newtonian Mechanics. The SP technique represents the movement of sentient beings by artificial forces between an individual and the environment in the same way Newtonian Mechanics represents movements via physical forces. The SP technique was originally developed as a way of modeling individuals' decisions. One of the earliest uses was in modeling flocks of birds and schools of fish (Reynolds, 1993). Then the techniques were applied to robotic movement and path planning (Herbert, 1998; Lee, 2003; Reif, 1999). The technique is set up to allow the behavior of an individual to be defined through a collection of simple force-like rules. These artificial forces sometimes relate the social interaction between individuals and therefore the name "Social Potential" was given to describe the modeling technique (Reif, 1999). The SP technique originally used potentials to calculate forces and then used these forces to determine the movement of the individuals. Research in the field has shown that forces other then potential based forces might be required (Helbing, 2002) to simulate the movement of some individuals. Therefore we will refer to any model which uses forces to determine the direction of movement as an SP model.

In order to use this technique, the causes of the movements must be identified and then artificial forces representing their effects must be designed. The appropriately designed forces will then define how the individual reacts to each of these causes. The final movement of an individual is then taken to be a superposition of all influencing forces. This separation into an individual force for each cause allows for a simple definition of the individual forces whose sum creates the specific movements in the individuals.

The SP technique treats each individual like a particle; these particles are attracted or repelled from points, obstacles, other individuals, and areas of interest. This technique creates interactions on a microscopic level by simulating the movement of each individual. Treating each individual as a particle allows the creator of the model to focus on what influences a given individual and how to define the reaction of the individual to these influences. This allows for simple definitions and relates these influences to a commonly used technique, Newtonian Mechanics. Groups of individuals can then be simulated by placing numerous individuals into a common area and allowing these individuals to interact. The sum total of all the individual movements then gives rise to the emergent behaviors which is referred to as macroscopic "Crowd Dynamics."

Simple Example

Consider searching for a place to eat when visiting a new location. This would have to be a place where you have never been before therefore you have no previous knowledge of the location of possible places to eat. Now assume that you intend to find a place by wandering around; in this way you will also get to know the area. What factors are going to be important to you?

1. Desire to stay close to the hotel, or where you are staying.
2. Attraction to visible restaurants.
3. Slight repulsion from other individuals.
4. Repulsion from crowded restaurants.

These four factors are identified as the causes for the movements of the individual. Assuming that there are no constraints on where you can walk (no walls or buildings) then there is a simple set of rules governing the movement. These rules are built as a set of forces representing the previously defined factors.

Since you want to stay near the hotel, the further you get away from the hotel, the larger the attraction to the hotel should be. The force keeping you near the hotel should have a form which increases as you get further away, like $f = -a \cdot r_{hotel}$ or $f = -a \cdot e^{b \cdot r_{hotel}}$ where r_{hotel} is the distance from the individual to the hotel and *a and b* are parameters. As you approach an eating establishment your attraction to the establishment should grow in the opposite manner, so the force should have something like the form

$$f = \frac{-c}{r_{restaurant}} \text{ or } f = -c \cdot e^{-d \cdot r_{restaurant}}$$

where *c and d* are parameters with $r_{restaurant}$ being the distance from the individual to the restaurant. Everyone has a certain amount of personal space they attempt to maintain, so they are generally repelled from nearby individuals by something like $f = \frac{g}{r_{individual}}$ or $f = g \cdot e^{-h \cdot r_{restaurant}}$. If you are currently hungry, you know that any crowd at a restaurant generally means a long wait time, so you should be repelled from crowded restaurants. You could represent this by using a force of the form $f = j \cdot (\#of\_individuals_{restaurant})$.

In general it is easiest to start out with the simpler polynomial type forces then try the exponential forces second. However these different forms could produce distinctly different behaviors. Choosing the simple polynomial functions to represent the forces will give a general idea of the movement, so we would take

$$f_1 = -a \cdot r_{hotel}$$

$$f_2 = \frac{-c}{r_{restaurant}}$$

$$f_3 = \frac{g}{r_{individual}}$$

$$f_4 = j \cdot (\#of\_individuals_{restaurant}).$$

The above illustrates the four general forces which one would use to begin simulating the above scenario. There are other forces which should be present, such as a small random force to start the individual looking for a restaurant, and to keep them from moving in perfectly straight lines. There could also be other interactions between individuals as well as interactions to prevent the individual from walking into buildings or other obstacles.

Generally, the approach taken in constructing an SP model involves the three broad steps discussed above. These are:

1. Define the important aspects which need to be modeled.
2. Decide on the types of forces and their functional form which would represent their causes.
3. Determine the appropriate values for the free parameters in the forces which would best represent the system you are trying to model.

If there is no driving reason for choosing a certain functional form for the forces then start as simple as you can. Begin with a simple polynomial and test the application to see if the individuals move in the general manner that you require. Get close approximations of the parameters then see if you need to adjust the types of forces, or possibly even add new forces. These three steps will be iterated numerous times before completing the construction of a model. Since this process can be very time consuming it can be helpful to start from an existing model.

CURRENT SP MODELS

There are only a few standard SP models being used to describe pedestrian movement. There are also models which have been developed for robotic movement (Khatib, 1985; Reif, 1999) which can also be used, but since each model for robotic movement is constructed for a specific goal, we will focus on the general models which are currently used for pedestrian movement. Each model has particular strengths as well as disadvantages, but they can be used as a starting point on which to build your model. These models already have the forces defined for basic movement and certain parameters have been set or bounded. This allows for a simple starting point and reduces the number of free parameter values which one would need to set (or determine) to represent a specific simulation.

Flocking/Herding

Flocking was one of the first recognized models using the SP technique. Craig Reynolds in the 1980s was trying to find a new way of defining movement of computer simulated individuals (Reynolds, 1987). Up to that time the movement of each individual was constructed by hand; this made simulating large numbers of individuals difficult and labor intensive. Reynolds found that he could represent these movements by four simple forces: cohesion, avoidance, flock centering, and a small random force. This simulation was called Boids and did an amazing job of representing both bird flocking and fish schooling

Of the social forces used in this model, cohesion is the force which causes the individuals to stick together; it is a mild attractive force toward other individuals within a local neighborhood of the individual. Avoidance is a repulsive force which balances the cohesive force so as to keep the individuals from running into one another. Flock centering is a force used to bring the individuals into a unified entity. The force representing the flock centering causes each individual to try to get into the center of the individuals it can see. This would give the individual the most protection from the surrounding elements and enemies. A small random force is necessary to prevent an individual from walking in a straight line. This randomness makes the simulation more accurate in portraying the life-like pattern of humans walking.

Flock centering is very noticeable in schools of fish. Since the fish on the edge of the school are most likely to be eaten, these fish constantly push themselves toward the center, thereby pushing the other fish out to the edges (Seghers, 1974). The constant pushing toward the center creates the shape of the school and causes the location of any individual in it to be constantly changing, not only in regard to its surroundings but also with regard to the school itself. Flock centering behavior is not as recognizable in flocks of birds, so in this case, this force is less important and can be given less influence. However, an exception to this is found in penguins. The emperor penguins guard their eggs over the long cold winter; the birds on the edge constantly move in towards the center causing the same cycle-type motion as mentioned above in fish. In this way the penguins keep the entire collection of birds at a reasonable temperature instead of leaving the edge to freeze (Gilbert, 2006).

Current implementations for pedestrian movement generally contain various forms of the above three types of social forces, excluding the random force. Since people do not generally have the need for protection whereby they would struggle to get toward the center, a centering concept as in flocking is not needed. In place of flocking a "consistency force" is added, keeping each individual moving in the direction he/she was generally moving.

A distinction in this model is that velocities are fixed and the forces are only used to determine the *direction* which the individuals will move. The collection of these forces is sometimes called a herding model, since the individuals loosely clump together and thereby act as a single collection, or herd. These forces would typically be of the form:

$$\vec{f}_{consistency} = c \cdot \vec{v}$$

$$\vec{f}_{avoidance} = \frac{\vec{r} \cdot a}{r^3}$$

$$\vec{f}_{cohesion} = -s \cdot \vec{r}$$

$$\vec{F} = \vec{f}_{consistency} + \vec{f}_{avoidance} + \vec{f}_{cohesion}$$

$$\Delta\vec{X} = \vec{F} \cdot \frac{v \cdot \Delta t}{\|F\|}$$

HMFV Model

Helbing, Molnar, Farkas and Vicsek realized that the representation of an individual's movements in a physical environment must consider standard physical forces because contact can occur with other pedestrians or objects (Helbing, 2002). In this model, an individual generally has both types of forces acting on him/her: the physical forces and the social forces. The physical forces are actual forces, like frictional and pushing forces, which occur when two individuals run into or otherwise contact each other, or when an individual collides with an obstacle. The social forces are those which represent how a self-determined individual would want to move. Both classes of forces are necessary in order to obtain realistic movement of individuals and realistic interaction between an individual and obstacles in the environment.

The HMFV model uses three primary forces: social, frictional, and pushing. The social force represents the personal space an individual wishes to keep open around them; it is modeled using exponential decay. The force of friction occurs when the individual contacts another individual or an obstacle. The frictional force on a pedestrian is tangential and opposite to the relative motion between them and the object or other individual. The pushing force occurs due to the fact that in a crowd, packed individuals are slightly compressible and therefore spring, or push back, when pressing on another individual or obstacle. The normal, pushing force is modeled by Hooke's law. The forms for these forces in the HMFV model are given below:

$$\vec{f}_{social} = \vec{r} \cdot a \cdot e^{-r/b}$$

$$\vec{f}_{friction} = \kappa \cdot \vec{N} \times \left[\vec{N} \times \vec{v}\right].$$

$$\vec{f}_{pushing} = \vec{N} \cdot c \cdot \delta r$$

LKF Model

Lakoba, Kaup, and Finkelstein modified the HMFV model by including more physically realistic parameter values in the physical forces (Lakoba, 2005). However when this was done, new issues arose, especially when dealing with different densities of individuals in the simulations. New social forces had to be included in order to create more physically realistic simulations for all densities. The new social forces dealt with the directionality of interactions between individuals as well as the "excitation level" of an individual. The physical forces kept the same basic form as in the original HMFV model (Lakoba, 2005), which are the first two equations listed below.

A description of the other forces introduced by LKF is in order. There are two different forms for the social forces: one for the social force acting on an individual (ind) due to any obstacle (obs) and another for the social force acting on an individual due to the presence of another individual. For the former, the force is a repulsive force along the line from the individual to the center of the obstacle. Its magnitude is given by $wF1(obs) \cdot \max faceToBack \cdot e^{\frac{}{B}}$, with the coefficient $wF1$ as an orientation factor. If the obstacle can be seen then $wF1$ is unity, otherwise as the angle increases from $\pi/2$, the value of the

Table 1. Variables for Flocking model

$\vec{r}$	The distance between two individuals, directed to the individual on which the force acts
r	The magnitude of the distance between two individuals
$\vec{v}$	The velocity of the individual of interest
v	The magnitude of the velocity
$\Delta\vec{X}$	The change in position of an individual
Δt	The time step used for the simulation
c, a, s	Free parameters to adjust strength of the individual forces

$wF1$ decreases to the value of b, which is defined below. The value of b is reached when the angle is π. The quantity max *faceToBack* represents the maximum value of this force when the individual is facing the back of an obstacle. For the social force between individuals, the form is the same except that the additional factor $wF3$ is included and has the effect of replacing max *faceToBack* with max *faceToFace* when the individual can see the other. The velocity of an individual is defined by the excitation of the individual, their current speed, and the average speed of nearby individuals. The excitation of the individual is also allowed to change over time, and this is based on the current excitation and the ability of the individual to move at their initial velocity. The function $wF3$ is defined as max *faceToBack* , or max *faceToFace* depending on if the entity in question can be seen. This notation given here is a change from the notation in Lakoba (2005), but without changing the value of any force they used. This only simplifies the definition of the social force acting on an individual, and causes $wF3$ to be nothing more than a switch from max *faceToFace* to max *faceToBack* .

$$\vec{f}_{friction} = \kappa \cdot \vec{N} \times \left[\vec{N} \times \vec{v}\right]$$

$$\vec{f}_{pushing} = \vec{N} \cdot c \cdot \delta r$$

$$\vec{f}_{social\_obs}(obs) = \vec{N} \cdot wF1(obs) \cdot \max faceToBack \cdot e^{\frac{-r}{B}}$$

$$\vec{f}_{social\_ind}(ind) = \vec{N} \cdot wF1(ind) \cdot wF3(ind) \cdot \left[\frac{e^{\frac{-r}{B}}}{m}\right]$$

where

Table 2. Variables for HMFV model

$\vec{N}$	The outward normal vector from the object or other individual, located at the point of contact
δr	The magnitude of overlap between an individual on the object of interest
a, b, c, k	Free parameters to adjust the strengths or ranges of the various forces

$$wF1(entity) = \begin{cases} 1 & canSee(entity) \\ 1 - \frac{2}{\pi}\left[\theta - \frac{\pi}{2}\right][1-b] & else \end{cases}$$

$$wF3(entity) = \begin{cases} \max faceToFace & canSee(entity) \\ \max faceToBack & else \end{cases}$$

$$\max faceToFace = F\left[1 + \frac{1}{e_{\max}}\right]\left[[1-\tilde{\rho}] + k2 \cdot \tilde{\rho}\right]$$

$$\max faceToBack = w_0 \cdot m \cdot [1 + e_{\max}] \cdot e^{\frac{-1}{B}\left[D - \frac{1}{\rho_{\max}}\right]}\left[k_0[1-\tilde{\rho}] + k1 \cdot \tilde{\rho}\right]\frac{1}{\tau[1-b]}$$

$$\vec{v}_0(t) = \vec{\theta}_g [1 + E(t)] v_{preferred} [1-p] + p \cdot \vec{v}_{local}$$

$$E(t) = E(t - \Delta t) - \Delta t \cdot \left[\frac{E(t-\Delta t)}{T} + \frac{e_{\max}}{T}\frac{1 - \|\vec{v}\|}{v_{preferred}}\right]$$

Some of these symbols have already been defined above. The new symbols introduced are given in Table 3 just below along with the values of the parameters used in the original LKF model.

COGNITIVE BEHAVIORS

Cognitive behavior forces are forces that can be added to the individual to create specific directional choices. These are things like wandering, seeking, following a path, or following a wall. They are considered cognitive behavioral forces due to the fact that the individual is making a decision using these forces; they are not purely reactive style forces.

Wander

Wander is sometimes referred to as a random walk. This type of force is generally needed in order to keep an individual from walking in a perfectly straight line. Basically it creates small deviations from the path the individual would otherwise take (Reynolds, 1999).

One method of applying this technique is to choose a small maximum angle (θ max) of deviation inside of which one would place an artificial attraction point and then add the force from the attraction point to the other forces acting on the individual (Figure 1). The strength of the force can be adjusted by choosing the distance (d) the artificial attraction point is placed from the center of the individual. For example:

$$\theta = (random * 2\theta \max) - \theta \max$$
$$f = (d * \cos\theta, d * \sin\theta)$$

where *random* is a randomly selected number between 0 and 1.

This force should never be so large that the individual will not follow the path at all; this is supposed to be small deviations in the movement of the individual as they follow the main path. The main path should still be selected by other forces.

The previous example is capable of creating a jittery movement in the individual. For a smoother movement, one could pick θ such that it would have a pattern instead of being purely random (Hebert, 1998). For example: $\theta(t) = \theta \max * \cos(t)$, would create a smooth, wave-like motion around the path instead of the jitter due to a random selection (Ueyama, 1993). The trigonometric function can be adjusted to modify the frequency of the wander.

Seek (Flee)/Pursue (Evade)

Seek (Flee)/Pursue (Evade) occurs when an individual either tries to head toward an individual of interest or away from an individual of interest. This is different from the standard attraction and repulsion between individuals in that it is a *selected* attraction or repulsion. If a man saw someone selling fruits when he was looking for an apple, then he would be attracted to that particular vendor, hence seeking or pursuing the vendor. If someone was being followed and was trying to not be caught, then they would be evading or fleeing. This is a technique used in predator/prey style simulations (Isaacs, 1999). The key feature to these behaviors is to predict where the individual (either following or being followed) will be at some point of time in the future. The point of attraction will actually be to the projected position and not the current position. If the pursuer goes to the point where the evader is currently at, then no matter how fast he is traveling he will never reach the evader. This is because the evader will have moved a little bit, and therefore will be just outside of the reach of the pursuer. This is why the pursuer must move to the projected location of the evader. For the evader, the force would be structured like

$$\vec{f} = \frac{a \cdot \vec{r}}{\|\vec{r}\|^b}$$

where

$$\vec{r} = \vec{X}_{evader(t)} - \vec{X}_{pursuer(t+\Delta t)}.$$

Similarly, for the pursuer the force would be

$$\vec{f} = \frac{-a \cdot \vec{r}}{\|\vec{r}\|^b}$$

where

$$\vec{r} = \vec{X}_{evader(t)} - \vec{X}_{pursuer(t+\Delta t)}.$$

Path Following

Path following is also sometimes referred to as "way-point based path planning". This is the ability to set up distinct way points to define a path that an individual will follow as he/she progresses to a destination point. In some ways, this goes against the idea of SP technique movement models in that the path is *not* determined by the forces. This technique can be very useful in planning out available routes that an individual can choose or to give an individual an idea of where movement should occur in an environment. The individual following the path must know the waypoints and the order in which to follow them. At the start, the individual gets an attraction force to the first waypoint. Once the individual gets close to the waypoint, the first attraction force is turned off and the attraction to the next waypoint in the list is turned on. This progression continues until the individual has passed all of the waypoints in the path. This is a way to create queues or lines in a simulation.

Wall Following

Wall following is a method which has been used for years to get out of mazes. Upon entering a maze, place a hand on one of the walls that touches the entry way, and continue to follow that wall. If you had started from the beginning of the maze then you are guaranteed to find the exit. On the other hand, if you were dropped into the middle of the maze, you could still use this principle. First, you would have to place a hand on a wall and mark where you are. Then follow that wall and if you found that you returned to the exact same spot, then you would move to the other wall and repeat the scenario. If you found that you returned once again to the exact same spot then both walls are interior walls and the technique fails because you are basically stuck inside a room with no doors. Otherwise, you will eventually find your way out.

Table 3. Variables for LKF model

Variable	Description
$b = 0.3$	Back-to-front ratio of perception
$B = 0.5m$	Approximate fall off length (personal space) for the social forces
$D \approx 0.7m$	The diameter of the individual
E	The excitation state of the individual
$e_{\max} = \frac{v_i}{w_0} \approx 1$	The maximum value allowed for the excitement parameter (E)
$m \approx 80kg$	The average mass of an individual
$p \in (0,1)$	The parameter representing the independence of an individual (does not change through the simulation)
ρ	The number of individuals inside a circle of radius B around the individual of interest, divided by the area πB^2
$\tilde{\rho} = \rho \cdot \frac{\pi D^2}{4}$	The non-dimensionalized density of individuals
$\rho_{\max} = 5.4 people / m^2$	Maximum allowable density of people per square meter (Weidmann, 1992)
$T = 2s$	The lag time for excitement to return to initial state when unaffected
$\tau = 0.2s$	The average reaction time of a person
θ	The angle between θ_g and r
$\vec{\theta}_g$	The vector representing the direction the individual is looking
$\vec{v}$	The velocity of the individual
$v_{preferred}$	The individual's preferred speed. The values used in LKF were 1.5, 3.0, and 4.5 m/s.
$\vec{v}_0$	The preferred velocity of the individual
$\vec{v}_{local}$	The average velocity of individuals in the local neighborhood
$w_0 = 1.34m/s$	Average walking speed of a non-panicked individual
$k_0 = 0.3$ $k_1 \in [1.2, 2.4]$ $k_2 = 1.5$	Parameters to adjust high density corrections for face-to-back orientation
c, F, k	Free parameters to adjust strength of the individual forces

Figure 1. Wander example

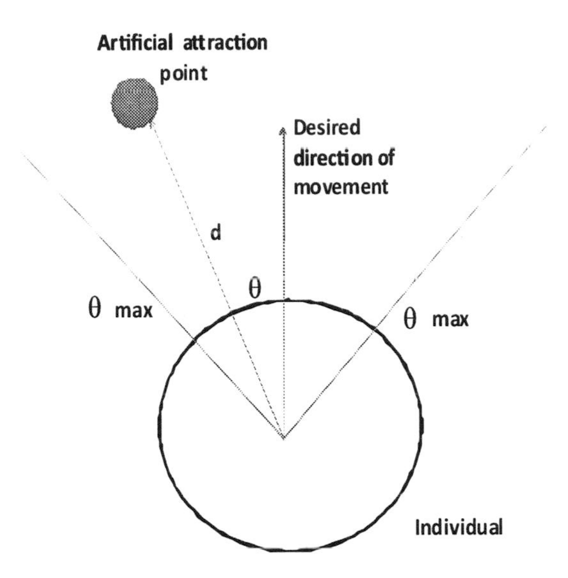

In simulations, wall following becomes useful because when one is using social forces to represent the movement, an individual can become stuck in closed areas and at corners. The individuals have to get out of these areas before they can reach their goal (e.g. there could be an obstacle between the individual and their goal that they would have to go around before they could reach their goal). Wall following can create the necessary break-out condition to move the individual out of these trapped situations and allow them to continue toward the intended goal.

One way to do this in a simulation is to set up an artificial attraction point which is parallel to the obstacle and in the direction of the individual's movement (Figure 4). This new point is there to pull the individual along the wall. This force should become active only when the individual is within a given distance of a wall and then should flip to repulsive once he/she gets too close to the wall. This will allow the individual to keep a given distance from the obstacle that the individual is walking along.

The second option (Figure 5) for applying wall following does not use an artificial point of attraction, but rather just modifies the calculated forces to cause the desired movement. First, the movement is calculated as originally defined to get a direction and magnitude; this is the calculated force vector. Next, the line is found which goes through the center of the individual of interest and is parallel to the obstacle of interest. Finally, the calculated force vector is projected onto the parallel line. This forces all movements to be parallel to the obstacle of interest. Using this approach, the individual will not follow the wall at all times, but will only follow a wall when the wall is impeding the individual's movements toward a given goal.

Both of these techniques can be very useful when trying to manuever around obstacles and explore environments. Some decisional logic must sometimes be included when two obstacles touch each other so that the individual will interact with the correct obstacle.

Figure 2. Seek Flee example

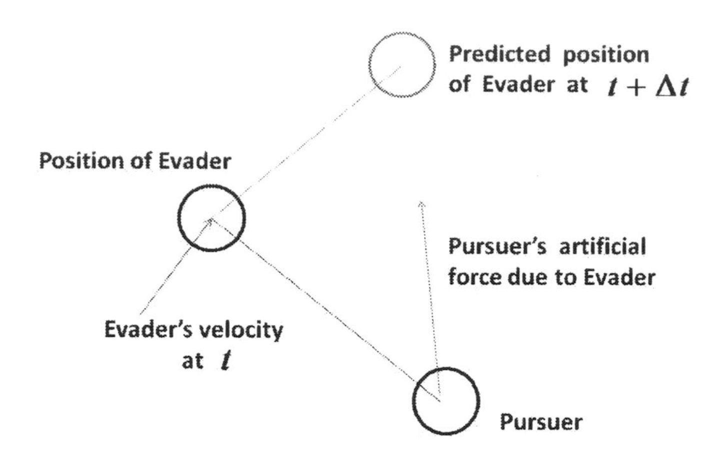

ENVIRONMENTAL FEATURES

The environment is a collection of geometric objects the individual must interact with, usually by avoiding them. The following are obstacles found in the simulation that define the environment in which individuals must maneuver.

Obstacles

An obstacle should have an external shape described in some manner such that the distance to points on it can be found. Also, obstacles should have a center. It is best to keep the definition of the obstacles to simple structures like rectangles and circles. Using pixilation principles defined for computer graphics, it is reasonably easy to represent all possible shapes by these two primitive structures (Pineda, 1988).

Walls

Walls are simply rectangular obstacles placed where a wall should occur in the simulation. There are some key points to consider though, primarily, what happens at the intersection of two walls. You do not want individuals walking between two connected walls, so make sure that there is no gap whatsoever between the two walls. Even a gap of a few centimeters could possibly be recognized and the individuals could attempt to squeeze between the two walls. This scenario can cause many problems in the simulation, and is sometimes very difficult to recognize. A simple solution to this is to always have the walls overlap slightly. This removes all possibility of an individual squeezing in between the walls.

Paths

Paths were described previously as a collection of waypoints the individual follows. Paths can be constructed as part of the environment and then handed to individuals when they need to use them. Consider an amusement park with five different rides. Each ride has a waiting line, and therefore each ride would have a path associated with it. These paths could reside as part of the environment. Once an individual decides to go on a given ride, a copy of the ride's associated path gets assigned to the individual. In this manner, the paths are part of the environment and the individuals only use these paths when they become of interest or are needed by the individuals being simulated (Lee, 2003).

Figure 3. Path following example

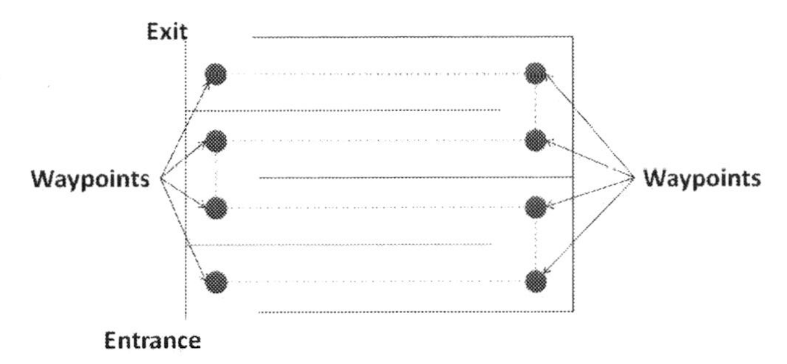

Moving Obstacles

There is nothing that restricts an obstacle from moving. It is possible to define a simulation where the obstacles move regularly, like a train at a train station, or with a more complicated description, like vehicles at an intersection. As long as the descriptions for the movement are defined on the same time step (Δt) as the SP models, the two different entities can interact simultaneously. Also, if any obstacles need to move, they could be defined as a different type of entity having a given movement pattern with all other obstacles being stationary. Either approach is valid; it depends on what is being modeled and which approach fits the scenario the best.

Regions

Sometimes there are areas, or regions, in an environment where certain events should happen or where certain effects occur on an individual. These can be constructed in a manner similar to the technique used in video games where a region of effect is created and all individuals within that region are affected. To do this, define a region as an obstacle in the environment which has no attraction or repulsion. Associate a given effect with this obstacle. This effect could be a speed reducer to represent tough terrain, or it could be a more mild repulsive force to represent an area where an individual would not like to enter. These regions could be associated with given individuals or all individuals to allow for a large variation in simulation scenarios.

Interactions

How an individual recognizes other entities and obstacles in the environment is a very important aspect to the simulation. There are a few different techniques used: centroidal, subdivision, force field, axial, and centroid with axial.

Centroidal

Traditionally the obstacles are treated as point masses (Reynolds, 1987) and are usually located at the center of the obstacle. This is similar to the way an introductory course in physics simplifies the features of Newtonian Mechanics. In progressing through the levels of physics, one learns that dealing with everything as only point masses is a drastic over-simplification to the system. This simplification can cause erroneous results or leave out important dynamics of the system.

Subdivision

Here, the environment is subdivided into small cells. Once the grid is developed, the obstacles are intersected with the grid and any cell of the grid intersecting with an obstacle is considered to be an obstacle. The grid divisions need to be

Figure 4. Wall following option 1

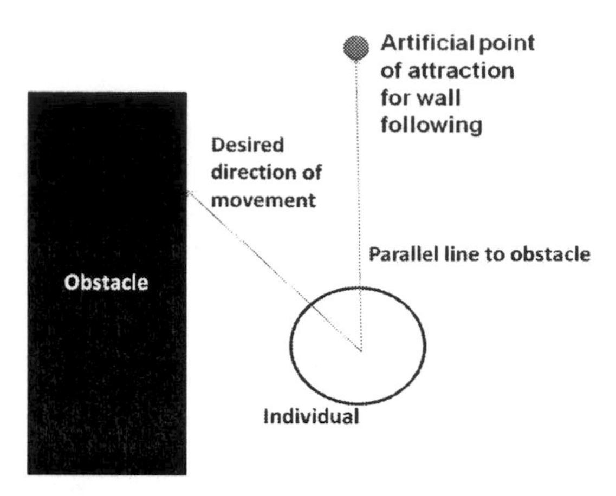

Figure 5. Wall following option 2

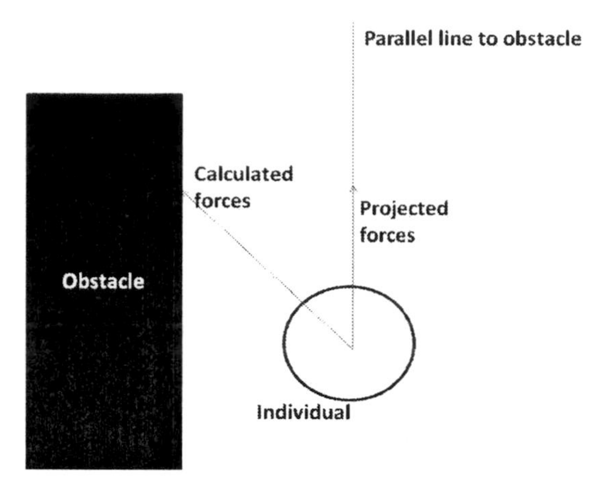

chosen according to the size of the obstacles and their general shapes. SP models are computationally dependent on the number of entities in the simulation since forces are calculated for each obstacle and for each individual. Because of these two factors, this grid-type division of the environment makes the calculations of movement for an individual in the simulation much more time consuming.

Force Field

If the environment is static, a force field can be generated from the environment definition. This technique can combine the information on strength of attraction/repulsion and overall shape of all obstacles in the environment (Gazi, 2005; Khatib, 1985). A map of the obstacles and their forces can then be used to determine the social forces on an individual due to the static environment. Once the map is generated, it can be referenced by the location of the individual, and the values for the social forces would be retrieved. The disadvantage here is that it takes a lot of work to define the environment and then to pre-calculate the necessary force fields to represent that environment.

Axial

This technique is based upon ray tracing concepts in computer graphics, but only a discrete number of rays are shot. This technique was used by Craig Reynolds who would shoot a single ray in the direction that the individual was moving and then check to see if it intersected any obstacles in the environment (Reynolds, 1993). The point of contact with the ray and the obstacle is the point used to calculate the interactions. Checking in the four axial directions gives a better idea of what was happening around an individual, instead of just what was occurring in front of the individual. The ray in the direction of movement could still be included but was not found to be that useful. This technique works reasonably well, but can miss a large number of obstacles which should be "seen" by an individual.

Centroid with Axial

Since all objects in a given vicinity of the individual are important, only checking in the axial directions is insufficient. The centroid with axial technique starts with gathering a collection of all obstacles in the known vicinity of the individual. The centroidal distance to the first obstacle is calculated. Next, one checks the four axial directions and calculates the distance to that obstacle. Then, one takes the minimum of the centroidal and axial distances and uses the point associated with that distance to calculate the social forces. Repeat this process for all obstacles in the vicinity of the individual. In this way, one is guaranteed to locate at least one point of interest for any obstacle near the individual.

Interacting with Various Models

A given SP model can be used simultaneously with a different SP model or even a different type of model altogether. SP models are continuous models discretized in time. The key point in ensuring that two continuous models work reasonably well together is they must have the same time step (Δt). In contrast, if the other model is a discrete model, like a Cellular Automata, the time step for the discrete model should be a multiple of the time step being used for the continuous models. If that is not possible, have the discrete model execute the first time step occurring after its execution should have occurred. Take care to make sure that the speeds and sizes of the individuals are in agreement between the models and then they can work reasonably well together.

CONCLUSION

SP techniques are very useful in describing the movements of individuals. The procedures described have been used to implement various models and to look at how individuals might be expected to react to given environments. However, it can easily be expanded and applied to individuals driving a car, riding a bicycle, etc. Recently Majid Ali Khan, Damla Turgut and Ladislau Bölöni (Khan 2008) have demonstrated the use of the SP technique for simulating trucks driving in highway convoys.

The mathematics of the models presented has been condensed, where needed, to allow for simpler implementations and easier understanding of the process of the SP techniques. These simplifications allowed relationships between interactions with obstacles and with other individuals to be apparent and quickly defined. Anytime individuals are in control of their movement and need to make decisions while simultaneously being constrained by the environment, SP models can be constructed to represent how individuals would tend to move.

Environments representing exiting rooms, walking in hallways, exiting gated areas, and wandering in a room have been visualized and simulated using this technique. By adding new parameters to existing models, ages and certain social characteristics were represented (Jaganthan, 2007; Kaup, 2006; Kaup, 2007). This has allowed the exploration of how environmental changes can affect different types of individuals. Differing exit strategies have been studied to see if environmental factors can be used to increase the efficiency of an exit. All of these results demonstrate the usefulness and applicability of the procedures described for the SP technique.

It also provides the possibility of eventually testing and validating social interaction theories. Given any theory, one could directly model that theory by programming a simulation so that the agents would respond per that theory. Then by running the simulation, one could observe what social structure(s) would arise.

FUTURE RESEARCH

Plans exist to continue to study additional parameters which can be included in current models to allow the design of better simulations for describing cultural and social differences. To do this correctly, one needs to have some reasonable measure by which one could determine whether or not two different simulations were sufficiently similar, as well as how close any one given simulation would compare to a real world event. Such methods need to be designed as quantitatively as possible.

As a first approach in this direction, videos have been created and gathered of various pedestrian movements in various venues, with the intention of gathering data from these videos which could be used for comparing simulations of these venues to real world videos of the same venue. A technique for doing this has been developed which is still in the testing phase. Preliminary results are encouraging.

REFERENCES

Gazi, V. (2005). Swarm aggregations using artificial potentials and sliding-mode control. *IEEE Transactions on Robotics* , (pp. 1208–1214).

Gilbert, C., Robertson, G., Le Maho, Y., Naito, Y., & Ancel, A. (2006). Huddling behavior in emperor penguins: Dynamics of huddling. *Physiology & Behavior , 88* (4-5), 479-488.

Hebert, T., & Valavanis, K. (1998). Navigation of an autonomous vehicle using an electrostatic potential field. *In Proceedings of the 1998 IEEE International Conference on Control Applications, 2*, 1328–1332.

Helbing, D., Buzna, L., Johansson, A., & Werner, T. (2005). Self-Organized Pedestrian Crowd Dynamics: Experiments, Simulations, and Design Solutions. *Transportation Science , 39*, 1-24.

Helbing, D., Farkás, I. J., Molnár, P., & Vicsek, T. (2002). Simulation of pedestrian crowds in normal and evacuation situations. In M. Schreckenberg, & S. D. Sharma (Eds.), *Pedestrian and Evacuation Dynamic* (pp. 21–58). Berlin, Germany: Springer.

Isaacs, R. (1999). *Differential Games: A Mathematical Theory with Applications to Warfare and Pursuit, Control and Optimization.* New York: Dover Publications.

Jaganthan, S., Clarke, T. L., Kaup, D. J., Koshti, J., Malone, L., & Oleson, R. (2007). Intelligent Agents: Incorporating Personality into Crowd Simulation. *I/ITSEC Interservice/Industry Training, Simulation & Education Conference.* Orlando.

Khan, M. A., Turgut, D., & Bölöni, L. (2008). A study of collaborative influence mechanisms for highway convoy driving. 5th Workshop on AGENTS IN TRAFFIC AND TRANSPORTATION. Estoril, Portugal.

Khatib, O. (1985). Real-time obstacle avoidance for manipulators and mobile robots. *In IEEE International Conference on Robotics and Automation , 2*, 500–505.

Kaup, D. J., Clarke, T. L., Malone, L., & Oleson, R. (2006). Crowd Dynamics Simulation Research. *Summer Simulation Multiconference.* Calgary, Canada.

Kaup, D. J., Clarke, T. L., Malone, L., Jentsch, F., & Oleson, R. (2007). Introducing Age-Based Parameters into Simulations of Crowd Dynamics. *American Sociological Association's 102nd Annual Meeting* . New York.

Lakoba, T., Kaup, D., & Finkelstein, N. (2005). Modifications of the Helbing-Molnr-Farkas-Vicsek Social Force Model for Pedestrian Evolution. *Simulation , 81*, 339–352.

Lee, J., Huang, R., Vaughn, A., Xiao, X., Hedrick, K., Zennaro, M., et al. (2003). Strategies of path-Planning for a UAV to track a ground vehicle. *AINS Conference* .

Pineda, J. (1988). A parallel algorithm for polygon rasterization. *SIGGRAPH '88: Proceedings of the 15th annual conference on Computer graphics and interactive techniques*, (pp. 17-20).

Reif, J., & Wang, H. (1999). Social potential fields: A distributed behavioral control for autonomous robots. *Robotics and Autonomous Systems, 27*(3), 171-194).

Reynolds, C. W. (1993). An Evolved, Vision-Based Behavioral Model of Coordinated Group Motion. *From Animals to Animats, Proc. 2nd International Conf. on Simulation of Adaptive Behavior.* Cambridge, MA: MIT Press.

Reynolds, C. W. (1987). Flocks, herds and schools: A distributed behavioral model. *SIGGRAPH Computer Graphics , 21*, 25-34.

Reynolds, C. W. (1999). Steering behaviors for autonomous characters. In A Yu (Ed.), *Proceedings of the 1999 Game Developer's Conference* (pp. 763 – 782). San Francisco, CA: Miller Freeman.

Seghers, B. H. (1974). Schooling Behavior in the Guppy (Poecilia reticulata): An Evolutionary Response to Predation. *Evolution , 28*, 486-489.

Ueyama, T., & Fukuda, T. (1993). Self-organization of cellular robots using random walk with simple rules. *In Proceedings of 1993 IEEE International Conference on Robotics and Automation , 3*, 595-600.

Weidmann, U. (1992). Transporttechnik der Fussgänger. *Zürich: Institutfür Verkehrsplanung.*

ADDITIONAL READING

Bachmayer, R., & Leonard, N. (2002). Vehicle networks for gradient descent in a sampled environment. *In Proceedings of the 41st IEEE Conference on Decision and Control, 1*, 112–117.

Barraquand, J., Langlois, B., & Latombe, J. (1991). Numerical potential field techniques for robot path planning. *In Fifth International Conference on Advanced Robotics, 2*, 1012–1017.

Breder JR., C. M. (1954). Equations Descriptive of Fish Schools and Other Animal Aggregations. *Ecology , 35*(2), 361-370.

Flacher, F., & Sigaud, O. (2002). Spatial Coordination through Social Potential Fields and Genetic Algorithms. *In In From Animals to Animats 7: Proceedings of the Seventh International Conference on Simulation of Adaptive Behavior* (pp. 389–390). Cambridge, MA: MIT Press.

Ge, S., & Cui, Y. (2000). New potential functions for mobile robot path planning. *IEEE Transactions on Robotics and Automation , 16*(5), 615–620.

Hamacher, H., & Tjandra, S. (2002). Mathematical Modelling of Evacuation Problems: A State of the Art. In M. Schreckenberg, & S. D. Sharma (Eds.), *Pedestrian and Evacuation Dynamcis* (pp. 228-266). Berlin: Springer.

Heigeas, L., Luciani, A., Thallot, J., & Castagne, N. (2003). A physically-based particle model of emergent crowd behaviors. *Graphikon* .

Helbing, D. (1991). A mathematical model for the behavior of pedestrians. *Behavioral Science* (36), 298–310.

Hoogendoorn, S., Bovy, P., & Daamen, W. (2002). Microscopic Pedestrian Wayfinding and Dynamics Modelling. In M. Schreckenberg, & S. Sharma, *Pedestrian and Evacuation Dynamics* (pp. 123-154). Berlin: Springer.

Kachroo, P., Al-nasur, S. J., Wadoo, S. A. & Shende, A. (2008). Pedestrian Dynamics: Feedback Control of Crowd Evacuation. Berlin: Springer.

Kirkwood, R., & Robertson, G. (1999). The Occurrence and Purpose of Huddling by Emperor Penguins During Foraging Trips. *Emu* , 40–45.

Kitamura, Y., T. Tanaka, F. K., & Yachida, M. (1995). 3-D path planning in a dynamic environment using an octree and an artificial potential field. *In International Conference on Intelligent Robots and Systems, 2*, 474–481.

Leonard, N. E., & Fiorelli, E. (2001). Virtual Leaders, Artificial Potentials and Coordinated Control of Groups. *Proceedings of the 40th IEEE Conference on Decision and Control*, (pp. 2968–2973).

Luce, R. D., & Howard, R. (1989). *Games and Decisions*. New York: Dover.

Millonig, A., & Schechtner, K. (2005). Decision Loads and Route Qualities for Pedestrians - Key Requirements for the Design of Pedestrian Navigation Services. In N. Waldau, P. Gattermann, H. Knoflacher, & M. Schreckenberg, *Pedestrian and Evacuation Dynamics 2005* (pp. 109-118). Berlin: Springer.

Rimon, E., & Koditschek, D. (1982). Exact robot navigation using artificial potential functions. *IEEE Transactions on Robotics and Automation*, *8*, 501–518.

Russo, F., & Vietta, A. (2002). Models and Algorithms for Evacuation Analysis in Urban Road Transportation Systems. In M. Schreckenberg (Ed.), & S. D. Sharma (Ed.), *Pedestrian and Evacuation Dynamcis* (pp. 315-322). Berlin: Springer.

Schreckenberg, M. (Ed.), & Sharma, S. D. (Ed.) (2001). Pedestrian and Evacuation Dynamcis. Berlin: Springer.

Thalmann, D., Musse, S. R., & Kallmann, a. M. (1999). Virtual humans behaviour: Individuals, groups, and crowds. In Proceedings of Digital MediaFutures. Bradford In.

Waldau, N. (Ed.), Gattermann, P. (Ed.), Knoflacher, H. (Ed.), & Schreckenberg, M. (Ed.) (2007). Pedestrian and Evacuation Dynamcis 2005. Berlin: Springer.

Vasudevan, C., & Ganesan, K.. (1996). Case-based path planning for autonomous underwater vehicles. *Autonomous Robots* , *3*(2-3), 79-89.

This work was previously published in Multi-Agent Systems for Traffic and Transportation Engineering, edited by A.L.C. Bazzan; F. Klügl, pp. 155-175, copyright 2009 by Information Science Reference (an imprint of IGI Global).

Chapter 7.23
Increasing Capital Revenue in Social Networking Communities:
Building Social and Economic Relationships through Avatars and Characters

Jonathan Bishop
Glamorgan Blended Learning Ltd. & The GTi Suite & Valleys Innovation Centre & Navigation Park & Abercynon, UK

ABSTRACT

The rise of online communities in Internet environments has set in motion an unprecedented shift in power from vendors of goods and services to the customers who buy them, with those vendors who understand this transfer of power and choose to capitalize on it by organizing online communities and being richly rewarded with both peerless customer loyalty and impressive economic returns. A type of online community, the virtual world, could radically alter the way people work, learn, grow consume, and entertain. Understanding the exchange of social and economic capital in online communities could involve looking at what causes actors to spend their resources on improving someone else's reputation. Actors' reputations may affect others' willingness to trade with them or give them gifts. Investigating online communities reveals a large number of different characters and associated avatars. When an actor looks at another's avatar they will evaluate them and make decisions that are crucial to creating interaction between customers and vendors in virtual worlds based on the exchange of goods and services. This chapter utilizes the ecological cognition framework to understand transactions, characters and avatars in virtual worlds and investigates the exchange of capital in a bulletin board and virtual. The chapter finds strong evidence for the existence of characters and stereotypes based on the ecological cognition framework and empirical evidence that actors using avatars with antisocial connotations are more likely to have a lower return on investment and be rated less positively than those with more sophisticated appearing avatars.

INTRODUCTION

The rise of online communities has set in motion an unprecedented power shift from goods and services vendors to customers according to Armstrong and Hagel (1997). Vendors who understand this power transfer and choose to capitalize on it are richly rewarded with both peerless customer loyalty and impressive economic returns they argue. In contemporary business discourse, online community is no longer seen as an impediment to online commerce, nor is it considered just a useful Web site add-on or a synonym for interactive marketing strategies. Rather, online communities are frequently central to the commercial development of the Internet, and to the imagined future of narrowcasting and mass customization in the wider world of marketing and advertising (Werry, 2001). According to Bressler and Grantham (2000), online communities offer vendors an unparalleled opportunity to really get to know their customers and to offer customized goods and services in a cost executive way and it is this recognition of an individual's needs that creates lasting customer loyalty. However, if as argued by Bishop (2007a) that needs, which he defines as pre-existing goals, are not the only cognitive element that affects an actor's behavior, then vendors that want to use online communities to reach their customers will benefit from taking account of the knowledge, skills and social networks of their customers as well.

According to Bishop (2003) it is possible to effectively create an online community at a click of a button as tools such as Yahoo! Groups and MSN Communities allow the casual Internet user to create a space on the Net for people to talk about a specific topic or Interest. Authors such as Bishop have defined online communities based on the forms they take. These forms range from special interest discussion Web sites to instant messaging groups. A social definition could include the requirement that an information system's users go through the membership lifecycle identified by Kim (2000). Kim's lifecycle proposed that individual online community members would enter each community as visitors, or "Lurkers." After breaking through a barrier they would become "Novices," and settle in to community life. If they regularly post content, they become "Regulars." Next, they become "Leaders," and if they serve in the community for a considerable amount of time, they become "Elders." Primary online community genres based on this definition are easily identified by the technology platforms on which they are based. Using this definition, it is possible to see the personal homepage as an online community since users must go through the membership lifecycle in order to post messages to a 'guestbook' or join a 'Circle of Friends'. The Circle of Friends method of networking, developed as part of the VECC Project (see Bishop, 2002) has been embedded in social networking sites, some of which meet the above definition of an online community. One of the most popular genres of online community is the bulletin board, also known as a message board. According to Kim (2000), a message board is one of the most familiar genres of online gathering place, which is asynchronous, meaning people do not have to be in the same place at the same time to have a conversation. An alternative to the message board is the e-mail list, which is the easiest kind of online gathering place to create, maintain and in which to participate (ibid). Another genre of online community that facilitates discussion is the Chat Group, where people can chat synchronously, communicating in the same place at the same time (Figallo, 1998). Two relatively new types of online community are the Weblog and the Wiki. Weblogs, or blogs are Web sites that comprise hyperlinks to articles, news releases, discussions and comments that vary in length and are presented in chronological order (Lindahl & Blount, 2003). The community element of this technology commences when the owner, referred to as a 'blogger', invites others to comment on what he/she has written. A Wiki, which is so named through taking the first letters from the axiom, 'what I know is', is a collaborative page-editing tool with which users may add or edit content directly through their Web browser (Feller, 2005). Despite their newness, Wikis could be augmented with older models of

Table 1. Advantages and disadvantages of specific online community genres

Genre	Advantages/Disadvantages
Personal Homepage	Advantages: Regularly updated, allows people to re-connect by leaving messages and joining circle of friends Disadvantage: Members often need to re-register for each site and cannot usually take their 'Circle of Friends' with them.
Message Boards	Advantages: Posts can be accessed at any time. Easy to ignore undesirable content. Disadvantages: Threads can be very long and reading them time consuming
E-mail Lists and News-letters	Advantages: Allows a user to receive a message as soon as it is sent Disadvantages: Message archives not always accessible.
Chat Groups	Advantages: Synchronous. Users can communicate in real time. Disadvantages: Posts can be sent simultaneously and the user can become lost in the conversation.
Virtual Worlds and Simulations	Advantages: 3-D metaphors enable heightened community involvement Disadvantages: Requires certain hardware and software that not all users have
Weblogs and Direc-tories	Advantages: Easily updated, regular content Disadvantages: Members cannot start topics, only respond to them
Wikis and Hypertext Fiction	Advantages: Can allow for collaborative work on literary projects Disadvantages: Can bring out the worst in people, for example, their destructive natures

hypertext system. A genre of online community that has existed for a long time, but is also becoming increasingly popular is the Virtual World, which may be a multi-user dungeon (MUD), a massively multiplayer online role-playing game (MMORG) or some other 3-D virtual environment. through taking the first letters from the axiom, 'what I know is', is a collaborative page-editing tool with which users may add or edit content directly through their Web browser (Feller, 2005). Despite their newness, Wikis could be augmented with older models of hypertext system. A genre of online community that has existed for a long time, but is also becoming increasingly popular is the Virtual World, which may be a multi-user dungeon (MUD), a massively multiplayer online role-playing game (MMORG) or some other 3-D virtual environment.

Encouraging Social and Economic Transactions in Online Communities

According to Shen et al. (2002), virtual worlds could radically alter the way people work, learn, grow, consume and entertain. Online communities such as virtual worlds are functional systems that exist in an environment. They contain actors, artifacts, structures and other external representations that provide stimuli to actors who respond (Bishop, 2007a; 2007b; 2007c). The transfer of a response into stimuli from one actor to another is social intercourse and the unit of this exchange is the transaction (Berne, 1961; 1964). A transaction is also the unit for the exchange of artifacts between actors and is observed and measured in currency (Vogel, 1999). Transactions can be observed in online communities, most obviously in virtual worlds, where actors communicate with words and trade goods and services. Research into how consumers trade with each other has considered online reputation, focusing on how a trader's reputation influences trading partner's trust formation, reputation scores' impact on transactional prices, reputation-related feedback's effect on online service adoption and the performance of existing online reputation systems (Li et al., 2007). According to Bishop (2007a), encouraging participation is one of the greatest challenges for any online community provider. There is a large amount of literature demonstrating ways in which online communities can be effectively built (Figallo, 1998; Kim, 2000; Levine-Young & Levine, 2000;

Preece, 2000). However, a virtual community can have the right tools, the right chat platform and the right ethos, but if community members are not participating the community will not flourish and encouraging members to change from lurkers into elders is proving to be a challenge for community providers. Traditional methods of behavior modification are unsuitable for virtual environments, as methodologies such as operant conditioning would suggest that the way to turn lurkers into elders is to reward them for taking participatory actions. The ecological cognition framework proposed Bishop (2007a; 2007c) proposes that in order for individuals to carry out a participatory action, such as posting a message, there needs to be a desire to do so, the desire needs to be consistent with the individual's goals, plans, values and beliefs, and they need to have to abilities and tools to do so. Some individuals such as lurkers, may have the desire and the capabilities, but hold beliefs that prevent them from making participatory actions in virtual communities. In order for them to do so, they need to have the desire to do so and their beliefs need to be changed. Traditional methods, such as operant conditioning may be able to change the belief of a lurker that they are not being helpful by posting a message, but it is unlikely that they will be effective at changing other beliefs, such as the belief they do not need to post. In order to change beliefs, it is necessary to make an actor's beliefs dissonant, something that could be uncomfortable for the individual. While changing an actor's beliefs is one way of encouraging them to participate in a virtual community, another potential way of increasing their involvement is to engage them in a state of flow which might mean that they are more likely to act out their desires to be social, but there is also the possibility that through losing a degree of self-consciousness they are also more likely to flame others (Orengo Castellá et al., 2000).

A CHARACTER THEORY FOR ONLINE COMMUNITIES

Kim's membership lifecycle provides a possible basis for analyzing the character roles that actors take on in online communities. Existing character theories could be utilized to explore specific types of online community (e.g., Propp, 1969) or explain to dominance of specific actors in online communities (e.g., Goffman, 1959). Propp suggested the following formula to explain characters in media texts:

$\alpha a^5 D^1 E^1 M F^1 T a^5 B K N T o Q W$

Propp's character theory suggests that in media texts eight characters can be identified; the villain who struggles against the hero; the donor who prepares the hero or gives the hero an artifact of some sort; the helper who helps the hero in their expedition; the princess who the hero fights to protect or seeks to marry; her father the dispatcher; and the false hero who takes credit for the hero's actions or tries to marry the princess. While Propp's theory might be acceptable for analyzing multi-user dungeons or fantasy adventure games, it may not be wholly appropriate for bulletin board-based online communities. Goffman's character theory according to Beaty et al. (1998) suggests that there are four main types of characters in a media text: the protagonists who are the leading characters; the deuteragonists who are the secondary characters; the bit players who are minor characters whose specific background the audience are not aware of; and the fool who is a character that uses humor to convey messages. Goffman's model could be useful in explaining the dominance of specific types of online community members, but does not explain the different characteristics of those that participate online, what it is that drives them, or what it is that leads them to contribute in the way they do.

Bishop's (2007a; 2007c) ecological cognition framework (ECF) provides a theoretical model for developing a character theory for online communities based on bulletin board and chat room models. One of the most talked about types of online community participant is the troll. According to Levine-Young and Levine (2000), a troll postsprovocative messages intended to start a flame war. The ECF would suggest that chaos drives these trolls, as they attempt to provoke other members into responding. This would also suggest there is a troll opposite, driven by order, which seeks to maintain control or rebuke obnoxious comments. Campbell et al. (2002) found evidence for such a character, namely the big man, existing in online communities. Salisbury (1965) suggests big men in tribes such as the Siane form a de facto council that confirms social policy and practices. Campbell et al. (2002) point out that big men are pivotal in the community as, according to Breton (1999), they support group order and stability by personally absorbing many conflicts. Actors susceptible to social stimuli activate one of two forces, either social forces or anti-social forces. Actors who are plainly obnoxious and offend other actors through posting flames, according to Jansen (2002) are known as snerts. According to Campbell, these anti-social characters are apparent in most online communities and often do not support or recognize any of the big men unless there is an immediate personal benefit in doing so. Campbell et al. (2002) also point out that the posts of these snerts, which they call sorcerers and trolls, which they call tricksters, could possibly look similar. Differentiating between when someone is being a troll and when they are being a snert although clear using the ECF, may require interviewing the poster to fully determine. Someone whose intent is to provoke a reaction, such as through playing 'devil's advocate' could be seen theoretically to be a troll, even if what they post is a flame. An actor who posts a flame after they were provoked into responding after interacting with another, could be seen theoretically to be a snert, as their intention is to be offensive. Another actor presented with the same social stimuli may respond differently. Indeed, Rheingold (1999) identified that members of online communities like to flirt. According to Smith (2001), some online community members banned from the community will return with new identities to disrupt the community. These actors could be labeled as e-venegers, as like Orczy's (1904) character the scarlet pimpernel, they hide their true identities. Driven by their emotions, they seek a form of personal justice. A character that has more constructive responses to their emotions exists in many empathetic online communities according to Preece (1998), and may say things such as "MHBFY," which according to Jansen (2002) means "My heart bleeds for you," so perhaps this character type could be known as a MHBFY Jenny. Using the ecological cognition framework there should be also those actors that are driven by gross stimuli, with either existential or thanatotic forces acting upon them. Jansen (2002) identified a term for a member of an online community that is driven by existential forces, known to many as the chat room Bob, who is the actor in an online community that seeks out members who will share nude pictures or engage in sexual relations with them. While first believed to be theoretical by Bishop (2006), there is evidence of members of online communities being driven by thanatotic forces, as reported by the BBC (Anon., 2003). Brandon Vedas, who was a 21-year-old computer expert, killed himself in January 2003. This tragic death suggests strongly that those in online communities should take the behavior of people in online communities that may want to cause harm to themselves seriously. The existence of this type of actor is evidence for the type of online community member who could be called a Ripper, in memory of the pseudonym used by Mr Vedas.

There are two more characters in online communities, driven by action stimuli that results in them experiencing creative or destructive forces. Surveying the literature reveals a type of actor that uses the Internet that are prime targets for "sophisticated technical information, beta test software,

Table 2. A character theory for online communities based on the ecological cognition framework

Label	Typical characteristics
Lurker	The lurker may experience a force, such as social, but will not act on it, resulting in them not fully taking part in the community.
Troll	Driven by chaos forces as a result of mental stimuli, would post provocative comments to incite a reaction.
Big Man	Driven by order forces as a result of mental stimuli, will seek to take control of conflict, correcting inaccuracies and keeping discussions on topic.
Flirt	Driven by social forces as a result of social stimuli, will seek to keep discussions going and post constructive comments.
Snert	Driven by anti-social forces as a result of social stimuli, will seek to offend their target because of something they said.
E-venger	Driven by vengeance forces as a result of emotional stimuli, will seek to get personal justice for the actions of others that wronged them.
MHBFY Jenny	Driven by forgiveness forces, as a result of experiencing emotional stimuli. As managers they will seek harmony among other members.
Chat Room Bob	Driven by existential forces as a result of experiencing gross stimuli, will seek more intimate encounters with other actors.
Ripper	Driven by thanatotic forces as a result of experiencing gross stimuli, seeks advice and confidence to cause self-harm.
Wizard	Driven by creative forces as a result of experiencing action stimuli, will seek to use online tools and artifacts to produce creative works.
Iconoclast	Driven by destructive forces as a result of experiencing action stiumli, will seek to destroy content that others have produced.

authoring tools [that] like software with lots of options and enjoy climbing a learning curve if it leads to interesting new abilities" (Mena, 1999), who are referred to as wizards. There is also the opposite of the wizard who according to Bishop (2006) seeks to destroy content in online communities, which could be called the iconoclast, which according to Bernstein and Wagner (1976) can mean to destroy and also has modern usage in Internet culture according to Jansen (2002) as a person on the Internet that attacks the traditional way of doing things, supporting Mitchell's (2005) definition of an iconoclast being someone that constructs an image of others as worshippers of artifacts and sets out to punish them by destroying such artifacts.

These eleven character types, summarized in Table 2, should be evident in most online communities, be they virtual worlds, bulletin boards, or wiki-based communities.to be explored and frameworks to be developed which can be used by both practitioners and researchers and also means that researchers can deal with real situations instead of having to contrive artificial situations for the purpose of quasi-experimental investigations (Harvey & Myers, 1995). While Yang (2003) argues that it is not feasible to spend a year or two investigating one online community as part of an ethnography, this is exactly the type of approach that was taken to evaluate the proposed character theory, partially due to the author receiving formal training in this method. Yang's approach, while allowing the gathering of diverse and varied information, would not allow the research to experience the completeness of Kim's (2000) membership lifecycle, or be able to fully evaluate the character theory and whether the characters in it can be identified.

Location and Participants

An online community was selected for study, this one serving Wales and those with an interest in the geographical locations of Pontypridd and the Rhondda and Cynon Valleys in South Wales. Its members consist of workers, business owners, elected members, and expatriates of the area the online community serves. This online community, known to its members as 'Ponty Town', with 'Ponty' being the shortened term for Pontypridd, was chosen by the author due to his cognitive interest in the Pontypridd constituency and his belief that he would be a representative member of the community and fit in due to holding similar personal interests to the members. This is in line with Figallo (1998), who argues that similar interests is what convinces some members of online communities to form and maintain an Internet presence. The members of the community each had their own user ID and username that they used to either portray or mask their identity. They ranged from actual names, such as 'Mike Powell' and 'Karen Roberts' that were used by elected representatives, names from popular culture, such as 'Pussy Galore', to location-based and gendered names, such as 'Ponty Girl', 'Bonty Boy' and 'Kigali Ken'.

Equipment and Materials

A Web browser was used to view and engage with the online community, and a word processor used to record data from the community.

Procedure

The author joined the online community under investigation and interacted with the other members. The community members did not know the author personally, however, he utilized his real name. Even though the author could have posted under a pseudonym it would have made the study less ecologically valid and more difficult for the author to assess the reaction of the participants. The author carried out activities in the online community by following the membership life-cycle stages, which manifested in not posting any responses, posting a few responses on specific topics to regularly posting as an active member of the community. Additionally, data collected by Livingstone and Bober (2006) was used to understand the results.

Results

Undertaking the ethnography proved to be time consuming, though revealing about the nature of online communities and the characteristics of the actors that use them. Of the eleven characters identified in the proposed character theory, eight were found in the investigated online community.

Lurkers could be identified by looking at the member list, where it was possible to find that 45 of the 369 members were lurkers in that they did not post any messages.

The Troll

The troll was easily identified as an actor that went by the pseudonym Pussy Galore, who even managed to provoke the author.

This Bishop baiting is so good I'm sure there will soon be a debate in the Commons that will advocate abolishing it. – Pussy Galore, Female, Pontypridd

Some of the troll's comments may be flames, but their intention is not to cause offence, but to present an alternative and sometimes intentionally provocative viewpoint, often taking on the role of devil's advocate by presenting a position that goes against the grain of the community.

There is some evidence of the troll existing in the data collected by Livingstone and Bober (2006), as out of a total of 996 responses, 10.9% (164) of those interviewed agreed that it is 'fun to be rude or silly on the internet'.

The Big Man

Evidence of actors being driven by order forces was also apparent, as demonstrated by the following comment from a big man.

I don't think so. Why should the actions of (elected member) attacking me unprovoked, and making remarks about my disability lead to ME getting banned? I am the victim of a hate crime here. – The Victim, Male, Trefforest

The example above clearly demonstrates the role of the big man as absorbing the conflicts of the community and having to take responsibility for the actions of others. While the big man may appear similar to the snert by challenging the actions of others, their intention is to promote their own worldview, rather than to flame and offend another person. The big man may resemble the troll by continually presenting alternative viewpoints, but their intention is not to provoke a flame war based on a viewpoint they do not have, but to justify a position they do have.

The importance of the big man to the community was confirmed approximately a year after the ethnography was completed when during additional observations, a particular actor, ValleyBoy, appeared to take over the role from the big man that was banned, and another member, Stinky, called for banned members to be allowed to return, suggesting the community was lacking strong and persistent characters, such as the big man.

The Flirt

In the studied online community, there was one remarkable member who posted mostly constructive posts in response to others' messages, known by her pseudonym Ponty Girl who was clearly identifiable as a flirt. Her comments as a whole appear to promote sociability as she responds constructively to others' posts. The flirt's approach to dealing with others appears to differ from the big man who absorbs conflict as it seems to resonate with the constructive sides of actors leading them to be less antagonistic towards the flirt than they would be the big man.

Yes, I've seen him at the train station on quite a few occasions," "A friend of mine in work was really upset when she had results from a feedback request from our team - I'd refused to reply on the principle that she is my friend and I would not judge her, but a lot of the comments said that she was rude, unsympathetic and aloof. She came to me to ask why people thought so badly of her. – Ponty Girl, Female, Graig

The Snert

There were a significant number of members of the community that responded to posts in an anti-social manner, characteristic of snerts. While members like Stinkybolthole frequently posted flames, the online community studied had one very noticeable snert, who went by the name of JH, whom from a sample of ten of his posts, posted six flames, meaning 60% of his posts were flames.

Nobody gives a shit what you want to talk to yourself about. Get a life," "I'm getting the picture. 'Fruitcake Becomes Councilor' is such an overused newspaper headline these days," "Sounds like you've won the lottery and haven't told us. Either that or your husband is a lawyer, accountant or drug dealer,, "The sooner we start to re-colonize ooga-booga land the better, then we'll see Britains (sic) prosperity grow. Bloody pc wimps, they need to get laid. – JH, Male, Trallwn

The existence of the snert is evident. The data collected by Livingstone and Bober (2006) reveals that from a sample of 1,511, 8.5% (128) of individuals have received nasty comments by e-mail and 7% (105) have received nasty comments in a chat room. Of the 406 that had received nasty comments across different media, 156 (38.42%) deleted them straight away. A total of 124 (30.54%) tried to block messages from the snert, 84 (20.69%) told a relative, 107 (26.35%) told a friend, 74 (18.23%) replied to ask the snert to stop their comments, and 113 (27.83%) engaged in a flame war with the snert.

The e-Venger

Evident in the online community investigated was the masked e-venger, who in the case of this particular community was an actor who signed up with the pseudonym elected member, claiming to be an elected representative on the local council, who the members quickly identified to be someone who had been banned from the community in the past. This user appeared to have similar ways of posting to the snert, posting flames and harassing other members. The difference between the e-venger and the snert is that the former is driven by wanting to get even for mistreatment in the past whereas the latter responds unconstructive to the present situation.

Poor sad boy, have you met him? He's so incompitent (sic). His dissabilty (sic) is not medical it's laughable. The lad has no idea about public perception," "Don't give me this sh#t, she and they cost a fortune to the taxpayer, you and I pay her huge salary. This is an ex-education Cabinet Member who was thrown out by the party, unelected at the next election and you STILL pay her wages!", "I'll see you at the Standards meeting [Mellow Pike]. 'Sponsor me to put forward a motion'! Bring it on. – Elected Member

The member appeared to be driven by emotional stimuli activating vengeance forces, seeking to disrupt the community and even making personal attacks on the members including the author. As outlined above, the data collected by Livingstone and Bober (2006) reveals that 27.83% of people that are flamed will seek revenge by posting a flame back.

The MHBFY Jenny

Sometimes the remarks of members such as flirts and big men are accepted, which can lead other actors to experience emotional stimuli activating forgiveness forces as was the case with Dave, the investigated online community's MHBFY Jenny.

Mind you it was funny getting you to sign up again as 'The Victim'. – Dave, Male, Pontypridd

While many of the MHBFY Jenny's comments are constructive like the flirt, they differ because the former responds to their internal dialogue as was the case with Dave above, whereas the flirt responds to external dialogue from other actors, as Ponty Girl clearly does.

The Chat Room Bob

The online community investigated, like many, had its own chat room Bob. The actor taking on this role in the investigated online community went by the name of Kigali Ken, and his contributions make one wonder whether he would say the same things in a real-world community.

Any smart women in Ponty these days? Or any on this message board? I've been out of the country for a while but now I'm back am looking for some uncomplicated sex. Needless to say I am absolutely lovely and have a massive... personality. Any women with an attitude need not apply. – Kigali Ken, Male, Pontypridd

While their action of seeking out others may appear to be flirting using the vernacular definition, the intention of the chat room Bob differs from the flirt who based on Heskell's (2001) definition is someone who feels great about themselves and resonates this to the world to make others feel good, as they will make pro-social comments about others and in response to others. The chat room Bob on the other hand, appears to be only after their own gratification, responding to their physical wants.

The existence of the chat room Bob is evident. The data collected by Livingstone and Bober (2006) reveals that 394 people from a sample of 1,511 have reported that they have received sexual comments from other users. Of these 238 (60.4%) deleted the comment straight away, 170 (43.14%) attempted to block the other person, 49 (12.44%) told a relative, 77 (19.54%) told a friend, and 75 (19.04%) responded to the message. This suggests that the chat room Bob is an unwanted character in online communities whose contributions people will want to delete and whom they may try to block.

The Ripper, Wizard, and Iconoclast

Despite studying the online community for over a year, there was no evidence of there being a ripper, a wizard or an iconoclast in the community, beyond the administrators of the site posting and deleting content and adding new features, such as polls. The closest an actor came to being a ripper was an actor called choppy, who faked a suicide and then claimed a friend had hijacked their account. Fortunately, it might be argued that a true ripper who was seeking to cause self-harm was not present, but the existence of this type of online community member should lead online community managers to show concern for them, and members should not reply with comments such as "murder/suicide" when they ask for advice, as happens in some online communities.

While visual representations are often absent from bulletin board communities, actors will often make their first interpretations of others in virtual worlds when they book at another's avatar and evaluates them based on their worldview, which may provoke a relation leading to the actor developing an interest in the other actor. In the context of online communities, an avatar is a digital representative of an actor in a virtual environment that can be an icon of some kind or an animated object (Stevens, 2004). According to Aizpurua et al. (2004), the effective modeling of the appearance of an avatar is essential to catch a consumer's attention and making them believe that they are someone, with avatars being crucial to creating interaction between customers and vendors. According to Puccinelli (2006), many vendors understand that customers' decisions to engage in economic transactions are often influenced by their reactions to the person who sells or promotes it, which seems to suggest that the appearance of an avatar will affect the number of transactions other actors will have with it.

A STEREOTYPE THEORY FOR INTERPRETING AVATARS IN ONLINE COMMUNITIES

Technology-enhanced businesses led by business leaders of a black ethnicity have been some of the most innovative in the world, with companies like Golden State Mutual ending the 1950s with electronic data processing systems in place, $133 million of insurance in force and $16 million in assets (Ingham & Feldman, 1994). Representations of black actors have also been some of the most studied, with potential applications for studying avatars in online communities. Alvarado et al. (1987) argue that black actors fit into four social classifications: the exotic, the humourous, the dangerous and the pitied. Furthermore, Malik (2002) suggests that male black actors are stereotyped as patriarchal, timid, assiduous, and orthodox. Evidence for these can easily be found in contemporary print media, such as Arena magazine (Croghton, 2007) where an advertisement for an electronic gaming system displays a black individual as pitied. In the same publication Murray and Mobasser (2007) argue that the Internet is damaging relationships, where images of women are of those in 'perfect' bodies, and although they do not define what a perfect body is, it would be safe to assume they mean those depicted in the publication, such as Abigail Clancy and Lily Cole, the later of which described herself as 'hot stuff', and perhaps depictions of this sort could be iconographic of an exotic avatar. Alvarado et al. (1987) supported by Malik (2002) have argued that these stereotypes have been effective in generating revenue for advertisers and not-for-profit organizations. While these stereotypes may be useful for developing an understanding of avatars and how they can generate both social and economic capital for individuals, they need to be put into the context of a psychological understanding of how actors behave and interact with others.

Utilizing the ecological cognition framework (Bishop, 2007a; 2007c); it can be seen that the visual appearance of an actor's avatar could be based on the five binary-opposition forces, with some of the stereotypes identified earlier mapping on to these forces. The image of actors as orthodox and pariahs can be seen to map onto the forces occupied by the flirts and snerts, respectively; the assiduous and vanguard stereotypes appear to be in harmony with the forces occupied by the wizard and iconoclast, respectively, the dangerous and timid stereotypes are consistent with the forces connected with the e-venger and MHBFY Jenny, the exotic and pitied stereotypes can be seen to map on to the forces used by the chat room Bob and ripper, respectively, and the patriarchal and humourous stereotypes appear to be consistent with the forces, respectively, used by the big man and troll. The stereotype theory provides a useful basis for investigating the role of avatars in online communities and the effect they have on social and economic transactions.

Location and Participants

A study was carried out in the second life virtual world and involved analyzing the avatars used to create a visual representation of the actor and profile pages displaying their personal details and avatar of 189 users, known as residents, of the community who met the criteria of having given at least one rating to another actor, a feature that has since been discontinued in the system despite it showing how popular a particular actor was.

Equipment and Materials

The Second Life application was used to view and engage with the online community, and a word processor and spreadsheet was used to record data from the community in the form of the number of times a person had received a gift or response from another.

Procedure

The author became a member of the online community under investigation and interacted with the other members over a period of three months. The members of the community did not know the author, especially as a pseudonym was adopted, as is the norm with Second Life. The author carried out activities in the online community by following the membership lifecycle stages that each individual member of an online community goes through as discussed earlier in the chapter. A search was done for actors and possible locations and groups of specific avatars identified. After an avatar was categorized, data from their profile was recorded and the return on investment (ROI) calculated. According to Stoehr (2002), calculating the ROI is a way of expressing the benefit-cost ratio of a set of transactions, and can be used to justify an e-commerce proposal.

Table 2. Mean (M) dollars ($) given and received by actors of specific avatars and their ROI (%)

Stereotype	Character	N	M Given $	M Received $	M ROI %
Exotic	Chat Room Bob	30	1171.67	1731.67	237.26
Pitied	Ripper	11	1743.18	2534.09	141.19
Humourous	Troll	48	362.5	446.35	120.69
Patriarchal	Big Man	17	4500	4588.24	428.43
Orthodox	Flirt	16	4393.75	4575	49.48
Pariah	Snert	26	149.04	107.69	-1.24
Assiduous	Wizard	16	267.19	159.38	-12.63
Vanguard	Iconoclast	4	75	162.5	233.33
Dangerous	E-venger	6	2587.5	2095.83	0.62
Timid	MHBFY Jenny	15	6150	4395	101.82

Results

The results, as summarized in Table 2, reveal that the avatar with greatest return on investment was the patriarchal stereotype with a 428.43% return and the one with the least ROI was the assiduous with a 12.63% loss. The most common avatar was the humorous, followed by exotic and pariah. The least common avatar was the vanguard, followed by the dangerous and pitied.

An independent samples test using the Mann-Whitey method was carried out on one of the highest ROI avatars, the patriarchal, with one of the lowest, the pariah. It revealed, as expected, a significant difference in the return on investment ($Z=-3.21$, $p<0.002$). Also interesting was the difference between the specific attributes rated. The mean appearance rating for the patriarchal stereotype was 29.24 compared to 17.27 for the pariah ($Z=-3.10$, $p<0.003$), the mean building rating for the patriarchal was 29.03 compared to 17.40 for the pariah ($Z=-3.06$, $p<0.003$), and the mean behavior rating was 30.62 for the patriarchal stereotype and 16.37 for the pariah ($Z=-3.68$, $p<0.001$). This would seem to suggest that as well as not getting as high a return on investment, other actors will not judge the more antisocial-looking pariah as well as they judge the more sophisticated-looking patriarchal avatar. Examples of the avatars are presented in Figure 3. Studies such as those by Zajonc (1962) and Goldstein (1964) have demonstrated that actors will seek to avoid the uncovering of beliefs and other thoughts thatcome about when an actor experiences threatening behavior from others or uncomfortable emotions. This being the case, it could be that when an actor is presented with an avatar that causes them discomfort or 'dissonance', then they will seek to resolve the conflict created by avoiding that particular avatar. This would explain why the pariah stereotype produces a limited number of economic transactions and has the one of the worst returns on investment, which would seem to support the findings of Eagly et al. (1991) that people that appear less discomforting are more popular with peers and receive preferential treatment from others.

THE FUTURE OF SOCIAL NETWORKING COMMUNITIES

In science fiction, the future is often portrayed as utopian or dystopian, where possible future outcomes of social trends or changes that are the result of scientific discoveries are depicted and the implications of them assessed (Csicsery-Ronay, 2003). In the cyberpunk genre of science fiction,

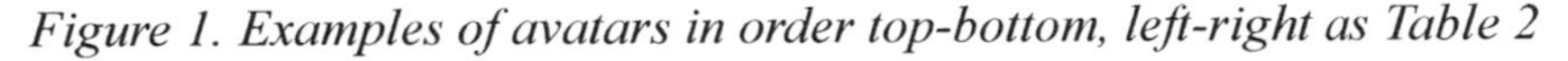

Figure 1. Examples of avatars in order top-bottom, left-right as Table 2

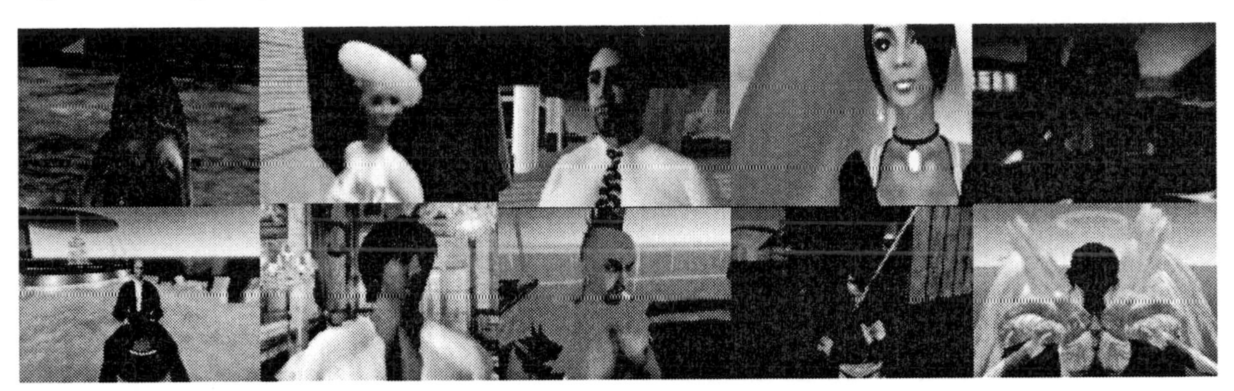

the dystopian future is often made up of corporations, who ruthlessly corrupt, corrode, exploit and destroy (Braidotti, 2003). Social networking communities are quickly being subsumed into corporate structures. In July 2005, News Corporation bought Myspace.com, which is a social networking service that integrates message boards with personal homepages and utilizes the Circle of Friends social networking technique, and in December of that year, the British broadcaster ITV bought the old school tie-based Friends Reunited social networking service (BBC, 2005; Scott-Joynt, 2005).

The ecological cognition framework has the potential to radically transform minor Web sites into highly persuasive and engaging communities where relationships between vendors and customers can be enhanced and the goals of each can be met. While there is also the possibility that a corporation that understands online communities can manipulate its members in such a way that it can easily exploit them, the model could be used by vendors with more of an interest in helping customers meet their goals to market their products and services effectively. Vendors that understand the stage of Kim's (2000) lifecycle they are at and the stage the consumer is at can more effectively target their messages in such a way that they are persuasive. Using the model, vendors can design avatars that provoke the particular responses they want from customers and continue that initial appeal by adopting the appropriate character type. This works well in some media texts where according to Kress (2004) media producers can use the appearance of their characters to convey that character's personality and build on that throughout the text.

DISCUSSION

The rise of online communities in Internet environments has set in motion an unprecedented shift in power from vendors of goods and services to the customers who buy them, with those vendors who understand this transfer of power and choose to capitalize on it by organizing online communities and being richly rewarded with both peerless customer loyalty and impressive economic returns. A type of online community, the virtual world, could radically alter the way people work, learn, grow consume, and entertain. Understanding the exchange of social and economic capital in virtual worlds could involve looking at what causes actors to spend their scarce resources on improving someone else's reputation. Actors' reputations may affect how willing others are to trade with them or even give them gifts, and their reputation is in part influenced by their appearance and how they interact with other actors and often feedback from other actors are displayed on their profile.

The ecological cognition framework provides

a theoretical model for developing a character theory for online communities based on bulletin board and chat room models. The five forces and their opposites can be used to develop the types, and the judgments of ignorance and temperance can be used to explain the behavior of those that do not participate, namely lurkers, which were accounted for in the investigated online community where it was possible to find that 45 of the 369 members were lurkers for the reason that they did not post any messages. The ECF would suggest that chaos forces drive trolls, as they attempt to provoke other members into responding as a result of experiencing mental stimuli. The troll was easily identified in the investigated online community as an actor that went by the pseudonym Pussy Galore, who even managed to provoke the author. Order forces can be seen to drive the big man and was represented in the investigated online community by the victim. Those actors who are plainly obnoxious and offend or harass other actors through posting flames are known as snerts, who were most obviously represented in the investigated online community by a user called JH. Flirts are members that respond to the text posted by other members as social stimuli, and will respond to it after activating their social forces and in the studied online community there was one remarkable member who posted mostly constructive posts in response to others' messages, know by her pseudonym Ponty Girl. There are actors driven by their vengeance forces, which could be labeled as e-venegers, represented in the investigated online community by elected member and those actors driven by forgiveness forces could be called MBHFY Jenny, represented in the studied online community by Dave. An actor in an online community that is driven by existential forces, known to many as the chat room Bob, who seeks out members who will share nude pictures or engage in sexual relations with them, was apparent in the investigated online community using the name Kigali Ken. There is evidence for an online community member driven by thanatotic forces, who could be called a ripper, but this member was not found in the investigated online community beyond an actor called Choppy. There are also theoretically two more characters in online communities, driven by action stimuli that results in them experiencing creative or destructive forces, with the one driven by creative forces being the wizard, and the opposite of iconoclast being the one that seeks to destroy content in online communities.

These character types are particularly evidenced in bulletin board communities, but in the virtual world it is likely that an actor's avatar will have some effect on how others perceive them before they are spoken to. The extent to which an actor is able to sustain an appeal to another could be analyzed as seduction. An actor's avatar forms an important part of the intimacy stage of seduction, as the visual appearance of an actor could possibly have an impact on how others perceivedthem, and an actor may construct an image based on their identity or the image they want to project and the relationship between an actor's avatar and their identity can be understood as elastic as even the best and strongest elastic can break, with there being the possibility that avatars can develop to the point where connection between them and the identities of the actors using them can be stretched so far that they cease to exist. There has been a debate over whether identity is unitary or multiple with psychoanalytic theory playing a complicated role in the debate. If there is a lifecycle to an actor's membership in an online community, then it is likely that they will develop different cognitions, such as beliefs and values at different stages that may become 'joindered'. This would mean that an actor's behavior would be affected by the beliefs and values they developed when joining the community when they are at a more advanced stage in their membership of the community. Utilizing the ecological cognition framework, it can be seen that the visual appearance of an actor's avatar could be based on the five binary-opposition forces, with some of the stereotypes identified earlier mapping on to these forces. The investigation found that the avatar with the greatest return on investment was the patriarchal stereotype with a 428.43% return and the one with the least ROI was the assiduous with a 12.63% loss. The most common avatar was the humorous, followed by exotic and pariah. The least common avatar was the vanguard, followed by the dangerous and pitied. An independent samples test revealed, as expected, a significant differences between the pariah and the patriarchal stereotype with the later having a greater return on investment, and higher ratings on appearance, building and behavior, suggesting that as not getting as high a return on investment, other actors will not judge the more antisocial-looking pariah as well as they judge the more sophisticated-looking patriarchal avatar.

The research methods used in this study were an ethnographical observation and document analysis. These methods seem particularly suited to online communities, where behavior can be observed through participation and further information can be gained through analyzing user profiles and community forums. The study has demonstrated that online communities, in particular virtual worlds, can be viewed as a type of media, and traditional approaches to media, such as investigating stereotypes, can be applied to Internet-based environments.

ACKNOWLEDGMENT

The author would like to acknowledge all the reviewers that provided feedback on earlier drafts of this chapter. In addition the author would like to thank S. Livingstone and M. Bober from the Department of Media and Communications at the London School of Economics and Political Science for collecting some of the data used in this study, which was sponsored by a grant from the Economic and Social Research Council. The Centre for Research into Online Communities and E-Learning Systems is part of Glamorgan Blended Learning Ltd., which is a Knowledge Transfer Initiative, supported by the University of Glamorgan through the GTi Business Network of which it is a member.

REFERENCES

Aizpurua, I., Ortix, A., Oyarzum, D., Arizkuren, I., Ansrés, A., Posada, J., & Iurgel, I. (2004). Adaption of mesh morphing techniques for avatars used in Web applications. In: F. Perales & B. Draper (Eds.), *Articulated motion and deformable objects: Third international workshop*. London: Springer.

Anon. (2003). *Net grief for online 'suicide'*. Retrieved from http://news.bbc.co.uk/1/hi/technology/2724819.stm

Armstrong, A., & Hagel, J. (1997). *Net gain: Expanding markets through virtual communities*. Boston, MA: Harvard Business School Press.

BBC. (2005). *ITV buys Friends Reunited Web site*. London: BBC Online. Retrieved from http://news.bbc.co.uk/1/hi/business/4502550.stm.

Beaty, J., Hunter, P., & Bain, C. (1998). *The Norton introduction to literature*. New York, NY: W.W. Norton & Company.

Berne, E. (1961). *Transactional analysis in psychotherapy*. New York, NY: Evergreen.

Berne, E. (1964). *Games people play: The psychology of human relationships*. New York, NY: Deutsch.

Bernstein, T., & Wagner, J. (1976). *Reverse dictionary*. London: Routledge.

Bishop, J. (2002). *Development and evaluation of a virtual community*. Unpublished dissertation. http://www.jonathanbishop.com/ publications/ display.aspx?Item=1.

Bishop, J. (2003). Factors shaping the form of and participation in online communities. *Digital Matrix*, *85*, 22–24.

Bishop, J. (2005). The role of mediating artefacts in the design of persuasive e-learning systems. In: *Proceedings of the Internet Technology & Applications 2005 Conference.* Wrexham: North East Wales Institute of Higher Education.

Bishop, J. (2006). Social change in organic and virtual communities: An exploratory study of bishop desires. *Paper presented to the Faith, Spirituality and Social Change Conference.* University of Winchester.

Bishop, J. (2007a). The psychology of how Christ created faith and social change: Implications for the design of e-learning systems. *Paper presented to the 2nd International Conference on Faith, Spirituality, and Social Change.* University of Winchester.

Bishop, J. (2007b). Increasing participation in online communities: A framework for human-computer interaction. *Computers in Human Behavior*, *23*, 1881–1893. doi:10.1016/j.chb.2005.11.004

Bishop, J. (2007c). Ecological cognition: A new dynamic for human computer interaction. In: B. Wallace, A. Ross, J. Davies, & T. Anderson (Eds.), *The mind, the body and the world* (pp. 327-345). Exeter: Imprint Academic.

Braidotti, R. (2003). Cyberteratologies: Female monsters negotiate the other's participation in humanity's far future. In: M. Barr (Ed.), *Envisioning the future: Science fiction and the next millennium.* Middletown, CT: Wesleyan University Press.

Bressler, S., & Grantham, C. (2000). *Communities of commerce.* New York, NY: McGraw-Hill.

Campbell, J., Fletcher, G., & Greenhil, A. (2002). Tribalism, conflict and shape-shifting identities in online communities. *Proceedings of the 13th Australasia Conference on Information Systems.* Melbourne, Australia.

Chak, A. (2003). *Submit now: Designing persuasive Web sites.* London: New Riders Publishing.

Chan, T.-S. (1999). *Consumer behavior in Asia.* New York, NY: Haworth Press.

Croughton, P. (2007). *Arena: The original men's style magazine.* London: Arena International.

Csicsery-Ronay, I. (2003). Marxist theory and science fiction. In: E. James & F. Mendlesohn (Eds.), *The Cambridge companion to science fiction.* Cambridge: Cambridge University Press.

Eagly, A., Ashmore, R., Makhijani, M., & Longo, L. (1991). What is beautiful is good, but…: A meta-analytic review of research on the physical attractiveness stereotype. *Psychological Bulletin*, *110*(1), 109–128. doi:10.1037/0033-2909.110.1.109

Feller, J. (2005). *Perspectives on free and open source software.* Cambridge, MA: The MIT Press.

Figallo, C. (1998). *Hosting Web communities: Building relationships, increasing customer loyalty and maintaining a competitive edge.* Chichester: John Wiley & Sons.

Freud, S. (1933). *New introductory lectures on psycho-analysis.* New York, NY: W.W. Norton & Company, Inc.

Givens, D. (1978). The non-verbal basis of attraction: Flirtation, courtship and seduction. *Journal for the Study of Interpersonal Processes*, *41*(4), 346–359.

Goffman, E. (1959). *The presentation of self in everyday life.* Garden City, NY: Doubleday.

Goldstein, M. (1964). Perceptual reactions to threat under varying conditions of measurement. *Journal of Abnormal and Social Psychology, 69*(5), 563–567. doi:10.1037/h0043955

Harvey, L., & Myers, M. (1995). Scholarship and practice: The contribution of ethnographic research methods to bridging the gap. *Information Technology & People, 8*(3), 13–27. doi:10.1108/09593849510098244

Heskell, P. *Flirt Coach: Communication tips for friendship, love and professional success.* London: Harper Collins Publishers Limited.

Ingham, J., & Feldman, L. (1994). *African-American business leaders: A biographical dictionary.* Westport, CT: Greenwood Press.

Jansen, E. (2002). *Netlingo: The Internet dictionary.* Ojai, CA: Independent Publishers Group.

Jordan, T. (1999). *Cyberpower: An introduction to the politics of cyberspace.* London: Routledge.

Kiesler, S., & Sproull, L. (1992). Group decision making and communication technology. *Organizational Behavior and Human Decision Processes, 52*(1), 96–123. doi:10.1016/0749-5978(92)90047-B

Kim, A. (2000). *Community building on the Web: Secret strategies for successful online communities.* Berkeley, CA: Peachpit Press.

Kress, N. (2004). *Dynamic characters: How to create personalities that keep readers captivated.* Cincinnati, OH: Writer's Digest Books.

Kyttä, M. (2003). *Children in outdoor contexts: Affordances and independent mobility in the assessment of environmental child friendliness.* Doctoral dissertation presented at Helsinki University of Technology, Espoo, Finland.

Li, D., Li, J., & Lin, Z. (2007). Online consumer-to-consumer market in China—a comparative study of Taobao and eBay. *Electronic Commerce Research and Applications.* doi:.doi:10.1016/j.elerap.2007.02.010

Lindahl, C., & Blount, E. (2003). Weblogs: Simplifying Web publishing. *IEEE Computer, 36*(11), 114–116.

Livingstone, S., & Bober, M. (2006). *Children go online, 2003-2005.* Colchester, Essex: UK Data Archive.

Malik, S. (2002). *Representing black Britain: A history of black and Asian images on British television.* London: Sage Publications.

Mann, C., & Stewart, F. (2000). *Internet communication and qualitative research: A handbook for Research Online.* London: Sage Publications.

Mantovani, F. (2001). Networked seduction: A test-bed for the study of strategic communication on the Internet. *Cyberpsychology & Behavior, 4*(1), 147–154. doi:10.1089/10949310151088532

Mena, J. (1999). *Data mining your Web site.* Oxford: Digital Press.

Mitchell, W. (2005). *What do pictures want?: The lives and loves of images.* Chicago, IL: University of Chicago Press.

Murray, S., & Mobasser, A. (2007). Is the Internet killing everything? In: P. Croughton (Ed.), *Arena: The original men's style magazine.* London: Arena International.

Orczy, E. (1905). *The scarlet pimpernel.* Binding Unknown.

Orengo Castellá, V., Zornoza Abad, A., Prieto Alonso, F., & Peiró Silla, J. (2000). The influence of familiarity among group members, group atmosphere and assertiveness on uninhibited behavior through three different communication media. *Computers in Human Behavior*, *16*, 141–159. doi:10.1016/S0747-5632(00)00012-1

Propp, V. (1969). *Morphology of the folk tale.* Austin, TX: University of Texas Press.

Puccinelli, N. (2006). Putting your best face forward: The impact of customer mood on salesperson evaluation. *Journal of Consumer Psychology*, *16*(2), 156–162. doi:10.1207/s15327663jcp1602_6

Rhiengold, H. (2000). *The virtual community: Homesteading on the electronic frontier.* London: The MIT Press.

Scott-Joynt, J. (2005). *What Myspace means to Murdoch.* London: BBC Online. Retrieved from http://news.bbc.co.uk/1/hi/business/4697671.stm.

Shen, X., Radakrishnan, T., & Georganas, N. (2002). vCOM: Electronic commerce in a collaborative virtual worlds. *Electronic Commerce Research and Applications*, *1*, 281–300. doi:10.1016/S1567-4223(02)00021-2

Smith, C. (2000). Content analysis and narrative analysis. In: H. Reis & C. Judd (Eds.), *Handbook of research methods in social and personal psychology.* Cambridge: Cambridge University Press.

Sternberg, R. (1986). A triangular theory of love. *Psychological Review*, *93*(2), 119–135. doi:10.1037/0033-295X.93.2.119

Stevens, V. (2004). Webhead communities: Writing tasks interleaved with synchronous online communication and Web page development. In: J. Willis & B. Leaver (Eds.), *Task-based instruction in foreign language education: Practices and programs.* Georgetown, VA: Georgetown University Press.

Stoehr, T. (2002). *Managing e-business projects: 99 key success factors.* London: Springer.

Turkle. (1997). *Life on the screen: Identity in the age of the Internet.* New York, NY: Touchstone.

Vogel, D. (1999). *Financial investigations: A financial approach to detecting and resolving crimes.* London: DIANE Publishing.

Wallace, P. (2001). *The psychology of the Internet.* Cambridge: Cambridge University Press.

Werry, C. (2001). Imagined electronic community: Representations of online community in business texts. In: C. Werry & M. Mowbray (Eds.), *Online communities: Commerce, community action and the virtual university.* Upper Saddle River, NJ: Prentice Hall.

Yang, G. (2003). The Internet and the rise of a transnational Chinese cultural sphere. *Media Culture & Society*, *25*, 469–490. doi:10.1177/01634437030254003

Zajonc, R. (1962). Response suppression in perceptual defense. *Journal of Experimental Psychology*, *64*, 206–214. doi:10.1037/h0047568

This work was previously published in Social Networking Communities and E-Dating Services: Concepts and Implications, edited by C. Romm-Livermore; K. Setzekorn, pp. 60-77, copyright 2009 by Information Science Reference (an imprint of IGI Global).

Chapter 7.24
Providing Mobile Multimodal Social Services Using a Grid Architecture

Stefania Pierno
Engineering IT, Italy

Massimo Magaldi
Engineering IT, Italy

Gian Luca Supino
Engineering IT, Italy

Luca Bevilacqua
Engineering IT, Italy

Vladimiro Scotto di Carlo
Engineering IT, Italy

Roberto Russo
Engineering IT, Italy

Luigi Romano
Engineering IT, Italy

ABSTRACT

In this chapter, we describe a grid approach to providing multimodal context-sensitive social services to mobile users. Interaction design is a major issue for mobile information system not only in terms of input-output channels and information presentation, but also in terms of context-awareness. The proposed platform supports the development of multi-channel, multi-modal, mobile context aware applications, and it is described using an example in an emergency management scenario. The platform allows the deployment of services featuring a multimodal (synergic) UI and backed up on the server side by a distributed architecture based on a GRID approach

DOI: 10.4018/978-1-60566-386-9.ch017

to better afford the computing load generated by input channels processing. Since a computational GRID provides access to "resources" (typically computing related ones) we began to apply the same paradigm to the modelling and sharing of other resources as well. This concept is described using a scenario about emergencies and crisis management.

INTRODUCTION

The penetration of mobile device in western countries is high and still increasing. At the same time new generation terminals feature ever increasing computing power, opening new possibilities for innovation, especially in service delivery.

One emerging trend in service evolution is for services to cater not only to individuals but also to communities of users. Communities are a social phenomenon where people with common interests, experiences, and objectives are brought together. They provide a social place where individuals exchange and share information, knowledge, emotions and jointly undertake activities. Managing the creation or deletion of flexible communities improves the user experiences in communities (NEM, 2006).

MoSoSo (Mobile Social Software), is a class of mobile applications that aims to support social interaction among interconnected mobile users (Lugano, G., 2007). While existing Internet-based services have already shown the growing interest in communication support for communities, *MoSoSo* adds additional dimensions to group communication by exploiting contextual data such as the user geographical position (Counts, S., 2006).

When designing MoSoSo applications, three important differences between desktop and mobile environments should be taken into account:

- The physical context of use is no longer static and poses some constraint to user attention;
- The social context is also dynamic: mobile communities member are tied up by common interest and contextual information, like location and time;
- MoSoSo applications are designed not just for communication but for usage in everyday life situations: users are always socially connected.

In our vision the MoSoSo concept could also benefit "public" (e-Government) services leading to innovative, more effective mobile services, able to leverage on dynamic management of *ad-hoc* communities, context-awareness (i.e. time and location), user profile management and multimodal interaction.

One domain where such benefits will matter most will be emergencies and crisis management. In fact, the response to such situations typically implies the coordination of physical resources, (emergency services personnel, often belonging to different organizations, or even possibly volunteers) in hardly predictable environments in situations where ineffective operations can cause the loss of lives.

From an IT standpoint, implementing such a vision requires coordination of services and sharing of resources among different organizations that typically operate heterogeneous hardware and software environments. The Virtual Organizations paradigm address this issue: "VOs enable disparate groups of organizations and/or individuals to share resources in a controlled fashion, so that members may collaborate to achieve a shared goal" (Foster I., 2001). In those circumstances dynamism, flexibility and interoperability become essential requirements.

Interoperability, in particular, is a key issue in the e-Government domain due to the increasing demand for integrated services. We aim to integrate MoSoSo users into a typical Grid resource management model. To this end, we designed an experimental platform to support the development of multimodal MoSoSo application, allowing an easy integration of mobile community users into a Grid based VO.

OGSA (Open Grid Service Architecture (Foster, I., 2006), a refinement of the SOA concept, allows the interoperability of "resources". In fact the OGSA specification allows each resource to be seen as a service with a standard interface. In the WS-Resource framework conceptual model, a Web service is a stateless entity that acts upon, provides access to, or manipulates a set of logical stateful resources (documents) based on the messages it sends and receives (WSRF, 2006) (Foster I, 2004).

Obviously, to evolve from SOA to OGSA, all architectural components must be extended to deal not only with services but also with resources. For example, as far as process execution is concerned, the workflow engine has to be able to compose both services and resources. Similarly a logical enhancement to the UDDI registry is required to store information about WS-Resources too.

Whereas there are interesting technology products (both commercial and open sources) dealing with multimodal client interfaces (EIF, 2004)(Frissen, V)(I.D.A.B.C)(Berners-Lee, Tim, 2001) and grid middleware (Foster, I., 2001) (Foster, I., 2006) (OASIS, 2006)(Mark Little, 2004), the innovative idea proposed in this article is to bring them together to enable innovative social services (multimodal emergency services being an example) and reduce digital divide.

The rest of this chapter is organized as follows. In the next section we will describe the multimodal part of the overall architecture (front-end). Then we will describe the back-end grid based architecture. Final remarks in the last section conclude the chapter.

MULTIMODAL ARCHITECTURE OVERVIEW

The multimodal part of the overall architecture (Figure 1, Figure 3), is composed by the following modules (Frattini G, 2006),(Frattini G, 2008):

- **Client side application (MMUIManager, MultiModal User Interface Manager):** shows an appealing graphical interface to the user, manages the output channels and collects the inputs from the user and context information.
- **Front end:** collects input from clients and routes it to distributed back-end recognizers (speech, sketch and handwriting recognizers);
- **Fusion:** semantically merges recognized input fragments coming from different channels;
- **Business Logic:** selects appropriate contents;
- **Fission:** sends the selected content to final users.

Building upon a fairly established conceptual multimodal reference architecture (W3C), we enriched it with Web 2.0 (Frattini, G, 2008), telecommunication & grid technologies, open to community model based interaction. We think the final result is interesting.

In the next sections we will expand on this evaluation. We will first describe our approach to assemble on-the fly mobile multimodal user interfaces using a thin client that exploits resources available on the network. Then, we will present the back-end architecture that deals with the induced computational load.

We focused on using commonly available mobile devices (PDAs or smart phones), and tried to avoid installing specific sw environments. Hence we had to develop a very light software

Figure 1. Architecture overview

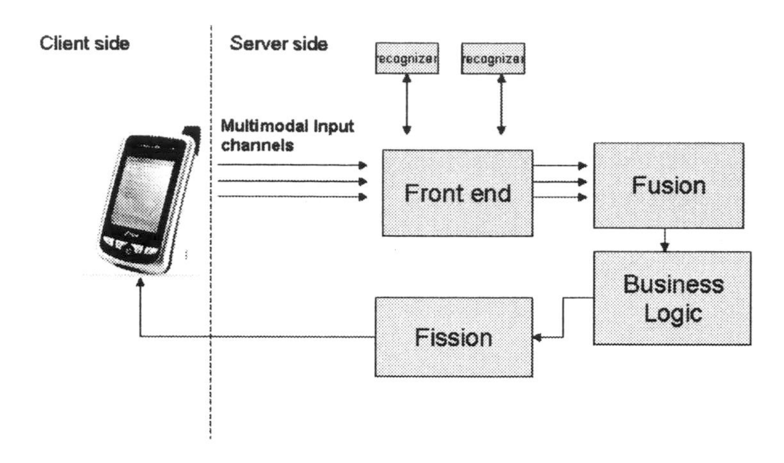

framework for building multimodal interfaces. The framework exploits several standard protocols, over commonly available networks, to interact with both local environment and remote servers (Frattini G, 2007).Being multimodal, such interfaces have to be able to:

- Collect multimodal input from different channels: speech, point, sketch, handwrite;
- Render outputs selecting the best possible channels (multimedia output);
- Exploit local resources to create an appropriate context for service fruition.

Since we chose to use light, standard "thin" terminals, all collected input fragments have to be routed to network based modal recognizers . All those functions are grouped in a small footprint application: the MMUIManager. The MMUIManager, operating in a fashion conceptually similar to an Internet browser, receives from servers the information describing the UI contents (images, text, audio/video resources), loading locally just the minimum software layer needed to support the desired user interaction.

Such an approach is similar to what is done fore other languages (XUL).

Implementing this concept was a not trivial task. Suffice it to say that, since a standard markup language to describe synergic mobile multimodal UI had not yet emerged, we had to develop our own: an XML based markup language for aggregating multimodal objects: LIDIM (from the Italian Linguaggio di Definizione Interfacce Multimodali, - language for designing multimodal interfaces).

LIDIM is a tag language that by exploiting alsoresources provided by a general multimodal framework is able to handle any potential combination of input modes and to compose multimodal output objects.

The thin client approach implies that the MMUIManager cannot carry out recognition tasks locally on the user terminal but instead it has to rely on back end recognition server distributed over the

Figure 2. Example of client interface

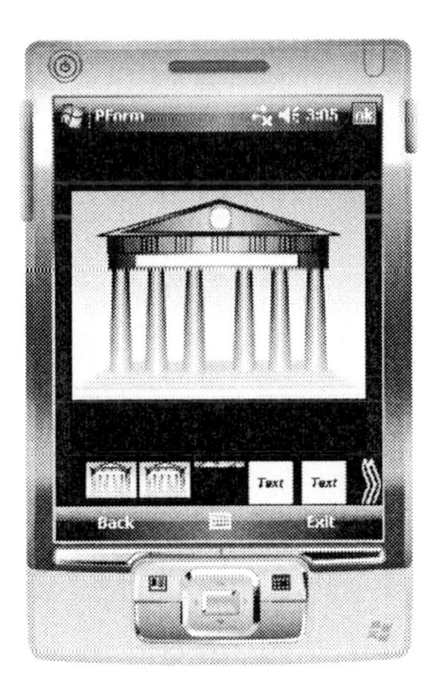

network, by streaming input fragments to them. To optimize this task, avoiding local buffering of input signals, we adopted the telecommunication protocols SIP/RTP and developed special multimodal objects able to manage streaming protocols in both input and output.

Mobile Interface: The Thin Client Approach

To reach such goals we have followed an XUL like approach. We defined the LIDIM language to describe the multimodal mobile interface built the support for LIDIM in our MMUIManager (MMUIManager acts as a browser for LIDIM building the multimodal interface on the fly and rendering it for user fruition). Such approach enables a very flexible interface design: a specific multimodal applicative frame can be dynamically built adapting to device capabilities, physical context, interaction type, user profile. Not only the graphical aspect, but also content fragments, need not to reside in the device. They reside on optimised content servers and are dynamically selected and streamed to the mobile device .

In short our MMUIManager had to be able to collect multimodal input fragments and send them to back end servers for recognition while at the same time managing multimedia output (text, image, audio, video) composed on the fly.

To be able to handle those tasks, rather than building a monolithic MMUIManager we chose to develop a multimodal framework: a rather comprehensive set of Multimodal elementar objects that can be composed into a coherent multimodal interface. At this moment available alements are buttons, textboxes, html browsers, imageboxes, audio/video panels (able to play audio/video resources) etc.

Those elements can be enriched with specific input modalities, like speech, stylus touch, or draw. For example it is possible to use an "image box" element, and make it sensible to stylus input. In this way the element can allow for an interface in

Figure 3. Sample of draw acquisition

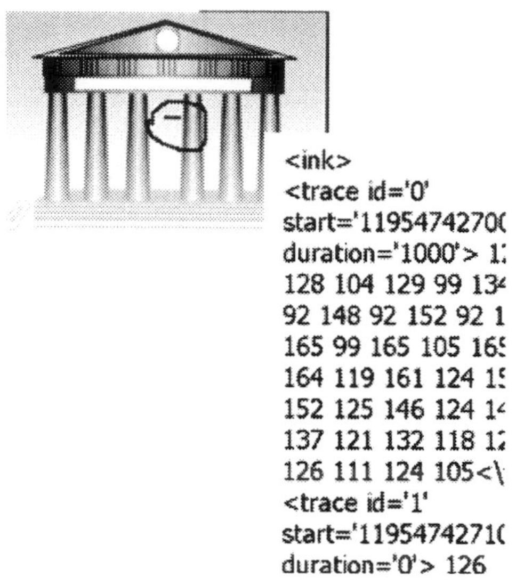

which the user can draw free hand sketches and/or handwriting on top of a picture.

In the specific example ink traces are saved in InkML standard format and sent over the network to appropriate recognition servers, as soon as possible (as soon as a recognition fragment is completed by the user).

Audio acquisition is different: since it is much more difficult to relate a speech fragment a specific interface object "a priori" (i.e. by not considering the semantic content of the speech utterance itself) such input is captured by the MMUIManager itself. This acquisition process is done continuously and the acquired utterances are continuosly streamed over RTP to backend speech recognizers.

As already anticipated we chose to introduce a specific language to describe modular, multimodal/multimedia interfaces: LIDIM.

An alternative choice could have been to extend an existing language to our requirements.

Since we wanted to be able to achieve synergic coordinated multimodal input and multimedia output in a modular interface, the existing technologies/standards proved overly difficult to extend: SMIL conceived for output was too difficult to adapt to input, while X+V and SALT showed the limits inherithed from their design that did not consider synergic multimodal input over simultaneously active input channels.LIDIM manages the following output channels:

- Graphic modular panel interface;
- Audio (pre-recorded sources and TTS);
- Text Messages (Pop up message).

For each resource of the modular interface one or more output media (image, audio, video, text) and input modalities (point, sketch) can be specified. Note that LIDIM just describes the output interface, while the MMUIManager builds the interface and manages the input/output channels. Let us give an example of how it is possible to build an object having point and sketch as input mode and media text, image and audio as possible output.

```
<object id="78678687">
        <output_media>
<text uri="http://...."/>
            <image uri="
http://...."/>
            <audio uri="
http://...."/>
        </output_media>
<input_mode>                    <point
isActive="true"/>
            <sketch
isActive="true"/>
        </input_mode>       </ob-
ject>
```

Like HTML, LIDIM does not include the resources (images, text, audio/video resources) which compose the interface; it just contains the addresses where the MMUIManager can get them.

The language is template based, hence leaving to the MMUIManager to dispose the objects on the screen. The current version cannot synchronize different output resources. Future releases are planned to remove such limitation.

We built two MMUIManager prototypes on top of J2ME and .NET Compact Framework software environments.This was motivated by two someway opposite requirements:

- The possibility to run on almost every commercial devices;
- The necessity to exploit all the hardware and OS capabilities of the devices.

The current .NET implementation is more advanced, for its support of streaming for audio input/output. The J2ME lacks this feature (handling only discrete speech utterances via HTTP) since J2ME lacks standard support for RTP protocol. The main MMUIManager tasks are:

- Build the graphic interface using an XML user interface language;
- Manage the input and output channels on the device;

The UI needs to be able to simultaneously collect multimodal input and show multimedia output according to instructions coming from the server.

This approach has a significant advantage: the very same MMUI manager can be used in very different applications contexts content or logic change will be only necessary on server side.

It is a crucial advantage for portable terminals whose users are not, and need not to become, used to installing and configuring local sw applications

According to our goals, the MMUIManager must be based on an engine for interpreting an XML language for creating multimodal/multimedia interface on the fly. Considering this, we have designed and developed a framework for aggregating multimodal objects. A composition of multimodal objects creates a complete multimodal user interface.The interaction modes enabled by our multimodal framework are, at the moment, the following:

- Point on specific buttons (ex. a "Back" button) and point on object;
- Draw/handwrite on the screen;
- speak.
- Output media supported are: image, text, Html, audio, video.

Input channels

To deliver synchronous coordinate synergic multimodality, the client has first to collect simultaneously different input channels, which currently are:.

- Speech
- Pointing
- Sketch
- Handwriting

To overcome the constraints posed by the somewhat limited hardware resources of mobile devices we chose to distribute the recognition process over back end recognisers.

To optimise this distribution we found useful to design an intermediate layer between the MMUIManager and the recognizers. This layer (that we call "Front End module") essentially manages the redirection of input from the MMUIManager toward the recognizers allowing for some location independence,failover and load balancing).

It also takes care of sending recognized input fragments to the fusion module in standard (EMMA) format, and for the TTS channel generate the output.

The Front-End was implemented as a set of cooperating Java servlets deployed on an application server (e.g. Tomcat, JBoss etc.).

The acquisition of the "pointing" modality is the simplest. Every time the user touches a sensible on screen object an HTTP call containing the object identifier is sent to the server to be merged with other concurrent modal fragments.

The acquisition of sketch/handwring is more complex. But still there is no need for a real time process.

The client side framework offers multimodal objects sensible to stylus inputs and traces. Acquired traces are buffered locally and then sent to the front-end server via HTTP using a standard format (InkML). The Front-end server, in turn, routes the acquired input to recognizer and fusion modules.

Speech Recognition is the most complex.

To recognize speech in real time by using distributed Speech Recognizers, an open streaming channel (RTP protocol) between the device and the recognizer (ASR, Automatic Speech Recognizer) has to be established.

This channel allows continuous recognition of the user speech and we chose to have it established through a direct negotiation process between the MMUIManager and the recognizer; such negotiations are made using the MRCP protocol (RFC 4463).

To trigger the recognition process start we introduced a special "start" keyword such as "computer"; so when a user utters "computer,.. do this" the recognition process starts. Obviously any other keyword can be chosen. A recognition fragment isidentified when the user pauses his speech.

Output process

In our architecture, the output process is managed by a business logic layer and by the Fission component. The former provides to thelatter application specific information (also taking into account factors such as user's preferences and behavioural patterns) while the Fission module, takes care of the technical coordination of the output. For example the Fission module guarantees synchronization between the visual presentation and the audio reproduction of the content. To do that the Fission module creates a specific LIDIM output file that is sent to the MMUIManager for rendering.

The Fission was designed to treat asynchronously clients' requests; when the business logic layer has new content for a specific client, it signals this availability to the Fission module by using the HTTP protocol and this may happen independently from client requests (for example because an element in the environment changesornew data becomes available). The Fission module notifies the MMUIManager that updated content issvailable by using the SIP (Session Initiation Protocol)

Figure 4. Fission module internal architecture

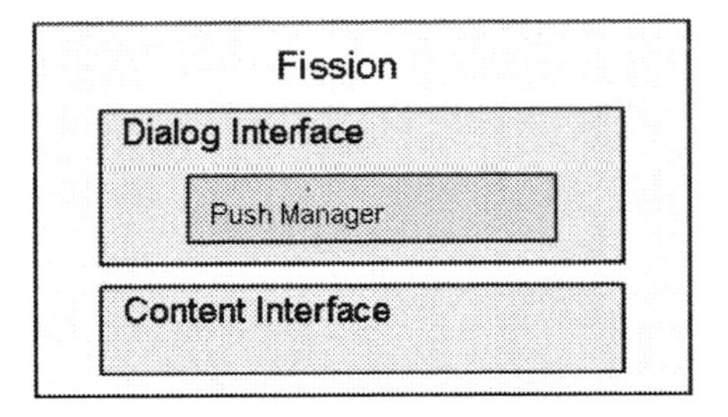

protocol. Then the MMUIManager receives from the Fission module the new content through an HTTP request.

SIP is the Internet Engineering Task Force's (IETF's) standard for multimedia conferencing over IP. SIP is an ASCII-based, application-layer control protocol (defined in RFC2543) that can be used to establish, maintain, and terminate calls between two or more end points.

The complete Fission architecture is illustrated in Figure 4.Figure 4 shows the following components:

- The *Dialog Interface* is the module by which the business logic layer talks with the Fission module for the notification of new contents.
- The *Push Manager* is responsible for notifying the MMUIManager about the availability of new content for a specific client.
- The *Content Interface* is the component by which the MMUIManager module communicates with the Fission module for retrieving new contents.

As previously mentioned, contents notification to the MMUIManager, is managed through SIP protocol. The Push Manager, part of the fission module, achieves this notification, after a registration process to the SIP server.

To complete contents notification, the PushManager performs a SIP call to the MMUIManager, that in turn hangs up and gets ready to receive the contents.

Figure 5 shows the steps just described.

The SIP server is not part of Fission module, but rather it is a component of the overall system. This component acts in proxy role: it is an intermediate device that receives SIP requests from a client and forwards them on the client's behalf.

The output channels that the MMUIManager can process are depicted in Figure 6.

The voice output channel can play either prerecorded audio streams or text streams converted to voice by a TTS engine.

The LIDIM file just contains information about resource location (an address). The MMUIManager fetches such content by using HTTP or RTP.

The text Message channel allows popup text

Figure 5. Asynchronous push

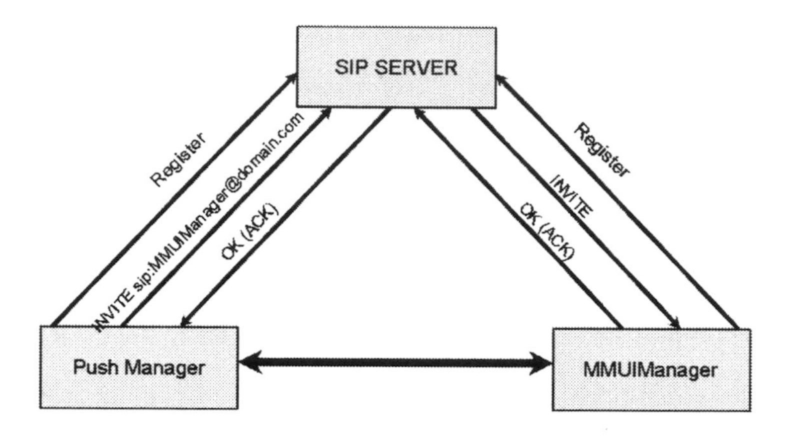

Figure 6. Output channels schema

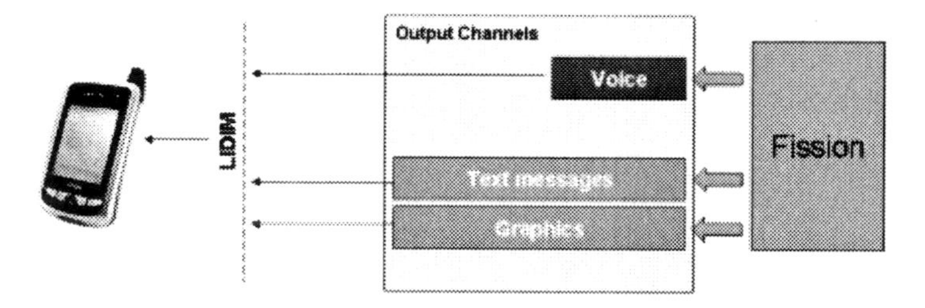

messages that can also be used for asking user confirmations.

The Graphics channel instructs the MMUIManager about the graphical interface layout to be shown. All the output information are included in a LIDIM page and the Fission module takes care of sending it to the MMUIManager.

It is worth noticing that the fission module is completely asynchronous. This is an important feature: real adaptive services must be able to react by adapting the user interface to context changes, even without an explicit user request, typically to react to usage context changes (noise or lightning conditions).

BACK-END ARCHITECTURAL OVERVIEW

Figure 7 shows the logical view of our back-end architecture (Pierno S, 2008).

The applications, being highly service specific, will not be discussed.

The Workflow Management System (WFMS) takes care of e-Government processes, dealing with process flow design and execution.

The Workflow Engine component belonging to WFMS, has to aggregate both WS and WS-Resources. Our architecture, as well as the functionalities of the components of the Grid oriented workflow system, are based on the model proposed by the Workflow Management Coalition (WfMC) (Yu, Jia, 2005)(Globus Alliance).

To enable processes combining both WS and WS-Resources, we selected, in the J2EE domain, JBOSS Jbpm as the workflow engine and bpel as the script language.

We are hence investigating how to extend them to fully support WSRF-compliant services (PVM) (Dörnemann T, 2004).

Processes that can be fully defined at design time do not pose significant research challenges, hence we concentrated our research efforts on processes that need to be planned dynamically at execution time. To this aim we are investigating different artificial intelligence techniques that will leverage semantic descriptions of both services and resources. This automatic process planning would either adapt an existing template (stored in a repository) or try to compose it ex novo.

A Service Flow Planner component in the WFMS (Pierno S, 2008) is devoted to this task while a Match-Maker component cooperates with it by finding the best service/resource available (early binding).

In more detail, the Service Flow Planner will be able to automatically plan a process from domain ontology in OWL and service descriptions in OWL-S into Planning Domain Definition Language PDDL.

In choosing the planner, we selected OWLS-Xplan because of its PDDL 2.1 compliance. The language PDDL2.1 is an extension of PDDL for temporal expressions.(Hoffmann, J, 2000)

Figure 7. Back-End (SIEGE) architecture

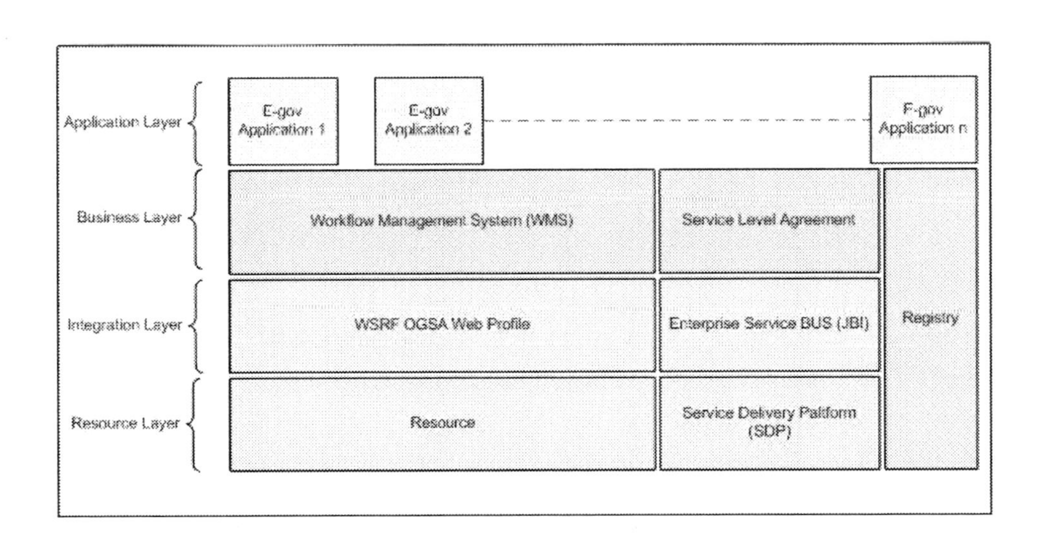

(Hoffmann, J, 2001)(Hoffmann, J, 2003)(Klusch Matthias). By exploiting this feature, the planner can chose the best service/resource according to QoS parameters.

To integrate the Service Flow Planner with the workflow engine previously described, we are developing a component that will be able to convert PDDL2.1 in both BPEL (OASIS) and JPDL languages.

The Match-Maker component (Pierno,S, 2008) can be used during process execution to find the best service at run time (process planning / re-planning) or, at build time, for discoverying service in the service registry. Our Match-Maker prototype does not depend on either language description, service registry, or matching strategies.

We are building upon the Match-Maker architecture proposed by Paolucci by extending it with the addition of other domain matching strategies such as e-Government QoS matching strategies. These extensions are very important in crisis management scenarios, where processes need to be highly dynamic.

The Service Flow Planner and the Match-Maker have the objective of increasing the capabilities of Workflow Management Systems for the autonomous self-management of composite processes, reducing human intervention to the minimum, primarily in the critical phases of process definition, execution and evolution.

In order to obtain such a result, we are starting from the body of research about Autonomic Web Processes. Autonomic Web Processes research aims to extend the Autonomic Computing model to the execution phase of Web Processes, with the objective of defining a model that extends autonomic self-management to the design, execution and post-execution phases.

In order to achieve such a result (that we define as Autonomic Workflow (Tretola Giancarlo, September 2007) we are proposing autonomic components for supervising a generic process,

Figure 8. Match-Maker

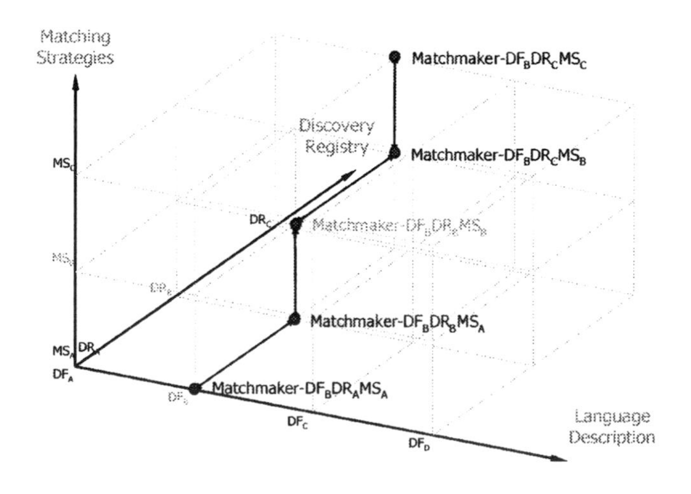

decreasing the need for human intervention, even during the design and supervision phases. (Tretola Giancarlo, April 25-29, 2006) (Tretola Giancarlo, February 7-9, 2007).

We are currently investigating how to integrate the JBOSS Process Virtual Machine (PVM) with Service Flow Planner and Match-Maker to fully support Autonomic Web Processing (Miers Derek (2005) (Tretola Giancarlo, March 26-30, 2007) (Tretola Giancarlo, June 19-20, 2007), (Tretola Giancarlo, July 9-13, 2007). While the Workflow Management System takes care of the high level coordination of services it does not need to concern itself with lower level details.

Functions such as data transformation, intelligent routing based on message content, protocol translation, message AAA (authentication, authorization, accounting) management, transaction management are best taken care of by a specialized component: the Enterprise Service Bus. We chose JBoss ESB because it is an ESB compliant with the Java Business Integration (JBI) standard.

WMS can invoke (via ESB) internal or external services and resources. The Service Delivery Platform (SDP), includes a library of loosely coupled services and it is responsible for their management and lifecycle control (Pierno, S, 2008).

The OGSA environment allows the use of resources and their lifecycle management Since the Globus Toolkit 4 (Globus Alliance) is one of OGSA most mature implementations (and one which is being evolved in parallel with OGSA specification), we selected it as a starting point for our investigations.

In distributed heterogeneous systems, mechanisms and technologies able to support a fast and effective services/resources discovery (an effective discovery mechanism has to deal with services/resources capabilities and resources status) are fundamental. In literature the use of semantic description is widely considered as the most promising approach about this. In particu-

Figure 9. Client application: Collecting useful information

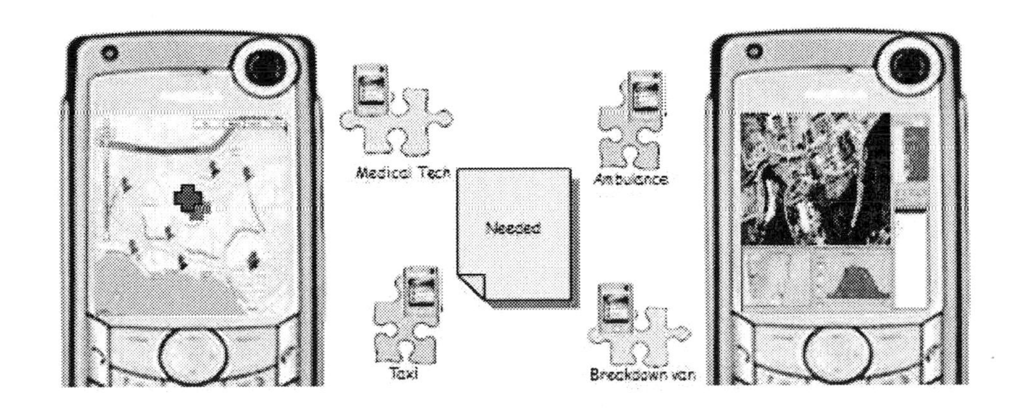

lar the Semantic-OGSA (Oscar Corcho, 2006) proposal treats metadata as a first class citizen while defining a set of services suited for metadata management (lifetime management, change notification, etc).

Semantic-OGSA is particularly interesting because it provides for a flexible introduction of semantic data in the architecture: grid resources extended with semantic descriptions may operate together with grid resources that do not receive such an extension. With this approach Semantic-OGSA semantically enables basic OGSA services.

The Registry is another key architectural element. A decentralized hierarchical structure is well suited to the e-Government domain where some kind of hierarchical topology of organizations (national, regional, local) often exists.

We hence started our studies by investigating the UDDI 3.0 registries federation. specification (federation of registries)and developed a framework that allows the navigation of its hierarchical structure. In other words, our framework enables *distributed queries*. It is based on an algorithm that, using configurable parameters, allows the best path for querying the federated registries to be selected.

We foresee future work to further extend our framework to enable *match-making* of services/resources in federated environments.

By so doing our framework will eventually enable *semantic distributed queries*.

Since we intend to use federated UDDI registries to discovery both WS and WS-Resources we need a mechanism to "refresh" resource status across the whole domain: whenever the status of a specific resource changes, the federated registry structure as a whole must be made aware of it.

For this function we are developing the architectural component named SLA (whose main functions are in fact retrieving the state of all services/grid resources and refresh the service registries).

To obtain a registry able to manage all the needed information, it is necessary to extend the registry with metadata annotations. We will follow

Figure 10. Server side: Information management

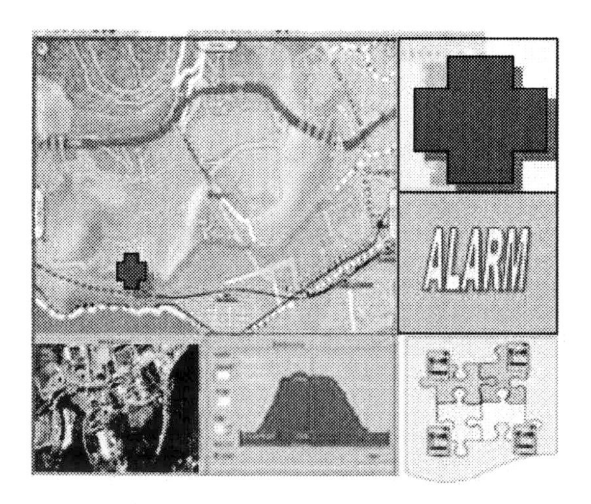

this approach to add semantic Web capabilities to UDDI registries (Paolucci, M., 2002).

Scenario

In a crisis, several organizations must work together as a virtual organization, sharing resources across organizational boundaries to deal with the complexities of such situations. In those scenarios resources are mainly physical ones: police cars, ambulances, emergency professionals and volunteers. Disaster response VO are hence characterized by resource heterogeneity and must rapidly reconfigure (structural and functional changes) to adapt to the changing communication and control demands which may be needed to handle such events (Sharad Mehrotra, 2008).

All this requires dynamic and adaptive workflows, able to coordinate fixed and mobile resources on the basis of their readiness, availability and capabilities.

In our scenario, local Emergency Operations Centers (ECOs) are in charge of collecting information and coordinating operations. To facilitate communication between VO members (EOC chief, workers and volunteer), our solution provides multimodal interaction support for the mobile devices of the operators involved in the operations, exploiting a "point and sketch" interaction mode, which is particularly useful in on-field mobile operation.

In case of emergency, mobile social community users, are asked by a Resource Planning Support (RPS) service to be involved in the operation and eventually integrated into an *ad-hoc* emergency VO created by the EOC. In this way, we may be able to augment the on-field operator team with additional resources. Furthermore, accessing community member profiles, the RPS service can organize operations considering user's skills and assigning the best task to each VO member.

Let's suppose, for example, that after an earthquake some teams are involved in on-site damage control on the affected area, may be requesting assistance.

Imagine that some team's member with medical skill identifies the symptoms of an heart attack for a citizen asking help. He can use his multimodal mobile device to request intervention of a properly equipped ambulance by indicating the location pointing to a map on the screen (Lugano G, 2007). Through support services, every organization in the VO that can provide ambulance resources, will be asked to return availability, location, capabilities (equipment and crew) and other relevant information.

The available resources will be discovered on the (federated) registry and the best matching one will be called upon to accomplish the task at hand.

CONCLUSION AND FUTURE WORK

So far the research activities already carried out show that the described approach is feasible, although it places high demands (in terms of computing power) to back end systems.

In particular the availability of suitable distributed input processing software for voice recognition on large scale is still elusive.

Future research activities will deal with solving those aspects and improving the processing of context sensitive but user unaware input. For example in case of an unexpected raise in temperature an environment sensor may signal the risk of a fire inception even if no user recognizes visually a fire yet.

REFERENCES

W3C. *Multimodal interaction activity.*

Berners-Lee, T., Hendler, J., & Lassila, O. (May). The Semantic Web. *Scientific American Magazine.* Retrieved on March 26, 2008.

Corcho, O., Alper, P., Kotsiopoulos, I., Missier, P., Bechhofer, S., & Goble, C. (2006). An overview of S-OGSA: A reference semantic grid architecture. *Web Semantics: Science, Services, and Agents on the World Wide Web, 4*(2).

Counts, S., Hofter, H., & Smith, I. (2006). Mobile social software: Realizing potential, managing risks. *Workshop at the Conference on Human Factors in Computing Systems (CHI '06)* (pp.1703-1706).

Dörnemann, T., Friese, T., Herdt, S., Juhnke, E., & Freisleben, B. (2007). Grid workflow modelling using grid-specific BPEL extensions. *German E-Science*.

EIF. (2004). *European interoperability framework for pan-European e-government services version 1.0*. Brussels.

Foster I., Frey, J., Graham, S., Tuecke, S., Czajkowski, K., Ferguson, D., Leymann, F., Nally, M., Sedukhin, I., Snelling, D., Storey, T., Vambenepe, W., & Weerawarana, S. (2004). *Modeling statefull fesources with Web services v. 1.1*.

Foster, I., Kesselman, C., Tuecke, S. (2001). The anatomy of the grid: Enabling scalable virtual organizations. *International J. Supercomputer Applications, 15*(3).

Foster, I., Kishimoto, H., Savva, A., Berry, D., Djaoui, A., Grimshaw, A., et al. (2006). The open grid services architecture, version 1.5. *Open Grid Forum*, Lemont, IL. GFD-I.080.

Frattini, G., Ceccarini, F., Corvino, F., De Furio, I., Gaudino, F., Petriccione, P., et al. (2008). *A new approach toward a modular multimodal interface for PDAs and smartphones*. VISUAL: 179-191. Frissen, V., Millard, J., Huijboom, N., Svava Iversen, J., Kool, L., & Kotterink, B. In D. Osimo, D. Zinnbauer & A. Bianchi (Eds.), *The future of e-government: An exploration of ICT-driven models of e-government for the EU in 2020*.

Frattini, G., Gaudino, F., & Scotto di Carlo, V. (2007). Mobile multimodal applications on mass-market devices: Experiences. *DEXA Workshops, 2007*, 89–93.

Frattini, G., Petriccione, P., Leone, G., Supino, G., & Corvino, F. (2007, September). Beyond Web 2.0: Enabling multimodal web interactions using VoIP and Ajax. In *Security and Privacy in Communications Networks and the Workshops, 2007. SecureComm 2007. Third International Conference* (pp. 89-97).

Frattini, G., Romano, L., Scotto di Carlo, V., Pierpaolo, P., Supino, G., Leone, G, & Autiero, C. (2006). Multimodal architectures: Issues and experiences. *OTM Workshops, (1)*, 974-983.

Globus Alliance. *Globus toolkit*.

Hoffmann, J. (2000). A heuristic for domain indepndent planning and its use in an enforced hill-climbing algorithm. In *Proceedings of 12th Intl Symposium on Methodologies for Intelligent Systems*. Springer Verlag.

Hoffmann, J. (2003). The metric-FF planning system: Translating ignoring delete lists to numeric state variables. *Artificial Intelligence Research (JAIR), 20*. BPEL.

Hoffmann, J., & Nebel, B. (2001). The FF planning system: Fast plan generation through heuristic search. *Journal of Artificial Intelligence Research*, (14): 253–302.

I.D.A.B.C. *Interoperable Delivery of European E-Government Services to Public Administrations, Businesses, and Citizens*.

JPDL.

Klusch, M., & Schmidt, M. *Semantic Web service composition planning with OWLS-Xplan*. Retrieved from www-ags.dfki.uni-sb.de/~klusch/i2s/owlsxplan-3.pdf

Little, M., Webber, J., & Parastatidis, S. (2004). Stateful interactions in Web services. A comparison of WSContext and WS-Resource framework. *SOA World Magazine*, April.

Lugano, G. (2007). Mobile social software: Definition, scope, and applications. *EU/IST E-Challenges Conference*, The Hague, The Netherlands.

Lugano, G., & Saariluoma, P. (2007). *Share or not to share: Supporting the user decision in mobile social software applications*.

Mehrotra, S., Znati, T., & Thompson, C. W. (2008). Crisis management. *IEEE Internet Computing Magazine*.

Miers, D. (2005). *Workflow handbook, workflow management coalition*. UK: Enix Consulting.

NEM. (2006, August). *Strategic research agenda, version 4.0*.

OASIS. (2006). *Reference model for service oriented architecture 1.0*.

OASIS. *Web services business process execution language*.

Paolucci, M., Kawamura, T., Payne, T., & Sycara, K. (2002). Importing the Semantic Web in UDDI. In *Web Services, e-business, and Semantic Web workshop*.

Paolucci, M., Kawamura, T., Payne, T., Sycara, R., & Katia, P. (2002). Semantic matching of Web services capabilities. *International Semantic Web Conference* (pp. 333-347).

Pierno, S., Romano, L., Capuano, L., Magaldi, M., & Bevilacqua, L. (2008). Software innovation for e-government expansion. In R. Meersman & Z. Tari (Eds.), *OTM 2008, Part I* (LNCS 5331, pp. 822–832). Springer-Verlag Berlin Heidelberg Computer Science.

PVM.

Tretola, G. (2007). *Autonomic workflow management in e-collaboration environment*. Department of Engineering University of Sannio, Benevento.

Tretola, G., & Zimeo, E. (2006, April 25-29). Workflow fine-grained concurrency with automatic continuations. In [th *International Parallel and Distributed Processing Symposium*, Rhodes Island, Greece.]. *Proceedings of the IEEE IPDPS, 06*, 20.

Tretola, G., & Zimeo, E. (2007, February 7-9). Activity pre-scheduling in grid workflow. In *Proceedings of the 15th Euromicro International Conference on Parallel, Distributed and Network-based Processing (PDP)*.

Tretola, G., & Zimeo, E. (2007, March 26-30). Client-side implementation of dynamic asynchronous invocations for Web services. In [st *International Parallel and Distributed Processing Symposium*, Long Beach, CA.]. *Proceedings of the IEEE IPDPS, 07*, 21.

Tretola, G., & Zimeo, E. (2007, June 19-20). Structure matching for enhancing UDDI queries results. In *Proceedings of the IEEE International Conference on Service-Oriented Computing and Applications (SOCA'07)*, Newport Beach, CA.

Tretola, G., & Zimeo, E. (2007, July 9-13). Extending Web services semantics to support asynchronous invocations and continuation. In *Proceedings of the IEEE 2007 International Conference on Web Services (ICWS)*, Salt Lake City, UT.

WSRF. (2006). *Web services resource framework 1.2 TC*. OASIS.

Yu, J., & Buyya, R. (2005). *A taxonomy of workow management systems for grid computing* (Tech. Rep.). Grid Computing and Distributed Systems Laboratory, University of Melbourne, Australia.

This work was previously published in Multimodal Human Computer Interaction and Pervasive Services, edited by P. Grifoni, pp. 315-330, copyright 2009 by Information Science Reference (an imprint of IGI Global).

Section VIII
Emerging Trends

This section highlights research potential within the field of social computing while exploring uncharted areas of study for the advancement of the discipline. Chapters within this section discuss social bookmarking, the emergence of virtual social networks, and new trends in educational social software. These contributions, which conclude this exhaustive, multi-volume set, provide emerging trends and suggestions for future research within this rapidly expanding discipline.

Chapter 8.1
Pedagogical Mash Up:
Gen Y, Social Media, and Learning in the Digital Age

Derek E. Baird
Yahoo!, Inc., USA

Mercedes Fisher
Milwaukee Applied Technical College, USA

ABSTRACT

In this chapter we outline how educators are creating a "mash up" of traditional pedagogy with new media to create a 21st Century pedagogy designed to support the digital learning styles of Gen Y students. The research included in this paper is intended as a directional means to help instructors and course designers identify social and new media resources and other emerging technologies that will enhance the delivery of instruction while meeting the needs of today's digital learning styles. The media-centric Generation Y values its ability to use the web to create self-paced, customized, on-demand learning paths that include using multiple platforms for mobile, interactive, social, and self-publishing experiences. These can include wiki, blogs, podcasts and other developing social platforms like Second Life, Twitter, Yackpack and Facebook. New media provides these hyper-connected students with a medium for understanding, social interaction, idea negotiation, as well as an intrinsic motivation for participation. The active nature of today's digitally connected student culture is one that more resourcefully fosters idea generation and experience-oriented innovation than traditional schooling models. In addition, we describe our approach to utilizing current and emerging social media to support Gen Y learners, facilitate the formation of learning communities, foster student engagement, reflection, and enhance the overall learning experience for students in synchronous and asynchronous virtual learning environments (VLE).

DOI: 10.4018/978-1-60566-120-9.ch004

INTRODUCTION TO WEB 2.0 & GENERATION Y

The basic idea of the Web is that an information space through which people can communicate, but communicate in a special way: communicate by sharing their knowledge in a pool. The idea was not just that it should be a big browsing medium. The idea was that everybody would be putting their ideas in, as well as taking them out.

— Tim Berners-Lee

Web 2.0: It's All About Relationships (and Interaction)

In the past social interaction required students and teachers to be tied to a physical space—such as a brick and mortar classroom. But as the Web has evolved, students and teachers have been able to utilize new media technologies to replicate face-to-face social interactions into Web-based learning environments.

This movement of using Web-based platforms for social interaction has been dubbed "Web 2.0." One of the main attributes of Web 2.0 is the transition of the user as passive participant to an active co-participant who creates both the content and context for their experience.

Web 2.0 (social media) is based on three very simple, yet often overlooked principles: 1) humans are inherently social creatures; 2) the continued viability of any social system is rooted in an individual's ability to trust the members of the group and control their level of interaction; and 3) social networking should be used in a situated and engaging context.

A 2005 study by the Pew Internet and American Life Project (Lenhart & Madden, 2005) reported that 48% of teens feel that using the Internet improves their relationships, and 74% report using Instant Messaging (IM) as the technology of choice when it comes to fostering and supporting social relationships with their peers.

In an educational context, social technologies, such as those outlined in the Pew Internet Study, have the potential to engage students in the learning materials and allow them to be included as active participants. Since Gen Y students are drawn to Web 2.0 tools, learning is facilitated by technology as they construct a learning landscape rooted in social interaction, knowledge exchange, and optimum cognitive development with their peers.

Meet Generation Y: Wired, Digital, and Always-On

Raised in the world of interactive, Web-based new media, today's student has different expectations and learning styles than previous generations. A key attribute valued by Gen Y is their ability to use the Web as a platform on which to create a self-paced, customized, interactive, on-demand experience, with plenty of opportunities for social networking with peers, and self-publish content to the Web.

A recent study conducted by the US-based Kaiser Family Foundation (Rideout, Roberts, & Foehr, 2005) found that teens routinely incorporate multiple forms of new and old media into

Table 1. What are the key attributes of Web 2.0?

Foundation Attributes **User-contributed value: Users make substantive contributions to enhance the overall value of a service.** **Network effect: For users, the value of a network substantially increases with the addition of each new user.**
Experience Attributes: **Decentralization:** Users experience learning on their terms, not those of a centralized authority, such as a teacher. **Co-Creation**: Users participate in the creation and delivery of the learning content. **Re-mixability**: Experiences are created and tailored to user needs, learning style, and multiple intelligences by integrating the capabilities of multiple types of social media. **Emergent systems**: Cumulative actions at the lowest levels of the system drive the form and value of the overall system. Users derive value not only from the service itself, but also the overall shape that a service inherits from user behaviors.

(Schauer, 2005)

their daily practice. For example, teens listen to music on their iPod, while simultaneously sending instant messages, watching TV, scouring the Web for information and writing a report for school. The end result is 8.5 hours of media consumption and multitasking squeezed into 6.5 hours a day (Rideout et al., 2005).

Moreover, although 90% of teen online access occurs in the home, the Kaiser Foundation Study (Rideout et al., 2005) found that many students access the Web via mobile devices such as a cell-phone (39%), portable game (55%), or other Web-enabled handheld device (13%).

Everyone involved in education needs to pay attention to these emerging sociological trends and design learning environments that will appeal to the "digital reality" of today's students. While the move from "Mass Media" to "My Media" is a shift in thinking for many, Gen Y views the world of virtual, social and always-on interactivity as their reality.

Understanding Gen Y & Digital Learning Styles

In the 21st Century classroom, the student wants to control the how, what, and when a task is completed. Social media and other Web-based technologies are well suited to provide avenues for students to engage in a social, collaborative, and active dialogue in the online learning environment with their peers and instructor.

A study conducted by the UK-based NESTA FutureLabs (2005) reported that the education "should be reversed to conform to the learner, rather than the learner to the system." Moreover, NESTA found that social media should be used to enable learners to study and be assessed according to their own learning style (BBC, 2005).

Digital learning theory and pedagogical practice also centers on the concept that learning needs to be situated in a social and collaborative context. Discussion among peers can make the often invisible community threads more visible and accessible, and may lead students to find others in the group who share the same interests.

Students are hard wired to look at the variety of available Web 2.0 technologies and then construct their own learning path, and content based on their intrinsic learning needs. As students go through process of choosing, utilizing, integrating and sharing content it provides opportunities for them to be actively engaged, provide and receive feedback, as well as acquire, share, and make use of community knowledge.

More importantly, this emerging digital pedagogy emphasizes providing students with a broad range of technology tools, then providing them with avenues to develop their own understanding and knowledge. As a result, students are highly motivated to discuss and create content, solve problems together, and apply new concepts which relate to their own practice. This approach also provides student's with access to flexible, self-paced, customizable content available on-demand for learning opportunities.

The use of social and new media provides students with an opportunity to self-assess their understanding (or lack of) of the current course topic with their peers. Moreover, as students utilize social technologies to share their thought processes and provide feedback to their learning community, they are able to help each other work through cognitive roadblocks, modify their perceptions, and negotiate their own views while simultaneously building a collaborative peer support system.

In addition, collaborative project-based learning environments help students develop critical thinking and problem solving skills—both essential skills for students to compete in today's global knowledge-based workforce.

Digital Disconnect: Sociological Trends and Implications

As you may expect, traditional academic institutions have generally resisted the influence and in-

Table 2. Gen Y digital learning attributes

Interactive	**Interactive, engaging content and course material that motivates them to learn through challenging pedagogy, conceptual review, and feedback. Students expect to find, use, and "mash up" various types of web-based media: audio, video, multimedia, edutainment and/or educational gaming/simulation.**
Student-Centered	Shifts the learning responsibility to the student, and emphasizes teacher-guided instruction and modeling. Customized, ability to use interactive and social media tools, and ability to self-direct how they learn.
Situated	Reconcile classroom use of social media with how technology is being used outside of the classroom. Use of technology is tied to both authentic (learning) activity and intrinsic motivation.
Collaborative	Learning is a social activity, and students learn best through observation, collaboration, and intrinsic motivation and from self-organizing social systems comprised of peers. This can take place in either a virtual or in-person environment.
On-Demand	Ability to multitask and handle multiple streams of information and juggle both short and long term goals. Access content via different media platforms, including mobile, PC based, or other handheld (portable) computer device.
Authentic	Active and meaningful activities based on real-world learning models. Industry driven problems and situations are the focus and require reflective elements, multiple perspectives and collaborative processes for relevant applicable responses from today's student.

(Baird, 2006)

creasingly pervasive presence of social networking activities in the life of their students, but recently the same institutions have had to look with new eyes at all of the aspects and consequences of these new modes of technological socialization sweeping the younger generations.

— Ruth Reynard

Gen Y students have grown up surrounded by new media and value the ability to choose how, what, where, and when they will learn. According to the 2005 Pew Internet study, teenagers are actively embracing the interactive capabilities of the Internet to create, publish, and share their own content (blogs, podcasts, Web pages, photographs, wiki, and/or video).

In fact, the Pew Internet Study concluded that fully half of all teens who use the internet could be considered *content creators* (Lenhart & Madden, 2005). Students report feeling a sense of growing disconnect between the authentic ways they use technology outside of the classroom and the ways they use it in the classroom (Levin, Arafes, Lenhart, & Rainie, 2002). This growing disconnect has resulted in many students feeling bored and constrained by traditional curriculum and pedagogical theory.

According to the *High School Survey of Student Engagement*, the majority of students interviewed reported they don't feel challenged in their coursework at school. Students also cited that they never or rarely received feedback from their teachers (Sanoff, 2005).

The expectation for highly interactive, flexible, collaborative, and desire to play a more active role in their own learning has already had an impact on the way colleges and even high schools educate students. The Michigan State Board of Education recently mandated that every high school student would have to take at least one online course to receive a diploma (Carnevale, 2005). Among the many reasons cited for adding the online course requirement was the realization that "much learning is going to take place in the 21st Century online."

The combination of social interaction with opportunities for peer support and collaboration creates an interesting, engaging, stimulating, and intuitive learning environment for students (Fisher & Baird, 2005). Effective course design will blend traditional pedagogy with the reality of the new media multitasking learner.

Digital Divide

The infusion of social and new media into the 21st Century pedagogy isn't without challenges. One of the key areas of concern is providing universal access to the Internet and bridging the digital divide between students and/or teachers who have technology and those who don't.

The issues around the digital divide, first raised in the 1990s, continue to be an area of concern. Even in countries where the Internet is widely accessible, there are still regions that remain digitally isolated.

According to 2008 *National Technology Scan*, a report conducted by Parks Associates, nearly 20 million American homes report being without Internet access and/or self-identified as lacking the technological expertise required to create content, search for information, or send an email.

At the 2008 ad:tech Miami Conference, Fabia Juliasz of ibope/NetRatings noted that Internet access in Latin America (Brazil 22%, Mexico 22%, Argentina 26%, and Chile 41%) continues to expand at a steady but slow pace. Since most student consumption of new and social media technologies occurs in the home, lack of at home access makes educational uses of the Internet problematic. This also makes it difficult for the teacher to assign projects or homework to students that require Internet access.

Increasingly there is a trend towards providing professional development online via Web-based platforms (Elluminate, Tapped-In, Facebook, Classroom 2.0, LearnHub etc.) all of which require participants to have Internet access. Teachers without access to the Internet at school or at home will miss out on these valuable opportunities to network with their colleagues and learn emerging and new best practices.

On a positive note, the divide has closed and the rapid adoption of mobile devices and broadband connections will continue to help shrink the divide and provide opportunities for students to participate in mobile learning (mLearning) environments.

NEW MEDIA, SOCIAL NETWORKS, & VIRTUAL LEARNING ENVIRONMENTS (VLE)

Learning requires more than just information, but also the ability to engage in the practice.

— Paul Duguid

The use of new media and social networks in a situated context provides both the structure and building blocks for interaction to take place. The end result is an environment which combines social media, Web-based information resources, and communities to provide a more diverse, active, and engaging learning experience.

Papert (1996) asserts that learning "is grounded in the idea that people learn by actively constructing new knowledge, rather than having information 'poured' into their heads." Moreover, he asserts that people learn with particular effectiveness when they are engaged in constructing personally meaningful artifacts", such as Weblogs, iPod, podcasting/audio blogs, wiki, social bookmarking, and other types of user-generated content (UGC).

How Social Media Supports Digital Learning Styles

The formation of an online learning community allows students to learn in a social context and turn to peers who are subject matter experts for immediate feedback and assistance. This approach also provides opportunities for students to learn through a cognitive apprentice with instructor and/or peers. In addition, opportunities should be provided for students to quantify their knowledge and skills in order to help them identify their place as well as other students with specific expertise. It's important to allow a community the freedom to discover where they fit in the learning community.

The collaborative and interactive aspects of projects undertaken in a course allow students to interact with other members of the class, allow students to identify who has a particular skill or expertise they want to acquire, and then provides opportunities for them to model and scaffold this knowledge from their peers. In addition the virtual learning environment allows students to explore and negotiate their understanding of the course content and find ways for the learner to develop a sense of intellectual identity (Papert, 1999).

When students collaborate they form social ties which motivate them to establish an identity within the group through active participation and contributions to the collective knowledge pool. Through this process learners become motivated on an individual level as well as fostering a sense of accountability to the group to continue to participate.

Anthropologist Lori Kendall, who spent almost two years researching the dynamics of social identity and community, concluded that members of virtual environments have intact social systems, and at times highly charged social relations. But unlike the electronic window of television, Kendall found that members of an online community feel that when they connect to an online forum, they enter a social, if not physical space (Kendall, 1999).

In this new digital age, we need to redefine our concept of what constitutes a legitimate "social system", "learning community" or "social interaction." In many ways, the effective use of new media to support instruction provides the same or better quality of socialization than a traditional classroom. If we are truly to expand educational opportunities via virtual learning environments and social networks, we will need to recognize and validate the existence of online communities, relationships, and interaction.

Teaching & Learning in Social Networks

Critics of e-learning often characterize online classrooms as neutral spaces devoid of human connection, emotion, or interaction with instructors or peers. However, effective use of social networking and new media technologies provides educators and students with the ability to interject emotion in the online space, thereby providing opportunities for peers to make emotional connections with classmates, and create a community of practice just as they do in the 'real time' world of the brick and mortar classroom. Social networks can also provide an outlet for students who are socially isolated or shy in the traditional classroom, a way connect, share ideas and collaborate with their peers.

Clearly, the key to a successful online user experience is to help students find ways to construct relationships with their peers, while simultaneously meeting their digital learning styles. A digital ethnographic study conducted by Goldman-Segall (1997) at the University of British Columbia pointed out how media tools create a constructivist learning environment which allows people to build interpretations of their data and utilize their individual life experience, multiple intelligences, while still working as part of a collaborative team.

Table 3. Education social networks

Platform	url
LearnHub	http://learnhub.com
Tutor Linker	http://tutorlinker.com/
Apple Learning Interchange	http://edcommunity.apple.com/ali/
Discovery Education Network	http://www.discoveryeducatornetwork.com/
Elgg Spaces	http://elggspaces.com/
Classroom 2.0/Ning	http://classroom20.com

Tosh and Werdmuller (2004) point out that students can use social networking to create their own learning and social communities. These self-directed learning communities could then provide resources, increase engagement in the course content, as well as provide a "network of knowledge transfer."

In the same vein as Vygotsky and other social learning theorist, their "power in the process" hypothesis states that the development of optimum cognitive development is rooted in the social exchange of information on both "the individual and collective levels" resulting in "opportunities to build one's learning instead of just being the recipients of information (Tosh & Werdmuller, 2004)."

Social networking media is an effective and authentic tool that engages the user in the content and allows them to be included as an active participant social interaction, knowledge exchange, and engagement with their peers.

Theory to Practice: Social Media in the Classroom

While teens have become increasingly hyper-connected and mobile, schools have been slow to respond to this cultural shift in the way students learn and communicate with each other. For the most part, educators, parents and school administrators have responded to the new digital reality by filtering, blocking, and restricting the use of digital devices, Web sites, new media and social networking in the classroom.

This growing tension between the digitally wired teens and their schools is reflected in a 2007 study by the US-based National School Board Association which showed that 96% of students use social networking technologies, such as chatting, text messaging, blogging, and participating in online communities such as Facebook, MySpace, and Webkinz, or Moshi Monsters (NSBA, 2007).

Adding to the growing sense of disconnect between wired teens and their schools, nearly 96% of districts participating in the NSBA study report that teachers are assigning homework that requires internet (NSBA, 2007).

Moreover, the NSBA study found that nearly 60% of online students report discussing education-related topics such as college or college planning, learning outside of school, and 50% of students reported that they use social networks to connect with peers to talk specifically about schoolwork. In short, today's Web-savvy students are stuck in text-dominated classrooms.

Preparing Teachers for 21st Century Learning

The other challenge is providing educators with the necessary professional development and training they need in order to effectively integrate new and social media technologies into their curriculum, as well as helping them develop a deeper understanding of the sociological shifts in students' learning styles.

In his book, "*Disrupting Class: How Disruptive Innovation Will Change the Way the World Learns*", Harvard University professor Clayton Christensen focuses on how education, technology, and innovation will impact the future of learning.

Among other things, Christensen predicts that by 2019 half of all high school courses will be taught online. If learning moves online as Christensen predicts, what are the implications for educators?

Teaching online with new and social media requires a different pedagogical approach from traditional teaching methods. Which raises an important question: Are educators getting the training and/or professional development required to teach our 21st Century students?

In the immediate future, teachers will need access to the correct pedagogical training for this shift — especially so they can realize the possibility that new and social media technology can truly improve learning.

Student Safety & Social Networks

Many educators face resistance from parents and school administrators about student use of the social Web. As a result, many schools use Web filters that block out large swaths of information. Understandably, the concern is that students will encounter inappropriate information or sexual predators.

However a recent study in the Journal of the American Psychologist (Wolak, Finkelhor, David, Mitchell, and Ybarra, 2008) found that many of the beliefs about sexual predators on the Web are overblown and, in some cases, not true. The study found that *"the stereotype of the Internet 'predator' who uses trickery and violence to assault children is largely inaccurate."*

While there isn't an easy solution when it comes to student Internet use, parents, teachers, and educators--need to take a less hyped, rational, measured approach on using social media in the classroom—and at home.

As a community, educators need to work on educating students to be more aware of the potential hazards and implications of disclosing too much personal information on social networking sites like MySpace and Facebook.

At a time when teens are constantly being reminded about the dangers lurking in social networks, it's always good to remind them that there are still plenty of dangers left in the non-digital world.

PEDAGOGICAL MASH UP: GEN Y & LEARNING IN THE DIGITAL AGE

Perhaps our generation focused on information, but these kids focus on meaning -- how does information take on meaning?

– John Seeley Brown

The variety of Web 2.0 tools are providing students with the opportunity to socialize around the context of the content, in terms of subject matter, production and commentary. These experiences become integrated into today's use of everyday devices in the everyday lives of the students for whom we design. As a result, the learning is embedded in and transferable to other contexts for the student.

Here we provide an overview of the current wave of Web-based tools and outline how social and new media can work together to support learning, and foster community in the frontline and offline classroom.

Social Bookmarking and Social Search

Social bookmarking provides students with a platform to exchange and share information found on the Web. As students search the Web they can save their search results, tag them with keywords, and then depending on whether they have marked the links private or public, share their pool of links and resources with their learning community and classmates.

Members of a community can also search, structure, and self-organize content via tags (keywords). You can then see what resources they are sharing with the community and add the ones you find most relevant to your tag list. And vice-versa. In this way, social bookmarking becomes an organic learning tool, evolving with the interests and needs of the community and the course.

Table 4. Social search/bookmarking resources

Platform	url
del.icio.us	http://del.icio.us
H20 Playlist	http://h2obeta.law.harvard.edu
Rollyo	http://rollyo.com
Blinklist	http://blinklist.com
Diigo	http://diigo.com

Using Social Bookmarking to Support Instruction

A teacher can place links in a community knowledge repository as a jumping off point for students. As students begin to research a topic, they can add content to and search the community pool. In this manner, students are scaffolding their own knowledge and the teacher is working as a facilitator, instead of a *"sage on the stage"*.

Weblogs/Blogs

Weblogs, more commonly known as "blogs", allow students to publish their thoughts and reflections while participating in a collective environment. As students reflect on their own Weblog entries, read their peers posts, receive feedback and network with their community of learners they are creating an environment for knowledge transfer to take place.

The user's ability to connect with members of their learning community via differing types of social media is an important consideration for today's learner. The interactive, collaborative, engaging nature of a blog combined with the ability to instantaneously publish content on the Web, enables students to use technology as a vehicle for presenting their own work as well as providing opportunities for feedback from their peers.

Moreover, blogs give students a chance to read, write, and expand their computing skills. For example, if one student reads another student's blog and sees a video in the blog, they want to learn how to complete that same skill. As a result, they collaborate with their peers to learn how to complete the same task (put video in a blog).

Vlogs or Movlogs are blogs which allow users to put video content on their blog. Platforms such as Flickr, contain mobile blogging tools, titled Moblogs, in which users post photographs or video taken from their camera enabled mobile phone.

Using Blogs to Support Instruction

The key feature of blogs is the author's ability to self-publish in an easy and quick manner on the Web. Students could be required to maintain a Web log (blog) or other Web-based journal throughout the program, as well as individual blogs for each course.

Reflection is a major component in online courses, and provides students with an avenue for expressing their own observed growth, and ability to make multiple connections within a course. Many students today use these types of blogs naturally and almost automatically.

In addition, unlike previous generations, today's digital student doesn't learn or consume information in a linear path. Rather they have

Table 5. Student perspectives on blogging

Source	Comment
• Synchronous Course Discussion	"One 'attitude' that might have changed for me regarding blogs, is that they don't necessarily have to be eloquently written (personal conversation, Mar 1, 2005)"
• Course Blog	"Other than using mandatory course-related academic discussion boards, I have never participated in this particular style of communication medium. It is necessary to become technologically informed and literate so thanks for providing this opportunity (personal conversation, February, 2005)."
• Synchronous Course Discussion	"I think if there is a focus or topic to blog on then the impact on a learning community would be tremendous—a guided blog. This type of journaling would offer a variety of POVs (point of views) and foster a culture of learning (personal conversation, March 1, 2005)."

Table 6. Weblog/blog resources

Platform	url
Blogger	http://www.blogger.com
Vox	http://vox.com
Squidoo	http://www.squidoo.com
Typepad	http://www.typepad.com
Wordpress	http://wordpress.com
Edublogs	http://edu.blogs.com
Gaggle	http://gaggle.net
Vlogs/Movlogs (Video Blog)	
Platform	**url**
Blip.tv	http://www.blip.tv
OurMedia	http://www.ourmedia.org
YouTube	http://youtube.com
Jumpcut	http://jumpcut.com
Vimeo	http://vimeo.com
Moblogs (Mobile Blog)	
Platform	**url**
Flickr	http://flickr.com
Shozu	http://shozu.com
Vox	http://vox.com

an "always-on" learning style that is driven by their intrinsic interests and looking at chunks of materials how and when they want.

By integrating a blog into your course, your class materials are available "on demand" thereby meeting the new digital learning styles of today's Gen Y student. In addition, students are able to utilize the latest in mobile technologies to access a myriad of information—including your course blog, right from their mobile phone.

Podcasts & Audio Tools

The Kaiser Family Foundation Study (Rideout et al., 2005) found that 65% of teens have a portable mp3 device. The ubiquitous use of these types of portable devices provides educators with a unique opportunity to use podcasts as a mobile content delivery tool. Not only will students and teachers will be able to use podcasting technology to locate and then download audio content, but it will also provide them with the software and tools to be able to create and share their own content in a podcast.

Teachers who incorporate podcasting into their curriculum cite many benefits, including an increased sense of student motivation stemming from community feedback, authentic and situated use of social technology in an instructional context, and the freedom to download the podcast content "on-demand."

Using Podcasts to Support Instruction

Podcasting will allow teachers to easily publish (or *podcast*) lectures, photos (perfect for the art history or architecture student), or foreign language accents pronunciations and drills, along with a myriad of other course content. Students

Table 7. Podcasting/audio resources

Platform	url
Kidcast	http://www.ftcpublishing.com/kidcast.html
Odeo	http://odeo.com
Education Podcast Network	http://epnweb.org
Yahoo! Podcasts	http://podcast.yahoo.com
iTunes U	http://www.apple.com/education/itunesu/
BBC MP3	http://www.bbc.co.uk/radio4/history/inourtime/mp3.shtml
Yahoo! Audio Search	http://audio.search.yahoo.com/audio/learnmore

will be able to subscribe to a course content feed and then automatically receive the content on their mp3 device.

YackPack

Developed by researchers at the Persuasive Technology Lab at Stanford University, YackPack is a social audio platform that allows users to record and send audio messages to friends inside privately formed groups. While there are other products that provide avenues for collaboration over the Web—most notably message boards, email, and instant messaging—YackPack is among the first social media tools to allow users to share both live and asynchronous voice messages.

The ability to interject voice into an online space is important because it provides opportunities for members of a community to convey the expression, emotion, and intimacy embedded in human speech. The ability to integrate human speech into the curriculum becomes even more important in pure online learning context where students and teachers only meet in a virtual environment.

Table 8. Yackpack resources

Yackpack	http://yackpack.com
StorytellingU	http://storytellingu.com
Yacklearning	http://yacklearning.net
Yackpack + PBWiki	http://www.blip.tv/file/196824

Using Audio Messaging to Support Instruction

An EFL (English Foreign Language) teacher (or Spanish, German, etc.) can post audio messages (verb conjugation, dialogue, etc) to an entire class. In turn, the students can respond to the teacher via a YackPack audio message. Instructors can also use YackPack as a tool to provide narrative feedback, assessment, and student support. In addition, you can also post Yackpack audio in PBWiki.

Wiki

A wiki is a collaborative Website where members can add, delete and change the content as needed. Wiki's can be used to brainstorm on ideas, create "work-in-progress" drafts, organize content, and provide participants with opportunities for interaction. Wikipedia is one of the most extensive and popular wiki's on the Web.

Many wiki clients allow you to create a mash up of rich media such as video, audio, PowerPoint, RSS feeds, widgets and other social media into your wiki. Not only does this make your wiki more interactive, but it also allows you to offer a variety of media that supports the multiple learning styles of students.

The WikiMedia Foundation is a non-profit organization that maintains several wiki's including one of the most well known, Wikipedia, a Web-based collaborative encyclopedia project.

Table 9. Wiki resources

Platform	url
PBWiki	http://pbwiki.com
Swicki	http://hnu.ida.liu.se/scwiki/Wiki.jsp
Wikimedia/Wikipedia	http://www.wikimedia.org
Zoho	http://wiki.zoho.com
Wetpaint	http://wetpaint.com
Social Text	http://www.socialtext.net/medialiteracy/index.cgi?wiki_resources
Miki (mobile wiki)	http://www.socialtext.com/node/75

Since WikiMedia is an open-source technology, students take can actively contribute to any of the WikiMedia projects.

Using a Wiki to Support Instruction

An instructor can have students form groups, conduct research on a topic of their choice, and then add their findings to the corresponding entry in Wikipedia. Or teachers could start a wiki to share teaching resources, curriculum ideas, or a forum for community support and interaction.

Wiki's are well suited to facilitate collaboration, communication and extend learning between peers. Most wiki clients provide privacy controls allowing you to choose which wiki pages you want to be public. Most importantly, wiki's provide a platform where everyone can contribute their ideas and extend the virtual boundaries of classrooms.

RSS

Really Simple Syndication (RSS) technology is an XML based format that provides the backbone for the distribution of Weblog, podcasting, and other content. RSS allows users to easily syndicate or publish their content for use by others.

There are several free RSS aggregators or news readers available, including *Bloglines*, *Feedburner*, *My Yahoo!*, *Google Reader* and *Yahoo! Pipes*. After a user subscribes to a RSS feed, the content (blogs, Websites, online community groups) automatically updates and is displayed in the feed reader. RSS readers also allow students to self-publish and share their content feed with members of their learning community.

Using RSS to Support Instruction

A key benefit is the user's ability to pick and choose (subscribe) to a particular RSS feed and then have the content updated in real time. In this manner, RSS is an important educational media tool to facilitate and support the "always on" learning styles of Gen Y.

RSS readers allow students to self-publish and share their content feed with members of their learning community. The use of RSS further supports multiple learning styles by allowing the user to select which content is relevant and then have it delivered directly to them for "*on demand*" viewing at their convenience.

As an assessment tool, RSS feeds provide teachers with several benefits. For example, instructors can subscribe to each students RSS feed and have their homework delivered directly into their aggregator, saving them the time consuming task of entering each student's URL in order to view their e-portfolio or blog.

Flickr

Sharing photos is an inherently social activity and Flickr, a Yahoo! company, was the first Web-based photo hosting service to successfully translate this experience into the online space. The key element that makes Flickr so unique is that active exploration and community are interwoven as main components of the design.

Flickr is important because its ease-of-use allows students to keep their focus on acquiring new skills, building on existing knowledge while at the same time developing writing, software, and strengthening social ties within their learning circle. This is especially important in geographically dispersed learning communities, where students may have limited face-to-face time to build a support network with their peers.

Using Flickr to Support Instruction

One of the unique features of Flickr is the ability of users to use their camera phones to take and upload pictures directly to their photoblog. Since most students already have access to a camera enabled cell phone, students can integrate Flickr into a mLearning activity. For example, students can use their camera phone fon a field trip to take pictures, and easily post them to their own Flickr photoblog. Later, students can write about their experiences on the field trip, reflect, and share their thoughts with their learning community via a Flickr group (Baird, 2005).

Flickr holds great potential as part of a multi-faceted approach that blends constructivist learning theory and mobile technologies in the curriculum. To be sure, Flickr and other mobile social media cannot, and should not, replace face-to-face communication between teachers and students; rather, it should be used as one of many digital tools that, when skillfully integrated into the curriculum, has the potential to open lines of dialogue, communication, and learning.

One of the challenges for educators is finding open copyright images and graphics they can use in their classroom. A partnership between Creative Commons, a non-profit that provides an alternative to copyright, Flickr and the generosity of the Flickr community has resulted in over a million photographs being made available for educators to use in their classroom.

Flickr provides educators with a powerful resource that can support differentiated instruction and support the multiple learning styles of their students. The visual and interactive nature of Flickr supports students who excel in learning activities that are centered on visual, kinthestic, and tactile learning activities.

Moreover, Flickr provides opportunities for students and instructors to create an engaging, open, and decentralized learning environment where ideas, creativity, and dialogue can be shared in an "always on" format that meets the needs of today's digital learner.

EDUCATION 2.0: MASH UP, REMIX, REUSE

A mash up is a Website, widget, or Web application that uses content from more than one source to create a completely new service (Wikipedia, 2006). They combine separate, stand-alone technologies into a new application.

The following chart illustrates how mash ups of new media platforms have been mashed up to create social and interactive learning activities that appeal to the digital and mobile sensibility of Gen Y students.

EMERGING EDUCATIONAL MEDIA

The fates guide those who go willingly; those who do not, they drag.

— Seneca

Looking towards the future, the next wave of learning will take place in the mobile space. The convergence of mobile technologies into student-centered learning environments requires academic institutions to design new and more

Table 10. RSS resources & tools

Platform	url
Yahoo! Pipes	http://pipes.yahoo.com/pipes
Google Reader	www.google.com/reader
Bloglines	http://www.bloglines.com
Feedburner	http://www.feedburner.com
New York Times RSS Generator	http://nytimes.blogspace.com/genlink

Table 11. Educational mash-up of Web 2.0 platforms

Platform	URL	About
Flickr		
delivr	http://www.delivr.net	Search Flickr tags to find photos and create postcard or greeting cards.
Slide	http://www.slide.com/flickr	Create embeddable slideshows using Flickr tag(s).
Spell with Flickr	http://metaatem.net/words	Use Flickr tags to enhance your spelling lists.
Huge Big Labs	http://bighugelabs.com/flickr	Several Flickr mashups including mosaic maker, slideshows, calendar & more.
Bubblr	http://pimpampum.net/bubblr/	Create comic strips using Flickr photos and/or tags.
Findr	http://www.forestandthetrees.com/findr	Use Findr to locate photographs by related tags and refining your tag search.
Spell with Flickr	http://www.krazydad.com/defacement/squirclescope.php	Create a kaleidoscope using Flickr tags.
North American Wildflower Guide	http://www.flickr.com/groups/wildflowers/	Search and discover hundreds of images of North American wildflowers.
Boardr	http://gallery.yahoo.com/apps/12356/locale/en	Create a storyboard using Flickr photos.
Google Maps		
Oral History of Route 66	http://maps.google.com	View landmarks & read narratives of the historic Route 66
Lit Trips	http://www.googlelittrips.org	Google Earth Maps mashed together with pictures, videos and other information tied to classic literature.
Google Mars	http://www.google.com/mars/	View topography, narratives of space explorers, and view spacecraft used to explore the Red Planet. Created in conjunction with NASA.
Jack Kerouac	http://maps.google.com	View landmarks and see pictures of the places in Kerouac's "On the Road."
Yahoo! Maps		
Exploring Shakespeare	http://gallery.yahoo.com/apps/5551/locale/en	Created by the Kennedy Center, this map plots the life and works of William Shakespeare.
Geologic Atlas of the United States	http://gallery.yahoo.com/apps/2490/locale/en	Map gallery of geologic features.
Life in San Francisco	http://gallery.yahoo.com/apps/1/locale/en	Watch videos mashed together with Yahoo! Maps explore San Francisco.
Disappearing Places	http://gallery.yahoo.com/apps/13126/locale/en	Archive and collective map of places that no longer exist.
Mile Calculator	http://gallery.yahoo.com/apps/11863/locale/en	Allows users to drag a path using the mouse on a mapped location and finds the miles or kilometers traversed over it.
Yahoo! Pipes (RSS)		
Yahoo! Pipes	http://pipes.yahoo.com/pipes/pipe.info?_id=hMj_M5_42xG0ZSPIJhOy0Q	Edublog mash up
WPR Science & Education	http://pipes.yahoo.com/pipes/pipe.info?_id=Drfl595U3BG5Ef_VouNLYQ	WPR interviews on science and education.
Second Life	http://pipes.yahoo.com/pipes/pipe.info?_id=qswEzwu92xGA5Up_lfXiAA	Mash up of RSS feeds on using Second Life in education.
CS Education	http://pipes.yahoo.com/pipes/pipe.info?_id=Hr9BCQTE2xGFf5qPmLokhQ	Mash up of K-12 CS education blogs.

Table 11, continued

Platform	URL	About
Wiki		
Yackpack + PBwiki	http://www.blip.tv/file/196824	Video showing how to embed Yackpack audio into a PBWiki.
WikiMapia	http://www.wikimapia.org/	Community generated content and Google Map mash up.
Musipedia	http://www.musipedia.org/	Collective musical encyclopedia & wiki platform.
Frappr	http://www.frappr.com/	Social map application; created with Google Maps. Users can create maps and embed on wiki, blog or web page.
MemoryWiki	http://memorywiki.org	Community generated collective encyclopedia of first-person narratives of historical events. Created using the Wikimedia suite of tools.
MS Office		
Blogger for Word	http://buzz.blogger.com/bloggerforword.html	Publish to Blogger via MS Word plug-in.
Creative Commons for MS Office	http://wiki.creativecommons.org/Microsoft_Office_Addin	Easily apply a Creative Commons license to your MS Word documents with this plug-in.

effective learning, teaching, and user experience strategies.

The rapid adoption of wireless, mobile and cloud computing by Gen Y learners will require educators to designl earning environments for wireless, mobile, or other portable Web-enabled devices (video iPod, PSP, Palm, iPhone). In addition to mLearning (mobile learning), Web applications like Twitter, Facebook and Second Life hold great promise as an educational platform.

Twitter

Twitter is an online microblogging application that is part blog, part social networking site, and part mobile phone/IM tool. It is designed to let users describe what they are doing or thinking at a given moment in 140 characters or less. As a tool for students and faculty, Twitter could be used academically to foster interaction and support metacognition (Educase, 2007).

Twitter also holds great promise as a way for seasoned educators to easily and quickly share their practice with novice or pre-service teachers. In this way, Twitter is being used as a digital legitimate peripheral participation or mentoring tool (Holahan, 2007).

Facebook

Facebook has taken an open source approach by releasing an API which allows developers to create Facebook Applications for the education community. The Facebook team has issued a call to action for the developer community to "create the applications that help people connect, track, and collaborate with their teachers, professors, and classmates (Moran, 2007)."

This open platform approach has resulted in an influx of new educational oriented Facebook Apps as well as a mash up of existing Web 2.0 tools. For example the wiki you created with Zoho can now be used in Facebook with a mash up between Zoho and Facebook. Other popular education tools like Slideshare, Flickr, Twitter, delicious and YouTube have all recently created Facebook applications.

Table 12. Second life teaching resources

Platform	url
Tutorial for Teen Second Life	http://wintermute.linguistics.ucla.edu/lsl
Salamander Project	http://www.eduisland.net/salamanderwiki/index.php?title=Main_Page
Second Life Tutorial	http://cterport.ed.uiuc.edu/technologies_folder/SL

Second Life

Second Life is an advanced virtual world simulation where users can create their own avatar (digital identity) and connect with other members of the Second Life community. Many higher education institutions, including ISTE, have already set up a virtual campus, classroom space and other learning environments within the Second Life grid.

This globally connected virtual learning environment (VLE) is also being used as a way to supplement traditional classroom activities, provide avenues for collaboration, as well as hosting distance-learning courses. There is also a new mobile version of Second Life that allows users to be connected anywhere they have an internet connection.

iPhone

The iPhone, a mobile device created by Apple, is getting a lot of buzz in educational circles as the next "killer app" for e-learning. In fact, since its release, several higher education institutions have started pilot programs to test the viability of using the iPhone as a mobile learning platform.

The App Store on iTunes has thousands of applications, many of them educational, that users can download to their iPhone. In addition, students can download podcasts and video from iTunes U and YouTube, created by their professors, onto the iPhone for on-demand viewing.

Learning 3.0: Mobile Learning

The use of mobile technologies continues to grow and represents the **next frontier for learning**. Increasingly we will continue to see academic and corporate research invest, design and launch new mobile applications, many of which can be used in a learning context. Learning 3.0 and beyond will be about harnessing the ubiquity of the mobile phone/handheld device and using it as an educational tool.

At the 2006 International Consumer Electronic Show, Yahoo! CEO Terry Semel outlined the explosive growth of mobile technology. According to Semel (2006), there are 900 million personal computers in the world. But this number pales in comparison to the 2 billion mobile phones currently being used in the world.

Even more astounding is how mobile devices are increasingly being used as the primary way in which people connect to the Internet. In fact, Semel notes that 50% of the Internet users outside the US will most likely never use a personal computer to connect to the Internet. Rather, they will access information, community, and create content on the Internet via a mobile device.

The convergence of mobile and social technologies, on-demand content delivery, and early adoption of portable media devices by students provides academia with an opportunity to leverage these tools into learning environments that seem authentic to the digital natives filling the 21st Century classroom. Clearly, the spread of Web-based technology into both the cognitive and social spheres requires educators to reexamine and redefine our teaching and learning methods.

Table 13. Other emerging educational technologies

Platform	url
Twitter (Mobile)	http://twitter.com
Second Life	http://secondlife.com/education
Red Halo (Mobile)	http://redhalo.com

The 2005 study conducted by the USA-based Kaiser Family Foundation (Rideout et al, 2005) found that, although **90% of teen online access occurs in the home**, most students also have Web access via mobile devices such as a **mobile phone (39%), portable game (55%), or other Web-enabled handheld device (13%).**

Palm estimates that mobile and handheld devices for public schools will be a **300 million dollar** market. A few progressive school districts in the USA, UK, and Ireland have already started using mobile devices in the classroom. Mobile devices are also seen by many as the solution bringing Internet access and information to students living in developing countries.

In order to create a better learning environment designed for the digital learning styles of Generation Y, there is a need to use strategies and methods that support and foster motivation, collaboration, and interaction. The use of mobile devices is directly connected with the personal experiences and authentic use of technology students bring to the classroom (Fisher & Baird, 2006).

The use of mobile technologies is growing and represents the next great frontier for learning. Increasingly we will continue to see academic and corporate research invest, design and launch new mobile applications, many of which can be used in a learning context.

CONCLUSION: IT'S ABOUT LEARNING, NOT TECHNOLOGY

With knowledge doubling every year or two, "expertise" now has a shelf life measured in days; everyone must be both learner and teacher; and the sheer challenge of learning can be managed only through a globe-girdling network that links all minds and all knowledge... We have the technology today to enable virtually anyone to learn anything... anywhere, anytime.

— Lewis Pereleman

Looking towards the future, perhaps the advice of management guru Peter Drucker provides educators with a mantra for teaching in the digital age: "*We now accept the fact that learning is a lifelong process of keeping abreast of change. And the most pressing task is to teach people how to learn.*"

The proliferation of old and new media, including the Internet and other emerging social and mobile technologies, has changed the way students communicate, interact, and learn. And a new digital pedagogy, based on authentic learning activities, observation, collaboration, intrinsic motivation, and self-organizing social systems, is emerging to meet the needs of Gen Y students filling our educational institutions.

In many cases, students spend as much (or more) time, receive more feedback, and interact with peers more in an online environment than they do with their teachers in the classroom. In fact, a 2002 Pew Internet Study (Levin, et al, 2002) found that 90% of student media consumption (8 hours worth) occurs *outside the classroom*.

Now more than ever, instructors must keep track of these sociological trends and learn how to effectively integrate social media into their curriculum as a means to meet both the learning goals and digital learning styles of their Gen Y students.

AUTHOR NOTE

Links for all of the resources, references, and services cited in this chapter can be found at http://del.icio.us/mashup.edu

REFERENCES

Baird, D. (2005). FlickrEdu: The Promise of Social Networks. *TechLEARNING, 4*(22).. San Francisco, CA: New Bay Media.

Baird, D. (2006). Learning 2.0: Digital, Social and Always-On. *Barking Robot*. Retrieved August 3, 2007 from http://www.debaird.net/blendededunet/2006/04/learning_styles.html

British Broadcasting Corporation. (2005). Make Lessons 'Fit the Learner'. *BBC News Education*. Retrieved November 29, 2005 from http://news.bbc.co.uk/1/hi/education/4482372.stm

Carnevale, D. (2005). Michigan Considers Requiring High-School Students to Take at Least One Online Course. *Chronicle of Higher Education*. Retrieved December 14, 2005 from http://chronicle.com/free/2005/12/2005121301t.htm

Christensen, Clayton. "*Disrupting Class: How Disruptive Innovation Will Change the Way the World Learns.*" Harvard University Press. Cambridge, MA.

Educase (2007). 7 Things You Should Know. *EDUCASE Learning Initiative*. Retrieved August 8, 2007 from http://www.educause.edu/7Things YouShouldKnowAboutSeries/7495

Fisher, M. & Baird, D. (2005). Online Learning Design that Fosters Student Support, Self-Regulation, and Retention. *Campus Wide Information Systems: An International Journal of E-Learning, 22*.

Fisher, M., & Baird, D. (2007). Making mLearning Work: Utilizing Mobile Technology for Active Exploration, Collaboration, Assessment and Reflection in Higher Education. *Journal of Educational Technology Systems, 35*(1).

Fisher, M., Coleman, P., Sparks, P., & Plett, C. (2006). Designing Community Learning in Web-based Environments. In B.H. Khan, (Ed.), *Flexible Learning in an Information Society*. Hershey, PA: Information Science Publishing.

Goldman-Segall. (1998). *Points of Viewing Children's Thinking: A Digital Ethnographer's Journey*. Mahwah, N.J.: Erlbaum.

Holahan, C. (2007). The Twitterization of Blogs. *Business Week*. Retrieved August 3, 2007 from http://www.businessweek.com/technology/content/jun2007/tc20070604_254236.htm

Kendall, L. (2002). *Hanging Out in the Virtual Pub: Masculinities and Relationships Online*. Berkeley, CA: University of California Press.

Lenhart, A., & Madden, M. (2005). Teen Content Creators and Consumers. *Pew Internet & American Life Project*. Retrieved November 4, 2005, from http://www.pewinternet.org/PPF/r/166/report_display.asp

Levin, D., Araheh, S., Lenhart, A., & Rainie, L. (2002). The Digital Disconnect: The Widening Gap Between Internet-Savvy Students and their Schools. *Pew Internet and American Life*. Retrieved January 5, 2006, from http://www.pewinternet.org/report_display.asp?r=67

Massachusetts Institute of Technology. (2006). *2005 Program Evaluation Findings Report*. Available at http://ocw.mit.edu/NR/rdonlyres/FA49E066-B838-4985-B548-F85C40B538B8/0/05_Prog_Eval_Report_Final.pdf (last accessed Sept. 19, 2006).

Moran, D. (2007). Goodbye Facebook Courses, Hello Facebook Platform Courses. *The Facebook Blog*. Retrieved August 11, 2007 from http://blog.facebook.com/blog.php?post=4314497130

National School Board Association. (2007). *Creating & Connecting: Research and Guidelines on Online Social and Educational Networking.* Retrieved August 14, 2007 from http://nsba.org/site/doc.asp?TRACKID=&VID=2&CID=90&DID=41336

Navidad, A. Potentially Useful Data on Latin American Internet Culture. *ad:tech Blog*. Retrieved June 3, 2008 from http://www.adtechblog.com/archives/20080603/potentially_useful_data_on_latin_american_internet_culture/

Papert, S. (1993). *The Children's Machine: Rethinking School in the Age of the Computer*. New York: Basic Books, Inc. Parks Associates. *National Technology Scan*. Retrieved May 13, 2008 from http://newsroom.parksassociates.com/article_display.cfm?article_id=5067

Pope, J. (2006). Some Students Prefer Classes Online. *ABC News*. Retrieved January 15, 2006 from http://abcnews.go.com/Technology/wireStory?id=1505420&CMP=OTC-RSS-Feeds0312

Reynard, R. (2008). Social Networking: Learning Theory in Action. *Campus Technology*. San Francisco, CA. Retrieved May 29, 2008 from http://www.campustechnology.com/articles/63319/

Richmond, T. (2006, September 15). OER in 2010 – Wither Portals? *Innovate Journal of Online Education, 3*(1), October/November. Online wiki article retreived Sept. 21, 2006 from http://www.nostatic.com/wiki/index.php/Main_Page Rideout, V, Roberts, D., & Foehra (2005). Generation M: Media in the Lives of 8-18 Year Olds. *Kaiser Family Foundation Study.* Retrieved November, 24, 2005, from http://www.kff.org/entmedia/7251.cfm

Sanoff, A. (2005). Survey: High School Fails to Engage Students. *USA Today.* Retrieved January 5, 2006, from http://www.usatoday.com/news/education/2005-05-08-high-school-usat_x.htm

Schauer, B. (2005). Experience Attributes: Crucial DNA of Web 2.0. *Adaptive Path*. Retrieved December 5, 2005 from: http://www.adaptivepath.com/publications/essays/archives/000547.php

Semel, T. (2006). *Yahoo! Keynote at 2006 International Consumer Electronics Show (CES)*. Retrieved January 6, 2006 from: http://podcasts.yahoo.com/episode?s=fa88e89d49dbbdbc77221b561570105a&e=15

Tosh, D., & Werdmuller, B. (2004). *Creation of a learning landscape: Weblogging and social networking in the context of e-portfolios*. Retrieved April 15, 2005 from: http://www.eradc.org/papers/Learning_landscape.pdf

Wikipedia, (2006). *Mashup (Web application hybrid)*. Retrieved (n.d.), from http://en.wikipedia.org/wiki/Mashup_(Web_application_hybrid)

Wikipedia, (2006). *RSS (file format)*. Retrieved (n.d.), from http://en.wikipedia.org/wiki/RSS_file_format

Wikipedia, (2007). *Multiple Learning Styles.* Retrieved (n.d.) from http://en.wikipedia.org/wiki/Multiple_intelligence

Wolak, J., Finkelhor, D., Mitchell, K. J., & Ybarra, M. L. (2008). Online "predators" and their victims: Myths, realities, and implications for prevention and treatment. *The American Psychologist, 63*(2), 111–128. doi:10.1037/0003-066X.63.2.111

KEY TERMS AND DEFINITIONS

Blog: A blog, short for "Weblog", is a Web site in which the author writes their opinions, impressions, etc., so as to make them public and receive reactions and comments about them.

Instant Messaging (IM): Instant messaging is the act of instantly communicating between two or more people over a network such as the Web.

Mash Up: A Web application that combines data from more than one source into an integrated experience.

Moblog (Mobile + Blog): A site for posting blog content from a mobile device, usually a cellular phone. Most often refers to photo sharing via a camera phone.

Palm: A handheld portable device or personal digital assistant.

Really Simple Syndication (RSS): Really Simple Syndication feeds provide Web content or summaries of Web content together with links to the full versions of the content. RSS is used by news Websites, Weblogs and podcasting to synch and deliver content.

SMS (Short Message Service): Written messages that you can send through a mobile phone.

Social Networks: A term used to describe virtual or online communities of shared practice.

Social Software, Social Media: Social software enables people to connect or collaborate through computer-mediated communication (wiki, Weblog, podcasts) and form online communities.

Text Messaging (TM): Another term used to describe SMS.

Web 2.0: Web 2.0 generally refers to a second generation of services available on the Web that lets people collaborate, and share information online.

Vlog: (Video + Blog): A Weblog using video as its primary presentation format.

Wiki: A collaborative environment where any user can contribute information, knowledge or embed rich media such as video, audio, or widget(s) (Adapted from Wikipedia and Wiktionary, 2006).

APPENDIX A: OVERVIEW OF SOCIAL LEARNING THEORY

Table 14.

Situated Learning (Lave/Wenger)	**Knowledge needs to be presented in an authentic context. Learning requires social interaction and collaboration with peers. As learners engage in social interaction they become involved in a community of knowledge and practice.**
Constructivist Theory (Bruner)	Learners construct new ideas based on their current or past knowledge or experiences. Learners acquire new knowledge by building upon what they have already learned. Understanding comes through "active dialogue" Learning takes place via collaboration and social interaction with peers.
Social Development Theory (Vygotsky)	Full cognitive development requires social interaction. The range of skills that can be developed with peer collaboration exceeds what can be attained alone.
Multiple Intelligences (Gardner)	Human intelligence is comprised of several faculties that work in conjunction or individually with each other to achieve full cognitive development.
Social Life of Information (Seeley/Duguid)	Become a member of a community of practice (CoP) Engage in its practice Acquire and make use of its knowledge
Cognitive Apprenticeship (Brown, Collins, and Duguid)	Cognitive apprenticeship is an instructional design and learning theory wherein the instructor through socialization, models the skill or task at hand for the student. Students may also receive guidance from their peers. The role of the teacher is to help novices clear cognitive roadblocks by providing them with the resources needed to develop a better understanding of the topic.
Legitimate Peripheral Participation (Lave/Wegner)	Theoretical description of how newcomers are integrated into a community of practice (CoP). Newcomers ability to observe experts in practice enables them to be integrated deeper into the community of practice.
Chart adapted from Wikipedia (www.wikipedia.com)	

APPENDIX B: ROLES OF STUDENTS IN VIRTUAL LEARNING ENVIRONMENTS/SOCIAL NETWORKS

Table 15.

Roles	**Task**	**Procedure**	**Group Value**
Organizer	**Provides an ordered way of examining information**	**Presents outlines, overviews, or summary of all information**	**Lead thinker**
Facilitator	**Moderates, keeps on task**	**Assures all work is done and/or all participants have opportunity**	**Inclusive**
Strategist	**Decides the best way to proceed on a task**	**Organization**	**Detail**
Analyst	**Looks for meaning within the content**	**Realizes potential of content to practical application**	**Analytical**
Supporter	**Provides overall support for an individual or group**	**Looks for ways to help members or groups**	**Helpful**
Summarizer	**Highlights significant points; restates conclusion**	**Reviews material looking for important concepts**	**Gives the overall big picture**

Roles	Task	Procedure	Group Value
Narrator	Generally relates information in order	Provides group with a reminder of order	Keeps group focused on goal
Elaborator	Relates discussion with prior learned concepts or knowledge	Presents previous information as a comparative measure	Application or expansion
Researcher	Supplies outside resources to comparative information	Goes looking for other information with which to compare discussion	Inclusiveness
Antagonist	Supplies contrasting ideas	Looks for opposing viewpoints and presents in a relative way	Opposing viewpoint
(Fisher, Coleman, Sparks, & Plett, 2006)			

APPENDIX C: GLOSSARY OF NEW/SOCIAL MEDIA TERMS

Table 16.

Term	Definition
Mash up	A Web application that combines data from more than one source into an integrated experience.
Social Software, Social Media	Social software enables people to connect or collaborate through computer-mediated communication (wiki, Weblog, podcasts) and form online communities.
Blog	A blog, short for "Weblog", is a Web site in which the author writes their opinions, impressions, etc., so as to make them public and receive reactions and comments about them.
Moblog (mobile + blog)	A site for posting blog content from a mobile device, usually a cellular phone. Most often refers to photo sharing via a camera phone.
Vlog (video + blog)	A Weblog using video as its primary presentation format.
SMS (Short Message Service)	Written messages that you can send through a mobile phone.
Palm	A handheld portable device or personal digital assistant.
Social Networks	A term used to describe virtual or online communities of shared practice.
Web 2.0	Web 2.0 generally refers to a second generation of services available on the Web that lets people collaborate, and share information online.
Instant Messaging (IM)	Instant messaging is the act of instantly communicating between two or more people over a network such as the Web.
Text Messaging (TM)	Another term used to describe SMS.
Really Simple Syndication (RSS)	Really Simple Syndication feeds provide Web content or summaries of Web content together with links to the full versions of the content. RSS is used by news Websites, Weblogs and podcasting to synch and deliver content.
Wiki	A collaborative environment where any user can contribute information, knowledge or embed rich media such as video, audio, or widget(s).
(Adapted from Wikipedia and Wiktionary, 2006)	

This work was previously published in Handbook of Research on New Media Literacy at the K-12 Level: Issues and Challenges, edited by L.T.W. Hin; R. Subramaniam, pp. 48-71, copyright 2009 by Information Science Reference (an imprint of IGI Global).

Chapter 8.2
Legal Issues Associated with Emerging Social Interaction Technologies

Robert D. Sprague
University of Wyoming, USA

ABSTRACT

This chapter focuses on legal issues that may arise from the increasing use of social interaction technologies: prospective employers searching the Internet to discover information from candidates' blogs, personal web pages, or social networking profiles; employees being fired because of blog comments; a still-evolving federal law granting online service providers sweeping immunity from liability for user-published content; and attempts to apply the federal computer crime law to conduct on social networking sites. The U.S. legal system has been slow to adapt to the rapid proliferation of social interaction technologies. This paradox of rapid technological change and slow legal development can sometimes cause unfairness and uncertainty. Until the U.S. legal system begins to adapt to the growing use of these technologies, there will be no change.

DOI: 10.4018/978-1-60566-368-5.ch031

INTRODUCTION

This chapter focuses on legal issues that may arise from the increasing use of two specific types of social interaction technologies: blogs and social networking sites. These two Web 2.0 applications are emphasized due to a particular paradox: while there has been tremendous growth of blogs and social networking sites during the first part of the twenty-first century, rules of law develop slowly. Within this gap, laws regulating online conduct are continuing to evolve, leaving the exact status of certain activities in limbo.

BACKGROUND

The promise of the Internet as an information sharing platform (Leiner et al., 2003) has been fulfilled to a large extent in the twenty-first century by the emergence of Weblogs or blogs and social networking sites. Blogs, which originated as online diaries in which authors published information of interest for themselves and their few readers, usually in reverse chronological order, now number over 70 million (Sifry, 2007), covering just about every conceivable topic. Blogs are interactive because they link to other content on the Internet and many have the capability for readers to post their own comments, creating the possibility for ongoing dialog.

Social networking sites allow individuals to create online profiles (also referred to as *pages*) providing information about themselves and their interests, create lists of users (often referred to as *friends*) with whom they wish to share information, and to view information published within the network by their friends (Boyd & Ellison, 2007). The two most popular social networking sites, Facebook and MySpace, together boast nearly 100 million users (Stone, 2007).

As blogs and the use of social networking sites have proliferated, so too have potential legal problems. Prospective employers are reviewing job applicants' social networking profiles to glean information not contained in résumés. Employees have been fired as a result of their personal blogs. Online services, including social networking sites, have been sued based on content provided by users. Criminal conduct has been partially extended to violating the terms of service required to join interactive sites. These situations present challenges to a legal system which historically has been slow to adapt to new technologies. As a result, many of these legal issues remain unsettled.

LEGAL ISSUES BROUGHT TO LIGHT BY EMERGING SOCIAL INTERACTION TECHNOLOGIES

Googling Job Applicants

Many employers wish to know more about job applicants than what can be discerned from a résumé and interview—and for good reason. Surveys have indicated that nearly half of job applicants mislead employers about their work history and education (How to ferret out instances of résumé padding and fraud, 2006). Employers also seek to find individuals who will work and perform well within the organization (Piotrowski & Armstrong, 2006).

Employers are compelled to investigate the backgrounds of prospective employees because of the negligent hiring doctrine, which will impose liability on an employer when it "places an unfit person in an employment situation that entails an unreasonable risk of harm to others" (Lienhard, 1996, p. 389). Negligent hiring occurs when, prior to the time the employee is actually hired, the employer knew or should have known of the employee's unfitness. Liability is focused on the adequacy of the employer's pre-employment investigation into the employee's background (Ponticas v. K.M.S. Invs., 1983). But employers' ability to investigate applicants is hindered by the reluctance of former employers to provide letters of reference in fear of defamation suits from former employees (Finkin, 2000). Traditional pre-screening techniques are also restricted by various laws. For example, an employer may not ask questions which would allow the employer to screen applicants based on a protected class (race, color, national origin, religion, or gender) under Title VII of the Civil Rights Act of 1964. The Equal Employment Opportunity Commission has issued a number of guidelines on what employers can and cannot ask in an employment interview to help ensure that employers do not discriminate on the basis of religion (29 C.F.R. § 1605.3), national origin (29 C.F.R. § 1606.6), or sex (29 C.F.R. § 1604.7).

Prospective employers are finding that with a quick search on Google, they can discover a substantial amount of additional information from a candidate's blog, personal web page, or social networking profile. Unfortunately for the candidate, some of that information may be embarrassing, or even frightening—eliminating the candidate from further consideration (Finder,

2006). More problematic, there is no control over how a prospective employer may use the Internet to discover additional information about candidates, allowing the employer to potentially base a hiring decision on information that would otherwise be prohibited from disclosure in a job interview or questionnaire.

Since most of the information a prospective employer may glean from an applicant's blog or social networking profile is published by the applicant himself or herself, there is no right to privacy that could protect the applicant. United States privacy laws presume there is no privacy interest in information one exposes to the public (Prosser, 1960). As one court has stated, it is "... obvious that a claim to privacy is unavailable to someone who places information on an indisputably, public medium, such as the Internet, without taking any measures to protect the information" (United States v. Gines-Perez, 2002, p. 225).

A person must assume, therefore, that any information he or she publishes on the Internet can and will be used by prospective employers when considering that person for an employment opportunity.

Fired for Blogging

The law also favors employers when employees have been fired based on content published in personal blogs. What, in the past, may have been simple "water cooler" griping easily becomes publicly-available criticisms when an employee complains about work on a blog, leading to dismissals even when the employee is publishing anonymously. There have been a number of instances in which employees have been dismissed because of blog comments about work after their true identity had been revealed to their supervisors (Blachman, 2005; Gutman, 2003; Joyce, 2005).

Most employees in the United States are "at-will" (Sprang, 1994), meaning either the employer or employee may terminate the employment relationship at any time, with or without cause. Taken to its extreme, the employment-at-will doctrine means that employers can dismiss employees for arbitrary or irrational reasons: "because of office politics, nepotism, preference for left-handedness, astrological sign, or their choice of favorite sports team" (Bird, 2004, p. 551). However, over time, exceptions to the employment-at-will doctrine have evolved.

Even if an employee is at-will, he or she cannot be fired if doing so would violate public policy. Traditionally, violations of public policy have prevented at-will employees from being fired after they refused to break a law on behalf of their employer, insisted on exercising a legal right (such as voting), or exposed the employer's illegal behavior to the pubic (e.g., whistle blowing) (Sprague, 2007). As part of this public policy exception, some courts have found that employers cannot fire employees based on information obtained through an invasion of the employees' privacy (Pagnattaro, 2004). But, as discussed above, since the information used to fire an employee comes from a blog published by the employee himself or herself, there would be no right to privacy in that information.

A number of states have adopted laws which prevent employers from considering certain off-site, off-duty conduct in employment decisions, such as the use of tobacco or other lawful products (Pagnattaro, 2004). Six states (California, Colorado, Connecticut, Massachusetts, New York, and North Dakota) restrict employers from considering off-site, off-duty conduct with minimal restrictions. These laws have been applied in only very limited circumstances, usually relating to romances between employees (Pagnattaro, 2004). However, in one case, a court upheld the dismissal of an employee who had written a letter which was published in the local newspaper complaining about management practices, concluding that the employee was wrongfully taking public a private employment dispute (Marsh v. Delta Air Lines, Inc., 1997). These laws also require that for the conduct in question to be protected, it must have no relationship to the employer's business. Therefore, if the employee's blog contains entries discussing his or her workplace, then the employee's off-duty conduct does relate to the employer's business, and would therefore not be protected.

Title II of the Electronic Communications Privacy Act, known as the Stored Communications Act ("SCA") makes it illegal to access stored electronic communications without authorization. The SCA can protect an employee's blog if access to the blog is restricted only to authorized users. For example, a pilot (Robert Konop) employed by Hawaiian Airlines maintained a website that contained commentary critical of the airline's management practices (Konop v. Hawaiian Airlines, Inc., 2002). Konop's website required users to have an assigned username and password to access the site, and Konop maintained a list of authorized users, which consisted primarily of other Hawaiian Airline employees. A Hawaiian Airlines senior manager (a vice president who did not have authorized access to the website) used other pilots' usernames and passwords (with their permission) to access the website. Because the manager had used other authorized users' accounts, with their permission, to access the website, it would appear the manager had not violated the SCA. However, the court concluded the manager did violate the SCA, but only because the accounts used by the manager had never been actually used by the authorized users—therefore, under a strict reading of the statute, the manager did not have the permission of an authorized "user." Konop v. Hawaiian Airlines, Inc. (2002) demonstrates that unauthorized access to a restricted website can be a violation of federal law.

Of course, in a profession which values blogging, a person's blog may lead directly to his or her being hired, rather than fired (Hammock, 2005). Most employees, though, may be jeopardizing their employment status, with no legal recourse available, if they choose to include in their blog any references to the workplace.

Liability for Online Content

Online service providers are granted sweeping immunity from liability for user-published content on their sites. This immunity is derived from § 230 of the Communications Decency Act ("CDA"), which provides two types of immunity: (1) that a provider or user of an interactive computer service will not be treated as the publisher or speaker of any information provided by another information content provider; and (2) websites that make good-faith attempts to screen objectionable material will also be immune from liability for the content of that material. In essence, the CDA provides immunity to Internet publishers of third-party content.

The case of Zeran v. Am. Online (1997) exemplifies the application of § 230 immunity. Kenneth Zeran was the victim of an anonymous prank in which someone published on America Online ("AOL") a fake advertisement for shirts containing offensive content, instructing users to contact Zeran for purchase. Zeran then received numerous derogatory and obscene phone calls, including death threats. When AOL did not immediately remove the advertisement, Zeran sued AOL for delay in removing the advertisement and failure to publish a retraction. In essence, according to the court, Zeran sought to hold AOL liable for defamatory speech initiated by a third party (Zeran v. Am. Online, 1997).

In refusing to hold AOL liable, the Zeran v. Am. Online (1997) court noted that the purpose of § 230 immunity under the CDA was "...to preserve the vibrant and competitive free market that presently exists for the Internet and other interactive computer services ..." (p. 330). With the vast amount of information published online by third parties, it would be impossible for interactive computer service providers to screen each message posted by users, and "[t]he specter of tort liability in an area of such prolific speech would have an obvious chilling effect" (Zeran v. Am. Online, 1997, p. 331).

Section 230 immunity is not absolute. As noted above, it protects Internet publishers from liability only for content provided by third parties. The immunity does not apply if the online publisher originates the content. For example, when the online auction site eBay was sued based on its claim, "Bidding on eBay Live Auctions is very safe[,]" a court refused to grant eBay § 230 immunity (Mazur v. eBay, Inc., 2008). If eBay's claim that its Live Auctions were safe was a material misrepresentation, then the CDA would not "... immunize eBay for its own fraudulent misconduct" (Mazur v. eBay, Inc., 2008, p. *28).

One concern is whether the publisher can make any editorial modifications and still preserve § 230 immunity. This issue was addressed in the case of Fair Housing Council of San Fernando Valley v. Roommates.com, LLC (2008). Roommates.com is an online roommate matching service. Before individuals can use the Roommates.com service, they must create profiles, which require disclosures as to sex, sexual orientation, and whether children would be brought into a household. California-based fair housing councils sued Roommates.com, alleging its business violated the federal Fair Housing Act and California housing discrimination laws. The court refused to grant Roommates.com CDA immunity because it both elicited the allegedly illegal content and made "aggressive" use of it in conducting its business (Fair Housing Council of San Fernando Valley v. Roommates.com, LLC, 2008, p. 1172). "Roommate[s.com] does not merely provide a framework that could be utilized for proper or improper purposes; rather, Roommate[s.com]'s work in developing the discriminatory questions, discriminatory answers and discriminatory search mechanism is directly related to the alleged illegality of the site" (Fair Housing Council of San Fernando Valley v. Roommates.com, LLC, 2008, p. 1172). In other words, Roommates.com was directly involved with developing and enforcing a system that subjected users to allegedly discriminatory housing practices.

What is not settled is the degree of involvement required for an information content provider to lose immunity. The Fair Housing Council of San Fernando Valley v. Roommates.com, LLC (2008) court did state that an editor's minor changes to the spelling, grammar and length of third-party content would not remove CDA immunity. However, "if the editor publishes material that he does not believe was tendered to him for posting online, then he is the one making the affirmative decision to publish, and so he contributes materially to its allegedly unlawful dissemination[,]..." and would thus be "deemed a developer and not entitled to CDA immunity" (Fair Housing Council of San Fernando Valley v. Roommates.com, LLC, 2008, p. 1171). Ultimately, the question may be whether the "the website did absolutely nothing to encourage the posting of defamatory content..." (Fair Housing Council of San Fernando Valley v. Roommates.com, LLC, 2008, p. 1171).

Shortly before the Fair Housing Council of San Fernando Valley v. Roommates.com, LLC (2008) case was decided, a different federal court followed this latter approach. In Chi. Lawyers' Comm. for Civil Rights Under Law, Inc. v. Craigslist, Inc. (2008), Craigslist, an online classified advertising service, was sued because of the discriminatory content of some of the real estate rental advertisements it published (e.g., ads proclaiming "NO MINORITIES") (p. 668). The Chi. Lawyers' Comm. for Civil Rights Under Law, Inc. v. Craigslist, Inc. (2008) court granted Craigslist § 230 immunity because it did not "cause" any discriminatory statements to be made (p. 671).

Section 230 immunity was specifically applied to a social networking site in Doe v. MySpace Inc. (2007). The mother of a thirteen-year-old female MySpace user sued MySpace for negligence after a nineteen-year-old male MySpace user befriended the girl and ultimately sexually assaulted her. Unlike Zeran v. Am. Online (1997), in which immunity was granted based on AOL's status as a publisher, the mother in Doe v. MySpace Inc. (2007) claimed MySpace was not subject to immunity because its negligence was based on its failure to take reasonable safety measures to keep young children off of its site, and not based on MySpace's "editorial acts" (p. 849). The court rejected this argument, noting it was the exchange of personal information through MySpace that led to the two individuals meeting (which ultimately led to the assault). The court reasoned that if MySpace had never published their communications, the victim and the perpetrator would never have met. As a result, the court concluded that the mother's claims against MySpace were directed toward "... its publishing, editorial, and/or [message] screening capacities[,]" for which MySpace is immune under § 230 of the CDA (Doe v. MySpace Inc., 2007, p. 849).

While an individual who publishes defamatory information on a social networking site can be held individually liable (Zeran v. Am. Online, 1997), one cannot conclude absolutely that as long as the complaint focuses on communications published on a website, the service provider itself will be immune from liability. While minor editing changes will not necessarily remove § 230 immunity, the courts have indicated that originating content (Mazur v. eBay, Inc., 2008), materially contributing to the creation of content or publishing material not originally intended for publication (Fair Housing Council of San Fernando Valley v. Roommates.com, LLC, 2008) can transform a website from merely a publisher to a developer of content.

Computer Crime

One trend that has developed with social media sites is *cyberbullying*, which has been defined as "an aggressive, intentional act carried out by a group or individual, using electronic forms of contact, repeatedly and over time against a victim who cannot easily defend him or herself" (Smith, Mahdavi, Carvalho, Tippett, 2005, p. 6). What has caught the attention of legislators and educators are suicides of young people who were reportedly the victims of cyberbullying (Norton, 2007; Steinhauer, 2008). Legislators and educators also say that cyberbullying has gone unchecked for years, with few laws or policies addressing it (Norton, 2007).

One notorious incident has caught the attention of not only the public but also legislators. In 2006, Lori Drew, a 48-year-old Missouri woman, created a MySpace account and posed as a teenage boy, Josh. Through this fictitious persona, Drew began corresponding with Megan Meier, a 13-year-old neighbor. Drew reportedly created the "Josh" profile to find out what Megan was saying online about her (Drew's) teenage daughter (Parents want jail time for MySpace hoax Mom, 2007). Initially, "Josh" showed a romantic interest in Megan, but after a few weeks, the tone changed with "Josh" telling Megan at one point "the world would be a better place" without her (Steinhauer, 2008). Shortly thereafter, Megan Meier committed suicide.

As abhorrent as Drew's actions may have been, the question quickly arose as to whether she had committed a crime. Missouri, where both Drew and Meier lived, has a criminal harassment law prohibiting statements made to frighten or intimidate someone or to cause emotional distress. However, at the time of Drew's conduct, the law only applied to statements made in writing or via telephone (Offenses Against the Person: Harassment, 2008). The Missouri attorney general believed online statements were not covered by the statute (Steinhauer, 2008).

Federal prosecutors believe Drew did, in fact, commit a crime. In 2008 they filed a federal felony indictment against Drew, accusing her of violating the Computer Fraud and Abuse Act ("CFAA"). In particular, prosecutors allege that by creating the "Josh" profile and communicating with Megan Meier, Drew violated the CFAA by intentionally accessing a computer (the MySpace servers) without authorization and in excess of authorized access, and obtained information from that computer to further tortious acts, namely, intentional infliction of emotional distress on Megan Meier (United States v. Drew, 2008).

If the language of the indictment sounds a bit odd, it is because of the language within the CFAA. When the CFAA was first enacted in 1984, no one envisioned the facts surrounding the United States v. Drew (2008) case. Initially, the scope of the CFAA was narrow, focusing on unauthorized access to certain governmental or financial computers (Andreano, 1999). The CFAA has been amended a number of times to address an expanding array of computer-related activities. Currently, the CFAA contains criminal and civil penalties for various forms of unauthorized access to computer systems: (1) intentionally accessing a computer without authorization or exceeding authorized access to obtain information from a computer; (2) knowingly and with intent to defraud, accessing a computer without authorization, or exceeding authorized access, and obtaining anything of value, which can include the use of the computer accessed; (3) knowingly causing the transmission of a program, information, code, or command, and as a result of such conduct, intentionally causing damage without authorization, to a computer; (4) intentionally accessing a computer without authorization, and as a result of such conduct, recklessly causing damage; or (5) intentionally accessing a computer without authorization with the intent to defraud, and trafficking in any password or similar information through which a computer may be accessed without authorization (Credentials Plus, LLC v. Calderone, 2002). Typical violations of the CFAA have involved: the introduction of an "Internet worm" that shut down computers across the country (United States v. Morris, 1991); taking customer login information from one website and using it to access another website (Creative Computing v. Getloaded.com LLC, 2004); and using computer programs to infiltrate hundreds of computers and steal encrypted data and passwords (United States v. Phillips, 2007).

But within the CFAA, the terms "access" and "authorization" are not defined, so it is left to the courts to determine what constitutes "unauthorized access" or "exceeding authorized access." One approach used by the courts has been to determine whether a user's access to a computer has gone beyond the norms of intended use. For example, introducing a computer program into a computer network that is designed to exploit weaknesses within the network is an unauthorized access because it was not in any way related to the network's intended function (United States v. Morris, 1991).

Courts have also found unauthorized access arising from a breach of contract. For example, in EF Cultural Travel BV v. Explorica, Inc. (2001), a court found that a former vice president of EF Cultural Travel BV ("EF") had used information that was subject to a confidentiality agreement between the vice president and EF to obtain pricing information from EF's website. The court concluded the vice president's use of the information breached the confidentiality agreement, constituting unauthorized access to EF's computers in violation of the CFAA. Following EF Cultural Travel BV v. Explorica, Inc. (2001), courts have held that violations of computer service terms of use agreements can constitute unauthorized access. For example, courts have found unauthorized access in violation of the CFAA in the following situations: accessing the WHOIS database by automated software to collect website registration information to use for mass marketing purposes was access with a means and purpose in violation of the end use agreement (Register.com, Inc. v. Verio, Inc., 2000); using an AOL account to harvest e-mail addresses and then send bulk e-mails ("spam") to AOL members violated AOL's Terms of Service (Am. Online v. LGCM, Inc., 1998); and sending spam from Hotmail accounts in violation of the Terms of Service agreement (Hotmail Corp. v. Van Money Pie Inc., 1998). Courts have also taken a rather expansive view of "access," holding that it is merely the ability to "make use of something" (Am. Online v. Nat'l Health Care Disc., Inc., 2000, p. 1373). For example, when someone sends an e-mail message from his or her own computer, and the message then is transmitted through a number of other computers until it reaches its destination, the sender is making use of all of those computers, and is therefore "accessing" them (Am. Online v. Nat'l Health Care Disc., Inc., 2000).

Based on these previous applications of the CFAA, one can discern the logic in the federal indictment of Lori Drew. She was accused of violating the MySpace Terms of Service by creating a fictitious profile, which therefore established her unauthorized access of MySpace computers. Commentators have referred to the United States v. Drew (2008) indictment as the "first of its kind" (Steinhauer, 2008), as well as "dangerously flawed" and "scary" (Zetter, 2008). In some corners of the blogosphere there is concern that merely pretending to be someone else on the Internet can lead to a federal felony indictment (User charged with felony for using fake name on MySpace, 2008). One blogger has expressed concern that government enforcement agencies are misreading and misusing website user agreements: "Most websites like MySpace include contractual restrictions like the ones at issue simply to preserve their ability to kick off troublesome users at their discretion—not to put every non-conforming user at risk of looking down the barrel of an FBI agent's .45" (Goldman, 2008).

But one must keep in mind that unauthorized access alone is not enough to violate the CFAA. The unauthorized access must be also accompanied by access to or theft of information, fraud, extortion, damage to computers by malicious software, trafficking in passwords, or the commission of a crime or tortious conduct. In the United States v. Drew (2008) indictment, the government alleges Lori Drew accessed MySpace computers without authorization, by creating a MySpace account under a fictitious name, and then used that unauthorized access to obtain information from Megan Meier, another MySpace user, to then "torment, harass, humiliate, and embarrass" her, which constituted the tort of intentional infliction of emotional duress (p. 6). As such, merely creating a fictitious Internet persona—assuming it violates a service's terms of use agreement—will not violate the CFAA without further nefarious or illegal conduct committed as a result of the unauthorized access.

Since the Drew/Meier incident, Missouri has amended its harassment statute to eliminate the requirement that the harassment must occur by writing or telephone, which means it could apply to cyberbullying (Missouri Senate Bills Nos. 818 & 795, 2008). Several states are also considering cyberbullying legislation (State action on cyberbullying, 2008), as is Congress (Megan Meier Cyberbullying Prevention Act, 2008). Should cyberbullying laws proliferate, it will eliminate the need to use the CFAA for purposes arguably beyond its original intent.

FUTURE TRENDS

As social interaction technologies become more ingrained in American society, the U.S. legal system will have to adjust. Traditional notions of employment, privacy, publishing responsibilities and liabilities, and computer crime will have to catch up with the use of new technologies. In the near term, the only safe prediction is that employees, current and prospective, will have to censor what they publish online about themselves to avoid either being fired or eliminated from consideration for a job. In the meantime, unless Congress amends § 230 of the Communications Decency Act, online service providers, such as MySpace, will enjoy near complete immunity from liability for content published by users on their sites. But, for MySpace as well as other websites, it is only "near complete" immunity, somewhere between minor editing and doing absolutely nothing to encourage wrongful content. And without legislation specifically written to address cyberbullying, the Computer Fraud and Abuse Act may continue to be applied to online activities beyond the original intent of the Act.

CONCLUSION

The U.S. legal system has been slow to adapt to the rapid proliferation of social interaction technologies. Consequently, employees are being fired as a result of blogging; employers are searching the Internet to learn as much as possible about job candidates—possibly too much; interactive websites enjoy substantial immunity from liability for content published by users on their sites; and the notion of computer crime has been extended to violating website terms of use agreements. As this chapter has discussed, these consequences, which arise from the paradox of rapid technological change and slow legal development, can sometimes cause unfairness and uncertainty. Until the U.S. legal system begins to adapt to the growing use of these technologies, the status quo will remain for the foreseeable future.

REFERENCES

Am. Online v. LGCM, Inc., 46 F. Supp. 2d 444 (E.D. Virginia 1998).

Am. Online v. Nat'l Health Care Disc., Inc., 121 F. Supp. 2d 1255 (N.D. Iowa 2000).

Andreano, F. P. (1999). The evolution of federal computer crime policy: The ad hoc approach to an ever-changing problem. *American Journal of Criminal Law*, *27*, 81–103.

Bird, R. C. (2004). Rethinking wrongful discharge: A continuum approach. *University of Cincinnati Law Review*, *73*, 517–579.

Blachman, J. (2005, August 31). Job posting. *New York Times*, p. A19.

Boyd, D. M., & Ellison, N. B. (2007). Social network sites: Definition, history, and scholarship. *Journal of Computer-Mediated Communication, 13*(1). Retrieved July 10, 2008, from http://jcmc.indiana.edu/vol13/issue1/boyd.ellison.html

Chi. Lawyers' Comm. for Civil Rights Under Law, Inc. v. Craigslist, Inc., 519 F.3d 666 (7th Cir. 2008).

Communications Decency Act, 47 U.S.C. § 230 (2008).

Computer Fraud and Abuse Act, 18 U.S.C. § 1030 (2008).

Creative Computing v. Getloaded.com LLC, 386 F.3d 930 (9th Cir. 2004).

Credentials Plus, LLC v. Calderone, 230 F. Supp. 2d 890 (N.D. Indiana 2002).

Doe v. MySpace Inc., 474 F. Supp. 2d 843 (W.D. Texas 2007), *aff'd* 528 F.3d 413 (5th Cir. 2008).

EF Cultural Travel BV v. Explorica, Inc., 274 F.3d 577 (1st Cir. 2001).

Electronic Communications Privacy Act, 18 U.S.C. §§ 2510-22, 2701-11 (2008).

Fair Housing Council of San Fernando Valley v. Roommates.com, LLC, 521 F.3d 1157 (9th Cir. 2008).

Finder, A. (2006, June 11). When a risqué online persona undermines a chance for a job. *New York Times*, p. 1.

Finkin, M. W. (2000). From anonymity to transparence: Screening the workforce in the information age. *Columbia Business Law Review, 2000*, 403–451.

Goldman, E. (2008, May 23). Lori Drew prosecuted for CFAA violations—some comments, and a practice pointer. Message posted to http://blog.ericgoldman.org/archives/2008/05/lori_drew_prose.htm

Gutman, P. S. (2003). Say what?: Blogging and employment law in conflict. *Columbia Journal of Law & the Arts*, *27*, 145–186.

Hammock, R. (2005, August 29). *Hired because of his blog.* Message posted to http://www.rexblog.com/2005/08/29/14586

Hotmail Corp. v. Van Money Pie Inc., No. C98-20064 JW, 1998 U.S. Dist. LEXIS 10729 (N.D. California April 16, 1998).

How to ferret out instances of résumé padding and fraud. (2006). *Compensation and Benefits for Law Offices, 06-0*, p. 1.

Joyce, A. (2005, February 11). Free expression can be costly when bloggers bad-mouth jobs. *Washington Post*, p. A1.

Konop v. Hawaiian Airlines, Inc., 302 F.3d 868 (9th Cir. 2002), *cert. denied*, 537 U.S. 1193 (2003).

Leiner, B. M., Cerf, V. G., Clark, D. D., Kahn, R. E., Kleinrock, L., Lynch, D. C., et al. (2003). A brief history of the Internet. *Internet Society*. Retrieved July 10, 2008, from http://www.isoc.org/internet/history/brief.shtml

Lienhard, R. (1996). Negligent retention of employees: An expanding doctrine. *Defense Counsel Journal, 63*, 389–395.

Marsh v. Delta Air Lines, Inc., 952 F.Supp. 1458 (D. Colorado 1997).

Mazur v. eBay, Inc., No. C 07-03967, 2008 U.S. Dist. LEXIS 16561 (N.D. California March 3, 2008).

Megan Meier Cyberbullying Prevention Act, H.R. 6123, 110th Cong. (2008).

Missouri Senate Bills Nos. 818 & 795. (2008).

Norton, J. M. (2007, February 23). *Some states pushing for laws to curb online bullying.* Retrieved July 10, 2008, from http://www.law.com/jsp/article.jsp?id=1172138587392

Offenses Against the Person: Harassment. Missouri Annotated Statutes, § 565.090 (2008).

Pagnattaro, M. A. (2004). What do you do when you are not at work?: Limiting the use of off-duty conduct as the basis for adverse employment decisions. *University of Pennsylvania Journal of Business and Employment Law, 6*, 625–684.

Parents want jail time for MySpace hoax mom. (2007, November 29). Retrieved July 10, 2008, from http://www.abcnews.go.com/GMA/story?id=3929774

Piotrowski, C., & Armstrong, T. (2006). Current recruitment and selection practices: A national survey of Fortune 1000 firms. *North American Journal of Psychology, 8*(3), 489–496.

Ponticas v. K.M.S. Invs., 331 N.W.2d 907 (Minn. 1983).

Prosser, W. L. (1960). Privacy. *California Law Review, 48*, 383–423. doi:10.2307/3478805

Register.com, Inc. v. Verio, Inc., 126 F. Supp. 2d 238 (S.D. N.Y. 2000), *aff'd*, 356 F.3d 393 (2d Cir. 2004).

Restatement (Second) of Torts. (1965). § 291.

Restatement (Second) of Torts. (1965). § 46.

Restatement (Second) of Torts. (1977). § 623A.

Sifry, D. (2007, April 5). *The state of the live Web, April 2007*. Retrieved July 10, 2008, from http://www.sifry.com/alerts/archives/000493.html

Smith, P., Mahdavi, J., Carvalho, M., & Tippett, N. (2005). An investigation into cyberbullying, its forms, awareness and impact, and the relationship between age and gender in cyberbullying. *Unit for School and Family Studies, Goldsmiths College, University of London*. Retrieved July 10, 2008, from http://www.anti-bullyingalliance.org.uk/downloads/pdf/cyberbullyingreportfinal230106_000.pdf

Sprague, R. (2007). Fired for blogging: Are there legal protections for employees who blog? *University of Pennsylvania Journal of Business and Employment Law*, *9*, 355–387.

Sprang, K. A. (1994). Beware the toothless tiger: A critique of the Model Employment Termination Act. *The American University Law Review*, *43*, 849–924.

State action on cyber-bullying. (2008, February 6). *USA Today*. Retrieved July 10, 2008, from http://www.usatoday.com/news/nation/2008-02-06-cyber-bullying-list_N.htm

Steinhauer, J. (2008, May 16). Missouri woman accused of driving girl to suicide is indicted in California. *New York Times*, p. A15.

Stone, B. (2007, May 25). Facebook goes off the campus. *New York Times*, p. C1.

Stored Communications Act, 18 U.S.C. §§ 2701-2711 (2008).

Title VII of the Civil Rights Act of 1964, 42 U.S.C. § 2000e-2(a) (2008).

United States v. Drew, Indictment, No. 08-cr-00582 (C.D. California May 15, 2008).

United States v. Gines-Perez, 214 F. Supp. 2d 205 (D. P.R. 2002).

United States v. Morris, 928 F.2d 504 (2d Cir. 1991).

United States v. Phillips, 477 F.3d 215 (5th Cir. 2007), *cert. denied*, 128 S. Ct. 119.

User charged with felony for using fake name on MySpace. (2008, July 7). Retrieved July 10, 2008, from http://yro.slashdot.org/article.pl?no_d2=1&sid=08/07/07/1824228

Zeran v. Am. Online, Inc., 129 F.3d 327 (4th Cir. 1997).

Zetter, K. (2008, May 15). *Experts say MySpace suicide indictment sets 'scary' legal precedent*. Retrieved July 10, 2008, from http://blog.wired.com/27bstroke6/2008/05/myspace-indictm.html

KEY TERMS AND DEFINITIONS

Blog: A personal diary published on the Internet, with entries appearing in reverse chronological order.

Defamation: A published false statement harmful to the interests of another (Restatement, 1977, § 623A). Defamation is a tort (a civil wrong) for which the victim may bring a lawsuit against the defamer to recover damages suffered as a result of the defamatory comments.

Employment-at-Will Doctrine: Legal doctrine applied to employment relationships of indeterminate length, allowing either the employer or employee to terminate the relationship at any time, with or without cause.

Intentional Infliction of Emotional Distress: Outrageous or extreme conduct which results in severe emotional distress. It does not apply to mere insults, indignities, threats, or annoyances (Restatement, 1965, § 46). However, as discussed in this chapter, tormenting or harassing an individual to the extent she commits suicide would constitute the requisite level of emotional distress.

Interactive Computer Service: A term defined by the Communications Decency Act, meaning any person or entity that is responsible, in whole or in part, for the creation or development of information provided through the Internet or any other interactive computer service.

Negligence: An act which a reasonable person would recognize as involving risk of harm to another (Restatement, 1965, § 291). Negligence is a tort (a civil wrong) for which the victim may bring a lawsuit against the negligent party to recover damages suffered as a result of the negligent conduct.

Negligent Hiring Doctrine: Legal doctrine holding an employer liable for harm caused to third parties by an employee, assuming the employer would not have placed the employee in a situation in which the third party was harmed had the employer, prior to hiring the employee, adequately investigated the employee's background.

Social Networking Site: Internet-based service which allows individuals to create online profiles and share information with other users. Popular examples include Facebook and MySpace.

Tortious Conduct: Causing harm to an innocent third party. The victim of tortious conduct seeks money damages through a private (civil) lawsuit brought against the tortfeasor.

This work was previously published in Handbook of Research on Social Interaction Technologies and Collaboration Software; Concepts and Trends, edited by T. Dumova; R. Fiordo, pp. 351-362, copyright 2010 by Information Science Reference (an imprint of IGI Global).

Chapter 8.3
Public Intimacy and the New Face (Book) of Surveillance:
The Role of Social Media in Shaping Contemporary Dataveillance

Lemi Baruh
Kadir Has University, Turkey

Levent Soysal
Kadir Has University, Turkey

ABSTRACT

In recent years, social media have become an important avenue for self-expression. At the same time, the ease with which individuals disclose their private information has added to an already heated debate about the privacy implications of interactive media. This chapter investigates the relationship between disclosure of personal information in social media and two related trends: the increasing value of subjective or private experience as a social currency and the evolving nature of automated dataveillance. The authors argue that the results of the extended ability of individuals to negotiate their identity through social media are contradictory. The information revealed to communicate the complexity of one's identity becomes an extensive source of data about individuals, thereby contributing to the functioning of a new regime of surveillance.

DOI: 10.4018/978-1-60566-368-5.ch035

INTRODUCTION

Since their first inception in 1997 (with SixDegrees.com), social network sites - such as *Facebook*, *Friendster*, *Orkut*, and *MySpace* - allowed users to create online profiles about themselves and connect with other users. Starting with MySpace, user profiles on social network sites were no longer limited by preset categories determined by the network owners (Boyd & Ellison, 2007). Today, the types of information that users can post on their social network accounts are virtually limitless. A few examples include: age, educational status, favorite music bands, movies or books, current mood, a detailed list of daily activities performed, relationship status, likes and dislikes, and hobbies.

According to Liu (2007), an important consequence of this characteristic of social media is that social network sites have become very suitable venues for self-expression and identity formation. By enabling users to list their own interests, hobbies, social preferences, among other forms of information, social network sites empower individuals to go beyond the traditional tokens of identity, such as profession and social class, to engage in what he calls "taste statements" (p. 253) and more freely communicate oneself to others. And according to Evans, Gosling, and Carroll (2008), what individuals have to say about themselves in social media does not fall on deaf ears: a person who views the online profile of another person usually forms impressions that are congruent with the profile owners.

However, the same feature that enables individuals to freely communicate their identity to their social networks also leaves traces of data in unprecedented detail. As such, the main purpose of this chapter is to discuss these two related trends and their implications for intimacy, social relations, privacy and identity in contemporary societies. Following a brief overview of social media, the chapter begins by arguing that increased transparency is one of the defining characteristic of the *new individual* in contemporary societies. Next, the chapter focuses on how social media, in a world of transparency, enable individuals to communicate their multiple identities to others. In the final sections, the chapter focuses on the privacy implications of this heightened transparency by discussing the characteristics of a regime of surveillance that increasingly relies on an automated collection, collation and interpretation of the data individuals reveal and by summarizing the role that social media play in this regime of surveillance.

BACKGROUND

According to Barnes (2006), *social media* is an all-encompassing term that describes loosely organized online applications through which individuals can create personas and communicate with each other. Especially since 2003, social network sites (such as MySpace, Orkut, Facebook, and LinkedIn) have become extremely popular. For example, in 2007, Facebook had close to 100 million and MySpace had more than 100 million unique visitors (Comscore.com, 2007). Weblogs or blogs are another form of widely used social media. By the end of 2007, there were an estimated 67 million blogs worldwide (Rappaport, 2007).

This rising popularity of social media, within which individuals reveal minute details of their lives, is closely related to the transformation of society's expectations about what constitutes an acceptable form of information. Noting this transformation in individuals' expectations about the type of truth that the media should make available, several commentators suggest that an important characteristic of current culture is the elevation of individualism around mid-1960s and the subsequent rise of the subjective and intimate experience of individuals as the guarantor of truth (Cavender, 2004; Corner, 2002). Commenting on this transformation, social theorist Beck (1994) points out that there has been a shift in individuals' relationship with institutions. Accordingly, whereas in early modernity, meaning and identity were grounded on somewhat loyal reliance on institutions and structures, starting with late 20th century, the locus of meaning shifted to the individual. The self became the primary agent of meaning.

Within this context, by aiding the circulation of the intimate, social media are quickly becoming a platform for self-expression and creation of meaning. However, the audiences for these attempts at self-expression via intimate disclosure are usually not limited to a few friends or potential friends. As such, the ease with which users reveal their personal information, while using social media, has triggered a heated debate over the privacy implications of social media in general and social network sites in particular (Solove, 2007; Viegas, 2005). Researchers have focused on a number of issues including: social media users' ability to limit who has access to identification information (Lange 2007); corporate snooping and intrusion (for marketing and employee recruitment) (Maher, 2007; Solove, 2007); data security and use of publicly accessible personal information for fraud (Gross & Acquisti, 2005; Jagatic, Johnson, Jakobsson, & Menezer. 2007); and protection of underage users' privacy (Barnes 2006; George 2006).

Despite their invaluable contribution to current debates regarding privacy in social media, most of the current studies in this area adopt a piecemeal approach. Within this approach, different privacy threats are considered in isolation from each other and from the greater framework of surveillance as an increasingly data-intensive risk management tool for institutions (government and private). Gary Marx (2004) argues that this newer form of surveillance has several important characteristics such as being continuous, automated, more intensive and extensive (because every individual is subjected to his/her data being collected in a data collection phase), invisible (as is the case when data about individuals is collected and subsequently disseminated to dispersed databases) and involuntary (partly as a result of this invisible nature of data intensive surveillance). The typical end-result that institutions seek from this process is to utilize the data to draw inferences about their identities and sort them into common types of so-called "unique" categories (Gandy, 1993; Lyon, 2001). Then, what could be ahead for individuals is a conflict between the personas and identities that they communicate in social media and the identity that they have been ascribed to as a result of this automated surveillance (also called *dataveillence*). The remainder of this chapter explores the relationship between disclosure of personal information in social media and two related trends: the increasing value of subjective/private experience as a social currency and the evolving nature of automated dataveillance.

PUBLIC INTIMACIES AND THE NEW SURVEILLANCE

Formations of the *New* Individual

Since the 1980s, the imaginary timeline of change, which can also be traced back to the anti-conservative upheavals of the 1960s in almost every facet of life, the individual has increasingly come to the center stage of social, economic, and technological order. Her rights have been significantly expanded, in particular with impetus generated by the hegemonic discourses of human rights (Benhabib, 2002; Turner, 1986; Soysal Y., 1994). The *new* individual, so to speak: (a) has rights to her identity and culture (in other words, she possesses identities as a member of a categorically cultured collectivity that is differentiated by gender, sexual preference, disability, ethnicity, religion or spirituality); (b) is extensively involved in financial and security markets as a rational actor (she is entrusted with security of her own self, family, and future under terms dictated by the market); and, (c) achieves intimacy in public (she lives her sociality and establishes her intimate relations primarily in public stages, enabled by institutionalized public discourses). She is at the center of multiple, and ever increasing, life spaces that enact synthetically modular lives.

In the globalizing world the new individual asserts herself, the lines that so preciously divide the time renowned cultural, social, and political categories into inside and outside, private and public are rapidly fading away under the duress of massive economies of circulation, imitation, and sociability. As sociability is amplified and externalized, and public and private become indistinguishable, intimacy (social, cultural, or personal) becomes displaced and public (Soysal L., 2007).

While intimacy as conventionally understood requires an inward movement toward the private sites of self, family, home, marriage, culture, and nation (Berlant, 1997, 1998; Herzfeld 1997), public intimacy suggests an outward move to locate the formations of intimacy. In public intimacy, the emphasis is on the shared discursive spaces of public engagement, rather than the shared, inviting spaces of the cultural or personal kind (Soysal L., 1999; Berlant, 1997, 1998; Wilson, 2004). Public discourses and expressions, even in their most formalized discursive modes, constitute and conjure intimate connections. They provide a vocabulary to engage and question prescribed techniques and "institutions of intimacy" (Berlant, 1997, 1998) as in romance, dating, and marriage, while suggestively constructing intimate attachments between persons.

Furthermore, in today's world, the "close association, privileged knowledge, deep knowing and understanding" (Jamieson, 1998) anticipated by proper definitions of intimacy are incomplete and temporary. When the engagement ends, the setting and conditions for organized intimacy simply cease to exist. Individuals leave behind their provisional partners in intimacy with whom they shared stories and sociality.

A corresponding transformation can be observed in how individuals live and enact sociality—in that sociality today is increasingly exogenous. Contemporary metropolitan spaces have become locations for year-round festivity. What's true for mega cities such as London, Paris, New York is true for most metropolitan centers. Festivals of all sorts and sizes mark the topography of culture in cities. City becomes unthinkable without its festivals—its impressive and expressive façade. Even cities not so famous for its carnivals, such as Istanbul and Berlin, are now year-round stages for spectacles—film, music, theater festivals, street parades, international sports meets, as well as commercial fairs, IGO and NGO meetings, state summits, and professional conferences. The moment one of them ends, another is given start (Soysal L., 2005). Add to this the fact that individuals spend more time and money on extra-home entertainment as epitomized by the proliferation of eating-out, fitness activities, shopping, and travel.

Said differently, as the sociality of the spectacular and extra-home entertainment—or the hold of what is anthropologically known as *expressive culture*—amplifies, the exogenous encircles the individual and the interior dissolves in the lives lived, enacted outside. Under circumstances of globalization, not only social lives increasingly happen outside the privacy of homes, offices, and selves. Gradually, but surely, sociality takes place in virtual worlds. In other words, intimacies are being carried into virtual worlds where privacy proper is not operational.

The new individual now lives, works, and shops in transparent interiors of buildings with glass facades (for example, Berlin's parliament with its transparent dome, Richard Meier apartments in New York). In fact, as Sternberg (2001) notes, the new individual now is occupied in a phantasmagoric workplace and is responsible to create a suitable persona to present her "iconographic capabilities" (p. 11). In other words, the labor of the new individual is a labor of self-presentation. Strangely enough, this labor of self-presentation, which used to be the domain of celebrities such as movie or rock stars, is now a full-time labor for many individuals, who, for example, wear their emotions on their t-shirts or sweatpants that read *Milf in Training*, *Jerk Magnet*, *Your Boyfriend Wants Me*, or *Juicy*.

In the digital realm, live webcam feeds through which individuals broadcast what transpired in their bedroom can be considered as an example to the trickling down of the act of self-presentation. And nowadays, the new individual has Facebooks, MySpaces, YouTube—the proliferated virtual worlds of sociality—where she not only displays but also actualizes intimacies in public. First, thanks to the modular structure of social media sites, e.g., Facebook or MySpace, the new individual can now determine what components of her own modular identity to display and prioritize (Donath & Boyd, 2004; Liu, 2007, Lampe, Ellison, & Steinfeld, 2007; Marwick, 2005). For example, she can choose to display information about her music taste at the top of the page whereas another might choose to share her travel experiences and the places she visited. Second, specialized social network sites allow the individuals to compartmentalize their personas by displaying information they see fit for different contexts. For social shopping, she can use VogueShopTV and go to StyleHive, FashionWalker. From the convenience of her cell phone, she can announce the course of her new love affair by minute to anyone who listens—actually to anyone who has Twitter, the "quick blogging" tool. If she is interested in networking to find new employment opportunities, she creates an account in LinkedIn to share her professional background. And let us not forget second lives and socialities she may enact in Second Life and its clones. Friends can even be determined via DNA matching by a visit to a social-networking website (https://www.23andme.com) to be unveiled by a new personal genomics start-up in Silicon Valley (Soysal L., 2007).

In this respect, what users of social media do by creating their online personas is to engage in what can be called "introversive publicity." In social psychology, individuals with introversive personality are characterized as retiring people who value introspective thinking and intimate relationships (Eysenck & Eysenck, 1975). The act of subjective expression on social media (introversive publicity), despite its public nature, is introspective. It requires careful consideration of how each modular component of one's identity works in coherence with each other. As such, the resulting persona is as intimate as it is public. It is as coherent as it is modular. However, the public presentation of the virtual modular self is not solely a self-publicity project. Rather, it is a crucial component of how individuals develop and negotiate their own identity. As Simmel (1922/1964) pointed out with respect to rational group memberships, each component added to this modular identity helps establish a unique identity. This is the world of amplified sociality, virtual intimacy, and actual simulacra we inhabit. And in this brave new world, as Google prophesizes in its newest slogan, "Social Will Be Everywhere" and intimacies that matter will be public (Soysal L., 2007).

Social Media: Reclaiming the Right to Privacy

What are the privacy implications of public intimacies on social media? Being able to create online profiles and communicate one's own identity in a manner that one prefers may be considered as exercising one's privacy rights. This perspective reflects a long-standing socio-legal understanding that defines *privacy* as one's ability to have control over when, how and to what extent information about them is known by others (Bing, 1972; Fried, 1984; Wacks, 1989; Weintraub, 1997). As such, by publicizing information about their subjective experiences and everyday lives, users may be exercising their privacy right to disseminate information about themselves.

Recently, several commentators suggested that individuals' voluntary submission to the gaze of other people (as is the case when Internet, users leave their Webcam turned on throughout the day) is not only an exercise of privacy rights but also an act of counter-surveillance (Dholakia & Zwick, 2001; Koskela, 2004). Accordingly, in an environment of extensive surveillance, self-disclosure is seen as the only viable way for individuals to actively participate in the creation of images about themselves (Groombridge, 2002).

The real situation of Hasan Elahi is a perfect illustration of this perspective. After being mistakenly put on the U.S. government terrorist watch list, Mr. Elahi decided that the best way to be free from government intrusion would be to document and publish online every single detail of his daily affairs. Today, Mr. Elahi's blog, sometimes visited by U.S. law enforcement officers, contains a slew of details including scanned images of the receipts of every transaction he enters and regularly updated GPS location images of his whereabouts.

The key to being left alone, Mr. Elahi says, is to give away one's privacy (Thompson, 2007).

Regarding such a conceptualization of privacy, Gavison (1980) argues that although knowing disclosure of information can be construed as an exercise of privacy rights, it is nevertheless a loss of control over information because after the act of disclosure, individuals will have little control in the subsequent dissemination of the information. The popular media frequently covers such mishaps. For example, recently Kansas University decided to penalize students after finding out that the photographs they uploaded on Facebook contained evidence that they violated an alcohol policy of the University (Acquisti & Gross, 2006). Similarly, Microsoft officials admit that they frequently peruse Facebook profiles of job candidates (Solove, 2007). However, as the remainder of this chapter will discuss, these incidents of corporate/institutional snooping may be the tip of the iceberg with respect to the problematic of privacy as control over personal information.

Uncertainty and Risk Externalization in the New Surveillance

Since Foucault's *Discipline & Punish: The Birth of the Prison* (1977), Bentham's panopticon—an architectural design that would allow the constant monitoring of prisoners from a central tower—has captured the imagination of many scholars studying contemporary surveillance. In perhaps one of the most influential studies of surveillance in the information age, Gandy (1993) used the panopticon metaphor to characterize the continuous surveillance of everyday transactions and sorting of populations into "consuming types" as a form of rationalization (in the Weberian sense) of inequality—through computers, which are for all practical purposes rationality incarnate.

An important characteristic of the new surveillance is that it relies on a machine based, automated collection of personal information. Even the most innocuous transactions leave data trail that can be stored for later analysis (Gandy, 2002; Marx, 2004). To some extent, the development of interactive media (e.g., the Internet, digital cable), which allow for a two-way flow of information between the content provider and the consumer, has added to the impetus for continuous tracking of individuals' behavior and creating profiles that can be used to categorize them into homogenous segments (Baruh, 2004; Turow, 2005b). Within this context, social media and social network sites add to what is already a very large pool of data about individuals. Private information that individuals voluntarily reveal in social media about their hobbies, favorite pastimes, music preferences, close friends and even changes in their mood can be used further to refine their profiles and categorize them into groups.

An important problem concerning the vast amount of information that institutions collect about individuals is to interpret the ensuing data. Just as with the collection phase, a process known as "data mining" increasingly allows the use of algorithms for automatic detection of patterns that can be used to predict future behavior and risk (Gandy, 2002; Zarsky, 2002, 2004). That the data collection and interpretation process is automated has important consequences in terms the uncertainty that surround individuals' interaction with contemporary surveillance. Clearly, uncertainty was an important component of the disciplining function of the panopticon. Whereas guards can observe prisoners at any time, prisoners have no way of knowing when they are being observed. The concept "chilling effect of surveillance" underlines an important consequence of this uncertainty regarding when one is being monitored. Accordingly, an individual will be less likely to express her controversial opinions in public if she suspected that any behavior she engages in can be recorded (Marx, 1988), or vice versa, individuals may reveal opinions, at times abundantly, as if it should matter to their listeners.

The prospect of fully automated analyses of data about individuals may introduce additional uncertainties. As Gary Marx (2004) argues, the new surveillance does not target suspected individuals. It is carried out superficially, with an intention to closely investigate later. As such, surveillance systems that rely on automated data mining are akin to a fishing expedition that starts by comparing each data-point to the population base. This comparison, done without human interpretation or prior hypotheses about what constitutes risk, has the potential to signal any deviation as risk, which could then invite further scrutiny (Andrejevic, 2007).

A related component of such uncertainty regarding what constitutes the automated risk categorization is the violation of the contextual integrity of information. As noted by several theorists, an important function of privacy in a world where information about us is abundant is to protect individuals from being (mis)identified out of context (Nissenbaum, 1998; Rosen, 2000). As we suggested before, the self is a modular and perennially evolving entity. This notion is perhaps best reflected in Erving Goffman's (1959) conceptualization of selfhood, which is comprised of multiple roles we play and masks we wear.

Each snapshot of the multi-modular self in a different context will provide factually correct information about that context. However, in automated data mining, rather than interpreting each photograph as an independent unit, the analysis is based on creating a collage without paying attention to the contextual background. A collage created from hundreds of independent snapshots of the same person will probably not contain factually incorrect information. Each bit of data is actually about the subject. However, depending on how the independent photographs are rearranged, the person may look overweight, underweight or right on scale.

The point that we seek to make with this discussion is not that the inferences made through automated data mining will always be factually inaccurate. Rather, this process largely diminishes individuals' ability to determine (and find ways to challenge) inferences that are made about them. This is partly due to an informational asymmetry between individuals and surveillant institutions. The concept of privacy, which supposedly protects individuals from undue attention, when combined with intellectual property rights, provides institutions with a high level of protection from external oversight regarding the accuracy of data, how the data are used, and whether the data are properly protected (either from individuals or from agencies representing individuals). Noting this trend, Andrejevic (2007) argues that *privacy* is now the keyword for increased surveillance with "diminished oversight and accountability" (p. 7).

FUTURE TRENDS

Considered from the perspective provided above, the new surveillance (if fully utilized) will be more Kafkaesque than Orwellian (Lyon, 2001; Solove, 2001). In Kafka's *The Trial* (1937) the main character, Joseph K. is subjected to a long judicial process without ever knowing what he was accused of.

In *The Trial* Joseph K.'s circumstances are particularly illustrative of two characteristics of the new surveillance this chapter discusses. First, the subject will not know when she is being surveilled, who uses the data, who wants the information, and what or who distinguishes acceptable behavior from risky behavior (Baruh, 2007; Solove, 2007). Second, the digital revolution (along with enhanced storage capacity) makes it increasingly difficult for our society to forget and move on, making it almost impossible for individuals to have a second chance (Solove, 2007): "We are losing control...because if what we do is represented digitally, it can appear anywhere, and at any time in the future. We no longer control access to anything we disclose" (Grudin, 2001, p. 11). Third, data mining rationalizes surveillance by removing humans from the interpretation process. The dehumanization of the analyses is important because it removes the so-called human bias from the interpretation process. As such, when combined with the fact that contemporary data mining relies on quantification of information (a seemingly dispassionate and objective method of interpreting the social world), this dehumanization projects an aura of objectivity, consequently making it even more difficult to challenge its premise (and the findings it provides).

In the end, data targets lose whatever control they used to have over the management of their multiple identities. Many scholars would argue that rather than being a loss of control over one's identity, what happens is increased accountability, which in turn reduces social costs associated with individuals' tendency to misrepresent themselves to others (Posner, 1978). However, it is very difficult to argue that just because individuals may occasionally misrepresent themselves, the inferences that institutions make about individuals should gain such an absolute credence over individuals' self-representations (Baruh, 2007).

CONCLUSION

The rise of social media coincides with shifting norms about what constitutes an acceptable form of private information in contemporary societies. Namely, an important characteristic of contemporary popular culture is the elevation of individualism (especially since the 1960's) and the subsequent rise of the subjective experience of the individual as an acceptable form of truth. Within this context, social media have become the loci of virtual public intimacies within which individuals communicate their multifaceted identities through their virtual personas.

Perhaps, then, virtual public intimacies can even be considered as enabling individuals to actively practice their privacy rights by giving them an opportunity to communicate the complexity of their identity. However, the paradoxical consequence of this ability to make active decisions regarding one's own immediate privacy through public intimacy is that the subjective information revealed in social media becomes the most extensive source of data about individuals, thereby contributing to the functioning of a new regime of surveillance. This new regime of surveillance is characterized by an expansion in the uncertainty that surrounds the criteria surveillance systems utilize to distinguish between prospects and threats. Each component of our modular online identity can be a potential factor that leads to discrimination. And in the end, the individual is left assigned to a category that may not only ignore the complexity of her modular identity but also is virtually (and practically) impossible to challenge because of its automated nature and consequent aura of objectivity.

REFERENCES

Acquisti, A., & Gross, R. (2006). *Imagined communities: Awareness, information sharing, and privacy on the Facebook*. Paper presented at the 6th Workshop on Privacy Enhancing Technologies, Cambridge, UK.

Andrejevic, M. (2007). *iSpy: Surveillance and power in the interactive era*. Lawrence, KS: University Press of Kansas.

Barnes, S. B. (2006). A privacy paradox: Social networking in the United States. *First Monday*, Retrieved November 11, 2007, from http://www.firstmonday.org/issues/issue11_9/barnes

Baruh, L. (2004). Audience surveillance and the right to anonymous reading in interactive media. *Knowledge Technology and Policy*, *17*(1), 59–73. doi:10.1007/BF02687076

Baruh, L. (2007). Read at your own risk: Shrinkage of privacy and interactive media. *New Media & Society*, *9*(2), 187–211. doi:10.1177/1461444807072220

Beck, U. (1994). The reinvention of politics: Towards a theory of reflexive modernization. In U. Beck, A. Giddens, & S. Lash (Eds.), *Reflexive modernization: Politics, tradition and aesthetics in the modern social order* (pp. 1-55). Cambridge, MA: Polity Press.

Benhabib, S. (2002). *The claims of culture: Equality and diversity in the global era*. Princeton, NJ: Princeton University Press.

Berlant, L. (1997). *The queen of America goes to Washington City: Essays on sex and citizenship*. Durham, NC: Duke University Press.

Berlant, L. (1998). Intimacy: A special issue. *Critical Inquiry, 24*, 281–288. doi:10.1086/448875

Bing, J. (1972). *Data banks and society*. Oslo, Norway: Universcitctsforlaget.

Boyd, D. M., & Ellison, N. B. (2007). Social network sites: Definition, history, and scholarship. *Journal of Computer-Mediated Communication, 13*(1), 210–230. doi:10.1111/j.1083-6101.2007.00393.x

Cavender, G. (2004). In search of community on reality TV: America's Most Wanted and Survivor. In S. Holmes & D. Jermyn (Eds.), *Understanding reality television* (pp. 154-173). New York: Routledge.

Corner, J. (2002). Performing the real: Documentary diversion. *Television & New Media, 3*(3), 255–269. doi:10.1177/152747640200300302

Dholakia, N., & Zwick, D. (2001). Privacy and consumer agency in the information age: Between prying profilers and preening Webcams. *Journal of Research for Consumers, 1*(1).

Donath, J., & Boyd, D. M. (2004). Public displays of connection. *BT Technology Journal, 22*(4), 71–82. doi:10.1023/B:BTTJ.0000047585.06264.cc

Evans, D. C., Gosling, S. D., & Carroll, A. (2008). *What elements of an online social networking profile predict target-rater agreement in personality impressions?* Paper presented at the International Conference on Weblogs and Social Media, Seattle, WA.

Eysenck, H. J., & Eysenck, M. W. (1975). *Manual of the Eysenck personality questionnaire*. San Diego, CA: Educational and Industrial Testing Service.

Foucault, M. (1975). *Discipline & punish: The birth of the prison*. New York: Vintage Books.

Fried, C. (1984). Privacy: A moral analysis. In F. D. Schoeman (Ed.), *Philosophical dimensions of privacy: An anthology* (pp. 203-222). Cambridge, MA: Cambridge University Press.

Gandy, O. H. (1993). *The panoptic sort: A political economy of personal information*. Boulder, CO: Westview.

Gandy, O. H. (2002). *Data mining and surveillance in the post-9.11 environment*. Paper presented at the annual meeting of the IAMCR, Barcelona, Spain.

Gavison, R. (1980). Privacy and the limits of law. *The Yale Law Journal, 89*, 421–471. doi:10.2307/795891

George, A. (2006). Living online: The end of privacy? *New Scientist*. Retrieved December 14, 2007, from http://technology.newscientist.com/article/mg19125691.700

Goffman, E. (1959). *The presentation of self in everyday life*. Garden City, NY: Doubleday.

Groombridge, N. (2002). Crime control or crime culture TV? *Surveillance & Society, 1*, 30–36.

Gross, R., & Acquisti, A. (2005). *Information revelation and privacy in online social networks (the Facebook case)*. Paper presented at the Privacy in the Electronic Society (WPES) Conference, Alexandria, VA.

Grudin, J. (2001). Desituating action: Digital representation of context. *Human-Computer Interaction, 16*(2), 269–286. doi:10.1207/S15327051HCI16234_10

Herzfeld, M. (1997). *Cultural intimacy: Social poetics in the nation-state*. New York: Routledge.

Jagatic, T. N., Johnson, N. A., Jakobsson, M., & Menezer, F. (2007). Social phishing. *Communications of the ACM, 50*(10), 94–100. doi:10.1145/1290958.1290968

Jamieson, L. (1998). *Intimacy: Personal relationships in modern societies*. Cambridge, MA: Polity Press.

Kafka, F. (1937). *The trial*. New York: A.A. Knopf.

Koskela, H. (2004). Webcams, TV shows and mobile phones: Empowering exhibitionism. *Surveillance & Society, 2*(2/3), 200–215.

Lampe, C., Ellison, N., & Steinfeld, C. (2007). *A familiar Face(book): Profile elements as signals in an online social network*. Paper presented at the Human Factors in Computing Systems Conference, New York.

Lange, P. D. (2007). Publicly private and privately public: Social networking on YouTube. *Journal of Computer-Mediated Communication, 13*(1), 361–380. doi:10.1111/j.1083-6101.2007.00400.x

Liu, H. (2007). Social network profiles as taste performances. *Journal of Computer-Mediated Communication, 13*(1), 252–275. doi:10.1111/j.1083-6101.2007.00395.x

Lyon, D. (2001). *Surveillance society: Monitoring everyday life*. Philadelphia, PA: Open University Press.

Maher, M. (2007). You've got messages: Modem technology recruiting through text-messaging and the intrusiveness of Facebook. *Texas Review of Entertainment & Sports Law, 8*, 125–150.

Marwick, A. (2005). *'I'm a lot more interesting than a Friendster profile': Identity presentation, authenticity, and power in social networking services*. Paper presented at the Association of Online Internet Researchers Conference, Chicago, IL.

Marx, G. T. (2004). What's new about the 'new surveillance'? Classifying for change and continuity. *Knowledge Technology and Policy, 17*(1), 18–37. doi:10.1007/BF02687074

Nissenbaum, H. (1998). Protecting privacy in an information age: The problem of privacy in public. *Law and Philosophy, 17*, 559–596.

Posner, R. (1978). An economic theory of privacy. *Regulation, 2*, 17–30.

Rappaport, S. D. (2007). Lessons from online practice: New advertising models. *Advertising Research, 47*(2), 135–141. doi:10.2501/S0021849907070158

Rosen, J. (2000). *The unwanted gaze: The destruction of privacy in America*. New York: Random House.

Simmel, G. (1922/1964). *Conflict and the web of group-affiliations*. New York: The Free Press.

Solove, D. J. (2001). Privacy and power: Computer databases and metaphors for information privacy. *Stanford Law Review, 53*, 1393–1462. doi:10.2307/1229546

Solove, D. J. (2007). *The future of reputation: Gossip, rumor, and privacy on the Internet*. New Haven, CT: Yale University Press.

Soysal, L. (1999). *Projects of culture: An ethnographic episode in the life of migrant youth in Berlin*. Unpublished doctoral dissertation, Harvard University, Cambridge, MA.

Soysal, L. (2005). Karneval als spektakel. Plaedoyer fuer eine aktualisierte perspektive. In M. Knecht & L. Soysal (Eds.), *Plausibe vielfalt. Wie der karneval der kulturen denkt, lernt und kultur schaft*. Berlin, Germany: Panama Verlag.

Soysal, L. (2007). *Intimate engagements of the public kind*. Paper presented at the annual meeting of the American Anthropological Association, Washington, DC.

Soysal, Y. N. (1994). *Limits of citizenship: Migrants and postnational membership in Europe*. Chicago, IL: University of Chicago Press.

Sternberg, E. (2001). Phantasmagoric labor: The new economics of self-presentation. *Futures*, *30*(1), 3–21. doi:10.1016/S0016-3287(98)00003-2

Thompson, C. (2007, June). The visible man: An FBI target puts his whole life online. *Wired Magazine, 74*.

Top global Web properties. (2008). Retrieved February 19, 2008, from http://www.comscore.com/press/data.asp

Turner, B. (1986). Personhood and citizenship. *Theory, Culture & Society*, *3*, 1–16. doi:10.1177/0263276486003001002

Turow, J. (2005b). *Niche envy: Marketing discrimination in the digital age*. Boston, MA: The MIT Press.

Viégas, F. B. (2005). Bloggers' expectation of privacy and accountability: An initial survey. *Journal of Computer-Mediated Communication*, *10*(3). Retrieved December 15, 2007, from http://jcmc.indiana.edu/vol10/issue3/viegas.html

Wacks, R. (1989). *Personal information: Privacy and the law*. Oxford: Clarendon Press.

Weintraub, J. (1997). The theory and politics of the public/private distinction. In J. Weintraub & K. Kumar (Eds.), *Public and private in thought and practice* (pp. 1-42). Chicago, IL: University of Chicago Press.

Wilson, A. (2004). *The intimate economies of Bangkok: Tomboys, tycoons, and Avon ladies in the global city*. Berkeley, CA: University of California Press.

Zarsky, T. Z. (2002). 'Mine your own business!': Making the case for the implications of the data mining of personal in the forum of public opinion. *Yale Journal of Law and Technology*, *5*, 1–54.

Zarsky, T. Z. (2004). Desperately seeking solutions: Using implementation-based solutions for the troubles of information privacy in the age of data mining and the Internet society. *Maine Law Review*, *56*, 13–59.

KEY TERMS AND DEFINITIONS

Contextual Integrity: Nissenbaum (1998) developed the concept of privacy as *contextual integrity* to propose a normative framework that evaluates the flow of information about individuals. Accordingly, given the multifaceted nature of individuals' identities, contextual integrity is violated when the informational norms associated with a specific social relationship are breached.

Data Mining: *Data mining* refers to a technologically driven process of using algorithms to analyze data from multiple perspectives and extract meaningful patterns that can be used to predict future behavior.

Dataveillance: The concept of *dataveillance* refers to the application of information technologies to monitor individuals' activities by investigating the data trail they leave through their activities.

Interactive Media: *Interactive media* is an umbrella term describing communication media that allow the two-way flow of information between content users and producers.

Public Intimacy: *Public intimacy* suggests an outward move to locate personal matters in the public domain. The emphasis is on the shared discursive spaces of public engagement, rather than inviting spaces of the cultural or personal kind. In other words, public discourses and expressions, even in their most formalized discursive modes, constitute and conjure intimate connections.

Social Media: The concept of *social media* refers to online applications and platforms through which individuals can create and distribute content and communicate with each other.

Social Network Sites: Social network sites are web-based systems that enable end-users to create online profiles, form virtual networks or associations with other users, and view other individuals' profiles.

This work was previously published in Handbook of Research on Social Interaction Technologies and Collaboration Software; Concepts and Trends, edited by T. Dumova; R. Fiordo, pp. 392-403, copyright 2010 by Information Science Reference (an imprint of IGI Global).

Chapter 8.4
Analysis of Content Popularity in Social Bookmarking Systems

Symeon Papadopoulos
Aristotle University of Thessaloniki, Greece Informatics & Telematics Institute, Thermi, Thessaloniki, Greece

Fotis Menemenis
Informatics & Telematics Institute, Thermi, Thessaloniki, Greece

Athena Vakali
Aristotle University of Thessaloniki, Greece

Ioannis Kompatsiaris
Informatics & Telematics Institute, Thermi, Thessaloniki, Greece

ABSTRACT

The recent advent and wide adoption of Social Bookmarking Systems (SBS) has disrupted the traditional model of online content publishing and consumption. Until recently, the majority of content consumed by people was published as a result of a centralized selection process. Nowadays, the large-scale adoption of the Web 2.0 paradigm has diffused the content selection process to the masses. Modern SBS-based applications permit their users to submit their preferred content, comment on and rate the content of other users and establish social relations with each other. As a result, the evolution of popularity of socially bookmarked content constitutes nowadays an overly complex phenomenon calling for a multi-aspect analysis approach. This chapter attempts to provide a unified treatment of the phenomenon by studying four aspects of popularity of socially bookmarked content: (a) the distributional properties of content consumption, (b) its evolution in time, (c) the correlation between the semantics of online content and its popularity, and (d) the impact of online social networks on the content consumption behavior of individuals.

DOI: 10.4018/978-1-60566-816-1.ch011

To this end, a case study is presented where the proposed analysis framework is applied to a large dataset collected from Digg, a popular social bookmarking and rating application.

INTRODUCTION

The emergence of Web 2.0 technologies and the widespread use of applications integrating such technologies have transformed the way people experience and act in online settings. In the first days of the Web, people were excited to browse through and consume online content (mostly static web pages) that was prepared and published by website owners or administrators. Nowadays, digital content consumption – e.g. online article reading, picture viewing and video watching – still appears to be one of the main activities for most internet users[1]. However, the advent of the Web 2.0 application paradigm has transformed the established "browsing-based" online content consumption behavior of users. This change was possible by means of offering users a host of rich interactivity features, such as content sharing, rating as well as online community building. Thus, users of today's Web 2.0 applications are empowered to share, organize, rate and retrieve online content. In addition, users are exposed to the content-related activities of other users and can even form online relations to each other. Consequently, online content consumption within a modern Web 2.0 application constitutes an overly complex phenomenon with interesting dynamics which have not been thoroughly investigated yet.

Social Bookmarking Systems (SBS) hold a prominent place among Web 2.0 applications with respect to content consumption since they provide a platform where users are provided with two significant features:

- Submitting and sharing bookmarks (links) to online resources, e.g. articles, photos or videos, which they consider interesting.
- Indicating their preference or disapproval to bookmarks submitted by other users, by voting for or against the interesting-ness/ appeal of online resources and by commenting on them.

In addition to these two features which are fundamental for an SBS, there are two other optional groups of features, namely Taxonomic and Social-Community features. Taxonomic features pertain to the possibilities offered to users for assigning bookmarks to a predefined topic-scheme or for tagging them with freely chosen keywords. Social-Community features enable the users to create "friendship" relations with each other (which can be unilateral or mutual) or to create groups of topical interests. A number of social bookmarking applications have been recently launched. Systems such as *delicious* are meant to be used as general bookmark organization and sharing applications, while there are also social bookmarking applications focused on online news such as *Digg* and *newsvine*, and even niche bookmarking services such as *CiteULike*, used only for bookmarking citations to research articles. Table 1 lists some of the most popular SBS along with their features.

There has been a recent surge in the usage of social bookmarking applications, many of which currently attract several million of unique users per month. An illustrative overview of usage statistics pertaining to the most popular SBS is provided in Table 2, where the top 10 social bookmarking applications are ranked based on the number of unique monthly visitors they attract. Several additional web popularity metrics are provided in the same table, namely the number of inbound links, the Google Page

Rank and the Alexa Rank. According to the table, *Digg*, *StumbleUpon*, propeller and *reddit* are the most popular social bookmarking applications in terms of the number of unique visitors per month. Although *delicious* is the oldest bookmarking application from this list and features the largest number of bookmarks, it appears to attract a smaller number of visitors than its main competitors.

The widespread and intense use of such applications is responsible for the ceaseless creation of massive data sets where the content consumption patterns of users are imprinted. The description and analysis of the intricate phenomena related to these patterns could provide answers to a series of interesting questions:

- What kinds of distributions emerge in the content consumption behavior of the masses?
- How does the attention attracted by a particular online resource evolve over time?
- Is there correlation between the content of a resource (semantics) and its popularity?
- What is the impact of social networking on the consumption of online content?

This chapter provides insights into such questions by investigating recent and ongoing research efforts in the area of Web 2.0 data mining as well as by carrying out an analysis of a large dataset collected from Digg. The next section provides an account of existing studies on the analysis of content consumption behavior within Web 2.0 applications, organized around four research tracks. The third section formalizes a framework for the analysis of content-related phenomena in SBS. The application of this framework is illustrated in the fourth section, with Digg used as a case study. The final two sections of the chapter present an outlook on future trends in this area and conclude the chapter respectively.

BACKGROUND

Considerable research interest has been recently developed in the analysis and modeling of the

Table 1. Taxonomic and social/community features of several social bookmarking services

Application name / URL	Type of Bookmarks	Taxonomic Feats		Social/Community Feats	
		Tags	Topics	Friends	Groups
Digg.com	News		+	+	
www.**StumbleUpon**.com	General	+		+	
www.**propeller**.com	News	+	+	+	+
www.**reddit**.com	News		+	+	
www.**fark**.com	News		+		
www.**newsvine**.com	News	+	+	+	+
delicious.com	General	+		+	
www.**blinklist**.com	General	+		+	
www.**clipmarks**.com	General	+		+	
furl.ne	General	+	+		+
www.**citeulie**.org	Ctation	+			+
bibsonomy.org	Citations	+			+

Table 2. Top 10 Social Bookmarking services ranked by averaging the number of unique visitors as provided by Compete[2] and Quantcast[3]. The number of inbound links was collected from Yahoo! Site Explorer. The data were collected in June 2008.

Site	Monthly Visitors		Inbound Links	Page Rank	Alexa Rank
	Compete	Quantcast			
Digg	**23,988,437** (1)	**19,906,963** (1)	27,589,161 (6)	8	**114**
StumbleUpon	2,338,242 (3)	1,331,110 (3)	160,863,707 (2)	8	372
propeller	1,521,706 (5)	6,055,679 (2)	8,590,778 (7)	7	876
reddit	2,489,583 (2)	1,115,655 (5)	4,859,451 (8)	8	4,074
fark	368,566 (7)	1,235,481 (4)	2,915,127 (9)	5	1,938
newsvine	625,115 (6)	986,471 (6)	66,154,514 (3)	8	6,923
delicious	1,632,204 (4)	420,043 (9)	**462,168,833** (1)	**9**	1,161
blinklist	307,673 (9)	510,838 (7)	43,226,590 (5)	7	5,613
clipmarks	322,011 (8)	448,437 (8)	460,278 (10)	6	6,208
furl	153,987 (10)	79,462 (10)	60,312,568 ()	8	15,914

content consumption behavior of Web 2.0 application users. Much of this research is focused on the mining of web log data where the user transactions are recorded. In our study, we have identified four major research tracks addressing the study of phenomena that arise through the online content consumption by masses of users:

1. *Statistical analysis:* The monitoring of the activities of large user masses enables the application of powerful statistical analyses in order to study the distributional properties of observed variables and to make inferences about the recorded data.
2. *Temporal data mining:* The analysis of content consumption patterns over time is crucial for in-depth understanding of the dynamics emerging in the phenomena that take place within social bookmarking applications. Discovering trends and periodic patterns as well as producing summaries of multiple data streams is the focus of this perspective.
3. *Content semantics:* The lexical and semantic analysis of the content that is consumed within SBS provides insights into the interests of the masses and has broad implications in the design of effective Information Retrieval (IR) systems.
4. *Social network effects:* The influence of the users' online social environment on their content consumption behavior is increasingly important for describing diffusion processes and viral phenomena arising in the SBS user communities.

In order to gain a high-level understanding of the phenomena emerging in complex systems such as SBS, researchers commonly employ statistical analysis techniques; more specifically, they inspect and analyze the distributional properties of the variables observed in the system under study. Previous studies of social, biological and computer systems have confirmed in a series of phenomena the emergence of highly-skewed distributions, frequently taking the form of a *power law*. Power-law distributions – commonly referred to as Zipf's laws or Pareto distributions – provide a statistical model for describing the "rich-get-richer" phenomena frequently appearing in complex systems. Two noteworthy survey studies on power laws are provided by Newman (2005) and Mitzenmacher (2004). In an attempt to explain the emergence of such distributions in complex systems, a series of generative models have been recently proposed; among those, one of the most prominent is the *preferential attachment* model by (Barabási & Albert, 1999). Later in the chapter, we will confirm that the interest attracted by online resources in Digg, as well as the voting patterns of SBS users follow highly-skewed patterns that can be often well approximated by a power law.

Furthermore, analysis of the temporal aspects of phenomena similar to the ones appearing within SBS-like environments provides further insights into the evolution of variables such as the intensity of user activity, or the number of votes that an online resource collects. For instance, the temporal analysis of the user posting and commenting activity in Slashdot, a social bookmarking and public discussion application focused on technology news, revealed that the time intervals between a post and its comments closely follow a log-normal distribution with periodic oscillatory patterns with daily and weekly periods (Kaltenbrunner et al., 2007a). Subsequently, the same authors managed to predict the future Slashdot user activity based on their past behavior by creating prototype activity profiles (Kaltenbrunner et al., 2007b). Another related study is provided by Cha et al. (2007) where the authors analyze the temporal video viewing patterns in YouTube[4] and Daum[5]. In line with these studies, we devote part of this chapter to analyzing the story popularity evolution in Digg as well as the temporal activity profiles of its users.

Another aspect of content popularity pertains to the correlation between the popularity of bookmarked online items (as quantified by number of votes or hits) and their semantics, which are conveyed by means of their textual features[6]. Considerable work has been carried out with the goal of separating between different classes of content based on machine learning methods that make use of features extracted from their text (Yang & Pedersen, 1997). For instance, automatic methods based on machine learning have been devised for differentiating between positive and negative online product reviews (Dave et al., 2003; Pang et al., 2002; Turney, 2002). Further text classification problems involve the automatic classification of textual items based on their utility (Zhang & Varadarajan, 2006) or their quality (Agichtein et al., 2008). In part of the case study presented in this chapter, we examine the potential of automatically predicting whether a given bookmarked item will become popular or not based on its textual content. Although this problem is very complex to tackle by means of the machine learning paradigm adopted in the aforementioned studies, we can establish significant correlations between the popularity of content items and their textual features.

Finally, the study of social network effects on the behavior of users constitutes another analysis perspective for content popularity in social bookmarking applications. In (Richardson & Domingos, 2002), evidence is provided supporting the significance of network effects on a customer's online purchase behavior. In an effort to exploit such effects, Song et al. (2007) propose an information flow model in order to exploit the different information diffusion rates in a network for improving on recommendation and ranking. On the other hand, an empirical study by Leskovec et al. (2007) based on an online recommendation network for online products (e.g. books, music, movies) indicated that there is only limited impact of a user's social environment on his/her purchasing behavior. Finally, the study by Lerman (2007) concludes that the users of Digg tend to prefer stories that their online friends have also found interesting. Here, we define two measures of social influence on (a) content popularity and (b) users' voting behavior, conceptually similar to the ones introduced by Anagnostopoulos et al. (2008). Then, we carry out a set of experiments to quantify the extent of the social influence effects on bookmarked content popularity and consumption in Digg.

SBS ANALYSIS FRAMEWORK

This section introduces the *Diggsonomy* framework, which aims at facilitating the study and the description of the phenomena arising in social bookmarking applications. This framework was originally presented in (Papadopoulos et al., 2008); here, we repeat the definition of the framework. The framework considers an SBS and the finite sets U, R, T, S, D, which stand for the sets of users, resources, timestamps, social relations and votes on resources respectively. Note that T is an ordered set.

Definition 1 (Diggsonomy): Given an SBS, its derived Diggsonomy $\boldsymbol{B}$ is defined as the tuple $\boldsymbol{B} = (U, R, T, S, D)$, where $S \subseteq U \times U$ is the social network of the SBS users, and $D \subseteq U \times R \times T$ is the users' voting set, modeled as a triadic relation between U, R, and T.

Definition 2 (Personomy): The Personomy $\boldsymbol{P}_u$ of a given user $\boldsymbol{u} \in \boldsymbol{U}$ is the restriction of $\boldsymbol{B}$ to u, i.e. $\boldsymbol{P}_u = (R_u, S_u, D_u)$ with $D_u = \{(r, t) \in R \times T \mid (u, r, t) \in D\}$, $S_u = \pi_U(S)$ and $R_u = \pi_R(D_u)$.

Definition 3 (Vote-history): The Vote-history for a particular resource (story) r, denoted as H_r is defined as the projection of the Diggsonomy D on $U \times T$ restricted on r, i.e. $H_r = \pi_{U \times T}(D \mid r) \subseteq U \times T$. The user u_0 for whom the statements $(u_0, t_0) \in H_r$ and $\forall\, t \in \pi_T(H_r)$, $t_0 < t$ hold is called the submitter of the story.

The framework is inspired by the Folksonomy definitions appearing in (Mika, 2005) and (Hotho et al., 2006a). The major difference is that the Diggsonomy formalism enables the description of the temporal aspects of content rating. The notation introduced by this framework will form the basis of the following discussion, which will be organized around the four analysis perspectives that were introduced in the background section:

- Analysis of statistical properties of variables measured in an SBS.
- Temporal analysis of content popularity and user behavior.
- Semantic aspects of popularity.
- Impact of social networks on popularity.

In the rest of this section, each of these perspectives will be discussed separately and existing analysis approaches will be reviewed.

The Heavy Tails of Social Bookmarking

A widely researched and empirically supported model for popularity (and variables generated by skewed distributions) is the power-law distribution. A comprehensive review of the properties observed in such distributions is provided in (Newman, 2005). According to this model, the probability density function of the skewed variable should be described by the following law:

$$p(x) = Cx^{-a} \tag{1}$$

In Equation (1), α is called the exponent or the scaling parameter of the power law (the constant C is part of the model in order to satisfy the requirement that the distribution sums to 1). A straightforward way to empirically identify a power-law is to plot its histogram. However, this might be tricky in practice since the tail of the distribution would appear very noisy (due to the regular histogram binning which is not appropriate for distributions of this nature). A potential solution to this problem would be to employ logarithmic binning; however, a more elegant way to deal with the problem is to calculate and plot the Cumulative Distribution Function (CDF), *P(x)*. Based on Equation (1), we get:

$$P(x) = \int_0^{\infty} p(x')dx' = C\int_0^{\infty} x'^{1-a}\, dx' = \frac{C}{a-1} x^{-(a-1)} \qquad (2)$$

It appears from Equation (2) that the cumulative distribution function *P(x)* also follows a power law, but with a different exponent (α-*1*). Since *P(x)* is derived by integrating over *p(x)*, the resulting curve has a much smoother tail (integration acts as a low-pass filter), thus rendering clear the power-law nature of the distribution. Also, an important consideration when modeling skewed distributions with power-laws (and other related models) is the range of values for which the power law approximates sufficiently well the real distribution. Typically, a value x_{min} can be identified such that Equation 1 is a reasonable approximation for the distribution only for $x \geq x_{min}$. The employed approach for estimating the parameters α and x_{min} of the power law will be described in the case study section.

A set of quantities measured within real-world complex systems have been reported to exhibit power-law behavior (Newman, 2005). Examples of such quantities are the frequency of words in the text of the *Moby Dick* novel, number of citations to scientific papers, number of calls received AT&T telephone customers and others (Newman, 2005). Recent research on social web data has confirmed the power-law nature of a series of Web 2.0 originating distributions, e.g. tag usage in delicious (Hotho et al., 2006a) and (Halpin et al., 2007), number of votes to questions/answers in the Yahoo! Answers system (Agichtein et al., 2008), video popularity in YouTube and Daum (Cha et al. 2007) and story popularity in Digg (Papadopoulos et al., 2008).

Apart from the classic power-law distribution of Equation (1) reported in the aforementioned works and used for the subsequent analysis of Digg popularity, a set of more elaborate models have recently been proposed in the literature for modeling skewed distributions. For instance, the Discrete Gaussian Exponential is proposed by Bi et al. (2001) as a generalization of the Zipf distribution (i.e. power-law) to model a variety of real-world distributions, e.g. user click-stream data. Furthermore, statistical analysis of the distribution of 29,684 Digg stories by Wu and Huberman (2007) resulted in a log-normal distribution model for the data. The truncated log-normal distribution was also found by Gómez et al. (2008) to accurately describe the in- and out-degree distributions of the Slashdot user network formed on the basis of their participation in the online discussion threads. Finally, the recently formulated Double Pareto log-normal distribution was presented by Seshadri et al. (2008) as an accurate model for a set of variables in a social network created by mobile phone calls.

There are significant benefits in recognizing and understanding the heavy-tail nature of skewed distributions. As pointed by Bi et al. (2001), typical statistical measures such as mean, median and standard deviation are not appropriate for summarizing skewed distributions. In contrast, parametric models, such as the power-law or the log-normal distribution, convey a succinct and accurate view of the observed variable. Furthermore, comparison of the observed variable with the fitted model may reveal deviant behavior (outliers). Similar benefits of employing a parsimonious model such as the power-law to summarize and mine massive data streams that depict skewed distribution were reported in (Cormode & Muthukrishnan, 2005). Finally, the work in (Cha et al., 2007) demonstrated the utility of understanding the heavy-tail content consumption patterns by demonstrating a potential for 40% improvement in video content consumption (in YouTube) by alleviating information delivery inefficiencies of the system (e.g. by recommending niche content lying in the long tail of the distribution). Later in the chapter, we will confirm the emergence of heavy-tail distributions in Digg.

Temporal Patterns of Content Consumption

The study of the temporal aspects of online content consumption and rating in the context of a Web 2.0 application has been beneficial for a series of tasks, e.g. planning an online campaign, anticipating voluminous requests for content items or detecting malicious user activities. For instance, by analyzing the temporal activity patterns of Slashdot users (i.e. posting and commenting on posts of others), the authors of (Kaltenbrunner et al., 2007b) could predict with sufficient accuracy the future comment activity attracted by a particular post. Similarly, the study by Cha et al. (2007) presents an analysis of the temporal video content popularity patterns observed in YouTube and demonstrates the potential for short-term popularity prediction. Furthermore, studies of the temporal aspects of story popularity in Digg were carried out by Lerman (2007) and Papadopoulos et al. (2008). In both studies, it was confirmed that Digg stories when moved to the front page of the site go through a staggering popularity growth phase, thus anticipating voluminous user requests for particular content items.

The temporal data appearing within a bookmarking system are usually generated by means of recording event timestamps; specifically, the set of instances when a story collects votes from users constitute the popularity timeline of the particular story (cf. $\pi_T(H_r)$ of Definition 3) and the instances when a user gives votes to stories form his/her activity timeline (cf. $\pi_T(D_u)$ of Definition 2). For convenience, we will denote the raw timestamp set comprising the event instances of object i (where i can either denote a story or a user) in an ordered fashion as $T_i = \{t_0, t_1 .... t_N\}$. The first step in analyzing such data is to select a small but sufficiently representative subset of stories or users and then to inspect their timestamp sets on an individual basis.

However, these timestamp sets are not time series in a typical sense, i.e. they are not the result of measuring the value of a variable at regular intervals. Thus, in order to visually convey the information contained in them in a meaningful way, we consider two kinds of time series based on these raw timestamp sets: (a) the time series of the aggregate count of events at time t, and (b) the time series of the count of events falling in the interval $[t-\Delta t, t+\Delta t]$. For ease of reference, we shall denote the aforementioned time series as $N(t)$ and $n(t)$ respectively. Figure 1 illustrates the characteristics of such time series for a small sample of Digg story popularity time series.

A complication arises when attempting to study the temporal behavior of numerous SBS entities (stories or users) in an aggregate manner: The entities of interest are active in different time intervals and have different activity rates. In order to overcome this complication, we consider the projection on T of the Vote-history set H_r, denoted by $T_r = \pi_T(H_r)$. For each story, we perform the following transformation:

Figure 1. Two alternatives for inspecting event-based time series: (a) cumulative number of Diggs, (b) number of Diggs per hour. Here, three sample Digg story popularity curves are shown

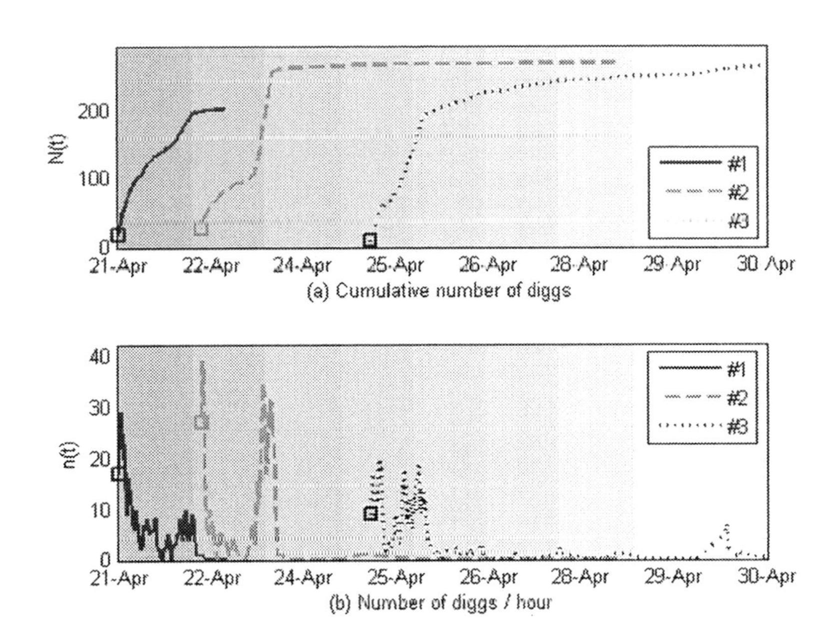

$$T_r = \frac{T_r - \min(T_r)}{\max(T_r) - \min(T_r)} \quad (3)$$

where *min(T)*, *max(T)* return the minimum and maximum timestamp values of the input timestamp set. In that way, it is possible to normalize the temporal activity of an SBS entity to an artificial temporal space spanning the interval [0, 1], which makes it possible to perform a set of operations between time series, e.g. addition, subtraction, averaging and so on. This possibility is of significance since we are particularly interested in deriving an "average" time series which is representative of hundreds or even thousands of time series. Figure 2 depicts the effect of the transformation on the set of time series of Figure 1 (under the representation *n(t)*, i.e. number of events per time interval). One should note that while the transformation removes the notion of temporal scale from the individual time series, it preserves their structural characteristics.

Finally, it is possible to apply the transformation of Equation 3 to a subsequence of a given time series and in that way to map a particular "phase" of a time series to the *[0, 1]* interval with the purpose of comparing the evolution of the phenomenon within this phase to its evolution within the full lifetime of the time series. Comparison between time series phases provides additional insights to the understanding of temporal phenomena, especially in cases where there is prior knowledge that a phenomenon takes place in more than one phases.

Semantic Aspects of Content Popularity

In a social media website where a stream of online stories is continuously flowing through the site's pages, one could argue that the popularity (number of votes) of a given resource (e.g. news article) will strongly correlate with the

Figure 2. The effect of the proposed time series normalization

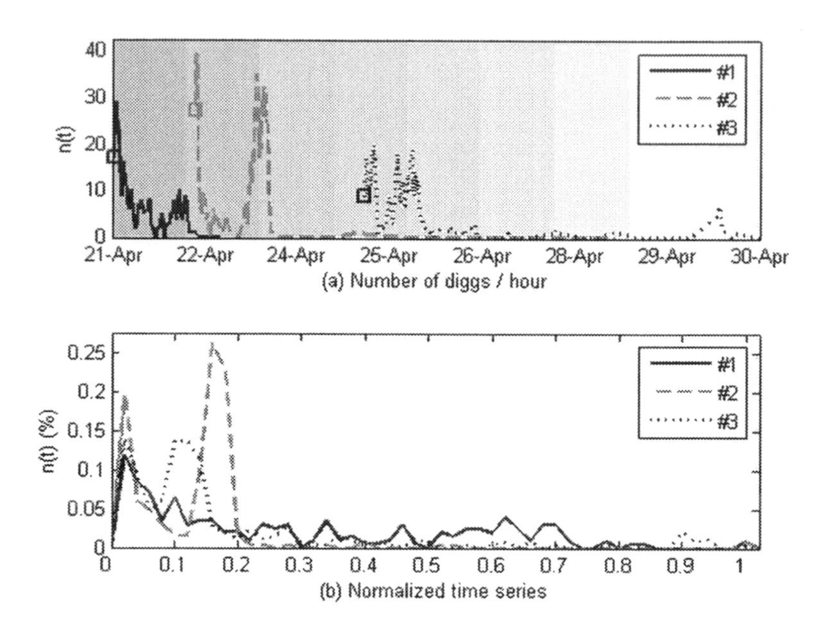

semantic content featured by the story as well as the linguistic style of the story text. Stories usually appear in the form of a short title and description which should provide sufficient incentive to readers to read the whole story and then vote in favor of it. Therefore, one could attempt to predict which stories will draw the attention of the masses by plain inspection of their textual content. In other words, one could cast the problem of popularity prediction as a text classification task, comprising a feature extraction and a training step (exploited by some machine learning algorithm).

Text classification usually takes place by processing a corpus of documents (bookmarked content items in our case), extracting all possible terms from them and considering them as the dimensions of the vector space where the classification task will be applied. This is commonly referred to as the Vector Space Model (VSM) in the Information Retrieval literature (Salton et al., 1975). However, due to the extremely large variety of vocabulary in large corpora (millions of unique terms), the dimensionality of such a vector space is frequently prohibitive for direct application of the model. The problem gets even worse when combinations of more than one term are considered as text features. For that reason, feature selection (or reduction) techniques are crucial in order to end up with a manageable set of dimensions that are sufficient for the classification task at hand.

In (Yang & Pedersen, 1997), a series of measures were evaluated regarding their effectiveness to select the "proper" features (i.e. these features that would result in higher classification performance). The simplest of these measures is the *Document Frequency* (DF), i.e. the number of content items where the particular feature (term) appears. *Information Gain* (IG) is a more sophisticated feature selection measure. It measures the number of bits of information obtained for class prediction (popular vs. non-popular) by knowing the presence or absence of a term in a document. Further, *Mutual Information* (MI) is an additional criterion for quantifying the importance of a feature in a classification problem; however, Yang & Pedersen (1997) found this measure ineffective for selection of discriminative features. In contrast, they found that the χ^2 *statistic* (CHI), which measures the lack of independence between a term and a class, was quite effective in that respect. Finally, another interesting feature selection measure introduced in the aforementioned paper is the *Term Strength* (TS), which estimates the term importance based on how commonly a term is likely to appear in closely-related documents.

Assuming that a subset of terms from the corpus under study have been selected as features, each content item is processed so that a feature vector (corresponding to the selected feature space) is extracted from it. Then, it is possible to apply a variety of machine learning techniques in order to create a model that permits classification of unknown pieces of texts to one of the predefined classes. Previous efforts in the area of sentiment classification (Dave et al., 2003; Pang et al., 2002) have employed Support Vector Machines (SVM), Naïve Bayes, as well as Maximum Entropy classifiers to tackle this problem.

In our case study, we investigated the potential of popularity prediction based purely on textual features. We conducted a series of feature extraction and text classification experiments on a corpus of 50,000 bookmarked articles. The feature selection measures used to reduce the dimensionality of the feature space were DF and CHI. For the classification task, three standard methods were used: Naïve Bayes (Duda et al., 2001), SVM (Cristianini & Shawe-Taylor, 2000) and C4.5 decision trees (Quinlan, 1993).

Social Impact on SBS Usage

It was previously argued that story popularity depends on the textual content of the particular story. However, it is widely recognized that readers do not select their content in isolation. Users of social bookmarking applications form online relations and are constantly made aware of the preferences and content consumption patterns of other SBS users. Therefore, one would expect the emergence of viral phenomena in online content consumption within an SBS. The value of understanding and exploiting viral phenomena has been already acknowledged in online knowledge sharing communities, for e.g. improving online advertising campaigns (Richardson & Domingos, 2002), understanding viral marketing dynamics (Leskovec et al., 2007) and identifying experts (Zhang et al., 2007). Therefore, we introduce here two measures in order to quantify the social influence on the content rating process. We name these two measures, the *user Social Susceptibility* (SS) I_u, and the *story Social Influence Gain* (SIG) I_r.

Definition 4 (Social Susceptibility - SS): The social susceptibility of a given user u denoted by I_u quantifies the extent to which his/her voting behavior (as expressed by his voting set D_u) follows the behavior of his/her friends' voting behavior (denoted by D_u').

$$I_u = \frac{|D_u'|}{|D_u|} \tag{4}$$

Definition 5 (Social Influence Gain - SIG): The social influence gain for a given story r denoted by I_r is a measure of the extent to which r has benefited from the social network of the story submitter.

$$I_r = \frac{|H_r^{'}|}{|H_r|}$$ where

$$H_r^{'} = \{(u,t_k) \in H_r \mid \exists (u_0,t_0) \in H_r, u_0 \in S_U, t_0 < t_k\} \quad (4)$$

and u_0 is the submitter of the story as defined in Definition 3.

SS and SIG are similar in nature to the concept of *Social Correlation* as discussed in (Anagnostopoulos et al., 2008). Social Correlation (SC) within an SBS can be defined either for two users, u_1, u_2 as the Jaccard coefficient of the sets R_{u1} and R_{u2} (cf. Definition 2) or for a single user u as the proportion of his/her stories that are common with the stories of the users of his/her social network. Here, we adopt the latter definition since it is directly comparable with the SS of a user, i.e. it can be derived by removing the temporal constraint from Equation 4. Note that SC may be attributed to a combination of the following: (a) an inherent tendency of friends to have similar interests (homophily), (b) some external factor causing two users to vote in favor of the same story (confounding) and (c) the possibility for users to see through the Digg interface which stories their friends have already dugg (influence). By imposing temporal constraints in Equations 4 and 5, we attempt to isolate the effect of (a) and (b) in order to use SS and SIG as measures of social influence rather than measures of generic SC.

CASE STUDY: SOCIAL BOOKMARKING PATTERNS IN DIGG

Since the case study for the discussion of the chapter is based on data collected from Digg, a short introduction on the specifics of the application will precede the presentation of the data analysis in order to facilitate the interpretation of the derived results. The basic rationale of Digg is the discovery of interesting web resources (or *stories* as they are commonly called in Digg-speak) by means of empowering simple users to submit and then collectively decide upon the significance (or interesting-ness) of the submitted web items (mostly news items, images and videos). In other words, Digg can be considered as an example of a Social Media application. When a story is submitted, it appears in the *Upcoming* section of the site, where stories are displayed in reverse chronological order. Users may vote on a story by "Digging" it. Digging a story a story saves it to a user's history (obviously a given user may Digg a particular story only once). There is also the possibility to "bury" a story, if one considers it to be spam or inappropriate material.

When a story collects enough votes, it is promoted to the *Popular* section, which is also the front page of the site. The vast majority of people who visit Digg daily or subscribe to its RSS feeds, mostly read the front page stories; thus the story exposure to the online public increases steeply, once it is transferred to the Popular section. The exact promotion mechanism is not disclosed to the public and is modified periodically in order to prevent malicious efforts of artificially promoting a story. The service moderators state that consensus by many independent users is required in order for a story to become popular.

Stories are categorized by media type (i.e. news, videos, images, and podcasts) and topic. Additionally, there is a predefined two-level topic hierarchy available for use, which is specified in Table 3. Apart from these story browsing capabilities, the application also features user-based story browsing, i.e. one can browse through the stories dugg by a particular user. Finally, it enables users to form social networks, by adding other users to their list of "friends". Such "friendship" relations are one-way (i.e. it takes

Table 3. Topic hierarchy of Digg

Container	Topic
Technology	Apple, Design, Gadgets, Hardware, Industry News, Linux/Unix, Microsoft, Mods, Programming, Security, Software
World & Business	Apple, Design, Gadgets, Hardware, Industry News, Linux/Unix, Microsoft, Mods, Programming, Security, Software
Science	Environment, General Sciences, Space
Gaming	Industry News, PC Games, Playable Web Games, Nintendo, PlayStation, Xbox
Lifestyle	Arts & Culture, Autos, Educational, Food & Drink, Health, Travel & Places
Entertainment	Celebrity, Movies, Music, Television, Comics & Animation
Sports	Baseball, Basketball, Extreme, Football (US/Canada), Golf, Hockey, Motorsport, Olympics, Soccer, Tennis, Other Sports
Offbeat	Comedy, Odd Stuff, People, Pets & Animals

both users to add each other to their friends list in order to have a mutual relation).

Apart from the basic SBS functionality mentioned above, the application offers a set of features for personalized reaction to content, more specifically it enables users to comment on stories, approve/disapprove of comments of others and blog on stories (i.e. write and submit a post to their own blog through Digg). Also, it is worth noting that a story recommendation engine has been recently launched (in July 2008) which recommends news items to users based on their past activity.

The bulk of the click stream generated by the masses of Digg users is made publicly accessible via a RESTful API. The analyses presented in the subsequent sections have been based on data downloaded via this API. The downloaded data contain information about the stories (title, description, container, topic, number of Diggs, number of comments, etc.), the users (username, subscription date, and friends) and the Diggs collected by stories (story, username, and timestamp). A first set of stories was collected by constantly monitoring the stories submitted to the site during the week between 24 and 30 April 2008. A set of 109,360 stories, S_0, were collected during this phase. This set of stories was submitted by a set of 34,593 users, U_0, who were used as seeds to collect data about a total of 354,150 users by requesting for the friends and the friends' friends of each user u belonging to U_0. Finally, the full Digging history for a random sample of these users was collected (both the Diggs and the stories that were dugg). In that way, a total of 98,034,660 Diggs given to 2,084,498 stories were collected. We will refer to the sets U_0 and S_0 of users and stories that were gathered in the first data collection phase as the *core* dataset, while the rest of the dataset will be referred to as *extended* dataset. It occasionally happened that stories or users were removed from the system (probably due to spamming behavior) in which case they were also removed from the local dataset.

Statistical Analysis of Digg Usage

The first step of our analysis involved the study of the heavy-tail nature of several variables of interest arising through the mass usage of Digg. The following distributions were examined:

- Diggs collected by stories.
- Comments collected by stories.
- Diggs given by users.
- Friends in the Digg social networks of users.

Figure 3: Four heavy-tail Digg Cumulative Distribution Functions (CDF) and their power-law approximations: (a) Diggs per story, (b) comments per story, (c) Diggs per user, and (d) Digg friends per user.

Figure 3 provides logarithmic plots of the aforementioned distributions. The figure renders clear the heavy-tail nature of the depicted distributions, by overlaying on top of the observed distributions their power-law fits according to the fitting method presented by Clauset et al. (2007). The proposed method employs an approximation to the Maximum Likelihood Estimator (MLE) for the scaling parameter of the power law:

$$\hat{a} \cong 1 + n\left[\sum_{i=1}^{n} \ln \frac{x_i}{x_{\min} - \frac{1}{2}}\right]^{-1} \quad (6)$$

This estimator assumes that the value x_{min} above which the power law holds is known. In order to estimate this value, the authors recommend the use of the Kolmogorov-Smirnov (KS) statistic as a measure of goodness-of-fit of the model with parameters (α, x_{min}) with the observed data. The KS statistic is defined as the maximum distance between the CDF of the data *S(x)* and the fitted model *P(x)*:

$$D = \max_{x \geq x_{\min}} | S(x) - P(x) | \quad (7)$$

Apparently, the plain power-law model is not sufficient for accurately fitting all of the observed distributions. For instance, the shape of the distribution of Diggs per user in Figure 3(c) indicates that a truncated log-normal distribution would be a better fit for the observed variable. Further, by inspection of the number of friends per user, a few conspicuous outliers can be identified that deviate significantly from the fitted power-law.

Figure 4. Two possible popularity evolution patterns. Note the two different phases in the popularity evolution for stories that are selected for the 'Popular' section of the site

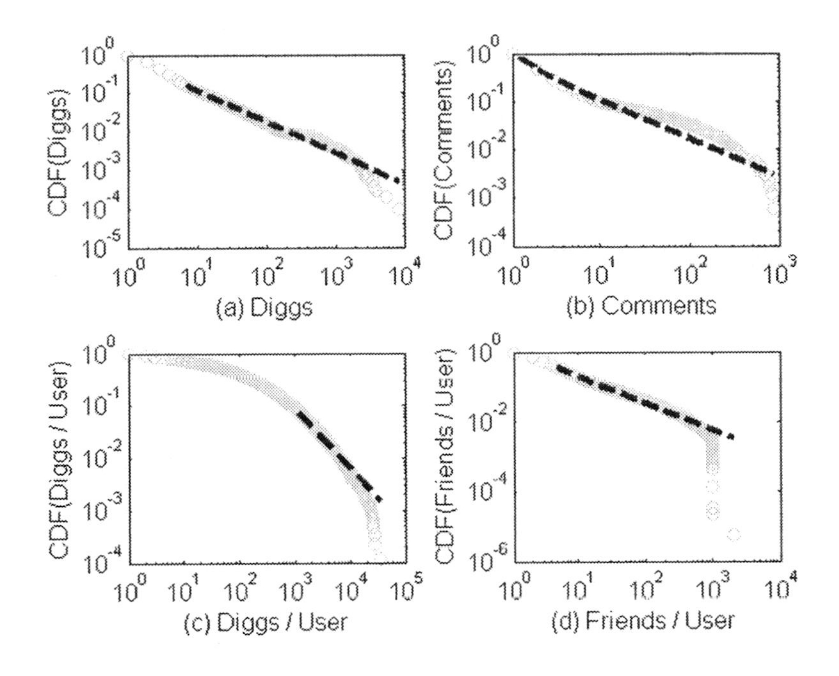

Temporal Analysis of Content Consumption

The temporal study of story popularity curves and user activity of Digg was carried out by means of the temporal representation and aggregation framework presented in the previous section. Based on that, a first noteworthy observation about the popularity of submitted stories in Digg is that they typically evolve in two ways: (a) they reach a plateau of popularity while in the 'Upcoming' section of the site and remain there until they are completely removed in case they do not receive any Diggs for a long time, (b) they attain the 'Popular' status after some time and they are moved to the 'Popular' section, where they undergo a second phase of popularity growth at a much higher intensity. Figure 4 depicts the cumulative number of Diggs, $N(t)$, collected by a sample popular and a sample non-popular story during their lifetime. For convenience, we denote the set of unpopular stories as R_U and the set of popular stories as R_P[7].

After establishing by inspection the difference of temporal evolution between popular and non-popular stories, we then proceed with comparing the distributions of their Digg arrival times. To this end, the time series, $n(t)$, of 5,468 non-popular and 852 popular stories, which were normalized by means of the transformation of Equation 5, were aggregated[8]. This resulted in the distributions of Figure 5. Together with the distributions we present the areas of confidence for their values; more specifically, around each instance $n(t)$, we draw the interval $[n(t)-\sigma_n/3, n(t)+\sigma_n/3]$. In 6(a), the time series of the number of Diggs $n(t)$ is normalized with respect to the total number of Diggs throughout the whole lifetime of each story. In that way, it is possible to directly compare the local temporal structures of popular stories to the ones of the non-popular stories. In Figure 5(b), we present the absolute number of Diggs per hour in order to provide a

Figure 5. Digg arrival time distributions of popular vs. non-popular stories

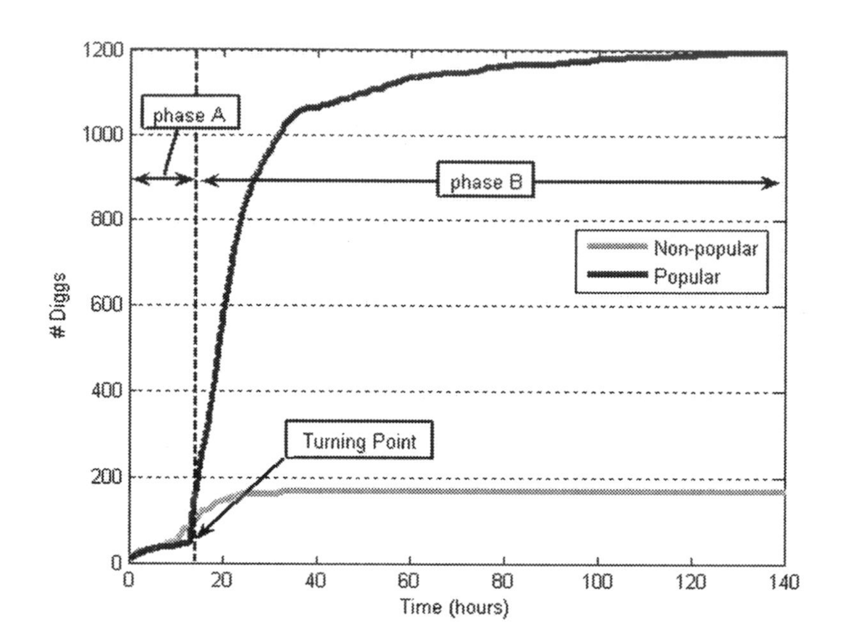

complete picture of the comparison between the popular and the non-popular stories.

Figure 5 clearly illustrates the fact that while the non-popular stories gather the majority of their Diggs during the very first moments of their lifetime (first two bins in the histogram), the popular ones are characterized by two growth stages: (a) a first growth stage which is similar to the full lifetime of the non-popular stories, i.e. it is characterized by a monotonically decreasing trend, (b) a second growth stage, which takes place once a story is moved to the 'Popular' section of the site and is characterized by a steep increase in the number of votes that a story receives.

The intensity of popularity growth for the stories that become members of R_p can be attributed to the high exposure that these stories get for a few minutes after they are moved to the 'Popular' section (which happens to be the front page of Digg). Also, one should note that big search engines regularly index and rank favorably the most popular stories of Digg (and stories coming from other SBS and social media applications) and thus they act as a secondary source of exposure for these stories, which contributes to sustain their popularity growth for some time.

After analyzing the temporal evolution of story popularity, we apply a similar analysis on the user Digging behavior. For such a study, we investigate the structure of the time series that are formed from counting the number of Diggs given by users to stories that fall within the interval $[t-\Delta t, t+\Delta t]$. Figure 6 presents the result of aggregating the user activity over 539 users drawn randomly from the set of users belonging to D_0^9. Inspection of the outer diagram reveals that Digg users are intensively active during the very first period after subscription to the service (note the extremely high number of Diggs occurring in the first 5-10% of the users' lifetime). This is not particularly surprising since users are more enthusiastic and eager to explore the service once they discover it. As time goes

Figure 6. Aggregate user Digging behavior. In subfigure 6(a), the embedded graph contains a zoomed view of the [0.1, 1.0] interval

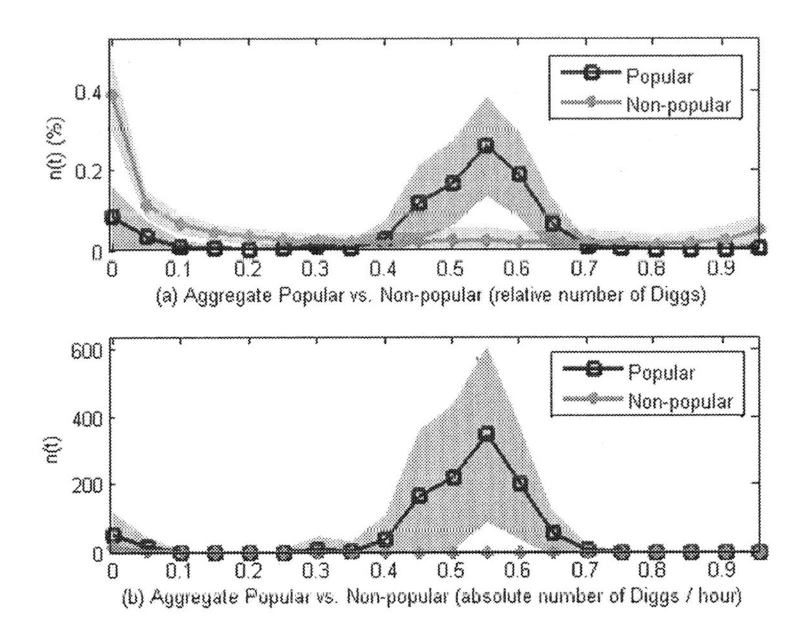

by, their enthusiasm wears off resulting in more stable usage patterns.

A further noteworthy observation can be made by comparing the aggregate user activity time of Figure 6(a) series with the three sample activity time series of Figure 6(b) which come from three individual users. It appears that the individual activity time series do not present any distinct pattern. On the other hand, the aggregate activity is quite stable (though the high variance indicated by the magnitude of the shaded area in 6(a) implies the instability of the individual time series). Thus, it appears that a set of independent behavior patterns of individuals leads to a stable mass behavior when aggregated.

Semantic Aspects of Digg Story Popularity

In this section, we are going to employ the previously discussed feature selection and text classification techniques in order to investigate the potential of popularity prediction based on text features. For that reason, we consider large samples of stories out of the dataset collected from Digg. The majority of Digg stories are written in English, with few stories being in German, Spanish, Chinese and Arabic. We filtered out such non-English text items by means of checking against characters or symbols that are particular

Table 4. Two-way contingency table of term t

	Popular	Non-popular
Term exists	A	B
Term doesn't exist	C	D

Table 5. Top 30 text features based on χ^2 statistic

#	Term	χ^2 (10^6)	#	Term	χ^2 (10^6)	#	Term	χ^2 (10^6)
1	see	61.5	11	seen	7.7	21	news	4.2
2	drive	60.4	12	nintendo	7.6	22	making	4.1
3	japanese	28.5	13	program	5.6	23	breaking	4.0
4	video	17.2	14	way	5.4	24	amazing	3.5
5	google	12.9	15	gets	5.2	25	say	3.4
6	long	11.7	16	computer	4.9	26	coolest	3.4
7	cool	11.0	17	need	4.8	27	release	3.3
8	term	9.9	18	want	4.7	28	right	.9
9	look	8.8	9	play	4.7	29	xbox	2.8
10	high	7.9	20	job	4.3	30	looks	.5

to those languages (e.g. characters with umlaut, non-ASCII characters, etc.).

First, a random sample of $N = 50{,}000$ English Digg stories was drawn from the extended dataset and was processed to extract the text features. For each story, the information on whether it became popular or not was available, so, after extracting all terms from the Digg stories, it was possible to create the two-way contingency matrix of Table 4 for each term t. According to this, A is the number of popular stories containing term t, B the number of non-popular stories containing t, and C and D are the number of popular and non-popular stories respectively that don't contain term t.

Then, the χ^2 statistic is calculated based on the following equation:

Table 6. Top 10 features ranked by χ^2 statistic for four Digg topics. The respective topic corpora consist of 50,000 documents each

	Technology		World & Business		Entertainment		Sports	
#	Term	χ^2	Term	χ^2	Term	χ^2	Term	χ^2
1	Apple	1215.88	Bush	2532.04	RIAA	1457.53	Amazing	1264.88
2	Windows	1080.11	president	1456.52	Movie	1346.79	Nfl	1181.59
3	Linux	995.02	Iraq	1182.86	movies	1015.56	game	1158.17
4	Google	945.72	house	1016.15	Industry	940.50	baseball	1146.77
5	Firefox	919.13	years	974.60	Time	727.12	time	1040.09
6	Just	784.83	administration	964.33	Just	672.92	history	975.46
7	Digg	773.09	congress	872.78	Show	671.09	year	972.57
8	Mac	756.55	officials	841.12	Says	647.37	just	894.85
9	OS	715.56	war	27.25	ilm	635.63	Top	892.82
10	check	643.97	federal	18.37	Lost	628.37	player	877.17

Table 7. Popularity prediction results, namely Precision (P), Recall (R) and F-measure (F). Classifier abbreviations used: NB → Naïve Bayes, SVM → Support Vector Machines and C4.5 → Quinlan's decision trees. Tests were carried out with the use of 500 features on a 50,000 story randomly selected corpus and by use of 10-fold cross validation to obtain the recorded measures

		Popular			Non-popular		
Classifier	**Accuracy (%)**	**P**	**R**	**F**	**P**	**R**	**F**
NB (CHI)	67.21	0.160	0.426	0.233	0.903	0.705	0.791
NB (DF)	88.32	1.000	0.001	0.003	0.883	1.000	0.938
SVM (CHI)	88.10	0.130	0.003	0.006	0.883	0.997	0.937
C4.5 (CHI)	88.18	0.222	0.004	0.008	0.883	0.998	0.937

$$\chi^2(t) = \frac{N \cdot (A \cdot D - C \cdot B)^2}{(A+C) \cdot (B+D) \cdot (A+B) \cdot (C+D)} \quad (8)$$

The χ^2 statistic naturally takes a value of zero if term t and the class of Popular stories are independent. The measure is only problematic when any of the contingency table cells is lightly populated, which is the case for low-frequency terms. For that reason, we filter low-frequency terms prior to the calculation of Equation 8. In addition, stop words and numeric strings were also filtered out of the feature selection process. In order to keep the experiments simple, no stemming or other term normalization was carried out. Table 5 lists the top 30 terms of this story set along with their χ^2 scores. Although not very informative to the human inspector, these keywords can be considered as the most appropriate (from a text feature perspective) for use in making the distinction between Popular and Non-popular stories.

In order to get a more fine-grained view of such keywords per topic, we also calculate the χ^2 scores on independent corpora that contain stories

Table 8. Popularity prediction results when features are selected per topic. Classification was done by means of the Naïve Bayes classifier with the use of 500 features selected based on CHI. For each topic, 50,000 stories were used as dataset and 10-fold cross-validation was used to obtain the performance results

		Popular			Non-popular		
Topic	**Accuracy (%)**	**P**	**R**	**F**	**P**	**R**	**F**
Entertainment	71.39	0.108	0.418	0.172	0.943	0.737	0.827
Gaming	70.26	0.152	0.389	0.218	0.910	0.740	0.816
Lifestyle	72.53	0.121	0.432	0.189	0.943	0.749	0.835
Offbeat	69.03	0.124	0.409	0.191	0.925	0.718	0.809
Science	64.71	0.141	0.444	0.214	0.909	0.672	0.560
Sport	65.79	0.079	0.538	0.138	0.964	0.664	0.787
Technology	67.53	0.159	0.415	0.230	0.902	0.710	0.794
World & B.	65.84	0.098	0.457	0.161	0.941	0.674	0.786

only from particular topics. Table 6 provides such a topic-specific χ^2-based ranking of terms.

After ranking the terms of each corpus based on their class separation ability, it is possible to select the top K of them and use them in an automatic text classification scheme. Table 7 presents the results achieved by such a classification scheme, i.e. the success of predicting the popularity of Digg stories, where three classifiers are compared, namely a Naïve Bayes classifier, an SVM and a C4.5 decision tree. The dataset used consists of 50,000 randomly selected stories and the performance metrics were calculated by use of 10-fold cross validation (i.e. repeated splitting of the dataset to 10 parts and usage of nine of them for training and one of them for testing).

Although in terms of accuracy, the combination of Naïve Bayes with DF-selected features appears to perform best, a closer examination of the Precision and Recall measures obtained separately for the classes Popular and Non-popular provides a different insight. Specifically, it appears that all classifiers have trouble achieving descent classification performance when the input stories are Popular. That means that classifiers can predict accurately that a story will remain Non-popular (when indeed that is the case), but they usually fail to identify Popular stories. The combination of Naïve Bayes with CHI-selected features is better in that respect. The aforementioned problem is related to the well recognized *class imbalance* problem in machine learning (Japkowicz, 2000). Indeed, in the 50,000 stories of the evaluation dataset the ratio of Popular to Non-popular stories is 0.132, i.e. there is almost only one Popular story for every ten Non-popular ones.

Similar results are also obtained for the case that the feature selection and classification process is applied separately per topic. The results of Table 8 provide the respective evidence.

Social Impact on Digg Story Popularity

The last part of our case study involved the estimation of the distributions for the social influence measures of Definitions 4 and 5, namely the SS of Digg users, as well as the story SIG values. These estimations were based on a subset of users and stories randomly selected[10] from the core dataset. From Equations 4 and 5, it is clear that these computations require data that fall outside the core data set (e.g. the Personomies

Figure 7. (a) Scatter plot of social correlation vs. social susceptibility, (b) Social susceptibility distributions for frequent vs. circumstantial users

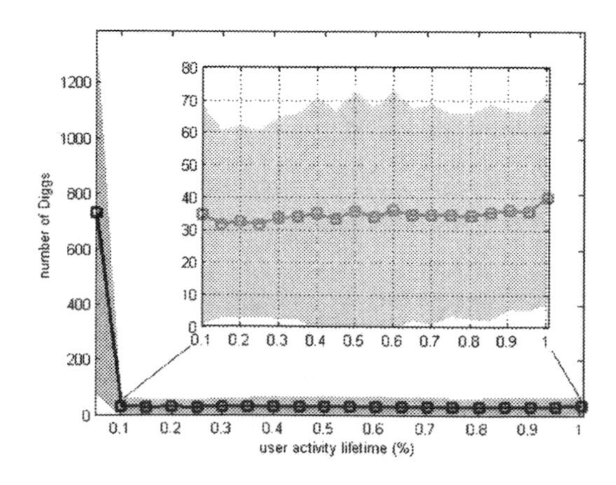

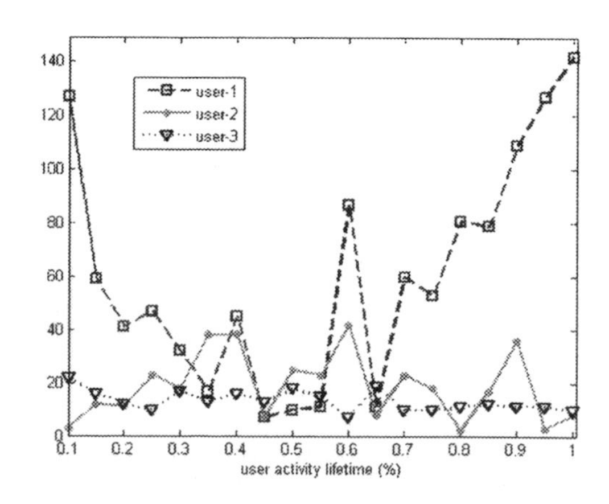

Figure 8. Distribution of story social influence gain (SIG) for popular and non-popular stories

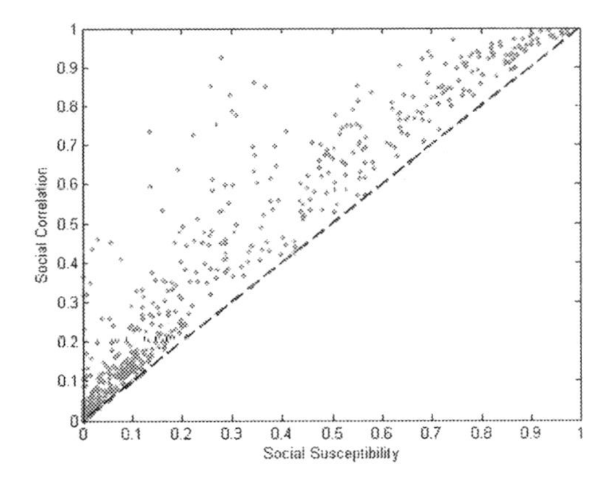

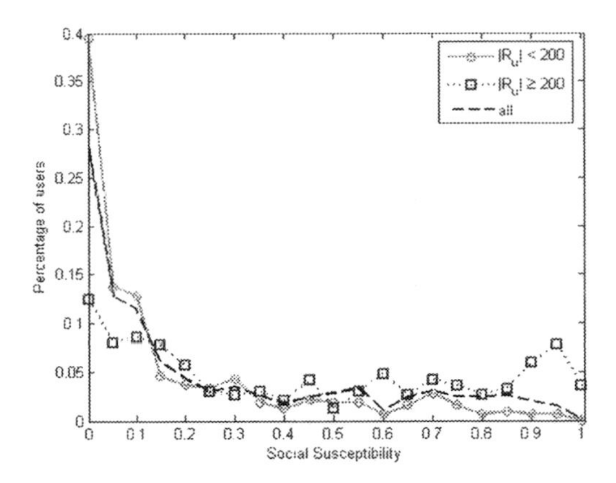

of the users' friends); it was for this reason that the extended dataset was collected.

The estimation of the SS distribution was based on a random sample of 672 users, of whom the SC and SS values were computed. Inspection of the scatter plot of the SC versus the SS of these users, cf. Figure 7(a), reveals that social susceptibility closely follows social correlation. That means that in most cases, when a story has been dugg by both a user and one or more of his/her friends, it is likely that the one or more of the user's friends dugg the story before the user. Few users deviate from this behavior and can thus be considered as the *opinion leaders* of the system (the more a point in Figure 7(a) deviates from the unitary straight line the more the user represented by this point can be considered an opinion leader).

Furthermore, a look into the user SS distributions as presented in Figure 7(b) indicates that the majority of circumstantial Digg users (these users that have dugg a story less than 200 times) present low social susceptibility. In contrast, Digg users with more intense activity tend to be influenced by their social network at a higher frequency. This may indicate that the more users are active within the system the more they rely on their online friends as a source of potentially interesting content.

Finally, insights into the mechanism of story promotion employed by Digg are provided through study of the story SIG patterns. Figure 8 clearly depicts the difference in the SIG distributions between the popular and the non-popular stories. These distributions were estimated by sampling 830 popular and 5076 non-popular stories and computing their respective SIG values. The histograms indicate that the event of a story with $I_r > 0.35$ becoming popular is highly unlikely while in contrast it is very common for non-popular stories to take SIG values in the interval *[0.2, 0.7]*. This could support the hypothesis that high SIG for a story implies low probability of becoming popular; however, instead of the above conclusion, we would rather speculate that Digg employs filters based on social measures similar to I_r to prevent groups of 'friends' from gaining control over which stories appear in the 'Popular' section[11].

FUTURE TRENDS

The startling success of social web applications may be just the preamble in the new era of information and communication technologies we are currently entering. Although it is extremely risky to attempt even rough predictions concerning the future of Social

Web, we would like to take this chance and provide a sketch of the future trends in systems involving the creation, discovery, rating and organization of digital content by masses of users.

In the short term, we anticipate the proliferation of existing social media applications such as Digg, reddit and newsvine, as well as the take-off of mobile social web applications, such as Twitter, used for micro-blogging through mobile devices (Java et al., 2007). Further, existing and new applications will incorporate personalization and context-awareness features, thus meshing with the daily reality of people. Thus, data will be recorded that will capture all aspects of people's lives, such that new insights into human behavior will be possible through "reality mining" (Eagle & Pentland, 2006), raising new concerns about privacy. By closely monitoring the actions of the individual, machine learning will be employed to predict future consumption patterns, e.g. in order to enable effective personalized advertising schemes (Piwowarski & Zaragoza, 2007) or to spawn social interactions between strangers based on profile matching (Eagle & Pentland, 2005).

There is already evidence that the analysis of data coming from social bookmarking usage can be beneficial for search engine tasks, such as new website discovery and authoritative online resource identification (Heymann et al., 2008). In the long run, advanced information extraction and semantic analysis techniques, e.g. automatic quality evaluation of user contributed content (Hu et al., 2007), are expected to be deployed in real-world applications (e.g. Wikipedia) and provide the basis for even more advanced collaborative applications, such as innovation management platforms (Perlich et al., 2007). The requirements of such applications will in turn instigate research into underlying technology disciplines, i.e. database systems for the support of online community services and collaborative platforms (Ramakrishnan, 2007), as well as frameworks for scalable knowledge discovery from streaming data (Faloutsos et al., 2007).

CONCLUSION

The widespread adoption of SBS has transformed online content consumption due to the powerful features that such systems offer to their users. The possibility for users to submit links to content of their interest, tag, rate and comment on online resources submitted by other users as well as to form relations with each other has stimulated intensive user activity in SBS, such that massive amounts of web activity data capturing the content consumption patterns of users are produced in a streaming fashion. This novel content consumption paradigm has spawned a series of interesting research questions related to the generation and evolution of such patterns. These questions pertain to the distributional attributes of online resource popularity, the temporal patterns of content consumption by users, the semantic as well as the social factors affecting the behavior of the masses with regard to their preferences for online resources.

This chapter presented an overview of existing research efforts that are germane to these questions and provided additional insights into the phenomena taking place in the context of an SBS by carrying out an analysis of a large dataset collected from Digg. The power-law nature of web resource popularity was established in accordance with previous studies of similar online systems (Cha et al; 2007, Hotho et al., 2006a; Halpin et al., 2007). Furthermore, a set of characteristic temporal patterns of content consumption were revealed which confirmed previous findings about social media content popularity evolution (Lerman; 2007). In addition, a preliminary investigation into the semantic elements of content popularity lent support to the hypothesis that popularity is affected by the semantic content and the linguistic style. What is more, it was empirically shown that users of SBS are socially susceptible, i.e. they tend to express interest for online resources that are also considered interesting by their online "friends".

Finally, the chapter provided an outlook on the exciting new prospects for online content publishing and mining on the massive amounts of data produced in the context of SBS and related applications.

ACKNOWLEDGMENT

This work was supported by the WeKnowIt project, partially funded by the European Commission, under contract number FP7-215453.

REFERENCES

Aberer, K., Cudré-Mauroux, P., Ouksel, A. M., Catarci, T., Hacid, M. S., Illarramendi, A., et al. (2004). *Emergent Semantics Principles and Issues.* (LNCS Vol. 2973, pp. 25-38). Berlin: Springer.

Agichtein, E., Castillo, C., Donato, D., Gionis, A., & Mishne, G. (2008). Finding high-quality content in social media. In *Proceedings of the international Conference on Web Search and Web Data Mining,* Palo Alto, CA, February 11 - 12, 2008 (WSDM '08, pp. 183-194). New York: ACM.

Anagnostopoulos, A., Kumar, R., & Mahdian, M. (2008). Influence and Correlation in Social Networks. In *Proceedings of the 14th ACM SIGKDD international Conference on Knowledge Discovery and Data Mining,* Las Vegas, Nevada, USA, August 24 - 27, 2008 (KDD '08, pp. 7 – 15). New York: ACM.

Barabási, A.-L., & Albert, R. (1999). Emergence of Scaling in Random Networks. [Washington, DC: AAAS.]. *Science, 286*(5439), 509–512. doi:10.1126/science.286.5439.509

Bi, Z., Faloutsos, C., & Korn, F. (2001). The "DGX" distribution for mining massive, skewed data. In *Proceedings of the Seventh ACM SIGKDD international Conference on Knowledge Discovery and Data Mining*, San Francisco, California, August 26 - 29 (KDD '01, pp. 17-26). New York: ACM.

Cha, M., Kwak, H., Rodriguez, P., Ahn, Y., & Moon, S. (2007). I Tube, You Tube, Everybody Tubes: Analyzing the World's Largest User Generated Content Video System. In *Proceedings of the 7th ACM SIGCOMM conference on Internet measurement*, San Diego, CA.

Clauset, A., Shalizi, C. R., & Newman, M. E. J. (2007). *Power-law distributions in empirical data.* Tech. Rep. submitted to arXiv on June 7, 2007 with identifier: arXiv:0706.1062v1.

Cormode, G., & Muthukrishnan, S. (2005). Summarizing and mining skewed data streams. In *Proceedings of the 2005 SIAM International Conference on Data Mining*, (pp.44-55). SIAM.

Cristianini, N., & Shawe-Taylor, J. (2000). *An Introduction to Support Vector Machines and other kernel-based learning methods*. New York: Cambridge University Press.

Dave, K., Lawrence, S., & Pennock, D. M. (2003). Mining the peanut gallery: opinion extraction and semantic classification of product reviews. In *Proceedings of the 12th international Conference on World Wide Web*, Budapest, Hungary, May 20 – 24, (WWW '03, pp. 519-528). New York: ACM.

Duda, R. O., Hart, P. E., & Stork, D. G. (2001). *Pattern Classification.* Chichester, UK: John Wiley & Sons, Inc.

Eagle, N., & Pentland, A. (2005). Social Serendipity: Mobilizing Social Software. *IEEE Pervasive Computing / IEEE Computer Society [and] IEEE Communications Society*, *4*(2), 28–34. doi:10.1109/MPRV.2005.37

Eagle, N., & Pentland, A. (2006). Reality Mining: Sensing Complex Social Systems. *Personal and Ubiquitous Computing*, *10*(4), 255–268. doi:10.1007/s00779-005-0046-3

Faloutsos, C., Kolda, T. G., & Sun, J. (2007). Mining large graphs and streams using matrix and tensor tools. In *Proceedings of the 2007 ACM SIGMOD international Conference on Management of Data*, Beijing, China, June 11 - 14, (SIGMOD '07, pp. 1174-1174). New York: ACM.

Golder, S., & Huberman, B. A. (2006). The Structure of Collaborative Tagging Systems. *Journal of Information Science*, *32*(2), 198–208. doi:10.1177/0165551506062337

Gómez, V., Kaltenbrunner, A., & López, V. (2008). Statistical analysis of the social network and discussion threads in slashdot. In *Proceeding of the 17th international Conference on World Wide Web,* Beijing, China, April 21 - 25, (WWW '08, pp. 645-654). New York: ACM.

Halpin, H., Robu, V., & Shepherd, H. (2007). The complex dynamics of collaborative tagging. In *Proceedings of the 16th international Conference on World Wide Web*, Banff, Alberta, Canada, May 08 - 12, 2007, (WWW '07, pp. 211-220). New York: ACM.

Heymann, P., Koutrika, G., & Garcia-Molina, H. (2008). Can social bookmarking improve web search? In *Proceedings of the international Conference on Web Search and Web Data Mining,* Palo Alto, CA, February 11-12, 2008, (WSDM '08, pp. 195-206). New York: ACM.

Hotho, A., Jäschke, R., Schmitz, C., & Stumme, G. (2006). Information Retrieval in Folksonomies: Search and Ranking. In *The Semantic Web: Research and Applications*, (pp. 411-426). Berlin: Springer.

Hotho, A., Jäschke, R., Schmitz, C., & Stumme, G. (2006). *Trend Detection in Folksonomies*. (LNCS 4306, pp. 56-70). Berlin: Springer.

Hu, M., Lim, E., Sun, A., Lauw, H. W., & Vuong, B. (2007). Measuring Article Quality in Wikipedia: Models and Evaluation. In *Proceedings of the Sixteenth ACM Conference on Conference on information and Knowledge Management*, Lisbon, Portugal, November 06 - 10, (CIKM '07, pp. 243-252). New York: ACM.

Japkowicz, N. (2000). The class imbalance problem: Significance and strategies. In *Proceedings of the 2000 International Conference on Artificial Intelligence* (ICAI 2000).

Java, A., Song, X., Finin, T., & Tseng, B. (2007). Why we twitter: Understanding the microblogging effect in user intentions and communities. In *Proceedings of WebKDD / SNAKDD 2007: KDD Workshop on Web Mining and Social Network Analysis, in conjunction with the 13th ACM SIGKDD International Conference on Knowledge Discovery and Data Mining* (KDD 2007, pp. 56-66).

Kaltenbrunner, A., Gómez, V., & López, V. (2007). *Description and Prediction of Slashdot Activity*. Paper presented at the LA-WEB 2007 5th Latin American Web Congress, Santiago, Chile.

Kaltenbrunner, A., Gómez, V., Moghnieh, A., Meza, R., Blat, J., & López, V. (2007). *Homogeneous temporal activity patterns in a large online communication space*. Paper presented at the 10th Int. Conf. on Business Information Systems, Workshop on Social Aspects of the Web (SAW 2007), Poznan, Poland.

Lerman, K. (2007). Social Information Processing in News Aggregation. *IEEE Internet Computing*, *11*(6), 16–28. doi:10.1109/MIC.2007.136

Leskovec, J., Adamic, L., & Huberman, B. A. (2007). The Dynamics of Viral Marketing. *ACM Transactions on the Web* (ACM TWEB), *1*(1).

Mika, P. (2005). Ontologies are us: A unified model of social networks and semantics. In *The Semantic Web – ISWC 2005*, (pp. 522-536). Berlin: Springer.

Mitzenmacher, M. (2004). A Brief History of Generative Models for Power Law and Lognormal Distributions. [Wellesley, MA: A K Peters, Ltd.]. *Internet Mathematics*, *1*(2), 226–251.

Newman, M. E. J. (2005). Power laws, Pareto distributions and Zipf's law. *Contemporary Physics*, *46*, 323–351. doi:10.1080/00107510500052444

Pang, B., Lee, L., & Vaithyanathan, S. (2002). Thumbs up? Sentiment classification using machine learning techniques. In *Proceedings of the Acl-02 Conference on Empirical Methods in Natural Language Processing - Volume 10*, (pp. 79-86). Morristown, NJ: ACL.

Papadopoulos, S., Vakali, A., & Kompatsiaris, I. (July, 2008). *Digg it Up! Analyzing Popularity Evolution in a Web 2.0 Setting.* Paper presented at MSoDa08 (Mining Social Data), a satellite Workshop of the 18th European Conference on Artificial Intelligence, Patras, Greece.

Perlich, C., Helander, M., Lawrence, R., Liu, Y., Rosset, S., & Reddy, C. (2007). Looking for great ideas: Analyzing the innovation jam. In *Proceedings of WebKDD / SNAKDD 2007: KDD Workshop on Web Mining and Social Network Analysis, in conjunction with the 13th ACM SIGKDD International Conference on Knowledge Discovery and Data Mining* (KDD 2007), (pp. 66-74).

Piwowarski, B., & Zaragoza, H. (2007). Predictive User Click Models based on Click-through History. In *Proceedings of the Sixteenth ACM Conference on Conference on information and Knowledge Management,* Lisbon, Portugal, November 06 - 10, 2007, CIKM '07, (pp. 175-182). New York: ACM.

Quinlan, J. R. C4.5 (1993). *Programs for Machine Learning*. San Francisco: Morgan Kaufmann Publishers.

Ramakrishnan, R. (2007). Community Systems: The World Online. In *Proceedings of CIDR 2007, Third Biennial Conference on Innovative Data Systems Research,* (pp. 341), Asilomar, CA.

Richardson, M., & Domingos, P. (2002). Mining knowledge-sharing sites for viral marketing. In *Proceedings of the 14th ACM SIGKDD International Conference on Knowledge Discovery and Data Mining*. Las Vegas, NV. New York: ACM Press.

Salton, G., Wong, A., & Yang, C. S. (1975). A vector space model for automatic indexing. *Communications of the ACM, 18*(11), 613–620. doi:10.1145/361219.361220

Seshadri, M., Machiraju, S., Sridharan, A., Bolot, J., Faloutsos, C., & Leskovec, J. (2008). Mobile Call Graphs: Beyond Power-Law and Lognormal Distributions. In *Proceedings of the Eighth ACM SIGKDD International Conference on Knowledge Discovery and Data Mining*, (pp. 61-70), Edmonton, Canada. New York: ACM Press.

Song, X., Chi, Y., Hino, K., & Tseng, B. L. (2007). Information flow modeling based on diffusion rate for prediction and ranking. In *Proceedings of the 16th international Conference on World Wide Web* (Banff, Alberta, Canada, May 08 - 12, 2007). WWW '07, pp. 191-200, ACM, New York, NY, USA.

Turney, P. D. (2001). Thumbs up or thumbs down?: Semantic orientation applied to unsupervised classification of reviews. In *Proceedings of the 40th Annual Meeting on Association For Computational Linguistics,* Philadelphia, PA, July 07 - 12, (pp. 417-424). Morristown, NJ: ACL.

Wu, F., & Huberman, B. A. (2007). Novelty and collective attention. *Proceedings of the National Academy of Sciences of the United States of America, 104*(45), 17599–17601. doi:10.1073/pnas.0704916104

Yang, Y., & Pedersen, J. O. (1997). A Comparative Study on Feature Selection in Text Categorization. In D. H. Fisher, (Ed.), *Proceedings of the Fourteenth international Conference on Machine Learning* (July 08 - 12, pp. 412-420). San Francisco: Morgan Kaufmann Publishers Inc.

Zhang, J., Ackerman, M. S., & Adamic, L. (2007). Expertise networks in online communities: structure and algorithms. In *Proceedings of the 16th international Conference on World Wide Web,* Banff, Alberta, Canada, May 08 - 12, WWW '07, (pp. 221-230). New York: ACM.

Zhang, Z., & Varadarajan, B. (2006). Utility scoring of product reviews. In *Proceedings of the 15th ACM international Conference on information and Knowledge Management,* Arlington, VA, November 06 - 11, (CIKM '06, pp. 51-57). New York: ACM.

ENDNOTES

1. An estimate of the allocation of internet users' time to different online activities is provided by the Online Publishers Association through the Internet Activity Index (IAI) in: http://www.online-publishers.org.
2. http://www.compete.com
3. http://www.quantcast.com
4. http://www.youtube.com
5. http://ucc.daum.net
6. Although the semantics of a digital resource can be also conveyed by other kinds of features (e.g. audio, visual), we restrict our study to the semantics conveyed by textual features of the content.
7. Whether a story jumps to the 'Popular' section or not does not solely depend on the number of Diggs it receives (although it is certainly taken into account). The Digg administrators make this decision on the basis of a set of proprietary criteria and heuristics, which they keep secret since sharing such knowledge would render the system prone to malicious attacks (e.g. to artificially boost the popularity of a story).
8. Only stories with $|H_r| > 20$ were studied to prevent the 'noisy' time series from distorting the resulting aggregate time series of the non-popular stories.

9. Users with $|R_u| < 20$ *were not considered in the sample selection in order to prevent users with sparse (and therefore noisy) activity to affect the aggregate activity time series.*
10. The filtering rules of $|R_u| > 20$ and $|H_r| > 20$ (same as above) were applied here too.
11. The post in http://blog.Digg.com/?p=106 provides further evidence in favor of this speculation.

This work was previously published in Evolving Application Domains of Data Warehousing and Mining: Trends and Solutions, edited by P. N. San-Bento Furtado, pp. 233-257, copyright 2010 by Information Science Reference (an imprint of IGI Global).

Chapter 8.5
Conceptualizing Codes of Conduct in Social Networking Communities

Ann Dutton Ewbank
Arizona State University, USA

Adam G. Kay
Arizona State University, USA

Teresa S. Foulger
Arizona State University, USA

Heather L. Carter
Arizona State University, USA

ABSTRACT

This chapter reviews the capabilities of social networking tools and links those capabilities to recent legal and ethical controversies involving use of social networking tools such as Facebook and MySpace. A social cognitive moral framework is applied to explore and analyze the ethical issues present in these incidents. Three ethical vulnerabilities are identified in the use of social networking tools: 1) the medium provides a magnified forum for public humiliation or hazing, 2) a blurring of boundaries exists between private and public information on social networking sites, and 3) the medium merges individuals' professional and non-professional identities. Prevalent legal and social responses to these kinds of incidents are considered and implications are suggested for encouraging responsible use. The chapter includes a description of the authors' current research with preservice students involving an intervention whereby students read and think about real cases where educators use social networking. The intervention was created to improve students' critical thinking about the ethical issues involved. Recommendations for applying institutional codes of conduct to ethical dilemmas involving online tools are discussed.

DOI: 10.4018/978-1-60566-729-4.ch002

INTRODUCTION

Social networking sites such as Facebook and MySpace have become ubiquitous. Whereas email was the electronic communication norm in the late twentieth century, social networking is rapidly replacing email as the most favored means of networking, connecting, and staying in touch. In fact, MySpace is the sixth most visited site on the Internet (Alexa, 2008) and Facebook is the world's largest and the fastest growing social networking site (Schonfeld, 2008). These tools are quite popular with teenagers, college-age students, and young professionals because they allow them to more easily stay connected. Using social networking sites, individuals can present themselves to others through an online identity that is tailored to their unique interests and desires, and participate in a variety of interconnected communication networks - personal, professional, creative, or informative. However, when individuals create a personal space online, they also create a digital footprint—the kind of footprint that can be permanent. And when a trail of personal information is left behind in a searchable and open format, notions of public and private information are challenged and the potential for liabilities may be high. This is of particular importance to those who wish to convey a professional image. An online profile that may have seemed innocuous and private during one stage of life may haunt an individual at the point in their life when they transition from student to professional.

For educational institutions, the widespread popularity of social networking sites as a means of communication, provide in-roads for experimenting with ways to connect with clientele. While innovative educators are quick to embrace and harness the learning potential of Web 2.0 tools, an understanding of the ethical issues in these unusual forms of social interaction has been slower to develop. Undoubtedly there are value-added features, many of which are yet to be discovered; but some institutions are refusing to innovate with this powerful technology tool due to the risks involved.

In order to design and endorse effective use of these tools, educators need socially responsible models and guidelines. What are the ethical considerations required of online social networking, and how can educational organizations capitalize on this innovative means of communicating while promoting responsible use? This chapter will highlight legal and ethical controversies surrounding social networking sites, identify ethical vulnerabilities associated with using the online tools through a social cognitive moral framework, and discuss implications for promoting socially responsible use of social networking tools.

BACKGROUND

Our inquiries into this topic began when one of the authors of this chapter encountered a situation involving social networking in her preservice teacher education class. What started as a class assignment turned into a moral and ethical dilemma for the instructor when a student revealed his MySpace profile as a part of a larger class assignment. Students were to create a homepage and provide three links to sites that a future teacher might use in the classroom as part of a lesson plan. Many students chose to link to their MySpace profiles as part of the assignment, but one particular link captured the attention of the instructor who was not prepared for what she saw—a MySpace profile showing a bloody machete stabbed into a hand with the caption that read, "Twist the hand that forces you to write." Other images and words on the profile were equally disturbing. The personal icon used to identify the profile owner was an image of a cut wrist with directions on how to commit suicide. The instructor wondered why a student would turn in what seemed to be a private and personal site as part of a class assignment. Perhaps Web 2.0 and online social networking caused this student to think differently than the instructor about revealing private thoughts in such a public forum. Because technology users in the Web 2.0 environment can be both consumers and creators of information, similar scenarios are occurring often. At what point is the boundary crossed when sharing information about self and others via social networking tools? And who draws that line? The ability to communicate personal, informational, or editorial information en masse—at the click of a mouse—poses new and different ethical dilemmas not as prevalent in the pre-Web 2.0 world. And this issue is compounded by the fact that as users share authored information with others, they invite countless people into their personal space. Social networking creates a window into users' lives that is much more immediate, permanent, and impactful than ever before.

Social Networking Tools and Their Capabilities

Social networking sites are Websites designed to bring together groups of people in order to communicate around shared interests or activities (Wikipedia, 2008). Because meeting places are virtual, the idea that any two people can be connected through several intermediaries, commonly known as "six degrees of separation," is magnified and expanded (Leskovec & Horvitz, 2007). This kind of interconnectivity among individuals would be impossible without the Web. The online manifestation of social networking typically refers to a minimum of three networking capacities, first popularized by Friendster in 2002. This includes publicly displayed profiles, publicly accessible lists of friends, and virtual walls for comments or testimonials (boyd, 2008). In any of the myriad social networking sites created since Friendster, individuals can join a Web service, and then design a profile to showcase and highlight personal information, hobbies, employment and any other topics they wish to be shared. Upon becoming a member of an online social network, the user can communicate with other members or groups, link their profile to others, and even invite those outside the social networking site to join the system and link to their profile. Thus, a network continually expands.

Revisiting the story of the student with the disturbing MySpace profile highlights the differences between in-person and online social networks. Prior to viewing the profile, the instructor had only the day-to-day interactions in the classroom to form an opinion of the student. The informal social network of this student was not revealed to the instructor prior to viewing the MySpace profile. However, it was the student's choice to share the public profile with the instructor, and when that happened the cloak over a previously invisible network was removed and a facet of this student's life that was markedly different than his in-class persona was revealed. The instructor had more information from which to form an opinion of the student, and could not help but think differently about the student from that day forth. However, the student was naively unaware of impact that his online profile had on interactions with the instructor.

More importantly though, with the lasting vision of a bloody hand, the instructor was in turmoil about the ethics behind what was brought to light. Unlike an incident in the classroom where a course of action would be clear, the most appropriate response to the personal information revealed through the online profile was not obvious to the instructor. Where did the instructor's authority end and the student's personal life begin? Was the instructor responsible for reporting the actions that took place outside of her classroom? Was the online information within or outside the classroom? How would this student interact with children during his field placement and in his future classroom? Was the profile indicative of a troubled individual or was it simply a manifestation of creative, albeit dark, expression? For these reasons, the instructor grappled with whether to report the student to campus authorities for the disturbing images and ideas relayed on his profile.

Legal Actions and Campus-Based Incidents

Professional or formal relationships may become tainted when people either purposefully or inadvertently share information about themselves via online social networking services, and some users fail to think about the consequences that may arise as a result. Furthermore, appropriate responses to online personal information by those in authority, such as instructors and supervisors, may not be clear.

In recent years, many campus-based incidents involving perceived student misuse of social networking sites have occurred in both K-12 and postsecondary institutions. For example, Elon University in North Carolina took disciplinary action against members of the baseball team after photos of players involved in hazing activities found their way onto a student's MySpace profile (Lindenberger, 2006). An academic institution's *in loco parentis* responsibilities are often interpreted by campus administrators as encompassing the cyberworld. Some universities including Penn State University and the University of California-Davis utilize information contained in social networking sites to investigate campus incidents of harassment, code of conduct violations, and criminal activity (Lipka, 2008).

Slightly different problems exist for younger students. K-12 students have been suspended or expelled for creating false and potentially libelous profiles of faculty or administrators. These suspensions have been met with an interesting reaction from parents; in some cases, parents have filed suit against the educational institution for impeding a student's right to free expression by nature of the discipline imposed for creating the profile (see Layshock v. Hermitage School District, 2007; Requa v. Kent School District, 2007; J.S. v. Blue Mountain School District, 2007; A. B. v. State of Indiana, 2008). These claims of free speech violations have largely been unfounded by the courts. However, in some cases decisions have held a student's right to free expression in the form of an Internet parody if no significant disruption to the educational institution has occurred. Faculty have also filed suit or pressed criminal charges against students for harassing, defaming, or intimidating speech online (see Wisniewski vs. Board of Education, Weedsport Central School District, 2007; WSBTV.com, 2006).

And faculty sanctions for perceived misuse of social networking are becoming commonplace as well. Regulation of faculty conduct outside of professional duties is embedded in institutional codes and social norms, and many cases exist where faculty have claimed that sanctions or dismissals are unconstitutional (Fulmer, 2002). However, with the advent of online social networking, traditional tests of rights vs. duty may not apply. In a conventional sense, educational institutions realize the boundaries of faculty behavior to be regulated. But the transparency of online social networking has somewhat eroded those boundaries. For example, do faculty have the *right* to free expression online even if it conflicts with the values of the institution? Or do institutions have the *duty* to ensure that the values of the institution are upheld online as they are in the physical world?

One particularly striking example of this dilemma is the Tamara Hoover case. Hoover, a high school art teacher in the Austin (Texas) Independent School District, was fired when nude photographs were discovered on her MySpace profile and on her photo-sharing website, Flickr (May, 2006). Hoover was fired based on "conduct unbecoming a teacher," even though the photographs displayed could be interpreted as artistic and professional. Hoover agreed to a cash settlement from the school district, and now uses her MySpace profile to promote teachers' free speech rights (Hoover, 2007). The case has attracted national media attention.

The Hoover case is not an isolated incident. Many other cases of faculty sanctions over social networking have occurred in recent years (Crawford, 2007; Phillips, 2007; Vivanco, 2007). Do education administrators have the right to screen potential employees by "Googling" them, or to monitor employees' electronic communications without evidence of inappropriate contact with students (Wheeler, 2007)? These and other questions concerning ethical conduct within social networking sites have been met with a variety of responses from teacher preparation programs, school districts, and universities. Some have warned faculty not to use the sites at all (e-School News, 2007) and others have taken an educational approach, encouraging users to critically think about what they post online (The Pennsylvania State University, 2007). Given the ubiquity of online social networking communities among youth (boyd, 2008) as well as the potential of these powerful tools to provide communications that would not otherwise be possible, barring their use strikes the authors of this chapter as an educational disservice. To best calculate the risks that could be incurred in leveraging the power of these innovative tools, we believe a careful analysis of the potential ethical issues involved in interactions in online social networks is necessary.

APPLYING A MORAL FRAMEWORK

The events described above begin to illustrate the confusing social and ethical landscape of communications in this changing time, especially for educators who are obligated by their professional standards to serve as role models. To add to the complexity, the multiple players, including faculty, students, administrators, and parents, appear to have vastly different points of view about what is appropriate and inappropriate conduct. This is partly due to the multiple and often competing social and moral concerns present in these types of incidents. To both investigate the ethical points of view involved in judging these incidents and to uncover the ethical vulnerabilities inherent in this new medium, we have applied a socio-moral framework with a legacy of describing moral and non-moral features of complex social interactions. Specifically, social cognitive domain theory (Turiel, 1983; Turiel, 2002) is an appropriate starting point for understanding the complexities in online social networking.

First, the theory provides an analytical framework that differentiates moral from non-moral concerns in social interactions. Prior research applying this framework has demonstrated that people consistently think about moral (such as notions of harm, fairness, and rights), conventional (such as social roles, institutional organization, and matters of social efficiency), and personal matters (such as tastes and choices) in different ways (see Smetana, 2006 for a review). From early childhood, individuals actively distinguish between these domains and make judgments specific to domains about these distinct categories of social interaction. These insights are critical because many real-world social interactions are multi-faceted in the sense that multiple social domains are involved. Judgments and actions based on social interactions often involve weighing and coordinating various moral and non-moral concerns. For example, a judgment about whether a teacher should be disciplined for approaching parents with alarming information acquired from a student's online profile involves the consideration of multiple issues. There are concerns for the student's welfare (moral), the limits of teacher authority (conventional), and the student's right to privacy (moral) when choosing to post information in a public forum (personal).

Second, the framework allows for analytical investigation rather than a prescribed approach to how one should behave in ambiguous situations. There are conflicting perspectives about whether a teacher should be disciplined in such a situation; the authors of this chapter do not pretend to be certain about the right or ethical course of action in complicated, multi-faceted events. Our purpose is not to prescribe a set of moral actions that fit under a wide variation of circumstances, but instead to better understand the issues involved and also discover the ways people weigh those various issues in their thinking as online social networks continue to grow. To this end we conducted an investigation into student ethical decision-making in online social networking communities. This study has implications for how instructors might develop ways to allow students to ponder their ethical reasoning while engaging in the use of these tools.

Research within this framework can provide insights and useful points of comparison for our topic because of multiple studies of reasoning about two relevant social issues: developing concepts of role-related authority (Laupa, 1991; Laupa & Turiel, 1993; Laupa, 1995) and thinking about rights and privacy issues with the use of modern technologies (Friedman et al., 2006; Friedman, 1997). For example, when asked about the limits of educators' authority and responsibility over students, older students are more likely than younger students to limit their influences to the concrete boundaries of the school context (Laupa & Turiel, 1993). As classroom and school boundaries become progressively virtual, limits on educators' responsibilities and authority are unclear for students and staff alike. While reasoning about moral issues in technology, one study demonstrated that many students who believe in property and privacy rights in non-technological arenas condone piracy and hacking activities on computers (Friedman, 1997). Interviews with these students revealed that this apparent contradiction had to do with fundamental aspects of technology: the perceived distance between the actor and potential victims, the indirect nature of the harmful consequences, the invisibility of the act, and the lack of established consequences for such behavior online.Therefore, social cognitive domain theory is a useful framework to guide an analysis of the kinds of issues that can arise in the use of social networking tools. It allows us to do so in a way that respects the complexity of these kinds of interactions. Finally, it enables us to connect directly with a body of research that informs investigations of the ways in which online social interaction might cultivate its own set of ethical vulnerabilities. For example, the studies highlighted above suggest at least three such vulnerabilities:

1. A magnified forum for public humiliation and hazing--Students might be more likely to engage in public humiliation through social networking tools because harmful consequences are not directly observed, in contrast to acts in physical public spaces such as the cafeteria or locker room. Furthermore, when hazing or humiliation is conducted online there is greater distance (and sometimes even anonymity) between the actor and the victim.
2. Privacy issues in public spaces--Online social networking has the power to re-frame the way we consider and apply traditional rights to privacy.
3. The merging of professional and non-professional identities--The classroom walls and school premises no longer frame the jurisdiction of the educational institution. How does this shift impact higher education? How can social networking tools be appropriately used by university programs, administrators, instructors, and students?

Using the social cognitive moral framework as a lens for analysis, these ethical vulnerabilities are described in detail in the following section.

ETHICAL VULNERABILITIES OF SOCIAL NETWORKING TOOLS REVEALED

A Magnified Forum for Public Humiliation and Hazing

It is not uncommon to see academic units represented on MySpace or Facebook, as they attempt to find ways to provide services and assistance to students who are familiar with online tools (Hermes, 2008). Additionally, academic units can use these spaces as an outreach tool for marketing (Berg, Berquam & Christoph, 2007; O'Hanlon, 2007). With this level of transparency it would be easy to witness students engaging in the kinds of online activities that could be characterized as public humiliation and hazing. Most university codes of conduct prohibit this sort of behavior. However, when the behavior is discovered online, questions may arise about whether academic officials should access online profiles at all. While conducting research recently on the feasibility of creating a presence for a university academic library in Facebook, one of the authors came across a student group called "Have you seen the homeless guy in the library?" The description of the group was, "He's always on the computers. He stinks really bad. And he has like 1,000's of plastic bags..." Updated in the *recent news* section of the page was a description of where he was sitting in the library that day as well as a Web site that he had viewed. There were seventeen students in this group (no doubt with various levels of engagement), creating and sustaining a public community with the sole purpose of ridiculing one specific human being.

Public humiliation is not a new phenomenon; but the power of the collective sentiment conveyed by this online "community" would be hard to fathom offline, as a group with a purpose such as this would take considerably more time and effort to create. But by accident this particular group became visible, and when the activity was reported to the library administration, the university student affairs department was alerted. The student affairs officer explained to the library representative that student issues involving online public humiliation and hazing are not uncommon, and that the department frequently engages in mediation to resolve disputes—student to student, student to instructor, and around potentially disturbing group behavior. The officer explained that she would talk to the students who had joined the Facebook group, and use the incident as a "teachable moment" to address ethical use of the social networking tool among the students.

More and more educational administrations are grappling with issues between and among students that have extended into online social networking. As these online forums become ubiquitous to the masses, they have developed into a natural extension of students' social and personal lives. In a recent survey the Pew Internet & American Life Project found that fifty-five percent of online teens (ages 12-17) have created a profile at a social networking site such as MySpace or Facebook (Lenhart, Madden, Macgill, & Smith, 2007). University students in particular frequent social networking sites. In fact, a recent survey of all first-year English students at the University of Illinois-Chicago (Hargittai, 2007), reported that 88% use social networking sites.

Incidents involving public humiliation or hazing such as the one that took place in the university library as described above, have become commonplace in both K-12 and institutions of higher learning. Through "cyberbullying," defined as "willful and repeated harm inflicted through the medium of electronic text," (Patchin & Hinduja, 2006, p.152) students and faculty can either become the targets or perpetrators of incidents that would be unacceptable in offline situations. Electronic humiliation and hazing of this nature can have lasting physical and psychological effects on the victims such as depression, insomnia, and anxiety disorders (Griffiths, 2002). In the interest of student and faculty welfare, educational institutions have responded to these incidents in a variety of ways. Some situations are minor and can easily be solved through mediation by administrators or student affairs professionals (Lipka, 2008) while others have warranted more serious disciplinary or even criminal action.

It is because of the sheer reach of electronic communication that the Internet and social networking have become a magnified forum for public humiliation and hazing. This activity occurs within all sectors of the educational spectrum. Almost one-third of teens who use the Internet say that they have been a victim of annoying or potentially threatening activities online, including others "outing" personal information via email, text messaging, or postings on social networking sites. Those who share personal identities and thoughts are more likely to be the targets of such activities (Lenhart, 2007). Additionally, in a recent survey conducted by the Teacher Support Network of Great Britain, 17% of K-12 teachers indicated that they had been a target of online humiliation or harassment (Woolcock, 2008). College-age students and faculty are not immune from online defamation of character. In a survey conducted at the University of New Hampshire 17% of students reported experiencing threatening online behavior, yet only 7% of those experiencing vicitmization reported it to campus authorities (Finn, 2004). Sites such as JuicyCampus.com and TheDirty.com, perhaps the most notorious Web sites aimed at college students, allow anyone to post humiliating or threatening messages and photos, some of which have cost students job opportunities and internships (O'Neil, 2008). Female law school students have reported sexual harassment and defamation on AutoAdmit.com, a message board about law school admissions (Nakashima, 2007). And Web sites like RateMyProfessors.com and RateMyTeachers.com allow students to be anonymous as they air their opinions about faculty, including rating an instructor's easiness and "hotness".

Such activity on social networking sites has led to several court cases over perceived defamation. For example, *Drews vs. Joint School District* (2006) described a situation where Casey Drews, a high school student, was the subject of rumors and gossip at school after a snapshot her mother took of Drews kissing a female friend was circulated on the Internet. Drews sued the school district for deliberately ignoring the harassment. Cases illustrating faculty harassment and defamation are evident as well. A Georgia teacher brought criminal charges against a high school student who created a fake MySpace profile about the teacher, claiming that the teacher "wrestled midgets and alligators" and stating that he liked "having a gay old time" (WSBTV.com, 2006). And a federal circuit court found that a student's distribution of a text message icon depicting a gun firing at a picture of his English teacher and the words "Kill Mr. Van der Molen" was threatening speech not protected by the First Amendment (Wisniewski vs. Board of Education, Weedsport Central School District, 2007).

Steve Dillon, director of student services at Carmel Clay Schools, states, "Kids look at the Internet as today's restroom wall. They need to learn that some things are not acceptable anywhere" (Carvin, 2006). If the Internet is a restroom wall, then it is a giant, unisex restroom open to all citizens. But unlike a restroom wall that can be painted over, the Internet can be a permanent archive of electronic communication. Localized or place-bound codes of conduct are clearly no longer adequate in a Web 2.0 environment. How do educational organizations come to terms that the school's four walls have been virtually obliterated, and craft appropriate responses in codes of conduct that protect the privacy and welfare of its students and faculty, while simultaneously honoring First Amendment rights?

Privacy Issues in Public Spaces

A number of recent controversies highlight our collective lack of clarity about how we can and should use personal information that is publicly available on the Web. English Education candidate Stacy Snyder of Millersville (Pennsylvania) University was denied her teaching certificate and given an English degree rather than an education degree after campus administrators discovered photos where she portrayed herself as a "drunken pirate" on her MySpace profile, even though she was of legal drinking age. The 27-year old filed a lawsuit against the university, and is asking for $75,000 in damages (Steiner, 2007). In another example (one which we have used to explore reasoning in our own research), a teacher revealed to students that she had a MySpace profile. The student consequently "friended" the teacher, giving the teacher access to the student's profile. In the process of exploring the student's profile, the teacher discovered information about activities such as underage drinking in which the student was engaged. Concerned about what she saw online about the student, the teacher contacted the parents. The parents contacted the school, outraged that the teacher was snooping into the student's personal life, and demanded that the teacher be disciplined. Scenarios such as these suggest that both producers and consumers of online information are unclear as to exactly what is public and exactly what is private.

Are producers aware of the extent to which their online disclosures are publicly accessible? Are consumers clear about producers' rights to privacy in this online community? Studies reveal that teens and adults alike underestimate who accesses their online submissions and how that information is used (Viègas, 2005). The nature of online social interaction allows for such vulnerabilities to protecting privacy. Palen and Dourish (2003) suggest that this illusion is perpetuated, at least in part, because of our reliance on non-virtual strategies for monitoring privacy, and that online interactions call for implications for better controlling privacy violations and for different ways of thinking about privacy rights in online environments.

According to this view, our mechanisms for managing privacy have traditionally been spatial and sensory: we know who our audience is (and can control it) because we can see, hear, and read traditional forms of communication. Online, these cues are distorted. Our audience is frequently unknown and underestimated (Viègas, 2005) and the boundaries between personal and professional domains are often easily crossed. Lastly, information shared online is often available for access at future times and for future audiences. Not only do these attributes impact our abilities to regulate our intended audience, they also weaken our control over how that information is interpreted and used. In an attempt to understand and improve privacy management in information technology, computer scientists have engaged in analysis of the concept of privacy and an examination of privacy online (Palen & Dourish, 2003). Conceiving privacy as a *boundary regulation process* (Altman, 1977), Palen & Dourish identify three boundary negotiation processes as essential to the management of privacy in a networked world: disclosure, identity, and temporality.

Issues involving disclosure dominate recent social networking controversies, including the two scenarios above. Some would argue that both teacher and student should have known better than to reveal personal information in a public forum -- both should simply have avoided disclosing information. But Palen & Dourish argue that this view undermines the true social interactive nature online. Disclosure is essential to online interaction. Effectively negotiating private and public spaces involves selective decisions about what to disclose and what kind of persona to display. Problems arise, they argue, in how we control who is targeted by this public display and consequently how the display is interpreted.

There is also an inherent tension in our attempts to control how others see us online. When creating and publishing our virtual identities we choose to affiliate with certain groups, networks, each with their own set of identity markers, language conventions and patterns of interaction (Yates & Orlikowski, 1992). Likewise, we modify our ways of interacting based on our perceptions of the identities and affiliations of our audience. Online, these various identities can appear quite fragmented and disconnected, such that viewing one facet of an online persona out of context (such as the "drunken pirate") can lead to distortions and errors in judgments of the person's character and personality, largely outside of the individual's control. Assessments can be made and then applied to the hiring and firing decisions within a variety of professions. Central to protecting privacy is "the ability to discern who might be able to see one's action" (Palen & Dourish, 2003, p.4).

Lastly, our attempts to control the information we share have a temporal quality. That is, in any moment of information sharing, we typically respond to the results of past attempts at information sharing and anticipate future consequences of information sharing. Moments of information sharing are connected historically and logically. However, online these moments can be viewed out of sequence, preserved for future use, and even reorganized into alternative sequences. The photos, stories, and conversations uploaded during college may return to contribute a completely different image of responsibility than the one conveyed by the professional resumé uploaded ten years later.

This analysis has implications for both education and ethics. First, these lenses offer useful entry points for developing awareness and understanding of the vulnerabilities to privacy online for both those sharing and those interpreting information shared online. Second, the review suggests a way of thinking about privacy rights online. In contrast to those who believe that privacy rights are surrendered when information is made public, this review suggests that rights to privacy might still be negotiated after information is publicly accessible. For example, individuals sharing the information might deserve the right to have that information understood in context or within its original logical sequence; that is, understood in a way which maintains the integrity of the sharer's initial intentions. There is some limited empirical evidence suggesting that people already do uphold privacy rights in public spaces (Friedman et al., 2006). In one study, when college students were asked to judge whether an installed video camera capturing video of them in a public place was an invasion of their privacy, the majority of students judged that it was. Furthermore, in this study, students' responses illustrated a complex construal of privacy issues in public. Judgments of privacy were mediated by a variety of factors such as the location of the camera, the perceived purpose of the video camera (safety vs. voyeurism), the audience viewing the footage, and the extent of disclosure about the video camera (from posted signs to informed consent). As new technologies continue to alter the nature of our social interactions in online communities, more studies are needed both to highlight the privacy vulnerabilities inherent in these types of social interaction and to capture the adaptive reasoning about privacy rights that are constructed through those experiences.

The Merging of Professional and Non-Professional Identities

Social networking tools can serve as both a rich resource and a potential liability. As with any powerful tool, there are far-reaching risks and potential disaster if use of the tool is not thoroughly calculated—but there are beneficial uses as well. Some online social networking tools go beyond a function of socialization to include professional communication functions. For example, Zinch.com helps students connect with the colleges and universities they are interested in attending. After students register with Zinch, they complete a personal profile and prepare an online digital portfolio illustrating their talents. Profiles are automatically private to those other than approved admissions officers. Recruiters across the nation use the network to connect with students whose profiles are of interest to their institutions. A similar site, Cappex.com, has the added feature of a calculator that estimates students' chances for admission to their desired institution. These tools take admissions criteria beyond testing, basic academics, and letters of recommendation, to allow those in non-local areas to connect with learning institutions, and provide a convenient opportunity for individual students and admissions officers to connect.

Once schooling is out of the way and the job hunt begins or a promotion is imminent, students in some professions face the news that a background check, also known as a consumer report, is required. Background checks can be conveniently conducted during the hiring process through third party investigation services to verify any level of candidate qualifications from education records to drug tests and credit records. The Fair Credit Reporting Act mandates that a consumer report should be conducted within compliance of the law to prevent discriminatory actions (Federal Trade Commission, 2004). To assure their compliance, most often employers conduct screenings by contracting with consumer reporting agencies that have access to specialized information sources. But consumer reports can also reveal information about a candidate that is irrelevant, taken out of context, or even inaccurate. This leaves room for concern for some applicants who have not paid attention or were unaware of how their prior behavior could be interpreted by employment agencies.

An online search of a person's name could also be conducted to obtain as much information as possible about a candidate's level of qualification. An Internet search can reveal a candidate's Web site or portfolio, professional accomplishments and awards, and other pertinent information. But use of an Internet search during the hiring process could also pull up a candidate's social networking profile or other Internet-based information outside a normal background check. Voluntarily-disclosed information on profiles that are public may reveal an applicant's sexual orientation, political affiliation, age, and marital status, and an employer who allows consideration to these factors could be acting in violation of workplace-discrimination statutes. But, it appears that use of information from social networking sites such as "drunken, racy, or provocative photographs" in order to make a judgment about the candidate's suitability is perfectly legal (Byrnside, 2008).

These scenarios play out over and over during the hiring process. For example, a Boston marketing recruiter was screening applications and noted one of particular interest. A member of the interview team asked the recruiter if she had seen the applicant's MySpace page, which included pictures of the recent college graduate "Jell-o wrestling." Based on more relevant factors the applicant was not interviewed, but the MySpace page remained in the recruiter's memory (Aucoin, 2007). Here again, misimpressions about privacy could pose life-changing implications if conscious actions were not promoted. How many emerging professionals have lost jobs or been denied opportunities because of the blurred boundaries between professional life and private life? The liabilities associated with social networking are potentially staggering, for both employers and employees.

Also consider the fact that individuals can experience a sort of identity theft through social networking sites. Cicero (Illinois) Town President, Larry Dominick, had two MySpace profiles. City officials found the sites, which were "replete with photos and questionable comments about his sexuality and ethics." But both sites were created by imposters (Noel, 2008). Imposter profiles are so prevalent that MySpace now has protocol and a division within the company to address cyber-bullying, underage users, and imposter profiles (MySpace, 2008). Cicero attorneys are asking MySpace to identify the anonymous users who created the profiles, and Dominick is planning on suing.

When social networking sites are made public, everyone in the world, including colleagues, has the capability to view the content. No doubt, we make judgments about a person's character based on what we see. The merging of professional and non-professional identities has implications for those who choose to create and display personal profiles.

IMPLICATIONS FOR SOCIALLY RESPONSIBLE USE OF SOCIAL NETWORKING TOOLS

Given the prevalence of incidents surrounding social networking within educational institutions, there is a need to embed socially responsible usage principles in academic programs rich in technological innovation. We have begun to implement and study such interventions. As instructors in a teacher education program we (Foulger, Ewbank, Kay, Osborn Popp, & Carter, 2008) investigated the use of case-based coursework (Kim et al., 2006; Kolodner, 1993) for encouraging change in preservice teachers' reasoning about ethical issues in Web 2.0 tools. As we have grappled with ethical dilemmas around social networking at our own institution, we were curious about the ways in which new technologies might alter traditional forms of social interaction. These circumstances gave rise to the following questions that drove our research: a) What are preservice teachers' perspectives regarding a social networking scenario that involves multiple ethical dilemmas? And b) In what ways does case-based coursework change preservice teachers' reasoning about social networking? In this study we assessed the effectiveness of a case-based intervention with a group university freshman-level education class. They participated in a homework assignment that was developed to help them better understand the features of social networking tools. It also helped them clarify their ethical positions about recent legal sanctions pertaining to the use of social networking tools by students and teachers.

Based on a review of the literature about case-based reasoning, we expected coursework that included case-based teaching about controversial social networking issues to (a) increase students' recognition and integration of multiple perspectives or viewpoints about the benefits and harms of teachers' use of social networking tools and (b) develop students' appreciation for the range of ethical vulnerabilities inherent in social networking media. Fifty students participated in a three-part assignment. First, they were asked to respond to online selections about the technological nature of social networking. Students then commented on cases where teachers used social networking tools for pedagogical purposes. Finally students responded to cases where teachers were disciplined or dismissed for inappropriate conduct as defined by educational institutions. Comparisons of perceptions before and after the assignment were examined to analyze the preservice teachers' reasoning about controversial social networking incidents. Some significant changes in student perceptions did occur, indicating that case-based coursework increases awareness of the ethical complexities embedded in social networking tools.

Several trends emerged from the analysis. The homework helped students develop more complex ethical reasoning to the scenarios posed and revealed a significant increase among students in the call for some form of teacher discipline. Additionally, the homework developed students' recognition of the complexities of social networking sites and the need to develop clearer protocols around their educational use. Finally, the assignment had an impact on students' understanding of the ethical vulnerabilities of social networking tools. A deeper exploration of one common set of responses revealed that the study participants grappled with the line between a teacher honoring a student's right to privacy and a teacher's responsibility of caring for the students' well-being.

It was apparent that the case-based coursework encouraged students to contemplate rights to privacy in a public online forum. This level of thought is important because of the unknown norms of social networking. Even though many students are immersed in technology every day, there is still room for education about social networking and professional ethics..

Future studies should include investigations about educator conduct and rights to privacy in online spaces. Those who are engaged in supporting future professionals should consider ways in which they can assist the development of thinking about these kinds of ethical dilemmas so that new professionals can anticipate and prevent potential problems, develop well-reasoned responses to ethical decisions, and participate in the construction of protocols that continue to harness the educational potential of social networking tools. Developing such awareness and protocols are initial steps toward encouraging responsible use of these tools.

THE FUTURE OF SOCIAL NETWORKING IN EDUCATION

News stories continue to surface about questionable social behaviors that occur online. Although some behave as though the faceless world of online communities is lawless territory and continue to test the waters, no firm legal precedents have been established to guide online codes of conduct.

Educational organizations have taken a variety of positions on this issue, some in response to real problems they have encountered, and some prompted by attempts to be proactive in light of the news events they hear. Lamar County School Board has taken a conservative position. Although no incidents led to the decision, the attorney of the southern Mississippi district recommended adoption of a policy to lessen liabilities. Now communications between teachers and students through social networking sites or through texting are prohibited (Associated Press, 2008).

But Tomás Gonzales, Senior Assistant Dean at Syracuse University School of Law and a nationally recognized speaker in the legal issues concerning on-line communities, believes that educational organizations should embrace collaborative technologies and explore appropriate uses even in the midst of much negative press about the drawbacks of social networking (2008). He also claims that current codes of conduct about appropriate face-to-face behavior are probably sufficient for providing online guidance for students and administrators.

Codes of conduct in educational institutions should be examined to determine whether protocols for online behavior are embedded within existing policies. At a minimum, institutions should consider how their existing codes of conduct would be applied in the event of a dilemma involving social networking tools. Additionally, education programs that result in awareness of both proactive behaviors as well as potential situations to be avoided in social networking would benefit students and faculty alike.

CONCLUSION

We must recognize the limitations of our own experience and expertise. This applies to the use of many Web 2.0 tools, including online social networking. With the use of social networking tools, as with any powerful tool, come many vulnerabilities. As society becomes more technologically advanced, it has become the responsibility of educational institutions to support the use of the kinds of technologies that might prove to strengthen and support the learning process. However, it is also the responsibility of policymakers to assure participation in a safe learning environment. Ironically, news broadcasts have been mostly negative press about the pitfalls of social networking tools, and have not showcased innovative and pedagogical uses of Web 2.0 features. In order for any beneficial uses of such tools to be realized and refined, and then incorporated into learning environments, the fear and apprehension surrounding them must be set aside long enough for real innovation to occur.

We encourage institutions to first create a safe place for ideas to percolate by revisiting existing codes of conduct to assure their policy and procedures embrace the idea of virtual connectivity, and to publish guidelines for acceptable and appropriate uses of technology. By providing guidance to their members, institutions can encourage them to utilize online tools in a socially responsible manner, without squelching innovative uses of technology. Just as institutions use codes of conduct to ensure the safety and rights of each member on campus, they can utilize those same codes in the online extension. By conceptualizing online spaces as an integral part of institutions' physical and temporal community, codes of conduct can be applied in a manner that respects privacy and individual rights, while allowing innovation and security for all participants.

REFERENCES

A. B. v. State of Indiana, 885 N.E. 2d 1223 (Ind., 2008).

Alexa. (2008). Traffic rankings for MySpace.com. *Alexa.com.* Retrieved April 25, 2008 from http://www.alexa.com/data/details/traffic_details/myspace.com

Altman, I. (1977). Privacy regulation: Culturally universal or culturally specific? *The Journal of Social Issues*, *33*(3), 66–84.

Associated Press. (2008). Mississippi school district bars teacher-student texting. *Yahoo News.* retrieved August 13, 2008 from http://news.yahoo.com/s/ap/20080721/ap_on_bi_ge/text_ban

Aucoin, D. (2007). MySpace or the workplace? *Boston Globe.* Retrieved April 25, 2008 from http://www.boston.com/news/globe/living/articles/2007/05/29/myspace_vs_workplace/

Berg, J., Berquam, L., & Christoph, K. (2007). Social networking technologies: A "poke" for campus services. *EDUCAUSE Review*, *42*(2), 32–44.

Boyd, D. (2008). Why youth (heart) social network sites: The role of networked publics in teenage social life. In D. Buckingham (Ed.), *Youth, identity, and digital media.* Cambridge, MA: The MIT Press.

Byrnside, I. (2008). Six clicks of separation: The legal ramifications of employers using social networking sites to research applicants. *Vanderbilt Journal of Entertainment and Technology Law, 445.*

Carvin, A. (2006, October 10). Is MySpace your space as well? *Learning.now.* Retrieved July 21, 2008 from http://www.pbs.org/teachers/learning.now/2006/10/is_myspace_your_space_as_well.html

Crawford, J. (2007, January 25). Teacher fired over MySpace page. January 25, 2007. *Tallahassee.com.* Retrieved December 3, 2007 from http://tallahassee.com/legacy/special/blogs/2007/01/teacher-fired-over-myspace-page_25.html

Drews vs. Joint School District, Not Reported in F.Supp.2d, 2006 WL 1308565 (D. Idaho).

E-School News Staff. (2007). Teachers warned about MySpace profiles. *e-School News.* Retrieved July 17, 2008 from http://www.eschoolnews.com/news/top-news/related-top-news/?i=50557;_hbguid=49a1babb-b469-4a85-a273-292a0514d91d

Federal Trade Commission. (2004). *The fair credit reporting act.* Retrieved August 13, 2008 from http://www.ftc.gov/os/statutes/031224fcra.pdf

Finn, J. (2004). A survey of online harassment at a university campus. *Journal of Interpersonal Violence*, *19*(4), 468–483. doi:10.1177/0886260503262083

Foulger, T., Ewbank, A., & Kay, A. Osborn Popp, S., & Carter, H. (2008). *Moral spaces in MySpace: Preservice teachers' perspectives about ethical issues in social networking* (Manuscript in progress).

Friedman, B. (1997). Social judgments and technological innovation: Adolescents' understanding of property, privacy, and electronic information. *Computers in Human Behavior, 13*(3), 327–351. doi:10.1016/S0747-5632(97)00013-7

Friedman, B., Kahn, P. H. Jr, Hagman, J., Severson, R. L., & Gill, B. (2006). The watcher and the watched: Social judgments about privacy in a public place. *Human-Computer Interaction, 21*(2), 235–272. doi:10.1207/s15327051hci2102_3

Fulmer, J. (2002). Dismissing the 'immoral' teacher for conduct outside the workplace-do current laws protect the interests of both school authorities and teachers? *Journal of Law and Education, 31*, 271–290.

Gonzales, T. (2008). *Facebook, Myspace, and online communities: What your college must know* (CD recording). Retrieved August 28, 2008 from https://www.higheredhero.com/audio/main.asp?G=2&E=1317&I=1

Griffiths, M. (2002). Occupational health issues concerning Internet use in the workplace. *Work and Stress, 16*(4), 283–286. doi:10.1080/02678370310000071438

Hargittai, E. (2007). Whose space? Differences among users and non-users of social network sites. *Journal of Computer-Mediated Communication, 13*(1), 14.

Hermes, J. (2008). Colleges create Facebook-style social networks to reach alumni. *The Chronicle of Higher Education, 54*(33), A18.

Hoover, T. (2008). *MySpace profile*. Retrieved August 26, 2008 from http://myspace.com/mshoover.

J.S. v. Blue Mountain School District, 2007 WL 954245 (M.D.Pa.).

Kim, S., Phillips, W. R., Pinsky, L., Brock, D., Phillips, K., & Keary, J. (2006). A conceptual framework for developing teaching cases: A review and synthesis of the literature across disciplines. *Medical Education, 40*, 867–876. doi:10.1111/j.1365-2929.2006.02544.x

Kolodner, J. (1993). *Case-based reasoning*. San Mateo, CA: Morgan Kaufmann.

Laupa, M. (1991). Children's reasoning about three authority attributes: Adult status, knowledge, and social position. *Developmental Psychology, 27*(2), 321–329. doi:10.1037/0012-1649.27.2.321

Laupa, M. (1995). Children's reasoning about authority in home and school contexts. *Social Development, 4*(1), 1–16. doi:10.1111/j.1467-9507.1995.tb00047.x

Laupa, M., & Turiel, E. (1993). Children's concepts of authority and social contexts. *Journal of Educational Psychology, 85*(1), 191–197. doi:10.1037/0022-0663.85.1.191

Layshock v. Hermitage School District, 496 F.Supp.2d 587 (W.D.Pa., 2007).

Lenhart, A. (2007). Cyberbullying and online teens. *Pew Internet & American Life Project.* Retrieved July 21, 2008 from http://www.pewInternet.org/pdfs/PIP%20Cyberbullying%20Memo.pdf

Lenhart, A., Madden, M., Macgill, A., & Smith, A. (2007). Teens and social media. *Pew Internet & American Life Project.* Retrieved July 18, 2008 from http://www.pewInternet.org/pdfs/PIP_Teens_Social_Media_Final.pdf

Leskovec, J., & Horvitz, E. (2007). *Planetary-scale views on an instant-messaging network* (Microsoft Research Technical Report MSR-TR-2006-186). Retrieved August 26, 2008 from http://arxiv.org/PS_cache/arxiv/pdf/0803/0803.0939v1.pdf

Lindenberger, M. (2006). Questions of conduct. *Diverse Issues in Higher Education, 23*(21), 36–37.

Lipka, S. (2008). The digital limits of "in loco parentis." . *The Chronicle of Higher Education, 54*(26), 1.

May, M. (2006, June 23). Hoover: Caught in the flash. *Austin Chronicle*. Retrieved December 3, 2007 from http://www.austinchronicle.com/gyrobase/Issue/story?oid=oid%3A378611

Myspace.com. (2008). Myspace.com safety and security. *Myspace.com*. Retrieved August 26, 2008 from http://www.myspace.com/safety

Nakashima, E. (2007, March 7). Harsh words die hard on the Web. *Washingtonpost.com*. Retrieved July 21, 2008 from http://www.washington-post.com/wp-dyn/content/article/2007/03/06/AR2007030602705.html

Noel, J. (2008, May 17). Cicero town president wants MySpace poser's identity revealed. *Chicagotribune.com*. Retrieved August 13, 2008 from http://www.chicagotribune.com/news/local/chi-myspaceimposters_bdmay18,0,3460074.story?page=1

O'Hanlon, C. (2007). If you can't beat 'em, join 'em. *T.H.E. Journal, 34*(8), 39–40, 42, 44.

O'Neil, R. (2008). It's not easy to stand up to cyberbullies, but we must. *The Chronicle of Higher Education, 54*(44), A23.

Palen, L., & Dourish, P. (2003). Unpacking "privacy" for a networked world. In *Proceedings of the ACM Conference on Human Factors in Computing Systems CHI 2003,* Fort Lauderdale, FL (pp. 129-136). New York: ACM.

Patchin, J., & Hinduja, S. (2006). Bullies move beyond the schoolyard: A preliminary look at cyberbullying. *Youth Violence and Juvenile Justice, 4*(2), 148–169. doi:10.1177/1541204006286288

Phillips, G. (2007, June 6). Teacher's blog draws probe from the system. *Southern Maryland Newspapers Online*. Retrieved December 3, 2007 from http://www.somdnews.com/stories/060607/rectop180341_32082.shtml

Requa v. Kent School District, 492 F.Supp.2d 1272 (W.D.Wash., 2007).

Schonefeld, E. (2008). Facebook is not only the world's largest social network, it is also the fastest growing. *Techcrunch.com*. Retrieved August 13, 2008 from http://www.techcrunch.com/2008/08/12/facebook-is-not-only-the-worlds-largest-social-network-it-is-also-the-fastest-growing/

Smetana, J. G. (2006). *Social-cognitive domain theory: Consistencies and variations in children's moral and social judgments*. Mahwah, NJ: Lawrence Erlbaum Associates.

Steiner, E. (2007, May 1). MySpace photo costs teacher education degree. *Washington Post.com*. Retrieved April 25, 2008 from http://blog.washingtonpost.com/offbeat/2007/05/myspace_photo_costs_teacher_ed.html

The Pennsylvania State University. (2007). *Student teaching handbook: The Pennsylvania State University college of education*. Retrieved July 17, 2008 from http://www.ed.psu.edu/preservice/things%20to%20update/2007-2008%20ST_HANDBOOK_August%202007.pdf

Turiel, E. (1983). *The development of social knowledge: Morality and convention*. Cambridge, UK: Cambridge University Press.

Turiel, E. (2002). *The culture of morality: Social development, context, and conflict*. New York: Cambridge University Press.

Viégas, F. B. (2005). Bloggers' expectations of privacy and accountability: An initial survey. *Journal of Computer-Mediated Communication, 10*(3). doi:.doi:10.1111/j.1083-6101.2005.tb00260.x

Vivanco, H. (2007, March 29). Teacher still posting on MySpace. *Inland Valley Daily Bulletin.* Retrieved on December 3, 2007 from http://www.dailybulletin.com/news/ci_5553720

Wheeler, T. (2007). Personnel pitfalls in the cyberworld. *School Administrator, 64*(9), 22–24.

Wikipedia. (2008). *Social networking*. Retrieved August 13, 2008 from http://en.wikipedia.org/wiki/Social_networking

Wisniewski v. Board of Education, Weedsport Central School District, 494 F. 3d 34, (2nd Cir. 2007).

Woolcock, N. (2008). Soaring number of teachers say they are 'cyberbully' vicitms. *The Times.* Retrieved July 21, 2008 from http://www.timesonline.co.uk/tol/life_and_style/education/article3213130.ece

WSBTV.com. (2006, May 16). Student faces criminal charges for teacher jokes. *WSBTV.com.* Retrieved July 21, 2008 from http://www.wsbtv.com/education/9223824/detail.html

Yates, J., & Orlikowski, W. J. (1992). Genres of organizational communication: A structurational approach to studying communication and media. *Academy of Management Review, 17*(2), 299–326. doi:10.2307/258774

This work was previously published in Collective Intelligence and E-Learning 2.0: Implications of Web-Based Communities and Networking, edited by H. H. Yang; S. C. Chen, pp. 27-43, copyright 2010 by Information Science Reference (an imprint of IGI Global).

Chapter 8.6
Modeling Cognitive Agents for Social Systems and a Simulation in Urban Dynamics

Yu Zhang
Trinity University, USA

Mark Lewis
Trinity University, USA

Christine Drennon
Trinity University, USA

Michael Pellon
Trinity University, USA

Phil Coleman
Trinity University, USA

Jason Leezer
Trinity University, USA

ABSTRACT

Multi-agent systems have been used to model complex social systems in many domains. The entire movement of multi-agent paradigm was spawned, at least in part, by the perceived importance of fostering human-like adjustable autonomy and behaviors in social systems. But, efficient scalable and robust social systems are difficult to engineer. One difficulty exists in the design of how society and agents evolve and the other difficulties exist in how to capture the highly cognitive decision-making process that sometimes follows intuition and bounded rationality. We present a multi-agent architecture called CASE (Cognitive Agents for Social Environments). CASE provides a way to embed agent interactions in a three-dimensional social structure. It also presents a computational model for an individual agent's intuitive and deliberative decision-making process. This chapter also presents our work on creating a

DOI: 10.4018/978-1-60566-236-7.ch008

multi-agent simulation which can help social and economic scientists use CASE agents to perform their tests. Finally, we test the system in an urban dynamic problem. Our experiment results suggest that intuitive decision-making allows the quick convergence of social strategies, and embedding agent interactions in a three-dimensional social structure speeds up this convergence as well as maintains the system's stability.

INTRODUCTION

In social environments, people interact with each other and form different societies (or organizations or groups). To better understand people's social interactions, researchers have increasingly relied on computational models [16, 40, 41, 42]. A good computational model that takes into consideration both the individual and social behaviors could serve as a viable tool to help researchers analyze or predict the complex phenomena that emerge from the interactions of massive autonomous agents, especially for the domain that often requires a long time to evolve or requires exposing real people to a dangerous environment. However, efficient, scalable, and robust social systems are difficult to engineer [3].

One difficulty exists in modeling the system by holding both the societal view and the individual agent view. The societal view involves the careful design of agent-to-agent interactions so that an individual agent's choices influence and are influenced by the choices made by others within the society. The agent view involves modeling only an individual agent's decision-making processes that sometimes follow intuition and bounded rationality [29]. Previous research in modeling theory of agents and society in a computational framework has taken singly a point of view of society or agent. While the single societal view mainly concentrates on the centralist, static approach to organizational design and specification of social structures and thus limits system dynamics [12, 16, 35], on the other hand, the single agent view focuses on modeling the nested beliefs of the other agents, but this suffers from an explosion in computational complexity as the number of agents in the system grows.

Another difficulty in modeling theory of agent and society exists in quantitative or qualitative modeling of uncertainty and preference. In the case of quantitative modeling, the traditional models like game theory and decision theory have their own limitations. Game theory typically relies on concepts of equilibria that people rarely achieve in an unstructured social setting, and decision theory typically relies on assumptions of rationality that people constantly violate [27]. In the case of qualitative modeling, there are three basic models: prescriptive, normative and descriptive [31, 37]. A prescriptive model is one which can and should be used by a real decision maker. A normative model requires the decision maker to have perfect rationality, for example, the classical utility function belongs to this category. Many normative theories have been refined over time to better "describe" how humans make decisions. Kahneman and Tversky's Prospect Theory [18, 34] and von Neuman and Morgenstein's Subjective Utility Theory [36] are noted examples of normative theories that have taken on a more descriptive guise. One of the central themes of the descriptive model is the idea of Bounded Rationality [29], i.e., humans don't calculate the utility value for every outcome; instead we use intuition and heuristics to determine if one situation is better than another. However, existing descriptive methods are mostly informal, therefore there is a growing need to study them in a systematic way and provide a qualitative framework in which to compare various possible underlying mechanisms.

Motivated by these observations, we have developed a cognitive agent model called CASE (Cognitive Agent in Social Environment). CASE is designed to achieve two goals. First, it aims to model the "meso-view" of multi-agent interaction by capturing both the societal view and the agent view. On one hand, we keep an individual perspective on the system assumed by the traditional multi-agent models, i.e. an agent is an autonomous entity and has its own goals and beliefs in the environment [5, 43]. On the other hand, we take into account how agent's decisions are influenced by the choices made by others. This is achieved by embedding agents' interactions in three social structures: group, which represents social connections, neighborhood, which represents space connections and network, which span social and space categories. These three structures reproduce the way information and social strategy is passed and therefore the way people influence each other. In our view, social structures are external to individual agent and independent from their goals. However, they constrain the individual's commitment to goals and choices and contribute to the stability, predictability and manageability of the system as a whole.

Our second goal is to provide a computational descriptive decision model of the highly cognitive process wherein an individual agent's decision-making. The descriptive theory assumes agents undergo two fundamental stages when reaching a final decision: an early phase of *editing* and a subsequent phase of *evaluation* [19]. In the editing phase, the agent sets up priorities for how the information will be handled in the subsequent decision-making phase and forms heuristics which will be used during the decision-making process, i.e. the agent only acts with bounded rationality. In the evaluation phase, there exist two generic modes of cognitive function: an *intuitive* mode in which decisions are made automatically and rapidly, and a controlled mode, which is *deliberate* and slower. When making decisions, the agent uses *satisfying* theory [30], i.e. it takes "good enough" options rather than a single "best" option.

The rest of the chapter is organized as follows. Section 2 introduces the related work. In Section 3, we give an overview to cognitive models for social agents, from both the societal view and individual agent view, and introduce preliminary contextual information. Section 4 presents CASE from the perspective of the societal view, i.e. how an agent's decision affects another. Section 5 presents CASE from the perspective of the individual agent's view, i.e. an intuitive and deliberative decision-making mechanism. Section 6 is a simulation supporting CASE agents that provides an integrated environment for researchers to manage, analyze and visualize their data. Section 7 reports the experiments and Section 8 concludes the chapter.

RELATED WORK

Multi-agent systems have been widely used to model human behaviors in social systems from the computational perspective. There have been many successful systems addressing this issue. Due to the lack of space, we limit this discussion to several of the most relevant systems. We review them from two categories: agent modeling and agent simulation.

Agent Modeling

COGnitive agENT (COGENT) [6] is a cognitive agent architecture based on Rasmussen's integrated theory of human information processing [28] and the Recognition Primed Decision (RPD) model [21]. It provides the decision-aiding at multiple levels of information processing, ranging from perceptual processing and situation feature extraction through information filtering and situation assessment, and not a direct process of real human social behaviors.

COgnitive Decision AGEnt (CODAGE) [20] is an agent architecture that derived its decision model from cognitive psychological theories to take bounded rationality into account. However, CODAGE does not consider an agent's influence on other agents and there is no communication between agents. We consider communication important since it permits individuals to expand their spheres of interest beyond the self. Moreover, CODAGE is a centralized system where only one decision maker makes decisions for each agent, while CASE is a distributed system where each agent makes their own decisions.

PsychSim [27] is a multi-agent simulation for human social interaction. In order to represent agents' influence on each other, PsychSim gives each agent full decision models of other agents. In PsychSim, bounded rationality is implemented as three limitations on agents' beliefs: 1) limiting the recursive nested-belief reasoning process to a certain level, 2) limiting the finite horizon of the agents' look-ahead, and 3) allowing the possible error in the agents' belief about others. However, we treat bounded rationality as a human tendency to anchor on one trait or piece of information when making decisions (the detail of this can be found in Section 5).

Construct-Spatial [24] combines an agent's communication and movement simultaneously. It aims to simulate many real world problems that require a mixed model containing both social and spatial features. They integrate two classical models: Sugarscape [7], a multi-agent grid model, and Construct [2], a multi-agent social model, and run virtual experiments to compare the output from the combined space to those from each of the two spaces. Our model is similar in that we also capture multi-dimensional interactions between agents. We embed agents' interactions in three social structures: group, which represents social connections, neighborhood, which represents space connections and network, which span social and space categories. These three structures reproduce the way information and social strategy is passed and therefore the way people influence each other.

Hales and Edmonds [14] introduce an interesting idea of using "tag" mechanisms for the spontaneous self-organization of group level adaptations in order to achieve social rationality. Their idea is to use agents that make decisions based on a simple learning mechanism that imitates other agents who have achieved a higher utility. This research reminds us that sometimes the simplest of techniques can have the most far-reaching results. However, agents in their system need a relatively large number of tag bits (32 tag bits for a population of 100 agents) for all agents to reach a socially rational decision. In this chapter we use a different approach for generating socially rational behaviors. We embed social interaction into three social structures and provide a model for diffusing one agent's strategy to others.

Jiang and Ishida [17] introduce an evolution model about the emergence of the dominance of a social strategy and how this strategy diffuses to other agents. Our model is similar in that it includes multiple groups and allows for diffusion of strategy. But our model differs in two aspects. First is in how the groups are defined. Jiang and Ishida define a one to one relationship between groups and strategies, i.e. for every one strategy there exists one group and each agent belongs to the group that has their strategy. However, in our model, an agent can belong to multiple groups at one time. The second difference exists in whether or not the group's strategy is dynamic. Because Jiang and Ishida define a one to one relationship between groups and strategies, there must always exist one group for every possible strategy. This means group strategies are static and will not change over time. In our model, we model the dynamics between the group and the agent, so both the group's strategy and the agent's may be changed with time.

Agent Simulation

RePast[1], perhaps one of the most feature filled packages, provides templates for easy construction of behavior for individual agents and integrates GIS (Geographic Information System) support, which is a feature that our simulations will need. It is also fully implemented on all systems. The multi agent system named MASON[2] is a lightweight system with a good amount of functionality. It has the ability to generate videos and snapshots as well as charts and graphs. JADE[3] is a project that we looked at for its ability to be distributed across multiple machines that do not need to be running on the same operating system, which is a feature of our system. It also allows configuration of a distributed model to be controlled by a remote GUI, which is also a feature we implement in our package. Cougaar[4], developed for use by the military is influential in that it allows for huge scaling of projects to simulate many agents working together, which is a very appealing feature for our simulations. JAS[5] (Java Agent based Simulation) library is a package that supports time unit management by allowing the user to specify how a system will operate in terms of hours, minutes or seconds. This is helpful for spatial modeling and simulations. SWARM[6], developed at the Santa Fe Institute, allows for users to write their own software. It also allows for development on a variety of systems. The package is open source and has a large community of developers as well. EcoLab[7] allows for the user to generate histograms and graphs but is only able to be implemented on a limited amount of systems. It provides a scripting language which can access the model's methods and instance variables, allowing experiments to be set up dynamically at runtime. This is good for the user without programming skills. Breve[8] allows the user to define agent behavior in a 3D world. It also allows for extensive use of plug-ins that fit seamlessly with user generated code.

COGNITIVE MODELS FOR SOCIAL AGENTS: AN OVERVIEW

According to social scientists, social behavior is behavior directed towards, or taking place between, agents of the same societies [26]. Understanding the emergence and nature of social behavior is necessary prior to the design of a computational framework. Social behaviors are complex phenomena, which may be better examined at two different levels: the society level and the individual agent level. These two levels are not independent but are intimately related and often overlap.

The Society

Any society is the result of an interaction between agents, and the behavior of the agents is constrained by the assembly of societal structures [9]. For this reason, a society is not necessarily a static structure, that is, an entity with predefined characteristics and actions. If societies such as public institutions or companies possess an individuality of their own which distinguishes them from the assembly created by the individualities of their members, it is not necessarily the same for simpler collective structures such as working groups or herds of animals. Even though societies are considered as being complex, such as colonies of bees or ants, they should not necessarily be considered as individuals in their own right if we wish to understand their organization and the regulation and evolution phenomena prevailing them there. Therefore, in our view, a society is the emergence of properties of individual interactions, without it being necessary to define a specific objective which represents such an outcome[9].

While decision rules were developed for the purpose of understanding decision-making on the individual level, it is not illogical to think that such a theory could be expanded to account for decision-making made by a group of individuals. There are generally two alternative methods of extending interest beyond the self. Both of these ways, however, present some problems.

The first method is to define a notion of social utility to replace individual utilities [4]. Such a concept is problematic, because, as put by Luce and Raiffa, the notion of social rationality is neither a postulate of the model nor does it appear to follow as a logical consequence of individual rationality [22]. Pareto optimality provides a concept of group interest as a direct attribute of the group, but this falls short of a viable solution for the concept of individually rational decision makers since no player would consent to reducing its own satisfaction simply to benefit another – it is not self-enforcing. Adopting this view would require the group to behave as a "super player", who can force agents to conform to a concept of group interest that is not compatible with individual interests. Therefore, there is a clear demand for keeping self-enforcement as the baseline of decision-making for agents behaving under social context [32, 33].

The second method is to incorporate the utility of other agents into the creation of individual utility, such as the RMM (Recursive Modeling Method) model [10]. The problem of this method is that the nesting of these agent models is potentially unbounded. Further, people rarely use such a deep recursive model although infinite nesting is required for modeling rational behavior [19]. Many multi-agent models of human decision-making made reasonable domain specific limitations to the number of nested levels and gains in computational efficiency [27]. But there is an inherent loss of precision. To better understand and quantify how people influence each other, Hogg and Jennings [15] introduce a framework for making socially acceptable decisions, based on social welfare functions which combine social and individual perspectives in a unified manner. It seems that the notion of the social welfare function, which represents the combination of individual and social interests, is especially useful for modeling social influence so that an individual agent's behavior is affected by others but is still able to maintain its individual goal and utility.

The Agent

From a human cognitive psychological perspective, a person's behaviors can be viewed as the outcome of his/her decision-making process [25, 11]. Kahneman and Tversky suggest that a person's decision-making processes follow intuition and bounded rationality [19]. Further, the knowledge that a person has learned through his/her life experience can be viewed as the extension of his/her intuitions [25]. In psychology, intuition has broad meaning encompassing both one's ability to identify valid solutions to problems and to quickly select a workable solution among many potential solutions. For example, the RPD model aims to explain how people can make relatively fast decisions without having to compare options [21], the Prospect Theory captures human intuitive attitudes toward risk, and the Multi-Attribute Decision-Making model [1, 39] draws intuition in terms of qualitative information.

First introduced by Simon, bounded rationality presents an alternative notion of individual optimization in multi-agent settings to the classic utility theory [29, 23]. Agents are only bounded rationally and use the satisfying theory to make decisions. The idea of the satisfying theory is to reconstruct utility around preferences, rather than actions. It basically states that the only information we can draw from are the preferences of individuals. This concept is an important one, since it reminds us not to ascribe spurious qualities to the individuals studied and abstracted by a utility function; such a function is a mere representation and may contain aspects that do not actually reflect the individual's nature. Stirling's satisfying game theory also shows that people do not judge the utility based off analysis of desired results, but based off their preferences [30].

AGENT-SOCIETY DUALITY

The agent/society duality, shown in Figure 1, characterizes the processes that take place between the agents and the societies which result from them. We are dealing with dynamic interaction, the logic of which depends simultaneously on the capabilities of individual agents and the dynamic interactions between them. On one hand, agents

Figure 1. Agent-Society Duality

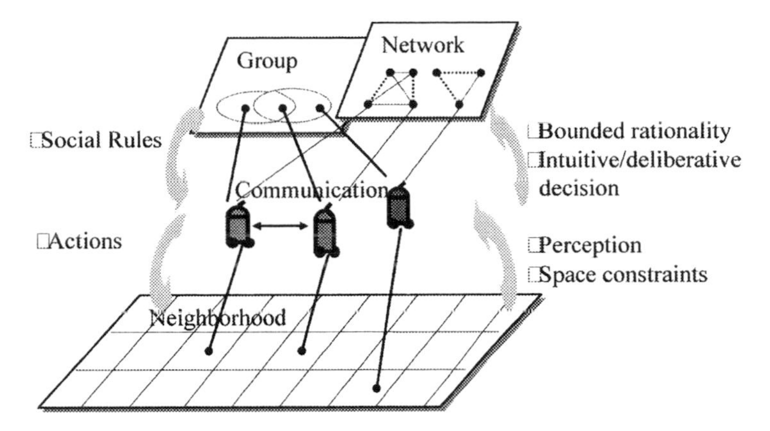

have their own goals and are capable of performing various actions. On the other hand, according to social scientists, agents interact with each other and the interactions are embedded in three social structures [13, 26]:

1. **Group:** represents social connections,
2. **Neighborhood:** represents space connections, and
3. **Network:** represents connections that span the social and space categories.

The purpose of these structures is to reproduce the way information and social strategy are passed and the way people influence each other. A group is a collection of agents who "think alike", or make similar decisions. It allows for a diffusion of social strategies through social space. An agent can belong to multiple groups at a time and can change groups over time. Each agent also has its own neighborhood and network. A neighborhood includes the agents whose behaviors this agent can observe within a predefined physical distance. It allows for a diffusion of information through physical space. One agent only has one neighborhood which includes the agents that it can directly observe. However, these neighbors can be different at every step because the agent moves. A network includes the agents that this agent chooses to communicate and interact with. Agents can't choose who is in their neighborhood but they can choose who they want to interact with. Therefore, the network connection allows for selective diffusion of information. Each agent also only has one network but its members can change over time. For example, if the agent has not heard from one agent over a certain step, the communication connection to that agent will be dropped.

Agent-Society Evolvement

Let agent $a \in A$ where A is the set of all agents. Each agent has a social strategy. This social strategy can be either ordinal or cardinal. We denote the social strategy for agent a by S_a.

Let $g \in G$ where G is the set of all groups. Each group also has a social strategy, denoted by S_g. Groups are formulated on the basis of a common preference. Each agent identifies itself with any group such that the agent's strategy falls within some threshold of the group's strategy.

$$\forall a \in A \textit{ and } g \in G,\ a \in g \textit{ if } \text{diff}(S_a, S_g) < \text{d} \quad (1)$$

where $\text{diff}(S_a, S_g)$ is the difference between the agent's strategy S_a and the group's strategy S_g, and d is the threshold. It can be seen that agent a can belong to more than one group at a time and can belong to different groups over time.

When an agent joins a group they are given a

rank in that group. An agent will have one rank for every group they belong to. The agent's rank can be evaluated based on the agent's importance, credibility, popularity, etc. Rank defines how much the agent will influence the group as well as how much the group will influence them. A high-ranking agent influences the group, and therefore its members, more than a low ranking agent and at the same time is influenced more than a low ranking agent. An agent's rank is specific to the domain and may change over time.

Each time step, every group will update their strategy. The update is determined by its members' strategy and the percentage of the total group rank they hold.

$$S_g = \sum_{a \in g} S_a \times \frac{R_a^g}{\sum_{b \in g} R_b^g} \quad (2)$$

where R_a^g denotes agent a's group rank. This allows for groups to be completely dynamic because both their members and their strategy can change each time step.

Just like an agent's rank in it groups, an agent also has a rank in its neighborhood and network. Each agent keeps track of the agents in its neighborhood and the agents it communicates with. Every time an agent observes another agent in his neighborhood, that agent's neighborhood rank will increase. Also, each time an agent communicates with another agent, that agents communication rank increases. Therefore every agent will have a rank value for every agent it interacts with, and a separate rank for every agent he communicates with. When an agent updates its strategy, it will take into account these ranks. Agents with a high rank relative to the other agents will have a stronger influence. Therefore the longer two agents are near each other, the more they will influence each other. The same is true for communications. Below is the update function for the neighborhood's strategy and the network's strategy.

$$S_n = \sum_{a \in n} S_a \times \frac{R_a^n}{\sum_{b \in n} R_b^n} \quad (3)$$

$$S_w = \sum_{a \in w} S_a \times \frac{R_a^w}{\sum_{b \in w} R_b^w} \quad (4)$$

where S_n is the strategy for neighborhood n, P_w is the strategy for network w, R_a^n is agent a's neighborhood rank and R_a^w is agent a's network rank.

Each time step, every agent also updates their strategy. An agent's update function is defined as:

$$S_a` = \alpha \times S_a + \beta \times S_g + \gamma \times S_n + \lambda \times S_w \quad (5)$$

where $\alpha, \beta, \gamma, \lambda \in [0, 1]$ and $\alpha+\beta+\gamma+\lambda = 1$. These values represent what percentage of influence the agent takes from itself, its group, its neighborhood and its network. This allows for multiple agent types. For example, (1, 0, 0, 0) represents a selfish agent because it cares nothing about the whole society, and (0, 0.33, 0.33, 0.34) represents a selfless agent who cares about the three social structures equally. Our system is fully distributed and uses discretized time. At each time step, every agent has an execution cycle, shown in Figure 2.

TWO-PHASE DECISION-MAKING PROCESS

Kahneman and Tversky suggest a two-phase decision model for descriptive decision-making: an early phase of editing and a subsequent phase of evaluation. In the editing phase, the decision-maker constructs a representation of the acts, contingencies and outcomes that are relevant to the decision. In the evaluation phase, the agent assesses the value of each alternative and chooses the alternative of highest value. Our decision model incorporates their idea and specifies it by the following five mechanisms:

Figure 2. Agent execution function

```
/* The function is executed independently by
   each agent, denoted agent a below.*/
execute(KB_a, S_a, env, mQueue, t)
  inputs: KB_a is the knowledge base for agent a
        S_a, the strategy of agent a
        env, the environment
        mQueue, the message queue for agent a
        t, the current step
  // making decision
        observation(env);
        update(KB_a);
        check(mQueue);
        M = situation_assess(KB_a);
        action = decision(KB_a, S_a, M);
  // performing the output of decision-making
        do(action);
        inform_resource_synchronize(env);
        update(env);
        inform_server_synchronize(masterserver);
  // updating society's strategy
        update_group_strategy();
        update_neighborhood_strategy();
        update_network_strategy();
  // updating agent's strategy
        update_agent_strategy();
  // moving to next step
        t++;
```

- Editing
  - **Framing:** The agent frames an outcome or transaction in its mind and the utility it expects to receive.
  - **Anchoring:** The agent's tendency to overly or heavily rely on one trait or piece of information when making decisions.
  - **Accessibility:** The importance of a fact within the selective attention.
  - Evaluation
  - **Two modes of function:** Intuition and deliberation.
  - **Satisfying theory:** Being good enough.

Figure 3 shows the two-phase decision-making process. Next we discuss each phase in a subsection.

Figure 3. Two-phase decision-making process

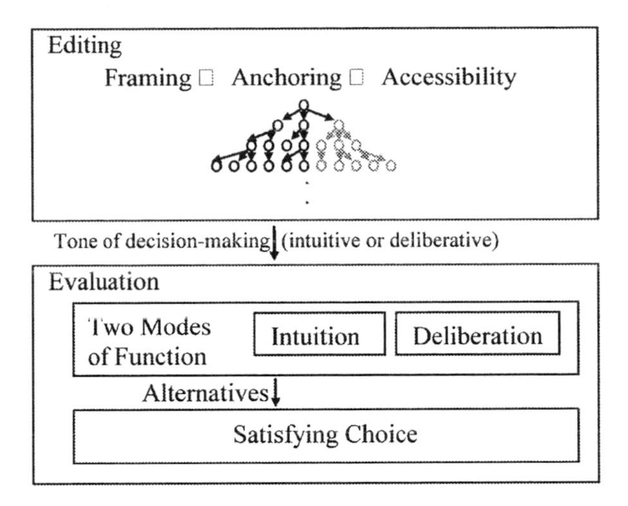

Editing Phase

One important feature of the descriptive model is that it is reference based. This notion grew out of another central notion called **framing** where agents subjectively frame an outcome or transaction in their minds and the utility they expect to receive is thus affected. This closely patterns the manner in which humans make rational decisions under conditions of uncertainty.

Framing can lead to another phenomenon referred to as **anchoring**. Anchoring or focalism is a psychological term used to describe the human tendency to overly or heavily rely (*anchor*) on one trait or piece of information when making decisions. A classic example would be a man purchasing an automobile, the client tends to "anchor" his decision on the odometer reading and year of the car rather than the condition of the engine or transmission.

Accessibility is the ease with which particular information come to mind. The concept of accessibility is applied more broadly in this research than in common usage. The different aspects and elements of a situation, the different objects in a scene, and the different attributes of an object, all can be described as more or less accessible for an individual agent exposed to a certain decision situation. As it is used here, the concept of accessibility subsumes the notions of stimulus salience, selective attention, and response activation or priming.

The editing phase gives us the main ideas that have been incorporated practically into the next phase, evaluation, to make quick intuitive decisions.

Evaluation Phase

In the evaluation phase, there exist two modes of cognitive function: an **intuitive** mode in which decisions are made automatically and rapidly, and a **deliberative** mode, which is effortful and slower. The operations of the intuition function are fast, effortless, associative, and difficult to control or modify, while the operations of the deliberation function are slower, serial, and deliberately controlled; they are also relatively flexible and potentially rule governed. Intuitive decisions occupy a position between the automatic operations of perception and the deliberate operations of reasoning.

Intuitions are thoughts and preferences that come to mind quickly and without much reflection. In psychology, intuition can encompass the ability to know valid solutions to problems and decision making. For example, the RPD model aimed to explain how people can make relatively fast decisions without having to compare options [21]. Klein found that under time pressure, high stakes, and changing parameters, experts used their base of experience to identify similar situations and intuitively choose feasible solutions. Thus, the RPD model is a blend of intuition and deliberation. Intuition is the pattern-matching process that quickly suggests feasible courses of action. Deliberation is a conscious reasoning of the courses of action.

We adopted a different approach from the RPD model to handle the intuitive and deliberative decision-making process. For our purpose, what becomes accessible in the current situation is a key issue in determining the tone of decision-making, i.e. intuitive or deliberative. Accessibility is determined in the editing phase by three factors.

- First, an agent utilizes prior knowledge of previous states to frame potential outcomes for its current state. In framing these potential outcomes, an agent ascribes reference based expected utility functions to them. Here, information anchoring or bias becomes a positive force as it leads to the agent's ability to make reference based utilities for each potential outcome.
- Second, when an agent makes decisions, it does not have to search all of its knowledge base. Instead, it concentrates on the relevant and important information.
- Third, a decision which was chosen before receives more attention (or high accessibility) than other alternatives and tends to be more positively evaluated before it is chosen again.

Based on the above analyses, we compile an information list. In addition to physical properties such as size and distance, the list keeps track of an abstract property called accessibility. The accessibility represents the relevance, similarity or importance of the information. At the beginning, this fact is known to the designer. It will be dynamically updated along with system processing. For example, the deliberation process may increase the accessibility because this information is important; if this information has been used in a previously successful decision, its accessibility will be increased, but if the previous decision was not successful, the accessibility will be decreased. The accessibility of all information is normalized and compared with a threshold for triggering the intuitive function for decision-making.

When making decisions, agents use the **satisfying theory**, i.e. they will take the good enough choice rather than the best one. We model the decision-making as a Multi-Attribute Decision-

Making problem [1, 39], which includes a finite discrete set of alternatives which is valued by a finite discrete set of attributes i. A classical evaluation of alternatives leads to the aggregation of all criteria into a unique criterion called value function V of the form:

$$V(\alpha) = w \cdot v(\alpha) = f_{i \in I}(w_i v_i(\alpha)) \quad (6)$$

where α is an action, $V(\alpha)$ is the overall value for action α, w_i is a scaling factor to represent the relative importance of the ith attribute, $v_i(\alpha)$ is a single attribute value with respect to attribute index $i \in I$ and f is the aggregation function. Function f normally is domain dependent, for example, it can be an additive value function for preference independence, a discounted value function when there is reward for different preferences, or a Constant Absolute Risk Aversion function for risk-averse decision-making. The action being finally selected is the first action whose value reaches a predefined desire value D:

$$\varepsilon(\alpha) = \exists \alpha \; s.t. \; V(\alpha) > D \quad (7)$$

Figure 4 shows the two-phase decision-making algorithm.

Figure 4. Two-phase decision-making algorithm

```
decision(KBa, Sa, M)
  inputs: KBa, the knowledge base for agent a
          Sa, the strategy for agent a
          M, the current situation

  //editing phase
    anchors = getAnchor(M);
    topAccessibleMemory = query(KBa, anchors);

  //evaluation phase
    //intuitive decision-making
    if (topAccessibleMemory is enough)
    rebuild individual tree
      return satisfyingDM(topAccessibleMemory, Sa);

    //deliberative decision-making
    else
      wholeMemory = query(KBa, M);
      if (wholeMemory is enough)
          return satisfyingDM(wholeMemory, Sa);
      else
          ask(network, anchors);
```

A SIMULATION FOR CASE AGENTS

We have developed a simulation for the above agent system. Our system is capable of scaling huge simulations, to be capable of being deployed on many machines with the ability to control what is happening in the simulation through the use of a single GUI running on one machine, and it is able to process large amounts of data and perform operations on that data.

Service-Oriented Architecture

The Service-Oriented Architecture (SOA) [8] is a design pattern commonly used in many large corporations, such as NASA. It provides a flexible and stable architecture for large scale software systems. For this reason we felt it would be a good fit for the distributed multi agent system we were developing.

SOA requires that services should be "loosely coupled", in other words encapsulated and have as few dependencies on other services as possible. This allows for services to be easily swapped and their implementation to be easily upgraded. One service could easily be swapped out for another service that accomplishes the same result.

For our system we use a three tier approach as seen in Figure 5, which consists of:

Figure 5. A three tier SOA diagram

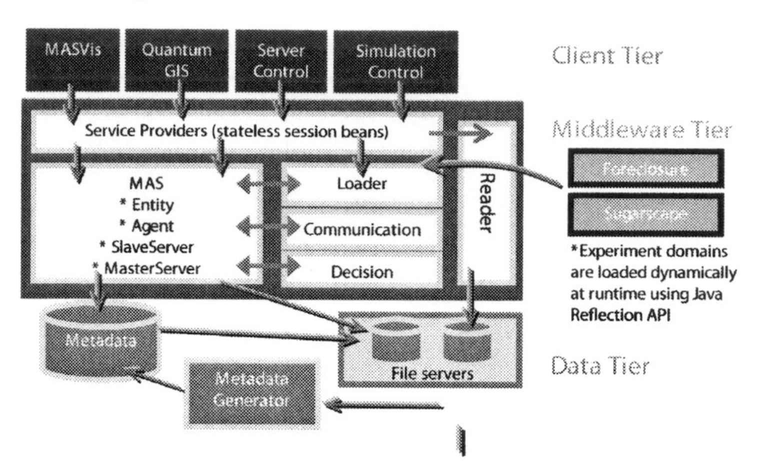

- The client tier
- The middleware tier
- The data tier

The *client tier* consists of the user interface. As defined by the nature of the SOA design pattern, the user interface maintains service encapsulation and could theoretically be used with any core system. All that would need to be ensured was that the core followed the service contract as far as how to communicate with the client tier. The first two services that the interface provides are server control and simulation control. This allows the user to set up their own system distribution and experiment. The other two services the interface provides are MASVis and Quantum GIS. MASVis is a data analysis and visualization package we developed for social simulations. It is built on the fundamental design of having data sources that connect through multiple filters to process elements of data and which can be plotted to display the data. Our simulation also supports GIS. The GIS system we use to display our data is Quantum GIS[10]. Quantum GIS is suitable for our needs because it allows for reprogramming. It is quite light and allows for plug-ins so it is easily expandable. It also requires few resources, being able to run on very little RAM and consume little processing power, which is necessary for our simulations since our simulations use the processor heavily. Further, it is completely compatible on all operating systems and has a very large community of active developers. Both MASVis and Quantum GIS are still works in progress.

The *middleware tier* consists of the core CASE agent system as described earlier. The core system is made up of the collection of master and slave servers as well as the CASE agents that occupy them. The middleware tier will also consist of the service providers that help the middleware tier communicate with both the client tier as well as the data tier. We have developed two experimental domains: Sugarscape and Foreclosure. Sugarscape is a classical test-bed for growing agent-based societies [7]. Foreclosure is a domain we developed for helping social scientists to analyze the nationwide "foreclosure crisis" problem (refer to section 4 for more details).

The *data tier* helps encapsulate the data, and stores it separately from the simulation running in the core system; this could entail having the data tier on a separate server. Not only does this help ensure the stability of the data, it also allows for the use of metadata. In our simulation, the data tier records its data using XML specifications specifying important meta information such as date, author, simulation, etc. This allows for a more robust data system, which can easily be searched using filters, such as range of dates or author.

Using these three tiers will help ensure that our simulation is robust enough to allow for any part of it to be updated with little or no difficulty. Also the modular nature of the SOA allows for other researchers to use only the parts of the system that they find to be appropriate for their own project.

Next, we explain in detail the middleware tier which focuses on the system distribution and the client tier which describes the user interface.

System Distribution

For the system to be effective at dealing with the large number of agents required for social research, it has to be highly distributed both in processing as well as in memory. To achieve this we developed a Master/Slave system, as showed in Figure 6.

The system has a single master server in charge of initializing and synchronizing the other servers (slaves). The master's responsibilities include initializing each step, facilitating agent communication and agent movement from one server to another, and most importantly load balancing all of the servers to ensure optimal performance. The slave is originally initialized with a given bounds and a set of agents. The slave's responsibility is to update all of its agents each step, this is a two step process:

- First it updates all of the agents' knowledge for what they perceived that step
- Secondly it has each of its agents implement its decided action for that step.

The agents are run in the maximum number of threads the server can handle; this ensures optimal performance through parallelization.

To ensure effective load balancing the master keeps track of the rest of the server by using a KD tree. The KD tree works by splitting the bounded region of the simulation on a different axis each time, until there is a single leaf per slave server, as seen in Figure 7. Splitting the region up in this manner allows the master server to go in and shift these splits one way or the other to balance the load between slaves. By shifting a KD split, the master server reduces the amount of the region and the amount of agents that a slave server is responsible for, and by doing so alleviates that server's load. Figure 8 is a screenshot of the server control interface.

Figure 6. Master/Slave relationship

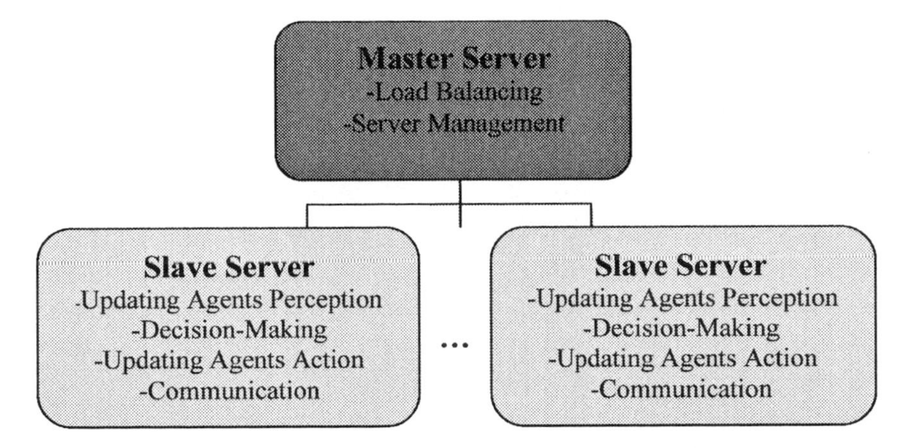

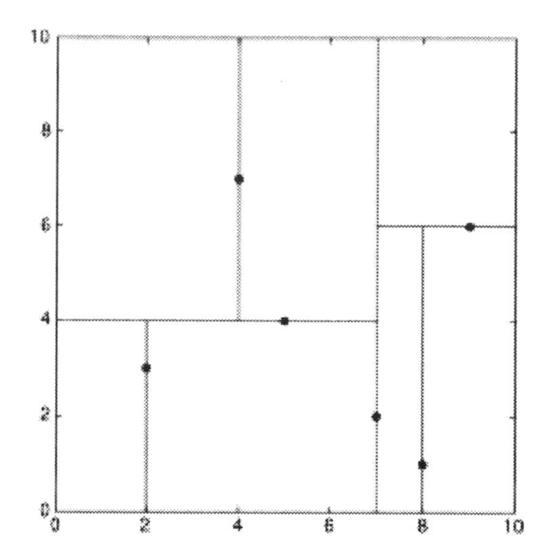

Figure 7. A graphical representation of the KD tree

User Interface

In creating the user interface, the primary consideration is that users will not be concerned with all of the system's features at once. Furthermore, there will not be enough screen space to conveniently and aesthetically portray all of these features, specifically the settings, server, data, graphical, and simulation windows, at once. To house each element of the system's features in detached windows, able to be hidden and then restored whenever needed, is therefore clearly plausible and arguably necessary for such an open-ended agent simulation. This can be argued for each individual element of the system's features.

Once a user calibrates the appropriate settings for his simulation implementation, he may not need to keep these settings in focus throughout the running of the simulation; rather, the user may be more focused on the system's output and the results of the simulation. To keep these settings on the screen for the duration of the simulation would force the user to work with cluttered space, constantly having to move windows around in order to see and collect data. To house all of the simulation settings in a detached window, able to

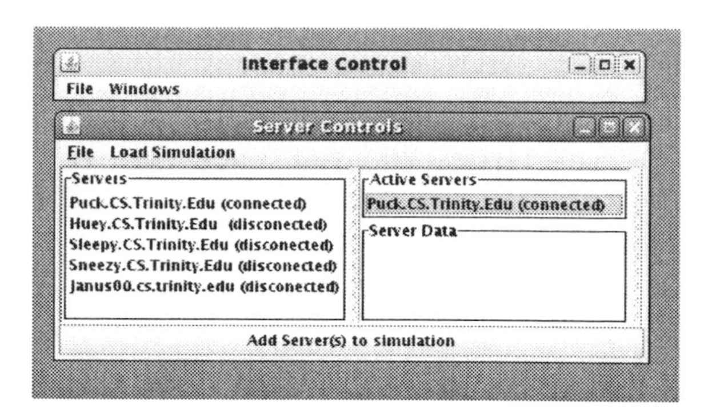

Figure 8. A screenshot of the server control interface

be hidden and then restored whenever needed, is, therefore, clearly plausible and arguably necessary for such a wide-breadth multi-agent simulation.

This same logic can apply to the server window, where server hosts can be added, managed, or booted. Clearly, server management is not the primary focus of any type of simulation. While servers need to be added at the beginning of a simulation and monitored and/or booted throughout the duration of the simulation, the server window need not be opened and visible the entire time. Again, the detached and hide-able window system works effectively in this situation.

This detached-window argument can also be applied to output-based elements of the simulation user interface. Consider when a user is focused on only the raw textual data output of a simulation. Clearly, any visual representation of the data in question, whether it is a graphical representation of this data or a step-by-step visualization of the individual agents in the system, would be extraneous and unneeded in the given situation.

Obviously, this scenario can be flipped into a situation in which the user is only focused on visuals, in which case the raw textual data output would be unneeded. Similar to the case of the settings window, housing both graphical data representations and textual data output in detached, escapable and restorable windows, is a necessary feature.

To keep almost every aspect of the simulation in separate, hide-able windows requires the use of a main menu, or a master controller, that cannot be closed. This controller handles not only the core, system-wide commands for MASVis but also the hide and restore functions for all other windows, while still remaining as small as possible and preserving screen space.

Without a doubt, users of MASVis have an advantage with a user interface system with separate, moveable, hide-able windows. Such a system allows for a maximization of the valuable resource of screen real estate and also puts the system's focus on what currently matters to the user. Figure 9 is a screenshot of the Sugarscape interface.

EXPERIMENT

Multi-agent systems are increasingly used to identify and analyze the social and economic problems of urban areas and provide solutions to these problems [9]. We tested the CASE architecture in the urban dynamics field. We did not use the classical benchmarks such as Santa-Fe Artificial Stock Market (SF-ASM) which is commonly used by the Agent-Based Simulation (ABS) community of social science and economics. This is because such classical testing domains were originally designed for game theory, which uses reactive agents and relies on the concept of equilibrium that is rarely seen in real-world environments, and decision theory, which was grounded on the level of a single individual with a lack of social interests.

Our simulation in urban dynamics focuses on the nationwide "foreclosure crisis" problem. With more than 430,000 foreclosure filings reported nationwide, the nation's rate of foreclosure is at an all time high[11]. This experiment simulated mortgage default in San Antonio, Texas. "San Antonio is dominated by predominately young Hispanic families. In the last decade the city's young and dynamic population has seen median household income grow significantly as childhood poverty declined rapidly. Unemployment remains fairly low, however, due to a lag in higher educational attainment the bulk of the city's households earn only low-to-meddle incomes. Despite this and other factors there was a considerable rise in homeownership as the growing population began to move into neighborhoods in both the metro and suburban areas of the city."[12] By the first quarter 2007, San Antonio's rate of foreclosures is nearly twice the national average and is among the top 20 cities with the highest foreclosure rates in the U.S.[13].

Based on the above observation, we choose to model San Antonio's "foreclosure crisis" as a

Figure 9. A screenshot of the Sugarscape Interface

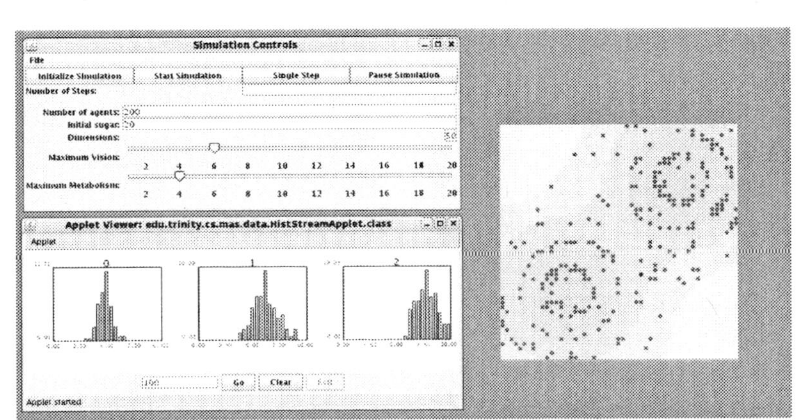

multi-agent system. This chapter reports our initial experiment. We modeled a simple housing market where agents purchase and sell homes. The agents need to carry out three tasks: selecting a home to purchase, obtaining a loan to purchase that home and making monthly payments on that loan. Each agent selects a home that is feasible for it to purchase based on its individual annual income, credit score and current interest rates. On the first time step each agent attempts to purchase a home and continues to do so each time step afterwards until it finds a suitable match. Throughout the simulation the interest rates and additional monthly expenses as signed to each agent are subject to change and no agent is given prior knowledge of the schedule or degree of these changes. Once an agent has purchased and occupied a home for one time step, that agent must begin making payments on the loan it initially took out to purchase that home. If during one time step an agent can not make its monthly mortgage payment (i.e. its annual income is less than the sum of its monthly expenses and mortgage payment) then that agent is forced into a situation of default. In remaining time steps that agent may attempt to purchase another home of equal or lesser value but this is made rather difficult as its credit score has dropped significantly. This is done to reflect the difficulty in reality of obtaining a new home after a foreclosure.

The purpose of this simulation is to examine the relationship between an individual agent's strategy, and the diffusion of those strategies to its society (groups, neighborhood and network), and finally the overall foreclosure rate of the city. Our initial simulation used only 100 homeowners (agents) and 100 homes that were both randomly generated within several carefully selected statistical parameters to reflect demographic realities of the city along the following lines: housing values, homeownership history, FICA (Federal Insurance Contributions Act) scores and household income. We ran the system for 100 steps with each single step simulating one month in reality.

We defined two groups of agents within the system: an *aggressive group* and a *conservative group*. Agents that are members of the aggressive group are more adapt to "over-extend" their credit lines and purchase a home that may or may not be within their budget. While agents who are members of the conservative group are more apt to purchase a home that is well within their budget. Initially each agent is given their own strategy that falls within some statistical range of one of the two groups. This initial strategy defines an agent's propensity to join one of the two groups. As time passes, agents are influenced in a variety of ways: either through members of their own groups, their neighbors who may or may not belong to the same group, and their extended social network. An agent's neighborhood is a direct function of its observability, i.e. the amount of agents it can observe at a given time. An agent's network is not a physical construct, but rather a meta-physical medium constructed from the agents it decides to frequently communicate with. Hence as the size of an agent's neighborhood, social network, and group membership grows its strategy becomes more dominant within society as a whole.

We use three teams as defined in Table 1. Except for the decision rules, conditions of all teams were exactly the same.

We report three experiments. Experiment 1 involves only the Reactive Team. These agents do not utilize any social structure (neighborhoods, groups or networks) and hence have no ability to learn through time or adapt to environmental changes. Figure 10 is the result of the reactive team. It shows that the foreclosure rate is extraordinarily high (~%60) and is rather unstable, moving up and down 15 points between some time steps. These large variations in the foreclosure rate are a product of both the agent's simplistic decision making mechanisms and their inability to diffuse successful strategies to other agents. Indicative of the fact that we do not see this high variable of foreclosure rates in reality is the underlying fact that humans utilize

Table 1. Three teams

Team	Feature	Purpose
Reactive Team	Reactive Agents	The base team
ID Team (Intuition + Deliberation)	Keep every condition in Reactive Team Replace Reactive rule by Intuitive & Deliberative decision rule	Test the effectiveness of pure I&D decision rule
IDS Team (Intuition + Deliberation + Society)	Keep every condition in I&D Team Add the three social structures to it (group, neighborhood and network)	Test the effectiveness of combined social structure and I&D decision rule

a much more complex and sophisticated means for making decisions.

Experiment 2 tests the ID Team. Figure 11 illustrates the results of adding only this intuition and deliberation capability to reactive agents. The upper line shows the number of agents who use intuitive decision-making. At the beginning of the simulation, this number is quite low. But agents are able to quickly learn from the decisions they made before and use them to make new decisions. The low line is the number of foreclosures. It is lower than the Reactive Team in Experiment 1 as the agents are able to recognize recurring instances of potential default and avoid them if possible. However, this adaptability is limited only to an individual agent and there are no mechanisms for one agent to diffuse its successful strategy onto other agents. This limitation is indicated by the presence still of variability in the foreclosure rate. The inability of individual agent leads to decreased stability of the system as a whole.

To further explore this, in experiment 3 we test the IDS Team. We add the three social structures (neighborhood, group and network) on top of the existing individual agent decision-making mechanisms. In doing so we allowed the agents to diffuse their successful strategies through physical (neighborhood) and non-physical (group and social network) space. As Figure 12 indicates, we see the same adaptability present in Experiment 2 as the knowledge base of the individual agents expands, allowing them to better predict future events based on past experiences, in addition to an increased level of stability (measured by less variability in the foreclosure rate) as the successful strategy is diffused through various social structures. Experiments 2 and 3 confirm our hypothesis that human decision-making is "embedded" in a social context.

Figure 10. The result of the reactive team

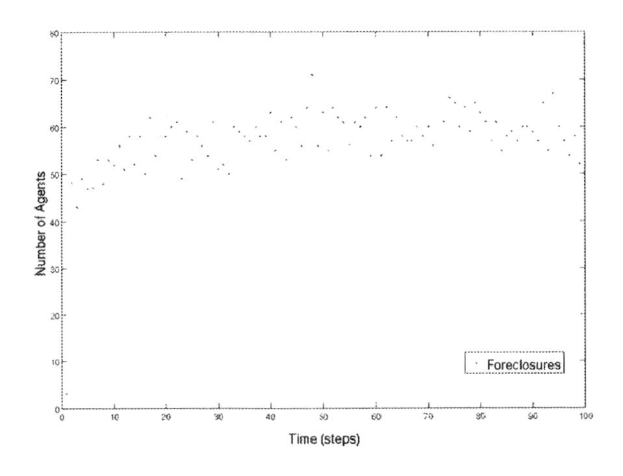

Figure 11. The result of the ID Team

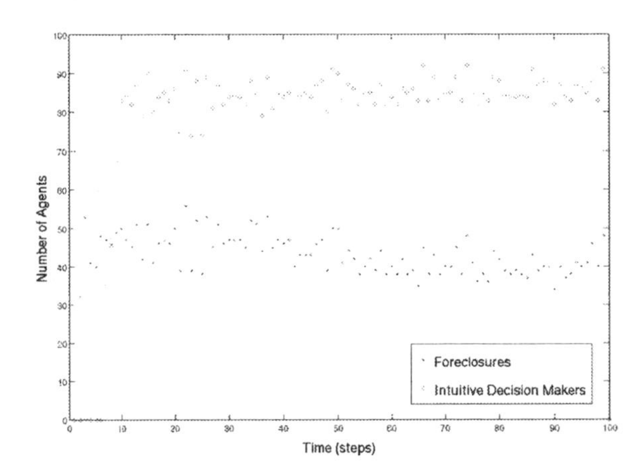

Figure 12. The result of the IDS team

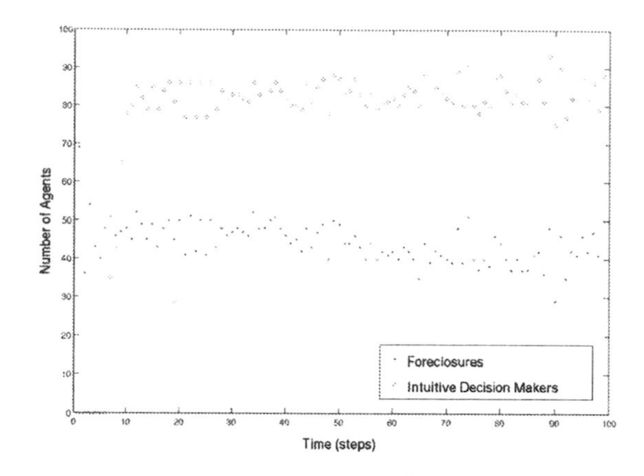

CONCLUSION

In this paper, we have presented CASE, a multi-agent architecture for supporting human social interaction as well as the intuitive and deliberative decision-making process. This approach allows us to observe a wide range of emergent phenomena of complex social systems and analyze their impact. The chapter also reports our first step in developing a robust and flexible multi-agent simulation that can be used by social scientists and economists.

Within CASE, there are many issues awaiting our research team. We are currently expanding CASE in four directions. The first direction is toward agent modeling. Our current agents use a simple memory structure for intuitive decision-making. The problem with this is it is not very adaptive to the dynamic environment. We are currently analyzing different situation assessment and learning technologies and will incorporate them into our agents. Another ongoing project is to theoretically study the emergent patterns in the interplay between the multi-dimensional relations (group, neighborhood and network) in the behavior of agents.

Second, we will add a range of technologies for the simulation. The purpose of the simulation is to provide an integrated environment for social scientists and economists to manage, analyze and visualize their data. Therefore it is important to provide a user friendly interface and a low programming-skill-required environment. We will continue simplifying the task of setting up different testing domains and develop a rich charting package for MASVis that will further simplify the creation of visualizations.

Third, we will continue investigating various data structures and algorithms of load balancing to better distribute CASE agents on clusters of computers or different types of grids. This development will further simplify and speed up the simulation by reducing the computing load and memory load in a single machine.

Fourth, we are also expanding the scale of the foreclosure experiment in order to draw more conclusions to the question regarding the social and economic factors that affect the "foreclosure climate" of a city. The questions that we will investigate include: 1) Where are the foreclosures occurring? 2) Are they clustered or isolated events? 3) Who is being foreclosed upon and do they share common characteristics? 4) What social and economic factors are driving an increase in the foreclosure rate? and 5) Could we predict the likelihood and location of future foreclosures?

REFERENCES

1Ahn, B. S. (2006). Multiattributed Decision Aid with Extended ISMAUT . *IEEE Transactions on Systems, Man, and Cybernetics. Part A, Systems and Humans*, *36*(3), 507–520. doi:10.1109/TSMCA.2005.851346

2Carley, K. M., & Hill, V. (2001). *Structural Change and Learning Within Organizations*, MIT Press/AAAI Press/Live Oak.

3Castelfranchi, C. (2000). Engineering Social Order . *Lecture Notes in Computer Science, 1972*, 1–18. doi:10.1007/3-540-44539-0_1

4Castelfranchi, C. (2001). The Theory of Social Functions: Challenges for Computational Social Science and Multi-Agent Learning . *Cognitive Systems Research, 2*(1), 5–38. doi:10.1016/S1389-0417(01)00013-4

5Coleman, P., Pellon, M., & Zhang, Y. (2007). Towards Human Decision-Making in Multi-Agent Systems, in Proceedings of the *International Conference on Artificial Intelligence*, Monte Carlo Resort, Las Vegas, Nevada.

6Das, S., & Grecu, D. (2000). COGENT: Cognitive Agent to Amplify Human Perception and Cognition. In *Proceedings of 4th International Conference on Autonomous Agents*, Barcelona, Spain, pp. 443-450.

7Epstein, J., & Axtell, R. (1996). *Growing Artificial Societies: Social Science from the Bottom Up*, The MIT Press.

8Erl, T. (2005). Service-Oriented Architecture:Concepts, Technology, and Design, Upper Saddle River: Prentice Hall PTR.

9Ferber, J. (1999). *Multi-Agent System: An Introduction to Distributed Artificial Intelligence*, Harlow: Addison Wesley Longman.

10Gmytrasiewicz, P. J., & Noh, S. (2002). Implementing a Decision-Theoretic Approach to Game Theory for Socially Competent Agents, in Parsons, S., Gmytrasiewicz, P., and Wooldridge, M. (eds.), *Game Theory and Decision Theory in Agent-Based Systems*, pp. 97-118, Kluwer Academic Publishers.

11Goldstone, R., Jones, A., & Roberts, M. E. (2006). Group Path Formation . *IEEE Transactions on Systems, Man, and Cybernetics. Part A, Systems and Humans, 36*(3), 611–620. doi:10.1109/TSMCA.2005.855779

12Goncalves, B., & Esteves, S. (2006). Cognitive Agents Based Simulation for Decision Regarding Human Team Composition. In *Proceedings of 5th International Conference on Autonomous Agents and Multi-Agent Systems* (AAMAS'06), pp. 34-41.

13Granovetter, M. (1985). Economic Action and Social Structure: the Problem of Embeddedness . *American Journal of Sociology, 91*(3), 481–510. doi:10.1086/228311

14Hales, D., & Edmonds, B. (2003). Evolving Social Rationality for MAS using "Tags". In *Proceedings of 5th International Conference on Autonomous Agents and Multi-Agent Systems* (AAMAS'03), pp. 497-503.

15Hogg, L. & Jennings, N. R. (2001). Socially Intelligent Reasoning for Autonomous Agents, *IEEE Transactions on Systems, Man and Cybernetics - Par t A*, 31(5):381-399.

16Hoggendoorn, M. (2007). Adaptation of Organizational Models for Multi-Agent Systems Based on Max Flow Networks . *IJCAI, 07*, 1321–1326.

17Jiang, Y., & Ishida, T. (2007). A Model for Collective Strategy Diffusion in Agent Social Law Evolution . *IJCAI, 07*, 1353–1358.

19Kahneman, D. (2002). Maps of Bounded Rationality: A Perspective on Intuitive Judgment and Choice, *Les Prix Nobel.*

18Kahneman, D., & Tversky, A. (1979). Prospect Theory: An Analysis of Decision under Risk . *Econometrica, 47*(2), 263–292. doi:10.2307/1914185

20Kant, J., & Thiriot, S. (2006). Modeling One Human Secision Maker with A Multi-agent System: The CODAGE Approach. In *Proceedings of 5th International Conference on Autonomous Agents and Multi-Agent Systems* (AAMAS'06), pp. 50-57.

21Klein, G. (1993). A Recognition-Primed Decision Making Model of Rapid Decision Making, in Klien, G., Orasanu, J., Calderwood, R. and Zsambok, C. (eds.), *Decision Making In Action: Models and Methods*, pp. 138-147.

22Luce, R. D., & Raiffa, H. (1957). *Games and Decisions*, New York, Wiley.

23March, J. (1994). *A Primer on Decision Making: How Decisions Happen*, Free Press, New York.

24Moon, I. I.-C., & Carley, K. M. (2007). Self-Organizing Social and Spatial Networks under What-If Scenarios. In *Proceedings of 6th International Conference on Autonomous Agents and Multi-Agent Systems* (AAMAS'07), pp. 1348-1355.

25Pan, X., Han, C. S., Dauber, K., & Law, K. H. (2005). A Multi-agent Based Framework for Simulating Human and Social Behaviors during Emergency Evacuations, *Social Intelligence Design*, Stanford University.

26Portes, A., & Sensenbrenner, J. (1993). Embeddedness and Immigration: Notes on the Social Determinants of Economic Action . *American Journal of Sociology*, *98*(6), 1320–1350. doi:10.1086/230191

27Pynadath, D. V., & Marsella, S. C. (2005). PsychSim: Modeling Theory of Mind with Decision-Theoretic Agents . *IJCAI*, *05*, 1181–1186.

28Rasmussen, J. (1986). *Information Processing and Human Machine Interaction: An Approach to Cognitive Engineering*, New York, North Holland.

29Simon, H. (1957). *A Behavioral Model of Rational Choice, in Models of Man, Social and Rational: Mathematical Essays on Rational Human Behavior in a Social Setting*, New York: Wiley.

31Stanovich, K. E., & West, R. F. (1999). Discrepancies between Normative and Descriptive Models of Decision Making and the Understanding/Acceptance Principle . *Cognitive Psychology*, *38*, 349–385. doi:10.1006/cogp.1998.0700

30Stirling, W. C. (2003). *Satisficing Games and Decision Making: with Applications to Engineering and Computer Science*, Cambridge University Press.

32Stirling, W. C. (2005). Social Utility Functions -part I: Theory . *IEEE Transactions on Systems, Man and Cybernetics. Part C, Applications and Reviews*, *35*(4), 522–532. doi:10.1109/TSMCC.2004.843198

33Stirling, W. C., & Frost, R. L. (2005). Social Utility Functions-part II: Applications . *IEEE Transactions on Systems, Man and Cybernetics. Part C, Applications and Reviews*, *35*(4), 533–543. doi:10.1109/TSMCC.2004.843200

34Tversky, A., & Kahneman, D. (1992). Advances in Prospect Theory: Cumulative Representation of Uncertainty . *Journal of Risk and Uncertainty*, *5*, 297–323. doi:10.1007/BF00122574

35Vazquez-Salceda, J., Dignum, V., & Dignum, F. (2005). Organizing Multiagent Systems . *Autonomous Agents and Multi-Agent Systems*, *11*(3), 307–360. doi:10.1007/s10458-005-1673-9

36von Neumann, J., & Morgenstern, O. (1947). *Theory of Games and Economic Behavior*, Princeton University Press, second edition.

37Weber, E. U., & Coskunoglu, O. (1990). Descriptive and Prescriptive Models of Decision-Making: Implications for the Development of Decision Aids. *IEEE Transactions on Systems, Man, and Cybernetics*, *20*(2), 310–317. doi:10.1109/21.52542

38Yoon, P. K., & Hwang, C. (1995). *Multiple Attribute Decision Making: An Introduction*, Sage Publications.

42Zhang, Y. Mark Lewis, Pellon, M. & Coleman, P. (2007). A Preliminary Research on Modeling Cognitive Agents for Social Environments in Multi-Agent Systems. *AAAI 2007 Fall Symposium, Emergent Agents and Socialities: Social and Organizational Aspects of Intelligence*, pp. 116-123.

39Zhang, Y., Ioerger, T. R., & Volz, R. A. (2005). Decision-Theoretic Proactive Communication in Multi-Agent Teamwork, in Proceedings of the *IEEE International Conference on Systems, Man and Cybernetics* (SMC'05), Hawaii, pp. 3903-3908.

41Zhang, Y., Pellon, M., & Coleman, P. (2007). Decision Under Risk in Multi-Agent Systems. In Proceedings of the *International Conference on System of Systems Engineering*, San Antonio, Texas, pp. 133-138.

40Zhang, Y., & Volz, R. A. (2005). Modeling Utility for Decision-theoretic Proactive Communication in Agent Team. In Proceedings of the 9th *World Multi-Conference on Systemics, Cybernetics and Informatics*, pp. 266-270, Orlando, FL, July 11-13.

ENDNOTES

1. repast.sourceforge.net
2. cs.gmu.edu/~eclab/projects/mason
3. jade.tilab.com
4. www.cougaar.org
5. http://jaslibrary.sourceforge.net/
6. http://www.swarm.org/
7. http://ecolab.sourceforge.net/
8. http://www.spiderland.org/
9. Conversely, this does not mean that it is impossible or useless to represent societies as entities in their own right. We can of course design a society in the form of an agent, and thus consider MASs as packages of agents and societies, like what has been done in [35].
10. http://www.qgis.org
11. Data provided by Realtytrac. Data Accessed May, 2007.
12. "San Antonio In Focus: A Profile from the Census 2000 " ©2003 Brookings Center on Urban and Metropolitan Policy.
13. Data provided by Realtytrac. Data Accessed May, 2007.

This work was previously published in Handbook of Research on Agent-Based Societies: Social and Cultural Interactions, edited by G. Trajkovski; S. Collins, pp. 104-124, copyright 2009 by Information Science Reference (an imprint of IGI Global).

Chapter 8.7
Embedding an Ecology Notion in the Social Production of Urban Space

Helen Klaebe
Queensland University of Technology, Australia

Barbara Adkins
Queensland University of Technology, Australia

Marcus Foth
Queensland University of Technology, Australia

Greg Hearn
Queensland University of Technology, Australia

ABSTRACT

This chapter defines, explores and Illustrates research at the intersection of people, place and technology in cities. First, we theorise the notion of ecology in the social production of space to continue our response to the quest of making sense of an environment characterised by different stakeholders and actors as well as technical, social and discursive elements that operate across dynamic time and space constraints. Second, we describe and rationalise our research approach, which is designed to illuminate the processes at play in the social production of space from three different perspectives. We illustrate the application of our model in a discussion of a case study of community networking and community engagement in an Australian urban renewal site. Three specific interventions that are loosely positioned at the exchange of each perspective are then discussed in detail, namely: Sharing Stories; Social Patchwork and History Lines; and City Flocks.

DOI: 10.4018/978-1-60566-152-0.ch012

INTRODUCTION

The intersection of urban and new media studies is a dynamic field of practice and research. There are a number of reasons why this is so. Technically these are both highly innovative domains, and the rate of change is significant and challenging. Urban life and media platforms are both in the midst of paradigm shifts. Theoretically, both fields can be understood as sites of signification and structuration of the social field—and because they both evidence such change they are potent laboratories for advancing understanding. The pragmatic corollary is that policy makers and corporate investors are also highly engaged in the intersection.

Apart from the complexity of maneuvering through the often differing agendas of researchers and practitioners and of private and public sector agencies that operate at this intersection, the objective of advancing understanding is also challenged by a plethora of different and sometimes differing theories. Yet, universally useful contributions to knowledge can be achieved if urban cultural studies, urban sociology, urban technology and human-computer interaction, urban architecture and planning, etc., overcome language and conceptual barriers. A cross-disciplinary approach requires effort to create models which help to overcome phenomenologically isolated attempts at explaining the city. Such models would ideally be cross-fertilised by the findings and insights of each party in order to recognise and play tribute to the interdependencies of people, place and technology in urban environments. We propose the notion of ecology (Hearn & Foth, 2007) as a foundation to develop a model depicting the processes that occur at the intersection of the city and new media.

In the context of the field of urban planning and development, the promise of digital content and new media has been seen as potentially serving new urbanist visions of developing and supporting social relationships that contribute to the sustainability of communities. As Carroll et al. (2007) have argued, recent critiques of assumptions underpinning this vision have pointed to the following outcomes as 'most in demand', and simultaneously most difficult to deliver:

- Community (Anderson, 2006; DeFilippis, Fisher, & Shragge, 2006; Delanty, 2000; Gleeson, 2004; Willson, 2006);
- Diversity (Talen, 2006; Wood & Landry, 2007);
- Participation (Hanzl, 2007; Sanoff, 2005; Stern & Dillman, 2006);
- Sustainability (Gleeson, Darbas, & Lawson, 2004; Van den Dobbelsteen & de Wilde, 2004);
- Identity (Al-Hathloul & Aslam Mughal, 1999; Oktay, 2002; Teo & Huang, 1996);
- Culture and History (Antrop, 2004; Burgess, Foth, & Klaebe, 2006; Klaebe, Foth, Burgess, & Bilandzic, 2007).

It is critical that the emergence of urban informatics as a multidisciplinary research cluster is founded on a theoretical and methodological framework capable of interrogating all these relationships and the assumptions that currently underpin them. As Sterne has warned in relation to research pertaining to the field of technology more generally,

> *the force of the 'preconstructed'—as Pierre Bourdieu has called it—weighs heavily upon anyone who chooses to study technology, since the choice of a technological object of study is already itself shaped by a socially organized field of choices. There are many forces in place that encourage us to ask certain questions of technologies, to define technology in certain ways to the exclusion of others, and to accept the terms of public debate as the basis for our research programs. (Sterne, 2003, p. 368)*

In this respect, if we are to promote an analytical focus on the capacities and possibilities of digital content and new media to meet the challenges of community, participation, sustainability, identity and so on, it is important to employ frameworks that permit systematic study of these relationships. We agree with Grabher (2004) who points to analytical advantages in resisting assumptions around passive adaptation to environments, and permitting a focus on networks, intricate interdependencies, temporary and permanent relationships and diverse loyalties and logics. We propose that these qualities are best integrated in an ecological model which we will develop in this chapter.

This chapter defines, explores and exemplifies our work at the intersection of people, place and technology in cities. First we introduce the notion of ecology in the social production of space to continue our response to the challenge of making sense of an environment characterised by different stakeholders and actors as well as technical, social and discursive elements that operate across dynamic time and space constraints (Foth & Adkins, 2006). Second we describe and rationalise our research approach which is designed to illuminate the processes at play from three different perspectives on the social production of space. We illustrate the application of our model in a discussion of a case study of community networking and community engagement in an Australian urban renewal site—the Kelvin Grove Urban Village (KGUV). Three specific interventions that are loosely positioned at the exchange of each perspective are then discussed in detail, namely: *Sharing Stories*; *Social Patchwork* and *History Lines*; and *City Flocks*.

THEORETICAL FRAMEWORK

Our theorisation is conscious of the extent to which technical, social and discursive elements of urban interaction work across (1) online and offline communication modalities; (2) local and global contexts; and (3) collective and networked interaction paradigms (Foth & Hearn, 2007; Hearn & Foth, 2007). The distinction between online and offline modes of communication—and thus, online and offline communities—is blurring. Social networks generated and maintained with the help of ICTs move seamlessly between online and offline modes (Foth & Hearn, 2007; Mesch & Levanon, 2003). Additionally, studies of Internet use and everyday life have found that the modes of communication afforded by Internet applications are being integrated into a mix of online and offline communication strategies used to maintain social networks (Wellman & Haythornthwaite, 2002).

Urban communicative ecologies operate within a global context increasingly dominated by Web 2.0 services (eg., search engines, instant messenger networks, auction sites and social networking systems). The notion of 'glocalization'—introduced by Robertson (1995) and later re-applied by Wellman (2001; 2002)—is useful here because it emphasises the need to develop locally meaningful ways of using this global service infrastructure rather than trying to compete with existing global sites and content. Studies have highlighted a range of opportunities for the development of local (and location-aware) services as well as locally produced and consumed content (Boase, Horrigan, Wellman, & Rainie, 2006; Burgess et al., 2006).

The similarly increasing ease with which people move between collective and networked community interaction paradigms hints at opportunities for new media services that can accommodate both kinds of interaction and afford the user a smooth transition between the two. Collective interactions ('community activism') relate to discussions about place; for example, community events, street rejuvenation initiatives and body corporate affairs. Networked community interactions ('social networking') relate to place-based sociability and features that for example, seek to raise awareness of who lives in the neighbourhood, provide opportunities for residents to find out about each other, and initiate contact.

In the introduction we pointed to the complexities of the relationships that need to be captured in understanding the role of information and communication technologies in the vision underpinning contemporary urban villages. Specifically it responds to recent calls for a more nuanced understanding of the patterns of relationships underpinning their uses in urban contexts. As Crang et al. (2006) observe, research in this area requires a more specific focus on the everyday uses of ICTs and the interactions of multiple technologies in everyday practices. In this respect, we argue that an ecological model enables a conceptualisation that opens up the possibility of diverse adaptations to a specific environment, and the different logics, practices and interdependencies involved. However, Crang is equally emphatic about the salience of social difference in the use of ICTs in urban contexts, pointing to research that underlines the episodic and instrumental use of them in contexts of deprivation and pervasive use in wealthier households (Crang et al., 2006). In this respect, we must provide for the possibility that different ecological configurations are related to the positioning of social agents in the field of urban life.

What is required, then, is a model that can capture ecological relationships in the context of the production of social difference in urban contexts. We extend previous conceptual work on ecological models (Adkins, Foth, Summerville, & Higgs, 2007; Dvir & Pasher, 2004; Foth & Adkins, 2006; Foth & Hearn, 2007; Hearn & Foth, 2007) by exploring Lefebvre's (1991) model in *The Production of Space*. It provides some conceptual tools to locate these ecologies as occurring in the context of different levels of space production. A central differentiation in his model is between 'representations of space' and 'representational spaces'. Representations of space refers to, "conceptualised space: the space of scientists, planners, urbanists, technocratic subdividers and social engineers [...] all of whom identify what is lived and what is perceived with what is conceived". Representational spaces on the other hand refers to, "space as directly lived through its associated images and symbols, and hence the space of the 'inhabitants' and 'users'" (Lefebvre, 1991, p. 38). In terms of this conceptual distinction the object of knowledge in the study of urban space,

> *is precisely the fragmented and uncertain connection between elaborated representations of space on the one hand and representational spaces [...] on the other; and this object implies (and explains) a subject—that subject in whom lived, perceived and conceived (known) come together in a spatial practice. (Lefebvre, 1991, p. 230)*

The model thus provides a context in which urban experience can be understood in terms of differential levels and kinds of power and constraint in the production of space and the relationships between them. Inequities can then be understood as involving different configurations of these relationships. Ecologies of ICT use in this framework are then understood in terms of the adaptations and interdependencies occurring in specific contexts underpinned by relationships of spatial use.

It is now possible to locate the vision of contemporary master-planning projects pertaining to, for example, 'community', 'identity' or 'culture and history', as produced at the level of the way space is conceived through planning and development practices, at the level of the way space is perceived through marketing material and spatial practice, and at the level of everyday experiences of residents as well as in the interrelationships between them (see Figure 1). This provides a framework in which the meaning of these visions can vary across different kinds of residents and can conflict with the conceived meanings of planners and developers. Studying ecologies of the use of new media and ICT from this perspective promises to illuminate the conceptual connections between these technologies and the field of urban life.

In the following part of the chapter, we discuss three initiatives that combine research and community development goals. They are positioned to facilitate a conceptual exchange between the three ecological domains of conceived, perceived and lived space, that is, *Sharing Stories*, *History Lines* and *City Flocks*. However, before we start, we introduce the emerging space of our study.

Figure 1. Embedding an ecology notion in Lefebvre's triad

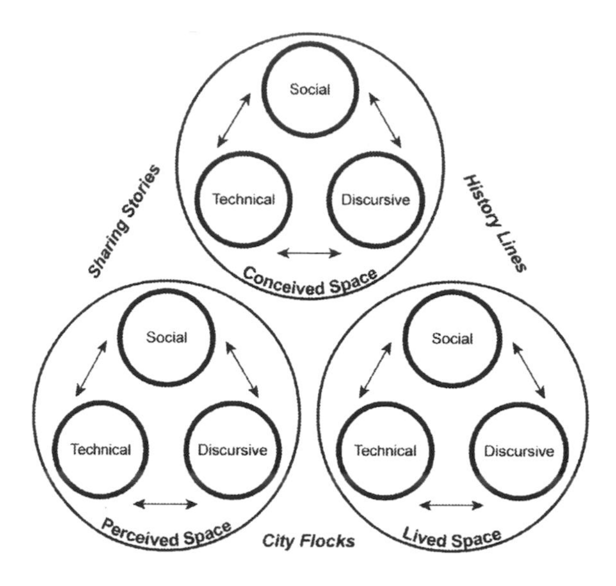

THE URBAN VILLAGE

The Kelvin Grove Urban Village (KGUV) is the Queensland Government's flagship urban renewal project. Through its Department of Housing, and in partnership with Queensland University of Technology, this 16 hectare master-planned community (see Figure 2) seeks to demonstrate best practice in sustainable, mixed-use urban development. By 'linking learning with enterprise and creative industry with community', the KGUV is designed to evolve as a diverse city fringe neighbourhood. Situated 2 km from Brisbane's CBD, it is based on a traditional village design, with a town centre and shops on the main streets. Since planning for the Village started in 2000 and construction started in 2002, AUD 1 billion had already been committed to deliver a heterogeneous design that brings together infrastructure with educational, cultural, residential, health, retail, recreational and business facilities within one precinct.

The following numbers and statistics illustrate the progress and development trajectory of the KGUV:

- When completed, there will be over 8,000 sqm (GFA) of retail space and in excess of 82,000 sqm (GFA) of commercial space located throughout KGUV.
- In 2007, there were 375 residential units (including 7 townhouses and 155 affordable housing units) in the KGUV. This is anticipated to exceed 1,000 two-bedroom equivalent

Figure 2. Aerial view courtesy of the Kelvin Grove Urban Village development group

units once the Village is complete (including student and senior accommodation).

- In 2007, there were 10,800 students and 1,800 staff based at the Kelvin Grove campus of QUT, and a total of 1,663 students and approx. 150 staff at Kelvin Grove State College.
- In 2006, 22,000 people attended exhibitions and performances at QUT's Creative Industries Precinct. Additionally, the KGUV-based theatre company has presented 137 performances to 40,000 patrons, plus there were various other events and productions in 2007.

Technical connectivity is established by a 'triple play' (ie., phone, TV, data) fibre-to-the-home and fibre-to-the-node network operated by a carrier within the KGUV. The services can include low or nil cost large bandwidths (for example, Internet Protocol at 100 Mbits/s) within and between points in the KGUV, fibre or wireless network access and quality of service management for multimedia over Internet Protocol. Internet and world wide web access are at commercial broadband speeds and prices. Wireless hotspots allow users to access the Internet in parks, cafés and other locations around the Village. The implementation of the AUD 700,000 infrastructure investment started in 2005. These pipes, wires, ducts and antennas provide the technical connectivity, yet the majority of the infrastructure and certainly the social effect is invisible or unnoticeable. The communication strategies and policies in the KGUV master plan call for ideas and strategies to enable, foster and showcase the social benefits of this infrastructure 'beyond access' (Foth & Podkalicka, 2007).

Our diverse research interests are positioned under the collective umbrella of *New Media in the Urban Village*. The Department of Housing acknowledges that the strategic design of the built environment and access to the ICT infrastructure are necessary but not sufficient neither to ensure 'effective use' (Gurstein, 2003) nor social sustainability. Therefore the master plan calls for the research and development of appropriate interventions, measures and systems which can provide mechanisms to help link the people and businesses that 'live, learn, work and play' at the KGUV, including residents of the KGUV and nearby areas (including affordable housing residents, seniors and students); university staff and students living or studying in the KGUV and nearby areas; businesses and their customers; and visitors. Our suite of research projects are aimed at responding to this call. We now introduce some of these projects, how they respond to the objectives of the KGUV master plan and how they are guided by and feed back into our theoretical underpinnings.

NEW MEDIA IN THE URBAN VILLAGE

The three initiatives discussed hereafter form part of a larger research project that examines the role of new media and ICT in place making efforts to ensure social sustainability of a master-planned urban renewal site. Apart from the projects presented in the following section, a number of affiliated studies are also underway, for example, an international exchange of experiences studying urban social networks (Foth, Gonzalez, & Taylor, 2006), an exploration of health communication to understand the link between the design of the built environment and residents' well-being (Carroll et al., 2007), and a design intervention to display visual evidence of connectivity (Young, Foth, & Matthes, 2007). New research trajectories are about to start examining ways to use new media to digitally augment social networks of urban residents (Foth, 2006) and the role of narrative and digital storytelling to inform urban planning (Foth, Hearn, & Klaebe, 2007).

Sharing Stories

What about just the ordinariness of everyday life not being thought of as important? People move on and then no one is there to take ownership for it. So there are no custodians of the evidence. How can we change that? (Gibson, 2005)

Sharing Stories (Klaebe & Foth, 2007) exemplifies how traditional and new media can work effectively together and in fact complement each other to broaden community inclusion. This multi-layered public history became a research vehicle used to expand the social interaction far beyond that of a community-based history project alone, so as to include the possibilities of global networks using new media applications. Leveraging opportunity, while negotiating and embracing a multidisciplinary new media approach proved rewarding for participants and local residents.

Kelvin Grove has always been a gathering point. While never densely populated, the land was once a meeting place for Indigenous clans, and in the last century, significant military and educational institutions were located there. In 1998 with the closure of the military barracks, the land was purchased by the Queensland Department of Housing. Together with QUT, planning began to transform the site into a creative urban village, lucratively located only two kilometres from the CBD that would include a mix of commercial, retail, university and residential land use. A triple bottom line approach was embraced as core to the master plan—incorporating economic, environmental and social sustainability.

Genuine creativity involves the capacity to think problems afresh or from first principles; to discover common threads amidst the seemingly chaotic and disparate; to experiment; to dare to be original; the capacity to rewrite rules; to visualise future scenarios; and perhaps most importantly, to work at the edge of one competences rather than the centre of them. (Landry, 2001)

The *Sharing Stories* history project was designed as a social engagement strategy that would become a longitudinal component of the development site from the outset. Its purpose was to collate the history of the site itself, to give a reference point and a context to future residents, to capture 'history in the making' as the ethos of the development was cutting edge. Furthermore, it was also essential to embrace the communities adjacent to the site, so they too were taken on the journey as their physical surroundings were to change so dramatically.

New technologies in communication are altering both the form and the content possible for historical discourse, with the processes of transmission arguably becoming less conventionally text-based, and instead offering visual and progressively more individuated options. Increasingly, visual life-story alternatives are being explored. *Sharing Stories* is predominantly an exercise in augmenting a participatory public history. This type of project and online visual display of the content created is still a relatively new approach to local community history and even more groundbreaking as a social engagement strategy to be incorporated in urban development. *Sharing Stories* engaged with the community using 'on the ground' interaction, seeking to be as inclusive as possible with its approach to participatory involvement. By accessing existing local social networks (schools, clubs, alumni groups, etc.) the profile of the project was raised and then promoted through public broadcasting coverage (local television and newspaper stories throughout the three year period)—all of which contributed to a ripple effect that continually drew more interest and contributions to the project as a whole.

Digital storytelling (Burgess, 2006; Klaebe et al., 2007; Lambert, 2002) was an innovative, alternative way of using new media to engage community in the *Sharing Stories* project. In the context of this project, digital storytelling is a combination of a personally narrated piece of writing (audio track), photographic images and sometimes music (or other such aural ambience) unified to produce a 2-3 minute autobiographical micro-documentary film. Traditionally, digital stories are produced in intensive workshops and this was the strategy employed for this project in 2004. Commonly, the thinking behind advocating this type of approach to create digital stories is to allow participants without access to new media the opportunity of using the technology in a hands-on way, so as to become part of the production team that produces their aesthetically coherent, broadcast quality story to a wider, public audience.

A one-to-many strategy included staging regular public events, so as to take the community on the journey of change in their locale, as the urban development process commenced. The primary aim of the *Sharing Stories* project was to capture the history and so a website was created to house content, as an evolving 'living archive' that would be accessible to interested parties locally and beyond. Fragments of the oral history and photographic collection, together with short story narratives produced from the historical research that was concurrently underway was seen as a strategic approach to keeping the community interested, informed and involved over the subsequent years of the development. Public events could be promoted online and afterwards, portions could be archived on the website. Public events held included: professional and local school visual art exhibitions, digital storytelling screenings and photographic exhibitions.

Throughout the *Sharing Stories* project it was noted that both 'real' and 'virtual' contact and exposure were critical in producing an effective participatory public history. The website stimulated interest in the public participating in the project and attendance at public exhibitions and similarly, the locally produced 'on the ground' activities created content and interest in the online representation. The project is thus an example of an initiative which combines opportunities for research data collection and analysis with community development outcomes. It was funded at the planning and development stages as a vehicle to inform the representations of space whilst preserving the history and heritage of KGUV. The local and historical knowledge that *Sharing Stories* produced has informed the marketing and public relations material and technical documentation of KGUV, but has also found its way into tangible representations of space in the form of for example, plaques embedded in the foot paths and other signage with historic anecdotes and citations.

History Lines

With the Web, you can find out what other people mean. You can find out where they are coming from. The Web can help people understand each other. (Berners-Lee, 2000)

An urban development does not occur overnight. A locale undergoing reformation rapidly changes shape. While the *Sharing Stories* phase concluded in December 2006, a new phase of rejuvenation is already beginning to occur. It was only in the latter months of 2006 for instance that new residents began to reside in the Kelvin Grove Urban Village. Thousands more are expected over the next few years, all of which will be living in apartment style accommodation. Research around social sustainability in urban developments turned its attention to capturing the migrational churn of the people coming into the neighbourhood. What are their stories? Where are they from and why are they moving to an inner-city location? And more importantly for some researchers, how can new media applications be used to engage the incoming population, so that the locale becomes *their* story and *their* emerging history?

As the new population grows, the Kelvin Grove Urban Village attracts researchers who are grappling with these issues. Their backgrounds diversely combine to possess sociology, anthropology, history, education, health, IT, media/communication, and cultural studies. These researchers work together informally to share ideas and data, as well as to conduct joint workshops and focus groups, reducing 'research fatigue' of new residents. Applications developed by some of these researchers (Klaebe et al., 2007) in the 'Social Patchwork' project, include *History Lines* and *City Flocks* as two examples which we discuss here.

The *Social Patchwork* project involves translating narratives into formats that are Web 2.0 amenable. Researchers aim to develop applications that are useful in the urban development, as well as build historically orientated prototypes that encourage socially sustainable community engagement, to build and strengthen local communities and identities within community. Narrative based applications for example that can support intercreativity, as opposed to conventional interactivity (Meikle, 2002). New media approaches, guided by interpretive narratives, can utilise shared networks, shared interests and can be linked to measure the migrational churn of the suburb.

History Lines brings a cross section of new residents together in an activity using narratives, digital maps and location markers pinpointing where they have lived during the course of their life (see Figure 3). Participants can map their life journey thus far by recording personal narratives of place, while also narrating their relationship with past communities. When stories and locations are collated together, overlapping and common lines of location emerge. The material can be used anonymously as content for an exhibition, as well as a link on a neighbourhood portal to stimulate interest in community networking.

Weyea & Geith, in chapter IV of this book, call for research to 'identify effective information tools to enable citizens to shape what their communities look like' and 'use community data, locative media and social software to enable effective local action'. Our objective with *History Lines* aligns with this research agenda in that it can be an aid in measuring the migrational churn of participating residents and become a tool for urban planners to give them a better understanding of why and where people have lived throughout their lives. Our experimental design is positioned to feed experiences of the lived space into the planning and development process of the conceived space. We think it can deliver a better understanding of

Figure 3. History lines

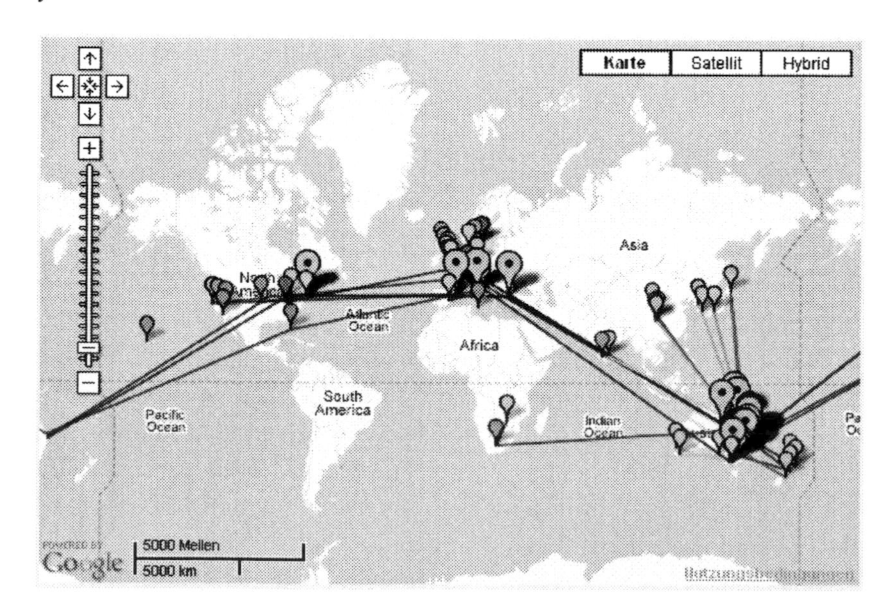

the role of narrative-based new media innovations in support of a more participatory urban planning process (Foth, Hearn et al., 2007; Foth, Odendaal, & Hearn, 2007).

City Flocks

City Flocks—developed by Mark Bilandzic (Bilandzic, Foth, & De Luca, 2008)—is a mobile information service for public urban places. The system is managed by local urban residents and is designed to tap into their tacit social knowledge. *City Flocks* (see Figure 4) allows participants to operate their mobile phones or computers to leave and access virtual recommendations about community facilities, making them easily accessible for other people employing user-generated tags. Residents can also voice-link to other residents, as participants can choose to nominate the mode of contact and expertise they are willing to share. *City Flocks* also allows users to plot their life's journey, but is primarily a rich resource for travellers and/or new residents to an urban location.

Both *History Lines* and *City Flocks* share networks and these shared interests can be linked using folksonomy tagging as opposed to traditional taxonomy directories (Beer & Burrows, 2007) so that they can be utilised by participants both globally and locally. Each application can also be used to encourage connections in the 'real' world. For instance *City Flocks* encourages users to contact fellow residents who have local knowledge of their neighbourhood and are happy to be contacted by newcomers either by email, SMS or by telephoning in order to gain first-hand tacit advice 'in person'. Groups that share 'history', for instance participants who have lived in Sydney and now live at Kelvin Grove, can find and contact each other to meet socially at a local café, thus using virtual connections creatively to meet in the 'real' world. This application is similar to many other social networking applications including *peuplade.fr*, *placebase.com*, *communitywalk.com*, *theorganiccity.com*, *urbancurators.com*

Figure 4. City flocks screen

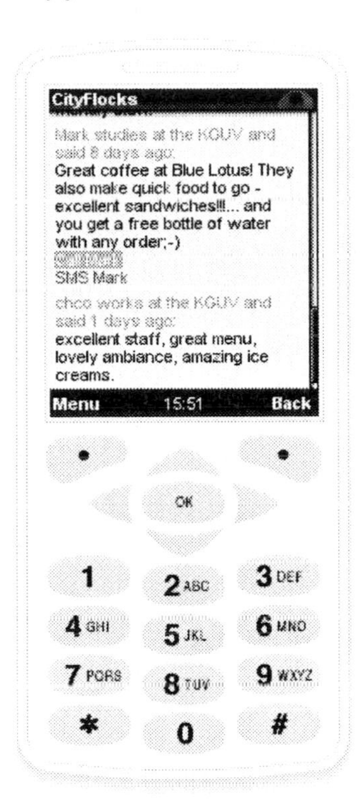

and *nearbie.com*. However, *City Flock*'s focus is to broaden the scope of the interface to include mobile devices and encourage users to interact directly with each other, rather than mediated via a website. Within the conceptual framework of the social production of space, we see *City Flocks* as a tool to link and balance the spatial practices in the perceived space of KGUV with the actual experiences of the lived space.

CONCLUSION

While both *History Lines* and *City Flocks* are in early stages of development, feedback from focus groups, urban developers and social planners has been encouraging. Whether this kind of online narrative-based testimony will be historically significant in the future is unclear; what is clear however is the fresh inter-creative way in which participants can freely leave their virtual 'mark' on a geographical location. *Sharing Stories* represents an exercise in interactivity, both on and offline. The content created offline could be later used and reflected upon in an online environment. *City Flocks* and *History Lines* comparatively demonstrate online and offline intercreativity, but in these examples the offline connection experience can augment more local, personal or face-to-face interaction. Communicative ecologies are both the enabler and outcome of the capacity to share interests, to 'find each other' and continue a virtual connection in reality. The use of these applications represents new configurations of online and offline relationships enabled by ICT that enhance the capacity of residents to appropriate represented space through their own representational practices. Together, however, they raise the question of the extent to which different kinds of users are able to assert their role in representational space and assert a level of agency over the represented space of the village. There is a key role for communicative ecologies to investigate the extent to which these services are amenable to use by people from different kinds of backgrounds—pervasive or episodic users, for example (Crang et al., 2006)—drawing attention to the configuration of relationships and interdependencies that enable people to assert a position in representational space.

We employed the notion of ecology to establish a holistic framework which allows us to differentiate interdependencies between forms of social interaction, technologies used by urban residents, as well as contextual and discursive aspects whilst at the same time keeping the bigger picture in mind. The design goal of the Urban Village to achieve and maintain a steady-state equilibrium of social sustainability requires further analytical and empirical work. Are there some features that are necessary and sufficient to maintain a socially sustainable equilibrium in a communicative ecology? Are there other factors which are detrimental and cause a gradual withering away of social relations and connective tissue? The interpretation of the various 'ingredients' which make up the Urban Village invokes a coming together of people, place and technology in an urban environment which is an inbetween, that is, not solely random, serendipitous, accidental, yet not solely master-planned and socially engineered development site either. The ecology notion points at an organic process which Gilchrist (2000) describes as 'human horticulture'.

In the introduction we chalked out a field of 'difficult to deliver' desires surrounding community, diversity, participation, sustainability, identity, culture and history. Reflecting back on these challenges, the ecology notion and the associated interventions which we trialled in this study prompt a critical and ongoing rethinking of the concept 'community' and its relation and contribution to the desired image of an 'Urban Village'. It is imperative to unpack the facets of the term 'community' and their individual meaning for different stakeholders and purposes at different times and places. One of the principles which guided the design of our interventions was the consideration of the diversity of the urban environment, both in socio-cultural as well as built environment terms. Rather than attempting an umbrella approach which would have regarded the residents of the Urban Village as a collective group united by their collocation, we tried to draw on their diversity, history and individual ability to express themselves creatively. Our preliminary observations and experiences allow us to argue that nurturing individual identities in this manner does support a local culture of participation, interaction and engagement conducive to engendering a sense of an urban village atmosphere. This culture of shared experiences combines some traditional place making efforts (Walljasper, 2007) with novel ideas employing new media, digital storytelling and social networking fosters the emergence of a socially sustainable urban development.

ACKNOWLEDGMENT

This research is supported under the Australian Research Council's Discovery Projects funding scheme (project number DP0663854) and Dr Marcus Foth is the recipient of an Australian Postdoctoral Fellowship. Further support has been received from the Queensland Government's Department of Housing. The authors would like to thank Mark Bilandzic, Jaz Choi, Aneta Podkalicka, Angela Button, Julie-Anne Carroll, Nicole Garcia, Peter Browning for supporting this research project and the anonymous reviewers for valuable comments on earlier versions of this chapter.

REFERENCES

Adkins, B., Foth, M., Summerville, J., & Higgs, P. (2007). Ecologies of Innovation: Symbolic Aspects of Cross-Organizational Linkages in the Design Sector in an Australian Inner-City Area. *The American Behavioral Scientist*, *50*(7), 922–934. doi:10.1177/0002764206298317

Al-Hathloul, S., & Aslam Mughal, M. (1999). Creating identity in new communities: case studies from Saudi Arabia. *Landscape and Urban Planning*, *44*(4), 199–218. doi:10.1016/S0169-2046(99)00010-9

Anderson, B. (2006). *Imagined Communities: Reflections on the Origin and Spread of Nationalism* (Rev. ed.). London: Verso.

Antrop, M. (2004). Landscape change and the urbanization process in Europe. *Landscape and Urban Planning*, *67*(1-4), 9–26. doi:10.1016/S0169-2046(03)00026-4

Beer, D., & Burrows, R. (2007). Sociology and, of and in Web 2.0: Some Initial Considerations. *Sociological Research Online*, *12*(5).

Berners-Lee, T. (2000). *Weaving the Web: The past, present and future of the World Wide Web by its inventor*. London: Texere.

Bilandzic, M., Foth, M., & De Luca, A. (2008, Feb 25-27). *CityFlocks: Designing Social Navigation for Urban Mobile Information Systems.* Paper presented at ACM SIGCHI Designing Interactive Systems (DIS), Cape Town, South Africa.

Boase, J., Horrigan, J. B., Wellman, B., & Rainie, L. (2006). *The Strength of Internet Ties*. Washington, DC: Pew Internet & American Life Project.

Burgess, J. (2006). Hearing Ordinary Voices: Cultural Studies, Vernacular Creativity and Digital Storytelling. *Continuum: Journal of Media & Cultural Studies*, *20*(2), 201–214. doi:10.1080/10304310600641737

Burgess, J., Foth, M., & Klaebe, H. (2006, Sep 25-26). *Everyday Creativity as Civic Engagement: A Cultural Citizenship View of New Media.* Paper presented at the Communications Policy & Research Forum, Sydney, NSW.

Carroll, J.-A., Adkins, B., Foth, M., & Parker, E. (2007, Sep 6-8). *The Kelvin Grove Urban Village: What aspects of design are important for connecting people, place, and health?* Paper presented at the International Urban Design Conference, Gold Coast, QLD.

Crang, M., Crosbie, T., & Graham, S. (2006). Variable Geometries of Connection: Urban Digital Divides and the Uses of Information Technology. *Urban Studies (Edinburgh, Scotland)*, *43*(13), 2551–2570. doi:10.1080/00420980600970664

DeFilippis, J., Fisher, R., & Shragge, E. (2006). Neither Romance Nor Regulation: Re-evaluating Community. *International Journal of Urban and Regional Research*, *30*(3), 673–689. doi:10.1111/j.1468-2427.2006.00680.x

Delanty, G. (2000). Postmodernism and the Possibility of Community. In *Modernity and Postmodernity: Knowledge, Power and the Self* (pp. 114-130). London: Sage.

Dvir, R., & Pasher, E. (2004). Innovation engines for knowledge cities: an innovation ecology perspective. *Journal of Knowledge Management, 8*(5), 16–27. doi:10.1108/13673270410558756

Foth, M. (2006). Research to Inform the Design of Social Technology for Master-Planned Communities. In J. Ljungberg & M. Andersson (Eds.), *Proceedings 14th European Conference on Information Systems (ECIS), June 12-14*. Göteborg, Sweden.

Foth, M., & Adkins, B. (2006). A Research Design to Build Effective Partnerships between City Planners, Developers, Government and Urban Neighbourhood Communities. *Journal of Community Informatics, 2*(2), 116–133.

Foth, M., Gonzalez, V. M., & Taylor, W. (2006, Nov 22-24). *Designing for Place-Based Social Interaction of Urban Residents in México, South Africa and Australia.* Paper presented at the OZCHI Conference 2006, Sydney, NSW.

Foth, M., & Hearn, G. (2007). Networked Individualism of Urban Residents: Discovering the communicative ecology in inner-city apartment buildings. *Information Communication and Society, 10*(5), 749–772. doi:10.1080/13691180701658095

Foth, M., Hearn, G., & Klaebe, H. (2007, Sep 9-12). *Embedding Digital Narratives and New Media in Urban Planning.* Paper presented at the Digital Resources for the Humanities and Arts (DRHA) Conference, Dartington, Totnes, UK.

Foth, M., Odendaal, N., & Hearn, G. (2007, Oct 15-16). *The View from Everywhere: Towards an Epistemology for Urbanites.* Paper presented at the 4th International Conference on Intellectual Capital, Knowledge Management and Organisational Learning (ICICKM), Cape Town, South Africa.

Foth, M., & Podkalicka, A. (2007). Communication Policies for Urban Village Connections: Beyond Access? In F. Papandrea & M. Armstrong (Eds.), *Proceedings Communications Policy & Research Forum (CPRF)* (pp. 356-369). Sydney, NSW.

Gibson, R. (2005, May 26). *Imagination and the Historical Impulse in Response to a Past Full of Disappearance.* Paper presented at the Centre for Public Culture and Ideas, Brisbane.

Gilchrist, A. (2000). The well-connected community: networking to the 'edge of chaos'. *Community Development Journal, 35*(3), 264–275. doi:10.1093/cdj/35.3.264

Gleeson, B. (2004). Deprogramming Planning: Collaboration and Inclusion in New Urban Development. *Urban Policy and Research, 22*(3), 315–322. doi:10.1080/0811114042000269326

Gleeson, B., Darbas, T., & Lawson, S. (2004). Governance, Sustainability and Recent Australian Metropolitan Strategies: A Socio-theoretic Analysis. *Urban Policy and Research, 22*(4), 345–366. doi:10.1080/0811114042000296290

Grabher, G. (2004). Learning in projects, remembering in networks? Communality, sociality, and connectivity in project ecologies. *European Urban and Regional Studies, 11*(2), 99–119. doi:10.1177/0969776404041417

Gurstein, M. (2003). Effective use: A community informatics strategy beyond the digital divide. *First Monday, 8*(12).

Hanzl, M. (2007). Information technology as a tool for public participation in urban planning: a review of experiments and potentials. *Design Studies*, *28*(3), 289–307. doi:10.1016/j.destud.2007.02.003

Hearn, G., & Foth, M. (Eds.). (2007). *Communicative Ecologies. Special issue of the Electronic Journal of Communication, 17(1-2)*. New York: Communication Institute for Online Scholarship.

Klaebe, H., & Foth, M. (2007). Connecting Communities Using New Media: The Sharing Stories Project. In L. Stillman & G. Johanson (Eds.), *Constructing and Sharing Memory: Community Informatics, Identity and Empowerment* (pp. 143-153). Newcastle, UK: Cambridge Scholars Publishing.

Klaebe, H., Foth, M., Burgess, J., & Bilandzic, M. (2007, Sep 23-26). *Digital Storytelling and History Lines: Community Engagement in a Master-Planned Development*. Paper presented at the 13th International Conference on Virtual Systems and Multimedia (VSMM'07), Brisbane, QLD.

Lambert, J. (2002). *Digital Storytelling: Capturing Lives, Creating Community*. Berkeley, CA: Digital Diner Press.

Landry, C. (2001, Sep 5-7). *Tapping the Potential of Neighbourhoods: The Power of Culture and Creativity*. Paper presented at the International conference on Revitalizing Urban Neighbourhoods, Copenhagen.

Lefebvre, H. (1991). *The Production of Space* (D. Nicholson-Smith, Trans.). Oxford: Blackwell.

Meikle, G. (2002). *Future active: Media activism and the Internet*. New York: Routledge.

Mesch, G. S., & Levanon, Y. (2003). Community Networking and Locally-Based Social Ties in Two Suburban Localities. *City & Community*, *2*(4), 335–351. doi:10.1046/j.1535-6841.2003.00059.x

Oktay, D. (2002). The quest for urban identity in the changing context of the city: Northern. *Cities (London, England)*, *19*(4), 261–271. doi:10.1016/S0264-2751(02)00023-9

Polanyi, M. (1966). *The Tacit Dimension*. Gloucester, MA: Peter Smith.

Robertson, R. (1995). Glocalization: Time-Space and Homogeneity-Heterogeneity. In M. Featherstone, S. Lash & R. Robertson (Eds.), *Global Modernities* (pp. 25-44). London: Sage.

Sanoff, H. (2005). Community participation in riverfront development. *CoDesign*, *1*(1), 61–78. doi:10.1080/15710880512331326022

Stern, M. J., & Dillman, D. A. (2006). Community Participation, Social Ties, and Use of the Internet. *City & Community*, *5*(4), 409–424.

Sterne, J. (2003). Bourdieu, Technique and Technology. *Cultural Studies*, *17*(3/4), 367–389.

Talen, E. (2006). Design for Diversity: Evaluating the Context of Socially Mixed Neighbourhoods. *Journal of Urban Design*, *11*(1), 1–32. doi:10.1080/13574800500490588

Teo, P., & Huang, S. (1996). A sense of place in public housing: A case study of Pasir Ris, Singapore. *Habitat International*, *20*(2), 307–325. doi:10.1016/0197-3975(95)00065-8

Van den Dobbelsteen, A., & de Wilde, S. (2004). Space use optimisation and sustainability: Environmental assessment of space use concepts. *Journal of Environmental Management*, *73*(2), 81–88. doi:10.1016/j.jenvman.2004.06.002

Walljasper, J. (2007). *The Great Neighborhood Book: A Do-it-Yourself Guide to Placemaking*. Gabriola Island, BC, Canada: New Society.

Wellman, B. (2001). Physical Place and Cyberplace: The Rise of Personalized Networking. *International Journal of Urban and Regional Research, 25*(2), 227–252. doi:10.1111/1468-2427.00309

Wellman, B. (2002). Little Boxes, Glocalization, and Networked Individualism. In M. Tanabe, P. van den Besselaar & T. Ishida (Eds.), *Digital Cities II: Second Kyoto Workshop on Digital Cities* (LNCS 2362, pp. 10-25). Heidelberg, Germany: Springer.

Wellman, B., & Haythornthwaite, C. A. (Eds.). (2002). *The Internet in Everyday Life*. Oxford, UK: Blackwell.

Willson, M. A. (2006). *Technically Together: Rethinking Community within Techno-Society*. New York: Peter Lang.

Wood, P., & Landry, C. (2007). *The Intercultural City: Planning for Diversity Advantage*. London: Earthscan.

Young, G. T., Foth, M., & Matthes, N. Y. (2007). Virtual Fish: Visual Evidence of Connectivity in a Master-Planned Urban Community. In B. Thomas & M. Billinghurst (Eds.), *Proceedings of OZCHI 2007* (pp. 219-222). Adelaide, SA: University of South Australia.

KEY TERMS AND DEFINITIONS

Communicative Ecology: As defined by Hearn & Foth (2007), comprises a technological layer which consists of the devices and connecting media that enable communication and interaction. A social layer which consists of people and social modes of organising those people—which might include, for example, everything from friendship groups to more formal community organisations, as well as companies or legal entities. And a discursive layer which is the content of communication—that is, the ideas or themes that constitute the known social universe that the ecology operates in.

Collective Interaction: Characterised by a shared goal or common purpose, a focus on the community rather than the individual. The interaction is more public and formal than private and informal, and resembles many-to-many broadcasts. The mode of interaction is often asynchronous, permanent and hierarchically structured. Technology that supports collective interaction includes online discussion boards and mailing list.

Digital Storytelling: Refers to a specific tradition based around the production of digital stories in intensive collaborative workshops. The outcome is a short autobiographical narrative recorded as a voiceover, combined with photographic images (often sourced from the participants' own photo albums) and sometimes music (or other sonic ambience). These textual elements are combined to produce a 2-3 minute video. This form of digital storytelling originated in the late 1990s at the University of California at Berkeley's Center for Digital Storytelling (www.storycenter.org), headed by Dana Atchley and Joe Lambert.

Local Knowledge: Knowledge, or even knowing, is the justified belief that something is true. Knowledge is thus different from opinion. Local knowledge refers to facts and information acquired by a person which are relevant to a specific locale or have been elicited from a place-based context. It can also include specific skills or experiences made in a particular location. In this regard, local knowledge can be tacitly held, that is, knowledge we draw upon to perform and act but we may not be able to easily and explicitly articulate it: "We can know things, and important things, that we cannot tell" (Polanyi, 1966).

Master-Planned Communities: Urban developments guided by a central planning document that outline strategic design principles and specifications pertaining to road infrastructure, building design, zoning, technology and social and community facilities. They are usually built on vacant land and thus in contrast with the type of ad-hoc organic growth of existing city settlements.

Networked Interaction: Characterised by an interest in personal social networking and a focus on individual relationships. The interaction is more private and informal than public and formal, and resembles a peer-to-peer switchboard. The mode of interaction if often synchronous, transitory and appears chaotic from the outside. Technology that supports networked interaction includes instant messengers, email and SMS.

Triple Play infrastructure: Combines broadband Internet access, television reception and telephone communication over a single broadband connection, usually a fibre optic network. It is a marketing term which refers to a business model offering a bundle package of all three services accessible over the same network infrastructure.

This work was previously published in Handbook of Research on Urban Informatics: The Practice and Promise of the Real-Time City, edited by M. Foth, pp. 179-194, copyright 2009 by Information Science Reference (an imprint of IGI Global).

Chapter 8.8
Affective Goal and Task Selection for Social Robots[1]

Matthias Scheutz
Indiana University, USA

Paul Schermerhorn
Indiana University, USA

ABSTRACT

Effective decision-making under real-world conditions can be very difficult as purely rational methods of decision-making are often not feasible or applicable. Psychologists have long hypothesized that humans are able to cope with time and resource limitations by employing affective evaluations rather than rational ones. In this chapter, we present the distributed integrated affect cognition and reflection architecture DIARC for social robots intended for natural human-robot interaction and demonstrate the utility of its human-inspired affect mechanisms for the selection of tasks and goals. Specifically, we show that DIARC incorporates affect mechanisms throughout the architecture, which are based on "evaluation signals" generated in each architectural component to obtain quick and efficient estimates of the state of the component, and illustrate the operation and utility of these mechanisms with examples from human-robot interaction experiments.

DOI: 10.4018/978-1-60566-354-8.ch005

INTRODUCTION

Effective decision-making under real-world conditions can be very difficult. From a purely decision-theoretic standpoint, the optimal way of making decisions – rational choice – requires an agent to know the utilities of all choice options as well as their associated likelihoods of succeeding for the agent to be able to calculate the expected utility of each alternative and being able to select the one with the maximum utility. Unfortunately, such rational methods are in practice often not applicable (e.g., because the agent does not have reliable or sufficient knowledge) or feasible (e.g., because it is too time-consuming to perform all necessary calculations).

Psychologists have long hypothesized that humans are able to cope with time, knowledge and other resource limitations by employing *affective evaluations* (Clore et al., 2001) rather than rational ones. For affect provides fast, low-cost (although often less accurate) mechanisms for estimating the value of an object, event, or situation for an agent, as opposed to longer, more complex and more computationally intensive *cognitive evaluations* (e.g., to compute the expected utilities) (Kahneman et al., 1997). Humans also rely on *affective memory*, which seems to encode implicit knowledge about the likelihood of occurrence of a positive or negative future event (Blaney, 1986). Finally, affect also influences human problem-solving and reasoning strategies, leading to global, top-down approaches when affect is positive, and local, bottom-up approaches when affect is negative (Bless et al., 1996).

For (autonomous) social robots that are supposed to interact with humans in natural ways in typically human environments, affect mechanisms are doubly important. For one, such robots will also have to find fast solutions to many of the same kids of difficult problems that humans ordinarily face, often with the same degree of uncertainty if not more. Hence, affect mechanisms in robotic architectures might help robots cope better with the intrinsic resource limitations of the real world. The second reason why affect mechanisms are essential for social robots is grounded in their intended role as *social agents* interacting with humans. For those interactions to be *natural* (and effective), robots need to be sensitive to *human affect*, both in its various forms of expression and in its role in human social interactions.

We have started to address affect mechanisms that can serve both functions in our DIARC architecture (Scheutz et al., 2006, Scheutz et al., 2007). DIARC is a "distributed integrated affect cognition and reflection" architecture particularly intended for social robots that need to interact with humans in natural ways. It integrates cognitive capabilities (such as natural language understanding and complex action planning and sequencing) (Scheutz et al., 2007, Scheutz et al., 2004, Brick and Scheutz 2007) with lower level activities (such as multi-modal perceptual processing, feature detection and tracking, and navigation and behavior coordination, e.g., see Scheutz et al., 2004, or Scheutz and Andronache 2004) and has been used in several human subject experiments and at various AAAI robot competitions (Scheutz et al., 2005, Scheutz et al., 2006, Schermerhorn et al., 2008, Schermerhorn et al., 2006). Most importantly, DIARC incorporates affect mechanisms throughout the architecture, which are based on "evaluation signals" generated in each architectural component, which effectively encode how "good" something (e.g., the current state of the world) is from the perspective of the component.

In this chapter, we will describe DIARC's mechanisms for affective goal and task selection, and demonstrate the operation of these mechanisms with examples from human-robot interaction experiments.

1. MOOD-BASED DECISION-MAKING

A perfectly rational agent with perfect information can make optimal decisions by selecting the action A with the highest expected utility

$$EU = \arg\max_{A}(p_A \cdot b_A - c_A)$$

where is the probability of action A succeeding, the benefit of A succeeding, and the cost of attempting A. If the agent knows the costs and benefits of each alternative and also the probabilities of each action succeeding, it cannot be wrong about which is the most profitable choice. In reality, however, costs and benefits are only approximately known. More importantly, real-world constraints can make it difficult to estimate accurately the probabilities of success and failure and, moreover, the dependence of the probabilities on other factors (e.g., past successes and failure).

Rational approaches probabilities that are often not available to robots. Without knowledge of the probabilities of failure and success associated with each potential alternative, it is not possible to calculate expected utility. Humans, on the other hand, are subject to the same kinds of real-world constraints, yet are able to make good evaluations, which are hypothesized to involve affective states ("gut feelings") in important ways (e.g., to help them prioritize goals).

Let an agent's overall affective state – its "mood" – be represented by two state variables, one which records positive affect (A_P), and the other of which records negative affect (A_N) (Sloman et al., 2005). and are reals in the interval [0,1] that are influenced by the performance of the agent's various subsystems (e.g., speech recognition). When a subsystem records a success, it increases the level of positive affect, and when it fails, it increases the level of negative affect. Specifically, success increases A_P by $\Delta A_P = (1-A_P) \times inc$ (failure updates analogously), where *inc* is a value (possibly learned) that determines the magnitude of the increase within the available range. This update function ensures that remains in the interval [0,1]. Both affective states are also subject to regular decay, bringing their activations in the absence of triggering events back to their rest values (i.e., 0): $\Delta A_P = (1 - A_P) \times dec$ (Scheutz 2001). Given that affective states can encode knowledge of recent events (e.g., the success or failure of recent attempts), they can be used to estimate probabilities (that take past evidence into account without the need for prior knowledge of the probabilities involved).

Consider, for example, a case in which the robot is deciding whether to ask for directions to some location. The robot does not know that it is in a noisy room where speech recognition is problematic. All else being equal (i.e., with both affect states starting at rest and no affect triggers from other sources), the value of *inc* determines how many failed communication attempts the agent will make of before giving up. With greater *inc*, the value of A_N rises faster, leading the agent to reduce its subjective assessment of the expected benefit (i.e., to become "pessimistic" that the benefit will be realized).

The agent makes online choices based on the expected utility of a single attempt, using the affect states A_P and A_P to generate an "affective estimate" of the likelihood of success . Examples presented below define *f* as follows:[2]

$$f(A_P, A_N) = \frac{1}{2} + \frac{(1 + A^{+2} - A^{-2})}{2}$$

This value is then used in the calculation of the expected utility of an action: $u = a \cdot b - c$.

The effect of positive and negative affect is to modify the benefit the agent expects to receive from attempting the action. When both A_P and A_N are neutral (i.e.,), the decision is based solely on a comparison of the benefit and the cost. However, given a history of actions, the agent may view the benefit more optimistically (if $A_P > A_N$) or pessimistically (if $A_P < A_N$), potentially making decisions that differ from the purely rational choice (overestimating true benefits or costs).

We can now demonstrate with a simple example of how overall mood states could be used in a beneficial way in the agent's decision making. Figure 1 depicts for the communication example the effect of various values of *inc* on estimates of utility: one that is too optimistic, willing to continue into the foreseeable future; one that is too pessimistic, stopping fairly early; and one that is more reasonable, stopping at about the point where the costs will outweigh the benefits. This suggests that the value of *inc* could be defined as a function of *b* and *c* to improve the likelihood that A_N will rise quickly enough to end the series of attempts before costs exceed potential benefits, for example. The agent could employ reinforcement learning to determine the value of *inc* for individual actions.

While the activation of each affective state is subject to decay, the rate of decay is slow enough

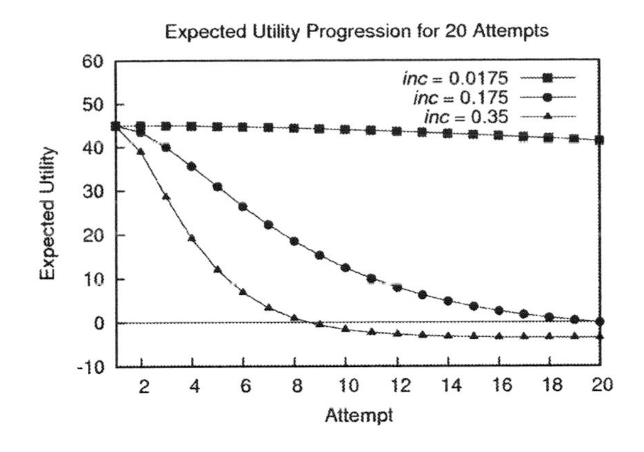

Figure 1. The expected utilities calculated at each attempt by the agent for various values of inc.

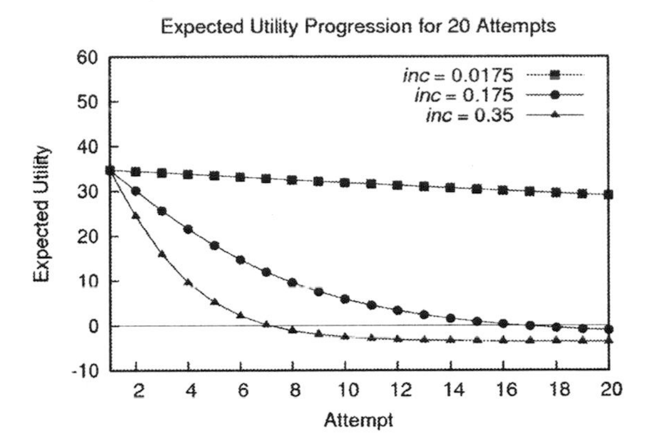

Figure 2. The expected utilities calculated at each attempt by agent for various values of inc, after an extended series of 20 failures and 100 decay cycles, demonstrating the role of affective states as memory

that they can serve as *affective memory*, carrying the subjective estimates of the likelihood of success and failure ahead for a period after the events that modified the states. Returning again to the robot example, after a series of failures leading to the agent deciding not to attempt to ask directions again, the activation of A_N begins to decay. If, after some period of time, the agent is again faced with the choice of whether to ask for directions, any remaining activation of A_N will reduce the likelihood that it will choose to do so. In this way, the agent "remembers" that it has failed recently, and pessimistically "believes" that its chances of failing again are relatively high (e.g., because it has likely not left the noisy room it was in). Figure 2 shows the expected utility of asking for directions calculated by an agent 100 cycles after a series of failed attempts (e.g., Figure 1). The increased "pessimism" leads the evaluation to drop below zero earlier, potentially saving wasted effort on fruitless attempts.

2. AFFECT REPRESENTATIONS IN ARCHITECTURAL COMPONENTS

We now show how the above decision-making process inspired by roles of human affect, where "affect states" are used to implicitly encode the history of positive and negative events from the agent's perspective, can be incorporated into an architecture at the level of functional components, where each component maintains its own "affective state". A primary determinant of the affective state of a component is its own performance, but in some cases the affective states of other functional components (e.g., those upon which it depends to function properly) or the occurrence of certain external events (e.g., a loud unexpected noise) can influence affect.

Specifically, we associate with each component of the architecture two state variables, one which represents *positive affect* (A_P), and the other which represents *negative affect* (A_N). A_P and A_N are reals in the interval [0,1] and define the "affective evaluation" of that component $a=f(A_P,A_N)$. Examples presented below define f as follows: $f(A_P,A_N)=1+A_P^{2}-A_N^{-2}$. The value of a is used by the component when making decisions about how to perform its function.

A component's affective state values can be passed on to other components to influence the calculation of their respective affect states. Associated with each affective state A_P is an increment variable inc^{+} that determines how much a positive event changes positive affect. Specifically, success increases A_P by $\sphericalangle A_P = (1 - A_P)\sphericalangle inc^+$ (this update function ensures that A_P remains in the interval [0,1]; failure updates A_N analogously). The value of inc^+ is computed based on the affective evaluation of connected components: $inc^+ = \sum_{i=1}^{n} w_i (A_i^{+2} - A_i^{-2})$, for $f(A_P, A_N) > 1$, where w_i is the weight assigned to the contribution of component *i*.

Similarly, for $inc^- = \sum_{i=1}^{n} w_i (A_i^{-2} - A_i^{+2})$, for $f(A_P, A_N) < 1$. Hence, positive affective evaluation of associated architectural components increases the degree to which positive outcomes influence positive affect A_P for a component, while negative affective evaluation of those components does the same for A_N. Affective states are also subject to regular decay, bringing their activations in the absence of triggering events back to their rest values (i.e., 0): $\sphericalangle A_P = (1 - A_P)\sphericalangle dec$.

The *affective goal manager* (AGM) prioritizes competing goals (i.e., those whose associated actions require conflicting resources) based on the expected utility of those goals and time constraints within which the goals must be completed. Each goal is assigned an *affective task manager* (ATM), which is responsible for action selection and dispatch. The AGM periodically updates the priority associated with each goal's ATM. These goal priorities are used to determine the outcome of conflicts between ATMs (e.g., resource conflicts, such as when each wants to move in a different direction). A goal's priority is determined by two components: its importance and its current urgency. The importance of a goal is determined by the cost and benefit of satisfying the goal. The affective evaluation *a* of the goal manager influences the assessment of a goal's importance: $u = a \cdot b - c$. The resulting *u* is scaled by the urgency component *g*, which is a reflection of the time remaining within which to satisfy the goal:

$$g = \frac{Time_{elapsed}}{Time_{allowed}} \cdot (g_{max} - g_{min}) + g_{min}$$

where g_{max} and g_{min} are upper and lower bounds on the urgency of that particular goal. The goal's priority *p*, then, is simply: $p = u \cdot g$. When there is a conflict over some resource, the ATM with the highest priority is awarded the resource. This formulation allows goals of lower importance, which would normally be excluded from execution in virtue of their interference with the satisfaction of more important goals, to be "worked in" ahead of the more important goals, so long as the interrupted goal has sufficient time to satisfy the goal after the less important goal completes (i.e., so long as the urgency of the more important goal is sufficiently low).

The ATM uses affect states similarly to select between alternative actions in service to a single goal. Each potential action has associated with it (in long-term memory) affect states A_P and A_N that result from positive and negative outcomes in past experience with that action, along with (learned) inc^+ and inc^- that determine how further experience influences the affect state that determine how further experience influences the affect states. The ATM makes online choices based on the expected utility of a single attempt of an action, using $a = f(A_P, A_N)$ as an "affective estimate" of the likelihood of success for the attempt in the utility calculation $u = a \cdot b - c$. The alternative with the highest expected utility is selected in service of the goal associated with the ATM.

The effect of positive and negative affect, then, is to modify the benefit the agent expects to receive from attempting the action. That is, the AGM/ATM implements a decision making process that can operate without exact knowledge of the prior and conditional distributions. When both and are neutral (i.e., $A_P = A_N = 0$), the decision is based solely on a comparison of the benefit and the cost. However, given a history of outcomes, the agent may view the benefit more optimistically (if $A_P > A_N$) or pessimistically (if $A_P < A_N$), potentially leading it to make decisions that differ from the purely "rational" decision strategy, as mentioned before.

The following two examples presented below focus on AGM and the ATM as they are currently implemented in our robotic architecture. For presentation purposes, simplified scenarios have been chosen to highlight the functionality and benefits of affect in decision-making.

2.1. Prioritizing Goals

The affective goal manager is responsible for prioritizing goals to determine the outcomes of resource conflicts. Priorities are recalculated periodically to accurately reflect the system's affect states and time-related goal urgencies. In this example, the AGM maintains priorities for two goals, *Collect Data* and *Report*. The *Collect Data* goal requires a robot to acquire information about a region by moving through the environment and taking readings (e.g., for the purpose of mapping locations of interest in the region). There is a limited time within which to gather the data before the robot needs to return with the data. The *Report* goal requires the robot to locate and report to the mission commander once the information is collected *or* when something goes wrong.[3]

One approach to accomplishing these two goals would be to explicitly sequence the *Collect* and *Report* goals, so that when the former was achieved, the latter would be pursued. The appropriate response to problems could similarly be explicitly triggered when problems were detected. However, the AGM allows for a more flexible unified approach in which both goals are instantiated at the start and the AGM's prioritization function ensures that the robot does the right thing at the right time. Figure 3 depicts the evolution of the two goals' priorities throughout a sample run of this scenario. Initially, the AGM's A_P=0 and A_N=0. The benefit associated with *Collect* (b_c) is 1800, while its cost (c_c) is 1200. The benefit associated with *Report* (b_r) is 200 and the cost (c_r) is 25. Both goals require the use of the robot's navigation system, but only one may do so at a time.

At the start, both goals have very low priorities due to the very low urgency (very little time had elapsed). *Collect* has a higher priority due to its greater net benefit ($b-c$); because the AGM's affect is neutral, there is no modification of the benefit component. As time passes, both priorities rise with the increasing urgency until an external event disturbs the system–the impact of an unknown object knocks out a sensor, causing a sharp increase in A_N for the AGM (this could be construed as a fear-like response to the impact event). The AGM output for time step 56 immediately preceding the impact was:

AGM A+: 0.0
AGM A-: 0.0
Collect PRIORITY 16.83
Report PRIORITY 5.89

Immediately following the impact event, the priorities have inverted:

AGM A+: 0.0
AGM A-: 0.5
Collect PRIORITY 1.71
Report PRIORITY 4.28

Both priorities were reduced due to the influence of A_N on the benefit component b, but because the reduction of relative to b_c was so much greater than relative to c_c, *Report* was given a higher priority. This allowed the robot to respond to the unexpected impact by seeking the mission commander, who would, presumably, be able to resolve the problem (e.g., by repairing the damage or redirecting the robot). Before the *Report* goal is achieved, however, the priorities were once again inverted (at time step 265), and *Collect* regained control of the navigation resources:

AGM A+: 0.0
AGM A-: 0.45
Collect PRIORITY 21.75
Report PRIORITY 21.73

Figure 3. Priorities calculated by the affective goal manager for the goals Collect Data and Report during a sample run

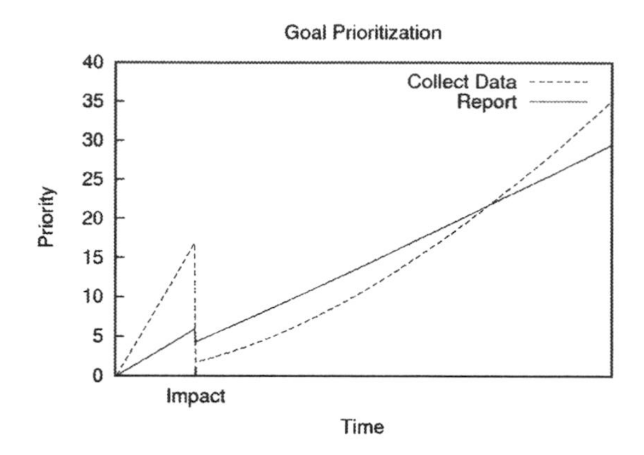

Figure 4. Expected utility of the two alternative actions Natural Language and Nonverbal Alert through a series of failed communication attempts

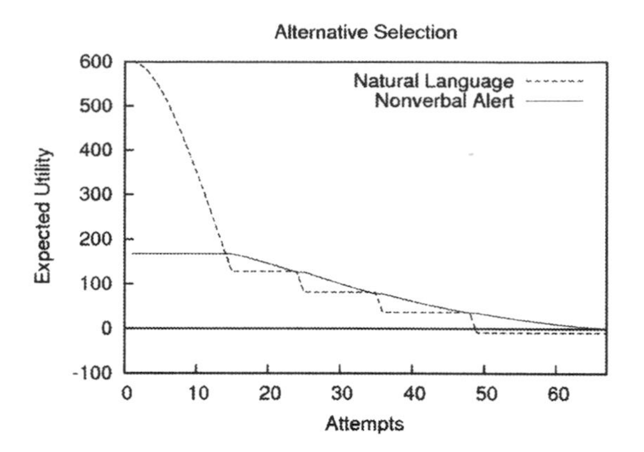

This switch is attributable to the decay of A_N in the AGM. No further impacts (or other negative events) occurred and the impact did not cause a catastrophic failure, so negative affect was gradually returning to zero. This (in addition to rising urgency) caused the priorities of both goals to rise, but the priority of *Collect* climbed faster, so that it eventually overtook *Report* and the robot was able to continue pursuing its "primary" goal.[4]

2.2. Choosing between Alternatives

The affective task manager (ATM) component selects and executes actions on behalf of a goal, as priority allows. When an action completes, the ATM is also responsible for updating the affect states associated with the completed action (based on its completion status, success or failure), in addition to updating its own affect states. The following example is extracted from a sample run in which the robot has noticed a problem and needs to communicate it to a human user. There are two modes of communication available: *Natural Language*, in which the robot attempts to explain the problem using natural language, and *Nonverbal Alert*, in which the robot uses "beep codes" to try to convey the message. *Natural Language* has a greater benefit (b_l=1800) than *Nonverbal Alert* (b_a=200), due to the ability to communicate more information about the problem, but also has a greater cost (c_l=1200 vs. c_a=25). Based on past experience, *Nonverbal Alert* has A_N (perhaps because of poor results trying to communicate failures using this method). This sample run depicts a series of failed attempts to communicate the problem to the human user (Figure 4).[5] At the beginning of the run, the ATM output is as follows:

Natural Language A-: 0.0
Natural Language UTILITY 600.0
Nonverbal Alert A-: 0.2
Nonverbal Alert UTILITY 167.0

The ATM selects *Natural Language* due to its higher expected utility. In the course of the next 14 attempts, u_l (the expected utility of *Natural Language*) falls, while u_a (the expected utility of *Nonverbal Alert*) remains unchanged:

Natural Language A-: 0.49
Natural Language UTILITY 173.70
Nonverbal Alert A-: 0.2
Nonverbal Alert UTILITY 167.0

After one more failure of *Natural Language*, $u_l<u_a$, so the ATM begins trying *Nonverbal Alert* instead:

Natural Language A-: 0.51
Natural Language UTILITY 127.54
Nonverbal Alert A-: 0.2
Nonverbal Alert UTILITY 167.0
Nonverbal Alert is repeated through attempt 23, and u_a is reduced:
Natural Language A-: 0.51
Natural Language UTILITY 127.54
Nonverbal Alert A-: 0.47
Nonverbal Alert UTILITY 130.96

After attempt 23, the increase in A_N for *Nonverbal Alert* causes its expected utility to fall below *Natural Language*, which is selected on attempt 24:

Natural Language A-: 0.51
Natural Language UTILITY 127.54
Nonverbal Alert A-: 0.50
Nonverbal Alert UTILITY 125.84

Natural Language is attempted only once before the ATM switches back to *Nonverbal Alert*, and the cycle begins again, with the robot occasionally attempting *Natural Language* before reverting to *Nonverbal Alert*, producing the "stair-stepping" effect seen in Figure 4.

3. HUMAN-ROBOT INTERACTION EXPERIMENTS WITH DIARC

Here we briefly give an example of an application of DIARC for studying affective human-robot interactions (Figure 5 shows the relevant components of the architecture for the given task).

In Scheutz et al., (2006), we reported an experiment that was intended to examine subjects' reactions to affect expressed by the robot. Subjects were paired with a robot to perform a task in the context of a hypothetical space exploration scenario. The task was to find a location in the environment (a "planetary surface") with a sufficiently high signal strength to allow the team to transmit some data to an orbiting spacecraft. The signal strength was detectable only by the robot, so the human had to direct it around the environment in search of a suitable location, asking it to take readings of the signal strength during the search and to transmit the data once a transmission

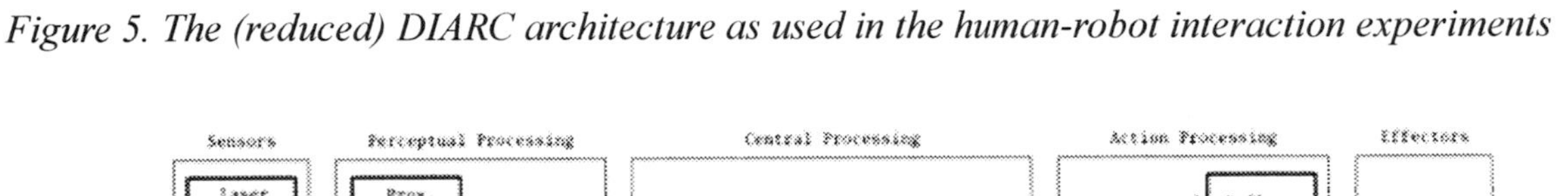

Figure 5. The (reduced) DIARC architecture as used in the human-robot interaction experiments

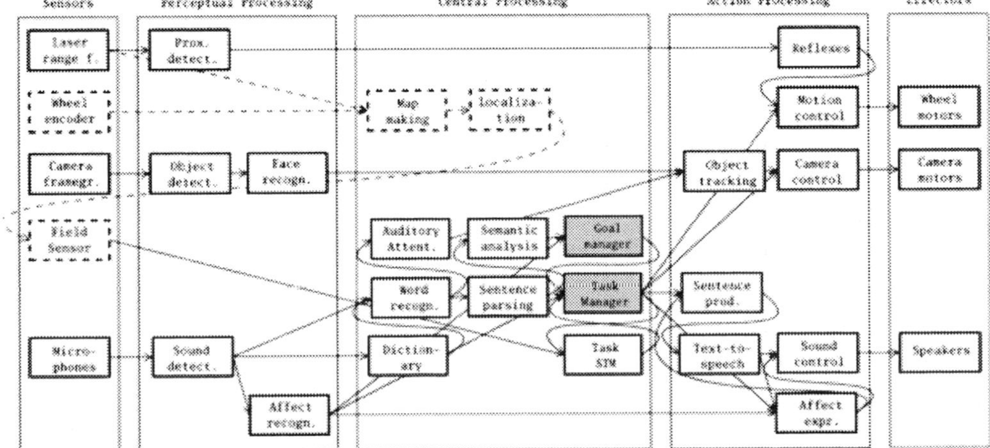

point was found. There was only one location in the room that met the criteria for transmission, although there were others that represented local peaks in signal strength; the "signal" was simulated by the robot, which maintained a global map of the environment, including all points representing peaks in signal strength. When asked to take a reading, the robot would calculate the signal based on its proximity to these peaks. The goal of the task was to locate a transmission point and transmit the data as quickly as possible; time to completion was recorded for use as the primary performance measure (see Scheutz et al.,, 2006 for further details).

Subjects were asked to respond to a series of survey items prior to beginning the interaction with the robot, in order to gauge their preconceived attitudes toward robots (e.g., whether they would think that it was useful for robots to detect and react to human emotions or whether they thought that it would be useful for robots to have emotions and express them). They were given a chance to interact with the robot for a short practice period before the actual experimental runs were conducted. The subjects and the robot communicated via spoken natural language. In order to evoke affective responses from subjects (and to impose an artificial time limit of three minutes on the task), a simulated battery failure was used. There were three points at which the robot could announce problems related to the battery, depending on whether the subject had completed the task or not. One minute into the experimental run, the robot announced that the batteries were "getting low." After another minute, it would follow with a warning that there was "not much time remaining" due to the battery problem. After three minutes (total), the robot would announce that the mission had failed.

We employed a 2x2 experimental design, with the first dimension, *affect expression*, being *affective* vs. *neutral* and the second, *proximity*, being *local* vs. *remote*. In the neutral affect expression condition, the robot's voice remained affectively neutral throughout the interaction, while in the affective condition, the robot's voice was modulated to express increasing levels of "fear" from the point of the first battery warning until the end of the task. Subjects in the local proximity condition completed the exploration task in the same room as the robot, whereas those in the remote condition interacted with the robot from a separate "control" room. The control room was equipped with a computer display of a live video stream fed from a camera in the exploration environment, along with a live audio stream of the robot's speech (using the ADE robot infrastructure, we were able to redirect the robot's speech production to the control station). Hence, the only difference between the two proximity conditions was the physical co-location of the robot and the subject. Most importantly, the channel by which affect expression is accomplished (i.e., voice modulation) was presented locally to the subject in both conditions–subjects in the remote condition heard the same voice in exactly the same as they would have if they had been next to the robot.

Subsequent analysis of the objective performance measure (i.e., time to completion) pointed to differences between the local and remote conditions with regard to the effect of affect. A 2x2 ANOVA for *time to completion* with independent variables *affect expression* and *proximity* showed no significant main effects ($F(1,46)=2.51, p=.12$ for affect expression and $F(1,46)=2.16, p=.15$ for proximity), but a marginally significant two-way interaction ($F(1,46)=3.43, p=.07$) due to a performance advantage in the *local* condition for *affect* over *neutral* ($\mu=123$ vs. $\mu=156$) that was not present in the *remote* condition ($\mu=151$ vs. $\mu=150$). The difference in the local condition between affect and no-affect groups is significant ($t(22)=2.21, p<.05$), while the difference in the remote condition is not significant ($t(16)=.09, p=.93$).

Affect expression provides a performance advantage in the local condition, but not in the remote condition. Given that the medium of affect expression (speech modulation) was presented identically in both proximity conditions, it seems unlikely that the remote subjects simply did not notice the robot's "mood" change. In fact, subjects were asked on a post-questionnaire to evaluate the robot's stress level after it issued the low-battery warning. A 2x2 ANOVA with with *affect expression* and *proximity* as independent variables, and *perceived robot stress* from the post-survey as dependent variable and found a main effect on affect ($F(1,44)=7.54, p<.01$), but no main effect on proximity and no interaction.[6] Subjects in the *affect* condition tended to rate the robot's behavior as "stressed" ($\mu=6.67, \sigma=1.71$), whereas subjects in the *neutral* condition were much less likely to do so ($\mu=5.1, \sigma=2.23$). Hence, subjects recognized the affect expression as they were intended to, and the lack of any effect or interaction involving proximity indicates that both conditions recognized the affect equally well. This, combined with the results of the objective performance task, strongly suggests that affect expression and physical embodiment play an important role in how people internalize affective cues.

A currently still ongoing follow-up study examines dynamic robot autonomy and how affect expression, as described above, influences subjects' responses to autonomy in the exploration task. The experimental setup is similar (Scheutz et al.,, 2006), but with an additional "distractor" measurement task included to induce cognitive load in the human team member concurrent to the exploration task. The measurement task consists of locating target "rock formations" (boxes) in the environment and "measuring" them (multiplying two two-digit numbers found on a paper in the box) to determine whether they were above a given threshold.

Dynamic robot autonomy is achieved via three goals in the AGM: *Commands*, which requires the robot to obey commands from the human team member, *Track*, which requires the robot to locate and stay with the transmission location, and *Transmit*, which requires the robot to gather and transmit data about the measurements. The priorities and costs were chosen to allow the tracking and transmission goals to overtake the commands goal at specific times. Obeying commands is originally given the highest priority, so that the other goals cannot acquire the resource locks for motion commands, etc. Hence, the autonomy condition starts out exactly as the non-autonomy control condition, with the robot taking commands from the subject related to searching the environment for the transmission location. Then, for example, when the tracking goal's priority surpasses the command goal's (Figure 6), the robot will no longer cooperate with commands that interfere with the robot's autonomous search for the signal peak. These transitions occur at approximately 150 seconds into the task for the tracking goal and 195 seconds for the transmission goal. This assumes that both and are in their rest states. Figure 7 shows the evolution of priorities for a case in which the robot begins the task with A_P =.25 and A_N=0 (e.g., as might be the case if the robot had recently detected positive affect in the voice of the human team member).[7] The elevated positive affect leads to an "optimistic" assessment of the benefit of following commands (relative to taking over and searching for the transmission location, for example), so the point at which the other goals take over is pushed back (by about ten seconds in either case). An analogous example of the impact of negative affect (A_P =0 and A_N=.25) is shown in Figure 8, which shows the "pessimistic" assessment hastening the takeovers by tracking and transmission by approximately 20 and 15 seconds, respectively.

The experimental design includes the *Autonomy* dimension and the *Affect Expression* dimension. This design allows us to explore the degree to which subjects are willing to accept robot dynamic autonomy, and how affect expression on the part of the robot influences the acceptability of autonomy. For example, it seems likely that the robot's expression of stress as a part of normal speech interactions will provide subjects with some context explaining *why* the

Figure 6. The priority evolution of the dynamic autonomy experiment with neutral starting affect

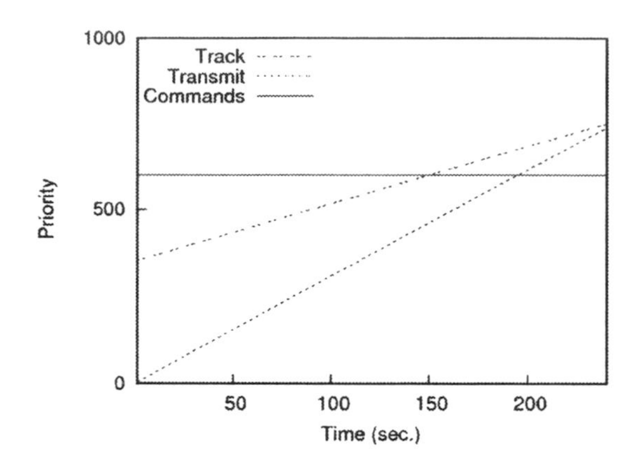

Figure 7. The priority evolution of the dynamic autonomy experiment with positive starting affect

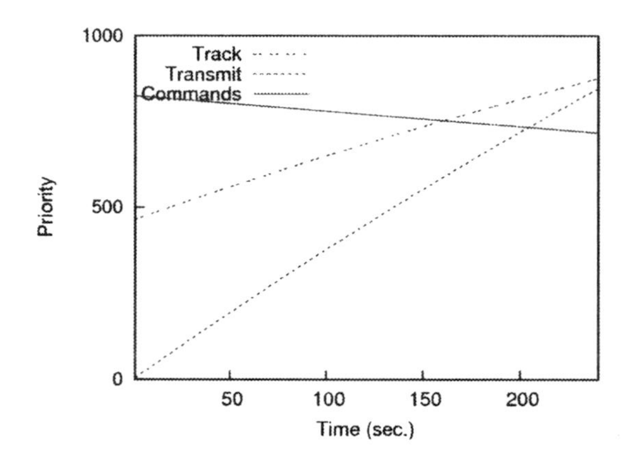

Figure 8. The priority evolution of the dynamic autonomy experiment with negative starting affect

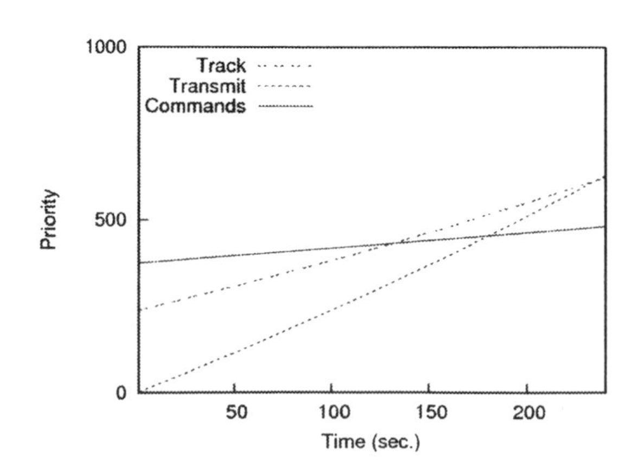

robot has stopped following commands, which could facilitate acceptance. We are currently conducting experiments and analyzing the results, having completed the first phase of experiments in a remote condition. As reported in (Scheutz and Schermerhorn under review), even without affect expression, subjects are positive with regard to dynamic autonomy in a robotic teammate, to the extent that they even characterize the robot in the autonomy condition as more cooperative than the robot in the non-autonomy condition, despite the fact that the autonomous version disobeyed in the later phase of the task, whereas the non-autonomous version obeyed throughout. We are currently analyzing the remaining data to determine if and how affect expression alters the picture.

4. RELATED WORK

While different forms of affective and deliberative processes (like reasoning or decision-making) have been in simulated agents (e.g., El-Nasr et al.,, 2000; Eliott, 1992; Gratch and Marsella, 2004), most robotic work has focused on action selection (e.g., Moshkina and Arkin, 2003, Murphy et al.,, 2002; Parker, 1998; Scheutz, 2002) using simple affective states, often times without explicit goal representations. Yet, complex robots (e.g., ones that work with people and need to interact with them in natural ways, Scheutz et al.,, 2007) will have to manage multiple, possibly inconsistent goals, and decide which to pursue at any given time under time-pressure and limited resources.

The two closest affective robotic architectures in terms of using emotion (a form of affective state) for internal state changes and decision-making on robots are (Murphy et al.,, 2002) and (Breazeal et al., 2004). In (Murphy et al.,, 2002) emotional states are implemented with fixed associated action tendencies (e.g., HAPPY–"free activate", CONFIDENT–"continue normal activity", CONCERNED–"monitor progress" and FRUSTRATED–"change current strategy") in a service robot as a function of two time parameters ("time-to-refill" and "time-to-empty" plus two constants). Effectively, emotion labels are associated with different intervals and cause state transitions in a Moore machine, which produces behaviors directly based on perceptions and emotional states. This is different from the explicit goal representation used in our architecture, which allows for the explicit computation of the *importance* of a goal to the robot (based on positive and negative affective state), which in turn influences goal prioritization and thus task and action selection.

The architecture in (Breazeal et al.,, 2004) extends prior work (Breazeal, 2002) to include natural language processing and some higher level deliberative functions, most importantly, an implementation of "joint intention theory" (e.g., that allows the robot to respond to human commands with gestures indicating a new focus of attention, etc.). The system is intended to study collaboration and learning of joint tasks. The mechanisms for selecting subgoals, subscripts, and updating priorities of goals are, however, different in our affective action interpreter, which uses a dual representation of positive and negative affect that is influenced by various components in the architecture and used for the calculation of the importance, and consequently the priority, of goals.[8]

5. CONCLUSION

In this chapter, we introduced the idea of integrating affect representations and processing mechanisms throughout a robotic architecture based on psychological evidence that affect permeates the human cognitive system. We present the specific mechanisms integrated in our DIARC architecture, with focus on DIARC's goal and task managers. We showed with several examples that these mechanisms can lead to effective decisions for robots that operate under time, computation, and knowledge constraints, especially given their low computational cost and knowledge requirements. As such, they can improve the functioning and level of autonomy of social robots. Moreover, we also demonstrated that DIARC can be used for systematic empirical studies that investigate the utility of affect mechanisms for social robots. Specifically, we described results from human-robot interaction experiments where affect expression by the robot in the right context could significantly improve the performance of joint human-robot teams. We also pointed at the potential of DIARC and its affective goal and task management mechanisms for further investigations of the interactions between affect and robot autonomy. Current experiments suggest that these interactions will be particularly important for robots that have to collaborate with humans.

While DIARC has already proven its robustness and applicability in real-world settings, it is still very much "work-in-progress". We investigating criteria for situations in which good values for some of the parameters (i.e., the increment and weight values in the affect update equations) can be found. We are also examining ways of making these parameters dependent on goal and task contexts, thus allowing for multiple context-dependent values (which can be learned using reinforcement learning techniques) to overcome the shortcomings of a single value.

REFERENCES

Blaney, P. H. (1986). Affect and memory: A review. *Psychological Bulletin*, *99*(2), 229–246. doi:10.1037/0033-2909.99.2.229

Bless, H., Schwarz, N., & Wieland, R. (1996). Mood and the impact of category membership and individuating information. *European Journal of Social Psychology*, *26*, 935–959. doi:10.1002/(SICI)1099-0992(199611)26:6<935::AID-EJSP798>3.0.CO;2-N

Breazeal, C., Hoffman, G., & Lockerd, A. (2004). Teaching and working with robots as a collaboration. In . *Proceedings of AAMAS, 2004*, 1030–1037.

Breazeal (2002). *Designing sociable robots*. MIT Press.

Brick, T., Schermerhorn, P., & Scheutz, M. (2007). Speech and action: Integration of action and language for mobile robots. In *Proceedings of the 2007 IEEE/RSJ International Conference on Intelligent Robots and Systems* (pp. 1423-1428).

Brick, T., & Scheutz, M. (2007). Incremental natural language processing for HRI. In *Proceedings of the Second ACM IEEE International Conference on Human-Robot Interaction* (pp. 263-270).

Clore, G. L., Gasper, K., & Conway, H. (2001). Affect as information. In J.P. Forgas (Ed.), *Handbook of Affect and Social Cognition* (pp. 121-144).

El-Nasr, M. S., Yen, J., & Ioerger, T. R. (2000). Flame – fuzzy logic adaptive model of emotions. *Autonomous Agents and Multi-Agent Systems*, *3*(3), 219–257. doi:10.1023/A:1010030809960

Eliott, C. (1992). *The affective reasoner: A process model of emotions in a multi-agent system*. PhD thesis, Institute for the Learning Sciences, Northwestern University.

Gratch, J., & Marsella, S. (2004). A domain-independent framework for modeling emotion. *Journal of Cognitive Systems Research*, *5*(4), 269–306. doi:10.1016/j.cogsys.2004.02.002

Kahneman, D., Wakker, P. P., & Sarin, R. (1997). Back to Bentham? Explorations of experienced utility. *The Quarterly Journal of Economics*, *112*, 375–405. doi:10.1162/003355397555235

Moshkina, L., & Arkin, R. C. (2003). On TAME-ing robots. In *IEEE International Conference on Systems, Man and Cybernetics, Vol. 4* (pp. 3949–3959).

Murphy, R. R., Lisetti, C., Tardif, R., Irish, L., & Gage, A. (2002). Emotion-based control of cooperating heterogeneous mobile robots. *IEEE Transactions on Robotics and Automation*, *18*(5), 744–757. doi:10.1109/TRA.2002.804503

Parker, L. E. (1998). Alliance: An architecture for fault-tolerant multi-robot cooperation. *IEEE Transactions on Robotics and Automation*, *14*(2), 220–240. doi:10.1109/70.681242

Schermerhorn, P., Kramer, J., Brick, T., Anderson, D., Dingler, A., & Scheutz, M. (2006). DIARC: A testbed for natural human-robot interactions. In *Proceedings of AAAI 2006 Robot Workshop*.

Schermerhorn, P., Scheutz, M., & Crowell, C. R. (2008). Robot social presence and gender: Do females view robots differently than males? In *Proceedings of the Third ACM IEEE International Conference on Human-Robot Interaction*, Amsterdam (pp. 263-270).

Scheutz, M. (2002). Affective action selection and behavior arbitration for autonomous robots. In H. Arabnia (Ed.), *Proceedings of the 2002 International Conference on Artificial Intelligence* (pp. 334-340).

Scheutz, M., & Andronache, V. (2004). Architectural mechanisms for dynamic changes of behavior selection strategies in behavior-based systems. *IEEE Transactions of System, Man, and Cybernetics Part B*, *34*(6), 2377–2395. doi:10.1109/TSMCB.2004.837309

Scheutz, M., Eberhard, K., & Andronache, V. (2004). A parallel, distributed, realtime, robotic model for human reference resolution with visual constraints. *Connection Science*, *16*(3), 145–167. doi:10.1080/09540090412331314803

Scheutz, M., McRaven, J., & Cserey, G. (2004). Fast, reliable, adaptive, bimodal people tracking for indoor environments. In *IEEE/RSJ International Conference on Intelligent Robots and Systems (IROS)* (pp. 1340-1352).

Scheutz, M., & Schermerhorn, P. (in press). Dynamic robot autonomy: Investigating the effects of robot decision-making in a human-robot team task.

Scheutz, M., Schermerhorn, P., Kramer, J., & Anderson, D. (2007). First steps toward natural human like HRI. *Autonomous Robots*, *22*(4), 411–423. doi:10.1007/s10514-006-9018-3

Scheutz, M., Schermerhorn, P., Kramer, J., & Middendorff, C. (2006). The utility of affect expression in natural language interactions in joint human-robot tasks. In *Proceedings of the 1st ACM International Conference on Human-Robot Interaction* (pp. 226–233).

Scheutz, M., Schermerhorn, P., Middendorff, C., Kramer, J., Anderson, D., & Dingler, A. (2005). Toward affective cognitive robots for human-robot interaction. In *AAAI 2005 Robot Workshop* (pp. 1737-1738).

Scheutz (2001). The evolution of simple affective states in multi-agent environments. In D. Cañamero (Ed.), *Proceedings of AAAI Fall Symposium* (pp. 123–128).

Sloman, A., Chrisley, A., & Scheutz, M. (2005). The architectural basis of affective states and processes. In J.M. Fellous & M.A. Arbib (Eds.), *Who needs emotions? The Brain Meets the Machine* (pp. 201-244). New York: Oxford University Press.

KEY TERMS

Affect: stating how emotions moves us in order to start actions action selection, Procedures to decide which action is more proper to a certain context goal management, Procedures and algorithms to sort the goals on a system according to the priorities that emerge from the context affective architecture, Constructing systems able to take decision based on simulations of the affective processes human-robot interaction Discipline that studies how humans and robots can interact, and finding ways to improve this interaction.

ENDNOTES

1. This material is based upon work supported by the National Science Foundation under Grant No. 0746950 and by the Office of Naval Research under MURI Grant No. N000140711049.
2. A_P and A_N are squared to amplify the difference between the two, which amplifies the effect of the dominant state on the agent's decision process.
3. This example is taken from the hypothetical space scenario that we have repeatedly used in human-robot interaction experiments (Scheutz et al., 2006), see also Section 3.
4. Note that there is nothing explicit in the architecture that makes *Report* primary; it is simply the relative costs and benefits of the two goals that make it the preferred goal in the zero-affect state.
5. Because there are no successful attempts, is not incremented for either action and remains zero throughout the run.
6. Two subjects had to be eliminated from the comparison since they did not answer the relevant question on the post-survey.
7. Note that the lines curve slightly due to the built-in decay of affect states.
8. The details for reprioritization of goals were not provided in (Breazeal et al.,, 2004).

This work was previously published in Handbook of Research on Synthetic Emotions and Sociable Robotics: New Applications in Affective Computing and Artificial Intelligence, edited by J. Vallverdú; D. Casacuberta, pp. 74-87, copyright 2009 by Information Science Reference (an imprint of IGI Global).

Chapter 8.9
Using Ambient Social Reminders to Stay in Touch with Friends

Ross Shannon
University College Dublin, Ireland

Eugene Kenny
University College Dublin, Ireland

Aaron Quigley
University College Dublin, Ireland

ABSTRACT

Social interactions among a group of friends will typically have a certain recurring rhythm. Most people interact with their own circle of friends at a range of different rates, and through a range of different modalities (by email, phone, instant messaging, face-to-face meetings and so on). When these naturally recurring interactions are maintained effectively, people feel at ease with the quality and stability of their social network. Conversely, when a person has not interacted with one of their friends for a longer time interval than they usually do, a situation can be identified in that relationship which may require action to resolve. Here we discuss the opportunities we see in using ambient information technology to effectively support a user's social connectedness. We present a social network visualisation which provides a user with occasional recommendations of which of their friends they should contact soon to keep their social network in a healthy state. [Article copies are available for purchase from InfoSci-on-Demand.com]

INTRODUCTION

When modelling the social interactions among a group of friends, certain recurring rhythms are identified between participants. Within this group, a single person may have a range of different rhythms with each of their friends, due to the similarity of their schedules, the differing strengths of those friendships, and a range of other social factors (Viegas et al., 2006). When these rhythms are maintained well—that is, the person interacts with each friend at the regularity that they normally do—the health of that friendship will feel natural. If on the other hand the friendship falls out of rhythm, through neglect or unfortunate circumstance, and the two people do not see each other or otherwise interact, this gap will be felt, though perhaps not always understood.

We refer to this as a person's social rhythm, and it describes the rate and regularity with which they interact with the various people they know. It is an intuitive, fuzzy metric; if asked how often you interact with a certain friend of yours, you may reply with "about twice a week" or "most days", not something more specific like "once every 37 hours." These frequencies will differ among subgroups of a social network: interactions with family members may have a different regularity than with work colleagues, and some friends may have special significance and be seen much more often. Still others may have a very low level of engagement—only being seen at annual events like birthdays or academic conferences.

A person's ability to effectively regulate their own social rhythm relies on their perception of time running like clockwork, but the human mind's perception of the passage of time is capricious at best (Harrison et al., 2007). Numerous studies have pointed to the fallibility of this ability, due to stress, anxiety, caffeine intake and a range of other factors (Chavez, 2003). Without external prompts, keeping up with friends—especially peripheral friends, who are not part of one's close social circle—can become a matter of chance and circumstance. Because social interactions are inherently vague and intuitive, there is no single point in time at which one is motivated to rekindle a relationship in decline. We believe that explicit cues based on historically observed rhythms will help alleviate this problem, just as they have been shown to support a user's health in other studies (Consolvo et al., 2008). We will discuss these issues in depth in the next section.

Intuitively, you may have experienced a digital or physical artefact that you come across arbitrarily which spurs you into thinking of a friend and then contacting them. For example, seeing a photograph of you and a friend may prompt you to send them a message to talk about a shared experience. Similarly, hearing a friend's name or reviewing past correspondences with them may remind you to contact them (Viegas et al., 2004). It is along these lines that we seek to provide subtle reminders of a friend at the right time, to induce a user into re-establishing contact. We have developed a visualisation for this purpose, which we present in section 4.

ATTENTION AND AWARENESS IN SOCIAL NETWORKS

One aspect of human memory is the remembrance of past experiences, known as "retrospective memory." A second form of memory, "prospective memory", works in the opposite direction and can be thought of as remembering to remember something (a task or object) at a certain time in the future (Winograd, 1988). For example, remembering to call a friend after work at 6 o'clock, or remembering to bring a book you have borrowed with you when going to visit a colleague.

Though the workings of prospective memory are not yet fully understood, the cognitive process is thought not to require external artefacts to trigger a memory (Meier et al., 2006), but can certainly be aided by such objects, like shopping lists. Setting an alarm on a phone or other device that is triggered at a certain time of the day is also effective, as it takes the burden of remembering when to do a task off the person's mind.

Facebook, a prominent social networking website[1], offer a feature they call a "news feed", which is a way to keep track of activity within your network of friends. The news feed presents a reverse-chronological list of events, such as photos being uploaded, public messages being exchanged, or status messages being updated. This gives the user a constant stream of activity and information about the members of their social network, though as the number of friends in your network increases, this list offers a view of a decreasing subset of your friends' activity on the website.

The first problem this brings up is that at any time, a user watching the activity on their news feed are mostly kept updated on the latest and loudest of their friends. Those friends who post status updates about every detail of their day will be featured much more frequently than those friends who broadcast information about themselves less frequently or not at all, even though it is quite likely that it is these quiet friends that are most likely to fall off a user's mental radar.

Second, none of the popular online social networks implement any concept of a friendship's inherent strength. All friends are presented equally, despite some presumably being more important than others to the person at the centre of the network. With many people having identified hundreds of friends on the website (Ellis et al., 2006), many of whom may be peripheral to them in everyday life (Fogg et al., 2008), it is easy for some more important friends to be lost among the throng. Indeed, a user may end up receiving many updates from friends whom they would be happy to hear from much less frequently.

Together, these factors have the effect of selectively emphasising a person's friends in proportion to their engagement with the social networking website, rather than in proportion to how frequently a user personally interacts with that friend. The phrase "out of sight, out of mind" describes the deleterious effects of this vicious cycle: as a friend is remembered less frequently, they become less likely to be contacted in future.

Previous studies have analysed social rhythms in socio-technical systems, although the focus of these studies was on the general trends of social rhythms apparent on a very large scale. Golder et al. studied interactions between college students on Facebook, and found that students' social calendars were heavily influenced by their school schedule (Golder et al., 2007). Leskovec et al. analysed all conversations conducted over Microsoft's Messenger instant messaging service in the month of June 2006, and concluded that users of similar age, language and location were most likely to communicate frequently (Leskovec et al., 2007).

Online social networking sites are used in part to maintain social connections which were originally forged offline (Ellison, 2007). "Dunbar's number" is a proposed upper bound to the number of people an individual can maintain stable social relationships with. Though we may suppose that this number would be a function of our circumstances and available free time, it is in fact related to the size of the neocortex. Among humans, this bound stands at approximately 150, and is due to the cognitive overheads involved in remembering and being able to meaningfully interact with others (Dunbar, 1992). Although social networking applications have long allowed users to have many more than this number of "friends" identified within the system (Boyd, 2007), it is unlikely that a user would report that they are friends with all of these people in the traditional sense (Boyd, 2006).

Despite the large number of friends identified by the average Facebook user, a person's social network cannot be described by data from any one source. Though the majority of a user's friends may indeed be present in an online social networking website, they will also have friends that they interact with purely offline, or mostly by phone or email. These ongoing social interactions are equally valid in characterising a user's circle of friends.

Because of the range of communication options available to us, reactivating links between people is relatively easy, if we are prompted at the right time. These social networking websites in particular present a low-cost way for people to evolve, maintain and reinforce a wide network of friends and acquaintances. The issue becomes one of identifying which friends are core to the network, and capturing information about the historical regularity of contact with all friends, so that we can accurately deliver helpful recommendations to the user.

Our application provides the user with suggestions of actions they can take to maintain the stability of their social network through a visual interface. This encourages users to contact their friends regularly, but also helps them to identify problems with certain friends early, so that they can take steps to correct a deviation before it becomes more pronounced. Thus, if a user tends their network well, they will have stronger ties with a wider and more diverse set of friends.

VISUALISING SOCIAL INTERACTIONS

There have been many visualisations generated of social networks, particularly since the rise of social networking websites and the rich data sets they present. Many visualisations use a familiar node-link diagram of a graph (Heer et al., 2005). Visualisations in this style will often present the graph from an "ego-centric" perspective, where the user being analysed is shown at the centre of the view, with their friends arrayed around them. In this project, because we are not interested in the network links that exist among friends, we can dispense with this network view, and focus on the strength of the connections between a user and their immediate network of friends, and therefore allow them to answer questions about the health of their network at a glance.

As with the social networking websites, a weakness we have identified with existing social network visualisations is that they frequently treat all edges in the network as being weighted identically. That is, an edge is either present or not present; there is no gradation to the strength of each link, and all links are drawn with equal length. In real life, we know that friendships do not behave like this. The social links between people become weaker over time and grow stronger through positive interactions. Representing this dynamism requires additional sources of data.

DATA SOURCES

Ambient systems can leverage the vast amounts of data available from the physical and virtual worlds. We now leave digital traces of most of our social interactions: all of our email is archived on a server somewhere, our instant messages are logged locally and remotely, posts to social network profiles are publicly visible, and so on. Even co-location data can be recorded if the users have a capable mobile device, allowing the identification of events like two people meeting in a bar or at a sports event.

Though all of this data is attractive, for this initial version of our application, we decided to focus on records of mobile phone interactions, which we are able to download as a spreadsheet from our telecommunication provider's website. These records gave us access to traces of a user's incoming and outgoing phone calls, as well as SMS text messages for the preceding month. We manually relate each phone number in the records to the corresponding friend from the user's Facebook network, which allows us to refer to each friend by name in the display.

The software has been built to be agnostic to the nature of the interactions, so adding support for emails in future, for example, is a matter of writing a client to connect to the user's inbox and find mails that they have sent or received from their friends. These, along with other discrete interactions like instant messaging conversations or comments left on eachother's Facebook pages, can then be entered into the system.

VISUALISATION OF SOCIAL RHYTHMS

Figure 1 shows a visualisation that we generated from the phone data combined with the user's

Figure 1. Our visualisation of a single user's social network, showing a record of their interactions with a subset of their friends over the course of six weeks. Blue circles are phone calls; the size of the circle reflects the length of the call. Red circles are SMS text messages. Suggested future social interactions are indicated by hollow circles on the right, giving the user time to act on those suggestions when it is convenient.

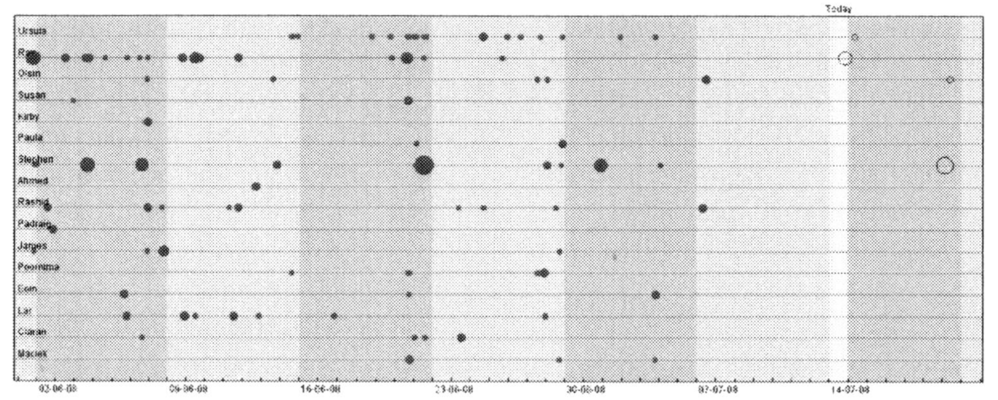

Facebook network information. Each row represents the social interactions a person has engaged in with one of their friends via their phone: blue dots indicate phone calls, with the size of the dot reflecting the length of the call, while red dots indicate text messages and are uniformly-sized. Weeks are delineated by differing background colours to provide users with an indication of their longer-term habits at a glance. Our visualisation is built using Processing (Reas et al., 2003), a Java-based visualisation framework.

The current day is highlighted, and the next week is visible on the right of the display. Cues for future interactions are displayed in this area as hollow circles. Their colour and size indicate the type of interaction suggested, based on a prediction algorithm that we have written for this purpose.

Predicted social interactions are drawn on the day that our algorithm has calculated to be most likely for them to occur, but the user can see them a week in advance. This gives the user several opportunities to act on the information being presented to them at an appropriate time. The intent is not to interrupt the user, but simply to plant the seed of memory so that they can act upon it when it is convenient.

If the user does not interact with their friend in any way before the suggested interaction , an "X" is marked at this position and this is counted as a "miss." The prominent marking of these events (or non-events, if you will) serve to draw the user's attention to these more critical cues. If the user then contacts the friend in question within a week of this event, the marker is removed and a regular blue or red marker is placed at this point.

PROPERTIES OF AMBIENT INFORMATION

Neely et al. have previously explored their hypothesis that some context sources are more applicable to being presented in an ambient manner than others (Neely et al., 2007). The reasons they described are precision, criticality, periodicity, interpretability and self-descriptiveness.

In addition, we propose three aspects of the reminders in our visualisation which make them appropriate for delivery by ambient information systems: they are passive, dynamic and simple. Passive means that changes in the information do not always require immediate attention; users can take note of reminders but choose not to act on them until later. Dynamic means that the data changes over time; if the display remains at the periphery of a person's attention, they can monitor for changes while concentrating on other activities. Simple means that the information can be digested easily; at a basic level, a reminder simply consists of the name of a friend whom they should contact soon. Other information may be present, such as a suggested contact time or medium, but this only serves to augment the primary information.

These three properties correspond well to the interaction, reaction and comprehension model proposed by McCrickard et al. (2003). Not all notification systems are as well-suited to an ambient implementation. Consider as a counterpoint the visualisation an air-traffic controller uses to direct planes at an airport, which satisfies none of the above criteria: the information requires immediate response, as planes must be given clearance to land or take off as quickly as possible; must be monitored constantly; and there are typically a huge number of variables to take into account for each notification, such as the plane's location, scheduled departure/arrival time, current velocity, etc. It would of course be possible to create an ambient display which delivers information about planes arriving and leaving an airport; while passengers might find this interesting and informative, air-traffic controllers would have no use for it.

AMBIENT APPLICATIONS

The implementation described above is used as both an interactive display, where a user filters the information processed by the system manually to achieve insights into their social trends, and as a passive information display, which allows a user to get a feel for the general health of their social environment in an instant. An implementation which more closely follows the traditional definition of an ambient display could adopt a similar presentation to the Whereabouts Clock developed at Microsoft Research (Sellen et al., 2006). This is a glanceable ambient display placed on a wall in a home, which displays the current location of all members of the family. A similar display which displays a collection of avatars representing some of the user's friends which harnesses the information traces we discussed previously would have an ideal marriage of these properties.

Since the critical information for the user—reminders indicating when a friendship is stagnating—is atomic and relatively simple, it could be used in conjunction with a number of lo-fi data delivery methods. The user could subscribe to receive suggestions as text messages on their mobile phone, or through email or twitter tweets, informing them of the person they need to catch up with.

One can imagine a future scenario where all devices in the home are connected, and digital photo frames could be updated on a frequency predicated by the requirement of the user to update that friendship. Facebook-enabled photo frames have already been released—it is simply the randomisation algorithm that needs to be more intelligent.

CONCLUSION

We have presented a visualisation of a user's interactions with members of their social network, and describe how this kind of information can help a user to keep their social network in a healthy state. Given sufficiently careful treatment, infrequent notifications can become a useful addition to an ambient information display. We have postulated that certain traits are desirable in an ambient reminder system; these are a long possible response time, variance in the timing and meaning of reminders, and simple, easy to interpret reminder information.

ACKNOWLEDGMENT

This work is supported by Science Foundation Ireland through an Undergraduate Research Experience and Knowledge grant (UREKA), and under grant number 03/CE2/I303-1, "LERO: the Irish Software Engineering Research Centre."

REFERENCES

Boyd, D., & Ellison, N. B. (2007). Social network sites: Definition, history, and scholarship. Journal of Computer-Mediated Communication, 13(1), 11.

Boyd, D. (2006). Friends, Friendsters, and MySpace Top 8: Writing community into being on social network sites. First Monday, 11(12).

Chavez, B. R. (2003). Effects of Stress and Relaxation on Time Perception. Masters thesis, Uniformed Services University of the Health Sciences, Bethesda, Maryland.

Consolvo, S., Klasnja, P., McDonald, D. W., Avrahami, D., Froehlich, J., LeGrand, L., Libby, R., Mosher, K., & Landay, J. A. (2008). Flowers or a robot army?: encouraging awareness & activity with personal, mobile displays. In UbiComp '08: Proceedings of the 10th international conference on Ubiquitous computing, NY, USA, 2008 (pp. 54–63). ACM.

Ellison, Nicole, C. S., & Lampe, C. (2006). Spatially Bounded Online Social Networks and Social Capital: The Role of Facebook. In Annual Conference of the International Communication Association, Dresden, Germany, 2006.

Fogg, B., & Iizawa, D. (2008). Online Persuasion in Facebook and Mixi: A Cross-Cultural Comparison. Persuasive Technology, 2008 (pp. 35–46).

Golder, S. A., Wilkinson, D., & Huberman, B. A. (2007). Rhythms of Social Interaction: Messaging within a Massive Online Network. In 3rd International Conference on Communities and Technologies (C&T 2007).

Harrison, C., Amento, B., Kuznetsov, S., & Bell, R. (2007). Rethinking the progress bar. In UIST '07: Proceedings of the 20th annual ACM symposium on User interface software and technology, New York, NY, USA, 2007 (pp. 115–118). ACM.

Heer, J., & boyd, d. (2005). Vizster: Visualizing Online Social Networks. In IEEE Symposium on Information Visualization (InfoVis 2005). Minneapolis, Minnesota, October 23-25.

McCrickard, D. S., Chewar, C., Somervell, J. P., & Ndiwalana, A. (2003). A model for notification systems evaluation- assessing user goals for multitasking activity. ACM Transactions on Computer-Human Interaction, 10(4), 312–338.

Meier, B., Zimmermann, T., & Perrig, W. (2006). Retrieval experience in prospective memory: Strategic monitoring and spontaneous retrieval. Memory, 14(7), 872–889.

Neely, S., Stevenson, G., & Nixon, P. (2007). Assessing the Suitability of Context Information for Ambient Display. In Workshop on Designing and Evaluating Ambient Information Systems at Pervasive 2007.

Reas, C., & Fry, B. (2003). Processing: a learning environment for creating interactive Web graphics. In SIGGRAPH '03: 2003 Sketches & Applications, NY, USA, 2003 (p. 1). ACM.

Sellen, A., Eardley, R., Izadi, S., & Harper, R. (2006). The whereabouts clock: early testing of a situated awareness device. In CHI '06: extended abstracts on Human factors in computing systems, New York, NY, USA, 2006 (pp. 1307–1312). ACM.

Viegas, F. B., boyd, d., Nguyen, D. H., Potter, J., & Donath, J. (2004). Digital artifacts for remembering and storytelling: posthistory and social network fragments. Proceedings of the 37th Annual Hawaii International Conference on System Sciences, IEEE Computer Society.

Viegas, F. B., Golder, S., & Donath, J. (2006). Visualizing email content: portraying relationships from conversational histories. In CHI '06: Proceedings of the SIGCHI conference on Human Factors in computing systems, New York, NY, USA, 2006 (pp. 979–988). ACM.

Winograd, E. (1988). Some observations on prospective remembering. Practical aspects of memory: Current research and issues, 1, 348–353.

ENDNOTE

[1] www.facebook.com

This work was previously published in International Journal of Ambient Computing and Intelligence, Vol. 1, Issue 2, edited by K. Curran, pp. 70-78, copyright 2009 by IGI Publishing (an imprint of IGI Global).

Chapter 8.10
Leveraging Semantic Technologies Towards Social Ambient Intelligence

Adrien Joly
Alcatel-Lucent Bell Labs, France
Université de Lyon, LIRIS / INSA, France

Pierre Maret
Université de Lyon, France
Université de Saint Etienne, France

Fabien Bataille
Alcatel-Lucent Bell Labs, France

ABSTRACT

These times, when the amount of information exponentially grows on the Internet, when most people can be connected at all times with powerful personal devices, we need to enhance, adapt, and simplify access to information and communication with other people. The vision of ambient intelligence which is a relevant response to this need brings many challenges in different areas such as context-awareness, adaptive human-system interaction, privacy enforcement, and social communications. The authors believe that ontologies and other semantic technologies can help meeting most of these challenges in a unified manner, as they are a bridge between meaningful (but fuzzy by nature) human knowledge and digital information systems. In this chapter, the authors will depict their vision of "Social Ambient Intelligence" based on the review of several uses of semantic technologies for context management, adaptive human-system interaction, privacy enforcement and social communications. Based on identified benefits and lacks, and on their experience, they will propose several research leads towards the realization of this vision.

DOI: 10.4018/978-1-60566-290-9.ch014

INTRODUCTION

These times, when the amount of information exponentially grows on the Internet, when most people can be connected at all times with powerful personal devices, users suffer from the growing complexity of the information society. Our use of technology is moving towards the vision of

"Ambient Intelligence", derived from the vision of "Ubiquitous computing" in which "the most profound technologies are those that disappear" (Weiser, 1991) . Thus, access to information is no longer limited to personal computers and the web browsing paradigm. This vision brings many technological and psychological challenges (Streitz & Nixon, 2005) that are considered in several research domains, including:

- Context-awareness: how to take one's context into account to improve his communication ?
- Multimodality: how to span user interfaces from a terminal into separate modal interfaces ? (e.g. various screens, input controllers, microphones, phones)
- Social networking: how to enhance and leverage social communication ?
- Privacy & Trust: how to ease one's life without delegating human control to machines ?

There is one transversal question yet to answer: is there a unified approach that could answer these challenges in a global way and that makes sense? Actually, a common approach exists that is considered in all these research domains, and in most corresponding works and has been shown as very promising. This approach is the use of semantic technologies.

In this chapter, we propose a review of research works relying on semantic technologies towards what we call "Social Ambient Intelligence", a social extension of ambient intelligence. The intention here is to identify the key technologies, approaches and issues that may be blended in order to build an optimal platform for a widescaled ubiquitous system that can support social applications. After defining the foundational terms of this chapter in the Background section, we will review several research works to identify their key technologies, approaches and issues in the State-of-the-Art section, then we will propose several research leads towards our vision of "Social Ambient Intelligence" in the Future Trends section, to finally conclude this chapter.

In this section, we propose and discuss the underlying definitions needed to set the foundations of this chapter: ubiquitous computing, context-awareness and semantic technologies.

Ubiquitous Computing, Ambient Intelligence and Context-Awareness

The phrase "ubiquitous computing" was proposed by Mark Weiser while working for the Xerox Palo Alto Research Center (PARC), to qualify a possible evolution of computers. "The Computer for the 21st century" (Weiser, 1991) has become a foundational paper for following works in this area. Indeed, it introduced a vision, in which "ubiquitous computers" are simple communicative devices and appliances that are suited for a particular task and are aware of their surrounding environment while fading into the background. For example, paper sheets could be replaced with flexible screens, bringing any information of the web as an independent element of a real desktop, an element that one could stack into piles, stick on a wall, lend to a colleague or take for lunch.

As depicted on Figure 1, the generation of ubiquitous computers has already arrived, as powerful and communicative computers are spread in many devices like watches, mobile phones, portable media players, game consoles, PDAs (Personal Digital Assistants), ticket machines, bike renting beacons and kids toys. Even though Mark Weiser's vision of interoperable and shared ubiquitous computers has not been reached yet, a significant research effort is done towards the vision of "Ambient intelligence". As such, "Ambient Intelligence" is considered as an evolution of "ubiquitous computing" in which networked devices can also be integrated in the environment (and thus not expecting any user intervention), can sense the environmental, personal and social situation to adapt the experience, and can anticipate

Figure 1. The evolution of computing, adapted from "Nano computing & Ambient intelligence" (Waldner, 2007), © 2007 Hermes Science Publishing. Used with permission.

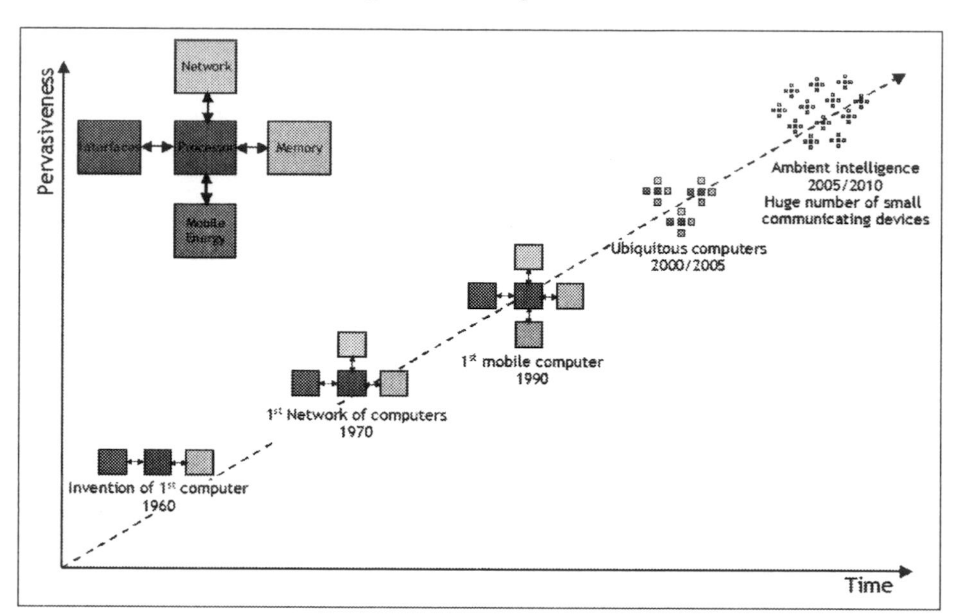

forthcoming situations or actions in order to ease and enhance people lives.

Firstly defined by (Schilit, Adams, & Want, 1994), context-awareness is a key research domain towards the vision of Ambient Intelligence. It consists in acquiring low-level context data (e.g. from sensors), inferring high-level knowledge from this data, and predicting context changes in order to clearly improve the user experience. As depicted on Figure 2, the low level of context contains current raw sensor data like GPS coordinates, IP address, surrounding Bluetooth MAC addresses or temperature. By combining and inferring on this knowledge, a meaningful position or activity like "in a meeting" or "watching TV" can be deduced to form a higher level of context. Then, after having learnt the habits of the user, predictions can be made about the actions that are probably going to happen next or about the exceptional cases that have occurred (e.g. the user is going to arrive late at work because he has not left home yet) in order to undertake relevant actions pro-actively (e.g. inform the colleagues that the meeting is delayed).

Context-awareness aims to make user interfaces automatically adapt to the user's environment and intents. It can enhance user inputs without requiring additional efforts from the user (Leong, Kobayashi, Koshizuka, & Sakamura, 2005) and also adapt outputs (Sadi & Maes, 2005). Although several works have been focusing on the implementation of context-awareness on mobile devices (Christopoulou, 2008; Korpipää, Häkkilä, Kela, Ronkainen, & Känsälä, 2004; Häkkilä & Mäntyjärvi, 2005), we will not specifically address mobile-based context-aware platforms in this paper.

Semantic Technologies: Ontologies, Knowledge Representation and Reasoning

In their study, (Strassner, O'Sullivan, & Lewis, 2007) define ontologies as « a formal, explicit specification of a shared, machine-readable vocabulary and meanings, in the form of various entities and relationships between them, to describe knowledge about the contents of one or more related subject

Figure 2. Levels of context (© Bell Labs. Used with permission.)

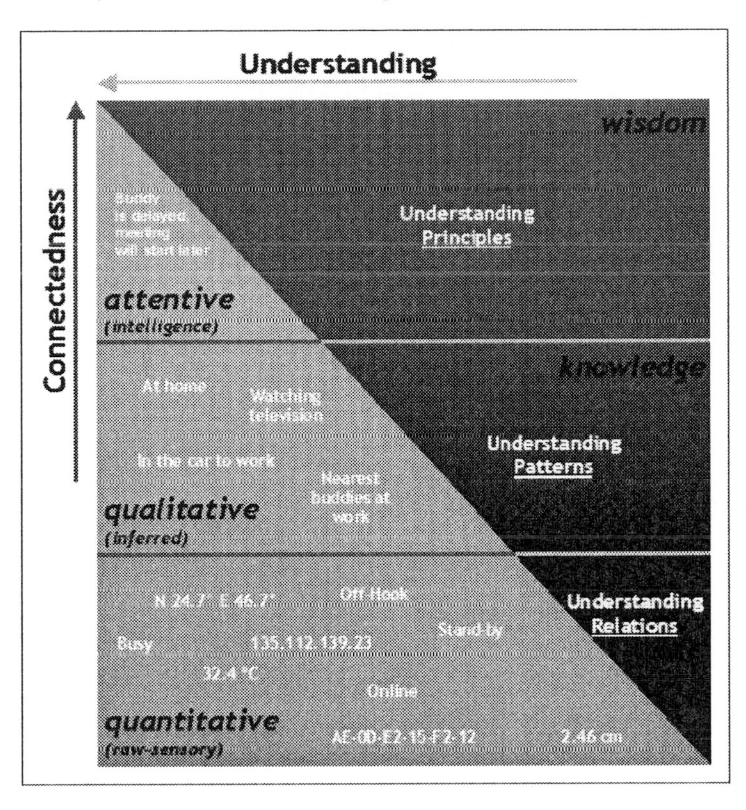

domains throughout the life cycle of its existence ». Semantic technologies, including ontologies and semantic description languages, are quite similar to human thinking and memorization: they allow the definition of concepts and instances (of these concepts) that are related with each other using semantically qualified links. They also allow to develop an inferred knowledge from the reasoning on this knowledge (Gruber, 1993). Applying such approach to information technologies enable machines to understand the actual meaning of data which is formulated using a distributed and evolving vocabulary. That way, ontologies fill the gap between ambiguous/fuzzy human thinking (e.g. in natural languages, a word can have different meanings) and formalized digital data (i.e. stored using specific formats and interpreted by specific applications for a specific purpose).

One of the benefits of using semantic languages is to allow progressive/incremental modeling of a system, reflecting the natural progression of conceptual understanding of domains. Ontologies can ease the communication between heterogeneous entities (i.e. using different languages/protocols) by matching similar portions of the semantic graph of the sender's knowledge with the recipient's knowledge.

On the other hand, we would like to prevent the reader to make the naïve assumption that semantic technologies are a magic solution to empower machines with autonomic intelligence. It may seem possible to model our universe as an ontology, allowing computers to understand the human world, but it is actually impossible. Indeed, modeling is always relative to a point of view, and integrating ontologies from experts of several domains would necessarily lead to inconsistencies. There is also a usual confusion about the so-called "Semantic Web" (Berners-Lee, Hendler, & Lassila, 2001). This expression does not mean that internet users will have to deal with semantic languages to communicate on the web, but it refers to a set of languages and tools that would allow web resources (i.e. web pages and services) to be described semantically in order to allow seamless processing of knowledge distributed among heterogeneous sites. Today, with the rise of the "Web 2.0" (O'Reilly, 2005), users are already able to create "mash-ups" relying on several components and data streams hosted on different sites. However, the next step is possibly to automatize (or, at least, to ease) the development of such mash-ups, assuming that web data and components are semantically described.

In the next section, we will investigate the use of semantic technologies in ubiquitous context-aware systems in order to identify the existing blocks that we will rely on to build our vision of "Social Ambient Intelligence".

STATE-OF-THE-ART

Previous studies (Strang & Linnhoff-Popien, 2004; Baldauf, Dustdar, & Rosenberg, 2007; O. Lassila & Khushraj, 2005) have identified ontologies as the most promising enabler for ubiquitous context-aware systems because they are heterogeneous and extensible by nature, and semantic technology enables « future-proof » interoperability. In this section, we will study the use of semantic technologies in four aspects of ambient intelligence: context management, human-system interactions, privacy enforcement, and social communications.

Semantic Context Management

According to (Dey, 2001), "a system is context-aware if it uses context to provide relevant information and/or services to the user, where relevancy depends on the user's task". By context, Dey means "any information that can be used to characterize the situation of an entity. An entity is a person, place, or object that is considered relevant to the interaction between a user and an application, including the user and applications themselves".

(Gu, Wang, Pung, & Zhang, 2004) gave an introduction to context-awareness by proposing the following requirements: « An appropriate infrastructure for context-aware systems should provide support for most of the tasks involved in dealing with contexts - acquiring context from various sources such as physical sensors, databases and agents; performing context interpretation; carrying out dissemination of context to interested parties in a distributed and timely fashion; and providing programming models for constructing of context-aware services. »

The use of ontologies to store and manipulate context have an impact on other aspects of the underlying system: context knowledge exchange, learning, user interactions, security and applications. In this section we will review several semantic-based approaches for context management platforms and identify the most successful approaches and current lacks.

Review of Major Context-Aware Platforms

One of the first semantic context modeling approaches was the **Aspect-Scale-Context (ASC)** model proposed by (Strang, Linnhoff-Popien, & Frank, 2003). Compared to non-semantic models, ASC enabled contextual interoperability during service discovery and execution in a distributed system. Indeed, this model consists of three concepts:

- Aspects are measurable properties of an entity (e.g. the current temperature of a room)
- Scales are metrics used to express the measure of these properties (e.g. Celsius temperature)
- Context qualifies the measure itself by describing the sensor, the timestamp and quality data

Contexts can be converted from a scale to another using Operations, also described semantically, and can be mapped to an implemented service. This model has been implemented as the CoOL Context Ontology Language. The CoOL core ontology can be formulated in OWL-DL (Dean & Schreiber, 2004) and F-Logic (object-oriented). The CoOL integration is an extension of the core to inter-operate with web services. OntoBroker (Decker, Erdmann, Fensel, & Studer, 1999) was chosen for semantic inference and reasoning, supporting F-Logic as knowledge representation and query language.

With EasyMeeting, (Chen et al., 2004) proposed a pragmatic application to demonstrate the benefits of their semantic context-aware system called **CoBrA**, for Context Broker Architecture. This application assists a speaker and its audience in a meeting situation by welcoming them in the room, dimming the lights, and displaying the presentation slides, either by vocal commands or automatically. The underlying prototype that they developed is a multi-agent system based on JADE (Java Agent DEvelopment Framework) [http://sharon.cselt.it/projects/jade/] in which a broker maintains a shared context model for all computing entities by acquiring context knowledge from various sensors and by reasoning on this knowledge to make decisions, as depicted on Figure 3. In the EasyMeeting application, this broker can deduce the list of expected participants and their role in the meeting by accessing their schedule, and can sense their actual presence when the bluetooth-enabled mobile phone declared in their profile is detected in the room. That way, the system can notify the speaker about their presence, decide to dim the lights and turn off the music when he arrives. These decisions are made possible by reasoning on the context knowledge using rules defined by the EasyMeeting application. The context knowledge is represented as RDF triples relying on the COBRA-ONT OWL ontology that includes vocabularies from the SOUPA ontology (Chen, Perich, Finin, & A. Joshi, 2004) covering time, space, policy, social networks, actions, location context, documents, and events, as depicted on Figure 4. Inferencing on the OWL ontology is handled by JENA's API [http://jena.sourceforge.net] whereas the JESS rule-based engine [http://herzberg.ca.sandia.

Figure 3. Overview of CoBrA, © 2003-2008 Harry Chen. Used with permission.

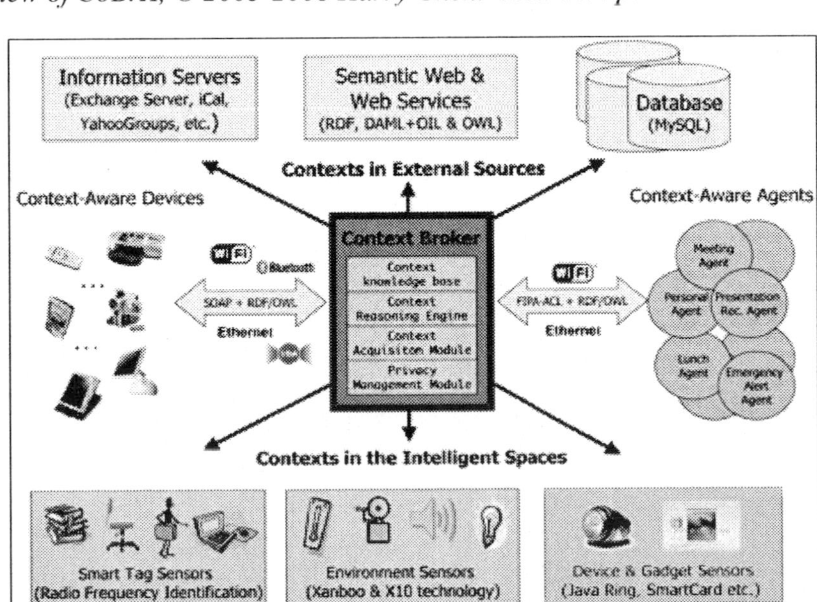

Figure 4. The SOUPA ontology, © 2003-2008 Harry Chen. Used with permission.

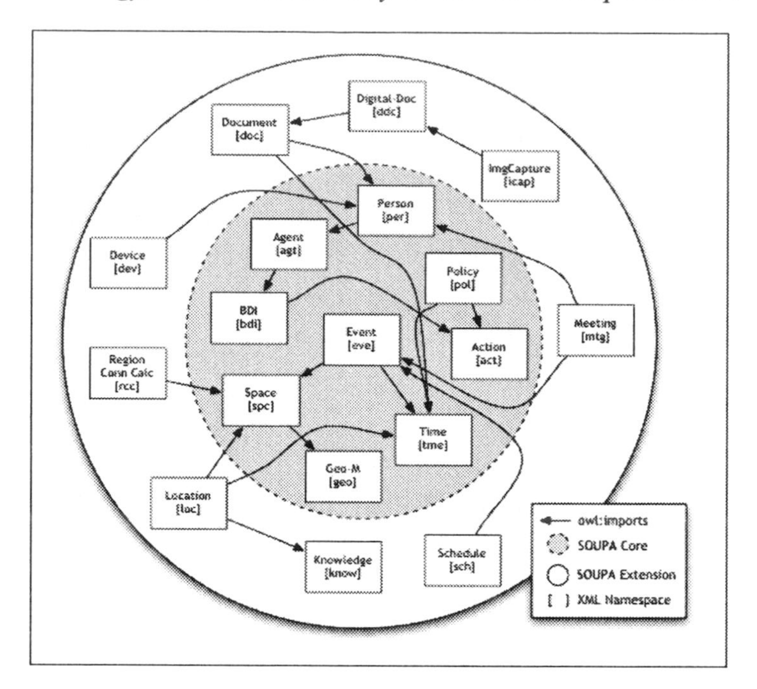

gov/] is used for domain-specific reasoning. The execution of rules (when results cannot be inferred from ontology axioms alone) uses the forward-chaining inference procedure of JESS to reason about contextual information. Note that, in this case, essential supporting facts must be extracted from RDF to JESS representation and the eventual results have to be injected in RDF to the knowledge base, which implies additional overhead in the process.

CoBrA's broker enforces privacy policies to define rules of behavior and restrict context communication. The enforcement of user-defined policies relies on the Rei role-based policy-reasoning engine (Kagal & T. A. Joshi, 2003) which does description logic inference over OWL. CoBrA also implements a meta-policy reasoning mechanism so that users can override some aspects of a global policy to define specific constraints at their desired level of granularity. However, they do not provide a tool for the user to express his/her privacy policy.

The **SOUPA** ontology proposed by (Chen et al., 2004) and used in CoBrA was a collaborative effort to build a generic context ontology for ubiquitous systems. Since 2003 it has been maintained by the "Semantic Web in Ubiquitous Comp Special Interest Group". The design of this ontology is driven by use cases and relies on FOAF, DAML-Time, OpenCyc (symbolic) + OpenGIS (geospatial) spatial ontology, COBRA-ONT, MoGATU BDI (human beliefs, desires and intentions) and Rei policy ontology (rights, prohibitions, obligations, dispensations). SOUPA defines its own vocabulary, but most classes and properties are mapped to foreign ontology terms using the standard OWL ontology mapping constructs (equivalentClass and equivalentProperty), which allows interoperability. In the core ontology in which both computational entities and human users can be modeled as agents, the following extensions are added: meeting & schedule, document & digital document, image capture and location (sensed location context of things).

Like CoBrA, **MOGATU** (Perich, Avancha, Chakraborty, A. Joshi, & Yesha, 2005) is a context-aware system based on the SOUPA ontology. However, this decentralized peer-to-peer multi-agent system implements several use cases covering automatic and adaptive itinerary computation based on real-time traffic knowledge, and commercial recommendation. In this approach, each device is a semi autonomous entity driven by the user's profile and context, relying on a contract-based transaction model. This entity is called InforMa and acts as a personal broker that handles exchanges with other peers. The user profile semantically defines his beliefs, desires and intentions, following the BDI model that is part of the SOUPA ontology. Beliefs are weighted facts depicting user's knowledge and preferences such as his schedule and food preferences, whereas desires express the user's goals. Intentions are defined as a set of intended tasks that can be inferred from desires or explicitly provided. However no clues are given by the authors about how these beliefs and intentions are defined by the user or the system, which let us assume that this is still a manual process yet to be enriched with profiling mechanisms and a graphical user interface to edit the profile. Moreover, this work being apparently focused on trusted peer-to-peer exchange of information according to the BDI user profile, details on the actual reasoning process on context knowledge are not given. InforMa is able to process queries that can possibly involve other peers and advertise information to these peers in vicinity, relying on graph search and caching techniques. However no details were given on how pro-activity is made possible. Another lack identified in the underlying BDI model is that the representation of pre-conditions and effects of intentions are left to the applications, but we have found no clues on how applications fill this issue. Facing an important cost of network transmissions in the exchange process, it seems that this research group is focusing on peer-to-peer networking optimization and trusted exchanges more than on the actual context management. However, they suggested that preparing purpose-driven queries in advance and caching intermediate query results could improve the performance of their system, which is an interesting approach that should be considered in distributed context-aware systems.

The CORBA-based **GAIA** platform proposed by (Ranganathan, Al-Muhtadi, & Campbell, 2004) focuses on hybrid reasoning about uncertain context, relying on probabilistic logic, fuzzy logic and Bayesian networks. In their approach, context knowledge is expressed using predicates which classes and properties are defined in a DAML+OIL ontology (Horrocks, 2002). Predicates can be plugged directly into rules and other reasoning and learning mechanisms for handling uncertainty. This choice reduces the overhead of the CoBrA system relying on RDF triples. Rules are processed by the XSB engine [http://xsb.sourceforge.net/], which is described as a kind of optimized Prolog that also supports HiLog, allowing unification on the predicate symbols themselves as well as on their arguments. HiLog's sound and complete proof procedure in first-order logic is needed to write rules about the probabilities of context.

GAIA's authentication mechanism demonstrates the usefulness of fuzzy/uncertain context reasoning. It allows users to authenticate with various means such as passwords, fingerprint sensor or bluetooth phone proximity. Each of these means have different levels of confidence, and some user roles may require that the user authenticates himself on two of them to cumulate their confidence level up to the required level.

Although GAIA proposes a common reasoning framework, application developers have to define the expected context inputs and specify the reasoning mechanism to be used by providing Prolog/HiLog rules (for probabilistic/fuzzy logic) or Bayesian networks. A graphical user interface is provided to help developers construct rules, whereas MSBN (Microsoft's Belief Network) can be used to create Bayesian nets. Although Bayesian networks are a powerful way to perform probabilistic sensor fusion and higher-level context derivation, they need to be trained. Moreover, inference with large networks (more than 50 nodes) becomes very

Figure 5. Partial definition of the CONON ontology extended with the home domain (Wang, Zhang, Gu, & Pung, 2004), © 2004 IEEE. Used with permission.

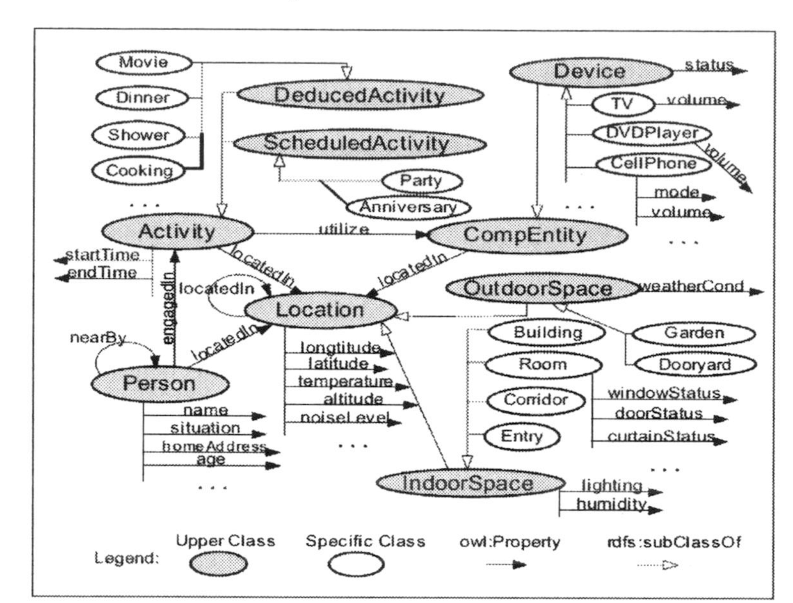

costly in terms of processing and can result in scalability problems.

Based on previous works, (Gu et al., 2004) propose **SOCAM** (Service-Oriented Context-Aware Middleware), another OWL-based context-aware framework with the aim to address more general use cases by adding more qualitative information on acquired context. The classifiedAs property allows the categorization of context facts as Sensed, Defined, Aggregated or Deduced. The dependsOn property allows the justification of a deduced context based on other context facts. Another contribution is the possibility to qualify context information with parameters such as accuracy, resolution, certainty and freshness. The SOCAM framework was proven (Gu, Pung, & Zhang, 2004) to reason successfully on uncertain contexts using Bayesian Networks, but no performance results were given. The same group of authors have also carried out a performance experiment of the **CONON** ontology (Wang, Zhang, Gu, & Pung, 2004) depicted on Figure 5, which is the name that was given to SOCAM's context ontology. Their results show that the duration of the reasoning process exponentially increases with the number of RDF triples stored in the context knowledge base, which reveals that this approach is not scalable for a widespread context-aware system. Therefore two leads were proposed to increase performance:

- to perform static, complex reasoning tasks (e.g., description logic reasoning for checking inconsistencies) in an off-line manner.
- to separate context processing from context usage, so that context reasoning can be performed by resource-rich devices (such as a server) while the terminals can acquire high-level context from a centralized service, instead of performing excessive computation themselves.

Later works of that team were focused on the peer-to-peer architecture for context information systems.

Basing on the CONON ontology, (Truong, Y. Lee, & S. Y. Lee, 2005) proposed the PROWL language ("Probabilistic annotated OWL") to generalize fuzzy/probabilistic reasoning from applications to domains by mapping Bayesian Networks to ontology classes and properties. This approach must be experimented with various context-aware applications to prove its feasibility.

The FP6 IST project **SPICE** (Service Platform for Innovative Communication Environment) brought a fresh approach to ubiquitous system, considering them in a wider scope centered on semantic knowledge management for improved ubiquitous end-user services (SPICE, 2006) (SPICE, 2007). On its Knowledge Management Layer, SPICE proposes two different implementations of the context provisioning subsystem: the IMS Context Enabler (ICE) (M. Strohbach, Bauer, E. Kovacs, C. Villalonga, & Richter, 2007) and the Knowledge Management Framework (KMF). In ICE, the SIP protocol (Session Initiation Protocol) is leveraged to control the parameters of the exchange sessions (e.g. data sets to communicate, update trigger, update frequency) and to flexibly adjust the communication path based on the changes in network structure and available context information. Both KMF and ICE rely on a shared ontology called the Mobile Ontology which is freely downloadable on the Internet [http://ontology.ist-spice.org/], the most important difference being the interfaces: ICE uses SIP whereas KMF uses OWL over SOAP Web Services for exchanging context information. However, gateways are also provided so that context data can be converted from a format to the other. Therefore we will abstract these implementations and focus on the common knowledge model. Embracing the recommendations of the W3C, SPICE Mobile Ontology is defined in OWL and the context data is expressed in RDF. Inspired from the Dutch project Freeband Awareness, SPICE's Physical Space ontology has a finer granularity than any previous context ontology: it notably defines properties for connections between rooms and floors. Following the approach of the « Doppelgänger User Modeling System » (Orwant, 1995), SPICE's User Profile ontology supports domain-specific and conditional (situation-specific) submodels. In this approach, the profile contains subsets which are considered on certain conditions expressed with the form: Context Type, Operator, Value. This allows variations of the profile, depending on the user's context and/or the targeted application/service.

The Knowledge Management Layer also contains a Knowledge Storage module, a Profile Manager, a Service and Knowledge Push and Notification module and three kinds of Reasoners: a Predictor, a Learner and a Recommender. The reasoners can request past knowledge directly from context sources or from an external knowledge storage source. Both feedback-based and observation-based learning are supported, generating LearntRule and LearntRuleSet instances in OWL. The results can be leveraged to propose Recommendations to the user. Experimental results on the use of different learning techniques are to be published. Another interesting contribution of SPICE in the context-awareness domain is the use of a KnowledgeParameter class that is used to qualify context information with values defining their probability, confidence, timestamp, temporal validity and accuracy. However we have not found any mechanism that is similar to the "dependsOn" property supported by SOCAM to justify high-level context with lower-level facts from which it was inferred.

Another part of the SPICE project called the Distributed Communication Sphere (Kernchen et al., 2007) allows dynamic discovery of users' surrounding devices, networks and services. This part includes components that leverage context knowledge to enable multimodal interaction, content delivery, data synchronization and dynamic widgets on terminals, requiring a lightweight rule engine to be deployed on every terminal. SPICE also provides the End User Studio, an Eclipse-based GUI (Graphical User Interface) shown on Figure 6 that allows end users to create custom trigger-action rules visually.

One of the biggest identified issues in previously reviewed semantic context-aware systems

Figure 6. Creating a rule-based service using SPICE's End User Studio (SPICE 2007) © 2008 SPICE. Used with permission.

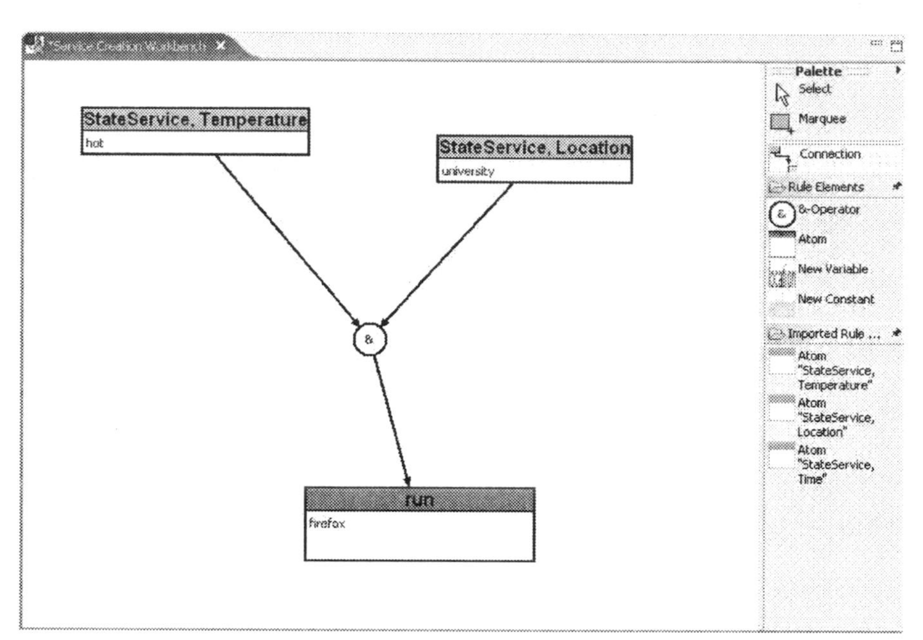

is the processing time required for reasoning on context knowledge. To answer this issue, (Ejigu, Scuturici, & Brunie, 2007) proposed an hybrid context management and reasoning system (HCoM) which relies on a heuristic-based context selector to filter the context data to be stored in the semantic context base for reasoning, the rest being stored in a relational database, as depicted on Figure 7. They report that this approach is more scalable than pure semantic context-awareness systems when the number of static context instances increases. (Lin, Li, Yang, & Shi, 2005) propose a similar approach but they filter context data according to their relevance to running applications instead of usage heuristics, in order to boost the reasoning performance.

(Tan, Zhang, Wang, & Cheng, 2005) propose to move from on-demand context reasoning to event-driven context interpretation so that reasoning on context data is processed as soon as it is received by the context management framework. However, in their distributed system, the performance is reduced because of increased communication overheads. Moreover, it does not support uncertainty yet.

Trends and Issues

In this section, we have reviewed several approaches addressing modeling, reasoning and distribution of contextual knowledge. Although semantic technologies have been shown as powerful tools to empower context-awareness, they also imply scalability problems, as the required processing time grows exponentially with the amount of knowledge, which is a major issue towards the realization of Ambient Intelligence. However, hybrid context management approaches leverage the assets of both relational and semantic context management, therefore they should be considered in the aim of building a powerful and scalable context-awareness system. Nevertheless, the selection/filtering of context data to be merged in the semantic database is not trivial and may need further research. Another track to consider is closer coupling or integration of rule

Figure 7. HCoM: Hybrid Context Management and reasoning system (Ejigu et al. 2007), © 2007 IEEE. Used with permission.

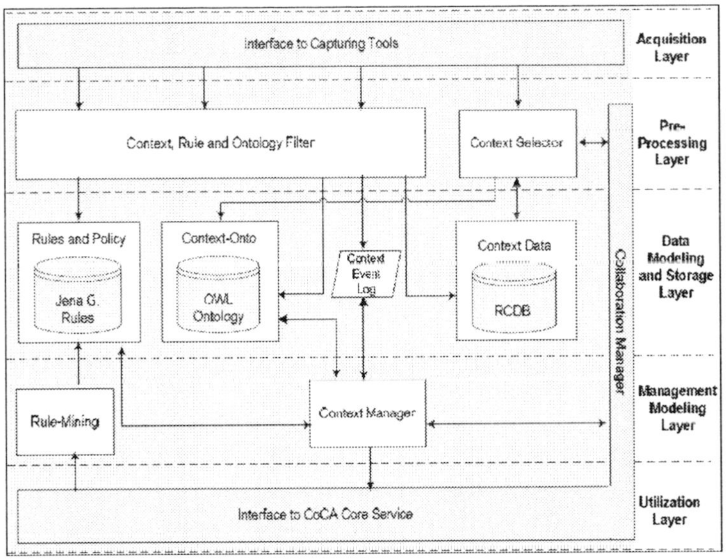

engines with knowledge bases in order to reduce processing overheads.

Semantics for Adapted Human-System Interactions

After context-awareness, another key aspect of Ambient Intelligence is how the user interacts with the digital world. Today, most internet-based interactions rely on the use of computers (i.e. a screen, a keyboard and a mouse). Whereas most people carry their own powerful mobile phone with them, most of the popular content and services are not adapted to general mobile devices with their constraints (small screen, no keyboard). Of course some of those have been adapted specifically to some popular platforms like the iPhone, but the vision of Ambient Intelligence is not only (i) to bring most of them to virtually any terminal according to its capabilities, but also (ii) to span various modalities of interaction over multiple interfaces (i.e. displays, inputs, speakers and other objects). Therefore, ambient services need to know the capabilities of every platform and interface they are used with, and they need to adapt the interaction to the user according to these capabilities. In this section we will review existing technologies for the discovery of devices and the description of their capabilities in order to enable rich user interactions and multimodality.

Semantic Discovery and Description of Interfaces

CC/PP (Composite Capabilities / Preferences Profile) (Klyne et al., 2004) is a recommendation from the W3C based on the Resource Description Framework (RDF) to create profiles that describe device capabilities and user preferences. It provides a syntax and tools to create terminal profiles and preference vocabularies, and thus can not be used as is. Indeed, the vocabulary of capabilities used for defining profiles is not in the scope of this recommendation and only structural rules and guidelines for interoperability are provided. However, the recommendation includes a pointer to the UAProf vocabulary as a referred example; we will review this vocabulary below. Among the features of the CC/PP syntax, allowed value types are listed, and the definition of default values is explained. The state-of-the-art of (SPICE, 2006) pointed out that conditional constraints are not supported in CC/PP. Moreover, the recommendation clearly informs that a CC/PP profile may include sensitive data, and delegates the enforcement of privacy to the application/system.

UAProf (User Agent Profile) (WAP Forum, 2001) is a CC/PP vocabulary for WAP (Wireless Application Protocol) enabled cell phones developed by the Open Mobile Alliance (OMA). The idea is that compliant cell phones have their capabilities described in a profile stored on a web repository so that adaptive services can gather this information in order to tailor content for embedded web browsers. This vocabulary is focused on software and hardware capabilities, and thus does not cover preferences.

WURFL (Wireless Universal Resource File) [http://wurfl.sourceforge.net/uaprof.php] is a collaborative effort to build an open XML file that describes device profiles based on fixes of their UAProf profiles. This promising initiative addresses several shortcomings of the original UAProf approach in which profiles can be inconsistent across providers, not up to date, or even do not exist.

The Foundation for Intelligent Physical Agents (**FIPA**) also proposed a device description ontology (FIPA, 2002) that can be used to reason and make decisions on the best device and modalities to create a user interface in multi-agent systems. Due to the nature of multi-agent systems, this approach differs from CC/PP in the manner of transmitting the profile. Instead of providing its complete profile on-demand, the terminal returns profile subsets adaptively to requests, allowing to set the granularity and scope of the required profile content in a gradual negotiation between agents. Whereas a CC/PP profile defines the capabilities for the software, hardware and the browser, FIPA Device Description supports the description of agent-related capabilities instead of the browser's. However, it is possible to use this ontology in a CC/PP profile, similarly to UAProf.

Even though this approach is not based on semantic technologies, the **UPnP** (Universal Plug and Play) discovery protocol (UPnP Forum, 2003) defines a XML language that can describe a physical device into a hierarchy of logical devices which map every hardware component of the device and thus its corresponding capability. A deeper study of UPnP is not in the scope of this chapter, but the modularity of this approach is interesting and should be considered in order to improve the re-usability of profiles, according to the fact that common hardware components are part of many devices.

Semantics for Multimodality

When devices and their capabilities are discovered, their use for multimodal interaction requires additional negotiation and synchronization so that user interaction constraints and preferences are respected for a rich user experience. The constraints to validate cover the quality of rendering/sampling, the robustness of the connectivity, the privacy of exchanges (e.g. displaying emails on a public screen should be avoided), and also the environmental context and user preferences.

The members of the W3C Multimodal Interaction Working Group propose their specifications of a **Multimodal Interaction Framework** (W3C, 2003) based on a central Interaction Manager that connects user inputs (e.g. audio, speech, handwriting, keyboard…) and outputs (e.g. speech, text, graphics, motion…) to applications and on two other components, as shown on Figure 8:

- The Session Component, which handles the state management for application sessions that may involve multiple steps, multiple modality modes, multiple devices and/or multiple users.
- The System and Environment component, which handles the changes of device capabilities, user preferences and contextual/environmental conditions.

Figure 8. The input process of the Multimodal Interaction Framework (W3C 2003), © 2003 World Wide Web Consortium

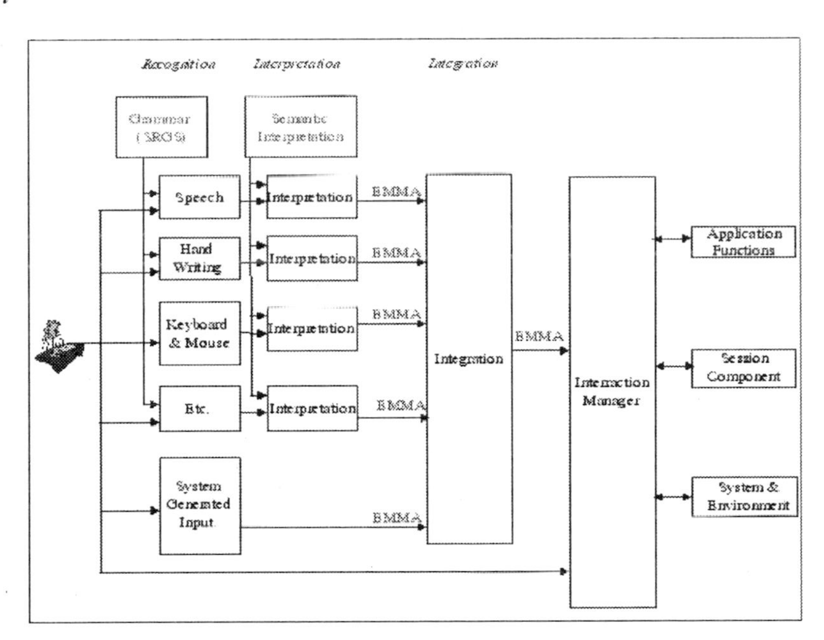

The Interaction Manager coordinates data and manages the execution flow from various input and output modality components. It combines various user inputs for submitting meaningful actions to applications (multimodal fusion) and dispatches responses to the user through various output interfaces (multimodal fission).

Also proposed by the W3C Multimodal Interaction Working Group, the EMMA (Extensible MultiModal Annotation) markup language (W3C, 2007) is a XML markup language for describing the interpretation of user inputs. An example of input interpretation is the transcription of a raw signal into words, for instance derived from speech, pen or keystroke input, or a set of attribute/value pairs describing a gesture. The interpretations of user's input are expected to be generated by signal interpretation processors, such as speech and ink recognition, semantic interpreters, and other types of processors for use by components that act on the user's inputs such as interaction managers. As shown on Figure 8, user inputs are processed in two layers to generate EMMA data which is integrated for submission to the Interaction Manager. The two layers of input processing consist of:

- Recognition components, which capture natural input from the user and translate them into a form useful for later processing. (e.g. speech to text, handwritten symbols and messages to text, mouse movements to x-y coordinates on a two-dimensional surface…)
- Interpretation components, which further process the results of recognition components by identifying the meaning/semantics intended by the user. (e.g. pointing somewhere on a map would result in knowing the name of the corresponding country, nodding or saying "I agree" would both mean acceptation from the user…)

Recommended by the W3C, EMMA is probably going to become a standard for annotation of multimodal inputs. It has shown its usefulness especially for speech-based dialog in extensible multimodal applications (Reithinger & Sonntag,

2005; Manchón, del Solar, de Amores, & Pérez, 2006; Oberle et al., 2006).

The IST project **Mobilife** proposed a solution (Kernchen, Boussard, Moessner, & Mrohs, 2006) to describe devices and modality services to form context-aware multimodal user interfaces. Their identified requirements include the deployment of a fission component implementing a rule-based algorithm on the device in order to adapt the user's mobile multimodal interface best to the current situation. In the SPICE project (Kernchen et al., 2007), the « Multimedia Delivery and Control System » depicted on Figure 9 has been developed as a part of the « Distributed Communication Sphere », is a multimodal platform relying on the W3C-recommended Synchronized Multimedia Integration Language (SMIL) (Ayars et al., 2000), that supports multimodal fusion and fission. First, the « resource discovery system » of the **MDCS** finds appropriate interfaces, then modalities are selected according to user preferences, context (e.g. Walking, driving), available resources in user's DCS and provision constraints. Modality, device and network recommendations are proposed by the knowledge management framework. This implementation is available as an open source project [https://sourceforge.net/projects/mdcs].

Trends and Issues

In this section, we have identified that semantic technologies have shown their usefulness to improve the discovery, description and exploitation of multimodal interfaces. Several collaborative efforts have been carried out to describe device capabilities. Besides, multimodal platforms are emerging with standardization support from the W3C. This progress leads to the interface-agnostic aspect of ubiquitous computing, but the state of the art of multimodality has still not been transferred from researchers to end-users.

The vision of ubiquitous computing, in which any screen can be used to display personal information, requires privacy enforcement mechanisms, especially if public screens are expected to be shared as well for this matter. In the next part, we will study how semantic technologies can help enforcing privacy in such systems.

Figure 9. SPICE Multimedia Delivery and Control System (Kernchen et al. 2007), © 2007 Ralf Kernchen. Used with permission.

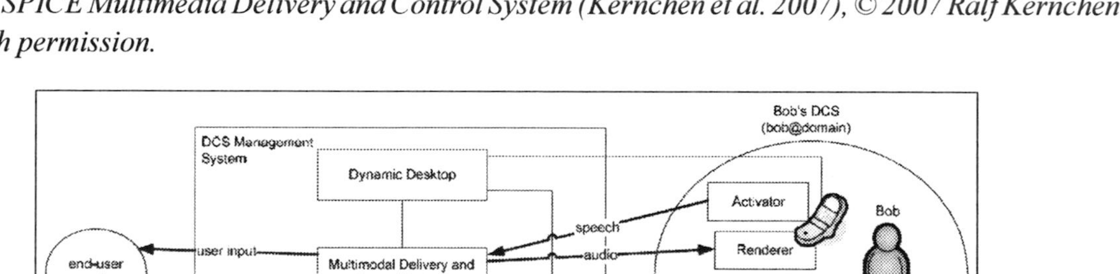

Semantics for Privacy

The vision of ubiquitous computing in which personal information flows in a highly networked ecosystem requires privacy enforcement mechanisms, especially if public screens are expected to be shared for displaying such information as well as personal terminals. Although privacy is a very rich and specific research domain, in this part, we will study how semantic technologies can help to enforce privacy in ubiquitous systems by reviewing a few approaches that must be considered to enforce privacy in Social Ambient Intelligence systems.

According to (Damianou, Dulay, Lupu, & Sloman, 2001), the use of policies is an emerging technique for controlling and adjusting the low-level system behaviors by specifying high-level rules. Policies enforced using semantic rule engines are implemented in most secure semantic context-aware platforms studied earlier in this chapter. In their review of semantic web languages for policy representation and reasoning, (Tonti et al., 2003) explain that "the use of policies allows administrators to modify system behavior without changing source code or requiring the consent or cooperation of the components being governed". **KAoS and Rei** are both semantic policy languages: KAoS is an OWL-based language and uses Java Theorem Prover to support reasoning whereas Rei uses Prolog and RDF-S. They also propose different enforcement mechanisms: KAoS requires the enforcers to be implemented and integrated in the system entities to control, whereas the Rei's actions are to be executed outside the Rei's engine.

(Shankar & Campbell, 2005) propose an extension to the ECA (Event-Condition-Action) rule framework, called Event-Condition-PreCondition-Action-PostCondition (**ECPAP**). In this framework, actions are annotated with axiomatic specifications that enable powerful reasoning to detect conflicts and cycles in policies.

(Brar & Kay, 2004) propose "secure persona exchange" (SPE), a framework based on W3C's Platform for Privacy Preferences (P3P) for secure anonymous/pseudonymous personal data exchange. This framework allows users to negotiate agreements with services that declare their privacy practices and request personal data. The P3P defines such a semantic service description format whereas privacy preferences are described using the APPEL language (A P3P Preference Exchange Language). SPE addresses the following identified end-user requirements: purpose specification, openness, simple and appropriate controls, limited data retention, pseudonymous interaction and decentralized control.

Trends and Issues

We have identified three semantic models that can be used to enforce privacy in ubiquitous systems: rule-based policies, ECA-based policies and secure exchange negotiation according to privacy preferences. Although the last one is the only one to address the issue of secure exchange of personal information, these approaches are complementary and promising to enforce privacy in Social Ambient Intelligence systems.

Semantics for the Social Communications and Activities

In the era of social networking and the participative web, of always-connected chat messengers and virtual worlds, people communicate and exchange more and more over the Internet. If computers are expected to disappear, the communication and exchange paradigms must be adapted to take the context of the users into account and to leverage the social knowledge held in web platforms in order to improve the awareness (and thus, intelligence) of Social Ambient Intelligence systems. One of the key points of such communications is user presence, because being online does not mean paying attention to any discussion at anytime. The second point that we will discuss covers user profiling techniques and the expression of the social graph. Finally, promising technologies for augmenting social activities with the Internet in an interoperable way will be discussed.

User Presence and Communication

The major context information in a synchronous communication network is presence, which is information on reachability, availability, and status across all communication channels (e.g., networks, applications, transports over Internet, wireless and wireline).

Two major presence exchange formats are considered here. The first one is **SIMPLE** (Session Initiation Protocol for Instant Messaging and Presence Leveraging Extensions), an extension of the SIP protocol recommended by the Open Mobile Alliance (OMA) that supports new features such as: voice, video, application sharing, and messaging. Leveraging the communication and security of the IMS (IP Multimedia Subsystem) platform, SIMPLE extends the user's presence to take into account the user's willingness, ability and desire to communicate across all different kinds of media types, devices, and places. Even though it is not a semantic language, the Dutch project Freeband Awareness (Bargh et al., 2005) chose the SIP/SIMPLE protocol for realizing a context-aware network infrastructure with the focus on secure and privacy-sensitive context exchange between a core network owner (e.g. a cell carrier) and external entities. In other projects, the use of SIP can be limited to exchanges that imply an interaction with the user: notifications, confirmations… In the SPICE project (M. Strohbach, E. Kovacs, & Goix, 2007), SIP is used to share presence information with the IMS platform and exchange data with the communicating user. On another hand, SPICE's Mobile Ontology includes a presence ontology based on **PIDF** (Presence Information Data Format) which allows definitions of the user's input, mood, contact relationship, place characteristics, current activity, and service. Transformation templates are provided to switch from the internal semantic representation in RDF into PIDF, and the other way round.

SIP has a wide range of possible uses but is not an optimal solution for all kinds of exchange. (Houri, 2007) criticized the weakness of SIP/SIMPLE in domain scaling. Furthermore it appears (Saint-Andre, 2005) that SIP/SIMPLE does not support advanced messaging mechanisms like workflow forms, multiple recipients, reliable delivery and publish-subscribe which are useful for context-aware systems. PIDF has shown to be suitable for the SPICE project.

Profiling and Social Graph

Considering the user's profile and social graph is important to personalize access to information and communication means. At a time when silo web-based social networking sites rapidly spread, many initiatives try to free our social data from these platforms using interoperable formats.

FOAF (Friend-of-a-Friend) (Brickley & Miller, 2007) is a RDF vocabulary based on an OWL ontology to describe people profiles, friends, affiliations, creations and other metadata related to people. FOAF's vision is a decentralized and extensible machine-readable social network based on personal profiles. The profile contains descriptions of personal user data, possibly his/her work history, and links to his/her contacts and affiliated services. Each person has a unique identifier, usually a hash of the email address. The community of FOAF users being principally made of researchers and semantic web enthusiasts, it does not compete with popular social networks like LinkedIn [http://www.linkedin.com], Myspace [http://www.myspace.com] or Facebook [http://www.facebook.com]. Many tools have appeared, including FOAFexplorer [http://xml.mfd-consult.dk/foaf/explorer/] which can be used to visualize FOAF profiles. However, there is a potential privacy issue with this language because selective privacy-aware views of a FOAF file are not addressed. It may be interesting to evaluate a mechanism similar to the conditional profiles proposed in the SPICE project or to enforce selective distribution of content using a policy-based system.

SIOC (Semantically-Interlinked Online Communities, http://sioc-project.org) represented on Figure 10 is an ontology-based framework aimed at interconnecting online community sites and internet-based discussions. The idea is to enable cross-platform interoperability so that conversation spanning over multiple online media (e.g. blogs, forums, mailing lists) can be unified into one open format. The interchange format expresses the information contained both explicitly and implicitly in internet discussion methods, in a machine-readable manner. A similar approach is proposed by the OPSN (Open Portable Social Network, http://www.opsn.net/) initiative which also covers notification and synchronization of contacts across platforms. However there is no existing implementation, and privacy control for personal published information seems not to have been addressed yet. DISO (distributed social networking, http://diso-project.org/), is yet another collaborative work to follow.

These initiatives would be a promising way to leverage consistent social relations, discussions and exchanges from various web platforms in order to build a more precise profile of user's interests, like with the **APML** language (Attention Profiling Mark-up Language, http://www.apml.org/), and qualify the types of relations in order to improve the social communication experience.

Social Interactivity

Social networking sites (SNSs) have become very popular communication platforms on the Internet, enabling new ways for people to interact with each other. Although the proposed interactions are similar on most SNSs, each of these sites were developed as silos, and thus their social graph (i.e. the list of "friends") and applications are not portable. We believe that consolidating SNS-based interactions is a key towards our vision of Social Ambient Intelligence, and that semantic technolo-

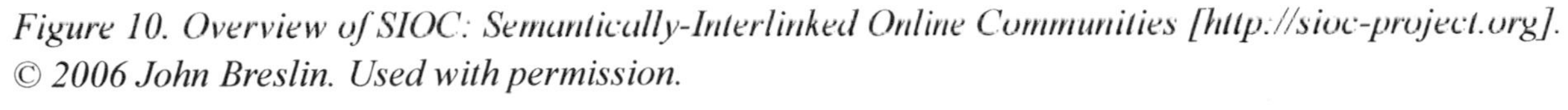

Figure 10. Overview of SIOC: Semantically-Interlinked Online Communities [http://sioc-project.org]. © 2006 John Breslin. Used with permission.

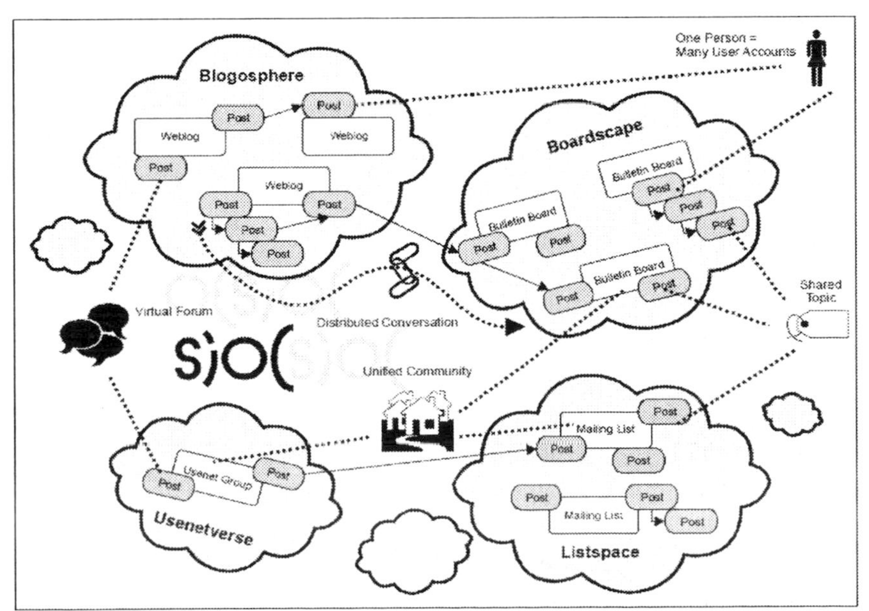

gies can help to solve this interoperability issue.

With its open application platform, the social networking site **Facebook** became a huge Internet player in a few months, attracting many service providers and increasing their population of users significantly. Indeed, Facebook made it easy for application developers to leverage the user's profile and social graph of the underlying platform, and thus bring user-friendly services with a social dimension. For example, as shown on Figure 11, the "Movies" application allows the user to rate movies so that his/her favorite movies are shown on his profile page. But the most interesting aspect of this application is the possibility for friends to compare their movie tastes to evaluate their compatibility.

Because there are many existing social networking sites on the Internet that are adopting the application platform approach à la Facebook, Google initiated the **OpenSocial** project, an interoperable framework to build applications on any compliant social networking site. However this framework implements basic contact management actions only and don't have access to all the information and capabilities of all social networking sites. For example, some of them are capable of exchanging "pokes", "gifts" and comments, but there is no interoperable way of invoking these capabilities from OpenSocial so far. This could be the opportunity to develop an ontology of social interaction which could be enriched by the platforms and gradually supported by applications without preventing them to work in degraded mode (e.g. by sending a comment instead of a gift, if this capability is not supported by the platform).

Trends and Issues

Despite the exponential popularity and value of Social Networking Websites (SNSs) on the Internet, the possible links between ubiquitous context-aware platforms and existing "Web 2.0" platforms (O'Reilly, 2005) have been neglected by academia, while Internet players are working together to build controlled interoperability. Although extraction of consistent knowledge from the Web 2.0 is not trivial (Gruber, 2006), there is a huge value in social networks sites (and user-generated content) that should be leveraged to extend the awareness of Social Ambient Intelligence systems, as we will explain in the next section. We believe that proposing a common SNS interaction ontology in current collaborative efforts such as the OpenSocial project is a good track for researchers towards our vision of Social Ambient Intelligence.

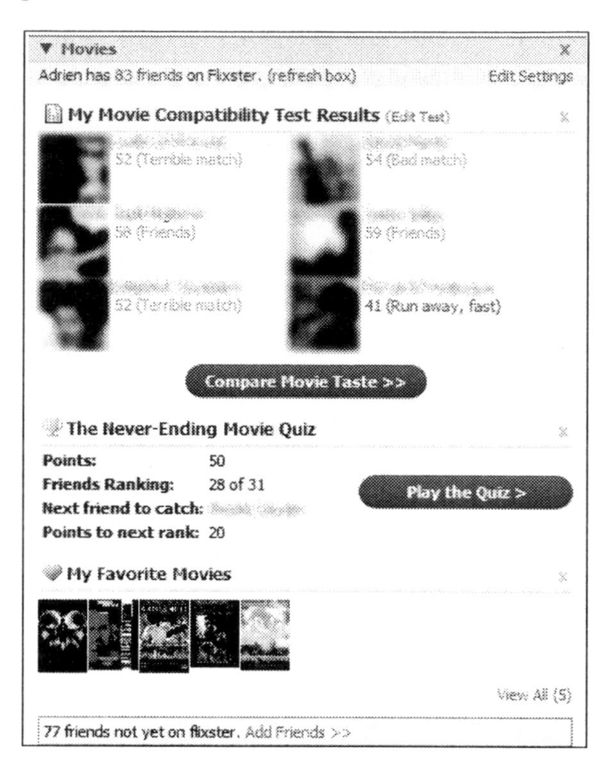

Figure 11. The "Movies" application on Facebook,© 2008 Flixster, Facebook. Used with permission.

Conclusions of the State-of-the-Art

In this State-of-the-Art part, we have depicted an overview of several past and current approaches for context-aware systems, adapted human-system interactions, privacy enforcement and social communications and activities. We have identified the assets of semantic technologies in all these domains, and several issues.

Whereas semantic technologies are a powerful tool to enable interoperability among heterogeneous entities, and to unify knowledge in a common model, we realise that existing research on Ambient Intelligence does not leverage the value of collective intelligence which has emerged with the Web 2.0 and its Social Networking Sites. In the next part of this chapter, we will respond to this paradox by defining our vision of "Social Ambient Intelligence" and proposing several research leads towards this vision.

REALIZING SOCIAL AMBIENT INTELLIGENCE

In this part, we define our vision of "Social Ambient Intelligence" and propose several research leads towards this vision, based on our previous study.

What is Social Ambient Intelligence?

As explained in the Background part of this chapter, the vision of "Ambient Intelligence" consists in leveraging new technologies and techniques (including context-awareness) to design applications that are user-centric, and thus more adapted to the user, his knowledge and his current environment/situation. As the Web 2.0 gave birth to the concept of Collective Intelligence, which consists in generating knowledge from user contributions and interactions on the Internet, it sounds like leveraging this knowledge would be extremely valuable to increase the awareness of "Ambient Intelligence" systems. Assuming that, for instance, recommendations coming from friends are necessarily given more confidence than recommendations coming from predictive statistics, adding a social dimension to "Ambient Intelligence" would result in more relevant results for users, and thus a better user-centricity, which was the rationale of "Ambient Intelligence".

Based on this analysis, we propose "Social Ambient Intelligence" (SocAmI) as an extension of "Ambient Intelligence" (AmI) that adds a social dimension in order to increase awareness, knowledge and intelligence of such systems. This social dimension would benefit from the "collective intelligence" of Web 2.0 platforms (such as Social Networking Sites), and therefore it will bring more relevance and confidence to users. The addition of this dimension also gives the opportunity to augment the user communication experience with new kinds of social interactions inspired by Social Networking Sites, without having to sit behind a computer.

As semantic technologies have been shown as an excellent framework to model, integrate and exchange formalized knowledge in a unified manner among heterogeneous agents/entities that constitute Ambient Intelligence systems, we believe in their capability to integrate the social knowledge gathered from the Collective Intelligence of the Web users.

In the following paragraphs, we will discuss the issues and challenges implied by the realization of Social Ambient Intelligence.

Converging with the Social Web

It is time for the social web, context awareness, and multimodal interfaces to converge into a Social Ambient Intelligence platform that enforces users' privacy. We believe that semantic technologies are the best enablers for interoperability, extensibility and intelligent exploitation of user, hardware and social web knowledge, in order to improve interactions between users and information. However, leveraging web knowledge in a semantic ubiquitous system may not be a trivial task according to (Strassner et al., 2007) who claimed that: in order for ontologies to be adopted by a system, this system should have a sufficient amount of semantic knowledge and minimal legacy information to carry. Indeed, the Semantic Web still being an unachieved vision (Berners-Lee, Hendler, & O. Lassila, 2001; Cardoso, 2007), most websites don't rely on semantic technologies to maintain their data. We have presented several initiatives that intend to create interoperable standards based on semantic technologies for universal use of user-generated content and communications kept in separate web platforms. Academics should get involved in this process, in order to take into account the requirements of Social Ambient Intelligence platforms that will leverage these standards. In the mean time, web platforms APIs (Application Programming Interface) can be used to build gateways between specific web social platforms and ubiquitous systems. For example, user feeds (e.g. Facebook's mini-feed, twitter, del.icio.us) could be analyzed as an additional source of context knowledge in the aim of identifying user activities and profile. On the other hand, ubiquitous systems could also be used to push content to these platforms, e.g. automatic presence information inferred from the context.

Bringing Ubiquitous Systems to People

Another issue that we want to address here by adding a Social dimension to Ambient Intelligence is the lack of integration and public visibility of research works related to Ambient Intelligence. The growing ubiquity of networks (infrastructures and ad-hoc), screens and mobile devices brings more exciting opportunities for people to communicate and exchange content but we lack interoperability standards, preventing people from experimenting state-of-the-art research results. In the meantime, innovative ubiquitous products appear on the market, such as electronic photo frames, widget displays, toys that can give weather reports and read emails, and powerful domestic management systems but they all work on their own because we lack common standards and platforms. One promising way of making people progressively adopt ubiquitous systems is to advertise them as applications on popular social platforms (e.g. Facebook), inviting users to deploy required software on their terminal to benefit from exciting new services that could possibly leverage users' context and social graph. Some people may be reluctant to use such systems at first, but we believe that there are solutions to make them accept them.

Gain Trust from Potential Users

Potential users of ubiquitous context-aware systems can be reluctant for the following reasons:

1. Privacy

Users will be concerned with the idea of provisioning private contextual knowledge (such as user positioning) to a "black-box" system which they may not trust, because they are afraid of loosing control of this information (Abowd & Mynatt, 2000), of being tracked or even spied (Bohn, Coroama, Langheinrich, Mattern, & Rohs, 2004). Moreover, most Internet users are already concerned with spam, and many already complain about profiling operated by web sites to improve the relevance of advertising; therefore sharing contextual knowledge can be seen as a major threat for privacy and control of personal information. We believe that advertising should be taken into account as the fair counterpart of a service, but it must be moderated by the system. E.g. a music recommendation service that advertises live performances and merchandising of one's favorite artists seems like an fair service that benefits both the user and the service provider, if the user is fond of music. Nevertheless, the user must constantly be in control of his private information, and confidentiality/security of exchanges must be enforced using mechanisms such as pseudonymity or cryptology. The transparency of the ubiquitous system's implementation and knowledge base can be a major source of trust for users, like it has been with open source software.

2. Intrusion

The subscription to many services that have access to extensive knowledge about users (e.g. their interests, their social network) and also privacy policy management can lead to digital pollution. Users could receive hundreds of recommendations, being asked hundreds of questions about their current situations and confirmations for proposed relevant actions to undertake. Research must be carried out to moderate explicit user interaction (i.e. requests and notifications) without compromising intended communications, user awareness and control. A promising approach for semi-autonomous control of user private data is the use of policies. However, as (O. Lassila, 2005) pointed out, we need a rich representation of policies so that users can define and visualize their privacy rules in a clear and easy way, and delegate their enforcement to the system.

In this section, we have sketched our vision of Social Ambient Intelligence. Main issues consist of the convergence of Ambient Intelligence with the Social web, the involvement of end-users with current research works, the definition of common standards, and the trust to be gained from users (regarding privacy and intrusion).

CONCLUSION

In this chapter, we have reviewed several uses of semantic technologies for context management, adaptive human-system interaction, privacy enforcement and social communications in the scope of Ambient Intelligence. Based on identified benefits and lacks, we defined our vision of "Social Ambient Intelligence" and proposed several research leads towards the realization of this vision based on the convergence of Ambient Intelligence, Collective Intelligence of the Social Web and Semantic Technologies. Through our involvement in several ongoing European, national and internal research projects, we will strive to focus our research on these points and to convey our position and trends to our collaborators.

ACKNOWLEDGMENT

The authors would like to thank Laure Pavlovic for her kind support to improve the quality of this chapter. We would also like to thank the authors of the figures included in this chapter for granting us permission to include them.

REFERENCES

W3C. (2003). W3c multimodal interaction framework. *W3C Note*. Retrieved from http://www.w3.org/TR/mmi-framework/.

W3C. (2007). Emma: extensible multimodal annotation markup language. *W3C Working Draft 14 December 2004*. Retrieved from http://www.w3.org/TR/emma/.

Abowd, G., & Mynatt, E. (2000). Charting Past, Present, and Future Research in Ubiquitous Computing. *ACM Transactions on Computer-Human Interaction*, *7*(1), 29–58. doi:10.1145/344949.344988

Ayars, J., Bulterman, D., Cohen, A., Day, K., Hodge, E., Hoschka, P., et al. (2000). Synchronized multimedia integration language (smil) 2.0 specification, *Work in progress. W3C Working Drafts are available at http://*www. w3. org/TR, 21.

Baldauf, M., Dustdar, S., & Rosenberg, F. (2007). A survey on context-aware systems . *International Journal of Ad Hoc and Ubiquitous Computing*, *2*(4), 263–277. doi:10.1504/IJAHUC.2007.014070

Bargh, M., Benz, H., Brok, J., Heijenk, G., Groot, S. H. D., Peddemors, A., et al. (2005, January 10). J. In Brok (Ed.), *Initial architecture for awareness network layer Retrieved from*http://awareness.freeband.nl.

Berners-Lee, T., Hendler, J., & Lassila, O. (2001). The semantic Web. *Scientific American, 284*(5), 28–37.

Bohn, J., Coroama, V., Langheinrich, M., Mattern, F., & Rohs, M. (2004). Social, Economic, and Ethical Implications of Ambient Intelligence and Ubiquitous Computing. *Ambient Intelligence. Springer-Verlag*.

Brar, A., & Kay, J. (2004). *Privacy and security in ubiquitous personalized applications*.

Brickley, D., & Miller, L. (2007, November 2). Foaf vocabulary specification, *Namespace Document*. Retrieved from http://xmlns.com/foaf/spec/.

Cardoso, J. (2007). The semantic web vision: where are we? *Intelligent Systems, 22*(5), 84–88. doi:10.1109/MIS.2007.4338499

Chen, H., Finin, T., Joshi, A., Kagal, L., Perich, F., & Chakraborty, D. (2004). Intelligent agents meet the semantic web in smart spaces. *IEEE Internet Computing, 8*(6), 69–79. doi:10.1109/MIC.2004.66

Chen, H., Perich, F., Finin, T., & Joshi, A. (2004). Soupa: standard ontology for ubiquitous and pervasive applications, *Mobile and Ubiquitous Systems: Networking and Services, 2004. MOBIQUITOUS 2004. The First Annual International Conference on*, (pp. 258-267).

Christopoulou, E. (2008). Context as a Necessity in Mobile Applications. *Handbook of Research on User Interface Design and Evaluation for Mobile Technology*.

Damianou, N., Dulay, N., Lupu, E., & Sloman, M. (2001). The ponder policy specification language. *Policies for Distributed Systems and Networks: International Workshop, Policy 2001*, Bristol, Uk, January 29-31, 2001: Proceedings.

Dean, M., & Schreiber, G. (2004). Owl web ontology language reference. *W3C Recommendation*.

Decker, S., Erdmann, M., Fensel, D., & Studer, R. (1999). Ontobroker: ontology based access to distributed and semi-structured information. *Database Semantics: Semantic Issues in Multimedia Systems*, (pp. 351–369).

Dey, A. K. (2001). Understanding and using context. *Personal and Ubiquitous Computing, 5*(1), 4–7. doi:10.1007/s007790170019

Ducatel, K., Bogdanowicz, M., Scapolo, F., Leijten, J., & Burgelma, J. C. (2001). *Scenarios for ambient intelligence in 2010 (ISTAG 2001 Final Report)*.

Ejigu, D., Scuturici, V., & Brunie, L. (2007). Semantic approach to context management and reasoning in ubiquitous context-aware systems In *The Second IEEE International Conference on Digital Information Management (ICDIM 2007), Proceedings of ICDIM'07*. (pp. 500-5005). Retrieved from http://liris.cnrs.fr/publis/?id=3242.

FIPA. (2002, December 6). *Fipa device ontology specification*.

Forum, W. A. P. (2001, October 20). Wag uaprof. Retrieved from http://www.openmobilealliance.org/tech/affiliates/wap/wap-248-uaprof-20011020-a.pdf.

Gruber, T. (2006). Where the social web meets the semantic web. *Lecture Notes in Computer Science, 4273*(994).

Gruber, T. R. (1993). A translation approach to portable ontology specifications. *Knowledge Acquisition, 5*(2), 199–220. doi:10.1006/knac.1993.1008

Gu, T., Pung, H. K., & Zhang, D. Q. (2004). A bayesian approach for dealing with uncertain contexts, *Proceedings of the Second International Conference on Pervasive Computing.*

Gu, T., Wang, X. H., Pung, H. K., & Zhang, D. Q. (2004). An ontology-based context model in intelligent environments, *Proceedings of Communication Networks and Distributed Systems Modeling and Simulation Conference*, 2004.

Häkkilä, J., & Mäntyjärvi, J. (2005). Combining Location-Aware Mobile Phone Applications and Multimedia Messaging. *Journal of Mobile Multimedia, 1*(1), 18–32.

Horrocks, I. (2002). Daml+oil: a description logic for the semantic web. *A Quarterly Bulletin of the Computer Society of the IEEE Technical Committee on Data Engineering, 25*(1), 4–9.

Houri, A. (2007, October 29). Draft-ietf-simple-interdomain-scaling-analysis-02 - presence interdomain scaling analysis for sip/simple, *Presence Interdomain Scaling Analysis for SIP/SIMPLE.* Retrieved November 13, 2007, from http://tools.ietf.org/html/draft-ietf-simple-interdomain-scaling-analysis-02.

Kagal, L. F., & Joshi, T. A. (2003). A policy language for a pervasive computing environment, *Policies for Distributed Systems and Networks, 2003. Proceedings. POLICY 2003. IEEE 4th International Workshop on*, 63-74.

Kernchen, R., Boussard, M., Hesselman, C., Villalonga, C., Clavier, E., Zhdanova, A. V., et al. (2007). Managing personal communication environments in next generation service platforms, *Mobile and Wireless Communications Summit, 2007. 16th IST*, 1-5.

Kernchen, R., Boussard, M., Moessner, K., & Mrohs, B. (2006). Device description for mobile multimodal interfaces In Mykonos, Greece.

Klyne, G., Reynolds, F., Woodrow, C., Ohto, H., Hjelm, J., Butler, M. H., et al. (2004). *Composite capability/preference profiles (cc/pp): structure and vocabularies 1.0. w3c recommendation, w3c, january 2004.*

Korpipää, P., Häkkilä, J., Kela, J., Ronkainen, S., & Känsälä, I. (2004). Utilising context ontology in mobile device application personalisation. *Proceedings of the 3rd international conference on Mobile and ubiquitous multimedia*, (pp. 133-140).

Lassila, O. (2005, August). *Using the semantic web in mobile and ubiquitous computing*. Jyväskylä (Finland).

Lassila, O., & Khushraj, D. (2005). Contextualizing applications via semantic middleware In . *Mobile and Ubiquitous Systems: Networking and Services, 2005*, 183–189. doi:10.1109/MOBIQUITOUS.2005.19

Leong, L. H., Kobayashi, S., Koshizuka, N., & Sakamura, K. (2005). CASIS: a context-aware speech interface system. *Proceedings of the 10th international conference on Intelligent user interfaces*, (pp. 231-238).

Lin, X., Li, S., Yang, Z., & Shi, W. (2005). Application-oriented context modeling and reasoning in pervasive computing, *Proceedings of the The Fifth International Conference on Computer and Information Technology*, (pp. 495-501).

Manchón, P., del Solar, C., de Amores, G., & Pérez, G. (2006). The mimus corpus In (pp. 56-59). Genoa, Italy.

O'Reilly, T. (2005). What is web 2.0: design patterns and business models for the next generation of software, *O'Reilly*. Retrieved from http://www.oreillynet.com/pub/a/oreilly/tim/news/2005/09/30/what-is-web-20.html.

Oberle, D., Ankolekar, A., Hitzler, P., Cimiano, P., Sintek, M., Kiesel, M., et al. (2006). Dolce ergo sumo: on foundational and domain models in swinto (smartweb integrated ontology), *Submission to Journal of Web Semantics (2006)*.

Orwant, L. (1995). Heterogeneous learning in the doppelganger user modelling system. *User Modeling and User-Adapted Interaction*, *4*(2), 107–130. doi:10.1007/BF01099429

Perich, F., Avancha, S., Chakraborty, D., Joshi, A., & Yesha, Y. (2005). *Profile driven data management for pervasive environments*. Springer.

Ranganathan, A., Al-Muhtadi, J., & Campbell, R. H. (2004). Reasoning about uncertain contexts in pervasive computing environments. *Pervasive Computing, IEEE*, *3*(2), 62–70. doi:10.1109/MPRV.2004.1316821

Reithinger, N., & Sonntag, D. (2005). An integration framework for a mobile multimodal dialogue system accessing the semantic web, *Ninth European Conference on Speech Communication and Technology*.

Sadi, S. H., & Maes, P. (2005). *xLink: Context Management Solution for Commodity Ubiquitous Computing Environments*. In Tokyo, Japan.

Saint-Andre, P. (2005, December 8). Xmpp-simple feature comparison, *XMPPSIMPLE Feature Comparison*. Retrieved November 13, 2007, from http://www.jabber.org/protocol/xmpp-simple.shtml.

Schilit, B. N., Adams, N., & Want, R. (1994). *Context-Aware Computing Applications*. Santa Cruz, CA, USA.

Shankar, C., & Campbell, R. (2005). A policy-based management framework for pervasive systems using axiomatized rule-actions. *Network Computing and Applications, Fourth IEEE International Symposium on*, (pp. 255-258).

SPICE. (2006). *Spice d4.1: ontology definition of user profiles, knowledge information and services*. Retrieved February 13, 2008, from http://www.ist-spice.org/documents/D4.1-final.pdf.

SPICE. (2007). *Spice unified architecture*. Retrieved from http://www.ist-spice.org/documents/SPICE_WP1_unified_architecture_Phase%202.pdf.

Strang, T., & Linnhoff-Popien, C. (2004). A context modeling survey. *Workshop on Advanced Context Modelling, Reasoning and Management, UbiComp*, (pp. 34-41).

Strang, T., Linnhoff-Popien, C., & Frank, K. (2003). Cool: a context ontology language to enable contextual interoperability, *Distributed Applications and Interoperable Systems: 4th Ifip Wg6. 1 International Conference, Dais 2003, Paris, France, November 17-21, 2003, Proceedings*.

Strassner, J., O'Sullivan, D., & Lewis, D. (2007). Ontologies in the engineering of management and autonomic systems: a reality check. *Journal of Network and Systems Management*, *15*(1), 5–11. doi:10.1007/s10922-006-9058-1

Streitz, N., & Nixon, P. (2005). The disappearing computer. *Communications of the ACM*, *48*(3), 32–35. doi:10.1145/1047671.1047700

Strohbach, M., Bauer, M., Kovacs, E., Villalonga, C., & Richter, N. (2007). *Context sessions: a novel approach for scalable context management in ngn networks*. Newport Beach, California, USA.

Strohbach, M., Kovacs, E., & Goix, L. W. (2007). Integrating ims presence information in a service oriented architecture.

Tan, J. G., Zhang, D., Wang, X., & Cheng, H. S. (2005). Enhancing semantic spaces with event-driven context interpretation, *Pervasive Computing: Third International Conference, Pervasive 2005, Munich, Germany, May 8-13, 2005, Proceedings*.

Tonti, G., Bradshaw, J. M., Jeffers, R., Montanari, R., Suri, N., & Uszok, A. (2003). Semantic web languages for policy representation and reasoning: a comparison of kaos, rei, and ponder. *The Semantic Web—ISWC*, (pp. 419-437).

Truong, B. A., Lee, Y., & Lee, S. Y. (2005). A unified context model: bringing probabilistic models to context ontology. *Lecture Notes in Computer Science*, 566–575. doi:10.1007/11596042_59

UPnP Forum. (2003, December 2). *Upnp device architecture 1.0*. Retrieved from http://www.upnp.org/.

Waldner, J. B. (2007). *Nano-informatique et intelligence ambiante: inventer l'ordinateur du XXIe siècle*. Lavoisier.

Wang, X., Zhang, D., Gu, T., & Pung, H. (2004). Ontology based context modeling and reasoning using owl. In *Pervasive Computing and Communications Workshops, 2004. Proceedings of the Second IEEE Annual Conference on* (pp. 18-22).

Weiser, M. (1991). The computer for the 21st century. *Scientific American, 265*(3).

This work was previously published in Context-Aware Mobile and Ubiquitous Computing for Enhanced Usability: Adaptive Technologies and Applications, edited by D. Stojanovic, pp. 311-337, copyright 2009 by Information Science Reference (an imprint of IGI Global).

Chapter 8.11
Virtual Social Networks: Toward A Research Agenda

Sunanda Sangwan
Nanyang Technological University, Singapore

Chong Guan
Nanyang Technological University, Singapore

Judy A. Siguaw
Cornell-Nanyang Institute, Singapore

ABSTRACT

Multi-user virtual communities have become an accepted fundamental component of communication whereby community members share information and knowledge for mutual learning or problem solving. Virtual communities in a multi-user virtual environment (MUVE) have evolved into active social networks, formulating an alternative social existence and this phenomenon warrants further investigations. In these virtual social networks (VSNs), member participation is essential for their success. Therefore, developing knowledge on how to manage and sustain participation of members in VSNs fills a gap in our academic understanding of the dynamics underpinning the processes of virtual community development. This article aims to address these issues by extending the theory of sense of community into a virtual context (SOVC) and by integrating it with other communication theory of U&G.

INTRODUCTION

Internet technologies have changed the methodology and content of communications in online environments (Koh, Kim, Butler & Bock, 2007). New forms of communication, such as collaborative communications, which enable people to communicate and interact with one another in the absence of face-to-face interactions, have emerged (Jepsen 2006). Such interactive collaborative communications have led to the formation of multi-user virtual communities for social networking known as virtual social networks (VSNs), and these have become an accepted fundamental component of communication (Koh et al. 2007). This proliferation of virtual communities worldwide provides an important form of communication whereby community members share information and knowledge for mutual learning or problem solving (Chen &Xie 2008). These communities in the multi-user virtual environment (MUVE) have expanded into a space to live and not just for occasional or casual participation (Kwai & Wagner 2008). These virtual social networks are essentially communities aiming to formulate alternative social existences and this evolving phenomenon warrants further investigation.

In this respect, the issue of what makes sense of a virtual community has captured the attention of both practitioners and researchers but so far research on this subject is limited (see Blanchard & Markus 2004, Fisher, Sonn & Bishop, 2002). According to communication theory, a sense of community (SOC) implies an emotionally positive effect which creates an intrinsically rewarding reason to continue participation in an online community (Blanchard & Markus 2004, Whitworth & DeMoor, 2003). When participants experience positive feelings in a virtual community, they are more likely to increase or maintain their membership, thereby contributing to its sustainability and success (Blanchard & Markus 2004, Sangwan 2005). We postulate that the sense of virtual community must be a principal construct in research designed to understand any form of virtual collaboration such as communication and participation in virtual communities-of-practice, or virtual social networks. Indeed this construct can facilitate our understanding of participation in virtual space where organizations are also increasingly using collaborative communication for various issues (Jepsen 2006, Koh & Kim 2003). Prior research has shown that virtual networks play a critical role in determining the way problems are solved, organizations are run, and the degree to which employees succeed in achieving their goals (see Kuk 2006, Kwai et al. 2008, Wietrz & Ruyter 2007). Similarly, virtual social network research has focused on how the structure of bonds affects members and their relationships in physically less-bounded social systems and global communities in a multi-user virtual environment (Wasserman & Faust, 1994). Most large-scale, virtual social networks that rely on members to contribute content or build services share the problem of how to increase participation across the board. The concentration of a small number of members who contribute both to postings and to other content shows that a larger number of participants do not feel motivated enough for regular participation (Koh et al. 2007). For example, in a study of a USENET forum, Franke and Hippel (2003) found that the most prolific 1% of members contributed 20% of the postings, and the top 20% of members contributed to 61% of the messages. Similarly, in a study of an online project, Koch and Schneider (2002) found that 17% of members made 80% of the contribution. Different levels of participation are related to the success of virtual communities (Kuk 2006). Therefore, developing knowledge of how to manage and sustain participation of members in VSNs fills a gap in our academic understanding of the dynamics underpinning the process of virtual community development.

This article aims to address these issues by extending the theory of SOC into virtual contexts (SOVC). First, we demonstrate the theoretical strength of SOVC from the members' perspective to develop a conceptual understanding of why members participate in virtual social networks. In accordance with previous research, we treat the social network context as a moderator (Venkatesh, Morris, Davis & Davis, 2003). Subsequently, we develop a research agenda with propositions for future research work. Implications relating to the satisfaction of members for sustainability of virtual social networks are also discussed. The article contributes toward an improved understanding of the SOVC construct on aspects contributing to the successful VSN for the interests of various stakeholders. VSN organizers can apply the theoretical understandings to increase participation by and satisfaction of their members (Koh & Kim 2003, Porter & Donthu 2008) for successful business models. An understanding of the dynamics of VSNs can also potentially be applied by enterprises and policy makers to facilitate virtual collaboration among their members, and to better transform activities of offline communities into an online context (Kuk 2006). These are important issues for business activities such as marketing and advertising of products and services where, through SOVC, promotion by word-of- mouth and other advertising influences and persuasions can be better exercised in virtual social networks (Chen & Xie 2008).

A COMMUNICATION THEORY PERSPECTIVE ON VIRTUAL SOCIAL NETWORKS

A community is characterized mainly by the relational interactions or social bonds that draw people together (Gusfield 1975). Relational communities are distinguished from territorial communities through their association with human relationships, irrespective of their location or being physically bounded together. In this sense, most virtual communities, including virtual social networks, are relational communities since these are based on common social interests such as social meetings, hobby clubs, religious groups, or fan clubs (Koh & Kim, 2003). Members of virtual social networks can be geographically dispersed while participating in common social actions.

Sense of community (SOC) is a feeling that members have of belonging; a feeling that members matter to one another and to their community; a shared belief that members' needs will be fulfilled through their commitment to be together; a feeling of being emotionally secure through bonding with other members (McMillan & Chavis 1986). This sense of belonging is about individual member's perception, understanding, attitudes, feelings toward the community and his or her relationship to it. At same time, this feeling extends to other members' participation for a complete and multifaceted community experience (Blanchard 2007, Forster 2004).

Similarly, in a virtual community or in a virtual settlement, a sense of community exists with a set of community-like behaviors and processes (Blanchard & Markus 2004, Forster 2004,). SOC is a vital component of a virtual community and is at the core of all efforts to strengthen and build a virtual social network (Chipuer & Pretty 1999, Obst 2002). More specifically, SOVC is defined as members' feelings of membership, identity, belonging or bonding, and attachment and connection to a group with common interests or goals (Koh & Kim, 2003) that interacts predominantly through electronic communication (Blanchard 2007). Affective or utilitarian bonds and connections manifesting a sense of virtual community are developed through social interactions between members using a technological medium. This bonding and connectivity is an important criterion that can be used to differentiate virtual social networks from other virtual settlements (Koh et al. 2007). With the ease of mobility and increased ability to communicate in MUVE, a sense of connection has evolved that ensures trust and value for members in order to build a virtual community (Forster 2004, Porter & Donthu 2008). Research demonstrates that SOVC relates to higher member participation and influence on other members in a virtual community for its success and sustainability (Andrews 2002, Blanchard & Markus, 2004).

Although there are differences between SOVC and SOC, researchers in various virtual settings have adapted the measure of SOC, as developed in the sense of community index (SCI) by Chavis et al. (1986), to study virtual groups (Blanchard 2007, Forster 2004, Koh & Kim 2003,). The SOC construct consists of four antecedents of feelings of membership, feelings of influence, integration and fulfillment of needs, and shared emotional connection (McMillan and Chavis 1986). Koh and Kim (2003) adapted two main dimensions of membership and influence and adjusted the construct for the virtual context. They also suggested immersion as a third dimension of SOVC to explain virtual communities' characteristics of anonymity, addictive behavior, and voluntary and total involvement.

A social network is a social structure made of individuals or organizations that are tied by one or more specific types of interdependency (Friedkin 1982). It reflects the ability to access information and exchange feelings through the links between network members (Friedkin 1982). Social network analysis views social relationships in terms of nodes (individual actors within the networks) and ties (relationships between the actors) (Wasserman & Faust 1994). Social communities can exist as personal and direct social ties that either link individuals who share values and beliefs or impersonal, formal, and instrumental social links (Mathwick, Wiertz & Ruyter, 2008). The likelihood and process of interaction within electronic social networks can be modeled by the nature of network size or degree of openess which is defined as the degree to which a member is near all other members in a network (directly or indirectly) (Constant, Sproull & Kiesler 1996, Mathwick et al. 2008). Prior research explores the capacity of the social networks as a communication channel and their role in influencing the spread of new ideas and practices (Davis 1989) in creating social capital (Mathwick et al. 2008) and other organizational usage (Constant et al. 1996). This approach can be useful for examining and explaining many online business and marketing phenomena and little research has focused on exploiting virtual communities from this perspective.

Building bonds in virtual communities relates to the various levels of participation contributing to the membership life cycle. Research in psychological SOC has revealed some effects on the individual level of participation in offline communities (Burroughs & Eby 1998) where members have a tendency to remain affiliated with others in their community (Fisher et al.2002). This trait of SOC can be extended to MUVE to examine the impact of SOVC on participation in social networks in order to explain members' exit or sustained membership, similar to participation in traditional or face-to-face communities.

Wenger (1999), using the principles of legitimate peripheral participation, illustrate a cycle of how users become incorporated into virtual communities. They suggest five types of trajectories within a learning community: peripheral (i.e. lurker), one who uses an outside, unstructured participation; inbound (i.e. novice), a newcomer who invests in the community and is heading towards full participation; insider (i.e. regular), one who is fully committed community participant; boundary (i.e. leader), one who sustains membership participation and brokers interactions; and outbound (i.e. elder) as one in the process of leaving the community due to new relationships, new positions, new outlooks (Wenger 1999). Kim (2000) reinforces this typology and explains that members of virtual communities begin their life as visitors, or lurkers. After breaking through the initial barriers of communication, members become novices and begin to participate in community life. After contributing for a sustained period, members become regular contributors. If they break through further barriers, members become leaders; and once they have contributed to the community for some time members become elders. This variance in participation can reflect on the building of SOVC and create value for members to join and stay in their virtual community for social networking. Other studies on communities in different research settings also demonstrate that participation is related to SOVC (Cho & McLeod 2007).

RESEARCH AGENDA

By the concept of virtual social networks we understand virtual communities which have a common social interaction interest, and these interactions generate and provide social capital for stakeholders. This social capital is an intangible resource embedded in and accumulated through a specific social structure (Mathwick et al. 2008). These specific social structures allow a higher level of member participation valuable to businesses such as electronic commerce, and organizational activities involving employees' collaboration.

In this article, we investigate SOC as a starting point and adapt it to the virtual community context. Thus, similar to SOC, the principles of SOVC have four antecedents: membership, influence, integration and fulfillment of needs, and shared emotional connection (McMillan 1996). Although the SOC construct has been reviewed, extended and refined in later research in both offline and online contexts (see Table 1 for an overview of selected studies), we used the original dimensions to develop our research propositions. The rationalization for this derives from the fact that none of the studies in online communities (see Blanchard 2007, Koh & Kim 2003, Long & Perkins 2003) has conducted any methodologically rigorous systematic analysis in MUVE. We envisage that, based on our propositions, future studies can provide both internal and external validation of the construct for sustained employment in further research. Moreover, in an offline environment, these dimensions have been established as the key to fostering a sense of connection among members in order to build a sense of community;

Table 1. An Overview of Selected Studies on SOC and its extension to SOVC

Authors	Context	Findings
McMillan and Chavis (1986).	Offline Communities	This study proposes four antecedent of psychological SOC: membership, influence, integration and fulfillment of needs and shared emotional connection.
Chavis, Hogge McMillan & Wandersman(1986)	Offline Communities	This study designs the widely used and broadly validated measure of Psychological Sense of Community from the Sense of Community Index (SCI) instrument (see Chipuer & Pretty 1999, Long & Perkins 2003).
Chipuer and Pretty (1999)	Offline Communities	The study measures SOC using SCI and shows some support for the existence of the four antecedents of the McMillan and Chavis (1986) SOC model in the SCI.
Long and Perkins, (2003)	Offline Communities	This study reassesses SCI using confirmatory factor analysis (CFA) and results yield a poor model fit for McMillan and Chavis's (1986) original theoretical formulation, and develops an eight-item, three-factor Brief SCI (BSCI).
Koh and Kim (2003)	Virtual Community	This study conceptualizes and operationalizes SOVC in three antecedents: membership, influence and immersion, and validates several other antecedents of SOVC: the enthusiasm of the community's leaders, offline activities available to members, and enjoyability.
Blanchard and Markus (2004)	Virtual Community	The paper explores whether online communities do have an SOVC. The antecedents of SOVC differ from physical SOC. The behavioral processes that contribute to SOVC- exchanging support, creating identities and making identifications, and the production of trust- are similar to those found in offline communities, but they are related to the challenges of electronic communication.
Blanchard (2007)	Virtual Community	This study develops an SOVC measure, building on the measure of SOC. Potential antecedents of SOVC, exchanging support, and identification determine that the newer SOVC measure is an improvement over the SCI.
Oh and Jeon (2007)	Virtual Community	This study investigates membership dynamics, patterns of interaction in the virtual community and how different network characteristics (i.e. network size and connectivity) influence the stability of the virtual community, providing insight into dynamic and reciprocal relations among community members.
Porter and Donthu (2008)	Virtual Community	By using membership dimension, this study shows that efforts to provide quality content and foster member embeddedness have positive effects on user attitude about the virtual community sponsor.
Mathwick et al. (2008)	Virtual Community	This study examines and supports the conceptualization of social capital as an index composed of the normative influences of voluntarism, reciprocity, and social trust in a virtual community.

thisrequires further research in different research settings (Fisher et al.2002).

MEMBERSHIP

Membership, defined as the experience of feelings of belonging to a community, includes five attributes: a sense of belonging and identification, emotional safety, boundaries, a common symbol system and personal investment (McMillan & Chavis 1986). It is an important component for a successful virtual community (Oh & Jeon 2007).

The first attribute, having a sense of belonging in a virtual community, is essentially about a member's expectation or faith that he or she will be able to fit into and gain acceptance by other participants of the social network (Chipuer & Pretty 1999) and be identified and recognized by the community (McMillan & Chavis 1986). This identification can be manifested in many ways in an online environment (Koh & Kim 2003), for example, by referring to each other by name in an online community of social networks (Blanchard & Markus 2004). This expression can create intimacy between members and a sense of belonging to the community.

The second attribute, emotional safety, is a feeling of security and willingness to reveal how one really feels in one's community (McMillan & Chavis 1986). Virtual community participants can feel a degree of emotional safety in their community whenever support is available, in terms of simple interaction or giving basic verbal support (Bagozzi and Dholakia 2002). Members herd together because the need for emotional safety reinforces them to do so and such reinforcement binds them into a cohesive virtual network (Oh & Jeon 2007).

The third attribute, boundaries, consists of elements such as language, dress, ritual and logistical time or place settings indicating who belongs or does not belong to the community (McMillan & Chavis 1986). Obst et al. (2002) argue that members are more aware of their membership within a common interest boundary, such as in a virtual community, than in a geographic community of common interest. Boundaries can also be criteria for membership (McMillan 1996). In virtual communities for social networks a list of frequently asked questions provides information about the community's character, core purpose, expectations of members, taboo topics and norms so that members with similar inclinations can identify with the community (Leimeister, Cesareni & Schwartz, 2005). Furthermore, some virtual communities are regulated by moderators and by criteria for membership that must be met prior to participation (McMillan 1996). Rituals such as "waving hello" when entering a particular chat room, or controlling access through an initiation ritual are also part of the boundaries separating members from non-members (see Wenger 1999).

The fourth attribute of membership, a common symbol system consists of elements such as special languages (from simple words to phrases that represent messages known only by the members of the community), or objects that have a special meaning for the members (for example, structures, paintings, sports equipment) (García, Giuliani & Wiesenfeld 1999). Understanding common symbol systems is a prerequisite to understanding the community and its members (McMillan & Chavis 1986). In the context of virtual communities, groups establish linguistic conventions such as signature styles, unique abbreviations, emoticons and specialized vocabularies or spellings which become a common symbol system of the communities that not only reflects their culture but also readily identifies regular participants (Blanchard & Markus 2004). Members sharing a common symbol system will be able to identify themselves with the community and develop a feeling of membership (Blanchard and Markus 2004,Chipuer & Pretty 1999).

The fifth attribute, personal investment, denotes the contribution that members make to the community. Personal investment, "material" or "non-material" (Blanchard 2007, García et al. 1999) can provide the feeling that one has earned a place in the community and its membership, which then becomes more meaningful and valuable (McMillan 1996). For the virtual context, members can show a willingness to invest in the community by offering unsolicited help without an obvious request or immediate benefit to themselves (Chipuer & Pretty 1999, Porter & Donthu 2008).

With regard to membership, McMillan and Chavis (1986) argue that the five attributes of membership: boundaries, emotional safety, a sense of belonging, personal investment and a common symbol system, fit together in a circular, self-reinforcing way (Oh & Jeon 2007). Thus, the greater the feeling of membership experienced by members of the community, the more it is likely to create an SOVC. Further, the greater the feeling of membership, the more likely that membership will be continued over a period of time.

Proposition 1a: *There is a positive relationship between membership and SOVC.*

Proposition 1b: *There is a positive relationship between a sense of membership and sustainability of membership over time.*

INFLUENCE

A bi-directional influence between a community and its members is needed to create an SOC (McMillan & Chavis 1986, Porter & Donthu 2008) and to cultivate trust and harvest value (Porter & Donthu 2008), and to create community cohesiveness (Massey, Montoya-Weiss & Hung, 2003). SOVC is contingent on a social network virtual community influencing its members and likewise a member must have the potential to influence the community. The ability of a virtual community to attract and simultaneously influence its members is crucial to maintaining cohesiveness (Chipuer & Pretty 1999, Massey et al. 2003). The salient element of influence is the development of trust (McMillan 1996). Being conceptualized as a form of belief, intention and behavior, trust refers to confident, positive expectations, the belief and willingness to act based on the community's conduct (Leimeister et al. 2005). Influence also reflects the cognitive aspects of a virtual community (Obst et al. 2002). Influence can be measured by three indicators: expressing a minority opinion, asking for or giving neutral, opposite or popular opinions, and asking for clarification (Ligorio et al. 2008, Massey et al. 2003). The influence expressed either by taking a stand on the divergent opinion or by asking for clarification shows a willingness to be influenced by another member, expressing both interest in another viewpoint and a willingness to expend extra effort to understand it (Massey et al. 2003).

Two elements exist regarding the level of dyadic influence the virtual community has on its members: first, the members' need for consensual validation; and second, the community's need for conformity (Bieber, Engelbart, Furuta, Hiltz, Noll, Preece, Stohr, Turoff, & VandeWalle 2002). One reason for conformity relates to the pressure on the members to validate the virtual community's views or values (Hung & Li 2007, Porter & Donthu 2008). This pressure can shift from the individual member to the community to consensually validate its members for creating common norms (McMillan & Chavis, 1986). Members with a stronger need for consensual validation tend to be more influenced by the virtual community because of the need for reassurances provided by other members who share a similar experience. Likewise, an individual will feel a stronger pressure to validate the community's view when the virtual community has a strong need for conformity (Hung & Li 2007).

Proposition 2: *There is a positive relationship between influence and SOVC.*

INTEGRATION AND FULFILLMENT OF NEEDS

Needs-fulfillment suggests that members of a community believe that the resources available in their community will meet their needs (McMillan & Chavis 1986). Prospective member attraction to a social networking virtual community depends on the perceived satisfaction the member may receive from fulfillment of his/her needs (Massey et al. 2003). For a stronger sense of community, the social networks virtual community can find ways to juxtapose and integrate members' needs and resources (see Oh & Jeon 2007). The extent to which integration and fulfillment of needs is achieved in a virtual community depends on the degree of ease with which an individual member's need fits among the community needs. In this sense, needs-fulfillment is similar to the "perceived usefulness" construct in the Technology Acceptance Model (TAM) (for example, Davis 1989), and "uses" dimensions of Uses and Gratification Theory (U&G) (Katz, Gurevitch & Hass 1973, Sangwan 2005).

The principle of integration and fulfillment of needs includes both the individual member's feeling of being supported by other members in the community while also supporting them (Blanchard & Markus 2002). The main types of satisfaction in being a member of a virtual community are status in the online social network, shared values, and meeting other's needs while having one's own needs fulfilled (Chipuer & Pretty 1999). Virtual social network resources such as content, in archives or other formats, represent aggregate collective expert resources and social capital of knowledge that can meet member needs (Mathwick et al. 2008). Member-generated content that derives from members' competencies and, through active participation, contributes to this social capital of knowledge, increases value for all members (Porter & Donthu 2008; Stohr, Turoff & VandeWalle, 2002).

Members look for support and reinforcement by either expressing a need, or offering to fulfill another member's needs (Ligorio 2008). The fulfillment of one's need helps members develop a sense of community (Bagozzi & Dholakia2002). We utilize the U&G theoretical framework which is grounded in the functionalist and communication research paradigm to extend the antecedent of integration and fulfillment of needs in the SOVC model in the virtual social network environment (Blumler 1979). This needs-gratification is concerned with what members do with the VSN, and suggests that members will be motivated to select a virtual community that best gratifies their needs. U&G theory proposes five categories of use-gratification, namely: cognitive, affective, personal integrative, social integrative and tension release needs (Katz et al.1973). Cognitive needs represent the intrinsic desire for information acquisition for knowledge and understanding in an increasingly information-rich society and in the context of VSN can be interpreted as functional or resource-based needs that are gratified by member-generated content (Stohr et al. 2002). Affective needs are related to emotional experiences and an intrinsic desire for pleasure, entertainment and aesthetics (Katz et al.1973). These emotive needs (Sangwan 2005) are served when a virtual community or VSN evolves beyond its functional or resource-based needs-fulfillment orientation. It then serves these emotive needs through their expression and exchange of social-emotional support (Ligorio 2008) or offers to help. These aspects also indicate that the members asking or offering emotional support believe in the mutual success of their VSN (see Blanchard and Markus 2004). Personal integrative needs are contextual needs (Sangwan 2005) and derive from an individual's desire to appear credible, be perceived as confident, and to have high self-esteem in a specific context. These needs are closely related to an individual's value system (Katz et al.1973). Social integrative needs are affiliation needs where members want to be part of a community and want to be recognized as part of it and develop a sense of belonging. These can be served in many ways; for instance, in the process of supporting other members' resource-based needs, one can also fulfill one's own need for recognition (Forster 2004). Tension-release needs relate to the need for escape and diversion from problems and routines (Blumler 1979, Ligorio 2008).

Prior research shows needs-gratification as one of the principal motivators for virtual community usage (Sangwan 2005, Stafford, Stafford & Schkade, 2004). We propose that functional, emotive and contextual needs-fulfillment affects the user's feeling of needs-fulfillment in a virtual social network.

Proposition 3: *There is a positive relationship between integration of needs and SOVC.*

Proposition 4a: *There is a positive relationship between cognitive needs-gratification and SOVC.*

Proposition 4b: *There is a positive relationship between affective needs-gratification and SOVC.*

Proposition 43c: *There is a positive relationship between personal integrative needs-gratification and SOVC.*

Proposition 4d: *There is a positive relationship between social integrative needs-gratification and SOVC.*

Proposition 4e: *There is a positive relationship between tension release needs gratification and SOVC*

SHARED EMOTIONAL CONNECTIONS

Shared emotional connections (SEC) is a commitment and belief that members have shared and will share history, common places and time together, and similar experiences in their community (McMillan & Chavis, 1986). SEC consists of the following mechanisms: frequency and quality of the interaction, a shared history, and the investment that people make in their community (Chipuer & Pretty 1999).

The frequency and quality of the interaction (for example, sending and reading messages) are important in creating positive experiences and bonds in virtual environments (Bieber et al. 2002). The frequency of reading messages has a greater influence than that of sending messages in forming close relationships amongst members (Hung & Li 2007). The quality of SEC is influenced by mechanisms such as common experiences of risk and values and traditions (McMillan 1996). In a virtual community or VSN, shared experience of crisis and threats from an external source can serve to increase connections through triggering of intense discussions and generating trust (Porter and Donthu 2008).

Shared history is another aspect that makes members more similar to each other, facilitating the feeling of belonging together (McMillan 1996). The longer members stay in a community, the more knowledge representing the community's values and traditions they will gather. In this sense, a common history creates a sense of continuity and stability (McMillan & Chavis, 1986). Participants who have a shared history or have a background in common, are more likely that a sense of community may develop and be shared *in* these interactive VCs than those sharing haphazard geographical communities with nothing in common (Hemetsberger 2002). Similarly, investment, such as contributing time and effort to organizing community-related activities, influences SEC by determining the importance of the community's success and status to individuals (Hemetsberger 2002).

Proposition 5: *There is a positive relationship between shared emotional connection and SOVC.*

STRUCTURE OF VIRTUAL SOCIAL NETWORKS

Virtual Social Networks or virtual communities exist for common use for social interaction, for sharing common places or space, and common bonds (Burroughs & Eby, 1998). These networks have varying levels of ties or bonds that form strong to weak interactions (Constant et al. 1996). Smaller, closed and tighter networks have stronger bonds. Nevertheless, these may provide a less positive experience for their members than open networks with lots of loose connections and weak ties (Constant et al. 1996). More open communication networks are more likely to introduce a wider range of innovative ideas and excitement to their members than closed networks (Friedkin 1982). In closed networks, only an exchange of information and feelings with each other exists while in open networks members can exercise an influence outside as well as within their subnetworks (Friedkin 1982). In larger groups it may be easier to take advantage of the benefits of living in a community without contributing to those benefits (Wasserman & Faust 1994). However, loosely-knit, open networks may limit the ability to recognize members or track emotional facts about all members of a group. The structure of a social network, virtual or not, affects the process of interaction between members within it. In this research, we expected the structure of social network would affect the development process of SOVC.

Proposition 6: *Virtual Social Network structures moderate the relationships between (a) membership, (b) feelings of influence, (c) integration, (d) fulfillment of needs, and(e) shared emotional connection and SOVC*

SENSE OF VIRTUAL COMMUNITY (SOVC) AND PARTICIPATION

Participation, treated as a dependent variable, reflects SOVC by implying motivation for member participation or a lack of it that influences sustainability of a virtual social network (Kuk 2006). A member who starts as a "lurker" by reading discussions but rarely actively participating may build up an SOVC by experiencing the specific culture and etiquette of the community gradually. There may always be some users with a lower SOVC, who will continue to habitually lurk on a forum, virtual community or VSN, and rarely contribute (Wenger 1999). But building SOVC can enforce higher participation from members. In this research, we expect SOVC to add to continuance and preference for a virtual social network.

Proposition 7: *There is a positive relationship between SOVC and the level of participation.*

CONCLUSION AND FUTURE RESEARCH

The framework offered in this article is designed to investigate SOVC by revisiting the theory of SOC and extending its antecedents into a virtual context of social networks. The structure of social networks is proposed as a moderator in the process. We demonstrate how SOVC may exert influences on participation in virtual social networks from the members' perspective. Subsequently, a research agenda with propositions for future research work is developed. Our major contribution is to develop a better theoretical understanding of SOVC within the context of virtual social networks and its impact on VSN participation. Our contributions to future research include presenting an integrated and improved research framework from communication theories perspectives and indicating avenues for further research by developing a set of propositions. We provide insights into the interaction between participation and social network structure. From a managerial practice perspective, our study also provides a rationale to explain various needs and levels of participation of members in virtual social networks and virtual communities which may have implications for businesses seeking to establish an online presence and trying to understand user behavior. Understanding virtual social networks is of interest to organizations that want to increase their content and other resources for a successful business model.

We recognize that our research framework suffers from some limitations. We focus on membership and participation in a VSN. However, there may be a difference between a member who logs on anonymously (such as a guest visitor) and a registered member logging on with a recognized name or pseudonym. This makes it difficult to establish the intensity of participation in terms of SOVC development. In addition, prior research finds that remaining anonymous encourages participation in a virtual community by ensuring personal privacy (Andrews 2002). This is an important distinction because of the possible effects of anonymity or identity may have on participation levels and on building SOVC.

Finally, in terms of suggestions for future research, a cross-cultural approach to the study of VSN and SOVC will be valuable. The knowledge of how cultural factors account for variance in members' behavior and participation will intensify the generalizability and validation of such research.

REFERENCES

Andrews, D.C. (2002). Audience-Specific Online Community Design, *Communications of the ACM*, 45(4), 64-68.

Bagozzi, Richard, Utpal Dholakia, (2002), Intentional social action in virtual communities, *Journal of interactive marketing,* 16(2) 2-21 .

Bieber, Michael; Engelbart, Douglas; Furuta, Richard; Hiltz, Starr Roxanne; Noll, John; Preece, Jennifer; Stohr, Edward A.; Turoff, Murray; Van de Walle, Bartel. (2002) Toward Virtual Community Knowledge Evolution. *Journal of Management Information Systems*, 18 (4), 11-35.

Blanchard, Anita L, Lynne Markus M.., (2004) The Experienced Sense of a Virtual Community: Characteristics and Processes, *Databases for Advances in Information Systems*, 35(1), 65.

Blanchard, Anita L..(2007) Developing a Sense of Virtual Community Measure. *CyberPsychology & Behavior*, 10(6), 827-830.

Blumler Jay G. (1979) The role of theory in uses and gratifications studies, *Communication Research* 6(1), 9-36.

Burroughs, Susan M., Eby, Lillian T., (1998). Psychological sense of community at work: A measurement system and explanatory framework. *Journal of Community Psychology*, 26, 509--532.

Chavis, D.M., Hogge, J.H., McMillan, D.W., Wandersman, A. (1986). Sense of community through Brunswick's lens: A first look. *Journal of Community Psychology*, 14(1), 24-40.

Chen Yubo; Xie Jinhong. (2008)Online Consumer Review: Word-of-Mouth as a New Element of Marketing Communication Mix. *Management Science*, Mar2008, 54 (3), 477-491.

Chipuer, Heather M.; Pretty, Grace M.H., (1999), A Review of the Sense of Community Index: Current Uses, Factor Structure, Reliability, and Further Development; *Journal of Community Psychology*; 27(6), 643-658.

Cho, Jaeho; McLeod, Douglas M.. (2007) Structural Antecedents to Knowledge and Participation: Extending the Knowledge Gap Concept to Participation. *Journal of Communication*, Jun2007, 57 (2), 205-228.

Constant, D., Sproull, L., Kiesler, S. (1996). The kindness of strangers: The usefulness of electronic weak ties for technical advice. *Organization Science*, 7(2), 119-135.

Davis, F.D. (1989) Perceived usefulness, perceived ease of use, and user acceptance of information technology. *MIS Quarterly*, 13, (3) (September 1989), 319–340.

Dueber Bill, Misanchuk Melanie (2001) Sense of Community in a Distance Education Course *Mid-South Instructional Technology Conference Murfreesboro*, TN, April 8-10.

Fisher A T, Sonn C C, Bishop B J.(2002)Psychological sense of community: Research, applications and implications. New York: Kluwer. Academic/Plenum Publishers.

Forster PM. (2004) Psychological sense of community in groups on the Internet. *Behaviour Change*, 21, 141–147.

Franke, N., E. von Hippel. (2003). Satisfying heterogeneous user needs via innovation toolkits: The case of Apache security software. *Research Policy*, 32, 1199–1215.

Friedkin, N. E. (1982). Information flow through strong and weak ties intraorganizational social networks. *Social Networks*, 3, 273-285.

García, I., Giuliani, F., Wiesenfeld, E. (1999). Community and sense of community: The case of an urban barrio in Caracas. *Journal of Community Psychology*, 27(6), 727—740.

Gusfield, J., (1975)The Community: A Critical Response, Harper Colophon, New York.

Hemetsberger, Andrea. (2002) Fostering Cooperation on the Internet: Social Exchange Processes in Innovative Virtual Consumer Communities. *Advances in Consumer Research*, 29(1), 354-356.

Hung, Kineta H.; Li Yiyan, Stella. (2007) The Influence of eWOM on Virtual Consumer Communities: Social Capital, Consumer Learning, and Behavioral Outcomes. *Journal of Advertising Research*, 47 (4), 485-495.

Jepsen, Anna Lund (2006) Information Search in Virtual Communities: Is it Replacing Use of OffLine Communication? *Journal of Marketing Communications*, 12 (4), 247-261

Katz, E., Gurevitch, M., Hass H. (1973), On the Use of Mass Media for Important Things, *American Sociological Review*, 38 (2), 164-181.

Kim, A.J. (2000). Community Building on the Web: Secret Strategies for Successful Online Communities. London: Addison Wesley.

Koch, S., G. Schneider. (2002). Effort, cooperation and coordination in an open source software project: GNOME. *Information Systems Journal*, 12, 27–42.

Koh, Joon; Kim Young-Gul, (2003) Sense of virtual community: A conceptual framework and empirical valuation, *International Journal of Electronic Commerce*, 8 (2), 75.

Koh Joon; Kim Young-Gul; Butler, Brian; Bock Gee-Woo (2007) Encouraging Participation in Virtual Communities. *Communications of the ACM*, 50 (2), 69-73.

Kuk, George (2006) Strategic Interaction and Knowledge Sharing in the KDE Developer Mailing List. *Management Science*, 52 (7), 1031-1042.

Kwai Rachael Fun IP, Wagner Christian (2008) Weblogging: A study of social computing and its impact on organizations, *Decision Support Systems*, 45(2), 242-250.

Leimeister, Jan Marco; Ebner, Winfried; Krcmar, Helmut. (2005) Design, Implementation, and Evaluation of Trust-Supporting Components in Virtual Communities for Patients. *Journal of Management Information Systems*, 21(4), 101-135.

Ligorio, M. Beatrice; Cesareni, Donatella; Schwartz, Neil.(2008) Collaborative Virtual

Environments as Means to Increase the Level of Intersubjectivity in a Distributed Cognition System. *Journal of Research on Technology in Education*, 40(3), 339-357.

Long, D. A., Perkins, D. (2003) Confirmatory factor analysis of the sense of. community index and development of a brief SCI. *Journal of Community Psychology*, 31(3), 279-296

Massey, A.P.; Montoya-Weiss, M.M.; Hung, Y-T, (2003), Because time matters: Temporal coordination in global virtual project teams, *Journal of Management Information Systems*,19 (4), 129-156.

Mathwick Charla, Wiertz, Caroline Ruyter Ko de. (2008) Social Capital Production in a Virtual P3 Community, *Journal of Consumer Research*. 34(6), 832.

McMillan, D.W., Chavis, D.M., (1986), Sense of community: A definition and theory, *American Journal of Community Psychology*, 14(1), 6.

McMillan D W. (1996)Sense of community. *Journal of Community Psychology*, 24(4): 315-325.

Obst P, Zinkiewicz L, Smith SG. (2002) Sense of community in science fiction fandom, Part 2: comparing neigh-borhood and interest group sense of community. *Journal of Community Psychology*; 30, 105–17.

Oh Wonseok, Jeon Sangyong. (2007) Membership Herding and Network Stability in the Open Source Community: The Ising Perspective, *Management Science*.53(7), 1086.

Porter Constance Elise, Donthu Naveen. (2008) Cultivating Trust and Harvesting Value in Virtual Communities, *Management Science*. 54(1), 113.

Sangwan, S. (2005). Virtual community success: a users and gratifications perspective. *Proceedings of the 38th HICSS Conference.*

Stafford, Thomas F; Stafford Marla Royne; Schkade, Lawrence L.(2004) Determining Uses and Gratifications for the Internet, *Decision Sciences Atlanta*, 35(2), 259-288.

Stohr, Edward A.; Turoff, Murray; Van de Walle, Bartel.Toward (2002) Virtual Community Knowledge Evolution. *Journal of Management Information Systems*, 18(4), 11-35.

Venkatesh, V., Morris, M. G., Davis, G. B., Davis, F. D. (2003). User acceptance of information technology: Toward a unified view. *MIS Quarterly*, 27(3), 425-478.

Wasserman, Stanley, Faust, Katherine. (1994). *Social Networks Analysis: Methods and Applications.* UK: Cambridge University Press.

Wenger Etienne (1999) Communities of Practice. Learning, meaning and identity. UK: Cambridge University Press.

This work was previously published in International Journal of Virtual Communities and Social Networking, Vol. 1, Issue 1, edited by S. Dasgupta, pp. 1-13, copyright 2009 by IGI Publishing (an imprint of IGI Global).

Chapter 8.12
Situated Evaluation of Socio-Technical Systems[1]

Bertram C. Bruce
University of Illinois at Urbana-Champaign, USA

Andee Rubin
TERC, USA

Junghyun An
University of Illinois at Urbana-Champaign, USA

ABSTRACT

This chapter introduces situated evaluation as an approach for evaluating socio-technical innovation and change. Many current evaluations simply identify the impacts of technology and deprecate alternate uses in their analysis. Situated evaluation instead calls for understanding how innovations emerge through use; this entails consideration of diverse uses, the contexts of use, and the reasons for the development of multiple realizations. The chapter presents a comparative study of different classroom uses of electronic Quill in order to demonstrate how this alternative evaluation can be conducted and to address the value of understanding and fostering diverse cultural appropriations of a socio-technical innovation.

What about the lay public as producers of technology and science? From the vernacular engineering of Latino car design to environmental analysis among rural women, groups outside the centers of scientific power persistently defy the notion that they are merely passive recipients of technological products and scientific knowledge. Rather, there are many instances in which they reinvent these products and rethink these knowledge systems, often in ways that embody critique, resistance, or outright revolt.

—Eglash, 2004, p.vii

INTRODUCTION

Implementing an innovation entails making changes to an existing system of social practices. People involved with that system naturally want to know what those changes mean and are, therefore, drawn to calling for some sort of an evaluation. Based on the results of the evaluation, practitioners, policy makers, and administrators make their practical decisions about the fate of the innovation. They often focus on evaluation outcomes alone, but the setting of evaluation questions and methods is as important

DOI: 10.4018/978-1-60566-264-0.ch045

as the outcomes. Evaluation processes embed evaluators' assumptions about the innovation and its relation to the relevant social contexts.

In this chapter, we raise questions about the basic assumptions and limitations that standard approaches to evaluations have, and introduce *situated evaluation* as an alternative approach that aims to uncover, not the way that an innovation interacts with practice, but rather the very emergence of innovations through practice. Through a study of Quill, an electronic composition system that was developed for teaching writing in the early 1980's, we demonstrate how this alternative evaluation can be conducted. We also discuss the values, challenges, and methodological issues related to using situated evaluation in supporting further understanding of socio-technical innovations. As new digital technologies increasingly pervade aspects of our daily lives, the innovations-in-use issues that arose in Quill implementations are even more relevant today.

QUESTIONING THE NATURE OF STANDARD EVALUATION

Standard evaluation practice tends to emphasize either formative or summative approaches. Formative evaluation is typically done during the development or improvement of a program and is conducted iteratively. Results are often informal and lead to recommendations for change. Summative evaluation provides information on the program's efficacy, such as improvement of student learning. In this chapter, we propose an alternative, which questions the basic assumption of "what" it is that is being evaluated.

In evaluating a new technology, researchers typically consider the innovation as a fixed object created by professional developers. They further assume that its benefits are somewhat fixed and known in advance with respect to social practice. For example, a program might be developed to help students learn a concept in science or to help a community engage in community building through better communication. Evaluation then becomes a way to improve that program or to assess its effectiveness. This is a reasonable approach, one that is fully in line with calls for reflective practice. But in its extreme form, the assumption that what the program actually is known prior to its integration into social practice becomes what Papert (1987) defines as *technocentrism*:

> *Egocentrism for Piaget does not, of course, mean "selfishness"—it means that the child has difficulty understanding anything independently of the self. Technocentrism refers to the tendency to give a similar centrality to a technical object—for example computers or Logo. This tendency shows up in questions like "What is THE effect of THE computer on cognitive development?" or "Does Logo work?" (p. 23)*

The problem here is that a technocentric perspective limits the scope of the evaluation, often making it difficult to see unexpected uses of an innovation. But, as any developer knows, technical innovations often result in unplanned uses and diverse readings of the innovation. Often, the variation in use is greater than the variation in programs, so that the claim to be evaluating a particular program becomes convoluted with discussions about faithfulness of implementation or effectiveness of the program per se versus effectiveness of its introduction.

One good example occurs in the discourse on online collaboration and learning systems. The early visions of new communication and information technologies asserted that their fundamental attributes could support innovative learning environments that promoted students' active participation, reflective thinking, attainment of self-discipline, and connections with the real world. However, this visionary perspective of educational computer-mediated communication has altered due to the unexpected effects of diverse teaching and learning practices.

For instance, Burniske (2001) designed and implemented several "telecollaborative" projects using e-mail, but eventually reported on the limitations of telecommunication for learning. Burniske's first project, "Project Utopia," used electronic mailing for having his students discuss utopia and dystopia with another colleague's students in a different location. Burniske judged that this project "had inspired a few constructive discussions, but many of them dissipated as students' imaginations, liberated from real-world concerns, took flight" (p. 36). Then he developed another project, "South African Elections' 94 Internet Project," which allowed e-mail exchanges among 11th and 12th grade students in South Africa and the U.S. However, he realized that students' discussions remained shallow and felt it difficult to improve the quality of the discourse. From these experiences, he started questioning the linear impact of new communication technology integration on student learning. Other scholars from critical perspectives have similarly questioned positivist views of technology's effects on practice (Bryson & De Castell, 1998; Bruce, Peyton, & Batson, 1993). These critical views have argued that new technologies do not generate social change, but are instead mutually constituted with social practice.

Standard (summative and formative) approaches have wide-ranging and important uses for evaluating socio-technical systems. But as they are usually carried out, they also have a crucial limitation related to examining the interaction of the technical innovation with the context in which the innovation is used. This makes it difficult to attend to the process of change, and consequently, to many of the concerns people have about innovations.

R. M. Wolf (1990) describes three key problems with standard evaluation. First, most evaluations do not identify the reasons for the observed phenomena. Thus, they do not say how the innovation can be improved, nor what aspect of it produced the measured effects. Second, not being able to account for why changes occur means that it is questionable to generalize to other settings in which the innovation might be used. Third, the development process often continues after the evaluation, so that most evaluations are effectively of innovations that no longer exist. Again, without knowing more about the situation and process or use, one cannot say whether initial results are still valid for the changed innovation.

Many researchers have proposed ways to attend more to the process of change. Some call for an emphasis on formative evaluation. Others call for broadening the range of measurement tools used for summative evaluation (Miles & Huberman, 1984). In *responsive evaluation*, evaluators become sensitive to the interests and values of the variety of participants involved with the innovation (Stake, 1990). Others call for multiple case studies across different settings to identify the variations and differences (Stenhouse, 1990). Each of these approaches makes a contribution to the study of socio-technical innovation and change. But often these methods fail to answer a basic question for a potential user: How can the innovation be recreated in one's own setting? Rather, they still designate which type of use is "acceptable" and which is "unacceptable." This leads us to raise a fundamental issue about the nature of evaluation: What is the "it" being evaluated?

SITUATED EVALUATION

Situated evaluation is an approach to articulating the emergence of innovations through practice, assuming that innovations are mutually constituted by social practice and some external input. It starts with the common finding that a program operates differently in different settings. But rather than postulating that there is one program used in different ways, it asserts that multiple programs come into being through use. This ontological shift leads to different ways of analyzing, describing, and conceptualizing alternate, or even non-uses. A

Table 1. Questions about innovations and change

Old Questions	New Questions
What can the innovation do?	What do people do as they use the innovation?
To what extent are the innovation's goals achieved?	How do social practices change, in whatever direction?
What constitutes proper, or successful, innovation?	What are the various forms of use of the innovation-in-use?
How should people or the context of use change in order to use the innovation most effectively?	How should the innovation be changed and how can people interact differently with it in order to achieve educational goals?
How does the innovation change the people using it?	How does the community fit the innovation into its ongoing history?

bibliography of situated evaluation studies can be found online at http://illinois.edu/goto/siteval.

A situated evaluation approach conceives technology users as active creators, rather than as "passive recipients of technological products and scientific knowledge" (Eglash, 2004). Users actively rethink the meaning and use of a technology and reinvent its practices by appropriating them within their situated, cultural contexts. Eglash (2004) calls this process *appropriating technologies*. We would go one step further to say *creating technologies*.

In these situations, we need a new type of evaluation that is open to new variables and sensitive to alternate uses and interpretations. This new concept of evaluation needs to focus on the *innovation-in-use*, and its primary purpose is to understand the different ways in which the innovation is realized and thus created. *Situated evaluation* then emphasizes the unique characteristics of each situation in which the innovation is used. With this approach, the object of interest is not the idealized form in the developer's specs, but rather, the realization through use. The "it" being evaluated is no longer the innovation (or even what we call the *idealization*), but the innovation-in-use, a situation-specific set of social practices. Recognizing the richness and the importance of the realization process also leads us to ask new sorts of questions for evaluation (see Table 1):

- What practices emerge as the innovation is incorporated into different settings?
- How well do the different uses of the innovation work?
- How can different realizations be improved?

KEY ELEMENTS OF SITUATED EVALUATION

Situated evaluation is a process of discovering relationships. Although it does not resolve into a simple, linear procedure, there are three major aspects of this process. First, it looks at the idealization of a technical system or program, in order to delineate as fully as possible what was intended by the developers. Second, it examines the settings in which a technology is used. Third, it analyzes the realization processes in different settings and generated hypotheses about how and why these realizations developed as they did.

The Idealization of the Innovation

We define the elements of the innovation as intended by developers as its *idealization.* An analysis of the idealization is part of a situated evaluation because it serves to characterize how participants in the setting of use might have perceived the innovation. It is also an index of the intentions of the developers, people who are often important participants not only in the initial creation of the innovation, but in its re-creation in context.

In contrast to the priorities for summative evaluation, the innovation is not privileged over any of its realizations; similarity to the idealization does not count as more successful, and non-use can be as important to consider as "faithful" use. Moreover, the innovation is not seen as an agent that acts upon the users or the setting, but rather as one more element added to a complex and dynamic system. It would be more correct to say that the users act upon the innovation, shaping it to fit their beliefs, values, goals, and current practices. Of course, in that process, they may themselves change, and their changes as well as those to the innovation need to be understood as part of the system.

The Setting in Which the Innovation Appears

The shift in perspective from the view that realizations are distortions of the ideal to one in which realizations are creations that result from active problem-solving has implications for the sorts of questions researchers need to ask in evaluating innovations. With this perspective, the social context in which the innovation is used becomes central. Questions relating to cultural, institutional, and pedagogical contexts need to be addressed. To answer these questions in full is a formidable task, but focusing on a few specific aspects may go far in providing what is needed for a situated evaluation.

The Realizations of the Innovation

The third aspect of a situated evaluation is to study the realizations of the innovation in different settings. This means, first, to examine the ways the innovation was used and search for the reasons that changes occur. This includes analyzing how the idealization was consonant or dissonant with existing social practices. It also includes studying how the innovation's use led to new social organizations. Second, is to look at the variety of uses across settings, treating each of these as an independent re-creation of the innovation, rather than as a data point for an aggregate statement about the innovation. Third, is to examine changes in the design of the innovation brought about by its use and the ways these changes relate to new practices.

Comparisons of Situated Evaluation with Standard Evaluations

A key difference between situated evaluation and the standard frameworks is that its purpose is to learn first how the innovation is used, not how it ought to be changed or whether it has claimed effects. Because it is concerned with actual use, it does not focus on the innovation or its effects, but rather on the social practices within the settings in which the innovation is re-created. This shift in focus has implications for the audience of the evaluation, the role of setting variability, the tools for evaluation, the time of assessment, and the presentation of results.

Focus

Standard evaluation is concerned either with properties of the innovation alone or with its "effects." In contrast, situated evaluation focuses on the way the innovation becomes social practices.

Audience

Situated evaluation results can be used by both users and developers. Users can make decisions not only about whether to use the innovation, but how to use it in their particular context. Developers can learn how to revise the innovation taking into account the variations in use.

Purpose

For situated evaluation, the audience is broad, as are the actions that follow from the findings. The results could lead to developers changing the innovation, to users changing their practices, to adoption of only parts of the innovation, or to deeper understanding of the process of use.

Variability of Settings

The central concern for situated evaluation is with characterizing the way an innovation comes into being in different contexts. Because the audience for the evaluation wants to know how to improve the use of innovation, it is useful to have a variety of contexts that they can compare to their own setting or to ones they might create. Thus, it is most appropriate when there are a variety of contexts of use, and differences across those settings.

Measurement Tools

With situated evaluation, the emphasis is on differences across contexts. This emphasis implies the use of qualitative tools, including observations and interviews that are structured to elicit information about recurring social practices in the setting and to draw out differences among realizations.

Time of Assessment

Situated evaluation can start once the innovation is developed enough to be placed in a classroom. This is in contrast to formative evaluation, which might start even earlier, in a laboratory setting. Situated evaluation can continue well after the developers have finished. It could be done before summative evaluation as a way to identify sites or issues to study, or afterwards as a way to study the process of change.

Results

Because a situated evaluation seeks to characterize alternate realizations, it requires multiple, detailed descriptions of specific uses. Changes need to be described using appropriate quantitative or qualitative representations, but more importantly, the reasons for changes need to be discussed and linked to characteristics of the settings of use. The process of change, including changes in the innovation, in the users, and in the setting, becomes paramount.

Situated Evaluation and Ethnographic Inquiry

Situated evaluation significantly differs from standard (summative and formative) evaluations that start with the given and ask how to improve it. Hence, evaluators who approach from a situated evaluation perspective would not simply identify the strengths and weaknesses of a technology and generalize the conditions for successful implementations. Situated evaluation also does not pursue wide and decontextualized dissemination of an innovation across different settings. Instead, through contrastive analyses and narrative accounts, evaluators seek to create a shared space for multiple technology users to reflect their values and practices so that they can continue re-creating their technology uses through practice. The audience for the evaluation would also want to compare to their own setting or to ones they might create.

Situated evaluation resembles the "sustained and engaged nature" of ethnography and extensively uses ethnographic methods, "long-term participant observation with in-depth interviewing" (Miller, Hengst, & Wang, 2003). To understand the process of change and to excavate different views or interpretations of socio-technical changes within contexts, situated evaluation demands evaluators' relatively long-term and ongoing engagement. An's study (2008) shows how ethnographic inquiry and methods have guided her situated evaluation of an alternative computer training practice implementing community service. According to her study, the methodological emphasis of situated evaluation needs to continuously create a dialectic between the "contextual" and "narrated" worlds in order to generate credible results throughout data collection, analysis and

Table 2. Comparisons among the three types of evaluation

	Formative	Summative	Situated
Focus	Innovation	Effects of the innovation	Social practices
Audience	Developer	User	User (but also developer)
Purpose	Improve the innovation	Decide whether to adopt innovation	Learn how the innovation is used
Variability of Settings	Minimized to high-light technology	Controlled by balanced design or random sampling	Needed for contrastive analysis
Measurement Tools	Observation/Interview/Survey	Experiment	Observation/ Interview
Time of assessment	During development	After initial development	During and after development
Results	List of changes to the technology	Table of measures contrasting groups	Ethnography

reporting. Different natural settings and uses of an innovation cannot be arbitrarily analyzed and compared in parallel. Rather, situated evaluation develops the researcher's continuous and meaningful construction of knowledge through sensitive use of multiple research methods.

Situated evaluation is also based on the idea that the researcher-participant relationship can significantly shape the researcher's understanding of the insiders' perspectives. What enables scientific inquiry is not the elimination of subject errors or biases, but the researcher's on-going, self-reflective learning to understand the multiplicity and complexity of modern social reality by carefully observing practice. Hence, evaluators weave possible interpretations about the phenomena on the basis of what they hear and observe. In this sense, conducting situated evaluation is a constructivist and historical process of learning for evaluators to make meaningful knowledge.

Briefly, situated evaluation requires an evaluator's sustained, extensive, and self-reflexive engagement. That effort is worthwhile if one wants to understand diverse cultural adaptations of technology and the process of technology design and use *in situ*.

A STUDY OF ELECTRONIC QUILL IN USE

Quill (Bruce, Michaels, & Watson-Gegeo, 1985; Bruce & Rubin, 1984; Liebling, 1984; Rubin & Bruce, 1985, 1986) was an approach to the teaching and learning of writing built around a software system that included both tools and environments for writing. From 1983 to 1987, it was used throughout the U.S. and Canada, primarily in upper-elementary and middle-school grades. Quill is no longer commercially available, but the Quill studies show extensive classroom data on its use. The studies examined how Quill was realized in different ways in diverse settings. They also looked at the details of the implementation processes to understand how the realization reflected the unique characteristics of Quill, as well as the particular classrooms in which Quill was used.

One of the Quill studies is described here in order to demonstrate how a situated evaluation can be conducted in a specific case. This study focused on the various ways that Quill's goal of purposeful writing was realized through the use of Mailbag, one component of the Quill software. Mailbag was a version of email used by the Quill students, years before many people became aware of it. The goal of the study was to understand

Table 3. Contrasts between QUILL and traditional classrooms

QUILL Classroom	Traditional Classroom
Prewriting	Sit and write
Topic choice	Designed topic
Multiple genres	Mostly narrative
Multiple real audiences	Teacher as audience
Real purposes	Writing for a grade
Conferencing	Red marks as response
Revision	Editing
Collaboration	Hidden papers
Sharing writing	Isolated writers
Writing across the curriculum	Writing in English class

how realizations of an innovation were created, and to use real classroom examples for insight into the process of integrating new technologies into teaching.

The following presents the findings in two major sections: the idealization of Quill and realizations of Quill. The latter describes alternate implementations of Mailbag and how the integration of students' and teachers' purposes and habits with the innovation produced different realizations. The data gathered include writing by the teachers about their own classrooms, student writing, electronic mail (both from Mailbag and from a network for teachers), and field notes from classroom observations.

The Idealization of Quill

Quill's design was based on research on composition, and encompassed prewriting, composing, revising, and publishing aspects of the writing process (Bruce, Collins, Rubin, & Gentner, 1982; Flower, 1981; Flower & Hayes, 1981; Graves, 1978, 1982; Newkirk & Atwell, 1982). It included a text storage and retrieval program (Library), a note-taking and planning program (Planner), and an electronic mail program (Mailbag), all supported by a text editor (*Writer's Assistant*; Levin, Boruta, & Vasconcellos, 1983).

In its software, accompanying curriculum (Quill *Teacher's Guide*; Bruce, Rubin, & Loucks-Horseley, 1984) and teacher workshops, Quill embodied a philosophy for teaching writing. Quill emphasized the process of writing, including the importance of both planning and revision. The contrast between Quill classrooms and traditional classrooms is highlighted in Table 3. On the left is a gloss of what we call the idealization of Quill, that is, the view of what Quill was supposed to become in classroom use. On the right are parallel descriptions of a more traditional writing class. Many teachers tried to integrate Quill with some of these discrepant practices. Although major changes in the teaching of writing have occurred since then, many classrooms still approach writing in the "traditional" way. Moreover, the issue of how classroom technology adoption is inseparable from pedagogy is still relevant (Mishra & Koehler, 2006)

A central element within the idealization of Quill was an emphasis on real audiences and purposes, which was expressed in the software, teacher's guide, and training. In the software, Mailbag, in particular, reified this emphasis on audience and purpose. Combining features of the post office, the telephone, and a bulletin board, it facilitated direct communication among students, groups of students, and teachers. With activities suggested in the Quill *Teacher's Guide*, it encouraged a variety of purposes for writing that students seldom experienced in school: "chatting," persuading, informing, instructing, and entertaining. It also motivated students to write more by introducing a personal element into the experience.

Many teachers introduced "writing as communication" to their students through Mailbag. Since they had used Mailbag extensively during training, teachers appreciated the differences between sending Mailbag messages and standard classroom writing assignments. They saw Mailbag as a way to help students understand writing as a communicative act through participation in writing activities that demanded a real audience and purpose.

Realizations of Quill

The realization of Quill in any real classroom was a re-creation that drew upon the idealization, but was usually more dependent upon characteristics of the situation of use, institutional forces, the teacher's goals and teaching style, the students, and idiosyncratic technical details, such as the number of computers or room layout. Thus, the many forms of Quill-In-Use differed markedly from the original conception.

Of course, each teacher understood the idealization of purposeful writing in Quill in his or her own way, and the variety of realizations were due in part to different teachers' interpretations of our message. What mattered was not just Quill's conception of purpose, but that of the people who used it: What did teachers and students think writing was useful for? How did they use writing to accomplish personal goals? What did teachers think students should learn about writing in school? What natural goals for writing existed in classrooms or community contexts?

In most classrooms, Mailbag use *did* lead to more purposeful writing. Students saw Mailbag as an unconstrained writing environment and were thus able to use it for their own purposes. But the specifics of this use took many different forms, often surprising both us and the teachers involved. A few teachers regarded the openness of the Mailbag environment as a pedagogical problem, and in these cases, little purposeful writing with Mailbag occurred.

For several teachers, Mailbag and its built-in assumptions were completely consistent with their current classroom practices and their attitudes toward teaching writing. These teachers firmly believed in "student-centered education" and in students' feeling ownership of the process and product of their work in school. They saw Mailbag as a welcome extension of the way they already taught writing. They were comfortable with students' deciding when, where, why, and on what topics to write. For instance, Bonnie's multigrade, village-school classroom reflects this symbiotic use of Mailbag. Students used the program frequently and enthusiastically from the beginning of the year. Bonnie offered the following comments about her class' early use of Mailbag:

Probably the best thing about Mailbag is communicating. The person at the keyboard is in complete control. I never made any Mailbag assignments. Students could use it or not, decide what they would say, to whom, when, how often, and why.

The Mailbag messages written in this class show their oral-language character. Students seemed to regard Mailbag as an environment in which they could carry out the same communicative functions for which they used oral language. Although many messages contained nonstandard grammar or spelling, Bonnie never corrected any student message. She considered Mailbag to be in the students' domain, where spelling and punctuation were secondary to just plain communicating.

In Bonnie's classroom, students expressed their control over Mailbag by deciding both when to use Mailbag and when to stop using it. Several other teachers also found that students' enthusiasm for Mailbag diminished as the year went on, but Bonnie's comment about this shift reflects again how her educational views easily encompassed such as change:

By springtime the Mailbag was hardly used at all. At first I was disappointed, then pleased. The students had learned that there were appropriate forms of communication for specific needs.

Especially in small classes where students knew each other well and saw one another frequently outside of school, the kind of communication Mailbag facilitated was mostly redundant. As Bonnie implies, students had become more sophisticated about audience and purpose and were not satisfied with a communicative situation that did not increase their access to real audiences.

In one class, however, interest in Mailbag remained strong during the entire year. Hans taught high school in Bonnie's village and used Mailbag with his class after learning about it from Bonnie. He designated one disk as the students' private Mailbag disk and promised the class that he would never read it. The students continued to send messages on the disk all year, and Mailbag remained the most popular Quill activity. As the year went on, Hans actually had to ration Mailbag's use because he wanted students to use the computer for other kinds of writing as well. Why did Mailbag remain so popular in this class? Certainly at least one influence was the unique audience Hans defined for Mailbag messages. It appears that the secrecy of the disk made the communication environment unusual enough that students did not consider it redundant with face-to-face communication.

Since many Quill classrooms had only a single computer, using Quill required some teachers to rethink their classroom management practices. How were they to integrate a free-form activity like Mailbag into a more structured day? Wilma, a fifth-grade teacher, invented a procedure to deal with the changes in her classroom structure. Wilma's students' excitement over Mailbag was particularly significant to her, since one of her goals for the year was to help her students learn to enjoy writing. While she was enthusiastic about Mailbag's effect on her students, she was troubled by its classroom management consequences:

When we started using Mailbag, I had a problem with my students wanting to be back at the computer constantly checking to see if they had any mail or not. We decided we needed to devise a system that would solve the problem. We talked about what we could do, and soon came up with a mailbox poster, which worked quite well. We each wrote our computer code name on a Library book card pocket, and glued the pockets to a piece of poster board. The poster board was then hung on the wall behind the computers. Another pocket was added to hold slips of red paper. When a student left a message on Mailbag for White Knight, he or she would put a red slip into White Knight's pocket. After While Knight read his messages, he returned the red slips to the extra pocket.

The classroom management issues were so central to teaching with Quill that Wilma's idea spread around the community via our technical assistance visits and the teachers' electronic mail network. The classroom management problem turned out to be a common one, and many teachers adopted Wilma's solution.

Not all integrations of purposeful writing with Mailbag into the classroom grew out of a symbiosis between Quill and a teacher's purposes. In one case, a teacher completely rejected Mailbag because it conflicted with her views of the appropriate way to teach writing. This teacher started out using Mailbag in the usual way, and students began sending messages according to their own purposes, such as love letters to one another. When the teacher discovered this, she immediately made Mailbag unavailable since she felt that the messages students had been exchanging were not appropriate classroom writing. The gap between her pedagogical assumptions and those underlying Quill was too great.

In a slightly different attempt at integration, a fourth-grade teacher tried to combine a fairly traditional writing assignment with Mailbag. The idea for her assignment came from the Quill *Teacher's Guide*, where we had described a "Classroom Chat" activity, based on a popular newspaper column called "Confidential Chat." In the newspaper prototype, writers send anonymous letters describing their personal problems; they usually adopt a pseudonym that refers to their situation (e.g., Hassled Mom or Concerned Commuter). Quill's variation had students sending anonymous messages to the Mailbag's Bulletin Board in order to discuss personal problems anonymously with others students in the class. Mixing the pseudonymous personal consultation idea of Classroom Chat with a more traditional teacher-directed writing assignment, the teacher sent the following message, complete with pseudonym:

Dear Classy Computer Kids,

There are five members in my family and only one shower. Because I'm the youngest member of our family, I'm the last one in line to take a shower. By then, there's usually no more hot water and not too much time for me to wash behind my ears! It's a horrible way to start a day. What can I do to solve this problem?

Cold, late, and dirty,

I. Needabath

The following tongue-in-cheek student response hovers between reality and fantasy, much as the original letter did:

Dear I. Needabath,

I think you should tell the first person that takes a shower you have to go to the bathroom. Then they should let you go before they take a shower. Quickly lock the door and take your shower. You will have enough of time to wash behind your ears.

Sneaky and Desperate,

Kerry N. and Jenny B.

An interesting problem emerged in this activity because of the conflict between the teacher's goals and the presuppositions of Mailbag. The form of the teacher's message mimicked that of the standard confidential chat letter, but the students in the class all knew who had sent the letter and, even more important, that it posed a fake problem. Thus, their assignment was to pretend they were answering a real letter from a needy person, while knowing it was an imaginary letter from their teacher. While students produced imaginative replies, we observed that students were confused about their audience (their teacher or I. Needabath) and their purpose (real or fantasy) while they were writing. This lack of clarity was most obvious when they were signing their names; many were not sure whether to use their own names or to make up clever pseudonyms. In this situation, the teacher's assignment worked only weakly as an attempt to integrate two inconsistent pedagogical goals.

Teachers were not the only ones for whom Mailbag offered new opportunities for integrating technology with personal goals. In several classrooms, students found in Mailbag, a new and unexpected way to pursue their own purposes in school. Students in Syd's fifth-grade class in Juneau discovered that Mailbag could serve an unexpected purpose in their relationships with others in the classroom. One of Syd's students "saw himself without friends"; Syd worried about both his academic and social development:

He chose late Friday for his time [on the computer] so he could miss it, not realizing that more often than not, late Friday was the easiest time for me to be his partner. The other children, in spite of their ugliness to one another, were able to sense his feelings and began writing [Mailbag] letters telling how much they liked him and that they wanted to be his friends. There is no way to describe the face of this handsome, brown-eyed boy as he read these notes, frequently slipped into his desk anonymously. He sat near me for obvious reasons and I would watch him remove one and literally clutch it to his chest.

Syd's students, having learned the power of writing, chose to use it to be kind to a troubled student with whom face-to-face communication was difficult.

Many students in field-test sites in Alaska used Quill to answer a pressing communicative need; they were unable to be in touch easily with people outside of their own villages and they had no way of meeting new people. Partly in response to their needs, the Quill project in Alaska instituted a long-distance network, implemented through a combination of human travel and U.S. mail (Barnhardt, 1984).

On one of our trips through Alaska to visit classrooms, we carried a disk called "Supermail." This was a very slow, but still effective, way to carry electronic messages from one village to the next, when even dialup connections were rare and unreliable. The Supermail disk facilitated communication for some students in Nikolai, as Don, their teacher, explains:

What made this activity fun for my class was the fact that Chip had just come from Telida and the most recent messages on the disk were from cousins and playmates upriver. This connection made the notion of sending hellos to strangers Outside seem less threatening.

Don reflects his students' view of the world by referring to the rest of the United States outside Alaska as Outside, to them a vast and little-known area. The Supermail disk provided an opportunity for the students to be in touch with the outside world; it made the transition gradual by allowing them to expand their understanding of communication from a familiar audience to a larger and unfamiliar audience Outside.

The crucial point for us here is that Supermail was nowhere envisioned in the original Quill design, or idealization. It didn't exist at all for most Quill classrooms and users. Instead, it emerged from the unique social and geographical situation of Alaskan village schools, and was thus as much a new technology as any other Quill component, although one created through use. For some in Alaska, Supermail became a salient part of the Quill experience. In a standard evaluation approach, we might footnote it as a user adaptation of the pre-existing program; with situated evaluation we describe it as an innovation created through practice.

It may be helpful to refer to Dewey's (1922) critique of the dualism of means and ends. He discusses how "means and ends are two names for the same reality"; that they are convertible, one into the other:

Only as the end is converted into means is it definitely conceived, or intellectually defined, to say nothing of being executable. Just as end, it is vague, cloudy, impressionistic. We do not know what we are really after until a course of action is mentally worked out. (p. 29)

Standard evaluations tend to assume a separation of means and ends: The program is a known, fairly well-defined means and the desired outcome is a known and somewhat fixed end. Situated evaluation, in contrast, assumes that means are created as much through use in a community or classroom as they are through development in the lab. Ends emerge as well, reflecting those new means. Supermail was an innovation created through use, because of ends that were unknown during development, or at best "vague, cloudy, impressionistic." Its creation defined new ends for the participants.

CONCLUSION

In the Quill study, the use of Mailbag for purposeful writing is only one area in which alternate realizations of Quill arose. In every case in which Quill raised significant pedagogical issues, teachers had to confront the relationship of their past practices to those implied by Quill. This resulted in a variety of solutions to the need to integrate Quill with sometimes disparate goals, values, and practices.

Our analysis views these as creative solutions to the complex and ill-defined problems teachers or, for that matter, anyone, must solve when presented with an opportunity to change. As we see through this study of Quill in use, an innovation is not an object that can be packed inside a box, but rather a set of practices that emerges from the social setting of its use. Thus, in a sense, the user does not accept or reject an innovation but instead creates it through action in the world.

The key notion about situated evaluation, as also shown in the Quill study, is that it does not postulate an *a priori* innovation to be used in various settings. Rather than investigating the practices or impact based on such an innovation (as formative or summative evaluation would do), it seeks to discover what innovation comes into being through practice.

Accordingly, situated evaluation highlights the power of the social context to affect the use of a new technology. How the features of the technology interact with human needs, expectations, beliefs, prior practices, and alternative tools far outweighs the properties of the technology itself. This does not mean that we ignore the influences of developers' visions and technical designs. Instead, we seek to develop a holistic understanding of an innovation as a mutual adaptation between technology and its situated social settings. This understanding of the idealization and various realizations of an innovation can help improve further re-creations of a socio-technical system.

REFERENCES

An, J. (2008). *Service learning in postsecondary technology education: Educational promises and challenges in student values development.* Unpublished doctoral dissertation, University of Illinois at Urbana-Champaign.

Barnhardt, C. (1984, April). The QUILL microcomputer writing program in Alaska. In R.V. Dusseldorp (Ed.), *Proceedings of the third annual statewide conference of Alaska Association for Computers in Education* (pp. 1-10). Anchorage: Alaska Association for Computers in Education.

Bruce, B., Collins, Rubin, A., & Gentner, D. (1982). Three perspectives on writing. *Educational Psychologist, 17*, 131–145.

Bruce, B., Michaels, S., & Watson-Gegeo, K. (1985). How computers can change the writing process. *Language Arts, 62*, 143–149.

Bruce, B., Peyton, J. K., & Batson, T. (1993). *Networked-based classrooms: Promises and realities*. New York: Cambridge University Press.

Bruce, B., & Rubin, A. (1984). *The utilization of technology in the development of basic skills instruction: Written communications* (Report No. 5766). Cambridge, MA: Bolt Beranek & Newman.

Bruce, B., Rubin, A., & Loucks-Horsley (1984). *Quill teacher's guide*. Lexington, MA: D. C. Heath.

Bryson, M., & De Castell, S. (1998). Telling tales out of school: Modernist, critical, and postmodern "true stories" about educational computing. In H. Bromley & M. W. Apple (Eds.), *Education/technology/power* (pp. 65-84). Albany, NY: State University of New York.

Burniske, S. W. (2001). Don't start evolution without me. In R. W. Burniske & L. Monke (Eds.), *Breaking down the digital walls* (pp. 30-58). New York: State University of New York Press.

Dewey, J. (1922). Habits and will. In J. A. Boydston (ed.). *The collected works of John Dewey; Middle works, 14*, 21-32. Southern Illinois University Press.

Eglash, R. (2004). Appropriating technology: An introduction. In R. Eglash, J. L. Croissant, G. Di Chiro, & R. Fouche, (Eds.), *Appropriating technology: Vernacular science and social power*, (pp. vii–xxi). Minneapolis, MN: University of Minnesota Press. Available at: http://www.rpi.edu/~eglash/eglash.dir/at/intro.htm

Flower, L. (1981). *Problem-solving strategies for writing*. New York: Harcourt Brace Jovanovich.

Flower, L. S., & Hayes, J. R. (1981). Problem solving and the cognitive process of writing. In C. H. Frederiksen, & J. F. Dominic (Eds.), *Writing: The nature, development and teaching of written communication* (pp. 39-58). Hillsdale, NJ: Erlbaum.

Graves, D. H. (1978). *Balance the basics: Let them write*. New York: Ford Foundation.

Graves, D. H. (1982). *Writing: Teachers and children at work*. Exeter, NH: Heinemann Educational Books.

Levin, J. A., Boruta, M. J., & Vasconcellos, M. T. (1983). Microcomputer-based environments for writing: A writer's assistant. In A. C. Wilkinson (Ed.), *Classroom computers and cognitive science* (pp. 219-232). New York: Academic Press.

Liebling, C. R. (1984). Creating the classroom's communicative context: How parents, teachers, and microcomputers can help. *Theory into Practice*, *23*, 232–238.

Miles, M. B., & Huberman, A. M. (1984). *Qualitative data analysis: A sourcebook of new methods*. Beverly Hills, CA: Sage.

Miller, P. J., Hengst, J. A., & Wang, S.-H. (2003). Ethnographic methods: Applications from developmental cultural psychology. In P. M. Camic, J. E. Rhodes, & L. Yardley (Eds.), *Qualitative research in psychology: Expanding perspectives in methodology and design* (pp. 219-242). Washington, D. C.: American Psychological Association.

Mishra, & Koehler, (2006). Technological Pedagogical Content Knowledge: A new framework for teacher knowledge. *Teachers College Record, 108*(6), 1017-1054.

Newkirk, T., & Atwell, N. (1982). *Understanding writing*. Chelmsford, MA: The Northeast Regional Exchange.

Papert, S. (1987, January-February). Computer criticism vs. technocentric thinking. *Educational Researcher*, *16*, 22–30.

Rubin, A. D., & Bruce, B. C. (1985). QUILL: Reading and writing with a microcomputer. In B. A. Hutson (Ed.), *Advances in reading and language research*. Greenwich, CT: JAI Press.

Rubin, A. D., & Bruce, B. C. (1986). Learning with QUILL: Lessons for students, teachers and software designers. In T. E. Raphael (Ed.), *Contexts of school based literacy* (pp. 217-230). New York: Random House.

Stake, R. E. (1990). Responsive evaluation. In H. J. Walberg & G. D. Haertel (Eds.), *The International encyclopedia of educational evaluation* (pp. 75-77). Oxford: Pergamon Press.

Stenhouse, L. (1990). Case study networks. In H. J. Walberg & G. D. Haertel (Eds.), *The International encyclopedia of educational evaluation* (pp. 644-649). Oxford: Pergamon Press.

Wolf, R. M. (1990). The nature of educational evaluation. In H. J. Walberg & G. D. Haertel (Eds.), *The International encyclopedia of educational evaluation* (pp. 8-15). Oxford: Pergamon Press.

KEY TERMS

Situated Evaluation: An approach to uncovering or articulating the emergence of innovations through practice, assuming that innovations are mutually constituted by social practice and some external input.

The Innovation-in-Use: Different ways in which the innovation is realized and thus created by diverse users. Situated evaluation, which is open to new variables and sensitive to alternate uses and interpretations, focuses on understanding *innovation-in-use*.

Idealization: The elements of the innovation as intended by developers.

Realization: The ways the innovation was used, modified, and re-created by users *in situ*.

Appropriating Technologies: Users actively rethinking the meaning and use of a technology and reinventing its practices within their situated, cultural contexts.

Technocentrism: The tendency to focus on technological artifacts or mechanisms to the exclusion of social, cultural or historical perspectives.

ENDNOTE

1. This chapter adapts portions of *Electronic Quills: A Situated Evaluation of Using Computers for Writing in Classrooms* (1993) by Bertram C. Bruce and Andee Rubin.

This work was previously published in Handbook of Research on Socio-Technical Design and Social Networking Systems, edited by B. Whitworth & A. de Moor, pp. 685-698, copyright 2009 by Information Science Reference (an imprint of IGI Global).

Chapter 8.13
Online Virtual Communities as a New Form of Social Relations: Elements for the Analysis

Almudena Moreno Mínguez
Universidad de Valladolid, Spain

Carolina Suárez Hernán
IE Universidad, Spain

ABSTRACT

The generalization of the new information technologies has favored the transformation of social structures and the way of relating to others. In this changing process, the logic of the social relationships is characterized by the fragility and the temporality of the communicative systems reciprocity which are established "online" in a new cybernetic culture. "Virtual communities" are created in which the interaction systems established by individuals exceed the traditional categories of time and space. In this manner the individuals create online social webs where they connect and disconnect themselves based on their needs or wishes. The new online communication technologies favor the rigid norms of the "solid society" that dilute in flexible referential contexts and reversible in the context of the "global and liquid society" to which the sociologists Bauman or Beck have referred to. Therefore the objective that the authors propose in this chapter is to try new theoretic tools, from the paradigms of the new sociology of technology, which let them analyze the new relational and cultural processes which are being generated in the cultural context of the information global society, as a consequence of the new communication technologies scope. Definitely the authors propose to analyze the meaning of concepts such as "virtual community", "cyber culture", or "contacted individualism", as well as the meaning and extent of some of the new social and individual behaviors which are maintained in the Net society.

DOI: 10.4018/978-1-60566-650-1.ch022

TOWARDS A MEANING OF "VIRTUAL COMMUNITY": A NEW STUDY OBJECT IN THE INFORMATION SOCIETY

The social dimension is one of the natural attributes of the human being, which must be understood as an individual person in interaction with a relational environment. Group, communities and culture are concepts which approach us to the man study in its complex network of social interaction as a social system. In this sense the plural evolution and interaction of the communities and the institutions as well as the forms of social relations and communication have shaped the history of humanity.

The community has been defined as a study object and subject from varied approaches, which comprise from the primitive forms of social association to the complex relationships of the post-industrial society in which the concept "virtual community" has emerged.

From the sociological point of view, community is a concept with a polysemous value, but as a global idea it responds to the anthropologic imperative of the social encounter and the need to create a sense and give shape to the human society. A possession feeling consolidates in it, understood as a psychological community feeling, in which one feels oneself as an active group member, which is decisive for the individuals own identity. Likewise, the feeling of a participation conscience and the link to a common territory are fundamental aspects (Gurrutxaga, 1991) (Pons Díez, J.; Gil Lacruz, M; Grande Gascón, J.M., 1996). The community is a network of social links, that can be based on a territory (a city), on common interests (associations, clubs) on similar characteristics of the individuals (Bar Association) or on an online platform (blogs, etc). Definitely, the community is an analytical category which defines human interaction as a constituent of the social reality, redimensioning the individual as a socialized person in a specific group, with social and symbolic representations and cultural values. Besides a social and anthropological approximation, we can consider the community as a context of action which contributes to the generation of realities based on symbolic truss.

The globalization and informationalization process has generated the transformation of our societies, including the space dimension. In such transformation the new space logic is characterized by the domination of a flow space, structured in electronic circuits that link themselves in the strategic nodes of production and management, which exceeds a space of places locally fragmented and the territorial structure as a way of daily organization. This new dimension takes us towards a Global City, understood as a "net of urban nodes of different levels and different functions, spreading all over the planet and that functions as the nervous center of Informational Economy. It is an interactive system to which companies, citizens and cities have to adapt themselves constantly. "(Castells, 1997: 2).

The Information Technologies are the fundamental instrument that allows the new logic of the social relationships to demonstrate themselves in reality. In them, Internet constitutes one of the most outstanding cases of the growing technological environment whose result has been the step from the Industrial Society to the Information Society (Cornella, 1997).

In the full expansion of Internet, the virtual communities are becoming a new social relationship format where the different communities turn to it to satisfy expectations and needs, to contribute its collaboration and to feel themselves as part of a great community. Unlike the Traditional Community, these impersonal spaces are characterized by the anonymity and the lack of human contact. These new relationship forms are giving way to a media society produced by a change of the social rules, by the capacity of transmitting ideology or by inducing behavior; definitely, by the generalization of a mass culture extensible to all social classes and communities. This turned Internet into a virtual community as means to unify the communications (Sánchez Noriega, 1997).

The first Virtual Communities were based mainly on the simple commerce or the sale of products through the Network, or on a web site where the users could place their personal webs freely. The current Virtual Community was born in this way, whose philosophy is based mainly on the leisure and recreation field, though it houses cultural societies or with certain scientific level: *Geocities*[1]. But the origin of these cyberspace centers is determined by the scientific communities that before the beginning of Internet formed a group and interchanged information (Cantolla, 2000). In 1985, the first Virtual Community of the history arose, *The Well*[2], created by a group of ecologists that "met" to debate about their issues.

Within the Network, new communication forms appear promoted by the use of the electronic mail, but the information exchange among the members was not immediate (deferred communication). Later, with the arrival of the World Wide Web as a hypermedia system that works on Internet, the communication is produced in real time, coming up the interactivity.

This new communication form obliges to distinguish between the Traditional Real Community and the Virtual Community (Iparraguirre, 1998).

Traditional Real Community

- Physical and temporal space for everyone.
- It is developed in the Real Society where the space-time nations and physical encounter determine its behavior and is limited by the territory.
- It is the material support of the Virtual Community.

Online Virtual Community

- The physical and temporal space is not a limitation any more.
- It is developed in the Virtual Society, the cyberspace territory, where there are no frontiers and is planetary.
- It appears when a Real Community uses the telematics to keep and widen the communication.

The breakage of the space-time barriers through the use of the new technologies allowed the development of numerous Virtual Communities. The permanence on the Network depends on basic elements such as the interactivity time and the emotional component among the members that form part of it, what refers us to the traditional sociability forms.

In order to understand the concept of "virtual community" properly one has to get back to the original concept of "community". For the sociologist Tönnies (1986:97-98), the concept of community (Gemeinschaft) (Community) is the stage on which the modern industrial society settles down (Gesellschaft) (Society). The community is characterized by the sort of relationships that prevail in it (based on the family, the homeland and the blood). He defines the community as a sort of social interaction in the emotional identification: he refers to the reciprocity that arises from sharing bonds based on the family, race or blood. The communication will is in the basis of this relationship which gives place to the social action. Therefore, for Tönnies, the community has its original root in the feelings. This definition is useful to refer to the community as a space of feelings and communication. Likewise, the members of a group share specific meanings and a collective vision that come forth from the shared experiences and that generates an own jargon at the end. (Prat, 2006: 29).

The sociological tradition considered the community as a group of persons who, besides exhibiting the social groups characteristics, has a territorial basis or a geographic territory that serves as seat. The first conceptualizations about the communities were carried out on the basis of territorial communities where a person could spend the whole life, as they were relatively self-sufficient. A city, a town, a village, a neighborhood constitute examples regarding this community concept. Under this concept it is present the idea that considers that the community implies more close bonds among the members that the ones existing among the members of a larger society

(Gesellschaft) (Society). A so called "communitarian feeling" exists among the members of a community.

However, nowadays, the use of the community concept is quite different according to the contexts and is used at present in a more varied and wide form. It is even tended to name community to groups that are not but conglomerates or social categories. Due to the urban sprawl, the social groups, the communities among them, went beyond the territorial frontiers. Those who emphasized the non territorial nature of the modern communities were the sociologists specialized in the analysis of the socials networks. (Scott, 1994; Wasserman and Faust, 1995). Besides studying the attributes of the group members, the sociologists of social networks analyze the relationships that are produced among them, their objective, intensity, quality as well as the structure and dynamic that arise from them. Wellman and Gulia, for example, have studied communities whose relations network goes beyond their geographic frontiers. Besides, these relationships tend to specialize being contextualized and not globalized at the same time; i.e. a person is related to other persons not in a total and integral manner but in certain specific contexts and will establish relationships with other persons if the context and objective of said relationship are different. According to Wellman and Gulia, the relationship network in which a person takes part might comprise a group of persons that are very far away in the geographic space and besides show time variations. This tendency is confirmed even more now in the cyberspace, where the sociability capacity of the persons is strengthened and creates the possibility of a new sociability manner among them. Wellman and Gulia have demonstrated that the virtual communities are also communities, although their members might not have physical nearness, similar bonds to the ones of the territorial communities are developed among them (Wellman, 1999). Definitely, the concept of virtual community seems to have its origin in the traditional concept of community and is clearly linked with the concepts of communication and socialization.

In fact, the Web allows now to integrate also communication functions and so the virtual communities arose which have a web site as coordination centre both of information and communication reservoirs. The web site became the "territory" of a virtual community. A non geographic territory as the communities studied by the sociologists during a social development stage, but an electronic territory, distributed in the new space that we call "cyberspace". Likewise, there are computer programs specialized in the construction and management of virtual communities, but, what is really a virtual community?

Howard Rheingold, to whom the term "virtual community" has been attributed, in his book, *The Virtual Community*, which became a classic work of literature about the cyberspace, defines the virtual communities as "... social aggregations that arise from the network when a sufficient quantity of persons enter into public discussions during a long enough period of time, with sufficient human feeling, to form networks of personal relationships in the cyberspace" (Rheingold, 1993: 5). We find three basic elements under this definition: the interactivity, the emotional component and the interactivity time, as conditions for a virtual community to exist and they are related to some of the characteristics of the communities in general.

According to Michael Powers, a virtual community is "an electronic place where a group of persons meet to interchange ideas in a regular manner... It is an extension of our daily life where we meet our friends, work mates and neighbors, at the park, at work or at the communitarian center". A more technical definition would be: "... a group of persons that communicate through a network of distributed computers, ... (the group) meets at an electronic locality, usually defined by a server software, while the customer software manages the information interchanges among the group members. All the members know the addresses of said localities and invest sufficient time in them in order to be considered a virtual community" (Powers, 1998: 3). The sociological concept of community as inclusive social group, with a territorial basis, is recreated in the virtual community, but the territory of the later is virtual and not geographic. The community does not occupy a space in the physical world but in cyberspace.

However, the essence of all these communities does not lie in the nature of the CMC (computer-mediated communication) that structures them, but in the fact that they are integrated by real individuals, flesh and blood and who for the sake to be included in the community, adopt a form of "online person"; a virtual identity that represents the self of the individual before the other, the totality of the social environment in which the individual is immersed (Turkle, 1995). Therefore, in order to understand the phenomenon reality it will always be necessary to take into account the dialectic of the self/other (identity/otherness), as the description of the virtual communities will always be connected to the recognition that they only exist when several individuals experience it as such, which refers us to the traditional concept of socialization once more.

Definitely, the social investigation about the virtual communities poses several serious queries on its relationship with the "real life" communities. Among them: what is lacking from the real communities for the human being to satisfy the social needs "virtually" through Internet? Do the virtual communities represent the beginning of the real communities' deterioration? Or simply, do they represent a new way of understanding and living the social relationships?

The theoretical and empirical works performed by the science and technology sociology have not been able so far to give an answer to said queries. During the last years, these studies became an alternative and useful theoretical, conceptual and methodological framework, in order to think about the analysis of the technical innovations and the incidence in the relationships and social behaviors, though the CMC technologies, or Internet itself, have not become yet a widely extended focus of attention among these investigators. In fact, it deals with the everlasting forgotten of the social sciences.

In order to reply to these queries in a general manner, we will adopt the theoretical paradigms of authors like Bauman, Beck or Castells to seize how the human needs of belonging to a community enlarge and adapt themselves in a relational and cultural environment, much more flexible with the advent of the new communication technologies. However, this interactive process that suggests new socialization and endoculturation forms is not exempt of contradictions, as the freedom and autonomy that allow the online communications generate certain frustrations regarding that innate human wish of belonging to an everlasting relational community. This has been named as the ambivalence of the new "connected individualism".

THE NEW SOCIAL/RELATIONAL AND CULTURAL DIMENSION OF THE "VIRTUAL COMMUNITIES"

The itinerary of the new communication technologies is closely entwined with the social and cultural changes as well as with the languages and narrative transformation. New concepts arise in order to seize new relational models as result of the convergence of these changes. One of these concepts that define the "information society", "network society" or the "connected society" is the "cyber-culture".

The investigation about the cyber-cultures started recently due to the accelerated advent process of the technological society. Bauman had already warned in the 90′ that it was necessary to break the individual-technology dichotomy and work on the reality/virtuality ambivalence as human product (Bauman, 1990). Therefore, it is essential to integrate the technology to the social and cultural environment (Feenberg, 1999), with the aim of knowing the dimensions plurality that characterize our existence in the information society.

In fact, the cyber-culture, understood as a set of social-technical-cultural systems that take place at the cyberspace (Lévy, 2007: XV), starts transforming the beliefs and speeches of the cybernauts through constant leaps and interactions between the "interface" and the "real world". The online practices start in this way to overwhelm the virtuality and to burst in the reality of the individuals beyond what has been imagined.

The transformations that we are witnessing in these stages of *attitudinal zapping* between the virtual and the real are configuring new ideas of the "being" and new expressions and representation manners of the individual and the community online.

Internet, more than a communication technology, constitutes itself in a par excellence representation technology of the new century. The initial fictional construction of the self is being replaced there by the reconstruction and recognition of the individual in the virtual practices.

The new forms of symbolical representation that are emerging from the Internet virtual space are giving place to new ways of privacy, personal and collective identity and, in brief, to new social relationships as manifests the use of spaces or virtual communities such as Youtube, Myspace, Hi5 or Facebook. Therefore the virtual and the real should not be understood as two opposed categories as the digital culture is, to a great extent, an extension of the culture concept, where the virtual really suggests "another" experience and another analysis of the real that compels us to a better understanding of the bonds and cores that link the realities and the appearances, the illusions and the symptoms, the images and the models. The virtual does not replace the real but it represents it; it is a laboratory of ontological experimentation that compels us to give up the support of the appearances and turns us hunters of the real in forests of symbols (Quéau, 1995: 79).

Therefore, we consider that it is not possible to separate technology, culture and society as autonomous and independent actors, as this would mean to understand the human independently from the material environment and the signs and images that give sense to life and world. "Therefore, the material world and less its artificial part of the ideas cannot be separated through the ideas with which the technical objects are harbored and used, nor from the human beings that invent them, produce them and use them" (Lévy, 2007: 6). The line that divides the real worlds from the virtual realities tends to blur with the progress of the simulation capacities that offer us the technology and its corresponding appropriation by the individuals, provoking new ideas and offering spaces to new experiences that would not be possible without the technological progress. In this sense to define the so called "cyber-culture" implies to understand how certain practices have been naturalized in the popular culture through the symbolic representation and the new communication forms that the individuals experience through the "virtual communities". Indeed and as Turkle points out, the computers by themselves would not have any value without the cultural representations and relationships which take place with the use the individuals make of them: "The computers would not be turning into powerful cultural objects if the people would not fall in love with his machines and with the ideas that the machines entail" (Turkle, 1997: 63).

One of the features that define the new social and communicative relationships that take place in the cyberspace cyber-culture is the simulation and the anonymity. Not everything in the cyberspace is simulation; however, the interfaces have caused the anonymity adoption from the beginning and the possibilities of constructing fictional personalities. The level of anonymity has a very important influence on our behavior as it leads to the lack of inhibitions or relaxation of the normal limits that the society imposes on us. Likewise, the anonymity becomes vital at the moment of experiencing on Internet with our personality; the falseness sensation gets lost and the adventure and exploration sensation is acquired. Therefore, it is interested to know what one feels while playing with the identity, experience different roles and see how the others react. This process changes the traditional sense of "role", "community" or "group" concepts defined by Durkheim, Weber or Mead in the classical sociology. In fact, the physical distance and the few social presence existing in the "virtual communities" make us feel less inhibited, safe from being discovered and less subdued to the command of our superego and the social structures.

Goffman referred to this process with the denomination "game information". From this definition we could say that the relationships in the network constitute a potentially infinite cycle of occultations, discoveries, false disclosures and rediscoveries through which we devote great efforts to produce and sophisticate the image we want to give to others without them knowing how much effort it requires (Goffman, 1959). On the Internet, the game information is much more flexible due to the opacity of the environment and due to the possibility of changing the interface if the game does not run well. The chats and forums offered the first experimentation windows at the beginning of the Network. Nowadays, with the development of the web 2.0, the new MUD as Second Life became the benchmarks in the role games and, therefore, the anonymity in the cyberspace.

This game information that enables Internet generated a new communitarian social idea through the socialization processes that are produced in the "cyber-culture". In fact, nowadays, the communitarian social idea is placed in the center of the theoretical debate: "The technological determinism is not any more a simple concept of intermittent appearance throughout the political thought of the XX century, to become, in fact, part of the communitarian idea about the technology. And it remains continuously corroborated when, curiously, both from technophobes and technophile positions; it is insisted on the inexorability of the technological development" (Aibar, 2002: 38).

The communitarian social idea of Internet, as a set of meanings and symbols, acts in the practice and in the everyday life contributing sense to the human behavior, to the social relationships and the human relationships with the objects, independently of its existence for the "conscious" of this society. According to Vayreda (2004), the technologies of the Computer-mediated Communication (CMC) are a component of the idea instituted of Internet that acquires sense in the daily practices of the individuals. The idea of the CMC is not a scheme of senses, even, without ruptures and fissures. On the contrary, its strength lies precisely in its capacity to adapt the diversity, and, even, the contradiction of the individual behavior of the persons who interact through these platforms (Vayreda, 2004). This is another example of the contradictions that characterize the liquid, individualized and globalized society referred to by Bauman, Beck and Giddens.

The new communication technologies offer us the possibility of connecting and disconnecting ourselves to the social relationships according to our will in the ontological need that the human being has to find protection in the community, either a real or virtual community. But for the communitarian idea that Internet recreates, new contradictions and relational uncertainties are generated that the citizens are not always able to distinguish. It has to do with the dichotomy between the community and the freedom, two opposed forces and equally powerful, two essential values, apparently incompatible and subject to a strain difficult to be placated, according to Bauman (2003).

The key to solve this contradiction is in what Castells calls "directed interconnection"; i.e., "the capacity of anyone to find the own destiny in the network, and in case of not finding it, create the own information arising the appearance of a new network" (Castells: 2001: 67). The electronic inter-connectivity, feature of the CMC technologies, becomes a customized connectivity, turns to be a self-management promise, of individual freedom. The only condition is not to switch off the computer, not to leave the network. According to Vayreda´s words (2004) the idea is to "change from forum, construct a new one, and invent an issue .... but never to switch off". This process responds to a new form of socialization in which the individual decides freely when to connect himself and how to manage the interaction with another individual, without anything being predetermined and defined beforehand as in the traditional socialization forms.

VIRTUAL COMMUNITY, "LIQUID SOCIETY" AND CONNECTED INDIVIDUALISM

The new context of a global society based on interactive communication favored by the information technologies boom is generating what we could call a "cyber cultural revolution". In the XVIII century series of phenomena converged which came to be called "industrial revolution" which meant a transformation of the social and production relationships with the market boom as a way of global interchange of material and cultural goods. Today we can talk about the "cyber cultural revolution" as a transformation process in which the new information technologies are transforming the social structures, the relational forms as well as the own cultural context in which those new forms which individuals adopt to relate to one another and with the environment acquire sense. It has to do with the new "online" environment in which social relationships are separated from the traditional time and space categories. The question one must pose is in what way are the social structures changing? The contemporary sociology does not have any answers to these new phenomena. The time of solid modernity certainties is giving place to another liquid modernity of uncertainties. Solid becomes liquid and with regard to the enigma of "social reality effect" as the one of "network effect" or the "Crowds" (Negri) and "Smart Mobs" (Rheingold, 2004), we only know that they exist but at the moment there is no paradigm that has the key to seize them in its totality. The only thing we can do so far is to learn to coexist and to know how to be in this new "liquid" context full of uncertainties, until we rebuild the concepts of these two basic categories (time and space) for any type of society.

In this interpretation line, one of the greatest descriptions about this new technologized age is given by the Polish sociologist Zygmunt Bauman, who in Liquid Love (2005) talks about a society that moves at a great speed through "liquid" individuals; in other words, people without lasting bonds who have the need to develop and establish ephemeral contact types based on the Internet connection, from Bauman´s point of view, implies an exercise of continuous connection and disconnection; in a virtual relationships network which have an easy access and output. Any resemblance with cinematographic Matrix is not pure coincidence.

In the passage from the solid world to the liquid phase of modernity captured by Bauman there is a fight between the globalizing power of Internet, based on the connection, and the local problematic of each individual or community. It is obvious that Internet is a global environment, but most of the investigators emphasize that the practices acquire a meaning in the local framework. The sites with most traffic in Europe and the United States are the search engines like Google, Yahoo or Windows Live), which are to access door to the navigation of individual and collective interests, big compartments of multiformat content (Fotolog and YouTube). Those are spaces of local information's (digital newspapers) and spaces to buy and sell products (e-Bay) whose usefulness only acquires meaning in the products and services exchange of local scope. Many talk about the kingdom of the "glob quality" in other words global sites due to its scope but with local focus to capture the attention of a specific audience.

The Internet revolution does not limit itself exclusively to the cyber space. In the "network society" (a definition of the Spanish sociologist Manuel Castells) converge the Web (the great generator of a paradigm change that allows, at least in papers, to overcome the temporal-space barriers of people who live in the planet), the globalization, and the institutions crisis in a new relational context that could be denominated as "connected individualism". In this context of the contemporary society, people live in networks not in groups. The groups assume that all the participants know and trust each other, while the essence of the networks is a set of interactions and exchange of information. Of course, this does not mean that groups do not exist, but rather that the life of an individual cannot be reduced nor to a concrete group, nor a fixed place, many times it is the blending of both interaction ways.

The new possibilities that the online interaction technological systems offer are not the reason of the transformation on the ways we connect ourselves. The technologies mainly are developed as an answer to the needs that we have to interact with others. Therefore, the social organization type and the technology that we use influence each other and start giving form to a social contemporary life.

The relationships which we create do not belong to a specific place but are at the same time local and global, product of the communication technologies development. In general, the traditional communities based on a concrete unit lose importance due to the relationships that we maintain with people who are physically in different places and in that way we participate of multiple social networks. The characteristics of modern life, more and more privatized and customized, are reflected in our ways of generating relationships which are more selective and voluntary than in the past. Although our contacts are global; that is to say, scattered in different areas, we continue connecting ourselves from some place, could be our house or our work, which means that we have globalized our relational network always having as reference a local context. (Ninova, 2008).

The new information technologies are changing the way in which we connect ourselves, as we do not necessarily have to be in a place to communicate with others. In fact the physical context becomes less important. The connections are among individuals and not among places, in that way technology offers a change: connects individuals wherever they are. The people become portable; they can be located for interaction through technology in wherever. In this way, the communication person to person becomes central and it supports the defragmentation of the groups and the communities turning them into "liquids". The individuals can "connect" and "disconnect" them to the social structures which even though they continue defining the social behaviors, they do so with more flexibility and liberty than in the past. They are the new "liquid times" the ones Bauman talks about, where the new technologies allow flexibility and fragmentation of the social relationships. Therefore the transition towards a customized world provides the connected individualism[3] where each person changes fast among bonds and networks. It is the person who defines how to operate to obtain information, support or collaborate in some project. We become more flexible when interacting in different spaces.

CONCLUSION

In this chapter it has been pointed out how the new information technologies are changing the traditional ways of communication and of relating with the immediate social environment. In fact in the "network society" or also called "information society" new concepts arise like the f"virtual community" or "cyber culture" associated with new social behaviors which are generated by the online communication programs. As a consequence of this, the traditional analytical categories used by the sociology to study the new social interaction systems generated in the information society are becoming obsolete. So, in this chapter the concepts used in sociology to explain the meaning and the scope of the new "online" communication cultural devices and its incidence in social relationships, in communication and definitely in the social and symbolic structure of the social groups have been checked and widened

Definitely, the virtual mobility that is being practiced in the last decades and which already forms part of our daily life demands a change in our ideas about the influence which the new technologies have, and at the same time, they make us assume that the online/offline dichotomy is a myth. The communication mediated by the computer offers flexibility and autonomy, and in no case, does it substitute the face to face communication but it supplements it and enlarges it. The online relationships fill in the empty spaces in our lives many times.. The proximity does not matter anymore; the communities and the groups are more disperse in time and space.

REFERENCES

Aibar, E. (2002). Contra el fatalismo tecnocientífico. *Archipiélago*, *53*, 37–42.

Bauman, Z. (1990). *Thinking sociologically*. Oxford, UK: Blackwell.

Bauman, Z. (2003). *Comunidad. En busca de seguridad en un mundo hostil*. Madrid, Spain: Siglo XXI.

Bauman, Z. (2005). *Amor líquido. Acerca de la fragilidad de los vínculos humanos*. Madrid, Spain: Fondo de Cultura Económica.

Beck, U. (1998). *La sociedad de riesgo. Hacia una nueva modernidad*. Barcelona, Spain: Paidós.

Beck, U., & Beck-Gernsheim, E. (2002). *Individualization. Institutionalized individualism and its social and political consequences*. London: Sage Publication.

Cantolla, D. (2000). *Comunidades Virtuales: ciudades en el ciberespacio*. Retrieved from http://www.ecommdigital.com

Castells, M. (1997). *La Era de la Información: Economía, Sociedad y Cultura. Vol. 1. La sociedad red. Vol. 2: El Poder de la Identidad*. Madrid, Spain: Alianza.

Castells, M. (2001). *La Galaxia Internet. Reflexiones sobre Internet, empresa y sociedad*. Barcelona, Spain: Areté.

Cornella, A. (1997). *La cultura de la información como institución previa a la sociedad de la información*. Barcelona, Spain: ESADE.

Cornella, A. (2000). *Infonomia com: La empresa es información*. Ediciones Deusto.

Feenberg, A. (1999). *Questioning technology*. New York: Routledge.

Feenberg, A., & Bakardjieva, M. (2004). Virtual communities: No "killer implication." *New Media & Society, 6*(1), 37-43. Retrieved from http://educ.ubc.ca/faculty/bryson/565/FeenbergVirComm.pdf

Fernández Sánchez, E., Fernández Morales, I., & Maldonado, A. (2000). *Comunidades virtuales especializadas: un análisis comparativo de la información y servicios que ofrecen al usuario*. In *Proceedings of the VII Jornadas Españolas de Documentación, 2000*.

Gálvez, A. (2005). Sociabilidad en pantalla. Un estudio de la interacción en los entornos virtuales. *AIBR. Revista de Antropología Iberoamericana, 2005*(Noviembre-Diciembre). Retrieved from http://www.aibr.org/antropologia/44nov/

Goffman, E. (1959). *The presentation of self in everyday life*. Garden City, NY: Doubleday.

Gurrutxaga, A. (1991). El redescubrimiento de la comunidad. *Reis*, *56*, 33–60.

Howard, P., & Jones, S. (Eds.). (2004). *Society online: The Internet in context*. London: Sage Publications.

Iparraguirre, J. (1998). *El taller de comunidades virtuales*. Retrieved from http://www.gpd.org/maig98/es/comvirtue.htm

Lévy, P. (2007). *Cibercultura. La cultura de la sociedad digital*. Barcelona, Spain: Anthropos.

Ninova, M. (2008). Comunidades, software social e individualismo conectado. *Athenea Digital, 13*.

Pons Díez, J., Gil Lacruz, M., & Grande Gascón, J. M. (1996). Participación y sentimiento de pertenencia en comunidades urbanas: Aproximación metodológica a su evaluación . *RTS*, *141*, 33–44.

Powers, M. (1997). *How to program a virtual community*. New York: Ziff-Davis Press.

Pratt Ferrer, J. (2006). Internet, hypermedia y la idea de la comunidad. *Culturas Populares. Revista Electrónica, 3*.

Quéau, P. (1995). *Lo virtual. Virtudes y vértigos*. Barcelona, Spain: Paidós.

Rheingold, H. (1993*). The virtual community*. Reading, MA: Addison-Wesley.

Rheingold, H. (2002). *Smart mobs: The next social revolution*. Cambridge, MA: Perseus Publishing.

Sánchez Noriega, J. L. (1997). *Crítica de la seducción mediática*. Madrid, Spain: Tecnos.

Scott, J. (1994). *Social networks analysis: A handbook*. London: Sage Publications.

Tönnies, F. (1986): *Comunidad y sociedad*. Buenos Aires, Argentina.

Turkle, S. (1997). *La vida en la pantalla. La construcción de la identidad en la era de Internet*. Barcelona, Spain: Paidós. Smith, M. A. (2003). La multitud invisible en el ciberespacio. El mapeado de la estructura social de usenet. In M. A. Smith & P. Kollock (Eds.), *Comunidades en el ciberespacio*. Barcelona, Spain: Editorial UOC.

Vayreda, A. (2004). Las promesas del imaginario Internet: Las comunidades virtuales. *Atenea Digital, 5*, 55-78. Retrieved from http://antalya.uab.es/athenea/num5/vayreda.pdf

Wasserman, S., & Faust, K. (1995). *Social network analysis: Methods and applications*. New York: Cambridge University Press.

Wellman, B. (2002). Little boxes, glocalization, and networked individualism. In M.Tanabe, P. Besselaar, & T. Ishida (Eds.), *Digital cities II: Computational and sociological approaches*. Berlin, Germany: Springer. http://www.chass.utoronto.ca/~wellman/netlab/PUBLICATIONS/_frames.html

Wellman, B., & Gulia, M. (1999). Virtual communities as communities: Net surfers don't ride alone. In M. Smith & P. Kollock (Eds.), *Communities in cyberspace*. London: Routledge.

KEY TERMS AND DEFINITIONS

Connected Individualism: It is about a term coined by Wellmann (2002) in which the individual operator of his/her network is important, rather than the household or work unit. It is called network individualism—where technology users are less tied to local groups and are increasingly part of more geographical scattered networks. From the sociology point of view it is an expression about the "new liquid and individualized society" referred to by Bauman and Beck, according to which the individuals are more and more determined for the great social structures and therefore they are more owners of their individual destinies thanks to the communication possibility and customized relationship that facilitates the new information technologies.

Cyber Culture: Is a new culture form (symbolic group of values, beliefs and rules that give sense to the social action) that is emerging, due to the use of the new communication technologies. Therefore, the cyber culture is an extension of the traditional concept of culture that brings together the set of human relationships mediated by the information control mechanisms through the different technological communication systems. This turns the communicative process into more fluid and flexible social relationships and in many cases distant from the traditional space-temporal categories.

Cyber Cultural Revolution: It is the social and cultural transformation process generated by the use of new communication technologies. The multimedia revolution has several ramifications in which Internet has a central place, but where other digital networks coexists. Under sociological terms it is about the coexistence and slow displacement of the Homo sapiens, product of the written culture, to the digital homo, who worships the new communication technologies. Definitely, it is about a transformation process of the social and communicative structures through which the daily consumption habits and the human relationships moved about gradually from the daily spaces of the social interaction to the virtual spaces of social relationships created by the new information technologies.

Smart Mobs: It is a collaborative social dynamics made up of persons able to act jointly in order to achieve common objectives when they even know each other. The persons who perform these strategies called Smart Mobs, collaborate in a new context and under circumstances where the collective action was not possible before, thanks to the use of new communication tools and data development.

Society Network: It is a sort of advanced social organization based on technological communication networks. The networks are made up of nodes and links that use a plurality of paths to distribute the information from one link to another. This society auto regulates itself through even governance hierarchies and power distribution. In this sense, Castells states that "we are passing from the information society to the networks society", where each one of the users is a node of different networks that exchange by means of the use of the information technology.

Online Socialization: It is the process through which the individuals internalize and learn the rules and values of a specific social and cultural context through the virtual relational spaces that are created on the online network. This concept is displacing the former socialization concept, as the new technologies create new virtual socialization spaces beyond the family, the educational system and the labor market. These new online socialization contexts do not refer to a specific space and time but in many cases they are created spontaneously and overlapped on the network arising virtual learning and socialization communities.

Virtual Communities: Virtual community is a community whose bonds, interactions and social relationships are not produced in a physical space but in a virtual one as Internet. Investigators like Rheingold define the virtual communities as social groups that emerge from the Network [Internet] when sufficient people establish social communication and interchange networks characterized by the relative space stay and based in a feeling of belonging to a group, to form links of personal relationships in the cyberspace. Three main elements of the social and communication relationships converge under this definition: the reciprocity, the relational affective component and the interactivity time.

ENDNOTES

1. See: www.geocities.com
2. See: www.well.com. The experience of this Virtual Community is gathered in Rheingold´s book. The author describes in a practical way, how this VC was formed, which was its development, etc., becoming one more member of it. We can point out a fragment in which it is described what the persons can do both on this VC and on others: "people who form part the VCs use the words that appear on the screen to interchange courtesies and to discuss, carry out commercial transactions, interchange information, supply emotional support, plan, have great ideas, fall in love, meet friends, etc.
3. "networked individualism" (Wellman, 2002)

This work was previously published in Handbook of Research on Social Dimensions of Semantic Technologies and Web Services, edited by M. M. Cruz-Cunha; E. F. Oliveira; A. J. Tavares & L. G. Ferreira, pp. 435-447, copyright 2009 by Information Science Reference (an imprint of IGI Global).

Chapter 8.14
Learning for the Future: Emerging Technologies and Social Participation

Guy Merchant
Sheffield Hallam University, UK

ABSTRACT

Over the last five years there has been a large scale shift in popular engagement with new media. Virtual worlds and massive multiplayer online games attract increasing numbers, whilst social networking sites have become commonplace. The changing nature of online engagement privileges interaction over information. Web 2.0 applications promote new kinds of interactivity, giving prominence and prestige to new literacies (Lankshear and Knobel, 2006). To date, discussion of the opportunities, and indeed the risks presented by Web 2.0 has been largely confined to social and recreational worlds. The purpose of this chapter is to open up discussion about the relevance of Web 2.0 to educational practice. A central concern in what follows will be to show how the new ways of communicating and collaborating that constitute digital literacy might combine with new insights into learning in ways that transform how we conceive of education (Gee, 2004).

DOI: 10.4018/978-1-60566-120-9.ch001

INTRODUCTION

The term Web 2.0 was originally coined by O'Reilly (2005) as a way of referring to a significant shift in the ways in which software applications were developing and the ways in which users were adopting and adapting these applications. New applications were tending to become easier for the non-expert to use and more interactive, thus widening the scope for participation in online communities - it was becoming possible for those with relatively unsophisticated technical skills to create and share content over the internet. The popularity of blogs as a medium for individuals and groups to publish and discuss their concerns, news, and interests (whether frivolous or serious) is testimony to the popularity and everyday currency of the Web 2.0 phenomenon (Davies and Merchant, 2007; Carrington, 2008). And so, the increased availability of broadband, together with the development of more responsive and user-friendly software has led to a greater recognition of the internet as a place for social interaction, a place for collaboration, and a place for strengthening and building social networks. Web 2.0 commentators have drawn our attention to the 'social' and 'participatory' nature of contemporary life online (Lessig, 2004) whereas innovators and early-adopters are just beginning to glimpse the educational possibilities of these new development.

Not only do educators need to understand and capitalize on these new ways of being and interacting, they also need to investigate the educational potential of social networking. In order to do this, there is a pressing need to conceptualize the difference between casual and frivolous online interaction and those kinds of communication that have the characteristics of 'learning conversations'. Whilst there has been considerable development in our knowledge about the characteristics of learning conversations in face-to-face interaction in classrooms (Mercer, 2005; Alexander, 2007) there is little equivalent work in the field of online social networking.

Can these new spaces for shared communication provide an arena for the more systematic and structured interactions that are associated with formal education? This chapter addresses this question by both drawing both on the literature and my own research and writing, highlighting how new kinds of software not only involve new literacies but also changing roles for teachers and learners. Most of the material is drawn from classroom studies with children in the 7-11 age range and includes email partnerships, literacy work in virtual worlds, educational blogging and wiki building.

TECHNOLOGY AND LITERACY

Children and young people are growing up in a rapidly changing social world - a social world that is marked by the spread of new digital technologies. The impact of these technologies on the toy and game industry, on mass entertainment and communication, and on the ways in which many of us live and work has been little short of transformative. In schools, despite a substantial investment in computer hardware and software, there is still unevenness of provision and access, and considerable professional uncertainty about how to integrate new technologies into the curriculum and how to develop appropriate pedagogies. Nowhere is this uncertainty more keenly felt than in the area of literacy. Literacy educators, it has been suggested, need to assess the significance of new communication technologies and the ways of producing, distributing and responding to messages that typify them (Lankshear and Knobel, 2003). This involves looking at new genres, emerging conventions of communication and the changes in language associated with them. In doing this, literacy educators will inevitably have to negotiate the tension between notions of correctness and the realities of linguistic change, as well as a whole host of other issues that emerge with the growth of peer-to-peer communication and digitally-mediated social networks. It is against a backdrop of rapid social change and professional uncertainty that the work on digital literacy and new communications technology described in this chapter is placed. The work focuses primarily on digital writing, but partly because of the multimodal nature of this communication, there is an inevitable overlap with the wider area of new media studies.

New trends in digital culture, collectively referred to as Web 2.0 (O'Reilly, 2005), have begun to emerge over the last few years. These have ushered in new kinds of social participation through user-generated content, exchange and playful interaction. Of particular note here are individual and group blogs; sites which are designed for collaborative authorship (such as wikis); sites for generating and exchanging media such as music, still and moving image; and 3D virtual worlds. These networking sites provide a context for affinity, and facilitate the development of ad hoc purpose- or interest-driven groups in which self-directed, informal learning can take place. They not only offer us alternative models for envisioning learning communities but also the opportunity, where appropriate, to modify existing practices to fulfill more explicitly educational goals.

Popular networking sites allow geographically dispersed groups and individuals to communicate, exchange information and develop ideas. They also serve to thicken existing social ties by opening new channels of communication for those who are already known to each other, such as family and friends. Furthermore they are places for rehearsing ideas, making new connections, and new meanings. As such, the practices of tagging and the creation of folksonomies are a powerful iteration of the new literacy practices involved (Marlow et al, 2006). For an increasing number of young people, social networking provides ways of communicating with friends and ways of making new friends. This sort of interaction lies at the very heart of online social-networking. As we know, computer systems can store and retrieve huge amounts of data in different media. Harnessing this capacity to enhance communication and collaboration is the life-blood of online social networking. At the same time it is important to recognize that social networking is almost exclusively mediated through written communication and as such constitutes a prime site for future research into digital literacy.

Similar observations could be made about the communicative spaces provided by virtual world technology. 3-D virtual worlds can provide life-like settings for multiple users to interact in real-time. Users are embodied as human (or non-human) avatars in order to explore a virtual environment and interact with each other. Again interaction and collaboration are normally achieved through digital writing – and this often resembles the synchronous conversations of chatrooms (Merchant, 2001) and instant messaging. The most popular of these virtual worlds, Second Life, is already being used for educational purposes, but more established providers, such as Active Worlds have designed purpose-built educational worlds (see: http//:www.virtually-learning.co.uk) as we see later.

Web 2.0 developments raise new questions about digital literacy. For instance: what should we teach children about kinds of online communication that are helpful to relationships and helpful to learning; how can teachers support and encourage peer-to-peer interaction without stifling it, and above all how can we help pupils to become critical readers and writers in online environments? My own research (Merchant, 2001; 2003; 2008) has begun to explore the characteristics of digital literacy and has helped in making sense of new forms of synchronous and asynchronous communication, the changing nature of literacy, and the skills, understandings and attitudes that we will need to encourage in our schools. I suggest that a clearer sense of what is involved in digital literacy will result in teachers and pupils being better prepared for digital futures (Merchant, 2007).

Gaps between real-world uses of technology and new technology in the classroom continue to be a cause for concern. At the centre of this concern is the sense that a whole range of cultural resources fail to be translated into cultural capital by the school system. What is needed now is innovative work in digital literacy and particularly in educational settings to investigate the implications of new forms of social networking, knowledge sharing and knowledge building. And finally, because of the pervasive nature of digital technology, the commercial interest that is invested in it and the largely unregulated content of internet-based sources we also need to begin to sketch out what a critical digital literacy might look like. There is, in short, plenty to be done if we are to prepare children and young people to play an active and critical part in the digital future.

GETTING STARTED: DIGITAL LITERACY AS CONNECTED LEARNING

Information and Communication Technologies (ICTs) provide new opportunities for children and young people in educational environments, and learners can now be connected with the world outside the classroom in interesting and productive ways. Even everyday applications such as email can be used to enhance learning and interaction, providing a significantly different kind of experience to traditional literacies. Listserv applications, which automatically update multiple email addresses, have proved to be a very successful tool for mobilizing individuals around a shared interest and in developing a sense of community. Although widely used in academic and professional circles and to a lesser extent with college students, they have made little impact on the education of younger learners. My own fieldwork in UK classrooms suggests that introducing even the simplest of email practices into the curriculum raises practical and structural challenges that are not always easy to resolve.

There is a growing body of work in the field of young children's uses of email communication in classroom contexts and this has raised a number of important issues. For example in a recent study, Harris and Kington (2004) report on a project which put ten year-olds in email contact with employees at a mobile phone factory some 30 miles away from the school. Those employees (or 'Epals') learnt about children's interests and in turn offered insights into the world of work. Teachers involved in the project commented on how they found out more about their pupils' lives and interests when reading the messages they exchanged. A more formal evaluation showed gains in pupil motivation and social skills.

McKeon's (1999) study of 23 children's email interactions with pre-service teachers looked at the balance between purely social exchanges and topic-focused exchanges (in this case book-talk). Roughly half of the exchanges of these nine and ten year olds fell into each category, leading McKeon to conclude that:

classroom e-mail partnerships may provide students with a new way to learn about themselves as they select information that defines who they are and send it via e-mail to another. (Mckeon, 1999)

From this it seems that digital literacy can provide useful opportunities for exploring identity and relationships whilst also providing a discursive form which depends on purposeful communication with audiences beyond the confines of the classroom. However, other commentators have expressed concerns about the use of e-communication in educational settings, suggesting that a medium that clearly works well for informal social interaction may not necessarily be an effective tool for learning. For example, (Leu,1996) suggests that digital literacy needs to do more than appeal to youngsters just because it is 'cool'.

In a close analysis of the frequency and content of email exchanges between 301 eleven year olds, van der Meij and Boersma (2002) draw attention to the inherently social nature of this communication. However, their work appears to be predicated on professional concerns that frivolous social interaction could undermine learning exchanges rather than blend with them. Nevertheless, this work emphasizes the importance of using email as a communicative tool rather than as an explicit focus in its own right (as is sometimes the case in skills-based ICT instruction). The researchers draw attention to the need for more work in this area, observing in passing that 'email is not yet the integrated communication tool that it is in business settings' (van der Meij and Boersma, 2002). In short, the ubiquity of interactive written discourse in work and leisure – and even in some educational settings - finds few parallels in most primary classrooms.

There is less work on the processes of digital writing. Matthewman and Trigg's (2003) report on children's use of visual features in onscreen writing. Their study suggests that visual elements (such as font size and colour, layout and use of image) may be significant at all stages of composition. Similar findings are reported by Merchant (2004) whose analysis of children's onscreen work focuses on the production of multimodal texts. This on-going attention to the visual appearance of text at all stages in its production contrasts with traditional models of writing which associate presentational features with the production of a final draft. These studies show some of the characteristics of children's digital writing and their use of e-communication and suggest some important lines of enquiry. A transformative approach would need to be both sensitive to these, as well as the literacy capital of the pupils themselves (Bearne, 2003). Importantly, previous analyses of children's onscreen writing have provided evidence of children's expertise, willingness to learn from each other and to solve problems through creative and playful interaction (Merchant, 2003; Burnett et al, 2004).

My own study of how teachers of 8-10 year olds set out to provide opportunities for pupils to explore digital literacy in ways which were meaningful to them involved setting up email links between children in geographically dispersed schools (Burnett et al, 2006). The project involved pupils from two very different primary schools emailing each other as a preparation for producing a joint PowerPoint presentation on children's views and interests to a group of trainee teachers. Although the focus was on pupils' use of digital literacy, there was a strong feeling from the class teachers involved that the social benefits - in terms of breaking down stereotypes and widening horizons - were positive by-products of the project. In order to facilitate an initial exploration of views and interests, pupils in both classes were provided with a shoebox to collect artifacts that were of significance to them (an idea first developed by Johnson, 2003). These children then used email to get to know their partners, attaching digital photographs of items from their shoebox as a starting point for their interaction. This use of image acted as a prompt for the receiver who responded by asking questions to find out more about the items and their significance.

This project illustrates one way of embedding the use of new communication in the primary classroom. It also suggested that email partnerships can be worthwhile and provide experience of an important medium of asynchronous communication. Furthermore, such partnerships can help to 'dissolve the walls of the classroom', and provide new purposes and audiences for children's writing. School and student partnerships provide opportunities for early exploration of two key characteristic of new media – interactivity, multimodality. But beyond this, the sort of work described here underlines the need to re-interpret the writing process in relation to the production of digital texts – and even more importantly, it suggests ways in which teachers may need to design and choreograph learning experiences that encourage meaningful and educational interaction between peers in different locations.

MOVING ON: WEB 2.0, PARTICIPATION AND LEARNING

As the earlier discussion of email and listserv applications suggested, the use of ICTs to promote learning through participation pre-dates Web 2.0. In fact, a number of commentators have observed that the term Web 2.0 is best seen as a way of describing a gradual change or evolution in online communication (for example: Elgan, 2006) Although not normally described as Web 2.0, listservs and discussion forums do display the characteristic of added value through participation - as user-generated content aggregates information and develops ideas. The development of learning platforms (Virtual Learning Environments and

Course Management Systems) shows how emerging technologies have been assimilated into online and blended learning. So, for example, many learning platforms now allow administrators to embed discussion boards, to create student blogs and wikis and to enable RSS feeds.

Web 2.0 applications allow users to create and share multimedia content over the internet with a relatively light demand on their technical knowledge. From the point of view of the end-user a commonly used phrase 'the read-write web' is useful in capturing the shift that O'Reilly (2005) describes as Web 2.0. The phrase suggests a change of emphasis - one in which web-based activity is no longer simply about storing and accessing information but more about interaction, providing a place in which individuals 'converse', react to each others' ideas and information, and thereby add to the stock of knowledge. User-generated content can vary enormously in topic and can exploit the affordances of different media from written text, to still image, moving image, and sound - and any combination of these. At the same time, user-interaction can be encouraged through applications that allow for such things as profile pages, messaging facilities, group formation, and category tagging. More sophisticated sites also allow you to see which of your friends are online, provide information on the latest changes to your favourite sites (through RSS feeds) and give users the choice of modifying or personalising their home pages, changing their look and the features included.

From this brief overview it should be clear that Web 2.0 pre-supposes a more active user – one who is encouraged to design an online presence (an identity, or a set of identities) and to participate, to a greater or lesser extent, in a community of like-minded users. Whether or not the social networks produced can be described as 'communities of practice' (Wenger,1999) and how we can best describe the informal learning that takes place in Web 2.0 environments is the subject of much current research activity.

The blog format offers a range of interactive and collaborative possibilities for individuals and teams. Some of these possibilities derive from features that are part of the architecture of blogs. During the last five years, a period in which the blogosphere has undergone a rapid expansion, diversification and innovation have been of central importance. So, for example, Lankshear and Knobel (2006) offer a provisional taxonomy of blogs identifying 15 different kinds of blogs, at the same time as recognising that blogs are an unstable form, as they continue to mutate and hybridise. There is clearly no standard way to blog. Arguably, the single defining feature of a blog is that of date-ordering (Walker, 2003). Although periodic updating is also a feature, some established bloggers post daily whilst others are less frequent. The sequential, chronological characteristics of the blog format suggest how it can be useful in capturing such things as the development of a narrative, the design and implementation of a project, the progress of research, emerging processes, the aggregation of links or references, and observations or reflections which develop over time (Davies and Merchant, forthcoming). Blogs, as multimodal texts, also allow us to represent these activities in written, still and moving image or audio format – and of course some of the most interesting blogs are a judicious combination of these modes. Educational blogging can capture learning as it unfolds over time and this has obvious benefits for both learners and teachers. In this most basic sense a blog can provide an analytical record of learning, or an online learning journal (Boud, 2001). Writing in 2003, Efimova and Fielder noted that alongside the 'diary-like format' blogs kept for family and friends there was a:

...growing cluster of weblogs used by professionals as personal knowledge repositories, learning journals or neworking instruments. (Efimova and Fielder, 2004:1)

They go on to suggest that these newer blogs not only serve the needs and interests of those writing them but also display emerging ideas in a public space. This suggests the development of more open learning journals which can be interlinked and commented upon within an emerging community of learners.

As Richardson (2006) points out, blogging can also involve users in an important and distinctive kind of learning; one that he characterises as: read- write- think -and -link. Richardson suggests that a blogger develops a kind of practice that he describes as 'connective writing' in which active reading, and involvement through comments and hyperlinks work alongside regular posting in the co-construction of meaning through social participation. This view accentuates the significance of a community of bloggers, either in the form of a cluster of related blogs or a team blog. From this point of view we can see blogging as a way of supporting a community of practice (Wenger, 1998) or an affinity space (Gee, 2004a). The growing number of educational blogs provide a variety of examples of how the perceived affordances of blogging can be used to support learning. For example, in my own work I describe how a teacher of 10 year olds used team blogs in the context of work on pollution in the environment (Davies and Merchant, forthcoming). The teams' initial posts were used to document their existing opinions on the topic. As the project developed, search results and hyperlinks provided a record of their learning and evaluation of web-based sources. Later, on a field visit, the students took digital photographs of environmental hazards such as fly-tipping, invasive non-native flora, and industrial effluent, and uploaded them to their blogs. Towards the end of the unit of work, students used their blogs to reflect on what they had learnt and share it with the wider school community. Where this project was based on the work of students in one particular school setting, providing a record of their learning over time, other projects have harnessed the potential of Web 2.0 to work collaboratively across settings.

Using wiki software, which allows multiple users to co-create interlinked pages, students in geographically dispersed locations can learn about each other and collaborate on shared interests. An example of such work is the partnership between the Helen Parkhurst School in Almere, in the Netherlands and the Gostivar Secondary School, in Macedonia (the MacNed Project). This project is developing intercultural understanding through the use of ICT as students share and analyse their own production and consumption of media. The MacNed Project illustrates how the new ways of communicating and collaborating that characterize Web 2.0 can be used to develop learning. Whilst it could be argued that the same kinds of understanding could be developed through more traditional approaches, the possibility of co-constructing text in different geographical locations, exchanging and commenting on work in different media creates a heightened sense of interactivity and a more overtly participatory space for learners. The work also begins to point to a changing role for educators who, in this case, needed to co-ordinate the work and provide the context for interaction – in short, to design a new kind of learning experience and to encourage participation and peer-to-peer dialogue.

FROM REAL CLASSROOMS TO VIRTUAL WORLDS

Similar issues of learning design are beginning to emerge from educational work that is based in virtual worlds, and in this section I explore and illustrate some of these issues. Schroeder (2002) describes a 3D virtual world as:

..a computer-generated display that allows or compels the user (or users) to have the feeling of being present in an environment other than one they are actually in... (Schroeder, 2002)

3D virtual worlds could well enhance or transform learning, but although recreational virtual world play continues to attract public attention, empirical research that investigates their learning potential in classrooms is still in its infancy. Although there are a number of claims about the high levels of learner engagement in gameplay (Squire, 2002) and the construction of 'powerful learning environments' in virtual worlds (Dede, Clarke, Ketelhut, Nelson and Bowman, 2006) there is clearly scope for more empirical evidence to back these claims. Despite the fact that some researchers have claimed that immersive environments may lead to a loss of focus and distraction (Lim, Nonis and Hedberg, 2006), there is, as yet, insufficient evidence to reach firm conclusions. Early studies such as those of Ingram, Hathorn and Evans (2000) focused on the complexity of virtual world chat. Fors and Jakobsson (2002) investigated the distinction between 'being' in a virtual world as opposed to 'using' a virtual world, but little rigorous attention has been given to their learning potential. The work of the Vertex Project (Bailey and Moar, 2001), which involved primary school children in the UK, makes some interesting observations on avatar gameplay, but placed its emphasis on the ICT learning involved in building 3D worlds rather than the learning and interaction which might take place within them.

An educational virtual world project, initiated by a UK local authority in Barnsley, aims to raise boys' attainment in literacy by an adventurous and innovative use of new technology that foregrounds digital literacy (Merchant, 2008). In partnership with the company Virtually Learning (http://www.virtuallylearning.com.uk), the project team – a group of education consultants and teachers - designed a literacy-rich 3D virtual world which children explore in avatar-based gameplay (Dovey and Kennedy, 2006). The children, in the 9-11 age range, work collaboratively to construct their own narratives around multiple, ambiguous clues located in the world and, as a result, engage in both on- and off-line literacy activities. The virtual world, called Barnsborough, is a three dimensional server-based environment which is explored from multiple but unique perspectives through local Active Worlds browsers. Navigational and communicational tools are built into the Active Worlds browser, enabling avatars controlled by the pupils to move around in virtual spaces such as streets, buildings and parks, to engage in synchronous written conversations, and in this particular example, to discover clues in order to build their own narratives.

Pupils in 10 different project schools have been using this 3D virtual world, interacting with each other using the Active Worlds' real time chat facility. The world itself consists of a number of interconnected zones which are life-like and familiar - in fact they are often modelled on real world objects. The zones include a town, complete with streets, alleyways, cafes, shops and administrative buildings some of which can be entered. There is also a park with a play area, bandstand, boating lake, mansion, woodland and hidden caves; a residential area with Victorian and contemporary housing, a petrol station and various local amenities and an industrial zone with old factories, canals and so on. In some of the connecting zones pupils may encounter other sites such as a large cemetery, a medieval castle and a stone circle. Rich media, tool-tip clues, hyperlinked and downloadable texts provide clues about the previous inhabitants of Barnsborough, suggesting a number of reasons why they have rather hurriedly abandoned the area. Some possible story lines include a major bio-hazard, alien abduction, a political or big business disaster or suggest something more mysterious. The planning team has seeded these clues throughout the Barnsborough environment, drawing on popular narratives such as Dr Who, Lost, Quatermass, the Third Man and Big Brother.

In this example, a 3D virtual world provides a stimulating environment for online exploration and interaction. Barnsborough is *designed* as a literacy-rich environment. To enter Barnsborough is to become immersed in a textual universe and to participate in what Steinkuehler (2007) has described as a 'constellation of literacy practices'. The following is a list of the main kinds of digital literacy encountered in the virtual world. These are not directly used for literacy instruction, with the exception of the hyperlinked texts, which are quite deliberately tied to national literacy objectives.

Environmental Signs and Notices

This material forms part of the texture of the 3D virtual world and is designed to create a real-world feel to the visual environment and also to provide children with clues. Examples of this include graffiti, logos, signs and notices, posters, and advertisements.

Tool Tips

These give additional explanations or commentaries on in-world artefacts and are revealed when 'moused over' with the cursor. Tool tip messages that draw attention to environmental features ('looks like someone's been here'); hold navigational information ('you'll need a code to get in'); or provide detail ('cake from Trinity's') are shown in text-boxes.

Hyperlinked Texts

Mouse-clicking on active links reveals a more extended text. Examples include an oil-drilling proposal (a Word document); a child's diary (a Flash document); and a web-page on aliens. Some of these links are multimedia (such as phone messages and music clips) whereas others provide examples of different text types, such as text messages and online chats.

Interactive Chat

This is the principle means of avatar interaction and involves synchronous chat between visitors to the world. Comments are displayed in speech bubbles above the avatars heads as well as in scrolling playscript format in the chat window beneath the 3D display.

The Barnsborough virtual world experience foregrounds some important dilemmas relating to engagement with digital literacy in the classroom. The most significant of these dilemmas stem from the fact that it introduces pupils and teachers to new ways of interacting with one another. So, for instance, in-world pupil-pupil interaction is not only conducted in the emerging informal genre of interactive written discourse (chat), but it also disrupts ideas of conventional spelling, turn-taking and on-task collaboration. New relationships between teachers, pupils from different schools and other adults have been significant in this work. Issues about authority and what kinds of behaviour are appropriate in a virtual environment were quick to surface, and this in turn has raised new issues for teachers who are understandably concerned about the safety of their pupils as well as how they might monitor children's online experiences and interactions. Onscreen digital practices can therefore give rise to uncertainty, particularly where these practices do not easily fit into established classroom routines. Squire, in an article on the educational value of video-gaming, suggests that:

...the educational value of the game-playing comes not from the game itself, but from the creative coupling of educational media with effective pedagogy to engage students in meaningful practices. (Squire, 2002)

This observation could apply equally to 3D virtual worlds as well as the other communicative spaces described in this chapter. In and of themselves, these technologies cannot *create* new forms of learning, but as educators become more familiar with their affordances, and the ways in which they are being used in recreational and work contexts, they can begin to experiment with educational uses, to design specific environments, and to envision new pedagogies.

CONCLUSION

In this chapter I have explored the ways in which the digital literacies that are central to new kinds of social practice can be incorporated into classroom settings. I have also shown how literacy continues to play a central role in social participation and knowledge-building – particularly in Web 2.0 environments - and how digital connection allows this to happen in ever more fluid and distributed ways. The question of whether the new communicative spaces described can provide an arena for the more systematic and structured interactions that are associated with formal education is not an easy one to answer. After all, classrooms are quite distinctive social contexts in which patterns of interaction and the availability of communicative tools are often restricted or carefully controlled (Kerawalla and Crook, 2002), and so, adopting and adapting digital literacies easily disrupts traditional classroom practices in ways that are unsettling to teachers. Indeed, as Carrington (2008) suggests alternative learning designs and pedagogies are required, and these may only be achieved through more far-reaching school reform.

There are also some important concerns about pupil safety that need to be addressed. Protecting school students from bullying, verbal abuse and inappropriate online behaviour cannot be passed over lightly. The tensions between adult supervision and surveillance, and trust and pupil autonomy become crucially important. Teachers and researchers involved in such work must ask themselves some key questions:

- How easy is it to leave the comfort zone of conventional, classroom-based pupil-teacher relationships and experiment with new and fluid online interactions? Teachers will have to take risks in using this sort of technology and both teachers and researchers need to document the new ways of working that emerge.
- What are the implications of working in an environment in which some pupils are more experienced or confident than the teacher? As in many other applications of new technology, children tend to be more experienced and more adaptable than their teachers. Although this is not always the case, teachers do need to be prepared to learn from pupils and to value their experimentation.
- How can this sort of work be justified and defended in an educational environment which regularly lurches back to a pre-occupation with 'the basics' and traditional print literacy skills? New and important digital literacies can be introduced through Web 2.0 work. Experience of these is likely to have a positive impact on learning in general, and on literacy in whatever form. Again more evidence is needed to support this case.
- How can the level of immersion and flexible online access required by such work be operationalised within the constraints of current resource and timetable structures? As others have observed (eg Holloway and Valentine, 2002) schools need to re-think the location, access and use of computer hardware. In common with other digital literacy practices, Web 2.0 work invites a more flexible approach to curriculum organisation and online access.
- What additional planning and co-ordination work is necessary to make the most of online work, to facilitate exchange between year groups and interactions between schools? One of the most important features of digital literacy is its potential to connect learners with others outside the immediate school environment. This will involve careful co-ordination and planning

between teachers in different locations.

- What real or perceived risks may be faced by engaging in Web 2.0 practices (eg: child protection; parental censure etc)? New projects need to pay careful attention to issues of online safety. Parents need to be kept informed, and teachers need a carefully rehearsed educational rationale for the work they undertake.

New literacy practices in the classroom contrast starkly with the educational routines of book-based literacy, as well as with the dominant ICT pedagogies. The former privilege print-based routines which, whilst still significant, are insufficient preparation for an increasingly digital future, whereas the latter reify centralised control through teacher-led use of whiteboards, instructional software, and highly structured learning platforms (VLEs and CMSs). Collaborative, peer-to-peer interactions, including communication with those not physically present in the classroom, suggest a very different set of resources and educational concerns. In short, everyday uses of new technology, and particularly recent Web 2.0 developments, raise new questions about digital literacy and its role in education.

Teachers concerns are for a safe and orderly space where controls, both subtle and gross are evoked to maintain a harmonious learning environment. Moreover, the classroom world is a world in which these relationships have traditionally been mediated by a set of schooled print literacy practices and instructional routines, powerfully structured by curriculum discourse. Disturbing this fragile ecology is a risky business – but experience shows that the use of emerging technologies can often destabilize. Consequently, strong support and sensitive professional development are required if we are to move beyond some of the curriculum constructs and pedagogical conventions that narrow our vision of learning through digital literacy. Teachers need not be the docile operatives of an outdated, centralised curriculum – as some of the work described in this chapter suggests - they can also be innovative in responding to the potential of powerful new technologies.

REFERENCES

Alexander, R. J. (2006) *Towards Dialogic Thinking: Rethinking classroom talk* (3rd ed.). York, UK: Dialogos.

Bailey, F. & Moar, M. (2001) The Vertex Project: children creating and populating 3D virtual worlds. *Jade 20*(1). NSEAD.

Bearne, E. (2003). Rethinking Literacy: communication, representation and text. *Reading Literacy and Language*, *37*(3), 98–103. doi:10.1046/j.0034-0472.2003.03703002.x

Burnett, C., Dickinson, P., Merchant, G., & Myers, J. (2004). Digikids. *The Primary English Magazine*, *9*(4), 16–20.

Burnett, C., Dickinson, P., Merchant, G., & Myers, J. (2006). Digital connections: transforming literacy in the primary school. *Cambridge Journal of Education*, *36*(1), 11–29. doi:10.1080/03057640500491120

Carrington, V. (2008). I'm Dylan and I'm not going to say my last name: some thoughts on childhood, text and new technologies. *British Educational Research Journal*, *34*(2), 1–16. doi:10.1080/01411920701492027

Davies, J., & Merchant, G. (2007) Looking from the inside out – academic blogging as new literacy. In M. Knobel & C. Lankshear (Eds.), *The New Literacies Sampler* (pp. 38 -46). New York: Peter Lang.

Davies, J., & Merchant, G. (in press). *Web 2.0 for Schools: social participation and learning*. New York . *Peter Lang.*

Dede, C., Clarke, J., Ketelhut, D., Nelson, B., & Bowman, C. (2006) *Fostering Motivation, Learning and Transfer in Multi-User Virtual Environments*. Paper given at the 2006 AERA conference, San Franscisco, CA.

Dickey, M. D. (2005). Three-dimensional virtual worlds and distance learning: two case studies of Active Worlds as a medium for distance learning. *British Journal of Educational Technology, 36*(3), 439–451. doi:10.1111/j.1467-8535.2005.00477.x

Dovey, J., & Kennedy, H. W. (2006) *Game Cultures: Computer Games as New Media.* Maidenhead, UK: Open University Press.

Elgan, M. (2006, September 14). Here's the skinny on Web 2.0. *Information Week.*Accessed 10th August, 2008 from: http://www.informationweek.com/news/software/open_source/showArticle.jhtml?articleID=193000630

Fors, A. C., & Jakobson, M. (2002). Beyond use and design: the dialectics of being in a virtual world. *Digital Creativity, 13*(1), 39–52. doi:10.1076/digc.13.1.39.3207

Gee, J. P. (2004) *What Videogames Have to Teach us About Learning and Literacy.* New York: Palgrave Macmillan.

Harris, S., & Kington, S. (2002) Innovative Classroom Practices Using ICT in England. *National Foundation for Educational Research,* Slough, UK. Accessed 27th February, 2005 at: http://nfer.ac.uk/research/down_pub.asp

Ingram, A. L., Hathorn, L. G., & Evans, A. (2000). Beyond chat on the internet. *Computers & Education, 35,* 21–35. doi:10.1016/S0360-1315(00)00015-4

Kerewalla, L., & Crook, C. (2002). 'Children's Computer Use at Home and at School: context and continuity.' *British Educational Research Journal, 28*(6), 751–771. doi:10.1080/0141192022000019044

Lankshear, C., & Knobel, M. (2003) *New Literacies: Changing Knowledge and Classroom Learning.* Buckingham, UK: Open University Press.

Lankshear, C., & Knobel, M. (2007) *New Literacies: Everyday Practices and Classroom Learning.* Buckingham, UK: Open University Press.

Lessig, L. (2004) *Free Culture: How Big Media Uses Technology and the Law to Lock Down Culture and Control Creativity.* New York: Penguin.

Leu, D. J. Jr. (1996). 'Sarah's secret: Social aspects of literacy and learning in a digital information age.' . *The Reading Teacher, 50,* 162–165.

Lim, C. P., Nonis, D., & Hedberg, J. (2006). Gaming in a 3D multiuser environment: engaging students in Science lessons. *British Journal of Educational Technology, 37*(2), 211–231. doi:10.1111/j.1467-8535.2006.00531.x

Marlow, C., & Naarman, M. boyd, d., & Davis, M. (2006) *HT06, Tagging Paper, Taxonomy, Flickr, Academic Article, ToRead.* Accessed 11th August, 2008 at: www.danah.org/papers/Hypertext2006.pdf

Matthewman, S., & Triggs. (2004). Obsessive compulsory font disorder: the challenge of supporting writing with computers. *Computers & Education, 43*(1-2), 125–135. doi:10.1016/j.compedu.2003.12.015

McKeon, C. A. (1999). The nature of children's e-mail in one classroom. *The Reading Teacher, 52*(7), 698–706.

Mercer, N. (2000) *Words and Minds: How We Use Language to Think Together.* London: Routledge.

Merchant, G. (2001). Teenagers in cyberspace: language use and language change in internet chatrooms. *Journal of Research in Reading, 24*(3), 293–306. doi:10.1111/1467-9817.00150

Merchant, G. (2003). E-mail me your Thoughts: digital communication and narrative writing. *Reading . Literacy and Language*, *37*(3), 104–110.

Merchant, G. (2007). Writing the future. *Literacy*, *41*(3), 1–19.

Merchant, G. (2008) Virtual Worlds in Real Life Classrooms. In V. Carrington & M. Robinson (Eds), *Contentious Literacies: Digital Literacies, Social Learning and Classroom Practices* (pp. 93-108). London: Sage.

O'Reilly, T. (2005) *What is Web 2.0? Design patterns and business models for the next generation of software.* Accessed 10 April, 2007 at: http://oreillynet.com/pub/a/oreilly/tim/news/2005/09/03/what-is-web-2.0.html

Schroeder, R. (2002) Social Interaction in Virtual Environments: Key Issues, Common Themes, and a Framework for Research. In R. Schroeder (Ed.) *The Social Life of Avatars: Presence and Interaction in Shared Virtual Environments,* (pp.1-19). London: Springer.

Squire, K. (2002) Cultural Framing of Computer/Video Games. *Game Studies.* Accessed 12th May, 2007 at http://gamestudies.org/0102/squire/

Steinkuehler, C. (2007). Massively Multiplayer Online Gaming as a Constellation of Literacy Practices. *E-learning*, *4*(3), 297–318. doi:10.2304/elea.2007.4.3.297

van der Meij, H., & Boersma, K. (2002). E-mail use in elementary school: an analysis of exchange patterns and content. *British Journal of Educational Technology*, *33*(2), 189–200. doi:10.1111/1467-8535.00252

Wenger, E. (1998) *Communities of Practice: Learning, Meaning and Identity.* Cambridge, UK: Cambridge University.

KEY TERMS AND DEFINITIONS

Digital Literacy: This term has been defined in different ways. I use it to describe written or symbolic representation that is mediated by new technology (Merchant, 2007). Whilst recognizing that many online texts are multimodal, digital literacy places the focus on the semiotic of written communication.

Folksonomy: Related to the term 'taxonomy', this describes the way in which participants in a Web 2.0 space have assigned tags or labels to content. These tags identify the prevalent themes, topics or areas of interest for individuals in that particular environment. Aggregating these tags creates a folksonomy. Visitors to the site can then search 'by tag' and see all the objects labelled by that specific tag.

Interactive written discourse: This is a term used to describe computer-mediated communication (CMC) that is based on two or more people 'taking turns'. These conversational exchanges range from email replies, to forum exchanges and synchronous chat.

Learning platform: A catch-all term for online learning environments designed for the education market. These are usually closed or controlled intranet systems. Alternative designations are Learning Management Systems or Virtual Learning Environments. Some Learning Platforms also include student-tracking and assessment data – sometimes these integrated systems are called Managed Learning Environments.

Multimodality: This term is used to describe the different modes of human communication (visual, verbal, gestural etc). In many web-based texts, meaning is communicated through a subtle interplay between different expressive modes.

This work was previously published in Handbook of Research on New Media Literacy at the K-12 Level: Issues and Challenges, edited by L.T.W. Hin & R. Subramaniam, pp. 1-13, copyright 2009 by Information Science Reference (an imprint of IGI Global).

Chapter 8.15
Metacognition on the Educational Social Software:
New Challenges and Opportunities

Margarida Romero
Université de Toulouse, France & Universitat Autònoma de Barcelona, Spain

ABSTRACT

In recent years, we have witnessed an information revolution. This revolution has been characterised by widespread access to the Internet and by the emergence of information which has been generated by end-users–the so-called user-generated content. The information thus generated has been supported by Web 2.0 applications or social software. These changes in the information society have had an important impact in education, with more and more adults enrolling on life-long learning programs; moreover, the availability of distance learning courses has grown in line with this increase in demand. In this emergent educational paradigm, the new 2.0 technology context implies new competencies for learners. These competencies include literacy in information and communication technology (ICT), learning autonomy, self-regulation and metacognition, while at the same time expanding the opportunities for metacognitive development. We will consider in this chapter these two perspectives of the 2.0 context; on the one hand, the new requirements provided by the environment and, on the other hand, the new learning opportunities which this environment brings.

DOI: 10.4018/978-1-60566-826-0.ch003

1. INTRODUCTION

The development of information and communications technology (ICT), and the Internet in particular, are producing a paradigm shift for the diffusion of information and the creation of knowledge. This revolution is even more pronounced in the context of the user-generated approach of Web 2.0. It is interesting to note that the "2006 Time Person of the Year" was none other than *You*. The reason for this surprising choice was the growth and influence of user-generated content on the Internet during the early years of the 21st century.

In education, this technological revolution is occurring in a context of globalisation, which challenges higher education organisations to structure and harmonise the length of their programs in order to improve quality standards and facilitate student mobility. In the context of the European Community, these changes are outlined in the Bologna Agenda, which sets targets for convergence within the European Higher Education Area (EHEA). This agenda imposes a greater responsibility on the learner – the learner's autonomy is to be developed by a shift from a system of teacher-driven provisions towards a student-centred approach.

From a teaching and learning perspective, we are experiencing two important changes. Firstly, the focus on knowledge construction by the learner, mostly through collaborative contexts; secondly, the willingness to globalise and standardise the information exchanges, the learning processes and the tools. In this new context, it is essential to become the manager of one's own learning. The challenge for the learner is about learning to learn; about developing individual skills for self-regulated learning and for enhancing metacognitive abilities. For this purpose, it is useful to take advantage of the new opportunities offered by Educational Social Software (ESS).

Metacognition, the knowledge about knowledge and how we learn, can be promoted by the social construction of knowledge and by its diffusion. This social construction and diffusion is facilitated by ESS, which provides new opportunities to develop inter-subjective awareness. However, the new facilities offered by Web 2.0 require learners to develop new competencies (ICT literacy, autonomy, metacognition, ...). At the same time, social software is expanding the opportunities for the development of the competencies of learners in the 21st century. Thus, we will consider in this chapter these two perspectives of the new 2.0 context – on the one hand, the new requirements provided by social software and, on the other hand, the new learning opportunities which this new context brings. Thus, in the first part we explore the changes in both learning and in social software. In the second part, we focus on the implications of Educational Social Software in regard to the needs created by ESS for the new competencies the learner must develop in order to succeed. Lastly, we will focus on the opportunities that ESS provides for the development of metacognitive learners.

2. TOWARDS A NEW LEARNER IDEAL: SELF REGULATION, METACOGNITION AND ENGAGEMENT IN LIFE-LONG LEARNING

The profound changes in the information society have also transformed the learner ideal. In a context of growing obsolescence of knowledge (Brandsma, 1998), subjects and organisations must learn throughout life (life-long learning), developing a willingness to learn continuously, encouraged by educational bodies at the highest national and international level (UNESCO, OECD, European Commission). In this context, autonomous learning is now regarded as a social issue (Moisan and Carré, 2002) fostered by ICT possibilities.

The rapid growth of the Internet in recent decades and the emergence of Web 2.0 have together developed a set of new learning opportunities. This has been achieved through a revolution in the social relations of learning, which can now take place at different times and in different places. These profound changes reshape the competencies of the 21st century learner; these competencies must now include a greater degree of autonomy. Learning skills need to be self-regulated; there is a requirement for a better metacognitive development and an increasing need for those following life-long learning courses to be able to use ICT. The latter implies a certain computer literacy. Learners in the 21st century should develop a complex set of skills before starting to take advantage of the learning potential offered by ICT solutions, and more specifically by social software. Next, we will continue the discussion by analysing four main ESS prerequisites, with special attention to the prerequisite for metacognitive competence.

2.1. Prerequisite I: The Use of ICT

Technology enhanced learning involves the use of technologies to aid learning, and a growing number of courses use Information and Communication Technologies (ICT). In the case of blended learning courses, ICTs are a technological complement to traditional face-to-face learning. With virtual distance courses, ICTs become the main learning environment. However, to take part in distance learning, learners must develop ICT skills in order to navigate through the information, choose the best sources, communicate with peers and professors and contribute to co-creation or information tagging. Computer literacy has become a basic skill for life-long learning and professional integration. In an ICT society, and even more in a 2.0 perspective, ICT skills are a prerequisite.

2.2. Prerequisite II: Learner's Autonomy and Self-regulation

Students engaged in distance learning courses are free to participate at the time and place of their choice. This freedom is both an opportunity and a heavy responsibility in terms of autonomy and self-regulation. Learner autonomy involves the ability to work alone without external regulation (teacher's instructions, timetables, etc.). Being a self-regulated student implies being able to manage one's own learning times and rhythms, planning one's own activities and their execution (revision, work, ...) and even managing emotions and motivations during the learning process.

2.3. Prerequisite III: Learner's Metacognitive Development

In order to manage his learning, the student must be aware of his current knowledge and should know his strong and weak points. This awareness of one's own cognition and the control of the cognitive processes is known as metacognition. For Marzano (1998) "metacognition is the engine of learning". A student with a good metacognitive development will be able to assess his own achievements, self-evaluate what he knows and what he still needs to learn or review, thus allowing him to plan and regulate the process. Having acquired the metacognitive skills to manage his own learning, he becomes manager of the task and gains a strategic approach to knowledge acquisition.

Learners with good metacognitive development will more than likely be "*people who possess self determination or autonomy in learning and problem solving. They will be able to refer to the what, how, when, where and why of learning when carrying out complex cognitive activities*" (Gordon, 1996). Metacognition has been defined as the "*conscious control of learning, planning and selecting strategies, monitoring the progress of learning, correcting errors, analysing the effectiveness of learning strategies, and changing learning behaviours and strategies when necessary.*" (Ridley et al, 1992), but also as *"cognitive strategies"*, (Paris and Winograd, 1990), *"monitoring of cognitive processes"* (Flavell, 1976), *"resources and self regulating learning"* (Osman and Hannafin, 1992) and *"evaluating cognitive states such as self appraisal and self management"* (Brown, 1987).

In other words, being a metacognitive student also means becoming a strategic learner who will choose a strategy to plan and regulate his learning. In this way, if I plan to master the English irregular verbs and know that I find it useful to read aloud, I will avoid the library, where silence is required, and find a place without such a restriction.

In a context of autonomous learning, metacognition is an important factor for learning performance (Osborne, 2000; Zimmerman and Martinez-Pons, 1990; Pintrich, 2002). In the context of distance learning, metacognition is also a key factor of success (Houssman, 1991). Learning with social software requires learners' metacognition to deal with the non-linear mass of information without wasting time and losing the learning objectives.

2.4. Prerequisite IV: Learner's Willingness to Engage in Life Long learning

However, there is a risk that the proliferation of learning opportunities will be meaningless unless participants and those offering such programs are willing to develop a new approach to learning. In a knowledge society, learning is an on-going requirement for citizens and businesses – both risk losing their competitiveness if they do not update their knowledge on a regular basis. This involves creating a new approach to learning – one not solely confined to secondary education. ICTs have contributed both to the consolidation of the knowledge society as well as to the development of e-learning and blended learning solutions. More informally, the emergence of social software helped to build learning communities (Lave and Wenger, 1991).

To sum up, a pre-requisite for the emergence of continuous learning and the use of ICT is a set of higher order skills – these skills must be acquired before the new e-learning opportunities can be exploited. Next, this chapter will focus on metacognition as a pre-requisite for effective learning in social software. In addition, Educational Social Software (ESS) will be stressed as a great opportunity to enhance metacognitive development activities.

3. METACOGNITION ON EDUCATIONAL SOCIAL SOFTWARE

3.1. The Emergence of New Learning Theorie Towards the ESS Approach

The 2.0 revolution opens up new educational opportunities based on collaborative learning and collective intelligence co-construction, going further than the mere transmission of information. For some authors, traditional pedagogical models no longer serve to explain learning in the context of social technologies. Instead they propose the introduction of a new learning model – connectionism.

From the earliest days of information technology, computers were considered as potential tools to facilitate learning and thus enhance human capabilities. The first approaches tried to implement the behaviourist approaches, creating DILL (digital and library learning) and practice programs (Bottino, 2004). These were followed in turn by MicroWorlds (Papert, 1980) and some educational interactive tools. With the development of the Internet, professors viewed the possibilities of the global network for publishing their learning resources, even if these resources were mostly designed for a traditional face-to-face use in the classroom, without the reengineering needed to fulfil web-based learning specificities (e.g. file formats and weight, human computer interaction and ergonomic rules, etc.). Broadly speaking, this consisted of PowerPoint presentations exported as pdf, and then uploaded into web sites or specific learning platforms. Fortunately, this approach has been modified and gradually enriched. Educational institutions and professors are beginning to reconsider the model in order to exploit the interactive possibilities of the Internet. The Internet is now seen as a tool for social construction of knowledge recommended by the socio-constructivist approach. Groupware has emerged and a new generation of Learning Management Systems (e.g. Moodle - *Modular Object-Oriented Dynamic Learning Environment*) has started to allow the development of constructivist learning through collaborative activities such as forums, co-constructed glossaries, wikis, or even, WebQuests. This decisive step has brought about the ESS revolution, breaking the learning and learners' atomization which had been imposed by traditional educational software and platform-based approaches. The activities and the learner community have been extended into a wider sphere. The World Wide Web as a global network promotes interaction among users and the collective intelligence generated by large communities of practices and learning. In this context, the emergence of social technologies is a great opportunity to develop the connectionist learning approach.

The connectionism (Siemens, 2004, inspired by Rumelhart and McClelland, 1986) considers learning as the result of connecting different sources of information, adopting a non-linear learning structure. In this approach, learning is a dynamic process – a process that requires our knowledge to be up-dated in order to take into account the evolution of information in the network. Consequently, the *"capacity to know more is more critical than what is currently known"* (Siemens, 2004). That places self-regulation and metacognitive abilities as a key factor for learning performance in a pervasive knowledge network which exhibits information in a continuous state of flux.

3.2. Metacognitive Requirements when Learning with ESS

The pre-requisites for 21st century learners to respond to the new opportunities for learning with ICTs were addressed at the beginning of this chapter, where ICT competence, autonomy and self-regulation and metacognitive development are identified as important. In this section we will explain the metacognitive needs related to ESS specificities: the complexity, the non-linear structure of knowledge and the continually changing network of people involved in the production and development of knowledge.

Metacognition is highly important for learning in complex situations. Students must manage their own learning in a context of continuous change where they need to regulate themselves as learners and take a large number of decisions concerning both their learning objectives and structure (what to learn and in which order) and the learning modalities (individual or collaborative learning, study time and rhythm, etc.). In urgent situations, metacognition could be supported by just-in-time context-aware mobile devices (Romero and Wareham, 2009), allowing flexible learning solutions adapted to the context.

The learning complexity with educational social software is due to the specificities of 2.0: a network-based structure of non-linear information and interactions. Already, in traditional educational software and web 1.0, interaction and information were limited by the software top-down approach of information workflow. In the bottom-up approach of social software, the number of participants and their contributions makes the system very complex.

The traditional web is already a complex metacognitive challenge for students (Veenman, Wilhelm and Beishuizen, 2004) because of the vast amount of information available and all the possible choices. This complexity is even more important in the case of Web 2.0, where we have an almost infinite number of knowledge contributors and their contributions. In this networked architecture, the information is not always displayed by its original author. Thus, we often reach information aggregations displayed in third sites. Websites like NetVibes or Google News can show a personalized information aggregation for each user. It is no longer a direct source of content, but a remix of external information sources. *"The content disappears behind the architecture. Speech is no longer anchored in a device (technology) but the device anchors speech."* (Ertzscheid, 2005). The information architecture as recovered by a search engine (e.g. Google, Kartoo ...) is also an information aggregation. In these levels of information remix, the original author of the content is often lost. In this sense, we can observe that the aggregation structure becomes even more important that the content structure (Saffer, 2005).

It is already difficult for learners to identify the structure of a linear content in an unknown area of knowledge. For this reason, identifying the information structure and the author in non-linear 2.0 aggregates of information presents a major challenge. Learners' metacognitive capabilities are required in social software more than in the traditional learning context, where the progressive linearity of teacher-led contents transmission, require fewer metacognitive capabilities.

Because of its complexity, ESS involves some learning risks that Thalheimer (2008) has summarized as follows:

1. Learners can learn bad information.
2. Learners can spend time learning low-priority information.
3. Learners can learn the right information, but learn it inadequately.
4. Learners can learn the right information, but learn it inefficiently.
5. Learners can learn at the wrong time, hurting their on-the-job performance.
6. Learners can learn good information that interferes with other good information.
7. Learners utilize productive time in learning. Learners can waste time learning.
8. Learners can learn something, but forget it before it is useful.
9. Previous inappropriate learning can harm learners' on-the-job learning.

To avoid these difficulties, learners need to be placed in the role of managers of their own learning. This implies taking into account the learning objectives, choosing what they will learn, then, planning and regulating their learning process. Learners who are not able to work in this level of complexity will have difficulties taking advantage of 2.0 learning opportunities.

Thus far, we have analysed the ESS learning requirements, and more especially, the metacognitive pre-requisite. From this point, we will address ESS proactively, as an opportunity to develop the metacognitive abilities of learners in order to better exploit the learning opportunities of this specialised software.

3.3 Metacognitive Development Opportunities in ESS

As metacognition has been a key issue in successful learning, and autonomous learning contexts in particular, many authors have studied how to develop learners' metacognitive capabilities. The nature of the learning interventions for the development of metacognitive learners is mainly based on the awareness of the learning process during the resolution of a specific task. Awareness implies the learner's ability to be conscious of the learning process. The learner must be conscious of his own knowledge – his metaknowledge (the knowledge about knowledge and the way we learn) – and of his own learning strategies. This awareness can be developed through metacognitive guidance, modelling activities or even collaborative dialogues.

In the field of ICT, the work of Jonassen (1990) introduced computers as cognitive tools and extended research has been done in the field. Recently, Azevedo (2005) introduced the potential of computer environments in the development of metacognitive learners. Until now, consideration of the use of computers for metacognitive development has held computers in a traditional perspective, without exploring the new possibilities of social software. This is probably due to the novelty of the 2.0 approaches.

Next in this chapter we will explore the various possibilities of ESS for metacognitive development. First, ESS could be considered as a metacognitive development opportunity because of its social aspects. Secondly, we will focus on activities that can be designed to develop the metacognitive abilities of the learner.

3.3.1 ESS: Social-Based Opportunities for Metacognitive Development

Collaborative activities play an important role in the development of metacognition. Social interaction is considered a pre-requisite in the learning of metacognition (Von Wright, 1992). According to Marzano (1988) metacognition is, initially, the result of a process of social interaction which, gradually, becomes internalized. According to Marzano, language and social relations play an important role in the interactions which will lead to metacognition. Manion and Alexander (1997) also demonstrated that collaborative work helps the development of metacognitive strategies.

We could consider three main social perspectives of metacognition. The first is the consideration of metacognition as an essential part of collaborative work (Salonen, Vauras, and Efklides, 2005). The second considers metacognition as a social interaction product (Goos et al, 2000). In the third perspective, metacognition is considered as socially distributed (socially shared metacognition) (Iiskala, Vauras and Lehtinen, 2004). The first two approaches are mainly accepted by the scientific community. However, a metacognition approach which is socially distributed faces the same criticism as that faced by distributed cognition; for example, the consideration of an external consciousness. For a large number of authors, cognition and metacognition are individual processes that could occur in a social context, but nevertheless remain individual functions because they are produced in one's brain.

Taking the same approach, Pata (2008) suggests that awareness or inter-subjective consciousness (inter-subjective awareness) could reflect the awareness of the individual cognitive and metacognitive process of others. Ligorio, Pontecorto and Talamo (2005) tried to develop metacognition during a distance learning collaborative activity. In this activity, Greek and Italian students were required to write a fairy tale. According to Ligorio and his colleagues the development of metacognition during this activity helped to foster interdependence among participants during the task.

3.3.2 Metacognitive Development through ESS-Based Learning Activities

The previous sections discussed the potential of ESS in metacognitive development because of its associated social aspect. However, despite the opportunities of their context, learners do not necessarily develop all their potential metacognition naturally (Hofer, Yu and Pintrich, 1998). We will now consider activities that could foment and support metacognition learning through the use of Educational Social Software.

A. Metacognitive Development through Tutored use of ESS: Metacognitive Dialogue and Modelling

During the learning activity, teachers and tutors could go beyond the traditional role of information transmitters and become metacognitive coaches, helping learners to develop their potential through metacognitive dialogue or modelling activities. Metacognitive development through dialogue helps to develop the awareness of learning strategies already used by the learner (Paris and Winograd, 1990).

Metacognitive modelling aims at a gradual internalization of metacognitive strategies through an initial explicit example given by the teacher or tutor. In this way, the teacher could show how to search for and to select information in Wikipedia or other 2.0 websites, making explicit the cognitive strategies he followed during the process of the planning, execution and regulation of this learning task. Thereafter, learners could try to reproduce the same cognitive and metacognitive strategies, describing their strategic behaviour in a written form or expressing it as a thinking aloud process. The explanation allows the teacher to check if the modelling activity served correctly to transfer the cognitive and metacognitive strategies to the learners.

B. Metacognitive Development through the use of ESS during Peer Tutoring Activities

In a certain way, peer tutoring could become a metacognitive development activity, assigning a *"metacognitive tutor"* role to some learners. That could be achieved by putting learners into pairs where one plays the *"metacognitive tutor"* while the other performs the task. This activity could have collaborative variations and be done by groups of learners or by assigning more specific roles (planning tutor, regulation tutor, ...) that could later be reassigned to another group member.

Metacognitive peer tutoring facilitates metacognitive development because of the closest zone of proximal development (ZPD) between peers. This ZPD allows a more effective metacognitive dialogue because of the close language and references employed between students of the same level. In such situations, learners may explain some examples of the ways they could use the social software that they already use (e.g. MySpace, Facebook) for achieving their learning tasks; they could also make explicit their strategic use of ESS solutions (e.g. Slideshare).

C. Metacognition Development through the Multiple and Complexity Perspectives of Social Software

Computers as cognitive tools allow outsourcing and expliciting knowledge in multiple perspectives (e.g. texts, static and interactive graphics, videos,...). For Clements and Nastasi (1999), using a microworld such as LOGO engenders a high-level type of conflict resolution involving coordination of divergent perspectives. In this situation, the teacher can explain the metacognitive experience. On the other hand, Witherspoon, Azevedo and Baker (2007) consider learning as the result of the confrontation of multiple representations, particularly effective in the context of external regulated learning (ERL). In the case of Web 2.0 applications, knowledge develops a myriad of different perspectives during a continuum of information which is continuously evolving. In this continually changing information, knowledge continues to evolve. For example, a definition in Wikipedia may change frequently during a single day. Thus someone who read the article at 9 am will get different information from that obtained by a person who reads it at 5 pm.

To foster learning through these multiple perspectives, and taking into account previous research, we suggest that the use of ESS as a source of multiple representations should be led by the teacher or tutor at primary and high school levels in order to make sure that the activity is correctly modelled. If not, there is a risk that the complexity of representations will render learning through multiple perspectives counter-productive.

D. Development of Metacognition through the Active Contribution to Web 2.0 Content Production

Participation in social software can contribute to the development of metacognition in different manners. The first way we could consider is the learner 2.0 participation that could be oriented to the sharing of metaknowledge about learning strategies of individual learners. During these activities, learners - individuals or teams - may communicate to the community (e.g. class, school, Internet) their learning process as they perform specific tasks. For example, how to find information by asking experts found in social networks. This metaknowledge sharing activity may also be achieved with distant learners. An example of this might be two students learning a foreign language and sharing their *"metacognitive tips"* or strategies for language learning. It may have even greater significance, and a real external impact, if students contribute to a Wikipedia article after previous work done in class or autonomously. In this set of metacognitive activities based on 2.0 participation and contribution, it is essential to ensure the explanation of metacognition strategies and the metaknowledge gained and developed during the activities. This will ensure a correct transfer to other learning situations.

CONCLUSION

The information society and the emergence of new social web approaches has changed our approach to teaching and learning and, more especially, changed the metacognitive strategies and metaknowledge that we engage when using social software for educational purposes. It has now become necessary to engage in a life-long learning approach, both at the individual level and at the organizational level. This implies a new relationship to knowledge – a relationship where we need to move to higher-level, becoming not just learners, but also managers of our own learning. This involves developing metacognitive skills in order to act more strategically when planning and regulating our learning. However, it also implies integrating social software as a potential life-long learning opportunity.

Throughout this chapter we addressed ESS from a metacognition point of view, first, analysing the metacognitive challenges of ESS, and secondly, considering ESS possibilities for learning and metacognition development. In a context where learning is no longer enough and we must learn to learn, the use of social software for educational purposes adds, firstly, new cognitive and metacognitive prerequisites to the learning process. At the same, ESS opens up new opportunities for metacognition development and life long-learning. Henceforth, learning passively will not be enough to maintain our competitiveness as knowledge workers or learners. We will need to develop our metacognition for learning strategically in an inter-connected world – a world where knowledge evolves permanently, but where we have a new universe of learning opportunities through social software.

REFERENCES

Azevedo, R. (2005). Computers environments as metacognitive tools for enhancing learning. [Special Issue on Computers as Metacognitive Tools for Enhancing Student Learning]. *Educational Psychologist, 40*(4), 193–197. doi:10.1207/s15326985ep4004_1

Brandsma, J. (1998). Financement de l'éducation et de la formation tout au long de la vie: Problèmes clés. In *Peut-on mesurer les bénéfices de l'investissement dans les ressources humaines. Formation professionnelle . Revue Européenne, 14*, 1–6.

Brown, A. L. (1987). Metacognition, executive control, self-regulation, and other more mysterious mechanisms. In F. E. Weinert & R. H. Kluwe (Eds.), *Metacognition, motivation, and understanding* (pp. 65-116). Hillsdale, NJ: Lawrence Erlbaum Associates.

Clarke, E., & Emerson, A. (1981). Design and synthesis of synchronization skeletons using branching-time temporal logic. *Logic of Programs, 1981*, 52–71.

Clements, D. H., & Nastasi, B. K. (1999). Metacognition, learning, and educational computer environments. *Information Technology in Childhood Education Annual, 1*, 5–38.

Ertscheid, O. (2005). Google a les moyens de devenir un guichet d'accès unique à l'information. *Le Monde*.

Flavell, J. H. (1976). Metacognition aspects of problem solving. In L. B. Resnick (Ed.), *The nature of intelligence*. Hilldale, NJ: Lawrence Erlbaum.

Goos, M., Galbraith, P., Renshaw, P., & Geiger, V. (2000, July 31-August 6). *Classroom voices: Technology enriched interactions in a community of mathematical practice*. Paper presented at the Working Group for Action 11 (The Use of Technology in Mathematics Education) at the 9th International Congress on Mathematical Education, Tokyo/Makuhari.

Gordon. (1996). Tracks for learning: Metacognition and learning technologies. *Australian Journal of Educational Technology, 12*(1), 46-55.

Hofer, B., Yu, S., & Pintrich, P. (1998). Teaching college students to be self-regulated learners. In D. Schunk & B. Zimmerman (Eds.), *Self-regulated learners: From teaching to self-reflective practice* (pp. 57-85). New York: Guilford.

Houssman, J. (1991). Self monitoring and learning proficiency. In *Computer classroom*. Hofstra University, EDD.

Hurme, T.-R., & Merenluoto, K. (2008, June 9-13). *Socially shared metacognition and feelings of difficulty in a group's computer supported mathematical problem solving*. Kesäseminaari pidetään Physicumissa, Helsingin yliopistossa. Retrieved on October 1, 2008, from http://per.physics.helsinki.fi/Tutkijakoulun_kesaseminaari_2008/Hurme.pdf

Iiskala, T., Vauras, M., & Lehtinen, E. (2004). Socially-shared metacognition in peer learning? *Hellenic Journal of Psychology, 1*, 147–178.

Jonassen, D. H., & Harris, N. D. (1990). Analyzing and selecting instructional strategies and tactics. *Performance Improvement Quarterly, 3*(2), 29–47.

Lave, J., & Wenger, E. (1991). *Situated learning: Legitimate peripheral participation*. Cambridge: Cambridge University Press.

Ligorio, M. B., Talamo, A., & Pontecorvo, C. (2008). Building intersubjectivity at a distance during the collaborative writing of fairytales. *Computers & Education, 5*, 357–374.

Manion, V., & Alexander, J. (1997). The benefits of peer collaboration on strategy use, metacognitive causal attribution, and recall. *Journal of Experimental Child Psychology, 67*, 268–289. doi:10.1006/jecp.1997.2409

Marzano, R. J. (1988). *Metacognition: The first step in teaching thinking. Professional handbook for the language arts*. Morristown, NJ: Silver Burdett and Ginn.

Marzano, R. J. (1998). *A theory-based meta-analysis of research on instruction*. Mid-continent Aurora, CO: Regional Educational Laboratory.

Moisan, A., & Carré, P. (2002). *L'autoformation, fait social? Aspects historiques et sociologiques*. Paris: L'Harmattan.

Osborne, J. W. (2000). *Assessing metacognition in the classroom: The assessment of cognition monitoring effectiveness*. Unpublished doctoral dissertation, University of Oklahoma.

Osman, M. E., & Hannafin, M. J. (1992). Metacognition research and theory: Analysis and implications for instructional design. *Educational Technology Research and Development, 40*(2), 83–99. doi:10.1007/BF02297053

Papert, S. (1980). Mindstorms: *Children, computers, and powerful ideas*. New York: Basic Books.

Paris, S. G., & Winograd, P. (1990). How metacognition can promote academic learning and instruction. In B. F. Jones & L. Idol (Eds.), *Dimensions of thinking and cognitive instruction* (pp. 15-51). Hillsdale, NJ: Lawrence Erlbaum Associates.

Pata, K. (2008). *Sociocultural and ecological explanations to self-reflection*. Retrieved on October 2, 2008, from http://tihane.wordpress.com/category/intersubjectivity/

Pintrich, R. (2002). The role of metacognitive knowledge in learning, teaching, and assessing. *Theory into Practice*, *41*(4), 219–225. doi:10.1207/s15430421tip4104_3

Ridley, D. S., Schutz, P. A., Glanz, R. S., & Weinstein, C. E. (1992). Self-regulated learning: The interactive influence of metacognitive awareness and goal-setting. *Journal of Experimental Education*, *60*(4), 293–306.

Romero, M., & Wareham, J. (2009). Just-in-time mobile learning model based on context awareness information. *IEEE Learning Technology Newsletter*, *11*(1-2), 4–6.

Rumelhart, D., & McClelland, J. (1986). *Parallel distributed processing*. MIT Press

Saffer, D. (2005). *The role of metaphor in interaction design*. Master's thesis, Carnegie Mellon University, Pittsburgh, PA.

Salonen, P., Vauras, M., & Efklides, A. (2005). Social interaction–what can it tell us about metacognition and coregulation in learning? *European Psychologist*, *10*(3), 199–208. doi:10.1027/1016-9040.10.3.199

Siemens, G. (2004). *Learning management systems: The wrong place to start learning. E-learnspace*. Retrieved on October 1, 2008, from http://www.elearnspace.org/Articles/lms.htm

Thalheimer, W. (2008, August 18). Evaluation e-learning 2.0: Getting our heads around the complexity. *Learning Solutions*.

Veenman, M. V. J., Wilhelm, P., & Beishuizen, J. J. (2004). The relation between intellectual and metacognitive skills from a developmental perspective. *Learning and Instruction*, *14*(1), 89–109. doi:10.1016/j.learninstruc.2003.10.004

Von Wright, J. (1992). Reflection on reflections. *Learning and Instruction*, *2*(1), 59–68. doi:10.1016/0959-4752(92)90005-7

Witherspoon, A., Azevedo, R., & Baker, S. (2007, July). *Learners' use of various types of representations during self-regulated learning and externally-regulated learning episodes*. Paper presented at a Workshop on Metacognition and Self-Regulated Learning at the 13th International Conference on Artificial Intelligence in Education, Los Angeles, CA.

Zimmerman, B. J., & Martinez-Pons, M. (1990). Student differences in self-regulated learning: Relating grade, sex, and giftedness to self-efficacy and strategy use. *Journal of Educational Psychology*, *82*, 52–59. doi:10.1037/0022-0663.82.1.51

This work was previously published in Educational Social Software for Context-Aware Learning: Collaborative Methods and Human Interaction, edited by N. Lambropoulos & M. Romero, pp. 38-48, copyright 2010 by Information Science Reference (an imprint of IGI Global).

Chapter 8.16
Digital Energy:
Clustering Micro Grids for Social Networking

Mikhail Simonov
Politecnico di Milano, Italy, and ISMB, Italy

Marco Mussetta
Politecnico di Milano, Italy, and Politecnico di Torino, Italy

Riccardo Zich
Politecnico di Milano, Italy

ABSTRACT

Since energy use is a type of consumer behavior reflecting the interests to maximize some objective function, the human being activities seen in energy terms might be used to create the social aggregations or groups. Electric energy generated from ecologic sources brings some unpredictability. Authors model the unpredictability of the distributed generation in order to create a tool for minimization. Authors propose the novel method to build real life smart micro grids in the distributed generation context characterized by zero emissions. The proposed tool becomes an instrument to create the social aggregation of users and negotiate locally the "social" energy in real time, strengthening and mastering a virtual neighborhood of the local community.

INTRODUCTION

Electric energy distribution networks were designed to distribute the centrally produced energy, considering the top-down structures and the Supervisory Control And Data Acquisition (SCADA hereafter) systems to manage them. One possible analogy with the human body gives the blood system, distributing oxygenated flows towards the cells (consumers) starting from the central heart/lung node (producer). The brain like any SCADA ensures the constant blood pressure (load balancing) analyzing the peripheral oxygenation thanks to the nervous lines. The system sends the stimuli to the heart/lung, keeping itself well functioning within certain limits even when infusions/drawings are administered.

The renewable energy, for instance the photovoltaic one, depends on the natural factors, such as clouds, wind, and ephemeredes. The alternative sources of energy introduce some entropy because their contributions have a certain extent of unpredictability. The photovoltaic power is generated during the daytime, and the flow intensity depends on the weather condition, adding some complexity to solve by modeling (Mellit, 2006). The eolic energy production is possible only during the windy condition, which is variable as well. Consequently the released power is uncertain, unpredictable in some extent, and it adds some instability to the power grid, simply because of the physical nature of the underlying phenomena, and might be seen as topic to deal with in load balancing perspective. Due to the higher unpredictability and uncontrollability an imbalance between supply and demand of electricity has to be settled by someone, leading to the extra costs, consequently these factors could finally result in a drawback for the integration of renewable energy sources (Frunt, 2006). The load balancing in the context of an unpredictable energy in power grids is ecologically expensive, because of the compensation coming form the instant, non eco-friendly, generators. The electric energy generated from the distributed renewable energy generators might be instantly consumed locally instead of flooding into the distribution network, eliminating specific controlling interventions. From the free energy market perspective, the best option appears also the immediate consumption of the locally produced energy following the well-known short chain "producer-to-consumer" because of the possible penalties for the Service Level Agreement (SLA hereafter) violations. Even if the ecologic alternative energy production today might be stimulated by incentives, however someone (the transmission system operator or the Society) has to pay for the imbalance induced by the unpredictable energy injection. In the liberalized market context the load balancing might become more expensive than the generated economic value, requiring the solution for the minimization of the impact induced. Moreover it becomes necessary adding the unpredictability/uncertainty indicators instead of characterizing nodes by power thresholds only.

The electric energy distribution network is notably storage free, requiring the adoption of the (a) real time load balancing, which is possible monitoring constantly the digital energy in real time (REMPLI project, 2008), or using the (b) user socialisation toolkit locally in order to establish the "social condominium", or (c) negotiating the energy locally, to avoid the accounting of the transits towards the main distribution backbones. The most interesting option from the social point of view is the creation of the local (virtual or not) community of human being, presenting common social and living patterns, e.g. those contributing to consume locally the renewable energy, consequently decreasing the energy demand.

An electric micro grid (Gellings, 1981) could be defined as a low voltage distribution network with distributed energy sources altogether with storage devices and loads, which could be operated, either interconnected to the main grid or either isolated from it, by means of a local management system with a communication infrastructure allowing control actions to be taken following any given strategy and objective. The social community of human being belonging to the above-mentioned topology might be organized to exploit better the possible synergies and benefits. Consequently we investigate on the possible optimization of the above-mentioned topology trying to minimize the uncertainty by the total local consumption of the produced energy. We create the cohesion between the human actors belonging to the said topology through the new concept of the "social" electric energy. Unlike traditional energy, the social one is locally produced and locally consumed by the micro power grid, thanks to the direct interactions among producers and consumers.

Human being is social by its nature and consumes energy to achieve the goals. The artifacts created by humans, such as houses, offices, electric vehicles, and other representative devices/appliances, might have some social meaning, because explicit the relationship with the community, a certain living standing and a social status. Local energy stakeholders reflect the membership in the modern society. Social group memberships might deliver some benefits also in energy terms, helping to save some energy or resources.

The human behavior is representative of the interests, goals, and the life style. Since energy use is a type of consumer behavior reflecting the interests to maximize some objective function, the human being activities seen in energy terms might be used to create the social aggregations or groups. Humans use the electric energy for heating, cooling, various household uses, devices and electric appliances, transportation, office activities, and industry. Humans influence the energy use indirectly through their purchase decisions, impacting on the amount of energy used to produce and maintain goods and services: let us remember the energy efficient appliances. To understand and manage energy demand one can analyze the energy-using behavior of users: individuals, households and other categories. Our analysis is not relying on a conception of energy as a commodity, but on the producer/consumer behavior. We have assumed that energy users are conscious and act in their own interests to maximize or minimize some objective function. The energy is not free introducing a budget optimization function for the demand side, there is the ecologic motivation to use the renewable energy, and we can envisage an additional function leading to the new aggregation of energy consumers to maximize some social objective function. It appears intuitive that energy users do not always take the actions maximizing the individual or collective benefits due to the incomplete information to guide decisions, or because of the market constraints.

MODELING

Researchers (Chan, Broehl, Gellings, 1981) have produced several load shaping models through the adoption of the econometric, statistical, engineering and combined approaches. Load shaping of the different categories of users, industrial, tertiary, and especially of the residential ones, is a highly complex task (EPRI, 1985), because it is linked to the lifestyle and related psychological factors, not precisely defined and subjective by their nature. It is known also that the definition of the standard behavior of the various types of customers through statistical correlations undertaken by the load management research in the domain of the electric engineering does not solve the problem, failing to consider the variability of the demand, which is the random factor. We start from the assumption that the human being is social by the nature, and all the activities undertaken daily needs some energy, consequently the load profile manifested reflects the lifestyle and the social inhabits appropriately. The manifestation of the common patterns among the portfolio might be used to cluster the social groups using the similarity principles.

The human actor behavior can be described by its **objective function** correlated with the goal's achievements. The set of resources used includes money, energy, vital resources, time, information, and knowledge. The determination to reach the above-mentioned goals becomes the motive and the sense, while the degree of the goal

$$\Theta(t) = \sum_{i=1,m} \gamma_i(t) \left(Fact_i(t) / Targ_i(t) \right) / m \rightarrow Max \quad (1)$$

$$Targ_i(t) = \varphi(t, x_i, y_i, z_i, g_i, \Omega_i), \; i \in [1,m] \quad (2)$$

$$Fact_i(t) = \psi(t, x_i, y_i, z_i, g_i, \Omega_i), \; i \in [1,m] \quad (3)$$

$$\sum_{i=1,m} x_i(t) < x_0(t), \; \sum_{i=1,m} y_i(t) < y_0(t), \; \sum_{i=1,m} z_i(t) < z_0(t), \; \sum_{i=1,m} g_i(t) < g_0(t) \quad (4)$$

achievements, reflected by the objective function, determines the personal satisfaction. We denote with i the index of the goal, m - the number of the goals, γ_i - the importance of the goal i, $Targ_i$ – the planned achievement level of the goal i, $Fact_i$ – the real achievement of the goal i, x_i – the amount of material resources (money, food etc.), y_i – the amount of energy, z_i – the amount of time available, g_i – the amount of the information/ knowledge available to materialize the goal i, Ω_i - the quality of the institute needed to achieve the goal i, while the φ and ψ represent the functions linking the vital resources with obtained results and describing the product/service. We assume an actor owns initially the asset $\{ x_0, y_0, z_0, g_0 \}$. The statement (1) defines an Objective function, with (2) and (3) representing the functional limitations, and (4) describing the resource limitations. The resources are consumed to achieve the goals, the activities undertaken by human being fit the model, and the resource limitations reflect the consumption dynamics.

Let us define the **Social Condominium** (SC hereafter) as an aggregation of the human actors presenting the common and complementary living patterns, and including the energy producers and consumers, which respective energy profiles – reflecting the living ones - consume totally the local energetic resources. The human actors being aggregated in the social condominium present the similar and complementary living patterns: those having a lunch at home from 1 PM to 2 PM, those looking after the children in the afternoon time, Small Office/Home Office Entrepreneurs (SOHO here after), and other similar cases. An interesting characteristic of the entity is the presence of the direct interactions among members and the explicit semantic descriptions labeling the aggregation, which contribute in the social cohesion. SC - being seen from the social point of view - includes a group of people showing similar patterns in their living activities. Geometrically, SC is a small sub-graph of the electric energy distribution network with qualified nodes embracing a number of energy stakeholders (Producers and Consumers) positioned in a given geographic area. The concept is useful for creating a virtualization of the real individual profiles through a new entity manifesting different load patterns.

The social network concept is well-known and simple: it is a number of individuals, or actors, becoming the nodes of the network because of the various relationships linking them. We use the social network concept to emerge and to show explicitly the existing relationships of the particular “society”, the electric energy market, analyzing the patterns and correlations in the digitized electric load streams. The concept of social network was proposed by J. Barnes in 1954, however till now it was never applied to the non-industrial distributed electric energy generation stakeholders.

The interactions among energy market participants in real time are apparently invisible and almost virtual. The Internet with its impersonalized contacts and the Internet of Things technology give a right paradigm to obtain a virtualized social network, and precisely the online one. The inclusion criteria can be elicited from the load data because of the user behavior contained as tacit knowledge, which can be made available by the soft computing methods. The above-defined SC social network is the real world entity of neighbors comprising a renewable energy producer plus some consuming members.

The energy profile is defined by the digital energy measured using digital meters $P(t_i)$, with the digitization which might happen every 1 sec., 1 min, 1 hour and so on: $\{ P(t_0), P(t_1), P(t_2), \ldots, P(t_i), \ldots \}$. In the past the manual energy measurement of the individual consumers was used for billing purposes only. The digital energy technology thanks to the time wrapping properties has enabled new features permitting to reveal the behavioral patterns, not detectable in the past. To understand the potential of the frequent sampling let us compare the traditional 25 frame per second video camera with the high-speed one filming a balloon explosion at 3000 frames per second rate. The first technology cannot detect the event, while the second one shows it. For example the residential meters in Italy produce 96 samples per day sets, being the data generated every 15 min. To simplify the analysis, in this study we consider 24 hourly samples per day reporting the aggregated values, however for the real time operations we suggest the 1 sec. sampling.

Fuzzy Set theory (Zadeh, 1965) permits the gradual assessment of the membership of elements in a set, described with the aid of a membership function valued in the real unit interval [0, 1]. Fuzzy Logic (Baldwin, 1981, Kruse, 1994, and Halpern, 2003) deals with reasoning that is approximate rather than precisely deduced from classical predicate logic, becoming a way of processing data by allowing partial set membership rather than crisp set ones, e.g. a problem-solving control system methodology providing a simple way to arrive at a definite conclusion based upon vague, ambiguous, imprecise, noisy, or missing input information. The model is empirically-based, relying on an operator's experience rather than their technical understanding of the system. For example imprecise terms like "IF (Consumption rises significantly) AND (lunch time) THEN (User has a lunch at home)" are very descriptive and meaningful of what must actually happen. Fuzzy Rules are linguistic IF-THEN constructions that have the general form "IF A THEN B" where A (*premise*) and B (*consequence*) are collections of propositions containing linguistic variables. In effect, the use of linguistic variables and Fuzzy IF-THEN rules exploits the tolerance for imprecision and uncertainty. In this respect, Fuzzy Logic mimics the crucial ability of the human mind to summarize data and focus on decision-relevant information. In a more explicit form, if there are *i* rules each with *k* premises in a system, the ith rule has the following form (5), where ***a*** represents the crisp inputs to the rule, and ***A*** and ***B*** are linguistic variables, while **Θ** is a Boolean operator AND, OR/XOR or NOT. Several rules constitute a Fuzzy rule-based system. Uncertainty is a fundamental and unavoidable feature of real life. In order to deal with uncertainty intelligently, we need to be able to represent it and reason about it. Further reading is available in (Halpern, 2003). Many of existing systems need the rules to be formulated by an expert.

$$\mathbf{IF}\,(a_1\,\mathbf{IS}\,A_{i,1})\,\Theta\,(a_2\,\mathbf{IS}\,A_{i,2})\,\Theta\ldots\Theta\,(a_k\,\mathbf{IS}\,A_{i,k})\,\mathbf{THEN}\; B_i \qquad (5)$$

Let us consider a LV/MV/HV segment of a power grid G (Figure 1) composed by nodes denoted as N_j with $j \in [0,m]$, where m is the total number of nodes in the topology.

The grid partitioning is possible combining nodes is many different ways, denoting resulting subsets or sub-topologies as micro grids M_i. Any M_i is the combination of some N_j. The grid G, being the union of N_j, becomes also an union of M_i, or $G = \mathbf{U}\, N_j = \mathbf{U}\, M_i$, where $i \in [0,k]$. Please note there are several possibilities to do it, depending on the practical needs and goals to pursue, consequently the number k is unknown a priori. The adoption of different clustering techniques would lead to the different automated partitioning schemes, demonstrating very different real life performances, especially those in terms of load balancing.

It is common to distinguish between the residential, industrial, tertiary and other types of nodes in electric energy distribution networks. In real life a commonly used approach in grouping is the aggregation of end-users behaving as Condominium - as a single entity – e.g. micro grid. Intuitive form the traditional flow dynamic viewpoint approach is not helpful in the distributed generation era: all customers have a very common energetic profile, excluding a chance to balance automatically the production from a local producer.

To enable a new electricity market segment – a local energy exchange – we need two components: an instrument to account the local exchanges, and

Figure 1. Power grid

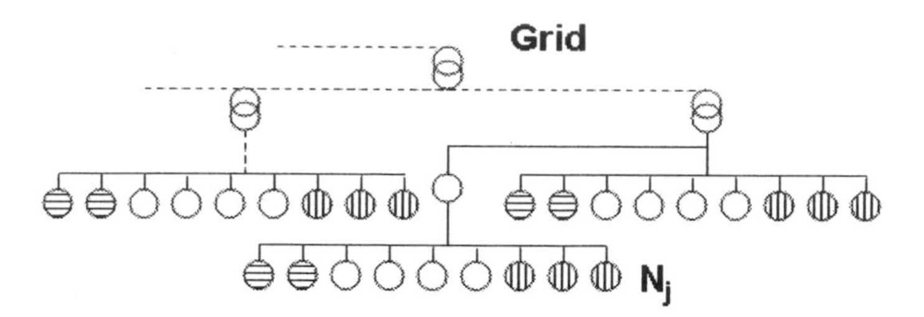

the tool to aggregate the local "market". The first instrument is a digital energy; the second one is a method to build the micro grid, while the additional one is the software framework controlling the new entity (like SCADA). After the adoption of the distributed generation, different classes of nodes are available in the power grid: Producers, Consumers and Prosumers - generating, consuming and releasing the exceeding energy into the grid. Traditional techniques for load management, forecasting and control should be complemented by new instruments to deal with such a complex and entropy charged components in order to simplify the grid's behavior and achieve some performance optimization, so authors propose a way to aggregate stakeholders in a way, ensuring the optimal consumption of the local renewable energy resources.

Let us consider a number of stakeholders (A) acting as nodes of a power grid, having the roles of Producer (P), Consumer (C) or pRosumer (R). Consequently $G = \mathbf{U}\, N_p(j_p)\, \mathbf{U}\, N_i(j_i)\, \mathbf{U}\, N_r(j_r)$, having/showing different user profiles. Each M_j represents one class of the topology, populated by actors with roles denoted as P, C and/or R.

Let take $M_j = \{ N_j \}$ a cluster of end user's nodes aggregated together and logically seen as an Unit. The research question we try to answer to is how to aggregate nodes $\mathbf{N}_j$ - in real life populated by physical neighbors - logically optimizing the cluster $\mathbf{M}_j$ in a way to make non positive the outgoing energy flows. Maximizing the local consumption of the energy generated at local nodes, it enables the local energy market and supports respective commercial models.

On each node in the real life topology a digital energy meter is installed, communicating with the utility company, generating "digital energy", e.g. real-time messages exchanged between the energy market actors, comprising producers, consumers, and Prosumers. Each N_j generates a stream of measurements AMR(N_j, t, date), becoming a right tool to undertake the data-warehousing for optimization purposes and/or similarity clustering. Typically there are 96 measurements a day coming every 15 min. To be more precise, we introduce in the model the uncertainty of the power expressed as Fuzzy function: $P(t) = \{ AMR(N_j, t, date), E(t)\}$, where E(t) is the uncertainty of the load profile for the point t. The AMR stream for M_j will be calculated as the sum of respective values for each N_j.

PROPOSITION

From the energy point of view the proposed approach delivers a new mathematical model describing the electric energy distribution network in terms of the transits through the qualified arches $\mathbf{M}_j$, instead of the nodes $\mathbf{N}_j$, enabling the minimization of the uncertainty because of the local energy consumption. Till now, the traditional models deal with the characterization of the nodes in energy terms referring a specific role: Producer or Consumer. The daily load consumption shape, which becomes an array of finite number of scalars after the digitization, is an attribute of the Prosumer entity. It becomes less important after the process clustering some nodes in one micro grid. The representative – virtual - nodes $\mathbf{N}_{j,1}$, $\mathbf{N}_{j+1,1}$, $\mathbf{N}_{j+2,1}$ shown on Figure 2 exemplify the transits towards the micro grids and automatically hide all the attributes of the nodes. In behavioral terms the new entity manifests the new social identity of the aggregation and the collective behavior in both social and energy terms.

We propose a method adding to the power grid one or more local smart micro grid components to reduce – or eliminate, whenever possible – the uncertainty. It is commonly understood that the aggregation of several nodes in a micro grid is beneficial because leveraging the behavioral phenomena and reducing the intermittent pumping between the micro and macro grids. It is less clear how to build the micro grids in order to achieve the above-mentioned result. The proposed algorithm to build a micro grid embeds local nodes in a way, leading to hide the disturbance effect introduced by alternative sources of energy.

We start from the N(j) populated by actors with roles denoted as P, C and/or R having ***k*** members: $j \in [0,k]$. Each N(j) is a set of ***l*** samples characterizing the energy. Initially we have to profile the users to define the possible similarity clusters. Producers (Figure 3A) add the positive values, while Consumers (Figure 3B, 3C) contribute by negative ones.

Let us undertake the similarity clustering (Figure 4) of each existing LV sub-net portfolio, obtaining the initial partitioning of the grid in "n" clusters by user profile's similarity { M_{jl} }. It might be done using evolutionary programming permitting to obtain the semantic characterizations. This partitioning should be done without considering the physical distance between the nodes. The outcome is the number of similarity clusters showing how many different behavioral models we have in the topology.

Second step is the characterization of each entity in terms of Fuzzy rules, understandable by human actor, deriving them from the manifestations of the lifestyle reflected by load patterns. It might be done empirically, on case-by-case basis, or using specific tools like GAMUT (Sammartino, 2000), or using any other technique. The outcome is a set of rule collections, annotating each cluster [M_{jl} , {FR_{jl}, FR_{j2}, … , }]. Every participant acquires now the role semantically characterized by the respective load profile, which might be done using the descriptive, Fuzzy or crisp, rules. Figure 4 exemplifies four different user models C_1, C_2, C_3 and C_4. The outcome is a set of rule collections, annotating each cluster.

Now we try leveraging among clusters in order to obtain more homogeneous aggregations compensating unbalancing, obtaining the micro

Figure 2. Power grid partitioning

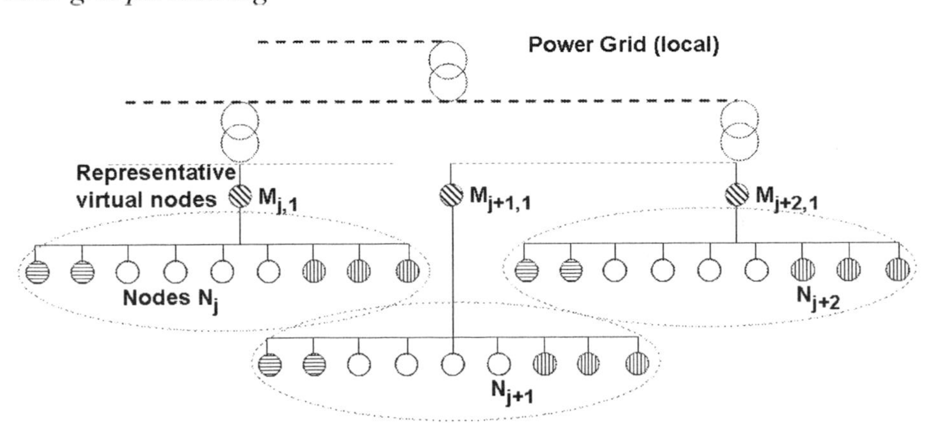

Figure 3. Load profiles

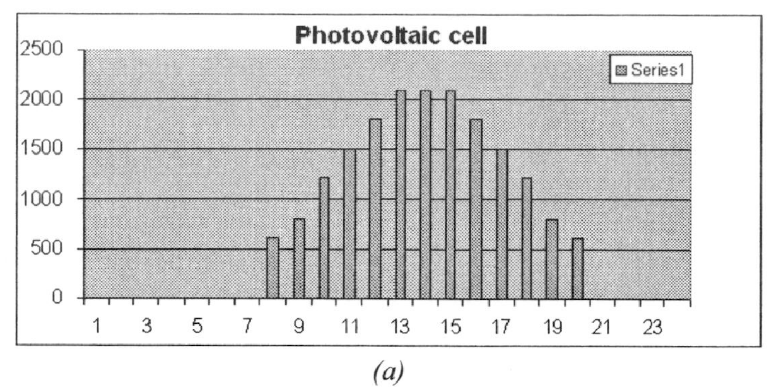

(a)

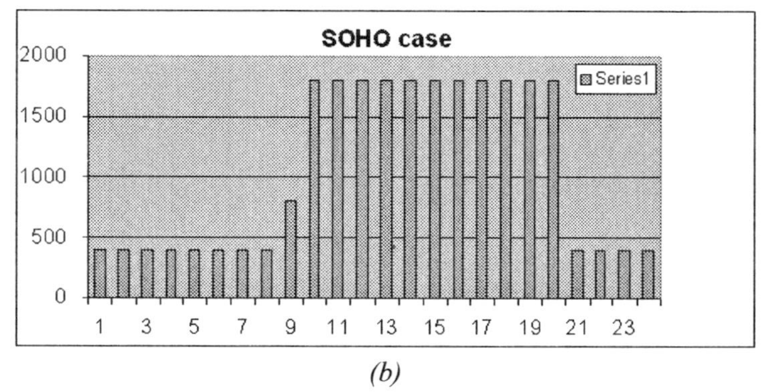

(b)

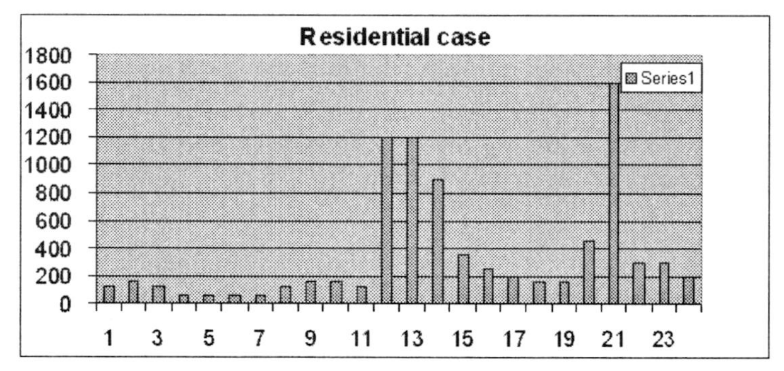

Figure 4. User profiling

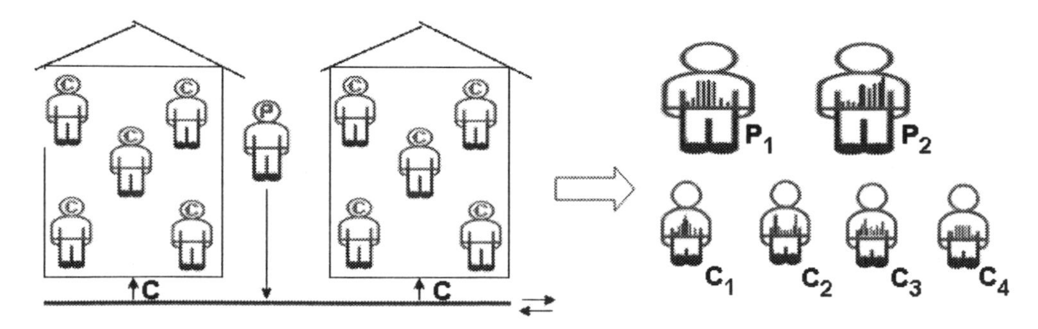

grids G_k. The objective now is to make non positive the outgoing flows generated by P_i grouping several C_j. for each G_k.

Third step is the characterization – by Fuzzy rules as well - of the Prosumer's nodes belonging to a given sub-net, which becomes the new gravity centers (of new partitioning). [M_{pl}, {FR_{pl}, FR_{p2}, … , }]. The number of Prosumers in the topology is the upper limit of clusters being created by the partitioning. It is possible to include more than one Prosumer in a micro grid being the nodes neighbors in real life. For illustrational purposes only let us build the simplest partitioning considering a micro grid G_l around each Prosumer R_m. To do it, we have to assess the similarity between {FR_{jl}, FR_{j2}, … ,} and {FR_{pl}, FR_{p2}, … ,} varying each index in [0,k], and build sub-groups to "neutralize" rules, e.g trying to minimize the number of resulting FR_j components in the micro grid characterization. We try to do it using advanced datamining techniques, applying the same soft computing tool. Once the optimal combination is found and fixed – by the terminal condition – we have the candidate technique releasing a criterion permitting the optimal composition of the micro grid.

Next step is to add consumers to the micro grid (Figure 5) until the maximum generated power becomes equalized by the sum of the consumptions. This operation considers all the samples of the digital energy (24 samples per day in our case), ensuring no one being positive: $\forall P_i < 0.001$. This step is different because now we have to consider the physical distance between the local nodes, in order to create a real life meaningful neighborhood. We have to fix the maximum distance between nodes, or consider simply the belongings to the same LV segment, or set up any other parameter we wish to use for grouping, simply labeling entities by an additional crisp value - becoming a new column in an array of data.

It is important to note that thanks to this procedure, the above-mentioned algorithm generates the micro grid behaving as a normal consumer, eliminating the hybrid role of the Prosumer. However the predictability of the load curve is not guaranteed suggesting to add an additional parameter ***E*** to the digital energy being exchanged – the uncertainty or entropy. From outside the micro grid, the virtual aggregated node will be always characterized by the same representation {AMR(N_j, t, date), E(t)}, as any real physical node, adding the value of the local automated leveraging of both maximum and minimum power.

Even if it is known that for a number of consumers lower than 10 – 20 the load power variations at any given time are significantly high and strongly depending on the number of consumers and the type and lifestyle of the family groups

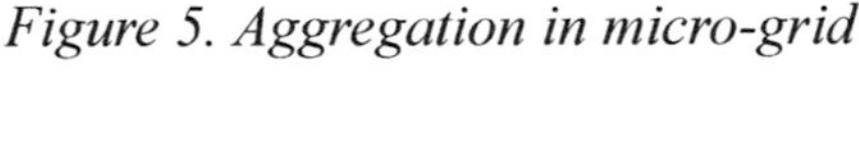
Figure 5. Aggregation in micro-grid

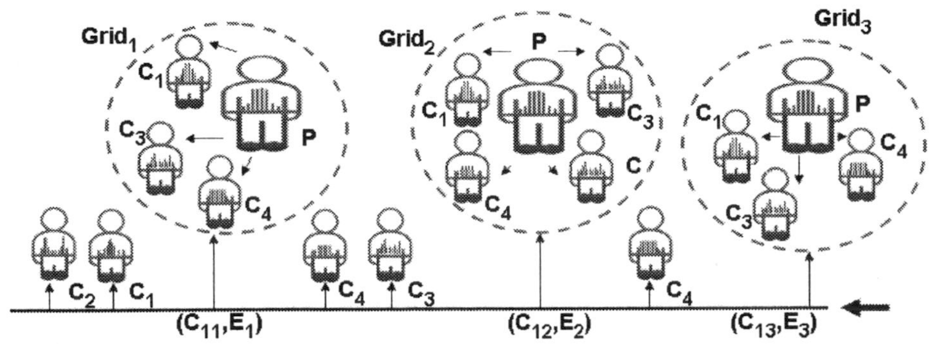

(Walker, 1985, Capasso, 1994, and Cagni, 2004), the minimization of the variations is not the aim of this study, because our objective function to solve is $P(t) = P_1(t) + C_{i1}(t) + C_{i2}(t) + \ldots < 0.001$, e.g. to ensure all the produced energy being consumed locally. The social aggregation created using the above-mentioned algorithms is meaningful because of the social cohesion capable to create the stable aggregation. The minimal micro grid, satisfying the above-mentioned equation might be complemented by the arbitrary number of other consumers to obtain a valuable social group, and in this configuration the redundant number of participants might lead also to the mitigation of the load power variations. Figure 6 shows an example of the unbalanced combination $P_1 + C_1$, $P_1 + C_2$, while the "balanced" one $P_1 + C_1 + C_2$. Aggregating an home user plus a SOHO gives the minimal micro grid we speak about: at any *t* it emits no energy outside consuming all the power locally.

Around each cluster produced by evolutionary algorithms we can expect a neighborhood populated by similar load patterns, following the normal distribution shown on Figure 7A. In our case the concrete family of patterns is shown on Figure 7B. More reading about the probabilistic characterization of the residential profiles is available in (Carpanetto, 2007).

EXPERIMENTATION

To implement the previously stated considerations, genetic procedure adaptively and dynamically analyzing and grouping together consumer profiles with a certain degree of similarity in their temporal evolution has been considered. In fact, the use of a direct approach can be very difficult if the data structure is very complex and with a relevant number of profiles to be extracted. Alternatively, one can use indirect procedures, based on a pseudo-casual algorithm, as the evolutionary algorithms are. The advantage of using evolutionary optimization algorithms is that it is possible to vary simultaneously several parameters, in order to optimize different quantities.

Evolutionary algorithms apply an indirect synthesis of Fuzzy parameters, by randomly choosing the parameters of interest and evolving their values towards a solution that is considered optimal; the fitness function, in particular, is the only link between the physical problem and the optimization procedure.

The most effective class of evolutionary algorithms developed until now is Genetic Algorithm (GA hereafter) that is now quite familiar to the scientific community and widely used (Goldberg, 1989). Genetic Algorithms simulate the natural evolution, in terms of survival of the fittest, adopting pseudo-biological operators such as selection, crossover and mutation. In GA, the set of parameters that characterizes a specific problem is called an individual or a chromosome and is composed of a list of genes. Each gene contains the parameter itself or a suitable encoding of them. Each individual therefore represents a point in the search space, and hence a possible solution to the problem. For each individual of the population a fitness function is therefore evaluated, resulting in a score assigned to the individual. Based on this fitness score, a new population is generated iteratively with each successive population referred to as a generation. The GA uses three basic operators - selection, crossover, and mutation - to manipulate the genetic composition of a population. Selection is the process by which the most highly rated individuals in the current generation are chosen to be involved as "parents" in the creation of a new generation. The crossover operator produces two new individuals (i.e. candidate solutions) by recombining the information from two parents. Crossover operation occurs in two steps. In the first one, a given number of crossing sites, along with the parent individual, are selected uniformly at random. In the second step, two new individuals are formed by exchanging alternate pairs of selection between the selected sites. The random

Figure 6. Balanced loads

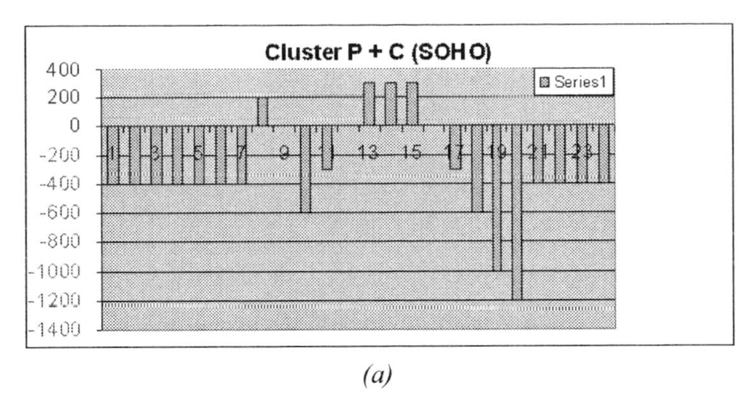

(a)

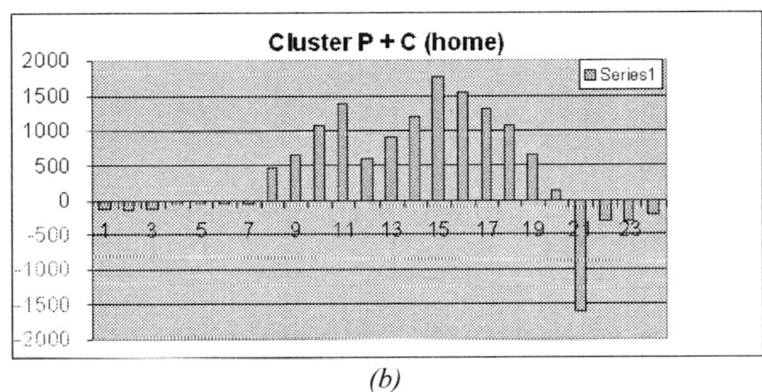

(b)

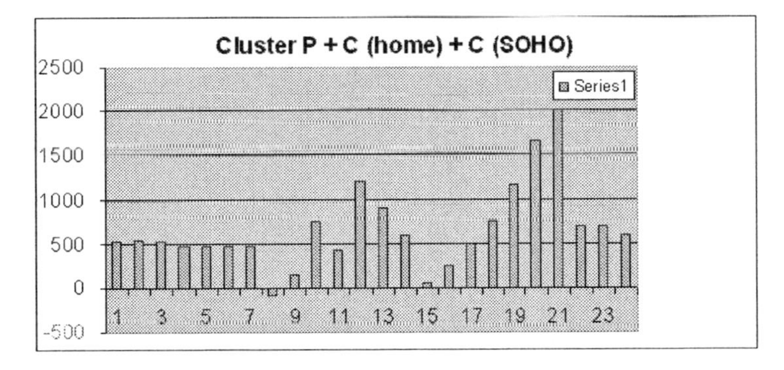

mutation of some gene values in an individual is the third GA operator. The allele of each gene is a candidate for mutation, and its function is determined by the mutation probability.

The Genetic Algorithm developed for this application is a standard implementation with real encoded genes (that represent the parameters of the specific problem, i.e. fuzzy parameters), since for high number of variables they result in faster than binary ones in convergence towards the maximum value. In this implementation, relevant operator parameters are the crossover probability P_c=0.8, the mutation probability P_m=0.1, the tournament selection and the elitism: these parameters have been used to speed up the convergence in the real coded GA.

To proceed with the evolutionary-assisted adaptive clustering, a pre-processing phase is performed first, where data are analyzed by a covariance matrix approach. Normalization and

Figure 7. Hyperplane of users

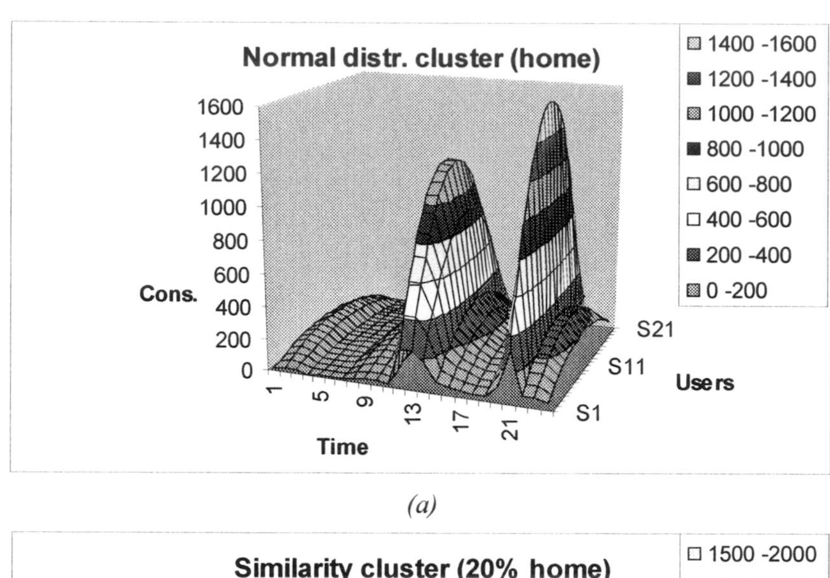

(a)

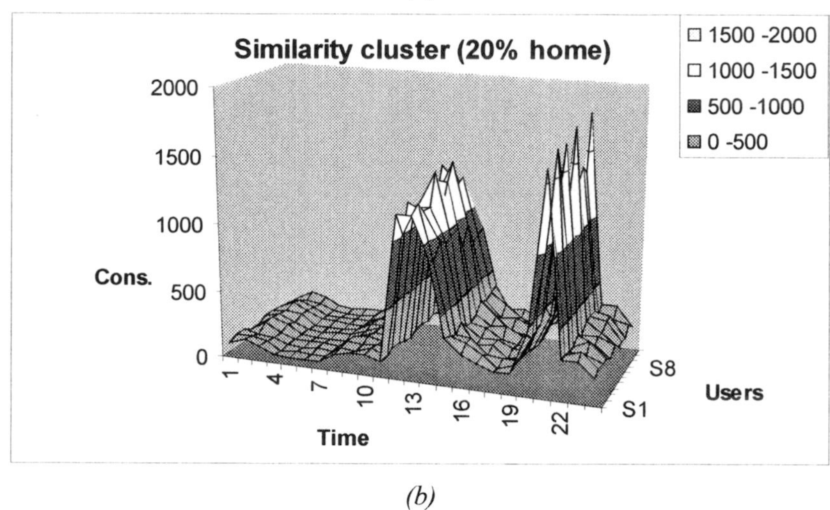

(b)

a set of Fuzzy threshold are applied to these results in order to roughly identify general behavior trends in several consumer profiles. Once this preliminary analysis is done, GA proposes a particular distribution of Fuzzy parameters to cluster together the different power consumption curves: the corresponding results are evaluated by the fitness function that, according with constrains defined before, assigns a score to the corresponding chromosome. The fitness score is the parameter that drives GA towards the optimum. After a certain number of evaluations, depending on the number of curves and of classes of consumers, a final classification is provided, with curves of different power consumption behavior being grouped together by the evolutionary algorithm and a Fuzzy set of rules for classification is provided too.

VALIDATION

To explore the performances of the proposed method for adaptively clustering consumer pro-

Figure 8. Results of data pre-processing

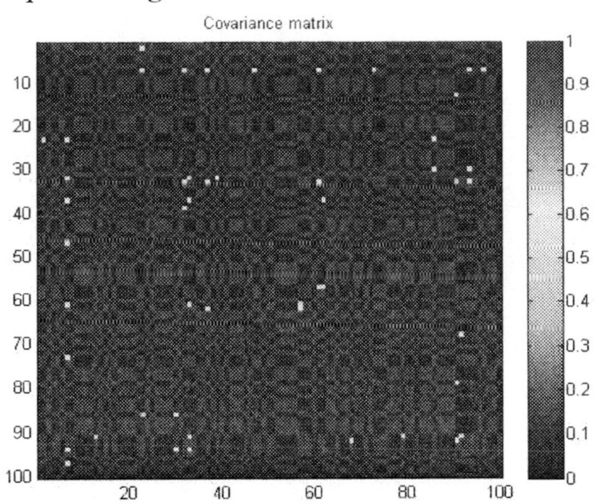

files, we chose to analyze a case-study considering - for the sake of simplicity- two possible typologies of (consumers) users: households and small offices. A set of randomized power consumption curves has been generated according to the predefined patterns previously described, in order to have a significant number of realistic consumer profiles. The results of the pre-processing phase are reported in Figure 8, where an image of the covariance matrix is shown: the values are normalized and filtered in order to show similarities between the considered randomly generated power consumption curves: red means that profile similarity is high, green that it is medium, blue means that profile similarity is low.

Once the similarities among different curves have been identified, the adaptive clustering phase takes place, according to the previously described evolutionary-assisted method. Results of this phase are reported in Figure 9, where a number of considered curves is reported in two separate panels: on the left side we have the residential-like set of curves, on the right panel we show the curves recognized as instances of the SOHO class. Two additional set of parameters have been reported in both plots: a continuous red line representing the class consumption behavior extracted by the evolutionary tool from the set of considered curves and a set of blue bars, to be used as the threshold for the Fuzzy logic classifier to be implemented. These very simple results are here reported for the sake of simplicity, but several other simulated evaluations have been conducted confirming that the proposed evolutionary-assisted procedure is capable to successfully identify and to cluster together different profiles presenting similar behavior.

CONCLUSION

Actors interacting on the energy market, e.g. individuals manifesting similar energy profiles can form a social network based on the cohesion criteria expressed in Fuzzy rules terms qualifying the electric load patterns (Figure 10). Authors have described the computational method, an algorithm, and a software tool helping to build a local virtual micro community optimizing the electric energy consumption locally, obtaining the equivalent of the short agricultural chain producer-to-consumer. There are three enablers: a smart real time metering giving the real life energy profile, the Fuzzy reasoning about the

Figure 9. Results of adaptive clustering operation

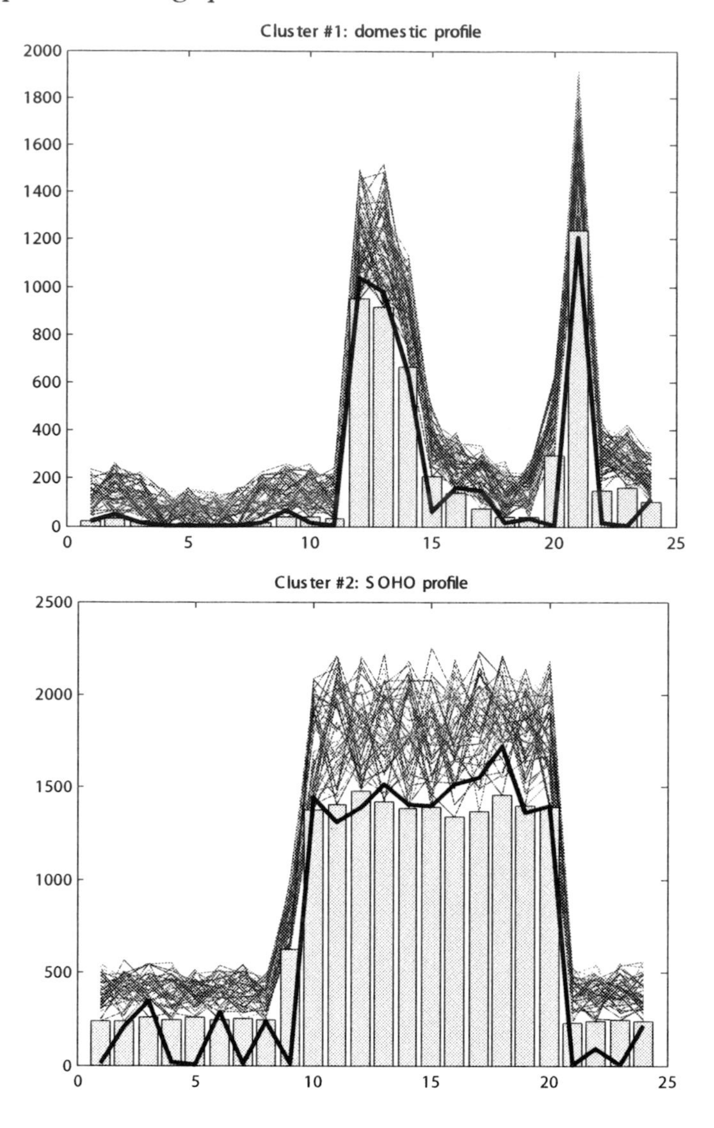

energy production forecast and the clustering technique creating the optimized local minimum. The Fuzzy system becomes criteria for the social cohesion of the newly proposed community, while the aggregated load profile represents the collective grid behavior.

The social aggregation proposed to physical persons is intentionally made virtual because of the nature of the transactions. However it might become real because of the motivation for grouping coming from the Fuzzy rules describing the objective function Θ (t) rewritten in lifestyle terms "people having the lunch at home" instead of the energy ones like "energy consumption rising between 1 and 2 PM". The minimal grouping ensuring the full local energy consumption can be easily extended adding an arbitrary number of new members to improve the social cohesion of the group. A minimal group achieves the main social goal of the total local energy consumption. Moreover an aggregation of a seniors looking after the children in the afternoon, a daytime working

Figure 10. Relationships and semantics

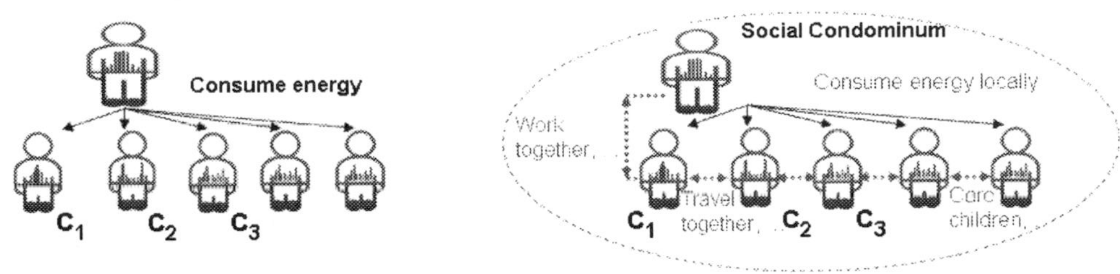

Figure 11. Software toolkit with energy features

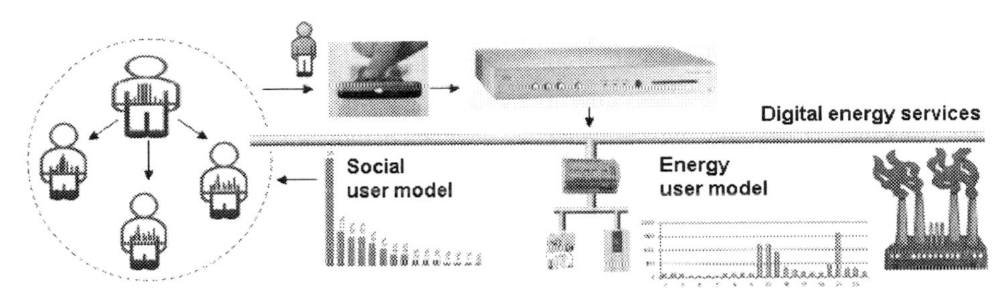

SOHO, plus some students coming back home to have a lunch, might create an efficient cooperative environment improving the collective safety of the Condominium. At the end, the above-mentioned communities become useful to leverage loads charging three phases lines inside buildings.

The most important finding confirmed in experimental settings, even if intuitive a priori, is the fact that each cluster M_j should comprise at least two different categories of users. The best legacy performance of the micro grid is achieved grouping the photovoltaic residential Prosumer with a small industrial user, fully confirming the initial assumption about the daytime generation – consumption. The resulting minimal threshold shown by the micro grid becomes very close to the traditional minimum of the residential load curve, if we add the correction coming form the local weather forecast for +1 day (tomorrow). The eolic component is less predictable; however the uncertainty might be reduced replacing 1 day forecast in photovoltaic case by shorter hourly interval.

Authors precise that no real time data elaboration was possible because of the lacking interface with the existing AMR devices: in Italy the Telegestore system is based on SITRED data exchange protocol, which is not publicly available. Authors have used the datasets of the measurements complemented by the randomly generated geo-referencing simulating the real life situation, also because of the privacy protection constraints. To set up the working vendor-independent interoperable solution, an alternative in-home device, like a media-server equipped by the digital Wattmeter, exchanging the data among the social condominium over IP (PLC) should be considered, which might become the future work.

The mathematical modeling of the energy micro grids as Internet of Things entities, designing the procedures ruling local markets extends the proposed research. Micro grids are the Internet of Things entities. A "long" ipv6 address would hide local intelligence exposing only the model MGrid = [{AMR(N_j, t, date), AMM(N_j, t, date), E(t), ID(t)}, M-class]. The micro grid behavior will be described by M-class in the remote Ontology.

The MGrid digital energy might bring only AMM and ID events, because of the encapsulation – all internal accountancy might be computed locally, discharging the net. MGrid becomes a valuable abstraction for the semantic management of micro grids.

Following the proposed approach, a new model describing the network in terms of the transits through the qualified arches $\mathbf{N}_j$, instead of the nodes $\mathbf{M}_j$, enables the minimization of the uncertainty because of the local energy consumption. A method building a micro-grid containing one Prosumer - the generator and the consumer of the renewable energy plus some other nodes - hide the complexity and eliminate the effect from the uncertainty. The above-mentioned toolkit might offer also the possibility to negotiate in real time the social electric energy among local consumers, instead of the transactions on the main energy market, proposing the further work (Figure 11), in which the Near-energy service becomes represented by the following event chain:

```
Notify _ event ( From, To, Expected _ power,
     Start _ At, End _ At, Now)
Sell _ energy  ( From, To, Amount _ Power,
     Start _ At, End _ At, Now)
Buy _ energy  ( From, To, Amount _ Power,
     Start _ At, End _ At, Now)
```

REFERENCES

Baldwin, J. (1981). Fuzzy logic and Fuzzy reasoning. In E. Mamdani & B. Gaines (Eds.), *Fuzzy Reasoning and its Applications*. London: Academic Press.

Broehl, J. (1981). An end-use approach to demand forecasting. *IEEE Transactions on Power Apparatus and Systems*, *6*(PAS-100), 2714-2718.

Cagni A., Carpaneto E., Chicco G., & Napoli R. (2004). Characterisation of the aggregated load patterns for extra-urban residential customer groups. *Proceedings from Melecon '04: 12th IEEE Mediterranean Electrotechnical Conference*, *3*, 951–954.

Capasso, A., Grattieri, W., Lamedica R., & Prudenzi A. (1994). A bottom-up approach to residential load modeling. *IEEE Trans. Power Syst.*, *9*(2), 957–964.

Carpaneto, E., & Chicco, G. (2007). Probabilistic characterisation of the aggregated residential load patterns, *IET Generation, Transmission & Distribution*, 2 (3), 373-382.

Chan, M.L., Marsh, E.N., Yoon, J.Y., Ackerman, G.B., & Stoughton, N. (1981). Simulation-based load synthesis methodology for evaluating load-management programs. *IEEE Transactions on Power Apparatus and Systems*, *4*(PAS-100), 1771- 1778.

EPRI, Electric Power Research Institute. (1985). *Combining engineering and statistical approaches to estimate end-use load shapes, Vol. 2: Methodology and results.* Palo Alto, California, USA: Report EA-4310.

Frunt, J., Kling, W.L., Myrzik, J.M.A., Nobel, F.A., & Klaar, D.A.M. (2006). Effects of Further Integration of Distributed Generation on the Electricity Market. *Proceedings from UPEC '06: The 41st Intl. Univ. Power Engineering Conf.*, 1, 1–5.

Gellings, C.W., & Taylor, R.W. (1981). Electric load curve synthesis – A computer simulation of an electric utility load shape. *IEEE Transactions on Power Apparatus and Systems*, *1*(PAS-100), 60-65.

Goldberg, D.E. (1989). *Genetic Algorithms in Search, Optimization and Machine Learning.* New York: Addison-Wesley

Halpern, J. (2003). *Reasoning about Uncertainty*. MIT Press.

Kruse, R., Gebhardt, J., & Klawonn, F. (1994). *Foundations of Fuzzy Systems*. Chichester: Wiley.

Mellit, A., & Kalogirou, S.A. (2006). Neuro-Fuzzy Based Modeling for Photovoltaic Power Supply System, *Proceedings from PECon '06: IEEE Power and Energy Conference*, (pp. 88–93).

REMPLI project, Retrieved January 20th, 2008, from http://www.rempli.org

Sammartino L., Simonov M., Soroldoni M., & Tettamanzi A. (2000). Gamut: A System for Customer Modelling based on Evolutionary Algorithms. In D.E. Goldberg & E. Cant (Eds.), *Proceedings from GECCO '00: Genetic and Evolutionary Computation Conference, 758*, Morgan Kaufmann.

Walker, C.F., & Pokoski, J.L. (1985). Residential load shape modeling based on customer behavior. *IEEE Trans. on Power Apparatus & Systems*, 7(PAS-104), 1703–1711.

Zadeh, L. (1965). Fuzzy Sets. *Information and Control*, 8, 338-53.

This work was previously published in International Journal of Virtual Communities and Social Networking, Vol. 1, Issue 3, edited by S. Dasgupta, pp. 75-93, copyright 2009 by IGI Publishing (an imprint of IGI Global).

Chapter 8.17
The Social Semantic Desktop:
A New Paradigm Towards Deploying the Semantic Web on the Desktop

Ansgar Bernardi
German Research Center for Artificial Intelligence (DFKI) GmbH, Kaiserslautern, Germany

Stefan Decker
National University of Ireland, Ireland

Ludger van Elst
German Research Center for Artificial Intelligence (DFKI) GmbH, Kaiserslautern, Germany

Gunnar Aastrand Grimnes
German Research Center for Artificial Intelligence (DFKI) GmbH, Kaiserslautern, Germany

Tudor Groza
National University of Ireland, Ireland

Siegfried Handschuh
National University of Ireland, Ireland

Mehdi Jazayeri
University of Lugano, Switzerland

Cédric Mesnage
University of Lugano, Switzerland

Knud Möller
National University of Ireland, Ireland

Gerald Reif
University of Lugano, Switzerland

Michael Sintek
German Research Center for Artificial Intelligence (DFKI) GmbH, Kaiserslautern, Germany

Leo Sauermann
German Research Center for Artificial Intelligence (DFKI) GmbH, Germany

ABSTRACT

This chapter introduces the general vision of the Social Semantic Desktop (SSD) and details it in the context of the NEPOMUK project. It outlines the typical SSD requirements and functionalities that were identified from real world scenarios. In addition, it provides the design of the standard SSD architecture together with the ontology pyramid developed to support it. Finally, the chapter gives

DOI: 10.4018/978-1-60566-112-4.ch012

an overview of some of the technical challenges that arise from the actual development process of the SSD.

INTRODUCTION

A large share of everybody's daily activities centres around the handling of information in one way or the other. Looking for information, digesting it, writing down new ideas, and sharing the results with other people are key activities both in work as well as in manifold leisure activities. The abundance of PCs and the Web in today's world result in new numbers and qualities of information exchange and interaction which are seen both as chance and as threat by the users. Supporting personal and shared information handling is thus a highly requested but yet unsolved challenge.

In traditional desktop architectures, applications are isolated islands of data – each application has its own data, unaware of related and relevant data in other applications. Individual vendors may decide to allow their applications to interoperate, so that, e.g., the email client knows about the address book. However, today there is no consistent approach for allowing interoperation and a system-wide exchange of data between applications. In a similar way, the desktops of different users are also isolated islands – there is no standardized architecture for interoperation and data exchange between desktops. Users may exchange data by sending emails or uploading it to a server, but so far there is no way of seamless communication from an application used by one person on their desktop to an application used by another person on another desktop.

The problem on the desktop is similar to that on the Web – also there, we are faced with isolated islands of data and no generic way to integrate and communicate between various Web applications (i.e., Web Services). The vision of the SW offers solutions for both problems. RDF[a] is the common data format which builds bridges between the islands, and Semantic Web Service technology offers the means to integrate applications on the Web.

The Social Semantic Desktop (SSD) paradigm adopts the ideas of the SW paradigm for the desktop. Formal ontologies capture both a shared conceptualization of desktop data and personal mental models. RDF serves as a common data representation format. Web Services – applications on the Web – can describe their capabilities and interfaces in a standardized way and thus become Semantic Web Services. On the desktop, applications (or rather: their interfaces) will therefore be modelled in a similar fashion. Together, these technologies provide a means to build the semantic bridges necessary for data exchange and application integration. The Social Semantic Desktop will transform the conventional desktop into a seamless, networked working environment, by loosening the borders between individual applications and the physical workspace of different users.

By realizing the Social Semantic Desktop, we contribute to several facets of an effective personal information handling:

- We offer the individual user a systematic way to structure information elements within the personal desktop. Using standard technology to describe and store structures and relations, users may easily reflect and express whatever is important in their personal realm.
- Standardized interfaces enable the integration of all kinds of available desktop applications into the personal information network. Investments in programs, data collections, and hard-learned working

styles are not lost but augmented and connected into a comprehensive information space.

- Based on the SW technology basis, all kinds of automated and semi-automatic support are possible, like, e.g., text classification services, image categorization, document relevance assessments, etc.
- The exchange of standard data formats between individual work spaces is supported not only on the technical level (e.g., standard communication protocols), but also on the semantic level (via sharing and alignment of ontologies and the corresponding annotated information elements). The integration with formal ontologies eases the sharing and understanding between different persons.
- Ultimately, we thus contribute to a solution for the initialization problem of the SW: As the individual user will receive immediate benefit from the semantic annotation within the personal workspace, the motivation is high to invest the necessary structuring and formalization work. As the standards used allow for an effortless sharing of such work, the amount of semantically annotated information which can be made available in the Web grows dramatically – which in turn makes it worthwhile to develop new SW-based services.

In this chapter we describe in detail the core components which are necessary for building a Social Semantic Desktop. We illustrate the necessary standard framework and describe the role and structure of the ontologies which support the spectrum from personal to social information handling. We outline the implementation decisions which need to be observed in order to realize a consequently ontology-oriented system, which is able to deal with the numerous flexibilities required within the Semantic Web. Finally, we show examples of the benefits obtained from the realization and use of an SSD.

The ideas and implementation principles presented in this chapter are distilled from our experiences in the NEPOMUK Project[b]. For each section we will describe the general motivation and principles and then give details on how the particular challenges have been solved in the NEPOMUK project.

BACKGROUND

The Social Semantic Desktop vision has been around for a long time: visionaries like Vannevar Bush (1945) and Doug Engelbart (1962) have formulated and partially realized these ideas. However, for the largest part their ideas remained a vision for far too long, since the necessary foundational technologies were not yet invented – figuratively speaking, these ideas were proposing jet planes when the rest of the world had just invented the parts to build a bicycle. Only in the recent years several technologies and research streams began to provide a foundation which will be combined and extended to realize the envisioned collaborative infrastructure of the SSD.

Figure 1 shows the highest-level architecture and connections between components of the SSD, i.e., the social networks, the P2P infrastructure, and the individual desktops. Traditional semantics, knowledge representation, and reasoning research are now interacting. While none of them can solve the problem alone, together they may have the explosive impact of the original Web:

The Semantic Web effort provides standards and technologies for the definition and exchange of metadata and ontologies. Available standard proposals provide ways to define the syntax (RDF) and semantics of metadata based on ontologies (Web Ontology Language – OWL(McGuiness

Figure 1. Component architecture of the Social Semantic desktop

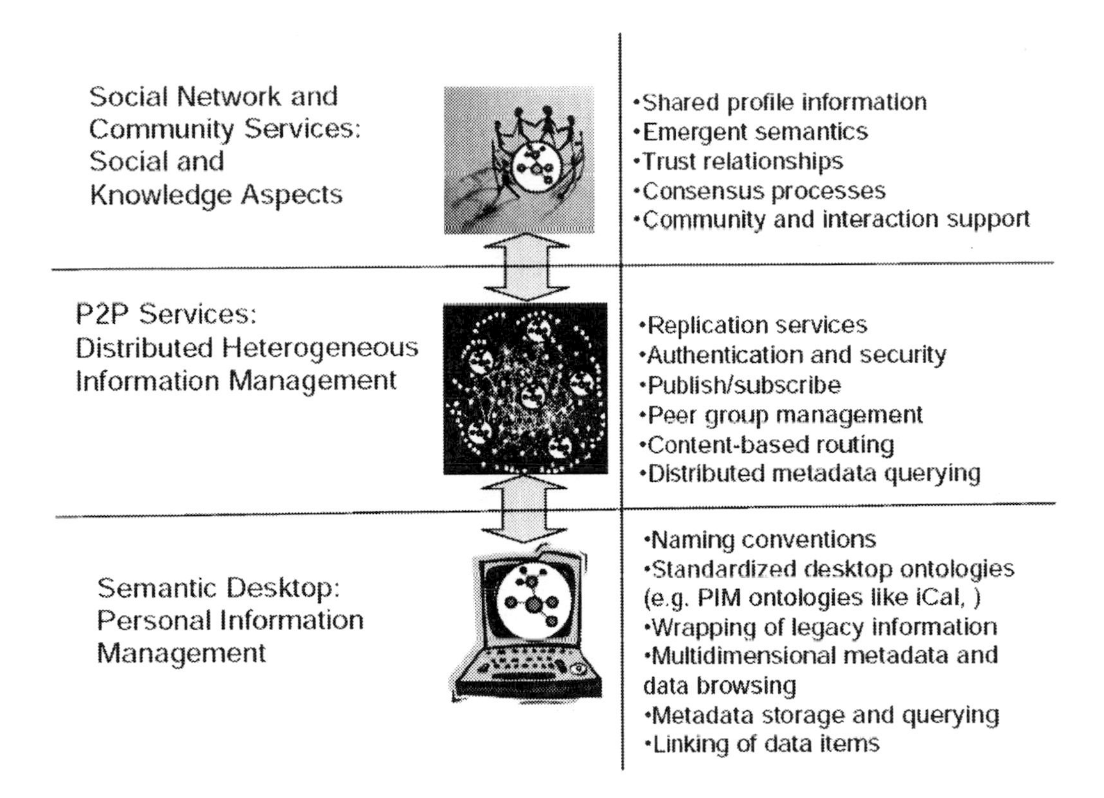

et. al, 2004), RDF Schema – RDFS). Research covering data transfer, privacy, and security issues is now also under development.

Social Software maps the social connections between different people into the technical infrastructure. As an example, Online Social Networking makes the relationships between individuals explicit and allows the discovery of previously unknown relationships. The most recent Social Networking Sites also help form new virtual communities around topics of interest and provide means to change and evolve these communities.

P2P and Grid computing develops technology to network large communities without centralized infrastructures for data and computation sharing. P2P networks have technical benefits in terms of scalability and fault tolerance, but a main advantage compared to central sites is a political one: they allow to build communities without centralized nodes of control, much as the Internet grew as fast as it did because it was based on reciprocity – it avoided political debate as to who gets to own big, expensive central facilities. Recent research has provided initial ways of querying, exchanging and replicating data in P2P networks in a scalable way.

By projecting the trajectory of current trends, we can simplify this picture by stating that next generation desktop applications will support collaboration and information exchange in a P2P network, connecting online decentralized social networks, and enabling shared metadata creation and evolution by a consensus process. The result of this process will be the Social Semantic Desktop. Figure 2 depicts the phases in which the relevant co-evolving technologies are combined to achieve the final goal, i.e., the realization of the Social Semantic Desktop.

Figure 2. Phases towards the Social Semantic desktop

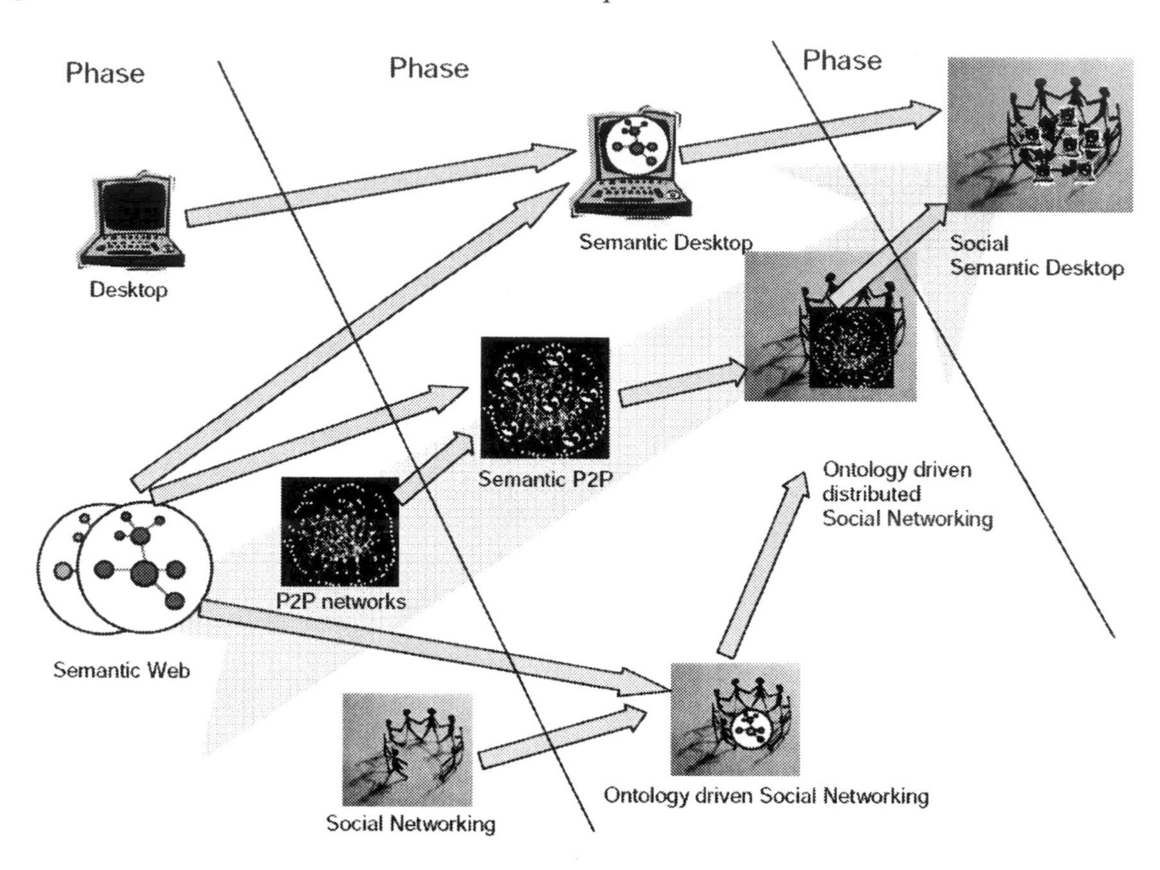

SCENARIOS

Before we move on to the specific functionalities that a Social Semantic Desktop supports and discuss how they are implemented, we will first present some scenarios that will illustrate what an SSD is, how it will be used, and how it will change the way we do knowledge work. We chose the scenarios such that they illustrate the different dimensions of an SSD: Sect. *The Semantic Dimension* describes example usage that shows the use of semantics on the desktop, and Sect. *The Social Dimension* will show the social dimension of an SSD, i.e., the interaction between desktops of different users. The scenarios give an overview of what is possible and how the SSD presents itself to the user.

The Semantic Dimension

A typical use of a single Semantic Desktop is to organize ones data: files, emails, pictures, etc. Users are able to tag those information resources with concepts from a network of ontologies. The ontologies also contain relations (or properties, to use RDF terminology), which can be used to link information resources on the desktop. Organizing information resources in this way helps users to find what they are looking for quicker, and makes it possible for the Semantic Desktop itself to aid the user in their daily work. When a user first begins using the Semantic Desktop, many often-used concepts and properties are already present. E.g., there are basic concepts such as **Person**, **Meeting** or **Place**, and properties such as **knows** or **located-in**. Also, we can assume that useful things like an ontology of all countries are already in place. Then, as the need arises, users can extend the existing ontologies – e.g., they can add concepts for a particular meeting they attend or people they know, such as **Executive-Committee-Meeting-07/06/07**, **Jane-Doe** or **Hans-Schmidt**. The following two scenarios give examples of this kind of Semantic Desktop usage. We will use two imaginary users (*personas*[c]) flesh them out: Dirk, who works in a research group for some company in Germany, and Claudia, who is the group leader and his boss. Both Dirk and Claudia work on a project called Torque.

Organizing pictures (*Annotation*). Dirk just got back from his holidays in Norway, where he took a lot of pictures. Using his Semantic Desktop, he now wants to organize them, so that he can later find the pictures he wants to easier, generate photo albums for particular topics, etc. A lot of the important concepts probably already exist on his desktop, such as Norway, or the cities he has visited: **Oslo**, **Bergen** and **Trondheim**. Other concepts will be added by Dirk himself, such as **Holidays-in-Norway-2007** and tourist sights like **Preikestolen** or **Holmenkollen**. Since these concepts are more than just tags, Dirk can also say things *about* them, e.g., that **Holidays-in-Norway-2007** was a **Trip** and took place in **2007**, and that **Preikestolen** is a **Location** in **Norway**. Dirk even managed to take a picture of Prince Håkon and Princess Mette-Marit, so he creates two more concepts **Håkon** and **Mette-Marit**. There are many ways in which Dirk can link (or tag) his pictures to the relevant concepts – however, part of the Semantic Desktop are intuitive user interfaces, which hide most of the intricacies that go on under the hood from the user. E.g., Dirk might have an application that shows all the concepts that he is interested in the form of a tag cloud. Linking the pictures would then simply require him to drag them onto the desired concept in the cloud.

Planning a trip (*Context*). Later, Dirk finds out that he has to go on a work trip: a conference in Oslo. The Semantic Desktop assists him in planning and organizing this trip, through the notion of *context*. Dirk can create a new **Trip** object **Trip-to-OOC2007-Oslo** and tell his desktop that he is now in the context of this trip. This means that everything he does from this moment on will be interpreted as happening in that context, until he quits the context again. When he books a flight in his Web browser, the destination field will automatically be filled in with "Oslo", similarly the departure field. Afterwards, when he books a hotel room, he will be assisted similarly. Dirk will receive a number of email confirmations, such as the flight itinerary and booking confirmation for his hotel. These emails and their attachments will automatically be filed as belonging to the **Trip-to-OOC2007-Oslo** context, so that Dirk can easily find them again later. Once he knows his exact flight dates and where his hotel will be, he enters this information into his calendar, which is also context-aware and will therefore remember that these entries belong to Dirk's trip.

The Social Dimension

Users will have a lot of benefit from just using the Semantic Desktop on their own. However, by connecting to others, a number of additional possibilities arise.

Assigning tasks in a group (*Social Interaction*). In the previous scenario, Dirk found out he had to go on a business trip. In fact, he found out about this because he was notified by his boss Claudia, who also uses a Semantic Desktop. Claudia plans to travel to the OOC2007 conference in Oslo to present a research prototype her group has developed as part of the *Torque* project. She does not want to travel alone, so she first needs to find out who of her group members are available while the conference runs. Through the network of Social Semantic Desktops, her calendar application has access to the calendars (or parts of them) of all her contacts. She can ask the calendar to give her a list of all people in her group (**My-Research-Group**) who are working on the *Torque* project (**Torque-**

Table 1. Functionalities of the Social Semantic desktop

Desktop	Annotation, Offline Access, Desktop Sharing, Resource Management, Application Integration, Notification Management
Search	Search, Find Related Items
Social	Social Interaction, Resource Sharing, Access Rights Management, Publish/Subscribe, User Group Management
Profiling	Training, Tailor, Trust, Logging
Data Analysis	Reasoning, Keyword Extraction, Sorting and Grouping

Project) and are free when OOC2007 is on. She finds out that Dirk is free at the desired time. Just like Dirk in the previous scenario, she creates a **Trip-to-OOC2007-Oslo** object and makes it her current context. She also links the trip to the **Torque-Project**. Now, she creates a new **Task** object **Dirk-Prepare-Trip-To-OOC2007**, with a subtask **Dirk-Prepare-Presentation-Slides** and afterwards sends an email to Dirk, asking him to accompany her to the conference, book flights and hotel rooms, and prepare slides for the conference presentation. Her email and the task will of course be automatically linked to the proper context. Also, in this version of the scenario, Dirk no longer has to create the **Trip-to-OOC2007-Oslo** object himself – instead, it will be added to his Semantic Desktop automatically when he gets Claudia's mail.

FUNCTIONALITIES

In this section we describe a list of functionalities that are needed to support the scenarios mentioned above, as well as other scenarios developed during the NEPOMUK project. The SSD is a platform used to develop different kinds of social and semantic applications. These applications share common functionalities which must be supported by the SSD. We have divided them into five groups, which can be considered different aspects of the SSD. Tab. 1 shows the five different aspects and the individual functionalities within each group. Below we briefly describe the use of each functionality.

Desktop. At the *desktop* level, the semantic functionality common to most applications is the ability to add information about any resource. **Annotation** comprises the facilities to store and retrieve semantic relations about anything on the desktop. When Dirk annotates his photos from his trip, he does it from his most favorite photo application (such as Picasa or iPhoto), the annotations are then stored by the SSD. We name this functionality **Application Integration**; applications interact with the SSD by means of different services. When Dirk got notified about the trip to Oslo, this was an example of **Notification Management**. The SSD handles different kinds of mechanisms such as emails, RSS, or text messaging. When Dirk creates a new concept or even a new file on the SSD, the application he uses interacts with the **Resource Management** facilities of the SSD, creating the needed semantics according to the current context and setup. Some of the information Dirk needs when booking his trip are stored on Claudia's desktop. If she is not connected to the network, the **Offline Access** facility exports the relevant information to another desktop. **Desktop Sharing** is the ability for different users of the SSD to work on the same resources. Claudia might write a report of the trip together with Dirk: the resource management is done on Dirk's desktop, but Claudia can access and edit it remotely.

Search. The semantic network created on the desktop unleashes a whole new way of searching on the SSD. **Search** uses the semantic relations as well as social relations to retrieve relevant items. Once an item is found a user can also **Find**

Related Items. For instance, when Dirk searches for a flight to Oslo, he can also search for related items and may find out that another company is actually cheaper, based on the experience of his social contacts.

Social. The SSD provides different means of **Social Interaction**, e.g., the embedding of semantic information in emails or text messaging, or the ability to annotate another user's resources. Some desktop level functionalities such as desktop sharing and offline access require the SSD to enable **Resource Sharing**. When Dirk and Claudia collaborate on the trip's report, Dirk might make it accessible to the whole group by adding it to a shared information space. When sharing resources or information on the network, the **Access Rights Management** of the SSD provides ways to define specific rights relations between users, groups and resources. The SSD's **User Group Management** system makes it easy for the rapid creation of new groups from a list of users. These groups can then be used to modify access rights or for resource sharing in a shared information space. E.g., some of Dirk's friends may have subscribed to get notifications of new pictures that Dirk annotates and makes available. The **Publish/Subscribe** mechanism of the SSD facilitates the creation of feeds of relevant information.

Profiling. If enabled, the **Logging** functionality of SSD logs user activity, which may help to detect the current user's context. The *profiling* of the SSD can be done automatically by **Training**: the SSD learns to predict the user's behavior. The user can still **Tailor** the SSD's intelligent behaviors: some learned contexts can become irrelevant and may need to be re-adapted or removed. The notion of **Trust** on the SSD between people or information sources is also a result of the profiling of the desktop. Dirk might define that he trusts Claudia's information, or Claudia's SSD might learn that Dirk is a trustworthy source of information regarding the *Torque* project.

Data analysis. To support the training behaviors of the SSD or querying related items, the SSD provides different *data analysis* mechanisms such as **Reasoning**. For instance, when Dirk tags a picture with **Preikestolen** and **Norway**, the SSD may infer that Preikestolen is in Norway. This information can later be reused for search. **Sorting and Grouping** supports applications that perform search. The SSD returns items from many sources and people and sorts and groups these items regarding different criteria, using the semantics defined on these resources. The **Keyword Extraction** from resources such as text resources is useful for automatically tagging or summarizing.

ONTOLOGIES

Ontologies form a central pillar in Semantic Desktop systems, as they are used to model the environment and domain of the applications. The common definition of an ontology is "a formal, explicit specification of a shared conceptualization" (Gruber, 1995)

We distinguish four levels of ontologies for the SSD: *Representational, Upper-Level, Mid-Level and Domain.* The main motivation for having these layers is that ontologies at the foundational levels can be more stable, reducing the maintenance effort for systems committed to using them. A core principle of the Semantic Desktop is that ontologies are used for personal knowledge management. Each user is free to create new concepts or modify existing ones for his *Personal Information Model*. This modeling takes place on the domain-ontology level, but the user is of course free to copy concepts from the other layers and modify them to fit his or hers own needs. In order of decreasing generality and stability the four layers are:

Representation(al) Ontology. Representational ontologies (i.e., ontology definition languages) define the vocabulary with which the other ontologies are represented; examples are RDFS and OWL. The relationship of a representational ontology to the other ontologies is quite special: while upper-level ontologies generalize mid-level ontologies, which in turn generalize domain ontologies, all these ontologies can be understood as instances of the representational ontology. Concepts that might occur in the Representational Ontology level include: classes, properties, constraints, etc.

Upper-Level Ontology. "An upper ontology [...] is a high-level, domain-independent ontology, providing a framework by which disparate systems may utilize a common knowledge base and from which more domain-specific ontologies may be derived. The concepts expressed in such an ontology are intended to be basic and universal concepts to ensure generality and expressivity for a wide area of domains. An upper ontology is often characterized as representing common sense concepts, i.e., those that are basic for human understanding of the world. Thus, an upper ontology is limited to concepts that are meta, generic, abstract and philosophical. Standard upper ontologies are also sometimes referred to as foundational ontologies or universal ontologies." (Semy et. al, 2004) In the upper-level ontology you will find concepts like: **Person**, **Organization**, **Process**, **Event**, **Time**, **Location**, **Collection**, etc.

Mid-Level Ontology. "A mid-level ontology serves as a bridge between abstract concepts defined in the upper ontology and low-level domain specific concepts specified in a domain ontology. While ontologies may be mapped to one another at any level, the mid-level and upper ontologies are intended to provide a mechanism to make this mapping of concepts across domains easier. Mid-level ontologies may provide more concrete representations of abstract concepts found in the upper ontology. These commonly used ontologies are sometimes referred to as utility ontologies." (Semy et. al, 2004). The mid-level ontologies may include concepts such as: **Company**, **Employer**, **Employee**, **Meeting**, etc.

Domain Ontology. "A domain ontology specifies concepts particular to a domain of interest and represents those concepts and their relationships from a domain specific perspective. While the same concept may exist in multiple domains, the representations may widely vary due to the differing domain contexts and assumptions. Domain ontologies may be composed by importing mid-level ontologies. They may also extend concepts defined in mid-level or upper ontologies. Reusing well established ontologies in the development of a domain ontology allows one to take advantage of the semantic richness of the relevant concepts and logic already built into the reused ontology. The intended use of upper ontologies is for key concepts expressed in a domain ontology to be derived from, or mapped to, concepts in an upper-level ontology. Mid-level ontologies may be used in the mapping as well. In this way ontologies may provide a web of meaning with semantic decomposition of concepts. Using common mid-level and upper ontologies is intended to ease the process of integrating or mapping domain ontologies." (Semy et. al, 2004). Domain ontologies consist of concepts like: **Group Leader**, **Software Engineer**, **Executive Committee Meeting**, **Business Trip**, **Conference**, etc.

Figure 3 shows how these four layers relate to the four ontologies created and used in the NEPOMUK Project. As detailed in Sect. "*Technology*", we were hesitant to make use of OWL for the representational ontology level in NEPOMUK, and in its place we developed the NEPOMUK Representational Language(Sintek et. al, 2007) (NRL). NRL defines an extension to the semantics offered by RDF and RDFS; the main contribution of NRL is the formalization of the semantics of named graphs. NRL allows multiple semantics (such as open and closed world) to coexist in the same application, by allowing each named graph to have separate semantics. The NEPOMUK Annotation Ontology (NAO) is a basic schema for describing annotations of resources, this is essentially a formalization of the tagging paradigm of Web2.0 applications. A specialized part of NAO is the NEPOMUK Graph Metadata schema (NGM)

Figure 3. NEPOMUK ontology pyramid

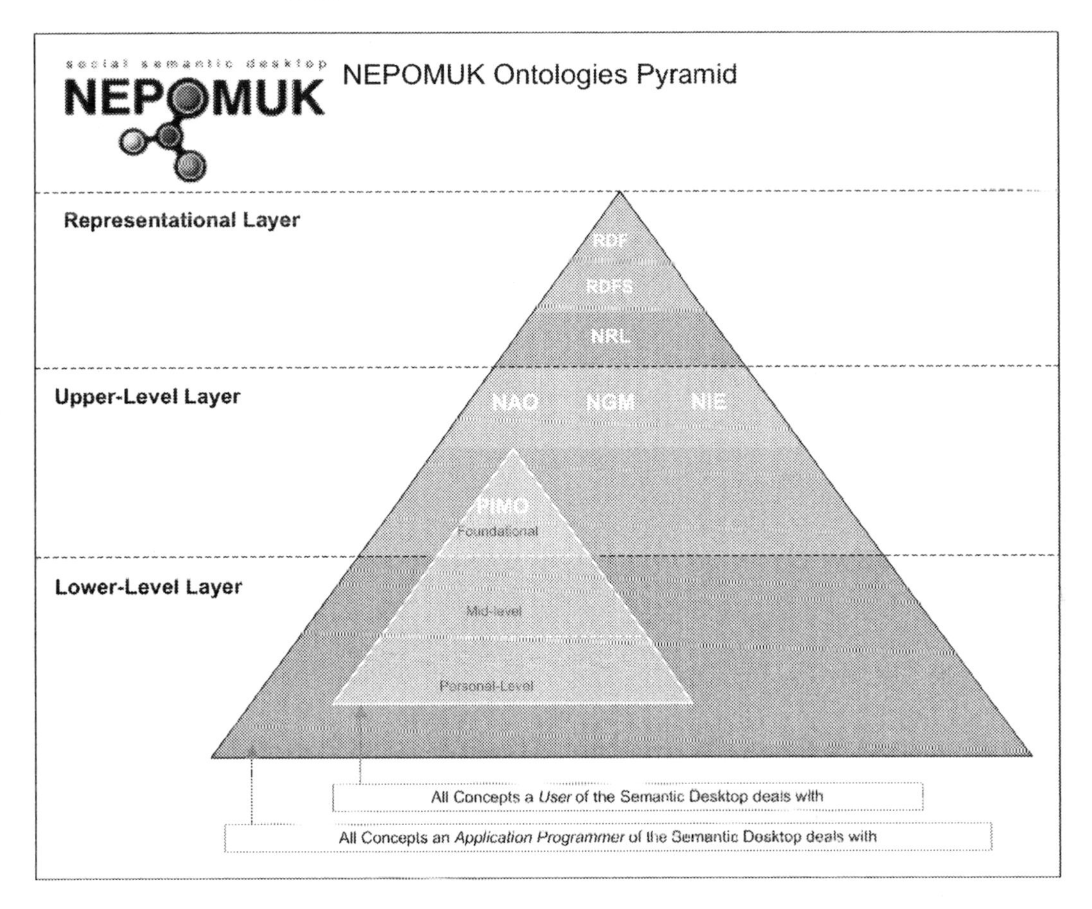

which allows the description of named graphs, defining meta-data properties such as the author, modification dates and version data.

Finally, the NEPOMUK Information Elements ontology (NIE) contains classes and properties for describing objects found on the traditional desktop, such as files (Word documents, images, PDFs), address book entries, emails, etc. NIE is based on existing formats for file meta-data such as EXIF for image meta-data, MPEG7 for multimedia annotations, ID3 for music files, iCal, and others.

TECHNOLOGY

The Social Semantic Desktop deploys the Semantic Web on the desktop computer. Therefore, the technology stack proposed for the Semantic Web (the famous "Layercake"[d] adapted in Figure 4) is adopted for the SSD as well.

However, there are some specific considerations for the desktop scenario: everything on the desktop should identifiable by URIs. This is partially solved for files, where RFC1738[e] specifies the form of file:// URIs, but requires considerable care for other applications which may not represent their data entities as individual files, such as address books or email clients.

Secondly, one can note that for the single desktop scenario there are fewer requirements on

Figure 4. The Semantic Web technology stack

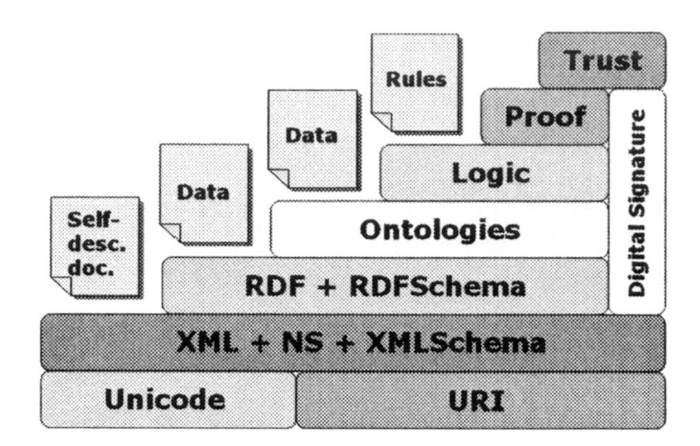

aspects such as trust, proof, and signatures. When one progresses to the Social Semantic Desktop, which involves interactions between many users, these aspects must be considered again.

In NEPOMUK we chose not to use the Web Ontology Language (OWL)(McGuiness et. al, 2004) as an ontology language, because of the challenge of dealing with (and implementing) OWL correctly; because our ontological modeling requirements were modest, and, most importantly, because OWL enforces an open-world view of the world, which did not seem to be appropriate for the (local) Semantic Desktop. In a World Wide Web context it is impossible for an application to read all available data, and an open-world assumption is natural, since additional data can be discovered at any moment. However, the open-world assumption makes it impossible to adopt negation as failure (Clark, 1978) which makes practical application development difficult and is also difficult to explain to the end-user. In the context of a local desktop application the situation is quite different, here it is perfectly possible to read all data available, and the closed world assumption makes much more sense. In place of OWL we developed our own ontology specification language called NRL, which uses the closed-world assumption. An additional RDF-based technology that we use widely, but which does not feature in the Semantic Web stack is the concept of named graphs(Caroll et. al, 2005). This allows one to divide a larger RDF store into sets of RDF statements (graphs), where each is identified with a URI (the name). In this way it is possible to make meta-statements about each graph, such as provenance information. Named graphs thus become an alternative to RDF reification, which also allows making statements about other statements, but is harder to implement and creates a significant overhead. NRL does also allows applying different semantics for different named graphs, thus allowing us to integrate the local closed-world with the open-world of the extended Semantic Web.

As noted previously, applications on the Semantic Desktop are analogous to services available on the Web. Each application will offer an interface for exposing the functionality it offers. Although a single desktop is not distributed, a network of SSDs is. It therefore suggests itself to adopt the Web Service stack of tools for inter-service communication for the Semantic Desktop: the Web Service Description Language (WSDL) [f] which is used for describing the interfaces of services offered, XML Schema (XSD)[g] which is used for primitive type definitions, and finally the Simple Object Access Protocol (SOAP)[h] which is used for the invocation of services. In Sect. "*Implementation and Engineering Principles*" we give further details on how these technologies work in relation to the Semantic Web technologies presented above.

Figure 5. Layered architecture of the Social Semantic desktop

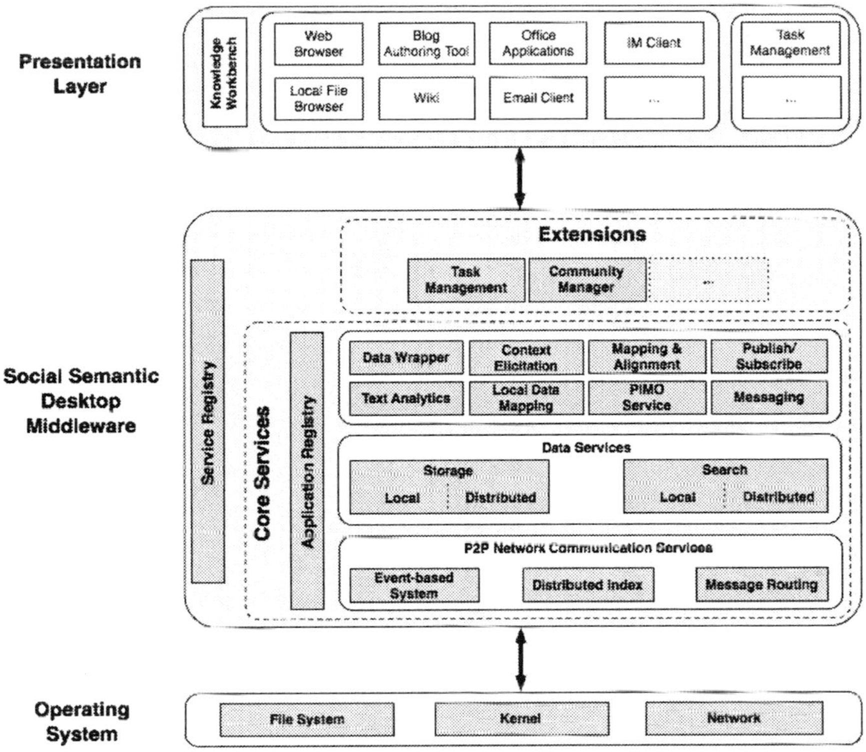

ARCHITECTURE

In our vision, the standard architecture comprises a small set of standard interfaces which allow application developers to easily extend it and ultimately lead to an evolving ecosystem. Figure 5 depicts this set of interfaces transposed into services, together with their placement in the NEPOMUK architecture. The architecture has to reflect the two aspects of the scenarios introduced in Sect. "*Scenarios*", i.e., the semantic (which can operate on a single desktop) and the social aspect (which is relevant in a network of desktops). To cover these requirements and the functionalities discussed in Sect. "*Functionalities*", the SSD is organized as a Service Oriented Architecture (SOA). Each service has a well defined WSDL interface and is registered at the *Service Registry*. The social aspect of sharing resources over the network is enabled by the peer-to-peer (P2P) infrastructure of the architecture. In the following we present the services of the SSD.

The architecture, as show in Figure 5, is organized in three layers. Like current desktop systems, the desktop environment builds on top of the *Operating System* core, such as the file system, kernel, and network environment. On the SSD the desktop environment is pooled in the *Social Semantic Desktop Middleware* Layer (SSD Middleware). The SSD Middleware groups the services of the SSD to be used in the Presentation Layer, which provides the user with SSD enabled applications that take advantages of the functionalities of the SSD.

The SSD is made up by individual desktops, which are organized in a P2P fashion. To support the communication between the peers, the SSD Middleware provides *P2P Network communication Services*. To enable information sharing between individual desktops, the RDF metadata of shared resources is stored in the *distributed index* of the P2P system. In NEPOMUK, the P2P system is based on GridVine (Aberer et. al, 2004), which in turn is built on top of P-Grid (Aberer et. al, 2003) and provides a distributed index with RDQL query supports.

Network Communication Services provide an *Event-based System*, which is responsible for the distribution of the events between the SSD peers. On the SSD, the event-based system is used to support the publish/subscribe system. Users as well as other services can use RDF to describe the kind of information they are interested in (e.g., new pictures of Norway become available, the status of a document changes to final, etc.). These subscriptions are stored in the distributed index of the P2P system. An event that was fired carries an RDF query as payload, which is matched against all subscriptions and triggers the notification of the subscriber. In addition, the *Messaging Routing* system uses RDF information to route messages to receiver.

The *Data Services* are responsible to control the insertion, modification, deletion, and retrieval of resources on the SSD. A resource can be a user, a document, a calendar entry, an email, and so on. It provides a service to store the RDF meta-data in the *Local Storage*. Resources and their RDF descriptions can either be added to the SSD manually, or the *Data Wrapper* or *Text Analysis* service extracts the information from desktop applications such as email clients or calendars. Data Wrappers are used to extract metadata form structured data sources (e.g., email headers, calendar entries, etc.). In NEPOMUK, data wrappers are implemented based on Aperture (Aperture, 2007). The Text Analysis service is used to extract metadata from unformatted text (e.g., email bodies, word processing documents, etc.). For local queries and for offline working the RDF metadata is stored in the *Local Storage*. If a resource is shared with other users in an information space, the metadata is also uploaded to the distributed index of the peer-to-peer file sharing system. The *Search* service can either issue a *local* search in the local storage or a *distributed* search in the underlying P2P system.

Before new metadata can be added to the repository, we have to check whether this metadata describes resources that are already instantiated (i.e., an URI has been assigned) in the RDF repository. In this case, the URI of the resource should be reused, rather then creating a new one. This process is known as information integration (Bergamaschi et. al, 2001). The *Local Data Mapper* service takes over this responsibility in the SSD Middleware. E.g., the Data Wrapping service extracts contact information from the address book and stores the metadata in the repository. Since this is the first time information about the contacts is added to the RDF repository, a new URI is generated for each person. If later the Data Wrapping service extracts information from an email header, the Local Data Mapping service is responsible to lookup whether information about the sender of the email is already in the repository and reuse the corresponding URI instead of creating a new one (Sauermann et. al, 2006).

Ideally only one ontology exists for a domain of interest such as contact data, calendar events. In reality, however, we are faced with many ontologies of (partly) overlapping domains (e.g., FOAF and vCard for contact data, or different personal information models). Since individual users share information over the SSD, it is likely to happen that they use different ontologies for their annotations even when talking about similar domains. Therefore, the SSD Middleware provides a *Mapping & Alignment* service that can be used by other middleware services and services in higher layers to translate RDF graphs from a source ontology to a target ontology.

The SSD Middleware logs the actions a user performs on the resources on his desktop. The logged data is stored in the Local Storage and analyzed by the *Context Elicitation* service to capture the current working context of the user. The context can for example be used to adapt the user interface or to suggest meaningful annotations to the users, depending on the task they are currently working on.

As discussed in Sect. "*Technology*", the services on the SSD use RDF to exchange data. Therefore, services need the capability to generate and process RDF graphs. To simplify the handling of the RDF graphs, the *Ontology Service* provides an easy way to create and manipulate concepts in RDF graphs.

The *Publish/Subscribe System* allows users or other SSD services to subscribe to events on the SSD. The subscriptions are stored as RDF graphs in the distributed index. If an event occurs, the RDF query of the event is matched against the subscriptions. When the subscription, i.e., the RDF query, matches the event, the *Messaging* service looks up the preferred notification media (e.g., email, instant messaging, SMS) and delivers the messages. The Messaging System is further used for synchronous and asynchronous communication between SSD users.

The *Core Services* of the SSD Middleware comprise the services which provide the basic functionalities of the SSD. These services can be accessed via the SSD Application Programming Interface (API). If a developer wants to exploit the SSD Core Services to build his domain-specific application, he can do this as an *extension* of the SSD Middleware. An example for such an extension is the *Task Management* which provides functionalities such as creating, delegating, and manipulating of tasks. Finally, the *Application registry* allows applications from the Presentation Layer to register call back methods at the SSD Middleware if they need to be notified by SSD services, e.g., when a message arrives and has to be displayed to the user in an Instant Messaging Client.

The top layer of the architecture is the presentation layer. It provides a user interface to the services provided by the SSD, and is built using the SSD API. Many desktop applications are possible sources for resources that should be managed by the SSD. Therefore, each desktop application should integrate support for the SSD Middleware. Since this assumption does not hold for most of the current off-the-shelf applications, we developed plug-ins and add-ons to enable a seamless integration with existing applications. These plugins for example extract email or calendar data and add them as resources to the SSD. However, within NEPOMUK we also develop dedicated applications that make use of the SSD API directly, such as a semantic *Wiki* or *Blogging Tools*. (Möller et. al, 2006)

In addition, the *Knowledge Workbench* is the central place to browse, query, view, and edit resources and their metadata. In this way the Knowledge Workbench aims to replace current file management tools such as the MS File Explorer. If the SSD is extended by usage extensions, the application programmer also has to provide the corresponding user interface in the Presentation Layer (e.g., for Task Management, Community Management, etc.).

IMPLEMENTATION AND ENGINEERING PRINCIPLES

As detailed in Sect. "*Architecture*". we deem a Service Oriented Architecture (SOA) to be most suitable for the SSD framework. Furthermore, we decided to use the industry standard SOAP (Simple Object Access Protocol) for exchanging messages between our components. For traditional applications the names and structure of SOAP messages is specified using the Web Service Description Language (WSDL), which in turn uses XML schema data-types to specify the form of the objects being exchanged. However, since the formal modeling of the target domain using ontologies is the core idea of a Semantic Desktop application, the best-practices for SOAs are slightly different. In this section we will discuss some important differences from a traditional SOA system.[i] Basing a system architecture on underlying domain ontologies is similar in nature to Model Driven Architectures (MDA)[j]. However, on the SSD, ontologies take the place of UML models.

Working with RDF

Sect. "*Ontologies*" described the substantial effort that went into the modeling of our domains as ontologies in a formal language. These ontologies give us a very powerful and flexible modeling language, although the structure of instances of such ontologies at first sight seem much more constrained than complex XML schema data-types, the flexibility of RDF introduces some additional requirements for developers of components that should handle RDF instances:

- The structure of the RDF instances received may not be fully known at design time. This means one must take great care that the code does not break when encountering unknown properties in the data, and these unknown properties must also be preserved. In general, programming services for the Semantic Desktop is more like programming services for the web, rather than for traditional desktop applications, and one should follow the general rule of web-programming: "Be strict in what you send and tolerant in what you receive."
- Conversely, other services might not be aware of all the properties the local service uses. Therefore each service must be programmed to be tolerant of missing data and do their best with the data that was provided.

Passing Instances in Messages

Normally, when using SOAP in connection with WSDL and XML schema for data modeling, some mapping is used that will convert the XML schema definition to class definitions in the programming language of choice. Furthermore, stubs and skeletons will be generated for the service themselves, so that the details of communication are hidden. Programming against remote services is then indistinguishable from programming against a local object. However, when using services that pass instances for which the structure is defined by ontologies, the mapping is not so straight forward. Although interaction with RDF data can always be done on a completely general level using basic RDF APIs we are interested in facilitating the job of programmers consuming our services, and allowing them to work on a higher level than RDF triples. We identify three alternatives for programming web services where parameters are instances from an ontology:

- Starting with the ontologies, a number of tools[k] can be used to create a set of Java classes from the ontologies. The service interface is written using parameters of these types, and another tool is used to generate the WSDL and associated XML schema types from these. By sharing the URIs of the concepts in the ontologies with the URIs of the XML schema types, the semantics of messages is retained. The benefit of this approach is that much of the SOAP technology is retained, existing tools may be reused. Also, developers who are not familiar with Semantic Web technology will find that developing and using these services is unchanged from a normal Java environment. The main problem with this approach comes from the fact that ontologies are in general more dynamic than Java class definitions. In particular, as noted in Sect. "*Ontologies*", we expect the personal information models to change frequently. This approach requires a complete re-run of the whole tool chain and a recompile of the system when an ontology changes, as well as introducing some constraints on the ontologies.

- On the other end of the spectrum it is possible to bypass the parameter passing of SOAP all together, and rely more on the Semantic Web technology. Each method offered by a service will take a single RDF document (possibly including several named-graphs), and all the details about the message are given in these RDF graphs. An additional ontology for messages and parameters must be constructed, and some named-graph aware serialization (e.g., TriG or TriX[l]) of RDF is used to construct the XML SOAP messages. This approach was, for instance, used in the SmartWeb project[m]. The benefit of this approach is that the effort that has gone into modeling the ontologies is not duplicated for modeling objects. Also, the full expressivity of RDF may be used when modeling, as it not required that the instances fit into another representation. The backside to this flexibility is that it is significantly harder to program with RDF graphs than with simple Java objects, and both service developers and consumers need good knowledge about RDF. One can of course envisage new tools that facilitate programming with such RDF messages, but since all the interesting details are hidden inside the RDF parameter, existing SOAP tools for development or debugging are no longer very useful.
- Finally, a hybrid approach of the two methods is possible. Here each method will retain multiple arguments, but each argument is represented by an RDF resource. We envisage two possibilities for doing this: either each parameter is given as a (*named-graph-uri, uri*) tuple pointing into an RDF document given as a special parameter; or, alternatively, each parameter is in itself an RDF graph plus the URI of the actual parameter (each RDF graph may contain several resources). The benefit of this method is that the changes in the ontology do no longer require a recompile of the system, while at the same time allowing slightly more compatibility with existing SOAP tools. The problem with this method remains that both client and server programmers need in-depth knowledge of RDF and the ontologies used.

Regardless of which of the three alternatives one chooses, it remains an important issue to make sure that the formal description of the services (i.e., the WSDL+XML Schema definitions) remain semantically correct and retain the pointers to the ontology concepts which the parameters represent. As mentioned, for the first approach this can be handled by well chosen URIs for the XMLSchema types. For the second and third approach the parameters have the form of simple string objects in both the WSDL definition and the SOAP messages, since the RDF serialization is represented as a string. However, both versions of WSDL available at the time of writing allow extensions to the WSDL format itself[n], and additional constraints on the type or form of the RDF instances contained inside the string parameters may be specified here. This is the approach taken by the *Semantic Annotation for WSDL and XML Schema* (SAWSDL) working group[o] and the NEPOMUK project makes use of their standard.

In this section we have considered a very lightweight approach to semantically enriching SOAP Web Services by passing RDF-based parameters. If a more powerful approach is required, the reader is advised to look into OWL-S[p] and the Web Service Modeling Language (WSML)[q], both defining much more sophisticated frameworks for Semantic Web Services.

RELATED WORK

In the following we review relevant research and development approaches for the Social Semantic Desktop. After providing a brief description, we discuss the lessons learned and state our conclusions.

Gnowsis (Sauermann, 2003) was among the first research projects targeting a Semantic Desktop system. Its goal is to complement established desktop applications and the desktop operating system with Semantic Web features, rather than replacing them. The primary focus of Gnowsis is on *Personal Information Management* (PIM). It also addresses the issues of identification and representation of desktop resources in a unified RDF graph. Gnowsis uses a Service Oriented Architecture (SOA), where each component defines a certain interface and it is available as an XML/RPC service.

The **Haystack** (Quan et. al, 2003) project presents a good example for an integrated approach to the SSD field. Inter-application barriers are avoided by simply replacing these applications with Haystack's own word processor, email client, image manipulation, instant messaging, etc. Haystack allows users to define their own arrangements and connections between views of information, thus making it easier to find information located in the personal space. The Haystack architecture can be split into two distinct parts: the Haystack Data Model (HDM) and the Haystack Service Model (HSM). The Data Model is the means by which the user's information space is represented, similar to what has been discussed in Sect. "*Ontologies*". The set of functionalities within Haystack is implemented by objects in the Haystack Service Model (HSM). Haystack has a standard three-tiered architecture, consisting of a user interface layer (the client), a server/service layer, and a database. Haystack was groundbreaking in terms of the dynamic creation of user interfaces, but the project ended before establishing any standards.

Another relevant personal information management tool is the **Semex System** (SEMantic EXplorer)(Dong et. al, 2005). Like other tools, it organizes data according to a domain ontology that offers a set of classes, objects and relationships. Semex leverages the Personal Information Management (PIM) environment to support on-the-fly integration of personal and public data. Information sources are related to the ontology through a set of mappings. Domain models can be shared with other users in order to increase the coverage of their information space. When users are faced with an information integration task, Semex aids them by trying to leverage data collected from previous tasks performed by the user or by others. Hence, the effort expended by one user later benefits others. Semex begins by extracting data from multiple sources and for these extractions it creates instances of classes in the domain model. It employs multiple modules for extracting associations, as well as allowing associations to be given by external sources or to be defined as views over other sets of associations. To combine all these associations seamlessly, Semex automatically reconciles multiple references to the same real-world object. The user browses and queries all this information through the domain model.

A similar idea is exploited by the **IRIS** Semantic Desktop(Cheyer et. al, 2005) ("Integrate. Relate. Infer. Share"), an application framework that enables users to create a "personal map" across their office-related information objects. IRIS offers integration services at three levels:

- Information resources (e.g., email messages, calendar appointments) and applications that create and manipulate them must be accessible to IRIS for instrumentation, automation, and query. IRIS offers a plug-in framework, in the style of the Eclipse architecture, where "applications" and "services" can be defined and integrated within IRIS. Apart from a very small, lightweight kernel, all functionality within IRIS is defined using a plug-in framework, including user interface, applications, back end persistence store, learning modules, harvesters, etc. Like Haystack, inter-application barriers do not exists, because all applications are made from scratch for IRIS.

- A Knowledge Base provides the unified data model, persistence store, and query mechanisms across the information resources and semantic relations among them. The IRIS user interface framework allows plug-in applications to embed their own interfaces within IRIS and to interoperate with global UI services, such as notification pane, menu toolbar management, query interfaces, the link manager, and suggestion pane.

DeepaMehta(Richter et. al, 2005) is an open source Semantic Desktop application based on the Topic Maps standard[r]. The DeepaMehta UI, which runs through a Web browser, renders Topic Maps as a graph, similar to concept maps. Information of any kind as well as relations between information items can be displayed and edited in the same space. The user is no longer confronted with files and programs. DeepaMehta has a layered, service oriented architecture. The main layer is the application layer, which offers various ways for the presentation layer to communicate with it via the communication layer (API, XML Topic Maps (XTM) export, EJB, SOAP). Finally, the storage layer holds all topics and their data either in a relational database or simply in the file system.

Other relevant projects include **Beagle++**(Brunkhorst et. al, 2005), a semantic search engine which provides the means for creating and retrieving relational metadata between information elements present on the destkop, **DBIN**(Tummarello et. al, 2006), which is similar to a file sharing client and connects directly to other peers, **PHLAT**(Cutrell et. al, 2006), a new interface for Windows, enabling users to easily specify queries and filters, attempting to integrate search and browse in one intuitive interface, or **MindRaider**[s], a Semantic Web outliner, trying to connect the tradition of outline editors with emerging SW technologies. The **MyLifeBits** project by Microsoft Research is a lifetime store of multimedia data. Though the system does not intent to be a SSD, one can learn from it how to integrate data, i.e., how to manage the huge amount of media and how to classify/retrieve the data(Gemmell et. al, 2002). It combines different approaches from HCI (Computer-Human Interaction) and information integration, while it lacks a conceptual layer beyond files. The **Apogée**[t] project deals with data integration in applications related to Enterprise Development Process (ECM). It aims at building a framework to create Enterprise Development Process-oriented desktop applications, independent from vendor or technologies. Finally, starting from the idea that everything has to do with everything, has a relationship with everything, **Fenfire**[u] is a Free Software project developing a computing environment in which you can express such relationships and benefit from them.

Although the systems we have looked at focus on isolated and complementary aspects, they clearly influenced the vision of the SSD presented here. Some of the architectural decisions made in the NEPOMUK project and presented in this chapter are similar to those of platforms like Haystack, IRIS, and DeepaMetha, e.g., in that we present a User Interface Layer, a Service and a Data Storage Layer. The modular architecture, also identified within the Haystack, SEMEX, and DeepaMetha systems, as well as the standardized APIs offer an easy way of introducing new components. Our approach guarantees that each component may be changed without affecting other components it interacts with. The interaction has to suffer only in the case in which the API of the component is modified. The NEPOMUK Architecture also provides service discovery functionalities: the

NEPOMUK Registry providing a proper support for publishing and discovering the existing NEPOMUK Services by using a standard interface.

CONCLUSION

We presented the Social Semantic Desktop as a comprehensive approach to information handling. Oriented at the needs of knowledge workers, this approach centers around supporting the main information-oriented activities: The articulation of knowledge and the generation of new information items; the structuring, relating, and organization of information, and the sharing of formal and informal information within networks of co-operating people. From this, we derived key functionalities of the desktop, but also for search, social interaction, profile building, and data analysis.

Building the SSD relies on basic principles: Whatever appears within the personal workspace is treated as an information item. Content, relations, special services all refer to formal annotations of such information items, which in turn link between information items and personal information models. Unifying the flexibility and personal liberty of expressing whatever concepts seem relevant with the commitment to socially shared conceptualizations results in a layered hierarchy of ontologies which allow the necessary differences in stability, sharing scope, and formality. Integrating the tools of everyday information processing asks for an easy and flexible integration of existing desktop applications. Finally, the adoption of Semantic Web standard technology for representation and communication enables the easy transgression from personal annotated information to shared Semantic Web content.

Consequently, the architecture of the SSD combines standards-based data repositories with a rich middleware, which in particular allows for manifold service integrations and communications. On top of that, various presentation clients and specific applications support whatever activities are performed on the desktop. Such applications may be highly domain-specific, although core functionalities of knowledge work trigger standard applications, e.g., for document processing, task management, communication, etc.

The design decisions presented result in particular implementation and engineering principles; we outlined the adaptation to RDF, the service integration, and the message passing mechanisms in particular.

In summary, the SSD offers the basic technology and tools for everyday information processing by knowledge workers. In order to reach the intended wide acceptance and broad uptake, care was taken to make the central software components available under open source licenses, and to encourage the development and contribution of application-specific enhancements and adaptations. The concept of the SSD is promising and relies on a number of techniques which reach their maturity right now – consequently, a number of research and development projects are under way and contribute to the overall evolution of the concept.

Following the realizations described in this chapter, we see the SSD as a basis for the self-motivated generation of semantically annotated information, which will not only help the individual by allowing multitudes of specific services and support, but will also initiate a wide movement to populate the Semantic Web.

FUTURE RESEARCH DIRECTIONS

Although the ideas of the Social Semantic Desktop are based on solid foundations as presented here, the research areas surrounding this topic are still in their infancies. We will briefly discuss some of the pre-dominant challenges in the coming years:

Trust and Privacy. As pointed out in the Semantic Web technology stack presented earlier, a crucial component for any high-level Semantic Web service is the issue of trust and privacy.

Trust touches on a wide range of issues, from the technical issues of cryptographic signatures and encryption, to the social issues of trust in groups and among individuals. These issues are all as valid for the Social Semantic Desktop as for the Semantic Web in general, or perhaps even more so, as people are less critical of putting personal data on their personal desktop.

User, group, and rights management. When a single personal Semantic Desktop allows easy sharing of information with the network of Social Semantic Desktops, determining access rights for this information becomes very important. The Social Semantic Desktop sets new requirements for distributed authentication, flexible group management, and fine-grained access rights, all the while remaining intuitive and unobtrusive for the end user.

Integration with the wider Semantic Web and Web 2.0. Currently we are talking about the Social Semantic Desktop as a network of Semantic Desktops built on the same standards. It is important to remember that the key benefit of Semantic technology is the easy access to integration with anyone using the same representational languages and ontologies. The growth of feature-rich Web applications is growing rapidly, and ensuring a strong bond between the Semantic Desktop and these services is a continuous challenge.

Ontologies and Intelligent Services. To date ontologies have been used to introduce a common vocabulary for knowledge exchange. On the Social Semantic Desktop ontologies are used to formalize and categorize personal information. This introduces many interesting issues around ontology versioning, ontology mapping, and ontology evolution. Furthermore, using ontologies with well-defined semantics will allow intelligent services to be built (e.g., using reasoning) that allow for much more than just browsing and (simple) searching.

User Evaluation. The underlying thesis of the whole (Social) Semantic Desktop effort is that the added semantics will improve productivity and enable new forms of cooperation and interaction which were not previously possible. In-depth empirical evaluation with real users of a Social Semantic Desktop systems are required to determine if this thesis really holds.

REFERENCES

Aberer, K., Cudré-Mauroux, P., Datta, A., Despotovic, Z., Hauswirth, M., Punceva, M., & Schmidt, R. (2003). P-Grid: A self-organizing structured P2P system. *SIGMOD Record, 32*(3), 29–33. doi:10.1145/945721.945729

Aberer, K., Cudré-Mauroux, P., Hauswirth, M., & Pelt, T. V. (2004). Gridvine: Building Internet-scale semantic overlay networks. In S. A. McIlraith, D. Plexousakis, F. van Harmelen (Eds.), *The Semantic Web – ISWC 2004: Third International Semantic Web Conference*, 107-121. Springer Verlag.

Aperture: A Java framework for getting data and metadata, Last visited March 2007. http://aperture.sourceforge.net/.

Bergamaschi, S., Castano, S., Vincini, M., & Beneventano, D. (2001). Semantic integration and query of heterogeneous information sources. *Data & Knowledge Engineering, 36*(3), 215–249. doi:10.1016/S0169-023X(00)00047-1

Berners-Lee, T., & Fischetti, M. (1999). *Weaving the Web – The original design and ultimate destiny of the World Wide Web by its inventor.* Harper San Francisco.

Berners-Lee, T., Hendler, J., & Lassila, O. (2001). The Semantic Web. *Scientific American.*

Brunkhorst, I., Chirita, P. A., Costache, S., Gaugaz, J., Ioannou, E., Iofciu, T., Minack, E., Nejdl, W., & Paiu. R. (2006). *The Beagle++ Toolbox: Towards an extendable desktop search architecture (Technical report)*, L3S Research Centre, Hannover, Germany.

Bush, V. (1945, July). *As we may think*. The Atlantic Monthly.

Carroll, J. J., Bizer, C., Hayes, P., & Sticker, P. (2005). Named graphs, provenance and trust, In A Ellis, T. Hagino (Eds.), *WWW 2005: The World Wide Web Conference*, 613-622.

Cheyer, A., Park, J., & Giuli, R. (2005, November 6). IRIS: Integrate. Relate. Infer. Share. In S. Decker, J. Park, D. Quan, L. Sauermann (Eds.), *Semantic Desktop Workshop at the International Semantic Web Conference*, Galway, Ireland, *175*.

Clark, K. L. (1978). Negation as failure. In J. Minker (Ed.), *Logic and Data Bases*, Plenum Press, New York, 293-322.

Cutrell, E., Robbins, D. C., Dumais, S. T., & Sarin, R. (2006, April 22-27). Fast, flexible filtering with PHLAT – Personal search and organization made easy. R. E. Grinter, T. Rodden, P. M. Aoki, E Cutrell, R. Jeffries, G. M. Olson (Eds.), *Proceedings of the 2006 Conference on Human Factors in Computing Systems,* CHI 2006, Montréal, Québec, Canada. ACM 2006, ISBN 1-59593-372-7.

Decker, S., & Frank, M. (2004, May 18). The networked semantic desktop. In C. Bussler, S. Decker, D., Schwabe, O. Pastor (Eds.), *Proceedings of the WWW2004 Workshop on Application Design, Development and Implementation Issues in the Semantic Web*, New York, USA.

Dong, X., & Halevy, A. Y. (2005). A platform for personal information management and integration. In M. Stonebraker, G. Weikum, D. DeWitt (Eds.), *Proceedings of 2005 Conference on Innovative Data Systems Research Conference*, 119-130

Engelbart, D. C. (1962). *Augmenting human intellect: A conceptual framework (Summary report)*, Stanford Research Institute (SRI).

Gemmell, J., Bell, G., Lueder, R., Drucker, S., & Wong, C. (2002, December 1-6). MyLifeBits: Fulfilling the memex vision. In *ACM Multimedia*, Juan-les-Pins, France, 235–238.

Gifford, D. K., Jouvelot, P., Sheldon, M. A., & O'Toole, J. W., Jr. (1991, October). Semantic file systems. In 13th *ACM Symposium on Operating Systems Principles*.

Gruber, T. R. (1995). Toward principles for the design of ontologies used for knowledge sharing. *International Journal of Human-Computer Studies*, *43*, 907–928. doi:10.1006/ijhc.1995.1081

Hendler, J. (2001, March/April). Agents and the SemanticWeb. *IEEE Intelligent Systems*, *16*(2), 30–37. doi:10.1109/5254.920597

McGuinness, D. L., & van Harmelen, F. (2004, February). *OWL Web Ontology Language Overview (Technical report)*. http://www.w3.org/TR/2004/REC-owl-features-20040210/.

Möller, K., Bojārs, U., & Breslin, J. G. (2006, June 11-14). Using semantics to enhance the blogging experience. In Y. Sure, J. Domingue (Eds.), *The Semantic Web: Research and Applications, 3rd European Semantic Web Conference*, ESWC 2006 Proceedings, Budva, Montenegro, 679-696.

Nelson, T. H. (1965). A file structure for the complex, the changing, and the indeterminate. In *ACM 20th National Conference Proceedings*, 84-100, Cleveland, Ohio.

Oren, E. (2006). An overview of information management and knowledge work studies: Lessons for the semantic sesktop. In S. Decker, J. Park, L. Sauermann, S. Auer, S. Handschuh (Eds.), *Proceedings of the Semantic Desktop and Social Semantic Collaboration Workshop (SemDesk 2006) at ISWC 2006*. Athens, GA, USA.

Quan, D., Huynh, D., & Karger, D. R. (2003). Haystack: A platform for authoring end user Semantic Web applications. In D. Fensel, K.P. Sycara, J. Mylopoulos (Eds.), *The Semantic Web – ISWC 2003: International Semantic Web Conference*, Proceedings, 738-753.

Richter, J., Völkel, M., & Haller, H. (2005). DeepaMehta – A Semantic desktop. In *Proceedings of the 1st Workshop on the Semantic Desktop - Next Generation Personal Information Management and Collaboration Infrastructure* at ISWC 2005, Galway, Ireland.

Sauermann, L. (2003). The Gnowsis – *Using Semantic Web Technologies to Build a Semantic Desktop*. Diploma Thesis, Technical University of Vienna, 2003.

Sauermann, L., Grimnes, G. A. A., Kiesel, M., Fluit, C., Maus, H., Heim, D., et al. (2006). Semantic desktop 2.0: The gnowsis experience. In I. Cruz, S. Decker, D. Allemang, C. Preist, D. Schwabe, P. Mika, M. Uschold, L. Aroyo (Eds.), *The Semantic Web – ISWC 2006: 5th International Semantic Web Conference*, Athens, GA, Proceedings.

Semy, S.K., Pulvermacher, M.K., & Obrst, L.J. (2004). *Toward the use of an upper ontology for U.S. Government and U.S. Military domains: An evaluation. (Technical report)*. MITRE, September 2004.

Sintek, M., van Elst, L., Scerri, S., & Handschuh, S. (2007). Distributed knowledge representation on the social semantic desktop: Named graphs, views and roles in NRL. In E. Franconi, M. Kifer, W. May (Eds.), *The Semantic Web – ESWC 2007: The 4th European Semantic Web Conference (ESWC 2007) Proceedings*.

Tummarello, T., Morbidoni, C., & Nucci, M. (2006). Enabling Semantic Web communities with DBin: An overview. In I. Cruz, S. Decker, D. Allemang, C. Preist, D. Schwabe, P. Mika, M. Uschold, L. Aroyo (Eds.), *The Semantic Web – ISWC 2006: 5th International Semantic Web Conference*, Athens, GA, Proceedings, 943-950.

ADDITIONAL READINGS

Current and recent research and development in the SSD domain has already been presented in Sect. "*Related Work*". However, one influence that has not been covered in this chapter so far, but is closely related to the idea of a Semantic Desktop is the concept of *Semantic File Systems* file systems in which files are not organized hierarchically, but rather according to their metadata. The concept and an early implementation are described in detail in (Gifford et. al, 2001).

Finally, as another entry point for additional reading, we would like to point the reader to the series of *Semantic Desktop Workshops* which were co-located with the International Semantic Web Conferences in 2005[v] and 2006[w].

From a historical perspective, the most important references in the Social Semantic Desktop domain are those by Vannevar Bush (1945) and Doug Engelbart (1962) which we mentioned in Sect. "*Background*". Another important early influence is certainly Ted Nelson's work on hypertext (Nelson, 1965). A modern vision of those ideas is a paper by Decker and Frank (2004), which also coined the term "Semantic Desktop". Of course, any work that is based on the ideas of the Semantic Web is not complete without references to seminal papers such as (Berners-Lee et. al, 2001) or (Hendler, 2001). In fact, the original vision of the World Wide Web itself already contained the idea of an information space that would reach from "mind to mind" (Berners-Lee, 1999); a thought that is central to the SSD.

Most of the references given in this chapter are of a technical nature. However, one has to keep in mind that the SSD is a tool for *information management* and *knowledge work*, and thus psychological and sociological research into the nature of knowledge work in any form are relevant as well. Oren (2006) provides a detailed overview of literature in this field, with the intention of applying the lessons learned to the development of the Semantic Desktop.

ENDNOTES

[a] RDF: http://www.w3.org/RDF/

[b] The NEPOMUK Project is supported by the European Union IST fund, grant FP6-027705

[c] Within the NEPOMUK Project, these personas were created by distilling typical users from a series of interviews and evaluations with our use-case partners.

[d] Tim Berners-Lee talk, XML and the Web: http://www.w3.org/2000/Talks/0906-xml-web-tbl/

[e] RFC1738: http://tools.ietf.org/html/rfc1738

[f] WSDL: http://www.w3.org/TR/wsdl

[g] XML Schema: http://www.w3.org/XML/Schema

[h] SOAP: http://www.w3.org/TR/soap

[i] In this chapter we make the assumption that a modern object-oriented programming language like Java will be used for implementation, but observations and solutions are equally valid for most other languages.

[j] MDA: http://www.omg.org/mda/

[k] RDFReactor: http://wiki.ontoworld.org/wiki/RDFReactor; RDF2Java: http://rdf-2java.opendfki.de; Elmo: http://openrdf.org, etc.

[l] TriG/TriX: http://www.w3.org/2004/03/trix/

[m] SmartWeb: http://www.smartweb-project.de/

[n] Language Extensibility in WSDL1: http://www.w3.org/TR/wsdl#_language and in WSDL2: http://www.w3.org/TR/wsdl20#language-extensibility

[o] SAWSDL: http://www.w3.org/TR/sawsdl/

[p] OWL-S: http://www.daml.org/services/owl-s/

[q] WSML: http://www.wsmo.org/wsml/

[r] ISO/EIC 13250:2003: http://www.y12.doe.gov/sgml/sc34/document/0129.pdf

[s] MindRaider: http://mindraider.sourceforge.org/

[t] Apogée: http://apogee.nuxeo.org/

[u] Fenfire: http://www.fenfire.org/

[v] SemDesk2005: http://tinyurl.com/yuxpld

[w] SemDesk2006: http://tinyurl.com/2hqfak

APPENDIX: QUESTIONS FOR DISCUSSION

Q: I prefer to handle my photo collection in a web 2.0 photo sharing environment. Is this compatible with the Social Semantic Desktop? May I keep the work I have invested here?

A: Yes. Every photo in your collection can be reached via a specific URI, thus it can be handled as a particular information item in the SSD. You might implement a suitable wrapper to transfer local annotations from your SSD onto the photo sharing platform, if you intend to disclose this information.

Q: The Social Semantic Desktop presupposes that everything is an information item. What about entities which are not information but real-world objects? Can I manage them in the SSD and add comments about them, e.g., about my friend's cat?

A: The solution is easy: Just create an appropriate description of the real world object within your SSD, thus creating an URI for the object in question. Let's say you create an instance of the class pet in your SSD (assuming you have this category within your SSD) and describe it as 'well-known house cat'. Then you can link this instance to, e.g., a photo of the animal, or you add an 'owns' link which connects it to the URI of your friend, and so on. Making an arbitrary object re-appear as a formal instance within the SSD models is often called 're-birthing', btw.

Q: Think about scenarios you encounter every day, and where the SSD can make your work easier.

A: The answer is of course a personal one, but for a typical knowledge worker (researchers, students, journalists, etc.) here are some example ideas:

• Show me related appointments when composing emails to a person, i.e., You also have lunch with Claudia next week.

• Show me previously viewed PDF documents on the same topic when researching on Wikipedia.

• Remember my meal and window preferences when booking flights.

• Remind me of my previous idea of combining topic A with topic B when reviewing my topic A notes.

• Let me connect an incoming email from a student to the colleague who introduced me to that student.

Q: What are the benefits of the Social Semantic Desktop compared to solution such as Microsoft Exchange server or the tight integration of applications on MacOSX? They also fulfil many of the functionalities required by the scenarios outline in this chapter.

A: The Social Semantic Desktop is different because of the standards used to build it. Firstly, by basing the representational layers of the Semantic Desktop on the existing (Semantic) Web standards we enable interoperability by a wide range of existing projects, and secondly, by creating new standards for desktop integration and data-formats we encourage future software developers to build on top of the Semantic Desktop. On the Semantic Desktop both the applications and the data encourages open access, and this exactly the opposite of the vendor lock-in that for instance Exchange Server aims for.

Q: Inspect the current state of the Semantic Web and the data available. What data-sources and/or ontologies do you think could be useful for integration with the Semantic Desktop?

A: The answer will of course change as the Semantic Web evolves, but at the time of writing relevant ontologies include:

- The Friend-of-a-Friend project – http://xmlns.com/foaf/spec
- The Description-of-a-Project schema – http://usefulinc.com/doap
- The Semantically Interlinked Online Communities project – http://siocproject.org/
- Dublin Core for basic meta-data – http://dublincore.org/

Useful data-sources and/or web-services include:

- GeoNames for (reverse) geocoding – http://www.geonames.org/
- DBpedia for a Semantic Web view of Wikipedia – http://DBpedia.org/

This work was previously published in Semantic Web Engineering in the Knowledge Society, edited by J. Cardoso; M. Lytras, pp. 290-314, copyright 2009 by Information Science Reference (an imprint of IGI Global).

Chapter 8.18
Social Media Marketing
Web X.0 of Opportunities

Lemi Baruh
Kadir Has University, Turkey

ABSTRACT

In recent years social media applications, which enable consumers to contribute to the world of online content, have grown in popularity. However, this growth is yet to be transformed into a sustainable commercial model. Starting with a brief overview of existing online advertising models, this chapter discusses the opportunities available for advertisers trying to reach consumers through social media. The chapter focuses on viral marketing as a viable option for marketers, reviews recent viral marketing campaigns, and offers recommendations for a successful implementation of social media marketing. In conclusion, the author examines future trends regarding the utilization of the emerging Semantic Web in marketing online.

INTRODUCTION

The brief history of the World Wide Web is filled with stories of unprecedented commercial success as well as shattered dreams of hopeful online entrepreneurs. It should not be surprising that, just as their predecessors, Web 2.0 and social media also bring about important questions regarding their sustainability. On the one hand, since 2006, social media sites have been growing in number and popularity (Boyd & Ellison, 2007). For example, according to comScore, a leading Internet information provider, as of December 2007 Facebook had close to 98 million unique visitors, and Fox Interactive Media, including MySpace, had more than 150 million. Similarly, recent years have seen a phenomenal growth in the popularity of weblogs (blogs): in 2007 every day, 175,000 new blogs were added to an estimated 67 million blogs that were already up and running (as cited in Rappaport, 2007). On the other hand, skeptics voice their belief that social media, despite their current popularity, may not have the staying power ("MySpace, Facebook and Other Social Networking Sites," 2006).

An important component of skeptics' concerns about the sustainability of social media pertains to the fact that there are no agreed upon ways of monetizing the rising popularity of social media (Allison, 2007; Hall, 2007). Perhaps, the most

DOI: 10.4018/978-1-60566-368-5.ch004

telling example of this problem is Facebook. Despite having a market value of around $15 billion, Facebook's 2007 revenue was $150 million (McCarthy, 2008) – a considerably small share of the $21 billion online advertising industry. Then, the question of whether social media will be more than just a fad boils down to advertisers' ability to utilize the unique opportunities presented by social media. Although advertisers and social media entrepreneurs are yet to agree on a marketing model for social media, recent discussions point to several important requirements that a successful model should accommodate. Given the decentralized architecture of the Internet in general and social media in particular, a central tenet of these recent debates concerns the relative merits of more conventional advertising methods and word of mouth (or word of "mouse") based marketing approaches that cede control to the consumers.

In the light of these debates, this chapter will start by summarizing online advertising methods. After this brief summary, the chapter will focus on the opportunities and challenges for online marketers that are brought about by the development of social media. Finally, the chapter will discuss viral marketing and integrated marketing communication principles to provide a roadmap for realizing the financial and marketing potential of Web 2.0.

BACKGROUND

Online Advertising

In its most traditional sense, *advertising* is defined as a paid form of communication appearing in media, usually with the purpose of reaching a large number of potential customers. Since 1993, when CERN announced that the World Wide Web would be available to anyone free of charge, advertisers experimented with different methods of reaching consumers online. Unsurprisingly, the first reaction of advertisers was to treat the World Wide Web as a natural extension of traditional media, such as newspapers and television. And, just as in conventional mass media, early online advertising methods, such as banners, pop-ups and interstitials, were characterized by intrusiveness and adoption of a one-way stimulus-response model within which information flows from the advertiser to the customer (McCoy, Everard, Polak, & Galletta, 2007; Rappaport, 2007).

However, even in the early years of online advertising, signs of what was to come in interactive marketing were revealed. Shortly after banners became a popular online advertising method in 1994, keyword-activated "smart banners" were introduced. What set smart banners apart from their predecessors was that the contents of the banners were personalized in response to the search words entered by the users. As such, smart banners were one of the first examples of how content variability in new media (Manovich, 2001) can be utilized to customize information to consumers' needs (Faber, Lee & Nana 2007).

Customization and Message Congruence in Interactive Media

As noted by several researchers, content variability and the consequent ability to customize content according to the needs of the consumer are made possible by the interactive capabilities of new media (Baruh, 2007; Faber et al., 2007). Two important characteristics of interactive media are the ability to facilitate a two-way flow of communication and the related ability to track and store every bit of information about how consumers use a system (McCallister & Turow, 2002). Real-time information about how consumers use a medium, especially when combined with other data through data mining, enables marketers to extract profiles about individuals that can then be used to tailor messages and products.

The ultimate aim of this process is to target different consumer groups with specific messages that can tie a product to their informational needs, lifestyles or predispositions. Extant literature on online targeting suggests that consumers will be more receptive to messages that are tailored as such (McCoy et al. 2007; Robinson, Wyscocka, & Hand, 2007). To a large extent, this higher receptivity is the result of being able to promote the "right" product, at the "right" time and place and the "right" tone. A case in point that supports these research findings is the success of Google's AdWords, which accounts for 40% of online advertising spending. The premise of AdWords is that the marketers can reach motivated consumers by providing them with contextual advertising messages congruent with their online keyword searches (and presumably, their interests). Similarly, a widely known feature of online vendors such as Amazon.com is their customized product recommendation systems. The recommendation system these online vendors utilize is based on a data mining system called *market-basket analysis* (also called *association discovery*). The premise of this system is that the marketer can create cross-selling opportunities by identifying the product types that a customer would be interested in (e.g., microwave popcorn) on the basis of other products that he or she has already purchased or is purchasing (e.g., a DVD movie). As such, what the market-basket analysis algorithm does is to identify product clusters that are purchased together or sequentially using the product purchasing history of customers whose tastes are similar to a specific customer.

ONE STEP FURTHER: WEB 2.0 OF OPPORTUNITIES

Customization and Data from Social Media

As can be inferred from the discussion above, collecting information about consumers is an important prerequisite of customizing advertising messages in accordance with the informational needs and lifestyles of consumers. Certainly, data about individuals' online media consumption and purchasing behavior, especially when combined with other sources of data such as credit history, provide marketers with an unprecedented capability to not only determine which customers to target (and avoid) but also when and how to target them.

Within this context, social network sites, such as Facebook, MySpace or LinkedIn, have a potential to extend what is already a large pool of data about consumers. Such social network sites are designed to allow users to create personal profiles and connect with other users, friends or strangers. And through the creation and perennial updating of their profiles, users of social network sites actively participate in the dissemination of information about themselves (Andrejevic, 2007; Solove, 2007). The types of information users of social network sites disclose include: information about their hobbies, interests, likes and dislikes, whom they associate with, a dinner they had a couple of days ago and, sometimes, disturbingly private details about their social and sexual lives. Blogs, another highly popular form of social media, are no different from social network websites. As Solove (2007) points out, any topic, any issue and any personal experience are fair game for more than 60 million bloggers around the world.

The massive quantities of data that social media users reveal online are not left untapped by media companies and marketers. For example, MySpace has recently begun an effort to mine data from its more than 100 million users in order to better target advertising messages. Named as MySpace HyperTargetting, the system initially began mining data about general interest categories, such as sports and gaming, and is now further dividing interests into thousands of subcategories (Morrisey, 2007).

The Community Touch

An important point to note with respect to the types of data available in social media is that the digital goldmine of information is not simply a more detailed version of data collected about consumers' interests and behaviors in other forms of interactive media. Rather, in social media, the available data contain unprecedented details about the network affinities of users. The data about the network affinities of users can be utilized at two levels. First, through the "tell me about your friends and I'll tell you about yourself" principle, marketers can make further refinements to consumers' profiles based on the interests shared by members of the communities they belong to. Secondly, information about the communities that an individual belongs to can be used to identify the paths through which they can be reached.

Recent marketing techniques devised by online vendors and social media outlets illustrate how information about social affinities can be used to reach consumers. For example, Amazon.com's iLike application, a music service that markets new music and concerts to interested listeners, works by scanning the music libraries of its subscribers. The service connects like-minded listeners and promotes new music to users through add-ons such as Facebook's iLike widget. Similarly, Facebook's own Beacon platform tracks purchases Facebook users make on partnering online vendors and then informs users' networks about the recent purchase (Klaassen & Creamer, 2007; Thompson, 2007; Tsai, 2008). In addition to leveraging existing social networks to disseminate marketing messages, some software applications, for example, Stealth Friend Finder automatically generate massive and targeted Facebook Friend Requests to directly connect with the consumers.

Web 2.0 of Opportunities: Viral Marketing in Social Media

These examples of social targeting pinpoint the direction that marketing in social media can take. Rather than being an advertising distribution system, Beacon is a viral marketing tool that lets community members know what their co-members have purchased. In other words, with the Beacon system, the consumer, through the publication of his/her purchasing decision, assumes the role of an influencer. Subramani and Rajagopalan (2003) suggest that consumers may assume such a role either passively or actively. In the passive form, the consumer spreads the word simply by using or purchasing a product (as is the case when an e-mail from a Blackberry user contains a message saying the e-mail was sent using a Blackberry account). On the other hand, active viral marketing requires that consumers participate in the message dissemination process by contacting other potential customers (Clow & Baack, 2007).

An important criticism of passive viral marketing systems in social media is that they fail to utilize an important characteristic of Web 2.0 in general and social media in particular. Instead of being a passive consumer of readily available content, Web 2.0 users are participants in both the creation and dissemination of content. Accordingly, despite utilizing social graphs to target messages more effectively, the "your friend just bought this book from Amazon.com" message is nevertheless an advertising method that affords the consumer very little power as a potential source of influence (Anderson, 2006; Windley, 2007).

Considered from this perspective, a more appropriate way of utilizing the viral potential of social media users is to invite them to actively participate in promoting the product. First, existing research shows that close to a quarter of users of online social networks, such as Facebook, use these sites to influence other users (Webb, 2007). Second, as evidenced by Facebook users' negative reaction to Beacon, social network sites are relatively intimate environments and advertising intrusion (especially given an overall mistrust for advertising messages) is not welcome (Clemons, Barnett, & Appadurai, 2007; Gillin, 2007; Hall, 2007). In contrast, 94% of online social network users find product recommendations from friends to be at least very worthwhile to listen to (MacKeltworth, 2007). This finding is not surprising since recommendations coming from friends, family members, or colleagues are more likely to be trustworthy and relevant to one's needs (Clemons et al., 2007). In fact, according to a recent survey, along with the reputation of the manufacturer, recommendations from friends and family members are the biggest factor that influences purchasing decisions made by individuals (Klaassen & Creamer, 2007). Third, thanks to synchronous connections between multiple users, a computer-mediated word of mouth can reach a larger number of people than word of mouth in the brick and mortar world.

As briefly mentioned before, in addition to these three important advantages of inviting social media users to actively disseminate marketing messages, product information, or recommendations, social media also provide marketers with an unprecedented capability to identify the individuals who would be the best candidates in a social network to act as viral marketers. Domingos (2005) suggests that in addition to actually liking a product, a suitable viral marketing candidate should have high connectivity and should be powerful as a source of influence. Using social network analyses (Hanneman & Riddle, 2005; Scott, 2000; Wasserman & Faust, 1995), data regarding personal affiliations and social network memberships can be utilized to identify opinion leaders ("hubs") who are central to and powerful in a given network.

Recently, there have been several attempts to apply social network analysis to social media to identify social network influencers. For example, Spertus, Sahami, and Büyükkökten (2005) used network data from Orkut.com to identify members who could be used to recommend new communities to users. Similarly, in a study of Flickr and Yahoo360 networks, Kumar, Novak and Tomkins (2006) were able to distinguish between passive users and active inviters that contributed to the extension of the network. And recently, MySpace announced that it is constructing an "influencer" option for advertisers who could be interested in reaching users with active and large networks. To identify potential influencers, MySpace will use data regarding users' group memberships and interests, their friends' interests, level of network activity in a given network and other factors (Morrissey, 2007).

The Integrated Marketing Communications Perspective

In 1976, Wayne DeLozier suggested that marketing communication was a process of creating an integrated group of stimuli with the purpose of evoking a set of desired responses. According to this integrated marketing communications perspective, which has been adopted by many companies since the 1980's, rather than being considered in isolation from one another, each component of the marketing mix should be coordinated to present a unified image to consumers.

Considered from this perspective, fulfilling the viral marketing promise of Web 2.0 and social media requires that the viral marketing effort be part of a greater scheme of corporate communications. In other words, rather than merely focusing on spreading the word, the viral marketing effort should fit the brand personality (Webb, 2007). A particular case illustrating this point is the "Top This TV Challenge" campaign of Heinz®. In this campaign, Heinz® invited consumers to produce 30-second TV commercials for Heinz® Ketchup and submit the commercials on YouTube. The winner of the contest, determined first by a panel of judges and then by the votes of consumers, was awarded $57,000 and a chance to get the commercial aired on national television. The premise of the campaign was not only that it fit the "fun" brand image of Heinz® Ketchup but also that the consumers would play a crucial role in disseminating Heinz Ketchup's name. Just as intended, many of the 4,000 qualified contestants who posted their videos on YouTube (as required) also created MySpace and Facebook pages promoting their own videos and consequently Heinz Ketchup.

Another example illustrating the connection between viral marketing and an integrated marketing communications approach that provides a fit between the marketing campaign and the organizational image is the "Download Day" organized by Mozilla Firefox in June 2008. Mozilla is a not for profit organization that is mostly known for its Firefox Web Browser (a challenger of the market leader, Internet Explorer). The organization is a self-proclaimed open source project that publicly shares the source codes of their own software for the development of new Internet applications. Unlike its major competitors, such as Internet Explorer and Safari, the Firefox Web Browser is positioned as an "organic browser" that has been developed through a collaborative process whereby thousands of software developers – the majority of which are not employed by Mozilla – contribute to the software. Likewise, the dissemination of Firefox largely relies on volunteers "spreading" the software.

In June 2008, Mozilla created a Download Day event to promote the third version of its Firefox Web Browser. The purpose of the Download Day was to set a world record in the number of downloads in 24 hours. To inform would-be users about the event, Mozilla heavily utilized social media and viral marketing. Following the initial announcement, the word of mouth about the Download Day first quickly spread through social news aggregators such as Digg™ and Reddit.com. Then, the links in the social news aggregators forwarded interested users to the Download Day homepage. In addition to asking individuals to pledge to download Firefox on Download Day and providing an update on the number of individuals who pledged to download, the homepage also invited them to engage in viral marketing by inviting their social networks to the event via Facebook, Bebo and MySpace, promoting the event on microblogging Twitter-like websites or organizing "Download Fests" on university campuses.

These two examples provide important insights regarding the criteria for a successful viral marketing campaign online (and in social media):

1. **Campaign-Organizational Image Congruence:** In the Download Day example, the event, the promoted goal (setting a world record) and the method of dissemination of the information of the event (through social media) were in line with Mozilla's overall image as a non-corporate, decentralized and innovative organization that relies on volunteers and users for its promotion as well as software development. Similarly, the "Top This TV Challenge" campaign of Heinz® fits the "fun" brand image of Heinz® Ketchup.
2. **Inciting Virality and Buzz:** This is the key for creating a pull rather than inducing a push in an organization's marketing campaign. An attractive event (in this case a world record setting event) or a message is a crucial component in developing an organic viral marketing process. The ability to create buzz through the event will also increase the chances that the viral marketing campaign will supplement other marketing communication goals: such as, providing material for other promotional efforts or getting coverage in traditional media—the latter being especially important for Firefox given that Mozilla does not have a centrally controlled advertising budget to spend on conventional media. For example, the overwhelming interest in the Top This TV Challenge (with 5.2 million views) also helped create publicity for the company in

the mainstream media and prompted Heinz® to repeat the challenge in 2008.

3. **Getting Consumers to be Personally Invested:** Mozilla's Download Day emphasized not only the possibility of a world record but that the consumers would be an integral part of this unique success. In this case, the prospects of being a part of a possible Guinness World Record-setting activity may have increased the chances that consumers identify with (and are invested in) not only the product or the brand but also the success of the campaign. Perhaps, for the contestants in the Heinz® Top This TV Challenge, the personal investment was even higher because their own success (in terms of getting enough votes to win the contest) partly depended on the popular votes they would get from other consumers.
4. **Creating Levels of Viral Involvement:** In terms of options available for viral marketing, social media not only expand the available options but also create the possibility of multiple levels of viral involvement. For example, in the Heinz® Top This TV Challenge, the level of viral activity of a contestant that promotes his/her video will naturally be higher than a regular YouTube user who happens to come across a challenger's video that is worth sharing with friends. The Mozilla's Download Day event, on the other hand, systematically utilized the social media (and other venues) to create tiers of consumer involvement. For example, an enthusiastic Firefox user could go as far as organizing a download festival whereas a regular user of Facebook or MySpace could invite friends to pledge for the download on the Mozilla's Download Day homepage.

FUTURE TRENDS

As discussed in the preceding sections, a central tenet of the debates regarding the marketing potential of social media pertains to the balance that needs to be struck between the efficiency of automatic recommendation systems and the organic involvement created by the real community touch of viral marketing campaigns that invite consumers to actively participate in the dissemination of the marketing messages. On the one hand, systems such as Facebook's Beacon platform and MySpace's "influencer" option promise to deliver large-scale, automated word of mouth that can expand the reach of viral marketing campaigns. However, the perceived intrusiveness of such systems, as well as their tendency to use consumers as passive hubs to automatically relay marketing messages, may call into question the true virality of such advertising efforts, consequently reducing their appeal for consumers.

Recent discussions regarding "Web 3.0" and the future of the Internet may point to the direction that this uneasy relationship between virality and automatic customization may take. Despite frequent disagreements regarding the definition of Web 3.0, an increasing number of commentators have started to use the concept interchangeably with the Semantic Web – a set of technologies that enable software agents to understand, interpret and extract knowledge from information, making it possible for them to complete "sophisticated tasks for users" (Berners-Lee, Hendler & Lassila, 2001). Michael Bloch provides a simple example explaining how the Semantic Web would work:

You want to go out to dinner...and your car is in the shop... You issue a command for the agent to search for a restaurant serving Indian food within a 10-mile radius... You want a restaurant that has a 4 star rating issued by a well-known restaurant critic. Furthermore, you want the table booked and a cab to pick you up from your place. Additionally you want a call to be made to your phone once that's all done; but you don't want to be disturbed by the call as you'll be in a meeting - just for the reservation details added to your phone organizer. (Bloch, 2007)

As this example suggests, the Semantic Web is more than a compilation of web pages. Rather, it is a network of systems and databases that can communicate with each other to perform tasks on an individual's behalf. Moreover, as recent developments suggest, the Semantic Web will have the potential for subtler customization of information in accordance with the cognitive (and perhaps emotional) styles/needs of consumers. For example, an article by Hauser, Urban, Liberali and Braun (forthcoming) from MIT's Sloan School of Management announces an algorithm that uses clickstream data to morph the website content and format to the cognitive style of its users.

As evidenced by recently developed semantic web advertising applications (such as *Semantic-Match*™ – a semantic advertising platform that utilizes a natural language processing algorithm to understand content and sentiments and target advertising accordingly), when applied to online advertising, semantic capabilities can enhance customization, decrease errors that are associated with keyword targeted advertising and provide a more conversational interaction between the advertiser and the consumer. With respect to viral marketing, such advancements in language processing and customization can address an important shortcoming of passive virality by making it more personal. Whereas social network analyses aid the identification of hubs that can act as active viral marketers, improvements in natural language processing can prove beneficial in terms of understanding the communicative processes and dynamics within a social network. This information can help the marketing organization create different strategies to reach various potential hubs, create levels of viral involvement depending on the depth and the context of the communicative processes between network members, and customize the webpage that potential customers arrive at as a result of the viral marketing effort.

CONCLUSION

In recent years, Web 2.0 applications that enable web users to contribute to the world of online content have grown in popularity. In 2008, the Top 10 most frequently visited web site list of Alexa Internet – a web traffic information service – consistently included several social media sites: namely, YouTube, MySpace, Facebook, Hi5, Wikipedia and Orkut.com (2008). Despite their popular appeal, however, many of the Web 2.0 initiatives are still struggling to turn their popularity into financial success.

What is important to note is that when it comes to monetizing social media, there are no magic formulas. However, as explained above, the interactive nature of social media, combined with consumers' participation in the creation and dissemination of information, make viral marketing a viable candidate to fulfill the promise of a Web 2.0 of opportunities. In contrast to impersonal advertising methods that consumers do not trust and find intrusive, viral marketing through social media has the potential to be a personal, personable, participatory and trustworthy. source of information. Nonetheless, this should not be taken for granted that any and all viral marketing efforts in social media would be successful.

Extant literature suggests that there are certain prerequisites to a successful implementation of a viral marketing campaign in social media. First, as Webb (2007) suggests, because the company is going to have to rely on consumers to push the message, the message (and the product) should be worth pushing. Second, as consumers grow more suspicious of traditional advertising methods, marketers engaging in viral marketing in social media should pay the utmost attention to keeping viral marketing free from centralized interference that can damage its credibility. For example, Coplan (2007) notes that to remain credible, consumer marketers should be "honest about their opinions good and bad, open about their affiliation – and unpaid" (p. 26). This second prerequisite of success in social media marketing is closely related to the third one: In the world of consumer marketers, companies should learn to "cede control to customers" (cited in Poynter, 2008, p. 12). Partially, this means that viral marketing may be mixed with negative word of mouth and backlash (Gillin, 2007; Giuliana, 2005). At the same time, both positive and negative word of mouth should be considered as an opportunity to engage in a conversation with customers. For example, recently Cadbury PLC decided to relaunch Wispa (a chocolate bar discontinued in 2003) as a response to demands from 14,000 Facebook members (Poynter, 2008). Finally, as evidenced by the recent negative public reaction to the inadequate privacy protection on Facebook, marketers should be aware of the relatively intimate nature of social network sites.

REFERENCES

Alexa.com. (2008). *Global top 500.* Retrieved July 6, 2008, from http://www.alexa.com/site/ds/top_sites?ts_mode=global&lang=none

Allison, K. (2007). Facebook set for a delicate balancing act. *Financial Times (North American Edition)*, 8.

Anderson, C. (2006). *The log tail: How endless choice is creating unlimited demand.* London: Random House Business Books.

Andrejevic, M. (2007). *iSpy: Surveillance and power in the interactive era.* Lawrence, KS: University Press of Kansas.

Baruh, L. (2007). Read at your own risk: Shrinkage of privacy and interactive media. *New Media & Society*, *9*(2), 187–211. doi:10.1177/1461444807072220

Berners-Lee, T., Hendler, J., & Lassila, O. (2001). The Semantic Web. *American Scientist.* Retrieved June 3, 2008, from http://www.sciam.com/article.cfm?id=the-semantic-web.

Bloch, M. (2007, July 28). *The Semantic Web–Web 3.0.* Retrieved June 3, 2008, from http://www.tamingthebeast.net/blog/online-world/semantic-web-30-0707.htm

Boyd, D. M., & Ellison, N. B. (2007). Social network sites: Definition, history, and scholarship. *Journal of Computer-Mediated Communication, 13*(1), 210–230. doi:10.1111/j.1083-6101.2007.00393.x

Clemons, E. K., Barnett, S., & Appadurai, A. (2007). *The future of advertising and the value of social networks: Some preliminary examinations.* Paper presented at the 9th International Conference on Electronic Commerce, Minneapolis, MN.

Clow, K. E., & Baack, D. (2007). *Integrated advertising, promotion, and marketing communications.* Upper Saddle River, NJ: Pearson Prentice Hall.

ComScore, Inc. (2008). *Top global Web properties.* Retrieved February 19, 2008, from http://www.comscore.com/press/data.asp

Coplan, J. H. (2007). Should friends pitch friends? *Adweek, 48*, 26–26.

DeLozier, M. W. (1976). *The marketing communications process.* London: McGraw Hill.

Domingos, P. (2005). Mining social networks for viral marketing. *IEEE Intelligent Systems, 20*(1).

Faber, R. J., Lee, M., & Nan, X. (2004). Advertising and the consumer information environment online. *The American Behavioral Scientist, 48*(4), 447–466. doi:10.1177/0002764204270281

Gillin, P. (2007). *The new influencers: A marketer's guide to the new social media*. Sanger, CA: Quill Driver Books.

Giuliana, D. (2005). *Alternative marketing techniques for entrepreneurs*. Retrieved January 3, 2008, from http://www.scribd.com/doc/35013/Alternative-Marketing-Techniques-for-Entrepreneurs

Hall, E. (2007). Study: Popularity of social networks hampers ad growth. *Advertising Age*, *78*(31), 18.

Hanneman, R., & Riddle, M. (2005). *Introduction to social network methods*. Retrieved December 17, 2007, from http://www.faculty.ucr.edu/~hanneman/nettext/C10_Centrality.html

Hauser, J. R., Urban, G. L., Liberali, G., & Braun, M. (forthcoming). Website morphing. *Marketing Science*. Retrieved July 4, 2008, from http://web.mit.edu/hauser/www/Papers/Hauser_Urban_Liberali_Braun_Website_Morphing_May_2008.pdf

Klaassen, A., & Creamer, M. (2007). Facebook to add shopping service to its menu. *Advertising Age*, *78*(44), 39–40.

Kumar, R., Novak, J., & Tomkins, A. (2006). *Structure and evolution of online social networks*. Paper presented at the 12th International Conference on Knowledge Discovery in Data Mining, New York.

MacKelworth, T. (2007). *Social networks: Evolution of the marketing paradigm*. Retrieved March 12, 2008, from http://www.amacltd.com/pdf/SocialNetworksWhitePaper.pdf

Manovich, L. (2001). *The language of new media*. Cambridge, MA: MIT Press.

McAllister, M. P., & Turow, J. (2002). New media and the commercial sphere: Two intersecting trends, five categories of concern. *Journal of Broadcasting & Electronic Media*, *46*(4), 505–515. doi:10.1207/s15506878jobem4604_1

McCarthy, C. (2008, February 1). *Report: Facebook raises '08 revenue projection*. Retrieved March 6, 2008, from http://www.news.com/8301-13577_3-9862792-36.html

McCoy, S., Everard, A., Polak, P., & Galletta, D. F. (2007). The effects of online advertising. *Communications of the ACM*, *50*(3), 84–88. doi:10.1145/1226736.1226740

Morrissey, B. (2007). Social network ads: Too close, too personal? *Adweek*, *48*, 11–11.

Poynter, R. (2008). Facebook: The future of networking with customers. *International Journal of Market Research*, *50*(1), 11–12.

Rappaport, S. D. (2007). Lessons from online practice: New advertising models. *Journal of Advertising Research*, *47*(2), 135–141. doi:10.2501/S0021849907070158

Robinson, H., Wyscocka, A., & Hand, C. (2007). Internet advertising effectiveness: The effect of design on click-through rates for banner ads. *International Journal of Advertising*, *26*(4), 527–541.

Scott, J. P. (2000). *Social network analysis: A handbook*. London: Sage Publications.

Solove, D. J. (2007). *The future of reputation: Gossip, rumor, and privacy on the Internet*. New Haven, CT: Yale University Press.

Spertus, E., Sahami, M., & Büyükkökten, O. (2005). *Evaluating similarity measures: A large-scale study in the Orkut social network*. Paper presented at the 11th International Conference on Knowledge Discovery in Data Mining, Chicago, IL.

Subramani, M. R., & Rajagopalan, B. (2003). Knowledge-sharing and influence in online social networks via viral marketing. *Communications of the ACM, 46*(12), 300–307. doi:10.1145/953460.953514

Thompson, R. J. (2007). Can't skip this: Consumers acclimating to Internet ads. *Brandweek, 48*, 5.
MySpace, Facebook and other social networking sites: Hot today, gone tomorrow? (2006, May 3). *Knowledge@Wharton*. Retrieved April 24, 2007, from http://knowledge.wharton.upenn.edu/article.cfm?articleid=1463

Tsai, J. (2008). Facebook's about-face. *Customer Relationship Management, 12*(1), 17–18.

Wasserman, S., & Faust, K. (1995). *Social network analysis: Methods and applications*. Cambridge, MA: Cambridge University Press.

Webb, G. (2007, October/November). A new future for brand marketing. *The British Journal of Administrative Management*, 13-15.

Windley, P. (2007). *The fine line between advertising and recommendations.* Retrieved December 12, 2007, from http://blogs.zdnet.com/BTL/?p=7134

Zarsky, T. Z. (2004). Desperately seeking solutions: Using implementation-based solutions for the troubles of information privacy in the age of data mining and the Internet Society. *Maine Law Review, 56*, 13–59.

KEY TERMS AND DEFINITIONS

Content Variability: *Content variability* refers to the notion that new media objects can exist in an infinite number of variations. This characteristic of new media is the result of the digital coding of content and consequently the modular nature of information.

Data Mining: Data mining is a technologically driven process of using algorithms to analyze data from multiple perspectives and extract meaningful patterns that can be used to predict future users behavior The market basket analysis system that Amazon.com uses to recommend new products to its customers on the basis of their past purchases is a widely known example of how data mining can be utilized in marketing.

Interactive Media: *Interactive media* is a catch-all term that is used to describe the two-way flow of information between the content user and the content producer. In addition to enabling consumers to actively participate in the production of content, interactive media also allow for the collection of real time data, which can later be used for content customization.

Semantic Web: The *Semantic Web* refers to a set of design principles, specifications, and web technologies that enable networked software agents to understand, interpret and communicate with each other to perform sophisticated tasks on behalf of users.

Social Network Analysis: Social network analysis is a research methodology utilized in research to investigate the structure and patterns of the relationship between social agents. Examples of sources of relational data include: contacts, connections, and group ties which can be studied using quantitative methodologies.

Social Network Sites: Social network sites are web-based systems that enable end-users to create online profiles, form associations with other users, and view other individuals' profiles. Examples of social network sites include: Match.com, MySpace, Facebook, Orkut, Hi5, Bebo and LinkedIn.

Viral Marketing: *Viral marketing* refers to a form of word of mouth marketing that relies on consumers relaying product information, a marketing message or a personal endorsement to other potential buyers.

Web 2.0: Introduced in 2004, during a conference brainstorming session between O'Reilly

Media and MediaLive International, *Web 2.0* refers to the second generation of web-based content. Rather than merely pointing to technological changes in the infrastructure of the Internet, the concept of Web 2.0 underlines the notion that end-users can do much more than consume readily available content: The user of Web 2.0 also plays a key role in the creation and the dissemination of content. Popular examples include: video-sharing and photo-sharing sites, such as YouTube and Flickr; social network sites, such as Orkut, MySpace and Facebook; and Weblogs (blogs).

This work was previously published in Handbook of Research on Social Interaction Technologies and Collaboration Software; Concepts and Trends, edited by T. Dumova; R. Fiordo, pp. 33-44, copyright 2010 by Information Science Reference (an imprint of IGI Global).

Chapter 8.19
Visualization in Support of Social Networking on the Web

J. Leng
Visual Conclusions, UK

Wes Sharrock
University of Manchester, UK

ABSTRACT

In this chapter the authors explore the contribution visualization can make to the new interfaces of the Semantic Web in terms of the quality of presentation of content. In doing this they discuss some of the underlying technologies enabling the Web and the social forces that are driving the further development of user-manipulable interfaces.

INTRODUCTION

The internet is a communication device. Its interface consists not just of static text but other static and dynamic constructs e.g., tables, images, animations and customised web applications. Some of these elements hold meaningful content while others are used for graphic reasons[1]. The rise of social networking in the form of Weblogs, discussion boards, wikis, and networking sites allows the general public to share content on the web. Such non-technical users require high level web-apps to help design and deliver their content with as little explicit dependence on the technicalities as possible.

Before talking about social networking on the web it is worth considering what this means. The expression 'Social Networking Sites' is used for sites with the primary purpose of supporting or creating sociable relationships, prominently friendships, but we use 'social networking' in a more inclusive way to include the formation of all kinds of networks such as those, for example, that form to collaborate on a task (as with an open source development).

DOI: 10.4018/978-1-60566-650-1.ch009

The naming conventions used within visualization can confuse, this relates to the difficulty in producing a visualization taxonomy which is discussed later. Visualization is a visual means to analyse data and is cross-disciplinary. The name of the discipline normally identifies a particular theme but where the name of the discipline is also the name of a technique the naming convention lacks clarity. For example social networking visualization could be the use of visual methods to show and analyse social networks or could be the use of visualizations in the support of social networking i.e., as a means to help people form and co-ordinate their activities. This chapter looks at the latter, at how visualization can aid communication on the web. Whilst this can include the visualization of social networks (because users of the web may like to understand the social networks they participate in) that is not the primary focus.

In this chapter we introduce visualization, its history and the two main visual paradigms in use, dividing the visualization community between those concentrating on scientific and information visualizations respectively. We survey the technologies that shape the web and the applications running on it. This allows us to look at how the technology shapes visualization systems (the visualization pipeline and the flow of data) and how these can be distributed to work efficiently in web environments. Finally we review some web applications that support social networking and consider what future trends may be.

WHAT IS VISUALIZATION?

The History of Visualization

There is no accepted definition of visualization but it can be adequately summarized for our purposes as using visual means to aid the communication and understanding of information. Modern visualization increasingly uses computer graphics technology to make information accessible. Visualization's long history predates the origin of computers by at least 8 thousand years. Maps are one of the oldest forms of graphical aid whose continued usefulness is demonstrated by the fact that mapping applications are amongst the most popular web-based applications. Before computers visualizations generally were not interactive, though there are exceptions to this as some scientists developed models and pop ups in books to explain their ideas but these were rare and expensive (Tufte, 1997; Tufte, 2001).

The roots of visualization are tangled into our history; a timeline of visualization is available on the internet (Friendly, 2008). Many historical breakthroughs were made possible through visualization, such as John Snow's use (in London in 1854), of maps to show the distribution of deaths from cholera in relation to the location of public water pumps. Visualization has never been an isolated discipline; it has been an integral element of scientific, intellectual and technical developments.

The timeline of visualization shows that the development of visualization has accelerated since 1975, since when important changes have depended upon advances in computing. Improved computer speed and capacity increasingly allow data to be visualized by increasingly intensive computational methods. Computers make visualizations more interactive and allow direct manipulation of data, e.g. selecting data by linking, brushing or using animation in grand tours. Also driving the development of visualization is the fact that visualization methods are being applied to and developed for an ever-expanding array of problem areas and data structures, including web applications that enable social networking.

Modern Visualization

Modern computer-based visualization developed through the accumulation of three specific areas (Schroeder, 1997). Scientific visualization was the first. It is a discipline stemming from computational science and started as an IT support activity. Computational simulations produced digital data representing natural phenomena, for example the weather forecast. Commonly the data has an inherent geometrical shape and the data represents continuous fields within this geometry. The data were produced as large arrays of values that were difficult to analyse by hand or eye. Instead, computers were used to produce images of the data within its native geometrical shape.

'Scientific visualization' is typically categorised by the dimensionality of the data values (number of dependant variables), and whether the data is scalar, vector, tensor, or multivariate (Brodlie, 1992; Schroeder, 1997). However another taxonomy distinguishes by technique between global, geometric and feature-based techniques (Post, 1999).

Subsequently, it was realised that techniques for scientific visualization could be applied to data from the humanities and social sciences. Since this data commonly looks at geographical distributions of populations this field focuses on statistical graphs and thematic cartography. Gradually, the field of data visualization diverged further and information visualization, the final area of visualization, emerged. This field was primarily aimed at visualizing computerised databases so that relationships could be found; commonly using tables, graphs and maps. Currently visual display and analysis of text in the form of academic papers, web pages, and patient records, are included in information visualization. 'Information visualization' can also be classified by data type, common types being multi-dimensional databases (with more than 3 dimensions), text, graphs and trees (Bohm, 2001; Schneiderman, 1996). Similar to scientific visualization, a technique classification system is based upon display styles which include table, or information landscape (Card, 1997; Chi, 2000). However unlike scientific visualization some classifications also look at tasks, for example gaining an overview or drilling down on detail (Schneiderman, 1996).

Over this period visualization has been transforming into an increasingly independent academic subject within the field of computer graphics and is commonly taught in computer science departments, and is divided between two main specialisms categorised as 'scientific visualization' (developing visualizations for purposes of aiding scientific research) and 'information visualization' (providing ways of communicating information generally e.g. in the design and presentation of charts, diagrams, graphs, tables, guides, instructions, directories and maps which can also be used to aid research whether scientific or not). However this division does not best reflect visualizations true nature, and there are attempts to improve on this classification. Tory (2004), complains that the division between scientific and information visualization encourages segregation of different kinds of activities from one another when many visualization problems cross that divide, proposing instead, a systematic scheme 'based on characteristics of models of the data rather than on characteristics of the data itself'. An easy-to-understand taxonomy would be useful to both users of visualization and researchers into visualization and if it existed it would have been presented here.

The application area of interest in this chapter falls mostly within the information visualization paradigm although map based visualization applications are also important but these techniques are on the border of the 2 paradigms drawing from both scientific and information visualization (Tory, 2004). However visual appearance is not the only theme in visualization research of importance to web 2.0. Visualization system design and the distribution of that system are relevant research themes. The web is a distributed computing environment, with users also distributed and if they are to work together then synchronization of their activities must be facilitated by the visualization system. These themes were originally titled 'remote visualization' and 'collaborative visualization', growing out of scientific visualization. To understand the underlying technology and social forces driving the development of the web it is necessary to understand the position of and the relationship between the information visualization and scientific visualization communities.

A Survey of Information Visualization Techniques

There are hundreds of techniques with thousands of low level elements that could be considered. This chapter is too short to take on this task so we aim to give the reader an idea of where to look for further information and an overview of a few relevant techniques. Many information visualization techniques rely on the use of colour, animation and interactive input from the user but in this book only static grayscale images are possible so where possible we refer the reader to websites for examples.

Visualization is a young subject without a definitive text but the selection of papers, books and websites in the references section should provide a suitable starting point. Friendly's timeline of visualization (Friendly, 2008) presents an online timeline of visualization with static images for each entry in the timeline. This site is a good starting point for examples of visualization techniques developed before 2004. Images and demos of information visualization techniques are on the websites of commercial vendors of general purpose information visualization software (although demos are not always available and users may require registration). There are 5 stable information visualization companies known to the authors: AVS (Advanced Visual Systems) has OpenVis (AVS, 2008); Spotfire focuses on business information (Spotfire, 2008); Tableau is similar to Spotfire (Tableau 2008); Steema has TeeChart and TeeTree (Steema, 2008); and finally aiSee is designed for large data (aiSee 2008).

Some key techniques from information visualization are:

- **Graphs:** a 2D plot showing the relationship between parameters. Typically orthogonal axes show the values of the parameters represented in the plot. Harris (1999) illustrates all the visual elements of graphs, tables and other information plots and recent research is available in the Journal of Graph Algorithms and Applications (Tamassia & Tollis, 2008).
- **Trees and Networks:** nodes are connected by lines showing relationships between nodes. If the nodes represent people then the plot represents a social network. These plots can be difficult to understand if many nodes are used. Networks do not occupy a real geography but are optimized by complex mathematics to look good in a 2D display. In information visualization these are considered to be graphs. Freeman (2000) and Bertini (2008) survey the history and applications in this area, and two case studies of social network visualization applications are in Brandes (2001) and Shen (2008).
- **Multiple Related Views:** a number of 2D plots are viewed in one problem solving environment. If the plots are graphs then the plots tend to be arranged so that the horizontal or vertical axis that relate to the same parameter are aligned and a user selection made in one graph can be seen to relate across to the other plots. The next 4 techniques use multiple related views but relationships are made by different user interactions.
- **Drilling:** if the plots are geographical then they may represent different levels of detail of a region the user selects from the higher resolution plot.
- **Brushing:** if a user is interested in the geographic distribution of different age groups within the US then using the histogram and selecting (brushing) regions from the

histogram this selection is displayed on a map.

- **Linking:** many elements are related so colour is used across multiple views to clearly identify each relationship.
- **Grand tours:** an animated tour is given through the data display space so that the user can get an overview of the data.

The Development of Visualization in Relation to Computer Technologies

The development of visualization is driven not only by the development of visualization techniques, but in response to the development of new computer technologies offering new possibilities for visual representation. Changes in computer hardware, graphics hardware, computer display devices, computer input devices, software design, collaborative working environments, remote visualization and visualization services all play an influential part in determining what can be visualized and how the user can manipulate it.

IN TRANSITION FROM WEB PAGES TO WEB APPLICATIONS

The Handbook is about the changes from Web 1.0 to Web 2.0, made possible by a mix of new technologies and approaches. The step changes in technology and user participation are not yet completed. We review the current state of the technology and how it might adapt to provide better graphical tools for the 'semantic web'.

From Static to Dynamic Web Pages

One of the main changes is the development of Content Management Systems (CMS) providing a web page delivery system that replaces static HTML web pages (North 2008). CMS separates two aspects of a web page, the content from the presentation. In CMS the content is separated from the web page as it no longer sits in an HTML file waiting for a user to access it, but is in a database so that the HTML page is dynamically reconstructed for the user when they select that web page.

The change from static to dynamic content has relied on these changes in technology:

1. **Static web pages:** the content and presentation are not only included in the same file but are mixed together throughout the text in the file
2. **Cascading Style Sheet (CSS) web pages:** the content and presentation are separated into separate text files. HTML (Hyper Text Markup Language) is the language encoding the information on a web page. It provides the information as text along with text that defines the style (presentation) of the informational text. The presentation information can be separated from the content information by the use of Cascading Style Sheets. Such CSS enable changes to the look and feel of a whole website through alterations to only one file, rather than requiring the individual rewriting of every page on the site.
3. **Dynamic web pages:** content and web pages are separated.

CMS are powerful because they separate out the responsibility for designing and developing a website from providing the content. CMS[3] build in tools to enable non-technical users to enter and manage their content (which is why CMS are popular for blogging sites). Many social networking websites give the user the option whether to produce their content as text or HTML, which contributes to the control that users now want over web content (manifested by the popularity of open source communities), giving them control over its style. There are in effect two interfaces to a website, one between the site and its non-technical users, the other between the site and those with technical competences. Site providers increasing encourage technically competent users to create and share service specific applications by releasing their API (Application Program Interface) and disseminating technical information which makes the addition of new modules easy.

HTML remains the underlying language that is recognized by web browsers so CMS must reconstruct HTML pages dynamically for the viewer. Even where web delivery technologies, e.g., wikis, have their own markup language or where websites allow plain text entry these inputs are still translated into HTML. Standardising interpretation of HTML across all browsers has taken considerable time and effort, improvements to HTML being coordinated between browser vendors by the World Wide Web Consortium (W3C). It would be too problematic to get the W3C to coordinate this effort and for browser developers to commit the resources for each brand of markup language, so wikis and comparable applications have developed their own markup with special functionality for their relevant formatting and semantic issues. Nonetheless, wiki applications have their own translator to convert that markup into more basic HTML components. HTML has limited functionality with regard to style, so the functionality that handles the graphical content and interactivity important to visualization is provided by supplementing HTML with other languages. These languages commonly provide applications (e.g., web-based games) or structural and navigational tools (e.g., menu systems) for a website. Depending on their specific functionality these languages can create applications that are stored in and served from a database in a CMS site, adding to the CMS functionality or providing style for a static web page. The main languages that contribute to visualization on the web were developed in the 1990's, including Java, JavaScript, VRML (Virtual Reality Modeling Language) and Flash.

The Java programming language was developed by Sun Microsystems (Gosling, 2005; Java Home Page, 2008; Wikipedia, 2008). It is an interpreted language meaning that a program written in Java can run on all the popular computer platforms without any adaption; "Write Once, Run Anywhere". Combined with its other features this makes it ideal for the web so web browsers quickly incorporated small Java programs (called applets) within web pages. Later Java was configured for particular platforms e.g. J2ME (Java 2 Micro Edition) for mobiles. Java Servlets, instead of being embedded in a web page as applets, are used to extend the functionality of the web server, allowing extra content to be added to web pages dynamically from the server. Whilst Applets and Servlets can be used to add important visual content to the web, Java is much more powerful for such a purpose as it has dedicated libraries that handle necessary functionality: the SWING library adds user interface functionality, the Drag and Drop library allows object manipulation by mouse, and the 2D and 3D libraries allow the graphics modeling of 2D and 3D objects.

JavaScript is a scripting language commonly used for client-side web development; the web browsers incorporate the ability to interpret JavaScript programs (W3 Schools, 2008; Wikipedia, 2008). JavaScript was designed to look like Java but has less functionality and is easier for non-programmers to work with. The primary purpose of JavaScript is to embed interactive functionality into web pages, typically to inspect or create content dynamically for that page. For example, a JavaScript may validate input values in a web form, control the opening of pop-up windows or change an image as the mouse passes over it. Because JavaScript code runs locally in a user's browser (rather than on a remote server), it can respond to user actions quickly, making an application feel more responsive. Furthermore, JavaScript code detects user actions which HTML cannot. JavaScript is heavily used in many web-based applications including CMS and those that support social networking through gmail and facebook.

VRML is a standard text file format similar to HTML used to represent 3D interactive objects (W3D Consortium, 2008; Wikipedia, 2008). Developed with the web in mind a browser can interpret VRML by installing the appropriate plug-ins. A number of small geometrical primitives are defined within the format and each of these primitives may have a number of properties that define the visual aspects of the object such as colour or transparency. The shape of large 3D objects are defined by combining the correct geometrical primitives e.g., 3D surfaces are defined as a mesh of triangular primitives. These models are interactive in a number of ways. Web links can be made by clicking with a mouse on a node, timers and external events can trigger changes in the scene and Java or JavaScript can be incorporated into the world (VRML files are called worlds, the term world is used where in other graphical systems the term scene would be used). The VRML format is an open format that has an ISO standard making it suitable for sharing geometrical model data which ensures its popular within academia. Successors to VRML include X3D and 3DMLW (based on XML).

VRML is useful in applications where the 3D shape of an object in important such as teaching anatomical structures to medical students. An early interactive web-based application using VRML was created to teach medical students to perform lumber punctures (John, 2001). A model of the external skin, spinal bones, spinal cord and cerebrospinal fluid (CSF) were combined in one world, enabling the student to manipulate the model in the viewer, viewing it from any angle and altering the transparency of all the elements. The student could then place the puncture needle in the model to simulate taking a sample of CSF. The student would be given feedback on their performance and could alter the transparency to see into the puncture site.

Flash (currently Adobe Flash, previously Shockwave Flash and Macromedia Flash) is a multimedia graphics program used to create interactive animations for web pages (W3 Schools, 2008; Wikipedia, 2008). Its features make it especially suited for a web environment. It uses vector graphics, which means that the graphics can be scaled to any size of display area without losing quality and it supports bi-directional streaming so that it can load into a web page more quickly than animated images. Most web browsers can interpret Flash but the Shockwave Player can be downloaded for free and used as a plug-in. Flash is being ported for use on mobiles and PDA. Flash applications are developed through an authoring tool.

From HTML to XML

HTML provides static content for the web and deals with formatting and style of text. While HTML allows the publisher to present their information in a particular style, Web 2.0 uses XML (Extensible Markup Language) to handle the data that is passed across the internet. XML has a user defined format to handle particular types of data which means that without standardization of XML file formats the variation in XML file content makes them difficult to handle. File formats based on XML and handling data relevant to visualization are developing e.g., the previously mentioned 3DMLW format. While that data does not have to be human-legible both HTML and XML are designed to be human-legible. Currently XML is rendered as raw text in a web browser with no unified display protocol for XML across all web browsers. In order to style the rendering of XML data the XML file must reference a style sheet that can not only give style but can also convert regions of the data into other data formats such as HTML.

Eventually XML may replace HTML and if it does good styling and visual display functionality may be included in the format. However currently the importance of XML is not in how it is rendered but as a way of formatting data to allow machine readable semantic information to be incorporated

Figure 1. An overview for the whole of the visualization pipeline originally given by Haber (1990a) in text and Domik (2008) diagrammatically and refers to computational science. "Computer representation of reality" referred to computer simulation but could refer to a database or any digital data

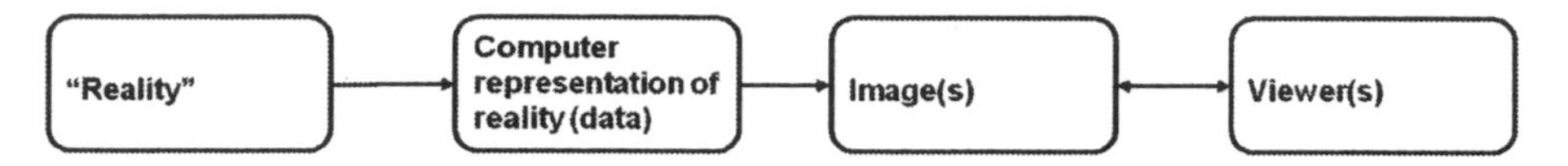

into the web. The best known function of XML is in price comparison websites, but it could be used to extend content to give information of interest (on when, what and the provenance of information) to social networking sites. Comparisons of large amounts of information where the structure, layout and interaction are important would benefit from visualization techniques.

VISUALIZATION SYSTEMS AND THE DEVELOPMENT OF TECHNIQUES

The Visualization Pipeline

The 'visualization pipeline' (Figure 1, from Haber and McNabb (1990a) gives an abstract presentation of the visualization computational process), indicating how scientific simulations are rendered in computerized form and thus made available to computational scientists. While this abstraction was developed for scientific visualization it can be expanded to encompass visualization more generally, and, some argue, covers computation generally. It shows (Figure 2) which parts of the pipeline are dependent on which computer hardware component. The performance of a pipeline is determined by the component with the worst performance called a "bottleneck". Developments in areas of computer technology on which the pipeline depends alter what is possible in the corresponding part of the visualization process. The graphics hardware and the display device are closely tied because the graphics card produces the images seen on the screen and this stream of images is a heavy load. The short dedicated cable connecting graphics processor and screen in a desktop computer is sufficient to enable the graphics card to service an acceptable refreshment of on-screen images—requiring at least 25 frames per second—but this works less well if the cable is replaced by a shared network for example the internet. While software on the pc comes with advice on minimum hardware requirements web-based applications do not, so users with low performance hardware will either have a poor experience of the application or the application must adapt to the user's system. The development of adaptive systems (Brodlie, Brooke, Chen, Chisnall, Hughes, John, Jones, Riding, Roard, Turner, & Wood, 2007) allows the detection of and adjustment for the nature of display systems and network speeds, allowing tailoring of the output so that, for example, a small mobile device could be fed lower resolution images to those fed into a wired pc or home entertainment system.

Physical interaction with a visualization is by two different types of input (Figure 3). Generally speaking interaction with the graphical object held on the graphics hardware is fast and uses a mouse or other device that feels natural. Interaction with the computation that occurs in the pipeline before the data reaches the graphics hardware tends to use application specific menu systems with mouse and keyboards input separated from the visual scene and disruptive of activity as the user must shift concentration from the graphical objects they are working with. Virtual Reality menu systems (Curington, 2001) appear to lessen disruption to the actions of the user but these can cause clutter in the visual scene and to reduce that clutter these menu systems have limited func-

Figure 2. Overview of the whole visualization pipeline in figure 1, but in this case the hardware that each element of the pipeline depends on is identified. Computer and graphics hardware can be part of the same machine but this is not necessarily so, but graphics hardware is normally closely tied to the display device

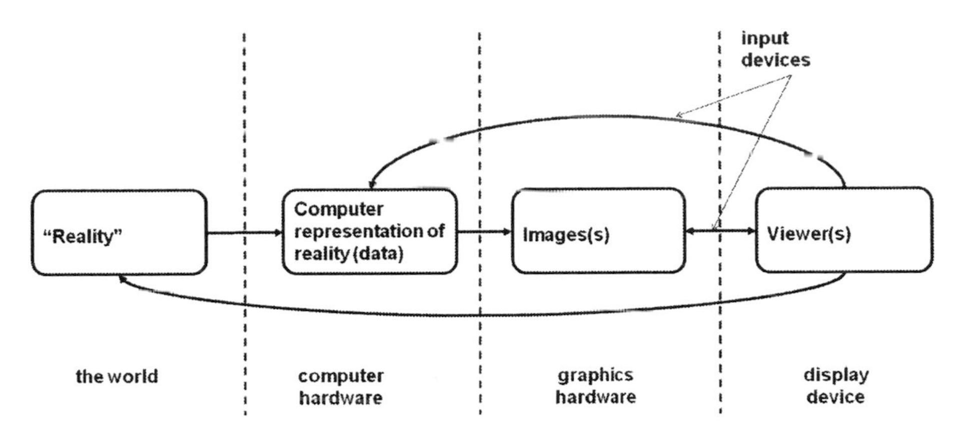

tionality. A third and dominant, but not physical, interaction is between the functionality of the application and the requirements of the user: if the visualizations do not appropriately represent reality then the user dependent on them may be misled when making their decisions, and if they are not confident that the results are useful they will abandon the system.

The visualization pipeline of the visualization system alone was further abstracted to categorise the interaction with the user and so exhibit the data-flow paradigm that is inherent in the software structure of visualization systems (Figure 4). The data is shown as flowing through the system while transformations are performed to deliver the final image. Haber (1990a) identified three types of transformation. The viewer is able to interact with the visualization and so affect each of these transformations, potentially gaining new insight into the data being displayed.

First, raw data is transformed by a data enrichment step into 'derived data'. Raw data can come

Figure 3. Overview of the whole visualization pipeline in figure 1, but with the 3 main types of user interaction given and numbered to match the order of discussion in the body of the chapter

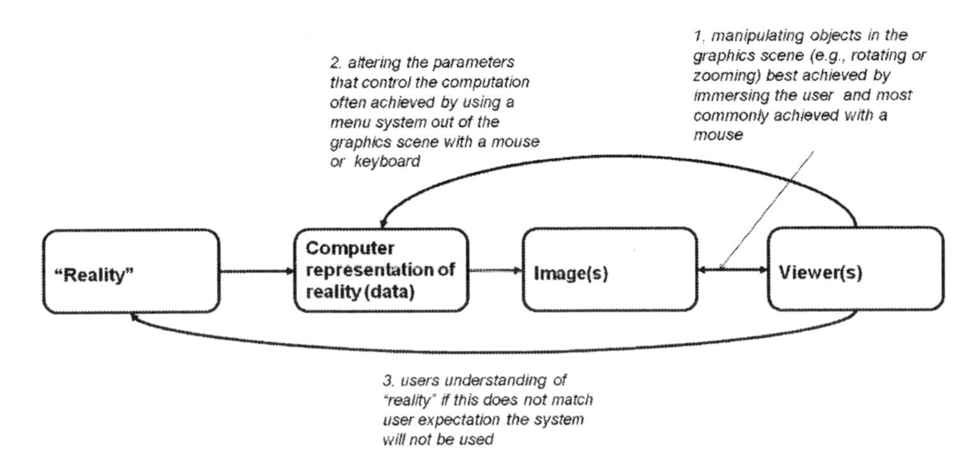

Figure 4. Overview of the visualization pipeline for just the visualization system showing how internally the visualization system handles data-flow (from left to right) and categorises types of interaction by data transformation type

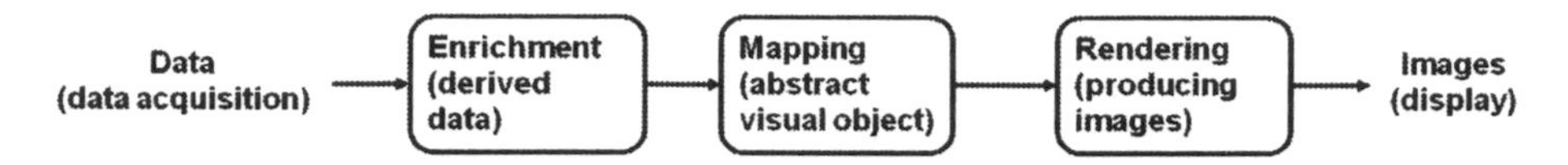

from any digital source such as simulations, data files and databases. Enrichment fits the data to the desired model perhaps including data manipulations such as smoothing, filtering, interpolating, or even hypothetical alterations. Preparing data for visualization can be time consuming and so deter the first time users.

The second transformation, mapping, converts data into an abstract visual object (AVO). Attributes of the data are converted to corresponding geometric features according to a set of classification rules. By adjusting these rules, the viewer can cause distinctions in the data to become perceptually obvious. This step establishes the representation of the data, but not necessarily the final visual appearance on the screen.

The third transformation, 'rendering', ultimately converts the AVO to a displayable image. This involves computer graphics techniques such as view transformations (rotation, translation, scaling), hidden surface removal, shading, shadowing. The transformations at this stage occur on the graphics hardware and alter the perspective on the display, contribute to the overall appearance and dictate what is or is not visible

This pipeline makes explicit the transformations the data must go through to produce images. These transformations must be performed in this sequence. The concept of the data passing successively through each transformation is built into visualization software as 'data-flow'. This pipeline is drawn to reflect generic features of data transformation but specific ones can be drawn for particular applications and used to optimize and/or split the process. This is a common strategy for complex visualization applications such as Google Earth which branch the pipeline and spread the computation across a number of computational resources in a way invisible to the user but designed to improve the functionality and interactivity for the user.

Interactive Rates

Interactive rates strongly constrain visualization strategies. There is no exact criterion for an appropriate interactive rate. In TV and film production there are three different frame rate standards, the slowest at 25 frames per second. Stereographic systems need at least twice this rate as one frame is produced for each eye. On a single unloaded machine[2] a visualization system will maintain a good frame rate up to a threshold in data size where the threshold is determined by the data's geometric composition, the complexity of the visualization techniques used, and the memory usage rather than by the disc space that the raw data occupies. Higher frame rates improve user satisfaction particularly for stereoscopic projection but in exceptional circumstances a rate may be acceptable down to about 10 frames per second, as when the user has no other way of viewing their data, however below this rate the user will prefer static images or animations produced in batch mode.

These issues affect the interactive rate of transformations occurring in the graphics hardware but changes to parameters in the earlier transformations (enrichment and mapping) alter the representation of the data output from a particular visualization pipeline without affecting the frame rate, so the rate that enrichment and mapping are updated is slower. There is no exact figure, but a lag of more than a few seconds causes delays that may confuse the user, who cannot be sure what transformations have already been applied and which are waiting to fed into the scene currently being viewed.

Visualization Systems

Scientific visualization software was initially developed before the idea of web services had taken hold. It mainly consisted in standalone software taking advantage of specialist graphics hardware and specialist display and interaction devices. This close tying of software to hardware improves performance and made this a suitable setting for the development of virtual environments. Scientific visualization started as a support activity for computational science when simulations were run on supercomputers that were administered in computing centers. The need for the visualization of simulations drove the development of computer graphics hardware and virtual reality environments by Silicon Graphics, a supercomputing vendor. Scientific visualization systems exploited local hardware using many processors and optimization strategies to allow the largest possible datasets to be visualized. Initially the display was to a simple screen but larger display areas were developed using projectors or tiled screens so that a number of users could explore and interact with the visualization scene at the same time. New input devices were needed in these environments such as the space mouse (a mouse designed for manipulating objects in a 3D environment by permitting the simultaneous control of all six degrees of freedom) as the users would walk close to the screen and away from the key board and mouse which were then wired into the computer terminal. The 3D scenes produced by the visualization software seemed flat; meaning that information such as the depth of an object was difficult to interpret so stereoscopic output was integrated into these environments to trick the eye into seeing these objects in their "real" 3D form.

When the visualization systems were physically tied to the hardware, the user needed to be in the computing center to take advantage of them. Thus users did not make intensive use of this specialist and very expensive equipment so the idea of remote visualization developed to allow users to stay in their office and make use of the graphics hardware over a network. The first remote visualization systems were developed before the web and were developed alongside the abstraction of the visualization pipeline (Haber, 1990b). Pioneering visualization web services used scientific visualization software on a supercomputer to produce static 3D images from medical scans served interactively through a web browser to medics (Jackson, 2000). Later more dynamic visualizations dependent on specialist graphics hardware again physically located in the computing centre using a product, Silicon Graphics VizServer were developed. However their use was restricted in practice to projects where the visualizations were produced on a proprietary Silicon Graphics supercomputer and a dedicated high bandwidth network was in place; the images produced on the graphics hardware were compressed before passing down a dedicated network to users who could interact with the objects in the images. The Op3D project delivered and projected3D medical visualizations onto the wall in operating theatres so surgeons could compare the patient to the 3D visualizations side by side (McCloy, 2003). Typically specialized research systems produced their software from scratch, in this case developing a user interface employing a joy stick that was easy to use and suitable for a sterile environment. Collaborative visualization allowed users to share their experience either by sharing the results of the visualization on a large display or by splitting the visualization pipeline at some point and distributing the output across the network where it is rendered and displayed for each collaborator. Collaborative environments used for teaching, consultations and group discussions enhance how knowledge is learnt and shared. The idea of adaptive visualization has been furthered by the Grid research initiative which aims to provide computer resources and services to academics across the web (Brodlie, et al. 2007). It is not a visualization system but it is a remote service that supplies users with visualization resources intelligently. In principal the user cannot choose which visualization software they use on which machine but instead define the scene that they wish to interact with and their local resources. As all visualization systems produce visually different results even for the same algorithm the user could find results from different softwares were not comparable. Visualization is a diverse and cross-disciplinary field so that the terminology within visualization and its applications is not consistent and this impacts on the ability to create an ontology or taxonomy for visualization making it difficult creating a visualization modeling language standardised across all applications (Brodlie & Mohd Noor, 2007). Technologies developed by the games industry are now the main driver for collaborative synchronous graphics environments.

Information visualization software followed a different path, developing later than scientific visualization software. Some early visualization systems for viewing social networks were adapted from scientific visualization software for molecular visualization using stereographic displays and VRML (Freeman 2000). However dedicated information visualization software was coded in Java rather than C or C++ and was suitable to integrate with web services from the start. Information visualization like scientific visualization displays large quantities of data often using different display techniques requiring less performance from the graphics hardware. Characteristically information visualization applications are more meaningfully displayed in 2D rather than 3D and use multiple related views that are easy to develop as web services. However such applications are computing intensive in the enhancement and mapping transformations rather than the rendering transformation (meaning that demands on the graphics card are contained). ManyEyes, a research application from IBM (Viegas 2007) is a web-based system that allows users to upload and visualize data through their browser. The visualizations are produced by java applets running asynchronously on the user's computer allowing each user to visualize and comment on uploaded datasets in a social networking environment similar to video and photo sharing in flickr or YouTube. Designed for visualization novices the system uses "ShowMe", a special interface to help users select suitable visualization techniques for their data, eliminating the need for a visualization modeling language. Its developers advocate good design web-based because it increases ease-of-use, as the authors of this chapter agree. Java applets are suitable for techniques that are not compute intensive.

Independently of the visualization community the Access Grid, an internet based video conferencing system, has developed. It is superior to video conferencing because the equipment is cheaper, using 'multicasting' technology that allows many sites to participate, facilitating on-line meetings, seminars and conferences. Access grid works by streaming video and audio over the internet, it is easy to stream visualization scenes across the internet and so include it in a session. This is interactive at the host site but not yet at other sites. Access grid sessions can be recorded and played back using visualization techniques designed to augment the analysis and playback of meetings (Buckingham Shum 2006; Slack 2006).

In Google Earth, a visualization web service combining satellite images, maps and other GIS information, vast amounts of computation are needed to serve the data to the user so that the best results are obtained for each user no matter what their graphics hardware (Jones, 2007). Web services will always be slower than locally run software running on high performance hardware. The distribution of computational resources needs to be balanced between the server and client, which are more noticeable for visualization than for non-graphical and non-interactive web applications. Currently, it is only by providing high performance data serving as in Google Earth that this complex visualization application becomes possible. There are other visualization applications that may be useful to social networking for example the production of large graph trees for particular members of a social networking website, perhaps including a famous politician who has an account on a social networking website. Large graph trees require high computational intensity and a low graphical intensity so the performance of an application aiming to produce such a tree would also require specialist hardware and code optimization, most likely to come from the proprietors of a social networking website rather than the open source community that is extending the functionality of the website. More generally such complex applications are most likely be developed within the commercial sector.

Visualization Design

Understanding visualization techniques is only part of the story. Knowing which techniques to pick, how to combine low level design elements and how to appraise a visualization are also important. Many elements of design are open to debate. Here we give the reader an idea of possible approaches to design and analysis.

The most scientific approach to design comes from psychology. Colin Ware, perceptual psychologist who for a short time worked as an artist before becoming interested in visualization, has studied visualization design from the perspective of the science of perception. His book (Ware, 1999) is the definitive guide to issues such as when to use words and/or images and the possibilities for visual programming languages. Lately techniques from psychology are used to test how "good" a newly developed visualization technique is by a method called 'user assessment'.

Another way to understand visualization design and analysis is through the history of how information has been visually represented. If a number of visual information representations can be understood from throughout history then it may be possible to use that understanding to analyse current visualizations and even to predict future developments (Friendly, 2008; Tufte, 1997; Tufte, 2001; Tufte, 2006; Wainer, 2005). By studying history it is possible to understand better the social context in which the techniques were developed and explain two types of negative outcomes. Firstly, the situation where techniques were developed but at that time it was disregarded only sometime later to become an accepted visual representation. This was the case with William Playfair who in the 16th Century developed the grounding for modern statistical plots 150 years before they were accepted (Playfair, 2005). Secondly, where the poor visual representation of information have contributed to the failure of projects. This approach has been applied to two NASA based projects involving the launch of the space shuttle Challenger and the space flight of the shuttle Columbia in 2003 (Tufte, 1997; Tufte, 2006). To improve his analyses Tufte shares information with a variety of experts who use visual methods to communicate information, creating a moderated web forum for this purpose.

Cross-disciplinary teams can not only be good for analyzing visualizations but also for designing and creating visual representations. The term "renaissance team" was coined in the mid 1980s by an artist, Donna Cox, who worked collaboratively with computer scientists and scientists to find novel visual ways to explore how the universe was formed. Collaborative cross-disciplinary teams including artists trained in design can be used to create novel and visually pleasing techniques. Art criticism is the discussion or evaluation of visual fine arts, and it has been suggested that it be extended for the analyses of the products of visualization i.e., images, animations or virtual environments.

The final approach to visualization design is not a formal design strategy or methodology but is mentioned only because it has the potential to affect the type of visualization applications available on the web. It involves the exploration and testing of technical possibilities for visualization through developing web-based open source applications by experienced graphics programmers who will probably have worked in creative environments reliant on computer graphics who create visualizations for fun, as does Jim Bumgardner who among other things has contributed open source software to flickr, an online photo management and sharing application (Bumgardner, 2008).

A SURVEY OF VISUALIZATION IN USE TO SUPPORT SOCIAL NETWORKING ON THE WEB

The current state of visualization to support social networking is poor. The web pages on social networking sites tend not to be visually novel or pleasing[4]. Good graphic design means some websites have a visually appealing header and menu system though there are arguments that web pages should have little visual clutter and that graphical headers are a distraction. We do not get involved in this argument as it is only how information is structured within the pages and the functionality of web-based applications that is really of relevance.

The layout of the web pages within social networking websites tend to be structured into tabular formats with information structured into long web pages that require scrolling, a typical artifact of CMS. Web pages in a blog style are placed in a list ordered by the date the blog entries were created. In websites primarily handling other types of data relevant to social networking such as images, video, film reviews or web bookmarks, the data are again tabulated but the entities are ranked by some property other than date. Often users of these websites will rate the data or other ways are used to calculate the popularity of the data. One of the features making these websites clumsy for social networking is the way different kinds of data are segregated within them i.e., the separation of images from video and the lack of support for other types of files such as pdf, powerpoint or even applets. While it can make sense to segregate data there are activities such as planning a wedding that require integration of a number of activities which may be carried out through social networking sites. Also, if file formats that can be uploaded and viewed cannot be annotated mean then their features of interest must be described in text that is isolated from the file.

Two web applications offering greater visual diversity are Google Earth/Maps (an online geographical application that combines maps and satellite images) and flickr (an online photo organization application).

Google Earth/Maps have advanced functionality allowing real time manipulation of large amounts of geographical information, with an open API to these applications so that further applications can be developed that extend the functionality for particular sets of users such as recreational runners who may use the MapMyRun website to share routes and to plot graphs of changing elevations over the run. Other information can be added for example 3D models of buildings can be attached to the maps or satellite images or outputs from simulations can be visualized within the Google Earth environment. The API is of commercial value to Google as the professional versions of these applications can be extended further than the non-commercial version.

Flickr also has advanced functionality but it really stands out for its graphical novelty. It has an open API but that is not supported by flickr. There are fewer stand-alone applications extending the functionality of flickr through the API than with Google Earth. One exception is the trip planner, using flickr's mapping functionality to link photos to a geographical location on a map, extended to allow planning trips in a shared environment. Flickr is graphically pleasing and especially within its open source community there is graphical experimentation, perhaps because this community is interested in photography and as such is aesthetically motivated.

FUTURE TRENDS

Though a great deal is happening in the field of Visualization, it is not yet a significant presence in the public life of the internet, though this situation will probably change drastically. With many fundamental problems of making information available over the internet now solved, demand for improvements to both presentational quality of information and to the appearance of websites and stand alone web-based applications will surely increase. This will bring the existing and evolving techniques of visualization to much wider attention. Google Earth and Google Maps, a visualization application, have 'introduced hundreds of millions of users to what have, in the years since Ivan Sutherland's Sketchpad, been concepts at the core of visualization research-projections, 3D user interaction, user feedback, motion models, level of detail, frame-rate management, view-dependent rendering, streamed textures, multimodal data composition, data-driven extensibility, and direct manipulation' (Rhyne, 2007).

Visualization may further facilitate user's access to and control over web content through contribution to the development of code writing. The development of techniques to support this task involves either the use of visualization to aid the writing of textual programs or the development of visual metaphors providing a visual language. In the first case the user must still understand the principals of programming, but techniques from text visualization may be appropriate (Card 2004) or functionality similar to that used in debugging packages could be applied. In the latter case the metaphors will only be helpful if they are apt but harmful if they are not and it is difficult to make them apt (Ware, 1999).

Using the web to aid academic research has been popular from the start but it seems that the semantic web will extend that popularity, aiding researchers in an ever increasing number of ways (Waldrop, 2008). Sharing textual information is the underlying means of communication. However other forms of communication will develop such as the access grid which allows video conference style communication but to multiple users is also popular for the delivery of seminars, conferences and meetings. Visualization could be used to structure information, for disciplines that have a geometrical or geographical dimension (such as engineering, anatomy or archeology) to share shape and location and in fields where there is intense cross-disciplinary activity which is not currently well supported. This trend could result in greater demands for more effective real time display across the internet, involving much greater visual access, and calling for shared visual environments. The semantic web also offers an important means of communication to hobbyists and intelligent lay people requiring similar functionality for different application areas such as football, patchwork quilting and dieting.

Visualization will become more relevant to domestic and leisure uses of the internet as the display and input devices available multiply in form, variety and functionality, with their presentational quality enhanced. With ever larger screens and projection systems used in home entertainment suites social groups and families could benefit from advances made in virtual environments (such as stereographic displays and the space mouse which has similar functionality to the Wii controller) and the access grid allowing groups to socialize through video streaming. Also the increasing array of input devices means that online gaming and training simulators can become more natural.

Comparable opportunities exist with mobile devices, in connection with problems of visualization for small screen spaces, the servicing of touch screens, the development of menu systems and the progressive improvement in interactivity. Google Earth can be accessed on mobile systems and this has the potential to turn any mobile device into a GPS (Global Positioning System).

Visualization can also contribute to the interfaces underlying social networking services. For example, Mayaviz (Roth, 2004) developed a system that uses web-based technology to help synchronize complex logistical efforts similar to planning social events. The ideas used in this product address many of the clunky features noticed in current social networking sites. Work sheet like constructs can be created by users and access can be limited appropriately. Data files of any type can be dragged and dropped anywhere on the sheet (they do not have to be in a tabular layout) and users can make clusters of elements holding related information. Data elements can be drawn on and annotated by the user and changes in data/information can be propagated into other related data without the user having to control these changes. An alternate example is improvements to techniques rather than systems such as to graphical representations of large social networks or to combine both social connectednesses in virtual space with geographic relations on a map.

CONCLUSION

The internet may be on the edge of a golden age for visualization as its techniques find use in many more application areas than those specialized niches within which it has developed. Much of the more complex functionality and varied applications will probably develop commercially but there is still a place for academic research into visualization and for the single open source developer to create interesting visual techniques.

With very powerful levels of functionality now available on the internet, there is the possibility of a rebalancing the emphasis toward more aesthetic and ease-of-use concerns. Future generations of technology will be integrated into the semantic web to provide superior audiovisual experiences. To this end visualization practices themselves will continue to evolve, taking advantage of the new devices and innovative functionality that continues to come on-stream. Not only will visualization become more commercially driven but more frontal emphasis to aesthetic quality is likely as it becomes more intensely cooperative with graphic designers and creative artists in forming the new user interfaces of the semantic web.

REFERENCES

W3 Schools. (2008). *Free training materials from the World Wide Web Consortium*. Retrieved June 16, 2008, from http://www.w3schools.com/

aiSee. (2008). *aiSee home page*. Retrieved September 25, 2008, from http://www.aisee.com/

AVS. (2008). *AVS (advanced visual systems) home page*. Retrieved June 16, 2008, from http://www.avs.com/

Bertini, E. (2008). *Social network visualization: A brief survey*. Retrieved June 16, 2008, from http://www.dis.uniroma1.it/~bertini/blog/bertini-socialnetvis-2.pdf

Bohm, C., Berchtold, S., & Keim, D. A. (2001). Searching in high-dimensional spaces: Index structures for improving the performance of multimedia databases. *ACM Computing Surveys*, *33*(3), 322–373. doi:10.1145/502807.502809

Brandes, U., Raab, J., & Wagner, D. (2001). Exploratory network visualization: Simultaneous display of actor status and connections. *Journal of Social Structure's*, *2*(4). Retrieved September 23, 2008, from http://www.cmu.edu/joss/content/articles/volume2/BrandesRaabWagner.html

Brodlie, K., & Mohd Noor, N. (2007b) Visualization Notations, Models and Taxonomies. In I. S. Lim & D. Duce (Eds.), *Theory and practice of computer graphics* (pp 207-212). Eurographics Association.

Brodlie, K. W., Brooke, J., Chen, M., Chisnall, D., Hughes, C. J., John, N. W., et al. (2007) Adaptive infrastructure for visual computing. In Theory and Practice of Computer Graphics 2007, pp 147-156, Eurographics Association.

Brodlie, K. W., Carpenter, L. A., Earnshaw, R. A., Gallop, J. R., Hubbold, R. J., Mummford, A. M., et al. (1992). *Scientific visualization: Techniques and applications*. Springer-Verlag.

Buckingham Shum, S., Slack, R., Daw, M., Juby, B., Rowley, A., Bachler, M., et al. (2006). Memetic: An infrastructure for meeting memory. In *Proceedings of the 7th International Conference on the Design of Cooperative Systems*, Carry-le-Rouet, France.

Bumgardner, J. (2008). *Krazy dad blog*. Retrieved June 9, 2008, from http://www.krazydad.com/blog/

Card, S. (2004). From information visualization to sensemaking: Connecting the mind's eye to the mind's muscle. In *Proceedings of the IEEE Symposium on Information Visualization*. IEEE Computer Society.

Card, S. K., & Mackinlay, J. (1997). The structure of the information visualization design space. In *Proceedings of the IEEE Symposium on Information Visualization*. IEEE Computer Society Press.

Chi, E. H. (2000). A taxonomy of visualization techniques using the data state reference model. In *Proceedings of the IEEE Symposium on Information Visualization*. IEEE Computer Society Press.

Curington, I. (2001). Immersive visualization using AVS/Express. In *Proceedings of the 2001 International Conference on Computational Science*. Springer-Verlag.

Domik, G. (2008). *Tutorial on visualization*. Retrieved June 15, 2008, from http://cose.math.bas.bg/Sci_Visualization/tutOnVis/folien.html

Freeman, L. (2000). Visualizing social networks. *Journal of Social Structure's, 1*(1). Retrieved September 23, 2008, from http://www.cmu.edu/joss/content/articles/volume1/Freeman.html

Friendly, M., & Denis, D. J. (2008). Milestones in the history of thematic cartography, statistical graphics, and data visualization. Retrieved April 24, 2008, from http://www.math.yorku.ca/SCS/Gallery/milestone/

Gosling, J., Joy, B., Steele, G., & Bracha, G. (2005). *The Java™ language specification third edition*. Addison-Wesley.

Haber, R. B., & McNabb, D. A. (1990a) Visualization idioms: A conceptual model for scientific visualization systems. In B. Shriver, G. M. Nielson, & L. J. Rosenblum (Eds.), *Visualization in scientific computing*. IEEE Computer Society Press.

Haber, R. B., McNabb, D. A., & Ellis, R. A. (1990b). Eliminating distance in scientific computing: An experiment in televisualization. *The International Journal of Supercomputer Applications, 4*(4).

Harris, R. L. (1999). Information graphics: A comprehensive illustrated reference. New York: Oxford University Press.

Jackson, A., Sadarjoen, I. A., Cooper, M., Neri, E., & Jern, M. (2000). Network oriented visualization in a clinical environment (NOVICE). In H. U. Lemke, M. W. Vannier, K. Inamura, A. G. Farman, & K. Doi (Eds.), *Computer assisted radiology and surgery, International Congress Series 1214* (pp. 1027-1027).

Java Home Page. (2008). *The official website for Java*. Retrieved June 16, 2008, from http://www.java.com/en/

John, N. W., Riding, M., Phillips, N. I., Mackay, S., Steineke, L., Fontaine, B., et al. (2001). Web-based surgical educational tools. In *Medicine Meets Virtual Reality 2001, Studies in Health Technology and Informatics* (pp. 212-217). Amsterdam: IOS Press.

Jones, M. T. (2007). Google's geospatial organizing principle. *IEEE Computer Graphics and Applications*, *27*(4), 8–13. doi:10.1109/MCG.2007.82

McCloy, R. F., & John, N. W. (2003). Remote visualization of patient data in the operating theatre during hepato-pancreatic surgery. In *Computer assisted radiology and surgery* (pp. 53-58).

North, B. M. (2008). *Joomla! A user's guide, building a Joomla! powered website*. Upper Saddle River, NJ: Prentice Hall.

Playfair, W. (2005). *Playfair's the commercial and political atlas and statistical breviary*. Cambridge, UK: Cambridge University Press.

Post, F. H., deLeeuw, W. C., Sadarjoen, I. A., Reinders, F., & van Walsum, T. (1999). Global, geometric, and feature-based techniques for vector field visualization. *Future Generation Computer Systems*, *15*(1), 87–98. doi:10.1016/S0167-739X(98)00050-8

Rhyne, T.-M. (2007). Editor's note. *IEEE Computer Graphics and Applications*, *27*(4), 8. doi:10.1109/MCG.2007.82

Roth, S. (2004). Visualization as a medium for capturing and sharing thoughts. In *Proceedings of the IEEE Symposium on Information Visualization*. New York: IEEE Computer Society Press.

Schneiderman, B. (1996). The eyes have it: A Task by data type taxonomy for information visualizations. In *Proceedings of the IEEE Symposium on Visual* Languages (pp. 336-345). New York: IEEE Computer Society Press.

Schroeder, W., Martin, K., & Lorensen, B. (1997). *The visualization toolkit an object-oriented approach to 3D graphics*. Upper Saddle River, NJ: Prentice Hall PTR.

Shen, Z., Ma, K., & Eliassi-Rad, T. (2006). Visual analysis of large heterogeneous social networks by semantic and structural abstraction. *IEEE Transactions on Visualization and Computer Graphics*, *12*(6), 1427–1439. doi:10.1109/TVCG.2006.107

Slack, R., Buckingham Shum, S., Mancini, M., & Daw, M. (2006). Design issues for VREs: Can richer records of meetings enhance collaboration? In *Proceedings of the Workshop on Usability Research Challenges for Cyberinfrastructure and Tools, ACM Conference on Human Factors in Computing Systems*, Montréal.

Spotfire. (2008). *Spotfire home page*. Retrieved June 16, 2008, from http://spotfire.tibco.com/index.cfm

Steema. (2008). *Steema home page*. Retrieved June 16, 2008 from http://www.steema.com/

Tableau. (2008). *Tableau home page*. Retrieved September 9, 2008, from http://www.tableausoftware.com/

Tamassia, R., & Tollis, I. G. (2008). *Journal of Graph Algorithms and Applications*. Retrieved September 23, 2008, from http://jgaa.info/

Tory, M. K., & Möller, T. (2004). Rethinking visualization: A high-level taxonomy. In *Proceedings of the IEEE Symposium on Information Visualization*. New York: IEEE Computer Society Press.

Tufte, E. R. (1997). *Visual explanations*. Cheshire, CT: Graphics Press.

Tufte, E. R. (2001). *Envisioning information* (2nd ed.). Cheshire, CT: Graphics Press.

Tufte, E. R. (2006). *Beautiful evidence*. Cheshire, CT: Graphics Press.

Viegas, F. B., Wattenberg, M., van Ham, F., Kriss, J., & McKeon, M. (2007). ManyEyes: A site for visualization at Internet scale. *Visualization and Computer Graphics, IEEE Transactions, 13*(6).

Wainer, H. (2005). *Graphic design: A Trout in the milk and other visual adventures*. Princeton, NJ: Princeton University Press.

Waldrop, M. M. (2008). *Science 2.0*. Scientific American Magazine.

Ware, C. (1999). *Information visualization, perception for design*. San Francisco, CA: Morgan Kaufmann Publishers.

Web3D Consortium. (2008). *Web-based community driving open standards in 3D modeling file formats*. Retrieved June 14, 2008, from http://www.web3d.org/x3d/vrml/index.html

Wikipedia. (2008). *Free online encyclopedia*. Retrieved June 16, 2008, from http://en.wikipedia.org/

KEY TERMS AND DEFINITIONS

API: (Application Program Interface) contains all the elements that a programmer needs to extend an application.

CMS: (Content Management System) a web delivery system separating content from presentation. These allow users to add content making them popular in social networking sites but the web pages have a tabular form that isolates the elements that make up the content.

Data visualization: The second area of visualization to emerge that focused on statistical plots and thematic cartography has now merged with information visualization.

Flash: is a multimedia web application adding animation and interactivity to web pages by using efficient streaming and vector graphics techniques.

Google Earth: The most popular visualization tool ever. It is a standalone web application combining maps, satellite and GIS information into a meaningful spatial context that gives the user direct manipulation of the applications elements.

Information visualization: The final area of visualization to emerge that aimed to show visually the relationships within databases.

Java: A powerful programming language that adds functionality to the web at the server side and the client (browser) side. Several libraries relevant to visualization are included within Java.

JavaScript: A scripting language that adds interactivity into web pages.

Scientific visualization: The first distinct area of visualization to have developed. Initially computer graphics technology was used to "view" the result of computer simulations which had an inherent geometry e.g., the flow of air over an aircraft.

Visualization: There are many definitions of visualization. In this chapter we use the term to cover the use of computer and computer graphics technology to present data to aid human understanding and communication. Today visualization is somewhat arbitrarily divided into scientific and information visualization.

VRML: is a file format that holds 3D models. Some animation and interactivity is encoded into the file.

ENDNOTES

1. This does not mean graphic design is unimportant or that the structure and layout of information is not part of the aim of visualization. However the original web language (HTML) made no allowance for design so that constructs such as html tables

have been adapted to take on a dual role i.e., one of enforcing a design.

[2] Single machine is stated here because the visualization system could run on a cluster where the process is spit over several machines. The 'unloaded' term is more important. On supercomputers there may be other users taking control of resources such as memory, processing power or I/O systems that affect the visualization system's ability to produce the optimum frame rate. On machines dedicated to the use of a single user this may still be a problem. If a machine is running background processes for example installing updates or if the user is using other software at the same time then there may also be a conflict in the sharing of resources affecting the frame rate.

[3] The authors have worked with three CMS Joomla!, Zope and BSCW.

[4] The authors could not view every possible website however they attempted to view websites that supported as many different activities as possible: YouTube (http://uk.youtube.com/), Flickr (http://www.flickr.com/), Fantasy Football (http://fantasyfootball.metro.co.uk/fantasy-games/), FaceBook (http://www.facebook.com/), LinkedIn (http://www.linkedin.com/), TopCoder (http://www.topcoder.com), ManyEyes (http://services.alphaworks.ibm.com/manyeyes/home), Del.ic.ious (http://delicious.com/), Jango (http://www.jango.com/), Wikipedia (http://www.wikipedia.org/), MapMyRun (http://www.mapmyrun.com/), Fetch Everyone (http://www.fetch-everyone.com/) and Now Public (http://www.nowpublic.com/). A blog was also developed on the blogspot website (https://www.blogger.com/start) that uses Google's blogger interface.

Chapter 8.20
On the Social Shaping of the Semantic Web

Paul T. Kidd
Cheshire Henbury, UK

ABSTRACT

Addressed in this chapter is the Social Shaping of the Semantic Web in the context of moving beyond the workplace application domain that has so dominated the development of both Information and Communications Technologies (ICTs), and the Social Shaping of Technology perspective. The importance of paradigms and the values that shape technology are considered along with the utility value of ICT, this latter issue being somewhat central in the development of these technologies. The new circumstances of ubiquity and of uses of ICT beyond mere utility, as a means of having fun for example, are considered leading to a notion of the Semantic Web, not just as a tool for more effective Web searches, but also as a means of having fun. Given this possibility of the Semantic Web serving two very different audiences and purposes, the matter of how to achieve this is considered, but without resorting to the obvious and rather simple conceptual formulation of the Semantic Web as either A or B. The relevance of existing Social Shaping of Technology perspectives is addressed. New thoughts are presented on what needs to be central to the development of a Semantic Web that is both A and B. Key here is an intelligent relationship between the Semantic Web and those that use it. Central to achieving this are the notions of the value of people, control over technology, and non-utility as a dominant design principle (the idea of things that do not necessarily serve a specific purpose).

DOI: 10.4018/978-1-60566-650-1.ch017

INTRODUCTION

When computing and communications technologies merged and moved from the industrial, commercial, academic and government settings in which the technologies initially developed, into society at large, something fundamental and quite profound happened. On achieving ubiquity, Information and Communications Technologies (ICTs) ceased to be the primary preserve of the professional developer and the work-based user, and became, in effect, public property. No more can the use of ICT be perceived as the domain of a select few. And the World Wide Web is the quintessential embodiment of this new circumstance. But it is not just the user community that has changed, for it is also the case that professional software developers also now operate in a world where anyone, potentially, can become a software or applications developer.

However, with the movement of ICT out from, and beyond, the workplace, into society at large, there is a need to look beyond traditional concerns with problem-solving, efficiency, utility, and usability. These are the criteria of the business world, of government departments, and the like, who are single-mindedly focused on delivering against targets set from high above. To these types of organization, computers, software, the Internet, and the World Wide Web are but functional utilitarian tools deployed in the service of profit, in the case of business, or policy as the embodiment of high political principles, in the case of governments.

Step beyond this well ordered world, into the life of an everyday citizen in modern society–the information society–and all these conventional concerns with problem-solving, efficiency, utility and the like, sit side-by-side, often uncomfortably, or are often substituted by, a second dimension representing a whole spectrum of other interests. The defining characteristics of this second dimension to the Web are typically: fun, enjoyment, happiness, fulfillment, excitement, creativity, experimentation, risk, etc. These in turn raise a much wider range of values and motivations than those of business, and embody notions such as freedom, individuality, equality, the human right to express oneself, and so forth. Moreover, these lead to a World Wide Web as a place, not of order, but of chaos, anarchy, subversion, and sometimes, even the criminal. And into this world moves the notion of semantic technologies, potentially transforming this chaotic and unpredictable environment into a place of order and meaning, taming this wild frontier, making it more *effective* and *useful*. But order, meaning, effectiveness, and usefulness for whom?

This paper addresses this rather simple, yet profound question. It is argued that the Semantic Web, as originally conceived, is not, as has been claimed, an improvement upon the first generation of the World Wide Web, but potentially a destructive force. What the Semantic Web may end up destroying is the unpredictability of interactions with the Web. This unpredictability, it is argued, is one of the Web's most appealing characteristics, at least to those who are *not* seeking to be more efficiency, or effective, and the like. And with the loss of unpredictability goes serendipity, which is something that has a value beyond quantification, for both the workplace user and the non-workplace user.

However, this outcome is not inevitable: there is no immutable law which states that the Semantic Web has to be such that all unpredictability in encounters with the Web has to be eliminated. The Semantic Web, like all technologies, can be shaped to produce outcomes (MacKenzie & Wajcman, 1985), and these do not have to be dominated by the need for efficiency and utility. In other words, insights form the social sciences can be used to design a technology that is different from one where purely technical and business considerations dominate.

To understand this perspective it is first necessary to understand how design paradigms influence the development of technology, and how these paradigms have failed to adapt to the age of computers and the information society (Kidd, 2007a). These paradigms, it is argued, are still predominately locked into a world dominated by inflexible electro-mechanical systems, these being technologies which severely limit design freedom and which do not allow designers to accommodate individualism. This, it is further argued, has implications, not just for social users of the Web, but also for business users.

Consideration is also given to values underlying science and technology, which still seem to embody Newton's Clockwork universe, with its belief in predictability and absoluteness, a perspective which sits in sharp contrast to the modern world view of a chaotic and relative universe. The journey will also consider notions of happiness and how this is connected to the concept of *meaning*. Meaning here does not refer to that which is commonly understood in the context of the Semantic Web. What *meaning* implies in this context is something that is strongly linked to the subjective, and key among this is emotions, which are a human trait that surpasses the domains of usability and conventional human factors. Emotions and meaning however are not well understood by software developers, but it has been argued that these are the very things that underlie the success of many of the modern world's most successful products (mobile phones being the classic example). Finally, the discourse will consider the concept of Ludic Systems, Ludic (Huizinga, 1970) being an obscure word meaning playful, but in a very wide sense, including learning, exploration, etc. This perspective does not just characterized people by thinking or achievements, but also by their *ludic* engagement with the world: their curiosity, their love of diversion, their explorations, inventions and wonder. Play is therefore not just perceived as mindless entertainment, but an essential way of engaging with and learning about the world and the people in it.

Bringing these perspectives together, the contribution will then consider the Social Shaping of the Semantic Web, namely the conscious design of the Web such that it accommodates the two very different, but important dimensions of the Web discussed in the chapter. It will be argued that the appropriate perspective is not to view these as competing and opposing views, a case of the Web as *A or B*, but rather the Web as *A and B*, in effect, order co-existing with chaos. And the key to achieving this lies in the development of a different sort of Semantic Web, one with different aims, and this in turn involves adopting a new design paradigm. This new perspective needs to explicitly acknowledge that usefulness, as defined in the classic sense, is not the beginning and end of matters, and that, it is legitimate to design the Web taking into account that not everything has to have a particular purpose. Put another way, as Gaver (2008a) has noted, some artifacts can just simply be *curious things for curious people*. The elements of such an approach will be described, including links to another Web development that goes under the label of *Web 2.0*; specifically context awareness.

BACKGROUND

The Semantic Web can be viewed as a business engendered ICT, that is to say, driven by and shaped by the needs of business. The promise is of a structured information resource from which information and knowledge can be more easily discovered than is the case with the first generation of the World Wide Web. A key requirement for creating the Semantic Web is the organization and classification of web content. In support of this requirement, knowledge-based technologies are needed that will provide a means of structuring this content, adding meaning to it, and automating the collection and extraction of information and knowledge.

Ultimately the goal appears to be transforming the World Wide Web from a chaotic and disorganized place, into an efficient information and knowledge source, providing a basis for the development of value added services based on this Semantic Web. One of the key elements in achieving this goal is the creation of ontologies (Benjamins *et.al.*, 2003), which are in effect agreed and shared vocabularies and definitions, that provide a basis for adding meaning to web content. These ontologies should provide a common understanding of the meaning of words used in different circumstances. One of the essential tasks in the development of the Semantic Web, is to research and define ontologies for specific applications. Another key research topic is that of content quality assurance, for without this there can be no trust that the data, information and knowledge derived are of any value.

There are of course other challenges of a technical nature, including scalability, multilingualism, visualization to reduce information overload, and stability of Semantic Web languages (Benjamins *et.al.*, 2003).

All these issues are recognized as being central to the successful development of the Semantic Web. However there is one more element, perhaps less well appreciated, that is also central to the success of the Semantic Web. This additional factor is the human and social dimension. It can be argued that the handling of this particular area could significantly affect, for better of worse, both the usefulness and the acceptability of the Semantic Web. To understand why this is so, it is necessary to understand something of the bigger picture that provides a contextual background to the development of the Semantic Web. This reflection begins with consideration of the future development of ICTs in general.

Future visions for ICTs are characterized by a belief in the ubiquity of these technologies (e.g. see Wiser (1991), Ductal *et.al.* (2001), Aarts & Marzano (2003)). Phrases such as *ubiquitous computing*, *pervasive computing*, and *Ambient Intelligence* represent visions of computing devices embedded into everyday things, thus transforming these items into intelligent and informative devices, capable of communicating with other intelligent objects. These intelligent and networked artefacts will assist people in their daily activities, whether these are associated with work, travel, shopping, leisure, entertainment, etc.

The prospect of embedded intelligence transforming the nature of everyday objects builds upon the already widespread use of ICTs in society at large. Personal computers, mobile telephones, laptop computers, personal digital assistants, MP3 players, games' consoles, and such forth are ubiquitous in society. And applications such as the Internet and the World Wide Web have provided a new access channel to existing services, and also opened up new activities, like social networking, blogs, instant messaging, personal web sites, and so forth, that were previously unforeseen and infeasible. Moreover, the ubiquity of computing and communication devices along with cheap (some times free) software downloadable from the Internet, has transformed the non-workplace user environment, for those with the inclination, into a *do-it-yourself* development environment. Many of the these developments in ICT can be described as socially engendered ICTs and applications, that is to say, driven by and shaped by social interests.

This is a very new circumstance and not surprisingly this has raised some concerns about the social effects of all this new and networked technology.

These concerns are wide ranging (Kidd, 2007b). They include worries over security, loss of privacy, and the rise of a surveillance society where the state, as well as some rich and powerful corporations, take on the features of the all seeing and all knowing *Big Brother* central to George Orwell's novel *Nineteen Eighty Four*. Other concerns include the fear that society is becoming too dependent upon computers, to the point where the computer becomes the defining feature of reality, and, if the computer says something is so, or not so, then this must be true, even if it is not. There are also worries about peoples' behavior. These range from increased detachment from the real world as people are increasingly drawn into the artificial world of cyberspace, through to concerns about lack of consideration for others, for example, by those using mobile telephones in public spaces without any regard for the affect on those around them (referred to as increasing civil inattention (Khattab & Love, 2008)).

In Europe there is also a worry that citizens will not accept the offered vision of *ubiquitous computing*. As a result of these concerns a different vision has been formulated for *ubiquitous computing*, one that goes by the name of *Ambient Intelligence*.

Specifically, the concept of *Ambient Intelligence* offers a vision of the information society where the emphasis is on greater user-friendliness, more efficient services support, user-empowerment, and support for human interactions (Ductal *et. al.*, 2001). People will be surrounded by intelligent intuitive interfaces that are embedded in all kinds of objects, and their environment will be capable of recognizing and responding to the presence of different individuals in a seamless, unobtrusive and often invisible way. This particular perspective is founded, at least in principle, on the notion of humans at the centre of all this technology, of the technologies serving people, of a human-centered information society.

But even the acceptance of this more human-centered vision by society at large is perceived to be a problem. This has led some senior representatives of the European ICT industry to propose that ordinary citizens become involved in the research, development and design processes that lead to the creation of so called intelligent everyday objects (ISTAG, 2004). This approach, as presented by ISTAG, is quite radical as it advocates design by, with, and for users, seeing these people not just as subjects for experiments, but also as a source of ideas for the development of new technologies and products.

In the USA there is also an appreciation of the importance of the human dimension with respect to ubiquitous computing (Abowd & Sterbenz, 2000). Key among the topics raised with respect to what are called *human-centered research issues*, is that of control. Control in this case means considering circumstances when an intelligent environment should initiate an interaction with a human, and vice versa, or when sharing of control is necessary between users and the system, and also flexibility in the level of control. Other issues raised are deciding which activities should be supported by these intelligent environments, what new capabilities they enable which go beyond current capabilities and activities, and support for task resumption following an interruption.

There is a danger however, that such matters might be seen as either usability issues or those relating to traditional human factors or ergonomics. And Abowd and Sterbenz (2000) do in fact raise matters that are very much in the sphere of traditional human factors. However there is more to *human-centred research issues* than these more traditional concerns. To understand why this is so, it is necessary to consider technologists' and engineers' perceptions of human-centredness.

Isomäki (2007) reports on a study of information systems designers' conceptions of human users. Of note is the observation that these designers occupy a continuum of perceptions. At one end there are those with very limited conceptions, right though to the other end where designers have more holistic and comprehensive perspectives. This tends to support the conclusion that different forms of responses are possible when taking into account people, ranging from ergonomics of human-computer interfaces, right through to designing technologies so that users have full control over the way that the technologies work and the way that they are used.

With different conceptions of human issues, it follows that designers will have different perceptions of the problems that need to be addressed. To those with limited conceptions, traditional ergonomics and human factors is all that is needed to address the human dimension. However, for those with more developed perceptions of the human dimension, of what human centeredness might imply, the scope of designing for the human, has, potentially, much greater scope.

These designer conception issues therefore raise matters of values and design paradigms concerning technologies and the relationships with people, and more broadly with society.

This then leads to the familiar ground of technological determinism (MacKenzie & Wajcman, 1985, page 4), where technology is perceived to lie outside of society, having effects upon society, but being neutral in the sense that the technology is not influenced by subjective elements within society (e.g. such as values). If there is a choice in relation to technology, then it is one of choice between competing inventions, and by virtue of rational assessment and judgments, the best technology can be selected. This chosen technology then has some effect upon society, and matters of human issues just reduce to finding the most appropriate way to interact with the technology.

An alternative view is to consider technology as the product of society, a result of the prevailing economic, political, social and value systems. Technology therefore is not neutral and independent of society, but is shaped by society and its dominant values. And if this is so, then it is possible to shape technology in different ways, depending upon the values that are dominant, which opens up the possibility for the *Social Shaping of Technology*, using insights from the social and psychological sciences to produce different technologies to those that would normally be produced when technologists are left to themselves.

This discussion about the factors that shape technology leads to considerations of design paradigms. Paradigms are comprised of core beliefs, shared values, assumptions, and accepted ways of working and behaving (Johnson, 1988). Technological determinism is part of the framework of beliefs that are held by technologists, and is therefore part of the paradigm of technology development.

It has been argued (Kidd, 2007a) that the shift to the knowledge age, to a world where Ambient Intelligent systems predominate, involves a paradigm shift, not just in technology, but also in terms of the values that technologists bring to the process of designing these technologies. The knowledge era is heralded as a new age for humankind, implying some sort of transition from the past, to a new and different future, one based on the value of information and knowledge. The old age that is being left behind, the industrial era, was, to a large extent, based on subjugation of human skills, knowledge, expertise, and purpose to the demands of a resource-intensive economic system based on mass production. This led to a relationship between people and machines, where the needs of machines were predominant, and technology was designed to, as far as possible, eliminate the need for human intelligence, or to move this need to a select group of people within organizations, such as engineers and managers.

In other ways also, technology design practices have been shaped by the limitations of past technologies. The age in which technology development and design practices emerged and were refined, was primarily characterized by rather inflexible electrical, mechanical, or (a combination) electro-mechanical systems, which severely limited what was possible. In many respects the advent of relative cheap and highly flexible computers, which are in effect universal machines, has done very little to change these established practices, especially when it comes to considering the human dimension. This is the power that paradigms hold over people, trapping them into avenues of thought and practice that no longer have relevance.

But the industrial age view of technology and machine, also reflected the dominant world view of the time, where the universe was seen as a machine, a majestic clockwork (Bronowski, 1973, pages 221-256) and where its workings are defined by causal laws (Newton's), and where, by using reductionist scientific methods, all in time could be understood and be known. Of course, the universe turned out to be a much stranger place than Newton imagined, a place of relativity, chaotic behavior within what were once perceived as predictable systems, and of the quirky behavior of matter at the quantum level.

But in many respects this clockwork view of the world still prevails in the world of technology, for it is far easier to conceive of people as being components of machines, since the alternative is to consider humans in their true light, with all their quirkiness, likes and dislikes, interests, emotions, wants, and so forth. Yet the advent of the information society, the knowledge era, the age of *Ambient Intelligence*, demands just such as shift, for no more can ICT be perceived primarily as technologies for the workplace, where the employees can be subjugated to the needs of employers and the purposes of the employing organization.

And the primary reason for this is the change, is that of a different context for the use of ICT, which is now just as much focused on the world outside of the familiar workplace environment, as it is on the workplace. But this not only has implications for technology and technology design practices, it also has implications for the social sciences. Traditionally, those concerned with the *Social Shaping of Technology* have developed their theories and concepts based on the use of ICTs, as well as early generations of automation technologies, in the context of workplace environments. In other words they have been motivated by business engendered ICTs. This is no longer the prevailing circumstance. ICTs now need to be designed for both the workplace user and the non-workplace user. Business engendered ICTs and socially engendered ICTs have to be accommodated, with an ill-defined boundary between the two, and this provides the *Social Shaping of Technology* perspective with new challenges.

MAIN FOCUS OF THE CHAPTER

Relevance of Existing Social Shaping of Technology Theories and Concepts

To begin to address the Social Shaping of the Semantic Web, it is first necessary to make a detour to consider some key theories and concepts of the *Social Shaping of Technology*, and then to address their potential relevance to the Social Shaping of the Semantic Web.

The *Social Shaping of Technology* has it roots in the understanding that technology is not neutral, and that technology is *shaped* by the values and beliefs of those that influence its development, these being mostly engineers, technologists, and scientists. Change these values and a different technology will result.

While the scientific and technical community tend (and like) to believe that technology is neutral and is not determined by values, many social scientists think differently. As a result there is a body of thought and knowledge in the social sciences, that provides a theoretical and conceptual basis for the *Social Shaping of Technology*. Key among the existing body of knowledge is *sociotechnical design*. Another important perspective is *interfacing in depth*.

For completeness, some of the key features of both will first be described. Then, the relevance of these to the Social Shaping of the Semantic Web will be addressed.

The Sociotechnical School developed in the United Kingdom shortly after World War II, in response to the introduction of new technology into the British coal mining industry. Its central tenant is that surrounding technology, which can be regarded as a sub-system, there is also a social sub-system. These two sub-systems can be designed to be compatible, either by changing the technology to match the social sub-system, or modifying the social sub-system to match the technology, or a mixture of both.

The Sociotechnical School of thought has been articulated in the form of principles, (Cherns, 1976, 1987) that embody the values and key features of sociotechnical design. These principles, of which there are 11, are: *Compatibility*; *Minimum Critical Specification*; *Variance Control*; *The Multifunctional Principle–Organism vs. Mechanism*; *Boundary Location*; *Information Flow*; *Support Congruence*; *Design and Human Values*; *Incompletion*; *Power and Authority*; and *Transitional Organization*.

Among the above there is a sub-set of principles that are primarily organizational in nature. These are: *The Multifunctional Principle–Organism vs. Mechanism*; *Boundary Location*; *Information Flow*; *Support Congruence*; and *Power and Authority*.

The *Multifunctional Principle–Organism vs. Mechanism* refers to traditional organizations which are often based on a high level of specialization and fragmentation of work, which reduces flexibility. When a complex array of responses is required, it becomes easier to achieve this variety if the system elements are capable of undertaking or performing several functions. *Boundary Location* is a principle that relates to a tendency in traditional hierarchical organizations to organize work around fragmented functions. This often leads to barriers that impede the sharing of data, information, knowledge and experience. Boundaries therefore should be designed around a complete flow of information, or knowledge, or materials, to enable the sharing of all relevant data, information, knowledge and experience. The *Information Flow* principle addresses the provision of information at the place where decisions and actions will be taken based on the information. *Support Congruence* relates to the design of reward systems, performance measurement systems, etc., and their alignment with the behaviors that are sought from people. For example, individual reward for individual effort, is not appropriate if team behavior is required. *Power and Authority* is concerned with responsibilities for tasks, and making available the resources that are needed to fulfill these responsibilities, which involves giving people the power and authority to secure these resources.

There is also another sub-set of principles that largely relate to the process by which technology is designed. These are: *The Compatibility Principle*; *The Incompletion Principle*; and *The Transitional Organization Principle*.

The *Compatibility Principle* states that the process by which technology is designed needs to be compatible with the objectives being pursued, implying that technologies designed without the involvement of users, would not be compatible with the aim of developing a participatory form of work organization where employees are involved in internal decision making. *Incompletion* addresses the fact that when workplace systems are designed, the design is in fact never finished. As soon implementation is completed, its consequences become more evident, possibly indicating the need for a redesign. The *Transitional Organization* principle addresses two quite distinct problems when creating new organizations: one is the design and start-up of new (greenfield) workplaces, the other relates to existing (brownfield) workplaces. The second is much more difficult than the first. In both situations the design team, and the processes it uses, are potentially a tool to support the start-up and any required transitions.

What remain from Cherns' set of 11 socio-technical design principles, is a sub-set that is significantly technology oriented, although the principles also have organizational implications. The principles in question are: *Minimum Critical Specification*; *Variance Control*; *Design and Human Values*.

The principle of *Minimum Critical Specification* states that only what is absolutely necessary should be specified, and no more than this, and that this applies to all aspects of the system: tasks, jobs, roles, etc. Whilst this is organizational in nature, it impacts technology as well. It implies that what has to be done needs to be defined, but how it should be done should be left open. In terms of features and functions of technology, the technology should not be over determined, but should leave room for different approaches. It implies a degree of flexibility and openness in the technologies. Turning now to *Variance Control*, this is a principle that, as its name suggests, is focused on handling variances, these being events that are unexpected or unprogrammed. Variances that cannot be eliminated should be controlled as near to the point of origin of the variance as possible. Some of these variances may be critical, in that they have an important affect on results. It is important to control variances at source, because not to do so often introduces time delays. Next on the list of principles is that of *Design and Human Values*. This is concerned with quality of working life. In the context of the working environment it manifests itself in issues such as stress, motivation, personal development, etc. This principle has both a social sub-system dimension and a technology sub-system dimension, in that both can be designed to reduce stress, and to enhance motivation and personal development.

The second approach mentioned as being of relevance to the Social Shaping of the Semantic Web, known as *interfacing in depth*, has its roots in technology design, specifically the design of computer-aided manufacturing systems.

This perspective on the *Social Shaping of Technology* rests on the observation (Kidd, 1992) of the importance of technology in influencing organizational choice and job design. There is a perspective (e.g. see Clegg, 1984) that suggests that technology is of secondary importance with respect to job design and organizational choice. However, as noted by Kidd (1992), technology is clearly not neutral and can close off options and choice in the design of organizations and jobs. Technology for example can be used to closely circumscribe working methods, to limit freedom of action and autonomy, and to determine the degree of control that users have over the work process.

This viewpoint, of technology shaping organizations, roles, and working methods, led to the notion of *interfacing in depth*. So, rather than just applying ergonomic and usability considerations to the design of human-computer interfaces, it was proposed that there is also a need to apply psychological and organizational science insights to the design of the technology behind the interface.

Kidd (1988) for example, describes a decision support system that was designed using this broader perspective. A key point about this decision support system is that the system characteristics were not achieved through the application of ergonomics or usability considerations to the design of the human-computer interface. Rather the characteristics arose from the technology behind the human-computer interface, where the technology refers to the algorithms, data models, architectures, and the dependency upon human judgment and skills that were built into the operational details of the software.

Kidd (1988) also points out that it is necessary to make a distinction between the surface characteristics of a system, as determined by the human-computer interface, and the deeper characteristics of a system, as determined by the actual technology. The surface characteristics are strongly related to ergonomics and usability, while the deeper characteristics relate more to the view of the user held by the designer, in that if values are driven by a desire to reduce user autonomy, this will be reflected in the details of the underlying technology. Likewise if values are such that autonomy is valued, then this will lead to a different type of underlying technology.

Consequently, good human-computer interface (surface) characteristics are necessary, but not sufficient. Attention must also be paid to the deep system characteristics, that is, the technology behind the human-computer interface. This is called *interfacing in depth*.

The relevance of both *sociotechnical design* and *interfacing in depth* to the *Social Shaping of the Semantic Web* has been addressed by Kidd (2008a). A key point in this consideration is the relevance of both approaches to non-workplace environments, for socially engendered ICTs, given that both the sociotechnical approach and *interfacing in depth* were developed in the context of workplace environments.

Kidd (2008a) specifically addresses the relevance of sociotechnical principles and *interfacing in depth* for the case of the Semantic Web, not just as a technology for the workplace, but in the context of the non-workplace user environment, for example in the home.

A key feature of any web technologies is that the outcome of the use of this type of technology is not known in advance. Consequently, to over determine how the technology is used, to over limit results based on semantics, could be incompatible with the purpose of the technology, as perceived by some people, and its value to users.

This implies that the sociotechnical variance control principle could potentially be very important in the design and development of the Semantic Web. One of the potential downsides of the Semantic Web is that it eliminates variances in web search results, thus destroying some of the value of the Web (the experience of discovering the unexpected). Consequently, enabling the user to decide how much variance to tolerate, in other words to place control of variances in the hands of users, could be an important attribute that needs to be designed into the Semantic Web, and for this reason therefore, variance control is potentially an important principle for the non-workplace environment. But it could also be an important principle in the workplace environment as well. The reason for this primarily lies in the competitive imperative for innovation, and in the need to be adaptive and responsive, especially in the face of structural changes in the business environment; changes that require agility, and corresponding organizational designs and operating principles that are open to bottom-up adaptation (Kidd, 2008b).

Control therefore is potentially important because not to have control over technology such as the Semantic Web, for users not to be able to decide which features of the technology should be employed, reduces the role of the Semantic Web to that of a vending machine for search results. This could be highly de-motivating to users of the Semantic Web.

This observation also arises from the *interfacing in depth* perspective. The whole philosophy of *interfacing in depth* is based on design of technologies where there is uncertainty and unpredictability in terms of outcomes. This approach provides a framework to counter the tendency to reduce human-computer encounters to circumstances where there is no uncertainty and unpredictability in outcomes. This theory is highly relevant to the Semantic Web, for this approach would seek to allow user autonomy and control to flourish, thus maintaining the potentially chaotic and serendipitous nature of the World Wide Web, but at the choice of the user.

Moving Towards the Social Shaping of the Semantic Web

Summarizing the central argument, the key issue is to shape the Semantic Web so that it does not reduce interactions with the Web to the circumstance where the Web becomes like a vending machine. But there is more involved here than a simple on-off switch that disables or enables, at will, the semantic features of the Web, although such as approach could be used. Ideally a circumstance should be created where the technology provides a more sophisticated approach. But the question remains how to do this? A key issue here relates to avoiding a circumstance where the Web is viewed in polar terms. The appropriate perspective therefore is not to see two competing possibility, two extremes, a case of the Web as semantic or the Web as non-semantic, but as a combination of both, as a continuum of infinite possibilities.

Such a perspective has been advanced before in connection with other technologies (Kidd, 1994, pages 301-303), but also more recently in relation to Ambient Intelligent Systems (Kidd, 2007a). The conceptual basis lies with *interfacing in depth* and variance control, and has been referred to as user defined human-computer relationships. This word *relationship* is important here as it implies more than just an interaction between person and machine. There has to be intelligence in this relationship, something that is often overlooked when the word intelligence is used in the context of computers. It is not, for example, very intelligent for a so-called *intelligent everyday artifact* to enforce a given way of working on users. An intelligent relationship with a semantically based World Wide Web would be built on control and understanding. Control comes from providing the technologies that will allow users to specify in some way, how semantic based searches operate, perhaps for example by including some form of *control knob* that would tone down the strength of the semantic dimension. But more than this, the Semantic Web needs to understand something of the context of the user.

This then touches upon a sensitive area, that of context awareness (Braun & Schmidt, 2006), which is a matter that arises in another area of World Wide Web development often referred to as *Web 2.0*. The sensitivity hinted at here relates to privacy, for to understand a user's context it is necessary to capture information, some of which may be of a personal nature, including patterns of usage and the like, much of which people may not want to have stored within a computer system. The information is also of the sort that commercial organizations, interested in marketing products and services, might be all too keen to lay their hands on.

However, putting aside for the moment these very serious issues, understanding the context under which users come to the Web, knowing something of their likes and dislikes, whether their are interests are broad or narrow, how much they value serendipity and how often they follow-up seemingly random and unrelated search results, could be a key factor in enabling the development of a more adaptive and responsive Semantic Web.

But this only defines the relationship. What about user motivation? Why should technology developers bother with such sophistication? This returns the discussion to the matter of paradigms and technologists' conceptions of the human dimension of the Semantic Web. Here, looking beyond utility, a factor that is such a dominant feature of the workplace, is critical. This involves addressing the non-utilitarian perspective, something that is perhaps an alien concept to the technologist. Put simply, not everything has to have utility.

Technologists need to understand that what makes people happy is not always something that is useful. Sometimes happiness comes from meaning, from emotional connection. This, it has been argued, is central to understanding why certain technologies are so successful, while others are less so (Lyngsø & Nielsen, 2007). For example, text messaging on mobile phones seems very much to be an example of a very useful tool, and it certainly has a very obvious utility. But text messaging is not just used for utilitarian purposes. Many young people use text messages to communicate with each other. But to adults the messages may seem to be pointless, like "where are you?" or "what are you doing?" or "I'm bored" and so forth. This is just chatter, which is meaningful to the younger generation, but not to adults. And the word *meaningful* is key here. It is the meaningfulness of the text messaging system that makes it so popular. And the same can be said for instant messaging, blogs, and social networking sites. They have very little in the way of utility. They are in fact just an extension of the face to face discussions that take place when people meet. But this point is important, because their motivation is not oriented to fulfilling a task, but to other more human inclinations, like for example, *having a good time*. These applications and many more, are examples of socially engendered ICT.

Socially engendered ICT points to a different type of driver for the development of the Semantic Web. The Semantic Web is not just a tool to undertake more efficient and effective searches of Web content, but can also be a means for people to *have a good time*. This is therefore links to the concept of Ludic Systems.

Ludic is something of an obscure word; it means playful (Huizinga, 1970), but in a very wide sense. Included are activities such as learning, exploration, etc. The Ludic perspective does not just characterized people by thinking or achievements, but also by their *ludic* engagement with the world: their curiosity, their love of diversion, their explorations, inventions and wonder. Play is therefore not just perceived as mindless entertainment, but an essential way of engaging with and learning about the world and the people in it (see Gaver (2008b) for an example).

Consequently, the Social Shaping of the Semantic Web needs to incorporate this perspective, which explicitly acknowledges that usefulness, as defined in the classic sense, is not the beginning and end of matters, and that, it is legitimate to design the Web taking into account that not everything has to have a particular purpose. Put another way, some artifacts can just simply be *curious things for curious people* (Gaver, 2008a). This could be of key importance in reshaping design paradigms, introducing a different dimension that explicitly recognizes that there is *life beyond mere utility*.

With this view in mind, Kidd (2008a) has proposed an additional sociotechnical design criteria to add to the 11 proposed by Cherns (1976,1987). This new criteria takes sociotechnical design out beyond the workplace environment, into the world of ubiquitous computing, of Ambient Intelligent systems, a world of the Web as used by a vast network of people seeking to *have a good time*, of a world of socially engendered ICT. The new principle embodies the mood of the age, as manifested in social networking Web sites, blogs, instant messaging, and so forth. The principle, referred to as the *Non-utility Principle*, is articulated as:

Non-utility Principle: *ICT in non-workplace contexts serve purposes beyond mere utility, and ICTs should therefore be designed to enable users to achieve emotional fulfillment through play, exploration, and several other dimensions, that are not traditionally associated with workplace environments.*

FUTURE TRENDS

Clearly during the early years of the 21st century there has been an emergence of, as well as a significant growth in, ICTs that are predominately focused on the world outside work. Many of these systems, while they also provide the workplace with useful tools that serve the utility oriented perspective of the working environment, were not conceived with this outcome in mind. It is more the case that they serve a purpose that is related to people as social creatures with a need to find meaning. Often these systems are used in what might be seen, when judged by the rational standards of work, as being nothing more than frivolous time wasting activities. But when looked at from a broader perspective, they seem to embody life, for life is made up of many activities such as enjoying oneself, socializing through small talk and casual chat, etc.

Further development and growth in these types of ICTs seems set to continue as social scientists and technologist begin to collaborate on the design and development of technologies that will make the experience of using these socially engendered systems, even better.

This collaboration between the social sciences and technologists is key to creating technologies that are better suited to the new circumstances of ICT and their use. With time, as technologists begin to realize that the value of technology does not just lie with utility, with making things more efficient, and so forth, and that it is quite legitimate to design technologies that will help people to find meaning through whatever activities (within reason) that they want to undertake, there should emerge a very different sort of technology to that which has already been developed.

What new delights lie ahead for the users of these systems is hard to foretell. What needs to be done to bring about these systems is however a little more predictable. Central will be the development of interdisciplinary design, and even the emergence of a new breed of technologist, with knowledge in social sciences as well as in technology subjects. Based upon this, the notion of a new breed of professional can be suggested, involving people who can operate in the spaces between the social sciences on one side, and engineering and technology on the other. Such people would be capable of taking into account both perspectives and would use their knowledge to design technologies more acceptable to society than those that might emerge from a more technology-oriented approach.

This in turn would lead to new research agenda, and in effect the implementation of *Social Shaping of Technology* in a world where technology is no longer perceived to lie outside of society, but to be an integral part of it. This development will in part be aided by research that is already underway looking at the development of complex systems science (European Commission, 2007) and its relevance and application in areas where ICT and society have already merged (social networking Web sites for example).

CONCLUSION

ICTs are beginning to develop along new paths, socially engendered and shaped to a significant extent by such concepts as instant messaging, chat rooms, social networking, and the like. These developments come from the world outside of work and are not based upon the notion of utility, but more on meaning, of doing things for fun, of explorations, etc. The Semantic Web on the other hand largely comes from business engendered thinking, from a world where the primary concerns are utility, effectiveness, efficiency, and usefulness.

These two worlds in many ways seem to clash, to be polar opposites. But this does not have to be so. The Semantic Web can be shaped in entirely new directions and does not have to become a tool for business, but could also be another means of having fun. To this end the paper has mapped out some preliminary possibilities, provided a conceptual basis for development, and highlighted a key guiding design principle. The challenge for the future is to take what is emerging from the domain of socially engendered ICT, and bring this to the area of business engendered ICT, to the benefit of both domains. For to do so would provide a means of preserving what is good about the Web, that is to say its unpredictability, providing a means by which meaning is found, while also delivering a Web that is also more useful for those with a more serious purpose. This new notion for the Semantic Web would provide a place for both work and play, adapting as required to the needs of users at specific moments in time.

REFERENCES

Aarts, E., & Marzano, S. (Eds.). (2003). *The new everyday: Views on ambient intelligence*. Rotterdam, The Netherlands: 010 Publications.

Abowd, G. D., & Sterbenz, J. P. G. (2000). Final report on the inter-agency workshop on research issues for smart environments. *IEEE Personal Communications*, *7*(5), 36–40. doi:10.1109/98.878535

Benjamins, V. R., Contreras, J., Corcho, O., & Gómez-Pérez, A. (2003). *Six challenges for the Semantic Web*. Retrieved July 2, 2008, from http://www.cs.man.ac.uk/~ocorcho/documents/KRR2002WS_BenjaminsEtAl.pdf

Braun, S., & Schmidt, A. (2006). *Socially-aware informal learning support: Potentials and challenges of the social dimension*. Retrieved July 2, 2008, from http://ftp.informatik.rwth-aachen.de/Publications/CEUR-WS/Vol-213/paper22.pdf

Bronowski, J. (1973). *The ascent of man*. London: BBC Publications.

Cherns, A. (1976). Principles of sociotechnical design. *Human Relations*, *29*(8), 783–792. doi:10.1177/001872677602900806

Cherns, A. (1987). Principles of sociotechnical design revisited. *Human Relations*, *40*(3), 153–162. doi:10.1177/001872678704000303

Clegg, C. W. (1984). The derivation of job design. *Journal of Occupational Behaviour*, *5*, 131–146. doi:10.1002/job.4030050205

Ducatel, K., Bogdanowicz, M., Scapolo, F., Leijten, J., & Burgelman, J.-C. (2001). *Scenarios for ambient intelligence in 2010*. Retrieved July 2, 2008, from ftp://ftp.cordis.europa.eu/pub/ist/docs/istagscenarios2010.pdf

European Commission. (2007). *Science of complex systems for socially intelligent ICT*. Retrieved July 2, 2008, from http://cordis.europa.eu/fp7/ict/fet-proactive/cosiict-ws-oct07_en.html#presentations

Gaver, W. (2008a). *Curious things for curious people*. Retrieved July 2, 2008, from http://www.goldsmiths.ac.uk/interaction/pdfs/36.gaver.curiousThings.inPress.pdf

Gaver, W. (2008b). *The video window: My life with a ludic system*. Retrieved July 2, 2008, from http://www.goldsmiths.ac.uk/interaction/pdfs/32gaver.videoWindow.3ad05.pdf

Huizinga, J. (1970). *Homo ludens: A study of the play element in culture*. London: Paladin.

Isomäki, H. (2007). Different levels of information systems designers' forms of thought and potential for human-centered design. *International Journal of Technology and Human Interaction*, *3*(1), 30–48.

ISTAG. (2004). Experience and application research: Involving users in the development of ambient intelligence. Retrieved July 2, 2008, from ftp://ftp.cordis.europa.eu/pub/ist/docs/2004_ear_web_en.pdf

Johnson, G. (1988). Process of managing strategic change. *Management Research News*, *11*(4/5), 43–46. doi:10.1108/eb027990

Khattab, I., & Love, S. (2008). Mobile phone use across cultures: A comparison between the United Kingdom and Sudan. *International Journal of Technology and Human Interaction*, *4*(2), 35–51.

Kidd, P. T. (1988). The social shaping of technology: The case of a CNC lathe. *Behaviour & Information Technology*, *7*(2), 193–204. doi:10.1080/01449298808901873

Kidd, P. T. (1992). Interdisciplinary design of skill based computer-aided technologies: Interfacing in depth. *International Journal of Human Factors in Manufacturing*, *2*(3), 209–228. doi:10.1002/hfm.4530020302

Kidd, P. T. (1994). *Agile Manufacturing: Forging New Frontiers*. Wokingham: Addison-Wesley.

Kidd, P. T. (2007a). Human-centered ambient intelligence: Human-computer relationships for the knowledge era. In P. T. Kidd (Ed.), *European visions for the knowledge age: A quest for new horizons in the information society* (pp. 55-67). Macclesfield, UK: Cheshire Henbury Publications.

Kidd, P. T. (Ed.). (2007b). *European visions for the knowledge age: A quest for new horizons in the information society*. Macclesfield, UK: Cheshire Henbury Publications.

Kidd, P. T. (2008a). Towards new theoretical and conceptual frameworks for the interdisciplinary design of information society technologies. *International Journal of Interdisciplinary Social Sciences*, *3*(5), 39–46.

Kidd, P. T. (2008b). Agile holonic network organizations. In G. D. Putnik & M. M. Cunha (Eds.), *Encyclopedia of networked and virtual organizations* (pp. 35-42). Hershey, PA: Information Science Reference.

Lyngsø, L., & Nielsen, A. S. (2007). Creating meaning: The future of human happiness. In P. T. Kidd (Ed.), *European visions for the knowledge age: A quest for new horizons in the information society* (pp. 217-228). Macclesfield, UK: Cheshire Henbury Publications.

MacKenzie, D., & Wajcman, J. (Eds.). (1985). *The social shaping of technology*. Milton Keynes, UK: Open University Press.

Sawyer, S., & Tapia, A. (2005). The sociotechnical nature of mobile computing work: Evidence from a study of policing in the United States. *International Journal of Technology and Human Interaction, 1*(3), 1–14.

Weiser, M. (1991). The computer of the 21st century. *Scientific American, 265*(3), 66–75.

KEY TERMS AND DEFINITIONS

Ambient Intelligence: A human-centered vision of the information society where the emphasis is on greater user-friendliness, more efficient services support, user-empowerment, and support for human interactions with respect to intelligent everyday objects and other ICT systems.

Business Engendered ICT: Information and Communications Technology, the development and use of which is driven and shaped by the needs of business.

Interfacing in Depth: Shaping the characteristics of a technology by considering the details of the technologies that lie behind the human-computer interface, where technology refers to the algorithms, data models, architectures, and the dependency upon human judgment and skills that are built into the operational details of the software.

Ludic Systems: Ludic refers to the play element of culture. Ludic systems are based on a philosophy of understanding the world through play, of play being primary to and a necessary condition for the generation of culture. Such systems therefore do not necessarily fulfill any particular purpose in the sense that most technological systems usually exist to fulfil a need, or have some useful function, or are a utility.

Non-Utility Principle: ICT in non-workplace contexts serve purposes beyond mere utility, and ICTs should therefore be designed to enable users to achieve emotional fulfillment through play, exploration, and several other dimensions, that are not traditionally associated with workplace environments.

Social Shaping of Technology: The philosophy that technology is not neutral and is shaped by the dominant social, political and economic values of society. As a result therefore, changes in values lead to different technological outcomes, and as a result, social science considerations can be used to shape technologies.

Socially Engendered ICT: Information and Communications Technology, the development and use of which is driven and shaped by social interests.

This work was previously published in Handbook of Research on Social Dimensions of Semantic Technologies and Web Services, edited by M. M. Cruz-Cunha; E. F. Oliveira; A. J. Tavares; L. G. Ferreira, pp. 341-356, copyright 2009 by Information Science Reference (an imprint of IGI Global).

Index

B

C

D

J

K

L

M

N

O

P

Q

R

S

U

V

W

Y

Z